$\|\mathbf{v}\|$	the length of the vector $\mathbf{v}$, pp. 182, 205
$\|\mathbf{u}-\mathbf{v}\|$	the distance between vectors $\mathbf{u}$ and $\mathbf{v}$, pp. 183, 208
$\cos\theta$	the cosine of the angle between two nonzero vectors $\mathbf{u}$ and $\mathbf{v}$, pp. 187, 206
$\mathbf{u}\cdot\mathbf{v}$	the standard inner product on R^2 or R^3, p. 186
$(\mathbf{u}, \mathbf{v})$	an inner product, pp. 186, 200
$\mathbf{u}\times\mathbf{v}$	the cross product operation, p. 192
$W^{\perp}$	the orthogonal complement of a subspace W of an inner product space, p. 227
$\operatorname{proj}_W \mathbf{v}$	the orthogonal projection of the vector $\mathbf{v}$ on the subspace W, p. 234
L, L_1, L_2	linear transformations, pp. 253, 291
$p'(t)$	the derivative of $p(t)$ with respect to t, p. 263 (Exercise 3)
$\ker L$	the kernel of the linear transformation L, p. 266
range L	the range of the linear transformation L, p. 268
$j_1 j_2 \cdots j_n$	a permutation of $S = \{1, 2, \ldots, n\}$, p. 315
$\det(A)$, $\|A\|$	the determinant of the matrix A, p. 316
$\det(M_{ij})$	the minor of a_{ij}, p. 331
A_{ij}	the cofactor of a_{ij}, p. 331
adj A	the adjoint of the matrix A, p. 339
λ_j	an eigenvalue of the linear transformation L or matrix A, pp. 352, 356
$\mathbf{x}_j$	an eigenvector of L or A associated with the eigenvalue λ_j, pp. 352, 356
$\det(\|\lambda I_n - A\|) = p(\lambda)$	the characteristic polynomial of A, p. 358
$\det(\|\lambda I_n - A\|) = p(\lambda) = 0$	the characteristic equation of A, p. 358
D	a diagonal matrix, p. 366
$\mathbf{x}^{(k)}$	age distribution at time k, p. 375
$g(\mathbf{x}) = \mathbf{x}^T A\mathbf{x}$	a real quadratic form in n variables, p. 406
$\mathbf{x}(t) = \begin{bmatrix} x_1(t) \\ x_2(t) \\ \vdots \\ x_n(t) \end{bmatrix}$	an $n \times 1$ matrix whose entries are functions of t, p. 437
$\mathbf{x}'(t) = \begin{bmatrix} x_1'(t) \\ x_2'(t) \\ \vdots \\ x_n'(t) \end{bmatrix}$	p. 437
$A(t)$	a matrix function, p. 455 (Supplementary Exercise 1)
e^{At}	the matrix exponential function, p. 456 (Supplementary Exercise 3)
$c = a + bi$	a complex number, p. 513
$\overline{c} = a - bi$	the conjugate of $c = a + bi$, p. 514
$\|c\| = \|a + bi\| = \sqrt{a^2 + b^2}$	the absolute value or modulus of the complex number $c = a + bi$, p. 516
$\overline{A} = \left[\overline{a}_{ij}\right]$	the conjugate of the matrix $A = \left[a_{ij}\right]$, p. 518

Elementary Linear Algebra

7th Edition

ELEMENTARY LINEAR ALGEBRA

Bernard Kolman
Drexel University

David R. Hill
Temple University

PRENTICE HALL, Upper Saddle River, New Jersey 07458

Library of Congress Cataloging-in-Publication Data

Kolman, Bernard
Elementary linear algebra. -- 7th ed. / Bernard Kolman and David R. Hill.
p. cm.
Includes index.
ISBN 0-13-085199-X
1. Algebras, Linear. I. Hill, David R. II. Title.
QA184.K668
512'.5--dc21 99-41239
CIP

Acquisitions Editor: George Lobell
Editor-in-Chief: Jerome Grant
Production Editor: Nick Romanelli
Senior Managing Editor: Linda Mihatov Behrens
Executive Managing Editor: Kathleen Schiaparelli
Assistant Vice President of Production and Manufacturing: David W. Riccardi
Marketing Manager: Melody Marcus
Marketing Assistant: Vince Jansen
Manufacturing Buyer: Alan Fischer
Manufacturing Manager: Trudy Pisciotti
Supplements Editor/Editorial Assistant: Gale Epps
Art Director: Ann France
Director of Creative Services: Paul Belfonti
Associate Creative Director: Amy Rosen
Assistant to Art Director: John Christiana
Art Manager: Gus Vibal
Art Editor: Grace Hazeldine
Interior Designer: Donna Wickes
Cover Image: Mikhail Matiushin, "Movement in Space," 1922/S A CORBIS/ The State Russian Museum

Printed in the United States of America

10 9 8 7 6 5 4 3 2 1

ISBN 0-13-085199-X

Prentice-Hall International (UK) Limited, *London*
Prentice-Hall of Australia Pty. Limited, *Sydney*
Prentice-Hall Canada Inc., *Toronto*
Prentice-Hall Hispanoamericana, S.A., *Mexico*
Prentice-Hall of India Private Limited, *New Delhi*
Prentice-Hall of Japan, Inc., *Tokyo*
Simon & Schuster Asia Pte. Ltd., *Singapore*
Editora Prentice-Hall do Brasil, Ltda., *Rio de Janeiro*

To Lisa, Stephen
and to the memory of Lillie

B. K.

To Suzanne

D. R. H.

CONTENTS

PREFACE

Linear algebra is an important course for a diverse number of students for at least two reasons. First, few subjects can claim to have such widespread applications in other areas of mathematics—multivariable calculus, differential equations, and probability, for example—as well as in physics, biology, chemistry, economics, psychology, sociology, and all fields of engineering. Second, the subject offers the student at the sophomore level an excellent opportunity to learn how to approach abstract concepts.

This book provides an introduction to the basic ideas and computational techniques of linear algebra at the sophomore level. In addition, it includes a few carefully selected applications. The book also introduces the student to working with abstract concepts. In covering the basic ideas of linear algebra, the abstract ideas are carefully balanced by the considerable emphasis on the geometrical and computational aspects of the subject. A new feature of this edition is the optional opportunity to use MATLAB or other software to enhance the pedagogy of the book.

WHAT'S NEW IN THE SEVENTH EDITION

We have been very pleased by the wide acceptance of the first six editions of this book throughout the 30 years of its life. Encouraged by the activities of the calculus reform movement that have been going on in this country for the last few years, progress has been made on developing ways to improve the teaching of linear algebra. *The Linear Algebra Curriculum Study Group* and others have made a number of recommendations for doing this. In preparing this edition, these recommendations, as well as faculty and student suggestions, have been carefully considered. Although a great many changes have been made to develop this major revision, our objective has remained the same as in the first six editions: *to present the basic ideas of linear algebra in a manner that the student will find understandable.* To achieve this objective, the following features have been developed in this edition:

- Greater use of linear-combinations-of-columns approach as a running theme throughout the book.
- Several sections in Chapters 2 and 4 have been moved to improve the organization, exposition, and flow of the material.

- Old Section 6.1, *Eigenvalues and Eigenvectors*, has been split into two sections to improve pedagogy.
- Section 6.3, *Stable Age Distribution in a Population*; *Markov Processes*, new to this edition, contains material dealing with applications of Sections 6.1 and 6.2.
- Section 6.5, *Spectral Decomposition and Singular Value Decomposition*, new to this edition, provides a generalization of eigenvalues and eigenvectors that is extremely valuable in applications.
- Section 7.2, *Dynamical Systems*, new to this edition, presents an introduction to the qualitative behavior of differential equations.
- More geometry has been added.
- More exercises at all levels have been added. Some of these are more open-ended, allowing for exploration and discovery.
- More material on linear transformations has been added.
- Chapter 8, MATLAB *for Linear Algebra*, new to this edition, provides an introduction to MATLAB .
- Chapter 9, MATLAB *Exercises*, new to this edition, consists of 147 exercises that are designed to be solved using MATLAB . However, we do not ask that users of this book write programs. The user is merely asked to employ MATLAB to solve specific numerical problems. The exercises in this chapter complement those given in Chapters 1–7 and exploit the computational capabilities of MATLAB . To extend the instructional capabilities of MATLAB we have developed a set of pedagogical routines, called scripts or M-files, to illustrate concepts, streamline step-by-step computational procedures, and demonstrate geometric aspects of topics using graphical displays. We feel that MATLAB and our instructional M-files provide an opportunity for a working partnership between the student and the computer that in many ways forecast situations that will occur once a student joins the technological workforce of the 21st century. The exercises in this chapter are keyed to topics rather than to individual sections of the text. Short descriptive headings and references to MATLAB commands in Chapter 8 supply information about the sets of exercises.
- The computer exercises that were present in the first seven chapters of the sixth edition were as software-neutral as possible. These exercises, with a few extra additions, have been retained in this new edition.

EXERCISES

The exercises form an integral part of the text. Many of them are numerical in nature, whereas others are of a theoretical type. The theoretical exercises as well as many numerical ones call for a verbal solution. In this technological age, it is especially important to be able to write with care and precision; exercises of this type should help to sharpen this skill. This edition contains nearly 200 new exercises. Computer exercises, clearly indicated by a special symbol , are of two types: in the first seven chapters there are exercises allowing for discovery and exploration that do not specify any particular software to be used for their solution; in Chapter 9, new to this edition, are 147 exercises designed to be solved using MATLAB. The answers to all odd-numbered exercises appear at the back of the book. An **Instructor's Solutions Manual**, containing answers to all even-numbered exercises and solutions to all theoretical exercises, is available (to instructors only) at no cost from the publisher.

PRESENTATION

We have learned from experience that, at the sophomore level, abstract ideas must be introduced quite gradually and must be based on some firm foundations. Thus we begin the study of linear algebra with the treatment of matrices as mere arrays of numbers that arise naturally in the solution of systems of linear equations, a problem already familiar to the student. Much attention has been devoted, from one edition to the next, to refining and improving the pedagogical aspects of the exposition. The abstract ideas are carefully balanced by the considerable emphasis on the geometrical and computational aspects of the subject.

MATERIAL COVERED

In using this book, for a one-quarter linear algebra course meeting four times a week, no difficulty has been encountered in covering eigenvalues and eigenvectors, omitting the optional material and Section 1.8. Varying the amount of time spent on the theoretical material can readily change the level and pace of the course. Thus, the book can be used to teach a number of different types of courses.

Chapter 1 deals with matrices and their properties. Methods for solving systems of linear equations are discussed in this chapter. In **Chapter 2**, we come to a more abstract notion, real vector spaces. Here we tap some of the many geometric ideas that arise naturally. Thus we prove that an n-dimensional, real vector space is isomorphic to R^n, the vector space of all ordered n-tuples of real numbers, or the vector space of all $n \times 1$ matrices with real entries. Since R^n is but a slight generalization of R^2 and R^3, two- and three-dimensional space are discussed at the beginning of the chapter. This shows that the notion of a finite-dimensional real vector space is not as remote as it may have seemed when first introduced. **Chapter 3** covers inner product spaces and has a strong geometric orientation. **Chapter 4** deals with matrices and linear transformations; here we consider the dimension theorems and also applications to the solution of systems of linear equations. **Chapter 5** introduces the basic properties of determinants and some of their applications. **Chapter 6** considers eigenvalues and eigenvectors and real quadratic forms. In this chapter we completely solve the diagonalization problem for symmetric matrices. Section 6.5, *Spectral Decomposition and Singular Value Decomposition*, new to this edition, highlights some of the very useful results in linear algebra. **Chapter 7** provides an introduction to the application of linear algebra in the solution of differential equations. It is possible to go from Section 6.2 directly to Section 7.1, showing an immediate application of the material in Section 6.2. Section 7.2, *Dynamical Systems*, new to this edition, provides an application of linear algebra to a new and exciting area of applied mathematics. **Chapter 8**, MATLAB *for Linear Algebra*, new to this edition, provides an introduction to MATLAB. **Chapter 9**,

MATLAB *Exercises*, new to this edition, consists of 147 exercises that are designed to be solved using MATLAB. **Appendix A** reviews some very basic material dealing with sets and functions. It can be consulted at any time as needed. **Appendix B**, on complex numbers, introduces in a brief but thorough manner complex numbers and their use in linear algebra.

MATLAB SOFTWARE

The instructional M-files that have been developed to be used for solving the exercises in this book, in particular those in Chapter 8, are available on the following website: **`http://www.prenhall.com/kolman`**. These M-files are designed to transform many of MATLAB's capabilities into courseware. Although the computational exercises can be solved using a number of software packages, in our judgment MATLAB is the most suitable package for this purpose. MATLAB is a versatile and powerful software package whose cornerstone is its linear algebra capabilities. This is done by providing pedagogy that allows the student to interact with MATLAB , thereby letting the student think through all the steps in the solution of a problem and relegating MATLAB to act as a powerful calculator to relieve the drudgery of tedious computation. Indeed, this is the ideal role for MATLAB (or any other similar package) in a beginning linear algebra course for, in this course, more than in many others, the tedium of lengthy computations makes it almost impossible to solve a modest-size problem. Thus, by introducing pedagogy and reining in the power of MATLAB , these M-files provide a working partnership between the student and the computer. Moreover, the introduction to a powerful tool such as MATLAB early in the student's college career opens the way for other software support in higher-level courses, especially in science and engineering.

MATLAB incorporates professionally developed, quality computer routines for linear algebra computation. The code employed by MATLAB is written in the C language and is upgraded as new versions of MATLAB are released. MATLAB is available from *The Math Works, Inc.*, 24 Prime Park Way, Natick, MA 01760, [(508) 653-1415], e-mail: info@mathworks.com. The Student version is available from *The Math Works* at a reasonable cost. This Student Edition of MATLAB also includes a version of Maple, thereby providing a symbolic computational capability.

STUDENT SOLUTIONS MANUAL

The **Student Solutions Manual**, prepared by Dennis R. Kletzing, Stetson University, contains solutions to all odd-numbered exercises, both numerical and theoretical.

ACKNOWLEDGEMENTS

We are pleased to express our thanks to the following reviewers of the first six editions: the late Edward Norman, University of Central Florida; the late Charles S. Duris, and Herbert J. Nichol, both at Drexel University; Stephen D. Kerr, Weber State College; Norman Lee, Ball State University; William Briggs, University of Colorado; Richard Roth, University of Colorado; David Stanford, College of William and Mary; David L. Abrahamson, Rhode Island College; Ruth Berger,

Memphis State University; Michael A. Geraghty, University of Iowa; You-Feng Lin, University of South Florida; Lothar Redlin, Pennsylvania State University, Abington; Richard Sot, University of Nevada, Reno; Raymond Southworth, Professor Emeritus, College of William and Mary; J. Barry Turett, Oakland University; Gordon Brown, University of Colorado; Matt Insall, University of Missouri at Rolla; Wolfgang Kappe, State University of New York at Binghamton; Richard P. Kubelka, San Jose State University; James Nation, University of Hawaii; David Peterson, University of Central Arkansas; Malcolm J. Sherman, State University of New York at Albany; Alex Stanoyevitch, University of Hawaii; Barbara Tabak, Brandeis University; Loring W. Tu, Tufts University; and of the seventh edition: Manfred Kolster, McMaster University; Daniel Cunningham, Buffalo State College; Larry K. Chu, Minot State University; Daniel King, Sarah Lawrence University; Kevin Vang, Minot State University; Avy Soffer, Rutgers University.

The numerous suggestions, comments, and criticisms of these individuals greatly improved the manuscript.

We thank Dennis R. Kletzing, Stetson University, who typeset the entire manuscript, the *Student Solutions Manual*, and the *Instructor's Solutions Manual*.

We also thank Nina Edelman, Temple University, for critically reading page proofs; and instructors and students from many institutions in the United States and other countries, for sharing with us their experiences with the book and offering helpful suggestions.

Finally, a sincere expression of thanks goes to Nicholas Romanelli, George Lobell, Gale Epps, and to the entire staff of Prentice Hall for their enthusiasm, interest, and unfailing cooperation during the conception, design, production, and marketing phases of this edition.

B.K.
D.R.H.

Chapter 1

LINEAR EQUATIONS AND MATRICES

1.1 SYSTEMS OF LINEAR EQUATIONS

One of the most frequently recurring practical problems in many fields of study—such as mathematics, physics, biology, chemistry, economics, all phases of engineering, operations research, and the social sciences—is that of solving a system of linear equations. The equation

$$a_1x_1 + a_2x_2 + \cdots + a_nx_n = b, \tag{1}$$

which expresses b in terms of the unknowns $x_1, x_2, \ldots, x_n$ and the constants $a_1, a_2, \ldots, a_n$, is called a **linear equation**. In many applications we are given b and must find numbers $x_1, x_2, \ldots, x_n$ satisfying (1).

A **solution** to linear Equation (1) is a sequence of n numbers $s_1, s_2, \ldots, s_n$ which has the property that (1) is satisfied when $x_1 = s_1, x_2 = s_2, \ldots, x_n = s_n$ are substituted in (1). Thus $x_1 = 2$, $x_2 = 3$, and $x_3 = -4$ is a solution to the linear equation

$$6x_1 - 3x_2 + 4x_3 = -13,$$

because

$$6(2) - 3(3) + 4(-4) = -13.$$

More generally, a **system of *m* linear equations in *n* unknowns,** $x_1, x_2, \ldots, x_n$, or a **linear system**, is a set of m linear equations each in n unknowns. A linear system can conveniently be written as

$$\begin{array}{ccccccccc} a_{11}x_1 & + & a_{12}x_2 & + & \cdots & + & a_{1n}x_n & = & b_1 \\ a_{21}x_1 & + & a_{22}x_2 & + & \cdots & + & a_{2n}x_n & = & b_2 \\ \vdots & & \vdots & & & & \vdots & & \vdots \\ a_{m1}x_1 & + & a_{m2}x_2 & + & \cdots & + & a_{mn}x_n & = & b_m. \end{array} \tag{2}$$

Note: Appendix A reviews some very basic material dealing with sets and functions. It can be consulted at any time, as needed.

Thus the ith equation is

$$a_{i1}x_1 + a_{i2}x_2 + \cdots + a_{in}x_n = b_i.$$

In (2) the a_{ij} are known constants. Given values of $b_1, b_2, \ldots, b_m$, we want to find values of $x_1, x_2, \ldots, x_n$ that will satisfy each equation in (2).

A **solution** to linear system (2) is a sequence of n numbers $s_1, s_2, \ldots, s_n$, which has the property that each equation in (2) is satisfied when $x_1 = s_1$, $x_2 = s_2$, $\ldots$, $x_n = s_n$ are substituted.

If the linear system (2) has no solution, it is said to be **inconsistent**; if it has a solution, it is called **consistent**. If $b_1 = b_2 = \cdots = b_m = 0$, then (2) is called a **homogeneous system**. Note that $x_1 = x_2 = \cdots = x_n = 0$ is always a solution to a homogeneous system; it is called the **trivial solution**. A solution to a homogeneous system in which not all of $x_1, x_2, \ldots, x_n$ are zero is called a **nontrivial solution**.

Consider another system of r linear equations in n unknowns:

$$\begin{aligned}
c_{11}x_1 + c_{12}x_2 + \cdots + c_{1n}x_n &= d_1 \\
c_{21}x_1 + c_{22}x_2 + \cdots + c_{2n}x_n &= d_2 \\
\vdots \qquad \vdots \qquad\qquad \vdots \quad &\quad \vdots \\
c_{r1}x_1 + c_{r2}x_2 + \cdots + c_{rn}x_n &= d_r.
\end{aligned} \tag{3}$$

We say that (2) and (3) are **equivalent** if they both have exactly the same solutions.

EXAMPLE 1

The linear system

$$\begin{aligned}
x_1 - 3x_2 &= -7 \\
2x_1 + x_2 &= 7
\end{aligned} \tag{4}$$

has only the solution $x_1 = 2$ and $x_2 = 3$. The linear system

$$\begin{aligned}
8x_1 - 3x_2 &= 7 \\
3x_1 - 2x_2 &= 0 \\
10x_1 - 2x_2 &= 14
\end{aligned} \tag{5}$$

also has only the solution $x_1 = 2$ and $x_2 = 3$. Thus (4) and (5) are equivalent. ■

To find solutions to a linear system, we shall use a technique called the **method of elimination**; that is, we eliminate some variables by adding a multiple of one equation to another equation. Elimination merely amounts to the development of a new linear system that is equivalent to the original system but is much simpler to solve. Readers have probably confined their earlier work in this area to linear systems in which $m = n$, that is, linear systems having as many equations as unknowns. In this course we shall broaden our outlook by dealing with systems in which we have $m = n$, $m < n$, and $m > n$. Indeed, there are numerous applications in which $m \neq n$.

EXAMPLE 2

Consider the linear system

$$\begin{aligned} x_1 - 3x_2 &= -3 \\ 2x_1 + x_2 &= 8. \end{aligned} \tag{6}$$

To eliminate x_1, we add (-2) times the first equation to the second one, obtaining

$$7x_2 = 14,$$

an equation having no x_1 term. Thus we have eliminated the unknown x_1. Then solving for x_2, we have

$$x_2 = 2,$$

and substituting into the first equation of (6), we obtain

$$x_1 = 3.$$

Then $x_1 = 3$, $x_2 = 2$ is the only solution to the given linear system. Thus the system is consistent. ■

EXAMPLE 3

Consider the linear system

$$\begin{aligned} x_1 - 3x_2 &= -7 \\ 2x_1 - 6x_2 &= 7. \end{aligned} \tag{7}$$

Again, we decide to eliminate x_1. We add (-2) times the first equation to the second one, obtaining

$$0 = 21,$$

which makes no sense. This means that (7) has no solution; it is inconsistent. We could have come to the same conclusion from observing that in (7) the left side of the second equation is twice the left side of the first equation, but the right side of the second equation is not twice the right side of the first equation. ■

EXAMPLE 4

Consider the linear system

$$\begin{aligned} x_1 + 2x_2 + 3x_3 &= 6 \\ 2x_1 - 3x_2 + 2x_3 &= 14 \\ 3x_1 + x_2 - x_3 &= -2. \end{aligned} \tag{8}$$

To eliminate x_1, we add (-2) times the first equation to the second one and (-3) times the first equation to the third one, obtaining

$$\begin{aligned} -7x_2 - 4x_3 &= 2 \\ -5x_2 - 10x_3 &= -20. \end{aligned} \tag{9}$$

This is a system of two equations in the unknowns x_2 and x_3. We multiply the second equation of (9) by $\left(-\frac{1}{5}\right)$, obtaining

$$\begin{aligned} -7x_2 - 4x_3 &= 2 \\ x_2 + 2x_3 &= 4, \end{aligned}$$

which we write, by interchanging equations, as

$$\begin{aligned} x_2 + 2x_3 &= 4 \\ -7x_2 - 4x_3 &= 2. \end{aligned} \tag{10}$$

We now eliminate x_2 in (10) by adding 7 times the first equation to the second one, to obtain

$$10x_3 = 30,$$

or

$$x_3 = 3. \tag{11}$$

Substituting this value of x_3 into the first equation of (10), we find that $x_2 = -2$. Then substituting these values of x_2 and x_3 into the first equation of (8), we find that $x_1 = 1$. We observe further that our elimination procedure has actually produced the linear system

$$\begin{aligned} x_1 + 2x_2 + 3x_3 &= 6 \\ x_2 + 2x_3 &= 4 \\ x_3 &= 3, \end{aligned} \tag{12}$$

obtained by using the first equations of (8) and (10) as well as (11). The importance of the procedure is that, although the linear systems (8) and (12) are equivalent, (12) has the advantage that it is easier to solve. ■

EXAMPLE 5

Consider the linear system

$$\begin{aligned} x_1 + 2x_2 - 3x_3 &= -4 \\ 2x_1 + x_2 - 3x_3 &= 4. \end{aligned} \tag{13}$$

Eliminating x_1, we add (-2) times the first equation to the second equation, to obtain

$$-3x_2 + 3x_3 = 12. \tag{14}$$

We must now solve (14). A solution is

$$x_2 = x_3 - 4,$$

where x_3 can be any real number. Then from the first equation of (13),

$$\begin{aligned} x_1 &= -4 - 2x_2 + 3x_3 \\ &= -4 - 2(x_3 - 4) + 3x_3 \\ &= x_3 + 4. \end{aligned}$$

Thus a solution to the linear system (13) is

$$\begin{aligned} x_1 &= x_3 + 4 \\ x_2 &= x_3 - 4 \\ x_3 &= \text{any real number.} \end{aligned}$$

This means that the linear system (13) has infinitely many solutions. Every time we assign a value to x_3 we obtain another solution to (13). Thus, if $x_3 = 1$, then

$$x_1 = 5, \quad x_2 = -3, \quad \text{and} \quad x_3 = 1$$

is a solution, while if $x_3 = -2$, then

$$x_1 = 2, \quad x_2 = -6, \quad \text{and} \quad x_3 = -2$$

is another solution. ■

These examples suggest that a linear system may have a unique solution, no solution, or infinitely many solutions.

Consider next a linear system of two equations in the unknowns x_1 and x_2:

$$\begin{aligned} a_1x_1 + a_2x_2 &= c_1 \\ b_1x_1 + b_2x_2 &= c_2. \end{aligned} \tag{15}$$

The graph of each of these equations is a straight line, which we denote by ℓ_1 and ℓ_2, respectively. If $x_1 = s_1$, $x_2 = s_2$ is a solution to the linear system (15), then the point (s_1, s_2) lies on both lines ℓ_1 and ℓ_2. Conversely, if the point (s_1, s_2) lies on both lines ℓ_1 and ℓ_2, then $x_1 = s_1$, $x_2 = s_2$ is a solution to the linear system (15). Thus we are led geometrically to the same three possibilities mentioned above. See Figure 1.1.

(a) A unique solution.

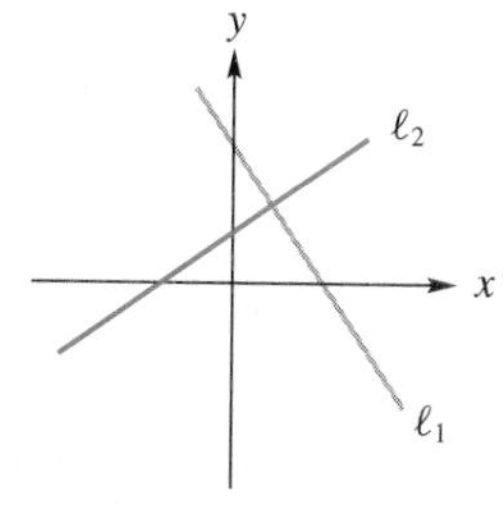

(b) No solution.

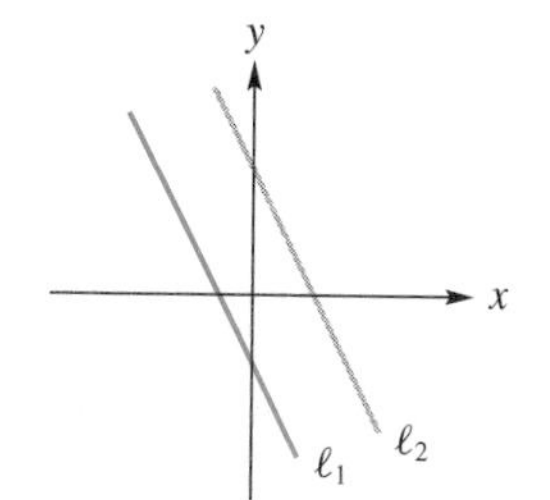

(c) Infinitely many solutions

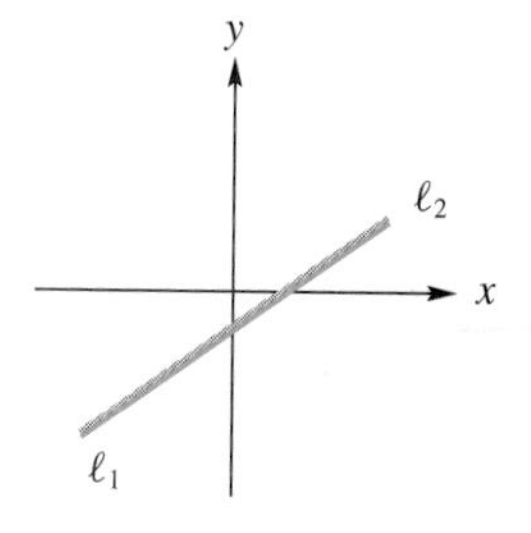

FIGURE 1.1

If we examine the method of elimination more closely, we find that it involves three manipulations that can be performed on a linear system to convert it into an equivalent system. These manipulations are as follows:

1. Interchange the ith and jth equations.
2. Multiply an equation by a nonzero constant.
3. Replace the ith equation by c times the jth equation plus the ith equation, $i \neq j$. That is, replace

$$a_{i1}x_1 + a_{i2}x_2 + \cdots + a_{in}x_n = b_i$$

by

$$(ca_{j1} + a_{i1})x_1 + (ca_{j2} + a_{i2})x_2 + \cdots + (ca_{jn} + a_{in})x_n = cb_j + b_i.$$

It is not difficult to prove that performing these manipulations on a linear system leads to an equivalent system. The next example proves this for the second type of manipulation. Exercises 22 and 23 prove it for the first and third manipulations.

EXAMPLE 6

Suppose that the ith equation of the linear system (2) is multiplied by the nonzero constant c, obtaining the linear system

$$\begin{aligned}
a_{11}x_1 + a_{12}x_2 + \cdots + a_{1n}x_n &= b_1 \\
a_{21}x_1 + a_{22}x_2 + \cdots + a_{2n}x_n &= b_2 \\
\vdots \qquad \vdots \qquad\qquad \vdots \quad &\quad \vdots \\
ca_{i1}x_1 + ca_{i2}x_2 + \cdots + ca_{in}x_n &= cb_i \\
\vdots \qquad \vdots \qquad\qquad \vdots \quad &\quad \vdots \\
a_{m1}x_1 + a_{m2}x_2 + \cdots + a_{mn}x_n &= b_m.
\end{aligned} \tag{16}$$

If $x_1 = s_1, x_2 = s_2, \ldots, x_n = s_n$ is a solution to (2), then it is a solution to all the equations in (16) except possibly to the ith equation. For the ith equation we have

$$c(a_{i1}s_1 + a_{i2}s_2 + \cdots + a_{in}s_n) = cb_i$$

or

$$ca_{i1}s_1 + ca_{i2}s_2 + \cdots + ca_{in}s_n = cb_i.$$

Thus the ith equation of (16) is also satisfied. Hence every solution to (2) is also a solution to (16). Conversely, every solution to (16) also satisfies (2). Hence (2) and (16) are equivalent systems. ■

As you have probably already observed, the method of elimination has been described, so far, in general terms. Thus we have not indicated any rules for selecting the unknowns to be eliminated. Before providing a very systematic description of the method of elimination, we introduce, in the next section, the notion of a matrix. This will greatly simplify our notational problems and will enable us to develop tools to solve many important applied problems.

1.1 Exercises

In Exercises 1 *through* 14, *solve the given linear system by the method of elimination.*

1. $$\begin{aligned} x_1 + 2x_2 &= 8 \\ 3x_1 - 4x_2 &= 4. \end{aligned}$$

2. $$\begin{aligned} 2x_1 - 3x_2 + 4x_3 &= -12 \\ x_1 - 2x_2 + x_3 &= -5 \\ 3x_1 + x_2 + 2x_3 &= 1. \end{aligned}$$

3. $$\begin{aligned} 3x_1 + 2x_2 + x_3 &= 2 \\ 4x_1 + 2x_2 + 2x_3 &= 8 \\ x_1 - x_2 + x_3 &= 4. \end{aligned}$$

4. $$\begin{aligned} x_1 + x_2 &= 5 \\ 3x_1 + 3x_2 &= 10. \end{aligned}$$

5. $$\begin{aligned} 2x_1 + 4x_2 + 6x_3 &= -12 \\ 2x_1 - 3x_2 - 4x_3 &= 15 \\ 3x_1 + 4x_2 + 5x_3 &= -8. \end{aligned}$$

6. $$\begin{aligned} x_1 + x_2 - 2x_3 &= 5 \\ 2x_1 + 3x_2 + 4x_3 &= 2. \end{aligned}$$

7. $$\begin{aligned} x_1 + 4x_2 - x_3 &= 12 \\ 3x_1 + 8x_2 - 2x_3 &= 4. \end{aligned}$$

8. $$\begin{aligned} 3x_1 + 4x_2 - x_3 &= 8 \\ 6x_1 + 8x_2 - 2x_3 &= 3. \end{aligned}$$

9. $$\begin{aligned} x_1 + x_2 + 3x_3 &= 12 \\ 2x_1 + 2x_2 + 6x_3 &= 6. \end{aligned}$$

10. $$\begin{aligned} x_1 + x_2 &= 1 \\ 2x_1 - x_2 &= 5 \\ 3x_1 + 4x_2 &= 2. \end{aligned}$$

11. $$\begin{aligned} 2x_1 + 3x_2 &= 13 \\ x_1 - 2x_2 &= 3 \\ 5x_1 + 2x_2 &= 27. \end{aligned}$$

12. $$\begin{aligned} x_1 - 5x_2 &= 6 \\ 3x_1 + 2x_2 &= 1 \\ 5x_1 + 2x_2 &= 1. \end{aligned}$$

13. $$\begin{aligned} x_1 + 3x_2 &= -4 \\ 2x_1 + 5x_2 &= -8 \\ x_1 + 3x_2 &= -5. \end{aligned}$$

14. $$\begin{aligned} 2x_1 + 3x_2 - x_3 &= 6 \\ 2x_1 - x_2 + 2x_3 &= -8 \\ 3x_1 - x_2 + x_3 &= -7. \end{aligned}$$

15. Given the linear system

$$\begin{aligned} 2x_1 - x_2 &= 5 \\ 4x_1 - 2x_2 &= t. \end{aligned}$$

(a) Determine a value of t so that the system is consistent.

(b) Determine a value of t so that the system is inconsistent.

(c) How many different values of t can be selected in part (b)?

16. Is every homogeneous linear system always consistent? Explain.

17. Given the linear system

$$\begin{aligned} 2x_1 + 3x_2 - x_3 &= 0 \\ x_1 - 4x_2 + 5x_3 &= 0. \end{aligned}$$

(a) Verify that $x_1 = 1, x_2 = -1, x_3 = -1$ is a solution.

(b) Verify that $x_1 = -2, x_2 = 2, x_3 = 2$ is a solution.

(c) Adding the corresponding x-values of the solutions in parts (a) and (b) gives $x_1 = -1, x_2 = 1, x_3 = 1$. Is this a solution to the linear system?

(d) Multiply each of the x-values in part (a) by 3. Are the resulting values a solution to the linear system?

18. Without using the method of elimination solve the linear system

$$\begin{aligned} 2x_1 + x_2 - 2x_3 &= -5 \\ 3x_2 + x_3 &= 7 \\ x_3 &= 4. \end{aligned}$$

19. Without using the method of elimination solve the linear system

$$\begin{aligned} 4x_1 \qquad\qquad &= 8 \\ -2x_1 + 3x_2 \qquad &= -1 \\ 3x_1 + 5x_2 - 2x_3 &= 11. \end{aligned}$$

20. Is there a value of r so that $x_1 = 1, x_2 = 2, x_3 = r$ is a solution to the following linear system? If there is, find it.

$$\begin{aligned} 2x_1 + 3x_2 - x_3 &= 11 \\ x_1 - x_2 + 2x_3 &= -7 \\ 4x_1 + x_2 - 2x_3 &= 12. \end{aligned}$$

21. Is there a value of r so that $x_1 = r, x_2 = 2, x_3 = 1$ is a solution to the following linear system? If there is, find it.

$$\begin{aligned} 3x_1 \qquad - 2x_3 &= 4 \\ x_1 - 4x_2 + x_3 &= -5 \\ -2x_1 + 3x_2 + 2x_3 &= 9. \end{aligned}$$

22. Show that the linear system obtained by interchanging two equations in (2) is equivalent to (2).

23. Show that the linear system obtained by adding a multiple of an equation in (2) to another equation is equivalent to (2).

24. For the software you are using, determine the command that "automatically" solves a linear system of equations.

25. Use the command from Exercise 24 to solve Exercises 3 and 4, and compare the output with the results you obtained by the method of elimination.

26. Solve the linear system

$$\begin{aligned} x_1 + \tfrac{1}{2}x_2 + \tfrac{1}{3}x_3 &= 1 \\ \tfrac{1}{2}x_1 + \tfrac{1}{3}x_2 + \tfrac{1}{4}x_3 &= \tfrac{11}{18} \\ \tfrac{1}{3}x_1 + \tfrac{1}{4}x_2 + \tfrac{1}{5}x_3 &= \tfrac{9}{20} \end{aligned}$$

using your software. Compare the computed solution with the exact solution $x_1 = \frac{1}{2}$, $x_2 = \frac{1}{3}$, $x_3 = 1$.

1.2 MATRICES; MATRIX OPERATIONS

If we examine the method of elimination described in Section 1.1, we make the following observation: Only the numbers in front of the unknowns $x_1, x_2, \ldots, x_n$ and the numbers $b_1, b_2, \ldots, b_m$ on the right side are being changed as we perform the steps in the method of elimination. Thus we might think of looking for a way of writing a linear system without having to carry along the unknowns. Matrices enable us to do this—that is, to write linear systems in a compact form that makes it easier to automate the elimination method on an electronic computer in order to obtain a fast and efficient procedure for finding solutions. Their use, however, is not merely that of a convenient notation. We now develop operations on matrices and will work with matrices according to the rules they obey; this will enable us to solve systems of linear equations and to handle other computational problems in a fast and efficient manner. Of course, as any good definition should do, the notion of a matrix provides not only a new way of looking at old problems but also gives rise to a great many new questions, some of which we study in this book.

Definition 1.1

A **matrix** is a rectangular array of numbers denoted by

$$A = \begin{bmatrix} a_{11} & a_{12} & \cdots & a_{1n} \\ a_{21} & a_{22} & \cdots & a_{2n} \\ \vdots & \vdots & & \vdots \\ a_{m1} & a_{m2} & \cdots & a_{mn} \end{bmatrix}.$$

▲

Unless stated otherwise, we assume that all our matrices are composed entirely of real numbers. The ***i*th row** of A is

$$\begin{bmatrix} a_{i1} & a_{i2} & \cdots & a_{in} \end{bmatrix} \qquad 1 \le i \le m,$$

while the ***j*th column** of A is

$$\begin{bmatrix} a_{1j} \\ a_{2j} \\ \vdots \\ a_{mj} \end{bmatrix} \qquad 1 \le j \le n.$$

If a matrix A has m rows and n columns, we say that A is an **m by n matrix** (written $m \times n$). If $m = n$, we say that A is a **square matrix of order n** and that the elements $a_{11}, a_{22}, \ldots, a_{nn}$ are on the **main diagonal** of A. We refer to a_{ij} as the **(i, j) entry** (entry in the ith row and jth column) or the **(i, j)th element** and we often write

$$A = \left[a_{ij} \right].$$

We shall also write A_{mn} to indicate that A has m rows and n columns. If A is $n \times n$, we merely write A_n.

EXAMPLE 1

The following are matrices:

$$A = \begin{bmatrix} 1 & 2 & 3 \\ 2 & -1 & 4 \\ 0 & -3 & 2 \end{bmatrix}, \quad B = \begin{bmatrix} 1 & 3 & -7 \end{bmatrix},$$

$$C = \begin{bmatrix} 2 \\ -1 \\ 3 \\ 4 \end{bmatrix}, \quad \text{and} \quad D = \begin{bmatrix} 0 & 3 \\ -1 & -2 \end{bmatrix}.$$

In A, $a_{32} = -3$; in C, $c_{41} = 4$. Here A is 3×3, B is 1×3, C is 4×1, and D is 2×2. In A, the elements $a_{11} = 1$, $a_{22} = -1$, and $a_{33} = 2$ are on the main diagonal. ■

Whenever a new object is introduced in mathematics, one must determine when two such objects are equal. For example, in the set of all rational numbers, the numbers $\frac{2}{3}$ and $\frac{4}{6}$ are called equal, although they are not represented in the same manner. What we have in mind is the definition that a/b equals c/d when $ad = bc$. Accordingly, we now have the following definition.

Definition 1.2

Two $m \times n$ matrices $A = \left[a_{ij} \right]$ and $B = \left[b_{ij} \right]$ are **equal** if they agree entry by entry, that is, if $a_{ij} = b_{ij}$ for $i = 1, 2, \ldots, m$ and $j = 1, 2, \ldots, n$. ▲

EXAMPLE 2

The matrices

$$A = \begin{bmatrix} 1 & 2 & -1 \\ 2 & -3 & 4 \\ 0 & -4 & 5 \end{bmatrix} \quad \text{and} \quad B = \begin{bmatrix} 1 & 2 & w \\ 2 & x & 4 \\ y & -4 & z \end{bmatrix}$$

are equal if and only if $w = -1$, $x = -3$, $y = 0$, and $z = 5$. ■

We next define a number of operations that will produce new matrices out of given matrices. When dealing with linear systems, for example, this will enable us to manipulate the matrices that arise and avoid writing down systems over and over again. These operations and manipulations are also useful in other applications of matrices.

Matrix Addition

Definition 1.3

If $A = \left[a_{ij}\right]$ and $B = \left[b_{ij}\right]$ are both $m \times n$ matrices, then the **sum** $A + B$ is an $m \times n$ matrix $C = \left[c_{ij}\right]$ defined by $c_{ij} = a_{ij} + b_{ij}$, $i = 1, 2, \ldots, m$; $j = 1, 2, \ldots, n$. Thus, to obtain the sum of A and B, we merely add corresponding entries. ▲

EXAMPLE 3

Let

$$A = \begin{bmatrix} 1 & -2 & 3 \\ 2 & -1 & 4 \end{bmatrix} \quad \text{and} \quad B = \begin{bmatrix} 0 & 2 & 1 \\ 1 & 3 & -4 \end{bmatrix}.$$

Then

$$A + B = \begin{bmatrix} 1+0 & -2+2 & 3+1 \\ 2+1 & -1+3 & 4+(-4) \end{bmatrix} = \begin{bmatrix} 1 & 0 & 4 \\ 3 & 2 & 0 \end{bmatrix}.$$

■

It should be noted that the sum of the matrices A and B is defined only when A and B have the same number of rows and the same number of columns, that is, only when A and B are of the same size. We now establish the convention that when $A + B$ is formed, both A and B are of the same size. The basic properties of matrix addition are considered in the following section and are similar to those satisfied by the real numbers.

Scalar Multiplication

Definition 1.4

If $A = \left[a_{ij}\right]$ is an $m \times n$ matrix and r is a real number, then the **scalar multiple** of A by r, rA, is the $m \times n$ matrix $C = \left[c_{ij}\right]$, where $c_{ij} = ra_{ij}$, $i = 1, 2, \ldots, m$ and $j = 1, 2, \ldots, n$; that is, the matrix C is obtained by multiplying each entry of A by r. ▲

EXAMPLE 4

We have

$$-2\begin{bmatrix} 4 & -2 & -3 \\ 7 & -3 & 2 \end{bmatrix} = \begin{bmatrix} (-2)(4) & (-2)(-2) & (-2)(-3) \\ (-2)(7) & (-2)(-3) & (-2)(2) \end{bmatrix}$$

$$= \begin{bmatrix} -8 & 4 & 6 \\ -14 & 6 & -4 \end{bmatrix}.$$

■

Thus far, addition of matrices has only been defined for two matrices. Our work with matrices will call for adding more than two matrices. Theorem 1.1 in the next section shows that addition of matrices satisfies the associative property: $A + (B + C) = (A + B) + C$.

If A and B are $m \times n$ matrices, we write $A + (-1)B$ as $A - B$ and call this the **difference between A and B**.

EXAMPLE 5

Let

$$A = \begin{bmatrix} 2 & 3 & -5 \\ 4 & 2 & 1 \end{bmatrix} \quad \text{and} \quad B = \begin{bmatrix} 2 & -1 & 3 \\ 3 & 5 & -2 \end{bmatrix}.$$

Then

$$A - B = \begin{bmatrix} 2-2 & 3+1 & -5-3 \\ 4-3 & 2-5 & 1+2 \end{bmatrix} = \begin{bmatrix} 0 & 4 & -8 \\ 1 & -3 & 3 \end{bmatrix}.$$

■

We shall sometimes use the **summation notation**, and we now review this useful and compact notation.

By $\sum_{i=1}^{n} r_i a_i$ we mean $r_1 a_1 + r_2 a_2 + \cdots + r_n a_n$. The letter i is called the **index of summation**; it is a dummy variable that can be replaced by another letter. Hence we can write

$$\sum_{i=1}^{n} r_i a_i = \sum_{j=1}^{n} r_j a_j = \sum_{k=1}^{n} r_k a_k.$$

Thus

$$\sum_{i=1}^{4} r_i a_i = r_1 a_1 + r_2 a_2 + r_3 a_3 + r_4 a_4.$$

The summation notation satisfies the following properties:

1. $\sum_{i=1}^{n} (r_i + s_i) a_i = \sum_{i=1}^{n} r_i a_i + \sum_{i=1}^{n} s_i a_i.$

2. $\sum_{i=1}^{n} c(r_i a_i) = c \sum_{i=1}^{n} r_i a_i.$

3. $\sum_{j=1}^{n} \left(\sum_{i=1}^{m} a_{ij} \right) = \sum_{i=1}^{m} \left(\sum_{j=1}^{n} a_{ij} \right).$

Property 3 can be interpreted as follows. The left side is obtained by adding all the entries in each column and then adding all the resulting numbers. The right side is obtained by adding all the entries in each row and then adding all the resulting numbers.

If $A_1, A_2, \ldots, A_k$ are $m \times n$ matrices and $c_1, c_2, \ldots, c_k$ are real numbers, then an expression of the form

$$c_1 A_1 + c_2 A_2 + \cdots + c_k A_k \tag{1}$$

is called a **linear combination** of $A_1, A_2, \ldots, A_k$, and $c_1, c_2, \ldots, c_k$ are called **coefficients**.

The linear combination in Equation (1) can also be expressed using the summation notation as

$$\sum_{i=1}^{k} c_i A_i.$$

EXAMPLE 6

The following are linear combinations of matrices:

$$3\begin{bmatrix} 0 & -3 & 5 \\ 2 & 3 & 4 \\ 1 & -2 & -3 \end{bmatrix} - \frac{1}{2}\begin{bmatrix} 5 & 2 & 3 \\ 6 & 2 & 3 \\ -1 & -2 & 3 \end{bmatrix},$$

$$2\begin{bmatrix} 3 & -2 \end{bmatrix} - 3\begin{bmatrix} 5 & 0 \end{bmatrix} + 4\begin{bmatrix} -2 & 5 \end{bmatrix},$$

$$-0.5\begin{bmatrix} 1 \\ -4 \\ -6 \end{bmatrix} + 0.4\begin{bmatrix} 0.1 \\ -4 \\ 0.2 \end{bmatrix}.$$

Using scalar multiplication and matrix addition, we can compute each of these linear combinations. Verify that the results of such computations are respectively

$$\begin{bmatrix} -\frac{5}{2} & -10 & \frac{27}{2} \\ 3 & 8 & \frac{21}{2} \\ \frac{7}{2} & -5 & -\frac{21}{2} \end{bmatrix}, \quad \begin{bmatrix} -17 & 16 \end{bmatrix}, \quad \text{and} \quad \begin{bmatrix} -0.46 \\ 0.4 \\ 3.08 \end{bmatrix}.$$

■

Matrix Multiplication

Definition 1.5

If $A = \begin{bmatrix} a_{ij} \end{bmatrix}$ is an $m \times n$ matrix and $B = \begin{bmatrix} b_{ij} \end{bmatrix}$ is an $n \times p$ matrix, then the **product** of A and B, $AB = C = \begin{bmatrix} c_{ij} \end{bmatrix}$, is an $m \times p$ matrix defined by

$$c_{ij} = \sum_{k=1}^{n} a_{ik}b_{kj} = a_{i1}b_{1j} + a_{i2}b_{2j} + \cdots + a_{in}b_{nj} \qquad i = 1, 2, \ldots, m;$$
$$j = 1, 2, \ldots, p.$$

▲

Note that AB is defined only when the number of columns of A is the same as the number of rows of B. We also observe that the (i, j) entry in C is obtained by using the ith row of A and the jth column of B. That is,

$$AB = \begin{bmatrix} a_{11} & a_{12} & \cdots & a_{1n} \\ a_{21} & a_{22} & \cdots & a_{2n} \\ \vdots & \vdots & & \vdots \\ a_{i1} & a_{i2} & \cdots & a_{in} \\ \vdots & \vdots & & \vdots \\ a_{m1} & a_{m2} & \cdots & a_{mn} \end{bmatrix} \begin{bmatrix} b_{11} & b_{12} & \cdots & b_{1j} & \cdots & b_{1p} \\ b_{21} & b_{22} & \cdots & b_{2j} & \cdots & b_{2p} \\ \vdots & \vdots & & \vdots & & \vdots \\ b_{n1} & b_{n2} & \cdots & b_{nj} & \cdots & b_{np} \end{bmatrix}$$

$$= \begin{bmatrix} c_{11} & c_{12} & \cdots & c_{1p} \\ c_{21} & c_{22} & \cdots & c_{2p} \\ \vdots & \vdots & c_{ij} & \vdots \\ c_{m1} & c_{m2} & \cdots & c_{mp} \end{bmatrix} = C.$$

EXAMPLE 7

Let

$$A = \begin{bmatrix} 1 & 2 & -1 \\ 3 & 1 & 4 \end{bmatrix} \quad \text{and} \quad B = \begin{bmatrix} -2 & 5 \\ 4 & -3 \\ 2 & 1 \end{bmatrix}.$$

Then

$$AB = \begin{bmatrix} (1)(-2) + (2)(4) + (-1)(2) & (1)(5) + (2)(-3) + (-1)(1) \\ (3)(-2) + (1)(4) + (4)(2) & (3)(5) + (1)(-3) + (4)(1) \end{bmatrix}$$

$$= \begin{bmatrix} 4 & -2 \\ 6 & 16 \end{bmatrix}.$$

■

The basic properties of matrix multiplication are considered in the following section. However, we note here that multiplication of matrices requires much more care than their addition, since the algebraic properties of matrix multiplication differ from those satisfied by the real numbers. Part of the problem is due to the fact that AB is defined only when the number of columns of A is the same as the number of rows of B. For instance, consider the following: if A is an $m \times n$ matrix and B is an $n \times p$ matrix, then AB is an $m \times p$ matrix. What about BA? Three different situations may occur:

1. BA may not be defined. This will take place if $p \neq m$.
2. If BA is defined (i.e., $p = m$), BA will be $n \times n$ and AB will be $m \times m$, and if $m \neq n$, AB and BA are of different sizes.
3. If BA and AB are of the same size, they may be unequal.

As in the case of addition, we establish the convention that when AB is written, it is defined.

EXAMPLE 8

Let A be a 2×3 matrix and let B be a 3×4 matrix. Then AB is 2×4 and BA is not defined. ■

EXAMPLE 9

Let A be 2×3 and let B be 3×2. Then AB is 2×2 and BA is 3×3. ■

EXAMPLE 10

Let

$$A = \begin{bmatrix} 1 & 2 \\ -1 & 3 \end{bmatrix} \quad \text{and} \quad B = \begin{bmatrix} 2 & 1 \\ 0 & 1 \end{bmatrix}.$$

Then

$$AB = \begin{bmatrix} 2 & 3 \\ -2 & 2 \end{bmatrix} \quad \text{while} \quad BA = \begin{bmatrix} 1 & 7 \\ -1 & 3 \end{bmatrix}.$$

Thus $AB \neq BA$. ■

One might ask why matrix equality and matrix addition are defined in such a natural way while matrix multiplication appears to be much more complicated. Only a thorough understanding of the composition of functions and the relationship that exists between matrices and what are called linear transformations would show that the definition of multiplication given above is the natural one. These topics will be covered later in the book.

It is sometimes useful to be able to find a column in the matrix product AB without having to multiply the two matrices. It is not difficult to show (Exercise 32) that the jth column of the matrix product AB is equal to the matrix product AB_j, where B_j is the jth column of B.

EXAMPLE 11

Let

$$A = \begin{bmatrix} 1 & 2 \\ 3 & 4 \\ -1 & 5 \end{bmatrix} \quad \text{and} \quad B = \begin{bmatrix} -2 & 3 & 4 \\ 3 & 2 & 1 \end{bmatrix}.$$

Then the second column of AB is

$$AB_2 = \begin{bmatrix} 1 & 2 \\ 3 & 4 \\ -1 & 5 \end{bmatrix} \begin{bmatrix} 3 \\ 2 \end{bmatrix} = \begin{bmatrix} 7 \\ 17 \\ 7 \end{bmatrix}.$$

■

It can also be shown (Exercise 33) that the jth column of the matrix product AB is the linear combination of the columns of A with coefficients the entries in B_j, the jth column of B.

EXAMPLE 12

If A and B are the matrices in Example 11, then the second column of AB, AB_2, can be written as the following linear combination of the columns of A:

$$AB_2 = \begin{bmatrix} 1 & 2 \\ 3 & 4 \\ -1 & 5 \end{bmatrix} \begin{bmatrix} 3 \\ 2 \end{bmatrix} = \begin{bmatrix} (1)(3) + (2)(2) \\ (3)(3) + (4)(2) \\ (-1)(3) + (5)(2) \end{bmatrix}$$

$$= \begin{bmatrix} (1)(3) \\ (3)(3) \\ (-1)(3) \end{bmatrix} + \begin{bmatrix} (2)(2) \\ (4)(2) \\ (5)(2) \end{bmatrix} = 3 \begin{bmatrix} 1 \\ 3 \\ -1 \end{bmatrix} + 2 \begin{bmatrix} 2 \\ 4 \\ 5 \end{bmatrix}.$$

■

Linear Systems

We now return to the linear system (2) in Section 1.1:

$$\begin{aligned} a_{11}x_1 + a_{12}x_2 + \cdots + a_{1n}x_n &= b_1 \\ a_{21}x_1 + a_{22}x_2 + \cdots + a_{2n}x_n &= b_2 \\ \vdots \qquad \vdots \qquad\qquad \vdots \quad &\quad \vdots \\ a_{m1}x_1 + a_{m2}x_2 + \cdots + a_{mn}x_n &= b_m \end{aligned} \tag{2}$$

and define the following matrices:

$$A = \begin{bmatrix} a_{11} & a_{12} & \cdots & a_{1n} \\ a_{21} & a_{22} & \cdots & a_{2n} \\ \vdots & \vdots & & \vdots \\ a_{m1} & a_{m2} & \cdots & a_{mn} \end{bmatrix}, \quad X = \begin{bmatrix} x_1 \\ x_2 \\ \vdots \\ x_n \end{bmatrix}, \quad B = \begin{bmatrix} b_1 \\ b_2 \\ \vdots \\ b_m \end{bmatrix}.$$

The matrix A is called the **coefficient matrix** of the system and the matrix

$$\left[\begin{array}{cccc|c} a_{11} & a_{12} & \cdots & a_{1n} & b_1 \\ a_{21} & a_{22} & \cdots & a_{2n} & b_2 \\ \vdots & \vdots & & \vdots & \vdots \\ a_{m1} & a_{m2} & \cdots & a_{mn} & b_m \end{array}\right]$$

is called the **augmented matrix** of the system. The coefficient and augmented matrices of a linear system will play key roles in our methods for solving linear systems. Since A is $m \times n$ and X is $n \times 1$, the matrix product AX is an $m \times 1$ matrix. Moreover,

$$AX = \begin{bmatrix} a_{11} & a_{12} & \cdots & a_{1n} \\ a_{21} & a_{22} & \cdots & a_{2n} \\ \vdots & \vdots & & \vdots \\ a_{m1} & a_{m2} & \cdots & a_{mn} \end{bmatrix} \begin{bmatrix} x_1 \\ x_2 \\ \vdots \\ x_n \end{bmatrix} = \begin{bmatrix} a_{11}x_1 + a_{12}x_2 + \cdots + a_{1n}x_n \\ a_{21}x_1 + a_{22}x_2 + \cdots + a_{2n}x_n \\ \vdots \qquad\qquad \vdots \qquad\qquad \vdots \\ a_{m1}x_1 + a_{m2}x_2 + \cdots + a_{mn}x_n \end{bmatrix}.$$

The entries in the product AX are merely the left sides of the equations in (2). Hence the linear system (2) can be written in matrix form as

$$AX = B. \tag{3}$$

EXAMPLE 13

Consider the following linear system:

$$\begin{aligned} 2x_1 + 3x_2 - 4x_3 + x_4 &= 5 \\ -2x_1 \qquad + x_3 \qquad &= 7 \\ 3x_1 + 2x_2 \qquad - 4x_4 &= 3. \end{aligned}$$

We can write this in matrix form as

$$\begin{bmatrix} 2 & 3 & -4 & 1 \\ -2 & 0 & 1 & 0 \\ 3 & 2 & 0 & -4 \end{bmatrix} \begin{bmatrix} x_1 \\ x_2 \\ x_3 \\ x_4 \end{bmatrix} = \begin{bmatrix} 5 \\ 7 \\ 3 \end{bmatrix}.$$

The coefficient matrix of this system is

$$\begin{bmatrix} 2 & 3 & -4 & 1 \\ -2 & 0 & 1 & 0 \\ 3 & 2 & 0 & -4 \end{bmatrix},$$

and the augmented matrix is

$$\left[\begin{array}{cccc|c} 2 & 3 & -4 & 1 & 5 \\ -2 & 0 & 1 & 0 & 7 \\ 3 & 2 & 0 & -4 & 3 \end{array}\right]$$

■

EXAMPLE 14

The matrix

$$\left[\begin{array}{ccc|c} 2 & -1 & 3 & 4 \\ 3 & 0 & 2 & 5 \end{array}\right]$$

is the augmented matrix of the linear system

$$\begin{aligned} 2x_1 - x_2 + 3x_3 &= 4 \\ 3x_1 \qquad + 2x_3 &= 5. \end{aligned}$$

■

From our remarks preceding Example 11, it follows that AX, the left side of Equation (3), can be written as a linear combination of the columns of A, whose coefficients are $x_1, x_2, \ldots, x_n$, the entries of X. Thus Equation (3) can be written as

$$\sum_{i=1}^{n} x_i A_i = x_1 \begin{bmatrix} a_{11} \\ a_{21} \\ \vdots \\ a_{m1} \end{bmatrix} + x_2 \begin{bmatrix} a_{12} \\ a_{22} \\ \vdots \\ a_{m2} \end{bmatrix} + \cdots + x_n \begin{bmatrix} a_{1n} \\ a_{2n} \\ \vdots \\ a_{mn} \end{bmatrix} = \begin{bmatrix} b_1 \\ b_2 \\ \vdots \\ b_m \end{bmatrix}, \tag{4}$$

where A_i is the ith column of A.

EXAMPLE 15

Write the linear system

$$\begin{aligned} 3x_1 \qquad\quad + 4x_3 - x_4 &= 2 \\ 2x_1 + x_2 + x_3 \qquad\quad &= -4 \\ -x_1 + 2x_2 + 3x_3 + x_4 &= 5 \end{aligned}$$

as a linear combination of the columns of A, as in Equation (4).

Solution We have

$$x_1 \begin{bmatrix} 3 \\ 2 \\ -1 \end{bmatrix} + x_2 \begin{bmatrix} 0 \\ 1 \\ 2 \end{bmatrix} + x_3 \begin{bmatrix} 4 \\ 1 \\ 3 \end{bmatrix} + x_4 \begin{bmatrix} -1 \\ 0 \\ 1 \end{bmatrix} = \begin{bmatrix} 2 \\ -4 \\ 5 \end{bmatrix}.$$

■

The next operation on matrices is one that is useful in a number of situations.

Definition 1.6

If $A = \left[a_{ij}\right]$ is an $m \times n$ matrix, then the **transpose** of A, $A^T = \left[a_{ij}^T\right]$, is the $n \times m$ matrix defined by $a_{ij}^T = a_{ji}$. Thus the transpose of A is obtained from A by interchanging the rows and columns of A. ▲

EXAMPLE 16

If

$$A = \begin{bmatrix} 1 & 2 & -1 \\ -3 & 2 & 7 \end{bmatrix}, \quad \text{then} \quad A^T = \begin{bmatrix} 1 & -3 \\ 2 & 2 \\ -1 & 7 \end{bmatrix}.$$

■

1.2 Exercises

1. Let

$$A = \begin{bmatrix} 2 & -3 & 5 \\ 6 & -5 & 4 \end{bmatrix}, \quad B = \begin{bmatrix} 4 \\ -3 \\ 5 \end{bmatrix}$$

and

$$C = \begin{bmatrix} 7 & 3 & 2 \\ -4 & 3 & 5 \\ 6 & 1 & -1 \end{bmatrix}.$$

(a) What is a_{12}, a_{22}, a_{23}?

(b) What is b_{11}, b_{31}?

(c) What is c_{13}, c_{31}, c_{21}?

2. If

$$\begin{bmatrix} a+b & c+d \\ c-d & a-b \end{bmatrix} = \begin{bmatrix} 4 & 6 \\ 10 & 2 \end{bmatrix},$$

find a, b, c, and d.

Consider the following matrices for Exercises 3 through 7.

$$A = \begin{bmatrix} 1 & 2 & 3 \\ 2 & 1 & 4 \end{bmatrix}, \quad B = \begin{bmatrix} 1 & 0 \\ 2 & 1 \\ 3 & 2 \end{bmatrix},$$

$$C = \begin{bmatrix} 3 & -1 & 3 \\ 4 & 1 & 5 \\ 2 & 1 & 3 \end{bmatrix}, \quad D = \begin{bmatrix} 3 & -2 \\ 2 & 5 \end{bmatrix},$$

$$E = \begin{bmatrix} 2 & -4 & 5 \\ 0 & 1 & 4 \\ 3 & 2 & 1 \end{bmatrix}, \quad \text{and} \quad F = \begin{bmatrix} -1 & 2 \\ 0 & 4 \\ 3 & 5 \end{bmatrix}.$$

3. If possible, compute:

(a) $C + E$. (b) AB and BA.

(c) $2C - 3E$. (d) $CB + D$.

(e) $AB + D^2$, where $D^2 = DD$.

4. If possible, compute:

(a) $DA + B$. (b) EC.

(c) CE. (d) $EB + F$.

(e) $FC + D$.

5. If possible, compute:

(a) $FD - 3B$. (b) $AB - 2D$.

(c) $3(2A)$ and $6A$. (d) $2F - 3(AE)$.

(e) $BD + AE$.

6. If possible, compute:

(a) $A(BD)$. (b) $(AB)D$.

(c) $A(C + E)$. (d) $AC + AE$.

(e) $3A + 2A$ and $5A$. (f) $A(C - 3E)$.

7. If possible, compute:

(a) A^T (b) $(A^T)^T$.

(c) $(AB)^T$. (d) $B^T A^T$.

(e) $(C + E)^T$ and $C^T + E^T$.

(f) $A(2B)$ and $2(AB)$.

8. Let $A = \begin{bmatrix} 1 & 2 & -3 \end{bmatrix}$, $B = \begin{bmatrix} -1 & 4 & 2 \end{bmatrix}$, and $C = \begin{bmatrix} -3 & 0 & 1 \end{bmatrix}$. If possible, compute:

(a) AB^T. (b) CA^T. (c) $(BA^T)C$.

(d) $A^T B$. (e) CC^T. (f) $C^T C$.

(g) $B^T C A A^T$.

9. Let $A = \begin{bmatrix} 2 & -3 & 1 \\ 1 & 2 & 4 \end{bmatrix}$ and $B = \begin{bmatrix} 3 \\ 5 \\ 2 \end{bmatrix}$.

(a) Verify that $AB = 3A_1 + 5A_2 + 2A_3$, where A_j is the jth column of A for $j = 1, 2, 3$.

(b) Verify that $AB = \begin{bmatrix} p_1 \\ p_2 \end{bmatrix}$, where $p_j = (\text{row}_j(A))B$, $j = 1, 2$.

10. Find a value of r so that $AB^T = 0$, where $A = \begin{bmatrix} r & 1 & -2 \end{bmatrix}$ and $B = \begin{bmatrix} 1 & 3 & -1 \end{bmatrix}$.

11. Find a value of r and a value of s so that $AB^T = 0$, where $A = \begin{bmatrix} 1 & r & 1 \end{bmatrix}$ and $B = \begin{bmatrix} -2 & 2 & s \end{bmatrix}$.

12. (a) Let A be an $m \times n$ matrix with a row consisting entirely of zeros. Show that if B is an $n \times p$ matrix, then AB has a row of zeros.

(b) Let A be an $m \times n$ matrix with a column consisting entirely of zeros and let B be $p \times m$. Show that BA has a column of zeros.

13. Let $A = \begin{bmatrix} -3 & 2 & 1 \\ 4 & 5 & 0 \end{bmatrix}$ with c_j = the jth column of A, $j = 1, 2$. Verify that

$$A^T A = \begin{bmatrix} c_1^T c_1 & c_1^T c_2 & c_1^T c_3 \\ c_2^T c_1 & c_2^T c_2 & c_2^T c_3 \\ c_3^T c_1 & c_3^T c_2 & c_3^T c_3 \end{bmatrix}.$$

14. True or false?

(a) $\sum_{i=1}^{n}(a_i + 1) = n + \sum_{i=1}^{n} a_i$.

(b) $\sum_{i=1}^{n}\sum_{j=1}^{m} 1 = mn$.

(c) $\sum_{j=1}^{m}\sum_{i=1}^{n} a_i b_j = \left[\sum_{i=1}^{n} a_i\right]\left[\sum_{j=1}^{m} b_j\right]$.

15. Let $A = \begin{bmatrix} 1 & 2 \\ 3 & 2 \end{bmatrix}$ and $B = \begin{bmatrix} 2 & -1 \\ -3 & 4 \end{bmatrix}$.
Show that $AB \neq BA$.

16. Consider the following linear system:

$$\begin{aligned} 2x_1 + 3x_2 - 3x_3 + x_4 + x_5 &= 7 \\ 3x_1 \qquad + 2x_3 \qquad + 3x_5 &= -2 \\ 2x_1 + 3x_2 \qquad - 4x_4 \qquad &= 3 \\ x_3 + x_4 + x_5 &= 5. \end{aligned}$$

(a) Find the coefficient matrix.

(b) Write the linear system in matrix form.

(c) Find the augmented matrix.

17. Write the linear system whose augmented matrix is

$$\left[\begin{array}{rrrr|r} -2 & -1 & 0 & 4 & 5 \\ -3 & 2 & 7 & 8 & 3 \\ 1 & 0 & 0 & 2 & 4 \\ 3 & 0 & 1 & 3 & 6 \end{array}\right].$$

18. Write the following linear system in matrix form.

$$\begin{aligned} -2x_1 + 3x_2 &= 5 \\ x_1 - 5x_2 &= 4. \end{aligned}$$

19. Write the following linear system in matrix form.

$$\begin{aligned} 2x_1 + 3x_2 &= 0 \\ 3x_2 + x_3 &= 0 \\ 2x_1 - x_2 &= 0. \end{aligned}$$

20. Write the linear system whose augmented matrix is

(a) $\left[\begin{array}{rrrr|r} 2 & 1 & 3 & 4 & 0 \\ 3 & -1 & 2 & 0 & 3 \\ -2 & 1 & -4 & 3 & 2 \end{array}\right]$.

(b) $\left[\begin{array}{rrrr|r} 2 & 1 & 3 & 4 & 0 \\ 3 & -1 & 2 & 0 & 3 \\ -2 & 1 & -4 & 3 & 2 \\ 0 & 0 & 0 & 0 & 0 \end{array}\right]$.

21. How are the linear systems obtained in Exercise 20 related?

22. Write each of the following linear systems as a linear combination of the columns of the coefficient matrix.

(a) $$\begin{aligned} 3x_1 + 2x_2 + x_3 &= 4 \\ x_1 - x_2 + 4x_3 &= -2. \end{aligned}$$

(b) $$\begin{aligned} -x_1 + x_2 &= 3 \\ 2x_1 - x_2 &= -2 \\ 3x_1 + x_2 &= 1. \end{aligned}$$

23. Write each of the following linear combinations of columns as a linear system of the form in (2).

(a) $x_1 \begin{bmatrix} 2 \\ 0 \end{bmatrix} + x_2 \begin{bmatrix} 1 \\ 3 \end{bmatrix} = \begin{bmatrix} 4 \\ 2 \end{bmatrix}$.

(b) $x_1 \begin{bmatrix} 1 \\ 2 \\ -1 \end{bmatrix} + x_2 \begin{bmatrix} 0 \\ 1 \\ 2 \end{bmatrix} + x_3 \begin{bmatrix} 3 \\ 4 \\ 5 \end{bmatrix} + x_4 \begin{bmatrix} 1 \\ 3 \\ 4 \end{bmatrix} = \begin{bmatrix} 2 \\ 5 \\ 8 \end{bmatrix}$.

24. Write each of the following as a linear system in matrix form.

(a) $x_1 \begin{bmatrix} 1 \\ 2 \end{bmatrix} + x_2 \begin{bmatrix} 2 \\ 5 \end{bmatrix} + x_3 \begin{bmatrix} 0 \\ 3 \end{bmatrix} = \begin{bmatrix} 1 \\ 1 \end{bmatrix}$.

(b) $x_1 \begin{bmatrix} 1 \\ 1 \\ 2 \end{bmatrix} + x_2 \begin{bmatrix} 2 \\ 1 \\ 0 \end{bmatrix} + x_3 \begin{bmatrix} 1 \\ 2 \\ 2 \end{bmatrix} = \begin{bmatrix} 0 \\ 0 \\ 0 \end{bmatrix}$.

25. Write each of the following as a linear system in matrix form.

(a) $x_1 \begin{bmatrix} 1 & 2 & 1 \end{bmatrix} + x_2 \begin{bmatrix} 3 & 0 & -1 \end{bmatrix} = \begin{bmatrix} 3 & 1 & 4 \end{bmatrix}$.

(b) $x_1 \begin{bmatrix} 2 & 1 & 0 & 1 \end{bmatrix} + x_2 \begin{bmatrix} 3 & -1 & 2 & 2 \end{bmatrix} + x_3 \begin{bmatrix} 0 & 1 & -1 & 3 \end{bmatrix} = \begin{bmatrix} 0 & 0 & 0 & 0 \end{bmatrix}$.

26. Is the matrix $\begin{bmatrix} 3 & 0 \\ 0 & 2 \end{bmatrix}$ a linear combination of the matrices $\begin{bmatrix} 1 & 0 \\ 0 & 1 \end{bmatrix}$ and $\begin{bmatrix} 1 & 0 \\ 0 & 0 \end{bmatrix}$? Justify your answer.

27. Is the matrix $\begin{bmatrix} 4 & 1 \\ 0 & -3 \end{bmatrix}$ a linear combination of the matrices $\begin{bmatrix} 1 & 0 \\ 0 & 1 \end{bmatrix}$ and $\begin{bmatrix} 1 & 0 \\ 0 & 0 \end{bmatrix}$? Justify your answer.

28. Let A be an $m \times n$ matrix and B an $n \times p$ matrix. What if anything can you say about the matrix product AB when:

(a) A has a column consisting entirely of zeros?

(b) B has a row consisting entirely of zeros?

29. If $A = \left[a_{ij}\right]$ is an $n \times n$ matrix, then the **trace** of A, $\text{Tr}(A)$, is defined as the sum of all elements on the main diagonal of A, $\text{Tr}(A) = \sum_{i=1}^{n} a_{ii}$. Show that:

(a) $\text{Tr}(cA) = c\,\text{Tr}(A)$, where c is a real number.

(b) $\text{Tr}(A + B) = \text{Tr}(A) + \text{Tr}(B)$.

(c) $\text{Tr}(AB) = \text{Tr}(BA)$.

(d) $\text{Tr}(A^T) = \text{Tr}(A)$.

(e) $\text{Tr}(A^T A) \geq 0$.

30. Compute the trace (see Exercise 29) of each of the following matrices.

(a) $\begin{bmatrix} 1 & 0 \\ 2 & 3 \end{bmatrix}$. (b) $\begin{bmatrix} 2 & 2 & 3 \\ 2 & 4 & 4 \\ 3 & -2 & -5 \end{bmatrix}$.

(c) $\begin{bmatrix} 1 & 0 & 0 \\ 0 & 1 & 0 \\ 0 & 0 & 1 \end{bmatrix}$.

31. Show that there are no 2×2 matrices A and B such that
$$AB - BA = \begin{bmatrix} 1 & 0 \\ 0 & 1 \end{bmatrix}.$$

32. (a) Show that the jth column of the matrix product AB is equal to the matrix product AB_j, where B_j is the jth column of B. It follows that the product AB can be written in terms of columns as

$$AB = \begin{bmatrix} AB_1 & AB_2 & \cdots & AB_n \end{bmatrix}.$$

(b) Show that the ith row of the matrix product AB is equal to the matrix product $A_i B$, where A_i is the ith row of A. It follows that the product AB can be written in terms of rows as

$$AB = \begin{bmatrix} A_1 B \\ A_2 B \\ \vdots \\ A_m B \end{bmatrix}.$$

33. Show that the jth column of the matrix product AB is a linear combination of the columns of A with coefficients the entries in B_j, the jth column of B.

34. (a) Show that

$$\sum_{i=1}^{3}\left(\sum_{j=1}^{2} a_{ij}\right) = \sum_{j=1}^{2}\left(\sum_{i=1}^{3} a_{ij}\right).$$

(b) Show that

$$\sum_{i=1}^{n}\left(\sum_{j=1}^{m} a_{ij}\right) = \sum_{j=1}^{m}\left(\sum_{i=1}^{n} a_{ij}\right).$$

35. For the software you are using, determine the commands for matrix addition, scalar multiplication, matrix multiplication, and the transpose of a matrix.

36. Use the matrices A and C in Exercise 3 and the matrix multiplication command in your software to compute AC and CA. Discuss the results.

37. For the software you are using, determine the command for obtaining the powers $A^2, A^3, \ldots$ of a square matrix A. Then for

$$A = \begin{bmatrix} 0 & 1 & 0 & 0 & 0 \\ 0 & 0 & 1 & 0 & 0 \\ 0 & 0 & 0 & 1 & 0 \\ 0 & 0 & 0 & 0 & 1 \\ 0 & 0 & 0 & 0 & 0 \end{bmatrix}$$

compute the matrix sequence A^k, $k = 2, 3, 4, 5, 6$. Describe the behavior of A^k as $k \to \infty$.

38. Experiment with your software to determine the behavior of the matrix sequence A^k as $k \to \infty$ for each of the following matrices.

(a) $A = \begin{bmatrix} \frac{1}{2} & \frac{1}{3} \\ \frac{1}{4} & \frac{1}{5} \end{bmatrix}$. (b) $A = \begin{bmatrix} 1 & -1 & 0 \\ 0 & 1 & -1 \\ -1 & 0 & 1 \end{bmatrix}$.

39. Using your software, compute $B^T B$ and BB^T for

$$B = \begin{bmatrix} 1 & \frac{1}{2} & \frac{1}{3} & \frac{1}{4} & \frac{1}{5} & \frac{1}{6} \end{bmatrix}.$$

Discuss the nature of the results.

1.3 ALGEBRAIC PROPERTIES OF MATRIX OPERATIONS

In this section we consider the algebraic properties of the matrix operations just defined. Many of these properties are similar to the familiar properties that hold for real numbers. However, there will be striking differences between the set of real numbers and the set of matrices in their algebraic behavior under certain operations, for example, under multiplication (as seen in Section 1.2). The proofs of most of the properties will be left as exercises.

Theorem 1.1 (Properties of Matrix Addition).

Let A, B, and C be $m \times n$ matrices.

(a) $A + B = B + A$.

(b) $A + (B + C) = (A + B) + C$.

(c) *There is a unique $m \times n$ matrix O_{mn} such that*

$$A + O_{mn} = A \tag{1}$$

for any $m \times n$ matrix A. The matrix O_{mn} is called the $m \times n$ **zero matrix**. *When $m = n$, we write O_n. When m and n are understood, we shall write O_{mn} simply as O.*

(d) *For each $m \times n$ matrix A, there is a unique $m \times n$ matrix D such that*

$$A + D = O. \tag{2}$$

We shall write D as $-A$, so that (2) can be written as

$$A + (-A) = O.$$

The matrix $-A$ is called the **negative** *of A. We also note that $-A$ is $(-1)A$.*

Proof

(a) Let

$$A = \begin{bmatrix} a_{ij} \end{bmatrix}, \quad B = \begin{bmatrix} b_{ij} \end{bmatrix},$$
$$A + B = C = \begin{bmatrix} c_{ij} \end{bmatrix}, \quad \text{and} \quad B + A = D = \begin{bmatrix} d_{ij} \end{bmatrix}.$$

We must show that $c_{ij} = d_{ij}$ for all i, j. Now $c_{ij} = a_{ij} + b_{ij}$ and $d_{ij} = b_{ij} + a_{ij}$ for all i, j. Since a_{ij} and b_{ij} are real numbers, we have $a_{ij} + b_{ij} = b_{ij} + a_{ij}$, which implies that $c_{ij} = d_{ij}$ for all i, j.

(c) Let $U = \begin{bmatrix} u_{ij} \end{bmatrix}$. Then $A + U = A$ if and only if[†] $a_{ij} + u_{ij} = a_{ij}$, which holds if and only if $u_{ij} = 0$. Thus U is the $m \times n$ matrix all of whose entries are zero: U is denoted by O_{mn}, or simply by O. ●

The 2×2 zero matrix is

$$O_2 = \begin{bmatrix} 0 & 0 \\ 0 & 0 \end{bmatrix}.$$

[†]The connector "if and only if" means that both statements are true or both statements are false. Thus (i) if $A + U = A$, then $a_{ij} + u_{ij} = a_{ij}$, and (ii) if $a_{ij} + u_{ij} = a_{ij}$, then $A + U = A$.

EXAMPLE 1

If

$$A = \begin{bmatrix} 4 & -1 \\ 2 & 3 \end{bmatrix},$$

then

$$\begin{bmatrix} 4 & -1 \\ 2 & 3 \end{bmatrix} + \begin{bmatrix} 0 & 0 \\ 0 & 0 \end{bmatrix} = \begin{bmatrix} 4+0 & -1+0 \\ 2+0 & 3+0 \end{bmatrix} = \begin{bmatrix} 4 & -1 \\ 2 & 3 \end{bmatrix}.$$

The 2×3 zero matrix is

$$O_{23} = \begin{bmatrix} 0 & 0 & 0 \\ 0 & 0 & 0 \end{bmatrix}.$$

■

EXAMPLE 2

If $A = \begin{bmatrix} 1 & 3 & -2 \\ -2 & 4 & 3 \end{bmatrix}$, then $-A = \begin{bmatrix} -1 & -3 & 2 \\ 2 & -4 & -3 \end{bmatrix}$. ■

Theorem 1.2 (*Properties of Matrix Multiplication*).

(a) *If A, B, and C are matrices, then*

$$A(BC) = (AB)C.$$

(b) *If A, B, and C are matrices, then*

$$A + B)C = AC + BC.$$

(c) *If A, B, and C are matrices, then*

$$C(A + B) = CA + CB. \tag{3}$$

Proof

(a) Suppose that A is $m \times n$, B is $n \times p$, and C is $p \times q$. We shall prove the result for the special case $m = 2$, $n = 3$, $p = 4$, and $q = 3$. The general proof is completely analogous.

Let $A = \begin{bmatrix} a_{ij} \end{bmatrix}$, $B = \begin{bmatrix} b_{ij} \end{bmatrix}$, $C = \begin{bmatrix} c_{ij} \end{bmatrix}$, $AB = D = \begin{bmatrix} d_{ij} \end{bmatrix}$, $BC = E = \begin{bmatrix} e_{ij} \end{bmatrix}$, $(AB)C = F = \begin{bmatrix} f_{ij} \end{bmatrix}$, and $A(BC) = G = \begin{bmatrix} g_{ij} \end{bmatrix}$. We must show that $f_{ij} = g_{ij}$ for all i, j. Now

$$f_{ij} = \sum_{k=1}^{4} d_{ik} c_{kj} = \sum_{k=1}^{4} \left(\sum_{r=1}^{3} a_{ir} b_{rk} \right) c_{kj}$$

and

$$g_{ij} = \sum_{r=1}^{3} a_{ir} e_{rj} = \sum_{r=1}^{3} a_{ir} \left(\sum_{k=1}^{4} b_{rk} c_{kj} \right).$$

Then, by the properties satisfied by the summation notation,

$$\begin{aligned} f_{ij} &= \sum_{k=1}^{4} (a_{i1}b_{1k} + a_{i2}b_{2k} + a_{i3}b_{3k})c_{kj} \\ &= a_{i1} \sum_{k=1}^{4} b_{1k}c_{kj} + a_{i2} \sum_{k=1}^{4} b_{2k}c_{kj} + a_{i3} \sum_{k=1}^{4} b_{3k}c_{kj} \\ &= \sum_{r=1}^{3} a_{ir} \left(\sum_{k=1}^{4} b_{rk}c_{kj} \right) = g_{ij}. \end{aligned}$$

(b) and (c) are left as exercises. ●

EXAMPLE 3

Let

$$A = \begin{bmatrix} 5 & 2 & 3 \\ 2 & -3 & 4 \end{bmatrix}, \qquad B = \begin{bmatrix} 2 & -1 & 1 & 0 \\ 0 & 2 & 2 & 2 \\ 3 & 0 & -1 & 3 \end{bmatrix},$$

and

$$C = \begin{bmatrix} 1 & 0 & 2 \\ 2 & -3 & 0 \\ 0 & 0 & 3 \\ 2 & 1 & 0 \end{bmatrix}.$$

Then

$$A(BC) = \begin{bmatrix} 5 & 2 & 3 \\ 2 & -3 & 4 \end{bmatrix} \begin{bmatrix} 0 & 3 & 7 \\ 8 & -4 & 6 \\ 9 & 3 & 3 \end{bmatrix} = \begin{bmatrix} 43 & 16 & 56 \\ 12 & 30 & 8 \end{bmatrix}$$

and

$$(AB)C = \begin{bmatrix} 19 & -1 & 6 & 13 \\ 16 & -8 & -8 & 6 \end{bmatrix} \begin{bmatrix} 1 & 0 & 2 \\ 2 & -3 & 0 \\ 0 & 0 & 3 \\ 2 & 1 & 0 \end{bmatrix} = \begin{bmatrix} 43 & 16 & 56 \\ 12 & 30 & 8 \end{bmatrix}.$$ ■

EXAMPLE 4

Let

$$A = \begin{bmatrix} 2 & 2 & 3 \\ 3 & -1 & 2 \end{bmatrix}, \quad B = \begin{bmatrix} 0 & 0 & 1 \\ 2 & 3 & -1 \end{bmatrix}, \quad \text{and} \quad C = \begin{bmatrix} 1 & 0 \\ 2 & 2 \\ 3 & -1 \end{bmatrix}.$$

Then

$$(A + B)C = \begin{bmatrix} 2 & 2 & 4 \\ 5 & 2 & 1 \end{bmatrix} \begin{bmatrix} 1 & 0 \\ 2 & 2 \\ 3 & -1 \end{bmatrix} = \begin{bmatrix} 18 & 0 \\ 12 & 3 \end{bmatrix}$$

and (verify)

$$AC + BC = \begin{bmatrix} 15 & 1 \\ 7 & -4 \end{bmatrix} + \begin{bmatrix} 3 & -1 \\ 5 & 7 \end{bmatrix} = \begin{bmatrix} 18 & 0 \\ 12 & 3 \end{bmatrix}.$$

■

Recall Example 10 in Section 1.2, which shows that AB need not always equal BA. This is the first significant difference between multiplication of matrices and multiplication of real numbers.

Theorem 1.3 (*Properties of Scalar Multiplication*).

If r and s are real numbers and A and B are matrices, then

(a) $r(sA) = (rs)A$.

(b) $(r + s)A = rA + sA$.

(c) $r(A + B) = rA + rB$.

(d) $A(rB) = r(AB) = (rA)B$.

Proof

Exercises. ●

EXAMPLE 5

Let

$$A = \begin{bmatrix} 4 & 2 & 3 \\ 2 & -3 & 4 \end{bmatrix} \quad \text{and} \quad B = \begin{bmatrix} 3 & -2 & 1 \\ 2 & 0 & -1 \\ 0 & 1 & 2 \end{bmatrix}.$$

Then

$$2(3A) = 2\begin{bmatrix} 12 & 6 & 9 \\ 6 & -9 & 12 \end{bmatrix} = \begin{bmatrix} 24 & 12 & 18 \\ 12 & -18 & 24 \end{bmatrix} = 6A.$$

We also have

$$A(2B) = \begin{bmatrix} 4 & 2 & 3 \\ 2 & -3 & 4 \end{bmatrix}\begin{bmatrix} 6 & -4 & 2 \\ 4 & 0 & -2 \\ 0 & 2 & 4 \end{bmatrix} = \begin{bmatrix} 32 & -10 & 16 \\ 0 & 0 & 26 \end{bmatrix} = 2(AB).$$

■

So far we have seen that multiplication and addition of matrices have much in common with multiplication and addition of real numbers. We now look at some properties of the transpose.

Theorem 1.4 (*Properties of Transpose*).

If r is a scalar and A and B are matrices, then

(a) $(A^T)^T = A$.

(b) $(A+B)^T = A^T + B^T$.
(c) $(AB)^T = B^T A^T$.
(d) $(rA)^T = rA^T$.

Proof

We leave the proofs of (a), (b), and (d) as exercises.

(c) Let $A = \begin{bmatrix} a_{ij} \end{bmatrix}$ and $B = \begin{bmatrix} b_{ij} \end{bmatrix}$; let $AB = C = \begin{bmatrix} c_{ij} \end{bmatrix}$. We must prove that c_{ij}^T is the (i, j) entry in $B^T A^T$. Now

$$c_{ij}^T = c_{ji} = \sum_{k=1}^{n} a_{jk} b_{ki} = \sum_{k=1}^{n} a_{kj}^T b_{ik}^T$$

$$= \sum_{k=1}^{n} b_{ik}^T a_{kj}^T = \text{the } (i, j) \text{ entry in } B^T A^T.$$

●

EXAMPLE 6

Let

$$A = \begin{bmatrix} 1 & 2 & 3 \\ -2 & 0 & 1 \end{bmatrix} \quad \text{and} \quad B = \begin{bmatrix} 3 & -1 & 2 \\ 3 & 2 & -1 \end{bmatrix}.$$

Then

$$A^T = \begin{bmatrix} 1 & -2 \\ 2 & 0 \\ 3 & 1 \end{bmatrix} \quad \text{and} \quad B^T = \begin{bmatrix} 3 & 3 \\ -1 & 2 \\ 2 & -1 \end{bmatrix}.$$

Also,

$$A + B = \begin{bmatrix} 4 & 1 & 5 \\ 1 & 2 & 0 \end{bmatrix} \quad \text{and} \quad (A+B)^T = \begin{bmatrix} 4 & 1 \\ 1 & 2 \\ 5 & 0 \end{bmatrix}.$$

Now

$$A^T + B^T = \begin{bmatrix} 4 & 1 \\ 1 & 2 \\ 5 & 0 \end{bmatrix} = (A+B)^T.$$

■

EXAMPLE 7

Let

$$A = \begin{bmatrix} 1 & 3 & 2 \\ 2 & -1 & 3 \end{bmatrix} \quad \text{and} \quad B = \begin{bmatrix} 0 & 1 \\ 2 & 2 \\ 3 & -1 \end{bmatrix}.$$

Then

$$AB = \begin{bmatrix} 12 & 5 \\ 7 & -3 \end{bmatrix} \quad \text{and} \quad (AB)^T = \begin{bmatrix} 12 & 7 \\ 5 & -3 \end{bmatrix}.$$

On the other hand,

$$A^T = \begin{bmatrix} 1 & 2 \\ 3 & -1 \\ 2 & 3 \end{bmatrix} \quad \text{and} \quad B^T = \begin{bmatrix} 0 & 2 & 3 \\ 1 & 2 & -1 \end{bmatrix}.$$

Then

$$B^T A^T = \begin{bmatrix} 12 & 7 \\ 5 & -3 \end{bmatrix} = (AB)^T.$$

■

We also note two other peculiarities of matrix multiplication. If a and b are real numbers, then $ab = 0$ can hold only if a or b is zero. However, this is not true for matrices.

EXAMPLE 8

If

$$A = \begin{bmatrix} 1 & 2 \\ 2 & 4 \end{bmatrix} \quad \text{and} \quad B = \begin{bmatrix} 4 & -6 \\ -2 & 3 \end{bmatrix},$$

then neither A nor B is the zero matrix, but $AB = \begin{bmatrix} 0 & 0 \\ 0 & 0 \end{bmatrix}$.

■

If a, b, and c are real numbers for which $ab = ac$ and $a \neq 0$, it follows that $b = c$. That is, we can cancel out the nonzero factor a. However, the cancellation law does not hold for matrices, as the following example shows.

EXAMPLE 9

If

$$A = \begin{bmatrix} 1 & 2 \\ 2 & 4 \end{bmatrix}, \quad B = \begin{bmatrix} 2 & 1 \\ 3 & 2 \end{bmatrix}, \quad \text{and} \quad C = \begin{bmatrix} -2 & 7 \\ 5 & -1 \end{bmatrix},$$

then

$$AB = AC = \begin{bmatrix} 8 & 5 \\ 16 & 10 \end{bmatrix},$$

but $B \neq C$. ■

We summarize some of the differences between matrix multiplication and the multiplication of real numbers as follows:

1. AB need not equal BA.
2. AB may be the zero matrix with $A \neq O$ and $B \neq O$.
3. AB may equal AC with $B \neq C$.

In this section we have developed a number of properties about matrices and their transposes. If a future problem either asks a question about these ideas or involves these concepts, refer to these properties to help answer the question. These results can be used to develop many more results.

1.3 Exercises

1. Prove Theorem 1.1(b).

2. Prove Theorem 1.1(d).

3. Verify Theorem 1.2(a) for the following matrices:

$$A = \begin{bmatrix} 1 & 3 \\ 2 & -1 \end{bmatrix}, \quad B = \begin{bmatrix} -1 & 3 & 2 \\ 1 & -3 & 4 \end{bmatrix},$$

$$\text{and} \quad C = \begin{bmatrix} 1 & 0 \\ 3 & -1 \\ 1 & 2 \end{bmatrix}.$$

4. Prove Theorem 1.2(b) and (c).

5. Verify Theorem 1.2(c) for the following matrices:

$$A = \begin{bmatrix} 2 & -3 & 2 \\ 3 & -1 & -2 \end{bmatrix}, \quad B = \begin{bmatrix} 0 & 1 & 2 \\ 1 & 3 & -2 \end{bmatrix},$$

$$\text{and} \quad C = \begin{bmatrix} 1 & -3 \\ -3 & 4 \end{bmatrix}.$$

6. Let A be an $m \times n$ matrix and $B = \begin{bmatrix} b_1 \\ b_2 \\ \vdots \\ b_n \end{bmatrix}$, an $n \times 1$ matrix. Prove that

$$AB = \sum_{j=1}^{n} b_j A_j,$$

where A_j is the jth column of A.

7. Let A be an $m \times n$ matrix and $C = \begin{bmatrix} c_1 & c_2 & \cdots & c_m \end{bmatrix}$ a $1 \times m$ matrix. Prove that

$$CA = \sum_{j=1}^{m} c_j A_j,$$

where A_j is the jth row of A.

8. Let $A = \begin{bmatrix} \cos\theta & \sin\theta \\ -\sin\theta & \cos\theta \end{bmatrix}$.

(a) Determine a simple expression for A^2.

(b) Determine a simple expression for A^3.

(c) Conjecture the form of a simple expression for A^k, k a positive integer.

(d) Prove or disprove your conjecture in part (c).

9. Find a pair of unequal 2×2 matrices A and B, other than those given in Example 8, such that $AB = O_2$.

10. Find two different 2×2 matrices A such that $A^2 = \begin{bmatrix} 1 & 0 \\ 0 & 1 \end{bmatrix}$.

11. Find two unequal 2×2 matrices A and B such that $AB = \begin{bmatrix} 1 & 0 \\ 0 & 1 \end{bmatrix}$.

12. Find two different 2×2 matrices A such that $A^2 = O_2$.

13. Prove Theorem 1.3(a).

14. Prove Theorem 1.3(b).

15. Verify Theorem 1.3(b) for $r = 4$, $s = -2$, and $A = \begin{bmatrix} 2 & -3 \\ 4 & 2 \end{bmatrix}$.

16. Prove Theorem 1.3(c).

17. Verify Theorem 1.3(c) for $r = -3$,

$$A = \begin{bmatrix} 4 & 2 \\ 1 & -3 \\ 3 & 2 \end{bmatrix}, \quad \text{and} \quad B = \begin{bmatrix} 0 & 2 \\ 4 & 3 \\ -2 & 1 \end{bmatrix}.$$

18. Prove Theorem 1.3(d).

19. Verify Theorem 1.3(d) for the following matrices:

$$A = \begin{bmatrix} 1 & 3 \\ 2 & -1 \end{bmatrix}, \quad B = \begin{bmatrix} -1 & 3 & 2 \\ 1 & -3 & 4 \end{bmatrix},$$
$$\text{and} \quad r = -3.$$

20. Determine a scalar r such that $AX = rX$, where

$$A = \begin{bmatrix} 2 & 1 \\ 1 & 2 \end{bmatrix} \quad \text{and} \quad X = \begin{bmatrix} 1 \\ 1 \end{bmatrix}.$$

21. Determine a scalar r such that $AX = rX$, where

$$A = \begin{bmatrix} 1 & 2 & -1 \\ 1 & 0 & 1 \\ 4 & -4 & 5 \end{bmatrix} \quad \text{and} \quad X = \begin{bmatrix} -\frac{1}{2} \\ \frac{1}{4} \\ 1 \end{bmatrix}.$$

22. Prove that if $AX = rX$ for $n \times n$ matrix A, $n \times 1$ matrix X, and scalar r, then $AY = rY$, where $Y = sX$ for any scalar s.

23. Determine a scalar s such that $A^2X = sX$ when $AX = rX$.

24. Prove Theorem 1.4(a).

25. Prove Theorem 1.4(b) and (d).

26. Verify Theorem 1.4(a), (b), and (d) for

$$A = \begin{bmatrix} 1 & 3 & 2 \\ 2 & 1 & -3 \end{bmatrix}, \quad B = \begin{bmatrix} 4 & 2 & -1 \\ -2 & 1 & 5 \end{bmatrix},$$
$$\text{and} \quad r = -4.$$

27. Verify Theorem 1.4(c) for

$$A = \begin{bmatrix} 1 & 3 & 2 \\ 2 & 1 & -3 \end{bmatrix} \quad \text{and} \quad B = \begin{bmatrix} 3 & -1 \\ 2 & 4 \\ 1 & 2 \end{bmatrix}.$$

28. Let

$$A = \begin{bmatrix} 2 \\ -1 \\ 3 \end{bmatrix}, \quad B = \begin{bmatrix} 3 \\ -2 \\ -4 \end{bmatrix}, \quad \text{and} \quad C = \begin{bmatrix} -1 \\ 5 \\ 1 \end{bmatrix}.$$

(a) Compute $(AB^T)C$.

(b) Compute B^TC and multiply the result by A on the right.

(c) Explain why $(AB^T)C = (B^TC)A$.

29. Determine a constant k such that $(kA)^T(kA) = 1$, where $A = \begin{bmatrix} -2 \\ 1 \\ -1 \end{bmatrix}$. Is there more than one value of k that could be used?

30. Find three 2×2 matrices, A, B, and C such that $AB = AC$ with $B \neq C$ and $A \neq O_2$.

31. Let A be an $n \times n$ matrix and c a real number. Show that if $cA = O_n$, then $c = 0$ or $A = O_n$.

32. Determine all 2×2 matrices A such that $AB = BA$ for any 2×2 matrix B.

33. Show that $(A - B)^T = A^T - B^T$.

34. Let X_1 and X_2 be solutions to the homogeneous linear system $AX = O$.

(a) Show that $X_1 + X_2$ is a solution.

(b) Show that $X_1 - X_2$ is a solution.

(c) For any scalar r, show that rX_1 is a solution.

(d) For any scalars r and s, show that $rX_1 + sX_2$ is a solution.

35. Show that if $AX = B$ has more than one solution, then it has infinitely many solutions. (*Hint*: If X_1 and X_2 are solutions, consider $X_3 = rX_1 + sX_2$, where $r + s = 1$.)

36. Show that if X_1 and X_2 are solutions to the linear system $AX = B$, then $X_1 - X_2$ is a solution to the associated homogeneous system $AX = O$.

37. Let

$$A = \begin{bmatrix} 6 & -1 & 1 \\ 0 & 13 & -16 \\ 0 & 8 & -11 \end{bmatrix} \quad \text{and} \quad X = \begin{bmatrix} 10.5 \\ 21.0 \\ 10.5 \end{bmatrix}.$$

(a) Determine a scalar r such that $AX = rX$.

(b) Is it true that $A^T X = rX$ for the value r determined in part (a)?

38. Repeat Exercise 37 with

$$A = \begin{bmatrix} -3.35 & -3.00 & 3.60 \\ 1.20 & 2.05 & -6.20 \\ -3.60 & -2.40 & 3.85 \end{bmatrix}$$

$$\text{and} \quad X = \begin{bmatrix} 12.5 \\ -12.5 \\ 6.25 \end{bmatrix}.$$

39. Let $A = \begin{bmatrix} 0.1 & 0.01 \\ 0.001 & 0.0001 \end{bmatrix}$. In your software, set the display format to show as many decimal places as possible, then compute

$$B = 10 * A,$$
$$C = \underbrace{A + A + A + A + A + A + A + A + A + A}_{10 \text{ summands}},$$

and $D = B - C$.

If D is not O_2, then you have verified that scalar multiplication by a positive integer and successive addition are not the same in your computing environment. (It is not unusual that $D \neq O_2$ since many computing environments use only a "model" of exact arithmetic, called floating-point arithmetic.)

40. Let $A = \begin{bmatrix} 1 & 1 \\ 1 & 1 \end{bmatrix}$. In your software, set the display to show as many decimal places as possible. Experiment to find a positive integer k such that $A + 10^{-k} * A$ is equal to A. If you find such an integer k, you have verified that there is more than one matrix in your computational environment that plays the role of O_2.

1.4 SPECIAL TYPES OF MATRICES AND PARTITIONED MATRICES

We have already introduced one special type of matrix O_{mn}, the matrix all of whose entries are zero. We now consider several other types of matrices whose structures are rather specialized and for which it will be convenient to have special names.

An $n \times n$ matrix $A = \begin{bmatrix} a_{ij} \end{bmatrix}$ is called a **diagonal matrix** if $a_{ij} = 0$ for $i \neq j$. Thus, for a diagonal matrix, the terms *off* the main diagonal are all zero. Note that O_n is a diagonal matrix. A **scalar matrix** is a diagonal matrix whose diagonal elements are equal. The scalar matrix $I_n = \begin{bmatrix} a_{ij} \end{bmatrix}$, where $a_{ii} = 1$ and $a_{ij} = 0$ for $i \neq j$, is called the $n \times n$ **identity matrix**.

EXAMPLE 1

Let

$$A = \begin{bmatrix} 1 & 0 & 0 \\ 0 & 2 & 0 \\ 0 & 0 & 3 \end{bmatrix}, \quad B = \begin{bmatrix} 2 & 0 & 0 \\ 0 & 2 & 0 \\ 0 & 0 & 2 \end{bmatrix}, \quad \text{and} \quad I_3 = \begin{bmatrix} 1 & 0 & 0 \\ 0 & 1 & 0 \\ 0 & 0 & 1 \end{bmatrix}.$$

Then A, B, and I_3 are diagonal matrices; B and I_3 are scalar matrices; and I_3 is the 3×3 identity matrix. ■

It is easy to show that if A is any $m \times n$ matrix, then

$$AI_n = A \quad \text{and} \quad I_m A = A.$$

Also, if A is a scalar matrix, then $A = rI_n$ for some scalar r.

Suppose that A is a square matrix. If p is a positive integer, then we define

$$A^p = \underbrace{A \cdot A \cdot \cdots \cdot A}_{p \text{ factors}}.$$

If A is $n \times n$, we also define

$$A^0 = I_n.$$

For nonnegative integers p and q, the familiar laws of exponents for the real numbers can also be proved for matrix multiplication of a square matrix A (Exercise 6):

$$A^p A^q = A^{p+q} \quad \text{and} \quad (A^p)^q = A^{pq}.$$

It should also be noted that the rule

$$(AB)^p = A^p B^p$$

does not hold for square matrices unless $AB = BA$ (Exercise 7).

An $n \times n$ matrix $A = \begin{bmatrix} a_{ij} \end{bmatrix}$ is called **upper triangular** if $a_{ij} = 0$ for $i > j$. It is called **lower triangular** if $a_{ij} = 0$ for $i < j$. A diagonal matrix is both upper triangular and lower triangular.

EXAMPLE 2

The matrix

$$A = \begin{bmatrix} 1 & 3 & 3 \\ 0 & 3 & 5 \\ 0 & 0 & 2 \end{bmatrix}$$

is upper triangular and

$$B = \begin{bmatrix} 1 & 0 & 0 \\ 2 & 3 & 0 \\ 3 & 5 & 2 \end{bmatrix}$$

is lower triangular. ■

Definition 1.7

A matrix A is called **symmetric** if $A^T = A$. ▲

Definition 1.8

A matrix A is called **skew symmetric** if $A^T = -A$. ▲

EXAMPLE 3

$A = \begin{bmatrix} 1 & 2 & 3 \\ 2 & 4 & 5 \\ 3 & 5 & 6 \end{bmatrix}$ is a symmetric matrix. ■

EXAMPLE 4

$$B = \begin{bmatrix} 0 & 2 & 3 \\ -2 & 0 & -4 \\ -3 & 4 & 0 \end{bmatrix} \text{ is a skew symmetric matrix.}$$ ■

We can make a few observations about symmetric and skew symmetric matrices; the proofs of most of these statements will be left as exercises.

It follows from the definitions above that if A is symmetric or skew symmetric, then A is a square matrix. If A is a symmetric matrix, then the entries of A are symmetric with respect to the main diagonal of A. Also, A is symmetric if and only if $a_{ij} = a_{ji}$, and A is skew symmetric if and only if $a_{ij} = -a_{ji}$. Moreover, if A is skew symmetric, then the entries on the main diagonal of A are all zero. An important property of symmetric and skew symmetric matrices is the following. If A is an $n \times n$ matrix, then we can show that $A = S + K$, where S is symmetric and K is skew symmetric. Moreover, this decomposition is unique (Exercise 21).

Partitioned Matrices

If we start out with an $m \times n$ matrix $A = \left[a_{ij}\right]$ and then cross out some, but not all, of its rows or columns, we obtain a **submatrix** of A.

EXAMPLE 5

Let

$$A = \begin{bmatrix} 1 & 2 & 3 & 4 \\ -2 & 4 & -3 & 5 \\ 3 & 0 & 5 & -3 \end{bmatrix}.$$

If we cross out the second row and third column, we obtain the submatrix

$$\begin{bmatrix} 1 & 2 & 4 \\ 3 & 0 & -3 \end{bmatrix}.$$ ■

A matrix can be partitioned into submatrices by drawing horizontal lines between rows and vertical lines between columns. Of course, the partitioning can be carried out in many different ways.

EXAMPLE 6

The matrix

$$A = \left[\begin{array}{ccc:cc} a_{11} & a_{12} & a_{13} & a_{14} & a_{15} \\ a_{21} & a_{22} & a_{23} & a_{24} & a_{25} \\ \hdashline a_{31} & a_{32} & a_{33} & a_{34} & a_{35} \\ a_{41} & a_{42} & a_{43} & a_{44} & a_{45} \end{array}\right]$$

is partitioned as

$$A = \begin{bmatrix} A_{11} & A_{12} \\ A_{21} & A_{22} \end{bmatrix}.$$

We could also write

$$A = \left[\begin{array}{cc:cc:c} a_{11} & a_{12} & a_{13} & a_{14} & a_{15} \\ a_{21} & a_{22} & a_{23} & a_{24} & a_{25} \\ \hdashline a_{31} & a_{32} & a_{33} & a_{34} & a_{35} \\ a_{41} & a_{42} & a_{43} & a_{44} & a_{45} \end{array}\right] = \begin{bmatrix} \widehat{A}_{11} & \widehat{A}_{12} & \widehat{A}_{13} \\ \widehat{A}_{21} & \widehat{A}_{22} & \widehat{A}_{23} \end{bmatrix}, \qquad (1)$$

which gives another partitioning of A. We thus speak of **partitioned matrices.** ■

EXAMPLE 7

The augmented matrix (defined in Section 1.2) of a linear system is a partitioned matrix. Thus, if $AX = B$, we can write the augmented matrix of this system as $\left[A \,\vdots\, B\right]$. ■

If A and B are both $m \times n$ matrices that are partitioned in the same way, then $A + B$ is obtained simply by adding the corresponding submatrices of A and B. Similarly, if A is a partitioned matrix, then the scalar multiple cA is obtained by forming the scalar multiple of each submatrix.

If A is partitioned as shown in (1) and

$$B = \left[\begin{array}{cc:cc} b_{11} & b_{12} & b_{13} & b_{14} \\ b_{21} & b_{22} & b_{23} & b_{24} \\ \hdashline b_{31} & b_{32} & b_{33} & b_{34} \\ b_{41} & b_{42} & b_{43} & b_{44} \\ \hdashline b_{51} & b_{52} & b_{53} & b_{54} \end{array}\right] = \begin{bmatrix} B_{11} & B_{12} \\ B_{21} & B_{22} \\ B_{31} & B_{32} \end{bmatrix},$$

then by straightforward computations we can show that

$$AB = \left[\begin{array}{c:c} (\widehat{A}_{11}B_{11} + \widehat{A}_{12}B_{21} + \widehat{A}_{13}B_{31}) & (\widehat{A}_{11}B_{12} + \widehat{A}_{12}B_{22} + \widehat{A}_{13}B_{32}) \\ \hdashline (\widehat{A}_{21}B_{11} + \widehat{A}_{22}B_{21} + \widehat{A}_{23}B_{31}) & (\widehat{A}_{21}B_{12} + \widehat{A}_{22}B_{22} + \widehat{A}_{23}B_{32}) \end{array}\right].$$

EXAMPLE 8

Let

$$A = \left[\begin{array}{cc|cc} 1 & 0 & 1 & 0 \\ 0 & 2 & 3 & -1 \\ \hline 2 & 0 & -4 & 0 \\ 0 & 1 & 0 & 3 \end{array}\right] = \begin{bmatrix} A_{11} & A_{12} \\ A_{21} & A_{22} \end{bmatrix}$$

and let

$$B = \left[\begin{array}{ccc|ccc} 2 & 0 & 0 & 1 & 1 & -1 \\ 0 & 1 & 1 & -1 & 2 & 2 \\ \hline 1 & 3 & 0 & 0 & 1 & 0 \\ -3 & -1 & 2 & 1 & 0 & -1 \end{array}\right] = \begin{bmatrix} B_{11} & B_{12} \\ B_{21} & B_{22} \end{bmatrix}.$$

Then

$$AB = C = \left[\begin{array}{ccc|ccc} 3 & 3 & 0 & 1 & 2 & -1 \\ 6 & 12 & 0 & -3 & 7 & 5 \\ \hline 0 & -12 & 0 & 2 & -2 & -2 \\ -9 & -2 & 7 & 2 & 2 & -1 \end{array}\right] = \begin{bmatrix} C_{11} & C_{12} \\ C_{21} & C_{22} \end{bmatrix},$$

where C_{11} should be $A_{11}B_{11} + A_{12}B_{21}$. We verify that C_{11} is this expression as follows:

$$\begin{aligned} A_{11}B_{11} + A_{12}B_{21} &= \begin{bmatrix} 1 & 0 \\ 0 & 2 \end{bmatrix}\begin{bmatrix} 2 & 0 & 0 \\ 0 & 1 & 1 \end{bmatrix} + \begin{bmatrix} 1 & 0 \\ 3 & -1 \end{bmatrix}\begin{bmatrix} 1 & 3 & 0 \\ -3 & -1 & 2 \end{bmatrix} \\ &= \begin{bmatrix} 2 & 0 & 0 \\ 0 & 2 & 2 \end{bmatrix} + \begin{bmatrix} 1 & 3 & 0 \\ 6 & 10 & -2 \end{bmatrix} \\ &= \begin{bmatrix} 3 & 3 & 0 \\ 6 & 12 & 0 \end{bmatrix} = C_{11}. \end{aligned}$$

■

This method of multiplying partitioned matrices is also known as **block multiplication**. Partitioned matrices can be used to great advantage in dealing with matrices that exceed the memory capacity of a computer. Thus, in multiplying two partitioned matrices, one can keep the matrices on disk and only bring into memory the submatrices required to form the submatrix products. The latter, of course, can be put out on disk as they are formed. The partitioning must be done so that the products of corresponding submatrices are defined. In contemporary computing technology, parallel-processing computers use partitioned matrices to perform matrix computations more rapidly.

Nonsingular Matrices

We now come to a special type of square matrix and formulate the notion corresponding to the reciprocal of a nonzero real number.

Definition 1.9

An $n \times n$ matrix A is called **nonsingular**, or **invertible**, if there exists an $n \times n$ matrix B such that $AB = BA = I_n$; such a B is called an **inverse** of A. Otherwise, A is called **singular**, or **noninvertible**. ▲

Remark In Theorem 1.19, Section 1.6, we show that if $AB = I_n$, then $BA = I_n$. Thus, to verify that B is an inverse of A, we need only verify that $AB = I_n$.

EXAMPLE 9

Let $A = \begin{bmatrix} 2 & 3 \\ 2 & 2 \end{bmatrix}$ and let $B = \begin{bmatrix} -1 & \frac{3}{2} \\ 1 & -1 \end{bmatrix}$. Since $AB = BA = I_2$, we conclude that B is an inverse of A. ■

Theorem 1.5

The inverse of a matrix, if it exists, is unique.

Proof

Let B and C be inverses of A. Then

$$AB = BA = I_n \quad \text{and} \quad AC = CA = I_n.$$

We then have $B = BI_n = B(AC) = (BA)C = I_nC = C$, which proves that the inverse of a matrix, if it exists, is unique. ●

Because of this uniqueness, we write the inverse of a nonsingular matrix A as A^{-1}. Thus

$$AA^{-1} = A^{-1}A = I_n.$$

EXAMPLE 10

Let

$$A = \begin{bmatrix} 1 & 2 \\ 3 & 4 \end{bmatrix}.$$

If A^{-1} exists, let

$$A^{-1} = \begin{bmatrix} a & b \\ c & d \end{bmatrix}.$$

Then we must have

$$AA^{-1} = \begin{bmatrix} 1 & 2 \\ 3 & 4 \end{bmatrix}\begin{bmatrix} a & b \\ c & d \end{bmatrix} = I_2 = \begin{bmatrix} 1 & 0 \\ 0 & 1 \end{bmatrix},$$

so that

$$\begin{bmatrix} a + 2c & b + 2d \\ 3a + 4c & 3b + 4d \end{bmatrix} = \begin{bmatrix} 1 & 0 \\ 0 & 1 \end{bmatrix}.$$

Equating corresponding entries of these two matrices, we obtain the linear systems

$$\begin{aligned} a + 2c &= 1 \\ 3a + 4c &= 0 \end{aligned} \quad \text{and} \quad \begin{aligned} b + 2d &= 0 \\ 3b + 4d &= 1. \end{aligned}$$

The solutions are (verify) $a = -2$, $c = \frac{3}{2}$, $b = 1$, and $d = -\frac{1}{2}$. Moreover, since the matrix

$$\begin{bmatrix} a & b \\ c & d \end{bmatrix} = \begin{bmatrix} -2 & 1 \\ \frac{3}{2} & -\frac{1}{2} \end{bmatrix}$$

also satisfies the property that

$$\begin{bmatrix} -2 & 1 \\ \frac{3}{2} & -\frac{1}{2} \end{bmatrix} \begin{bmatrix} 1 & 2 \\ 3 & 4 \end{bmatrix} = \begin{bmatrix} 1 & 0 \\ 0 & 1 \end{bmatrix},$$

we conclude that A is nonsingular and that

$$A^{-1} = \begin{bmatrix} -2 & 1 \\ \frac{3}{2} & -\frac{1}{2} \end{bmatrix}.$$

■

EXAMPLE 11

Let

$$A = \begin{bmatrix} 1 & 2 \\ 2 & 4 \end{bmatrix}.$$

If A^{-1} exists, let

$$A^{-1} = \begin{bmatrix} a & b \\ c & d \end{bmatrix}.$$

Then we must have

$$AA^{-1} = \begin{bmatrix} 1 & 2 \\ 2 & 4 \end{bmatrix} \begin{bmatrix} a & b \\ c & d \end{bmatrix} = I_2 = \begin{bmatrix} 1 & 0 \\ 0 & 1 \end{bmatrix},$$

so that

$$\begin{bmatrix} a + 2c & b + 2d \\ 2a + 4c & 2b + 4d \end{bmatrix} = \begin{bmatrix} 1 & 0 \\ 0 & 1 \end{bmatrix}.$$

Equating corresponding entries of these two matrices, we obtain the linear systems

$$\begin{aligned} a + 2c &= 1 \\ 2a + 4c &= 0 \end{aligned} \quad \text{and} \quad \begin{aligned} b + 2d &= 0 \\ 2b + 4d &= 1. \end{aligned}$$

These linear systems have no solutions, so our assumption that A^{-1} exists is incorrect. Thus, A is singular. ■

We next establish several properties of inverses of matrices.

Theorem 1.6

If A and B are both nonsingular $n \times n$ matrices, then AB is nonsingular and $(AB)^{-1} = B^{-1}A^{-1}$.

Proof

We have $(AB)(B^{-1}A^{-1}) = A(BB^{-1})A^{-1} = (AI_n)A^{-1} = AA^{-1} = I_n$. Similarly, $(B^{-1}A^{-1})(AB) = I_n$. Therefore, AB is nonsingular. Since the inverse of a matrix is unique, we conclude that $(AB)^{-1} = B^{-1}A^{-1}$. ●

Corollary 1.1

If $A_1, A_2, \ldots, A_r$ are $n \times n$ nonsingular matrices, then $A_1A_2 \cdots A_r$ is nonsingular and $(A_1A_2 \cdots A_r)^{-1} = A_r^{-1}A_{r-1}^{-1} \cdots A_1^{-1}$.

Proof

Exercise. ●

Theorem 1.7

If A is a nonsingular matrix, then A^{-1} is nonsingular and $(A^{-1})^{-1} = A$.

Proof

Exercise. ●

Theorem 1.8

If A is a nonsingular matrix, then A^T is nonsingular and $(A^{-1})^T = (A^T)^{-1}$.

Proof

We have $AA^{-1} = I_n$. Taking transposes of both sides, we obtain

$$(A^{-1})^T A^T = I_n^T = I_n.$$

Taking transposes of both sides of the equation $A^{-1}A = I_n$, we find, similarly, that

$$(A^T)(A^{-1})^T = I_n.$$

These equations imply that $(A^{-1})^T = (A^T)^{-1}$. ●

EXAMPLE 12

If

$$A = \begin{bmatrix} 1 & 2 \\ 3 & 4 \end{bmatrix},$$

then from Example 10

$$A^{-1} = \begin{bmatrix} -2 & 1 \\ \frac{3}{2} & -\frac{1}{2} \end{bmatrix} \quad \text{and} \quad (A^{-1})^T = \begin{bmatrix} -2 & \frac{3}{2} \\ 1 & -\frac{1}{2} \end{bmatrix}.$$

Also (verify),

$$A^T = \begin{bmatrix} 1 & 3 \\ 2 & 4 \end{bmatrix} \quad \text{and} \quad (A^T)^{-1} = \begin{bmatrix} -2 & \frac{3}{2} \\ 1 & -\frac{1}{2} \end{bmatrix}.$$

■

Suppose that A is nonsingular. Then $AB = AC$ implies that $B = C$ (Exercise 36), and $AB = O_n$ implies that $B = O_n$ (Exercise 37).

It follows from Theorem 1.8 that if A is a symmetric nonsingular matrix, then A^{-1} is symmetric (see Exercise 40).

Linear Systems and Inverses

If A is an $n \times n$ matrix, then the linear system $AX = B$ is a system of n equations in n unknowns. Suppose that A is nonsingular. Then A^{-1} exists and we can multiply $AX = B$ by A^{-1} on both sides, obtaining

$$A^{-1}(AX) = A^{-1}B,$$

or

$$I_n X = X = A^{-1}B. \tag{2}$$

Moreover, $X = A^{-1}B$ is clearly a solution to the given linear system. Thus, if A is nonsingular, we have a unique solution.

If A is a nonsingular $n \times n$ matrix, Equation (2) implies that if the linear system $AX = B$ needs to be solved repeatedly for different B's, we need only compute A^{-1} once; then whenever we change B, we find the corresponding solution X by forming $A^{-1}B$. Although this is certainly a valid approach, its value is of a more theoretical rather than practical nature, since a more efficient procedure for solving such problems is presented in Section 1.8.

EXAMPLE 13

Suppose that A is the matrix of Example 10 so that

$$A^{-1} = \begin{bmatrix} -2 & 1 \\ \frac{3}{2} & -\frac{1}{2} \end{bmatrix}.$$

If

$$B = \begin{bmatrix} 8 \\ 6 \end{bmatrix},$$

then the solution to the linear system $AX = B$ is

$$X = A^{-1}B = \begin{bmatrix} -2 & 1 \\ \frac{3}{2} & -\frac{1}{2} \end{bmatrix} \begin{bmatrix} 8 \\ 6 \end{bmatrix} = \begin{bmatrix} -10 \\ 9 \end{bmatrix}.$$

On the other hand, if

$$B = \begin{bmatrix} 10 \\ 20 \end{bmatrix},$$

then

$$X = A^{-1}\begin{bmatrix} 10 \\ 20 \end{bmatrix} = \begin{bmatrix} 0 \\ 5 \end{bmatrix}.$$

■

1.4 Exercises

1. (a) Show that if A is any $m \times n$ matrix, then $I_m A = A$ and $AI_n = A$.

(b) Show that if A is an $n \times n$ scalar matrix, then $A = rI_n$ for some real number r.

2. Prove that the sum, product, and scalar multiple of diagonal, scalar, and upper (lower) triangular matrices is diagonal, scalar, and upper (lower) triangular, respectively.

3. Prove: If A and B are $n \times n$ diagonal matrices, then $AB = BA$.

4. Let

$$A = \begin{bmatrix} 3 & 2 & -1 \\ 0 & -4 & 3 \\ 0 & 0 & 0 \end{bmatrix} \quad \text{and} \quad B = \begin{bmatrix} 6 & -3 & 2 \\ 0 & 2 & 4 \\ 0 & 0 & 3 \end{bmatrix}.$$

Verify that $A + B$ and AB are upper triangular.

5. Describe all matrices that are both upper and lower triangular.

6. Let p and q be nonnegative integers and let A be a square matrix. Show that

$$A^p A^q = A^{p+q} \quad \text{and} \quad (A^p)^q = A^{pq}.$$

7. If $AB = BA$ and p is a nonnegative integer, show that $(AB)^p = A^p B^p$.

8. If p is a nonnegative integer and c is a scalar, show that $(cA)^p = c^p A^p$.

9. Find a 2×2 matrix $B \neq O_2$ and $B \neq I_2$ such that $AB = BA$, where $A = \begin{bmatrix} 1 & 2 \\ 2 & 1 \end{bmatrix}$. How many such matrices B are there?

10. Find a 2×2 matrix $B \neq O_2$ and $B \neq I_2$ such that $AB = BA$, where $A = \begin{bmatrix} 1 & 2 \\ 0 & 1 \end{bmatrix}$. How many such matrices B are there?

11. Prove or disprove: For any $n \times n$ matrix A, $A^T A = AA^T$.

12. (a) Show that A is symmetric if and only if $a_{ij} = a_{ji}$ for all i, j.

(b) Show that A is skew symmetric if and only if $a_{ij} = -a_{ji}$ for all i, j.

(c) Show that if A is skew symmetric, then the elements on the main diagonal of A are all zero.

13. Show that if A is a symmetric matrix, then A^T is symmetric.

14. Describe all skew symmetric scalar matrices.

15. Show that if A is any $m \times n$ matrix, then AA^T and $A^T A$ are symmetric.

16. Show that if A is any $n \times n$ matrix, then:

(a) $A + A^T$ is symmetric.

(b) $A - A^T$ is skew symmetric.

17. Show that if A is a symmetric matrix, then A^k, $k = 2, 3, \ldots$, is symmetric.

18. Let A and B be symmetric matrices.

(a) Show that $A + B$ is symmetric.

(b) Show that AB is symmetric if and only if $AB = BA$.

19. (a) Show that if A is an upper triangular matrix, then A^T is lower triangular.

(b) Show that if A is a lower triangular matrix, then A^T is upper triangular.

20. If A is a skew symmetric matrix, what type of matrix is A^T? Justify your answer.

21. Show that if A is an $n \times n$ matrix, then $A = S + K$, where S is symmetric and K is skew symmetric. Also show that this decomposition is unique. (*Hint*: Use Exercise 16.)

22. Let

$$A = \begin{bmatrix} 1 & 3 & -2 \\ 4 & 6 & 2 \\ 5 & 1 & 3 \end{bmatrix}.$$

find the matrices S and K described in Exercise 21.

23. Show that the matrix $A = \begin{bmatrix} 2 & 3 \\ 4 & 6 \end{bmatrix}$ is singular.

24. If $D = \begin{bmatrix} 4 & 0 & 0 \\ 0 & -2 & 0 \\ 0 & 0 & 3 \end{bmatrix}$, find D^{-1}.

25. Find the inverse of each of the following matrices.

(a) $A = \begin{bmatrix} 1 & 3 \\ 5 & 2 \end{bmatrix}$. (b) $A = \begin{bmatrix} 1 & 2 \\ 2 & 1 \end{bmatrix}$.

26. If A is a nonsingular matrix whose inverse is $\begin{bmatrix} 2 & 1 \\ 4 & 1 \end{bmatrix}$, find A.

27. If

$$A^{-1} = \begin{bmatrix} 3 & 2 \\ 1 & 3 \end{bmatrix} \quad \text{and} \quad B^{-1} = \begin{bmatrix} 2 & 5 \\ 3 & -2 \end{bmatrix},$$

find $(AB)^{-1}$.

28. Suppose that

$$A^{-1} = \begin{bmatrix} 1 & 2 \\ 1 & 3 \end{bmatrix}.$$

Solve the linear system $AX = B$ for each of the following matrices B.

(a) $\begin{bmatrix} 4 \\ 6 \end{bmatrix}$. (b) $\begin{bmatrix} 8 \\ 15 \end{bmatrix}$.

29. Consider the linear system $AX = B$, where A is the matrix defined in Exercise 25(a).

(a) Find a solution if $B = \begin{bmatrix} 3 \\ 4 \end{bmatrix}$.

(b) Find a solution if $B = \begin{bmatrix} 5 \\ 6 \end{bmatrix}$.

30. Find two 2×2 singular matrices whose sum is nonsingular.

31. Find two 2×2 nonsingular matrices whose sum is singular.

32. Prove Corollary 1.1.

33. Prove Theorem 1.7.

34. Prove that if one row (column) of the $n \times n$ matrix A consists entirely of zeros, then A is singular. (*Hint*: Assume that A is nonsingular; that is, there exists an $n \times n$ matrix B such that $AB = BA = I_n$. Establish a contradiction.)

35. Prove: If A is a diagonal matrix with nonzero diagonal entries $a_{11}, a_{22}, \ldots, a_{nn}$, then A is nonsingular and A^{-1} is a diagonal matrix with diagonal entries $1/a_{11}, 1/a_{22}, \ldots, 1/a_{nn}$.

36. Show that if $AB = AC$ and A is nonsingular, then $B = C$.

37. Show that if A is nonsingular and $AB = O_n$ for an $n \times n$ matrix B, then $B = O_n$.

38. Let $A = \begin{bmatrix} a & b \\ c & d \end{bmatrix}$. Show that A is nonsingular if and only if $ad - bc \neq 0$.

39. Consider the homogeneous system $AX = O$, where A is $n \times n$. If A is nonsingular, show that the only solution is the trivial one, $X = O$.

40. Prove that if A is symmetric and nonsingular, then A^{-1} is symmetric.

41. Formulate the method for adding partitioned matrices, and verify your method by partitioning the matrices

$$A = \begin{bmatrix} 1 & 3 & -1 \\ 2 & 1 & 0 \\ 2 & -3 & 1 \end{bmatrix} \quad \text{and} \quad B = \begin{bmatrix} 3 & 2 & 1 \\ -2 & 3 & 1 \\ 4 & 1 & 5 \end{bmatrix}$$

in two different ways and finding their sum.

42. Let A and B be the following matrices:

$$A = \begin{bmatrix} 2 & 1 & 3 & 4 & 2 \\ 1 & 2 & 3 & -1 & 4 \\ 2 & 3 & 2 & 1 & 4 \\ 5 & -1 & 3 & 2 & 6 \\ 3 & 1 & 2 & 4 & 6 \\ 2 & -1 & 3 & 5 & 7 \end{bmatrix}$$

and

$$B = \begin{bmatrix} 1 & 2 & 3 & 4 & 1 \\ 2 & 1 & 3 & 2 & -1 \\ 1 & 5 & 4 & 2 & 3 \\ 2 & 1 & 3 & 5 & 7 \\ 3 & 2 & 4 & 6 & 1 \end{bmatrix}.$$

Find AB by partitioning A and B in two different ways.

43. What type of matrix is a linear combination of symmetric matrices? (See Section 1.2.) Justify your answer.

44. What type of matrix is a linear combination of scalar matrices? (See Section 1.2.) Justify your answer.

45. For the software you are using, determine the command(s) or procedures required to do the following.

(a) Adjoin a row or column to an existing matrix.

(b) Construct the partitioned matrix

$$\begin{bmatrix} A & O \\ O & B \end{bmatrix}$$

from existing matrices A and B using appropriate size zero matrices.

(c) Extract a submatrix from an existing matrix.

46. Most software for linear algebra has specific commands for extracting the diagonal, upper triangular part, and lower triangular part of a matrix. Determine the corresponding commands for the software that you are using, and experiment with them.

47. Determine the command for computing the inverse of a matrix in the software you use. Usually, if such a command is applied to a singular matrix, a warning message is displayed. Experiment with your inverse command to determine which of the following matrices are singular.

(a) $\begin{bmatrix} 1 & 2 & 3 \\ 4 & 5 & 6 \\ 7 & 8 & 0 \end{bmatrix}$. (b) $\begin{bmatrix} 1 & 2 & 3 \\ 4 & 5 & 6 \\ 7 & 8 & 9 \end{bmatrix}$.

(c) $\begin{bmatrix} 1 & 2 & 4 \\ -1 & 1 & -1 \\ 2 & -1 & 3 \end{bmatrix}$.

48. If B is the inverse of $n \times n$ matrix A, then Theorem 1.9 guarantees that $AB = BA = I_n$. The unstated assumption is that exact arithmetic is used. If computer arithmetic is used to compute AB, then AB need not equal I_n and, in fact, BA need not equal AB. However, both AB and BA should be close to I_n. In your software use the inverse command (see Exercise 47) and form the products AB and BA for the following matrices.

(a) $A = \begin{bmatrix} 1 & \frac{1}{3} \\ 0 & \frac{1}{3} \end{bmatrix}$. (b) $A = \begin{bmatrix} \frac{1}{2} & \frac{1}{4} \\ \frac{1}{4} & \frac{1}{2} \end{bmatrix}$.

(c) $A = \begin{bmatrix} 1 & \frac{1}{2} & \frac{1}{3} \\ \frac{1}{2} & \frac{1}{3} & \frac{1}{4} \\ \frac{1}{3} & \frac{1}{4} & \frac{1}{5} \end{bmatrix}$.

49. In Section 1.1 we studied the method of elimination for solving linear systems $AX = B$. In Equation (2) of this section we showed that the solution is given by $X = A^{-1}B$, if A is nonsingular. Using your software's command for automatically solving linear systems and its inverse command, compare these two solution techniques on the following linear systems.

(a) $A = \begin{bmatrix} \frac{1}{3} & \frac{2}{3} & \frac{4}{3} \\ 0 & \frac{2}{3} & \frac{4}{3} \\ 0 & 0 & \frac{5}{3} \end{bmatrix}$, $B = \begin{bmatrix} 2 \\ 2 \\ \frac{10}{3} \end{bmatrix}$.

(b) $A = \begin{bmatrix} a_{ij} \end{bmatrix}$, where

$$a_{ij} = \frac{1}{i + j - 1},$$

$i, j = 1, 2, \ldots, 10$, and $B =$ the first column of I_{10}.

1.5 ECHELON FORM OF A MATRIX

In this section we take the elimination method for solving linear systems, learned in high school, and systematize it by introducing the language of matrices. This will result in two methods for solving a system of m linear equations in n unknowns. These methods take the augmented matrix of the linear system, perform certain operations on it, and obtain a new matrix that represents an equivalent linear system (i.e., a system that has the same solutions as the original linear system). The important point here is that the latter linear system can be solved more easily.

To see how a linear system whose augmented matrix has a particular form can be readily solved, suppose that

$$\left[\begin{array}{ccc|c} 1 & 2 & 0 & 3 \\ 0 & 1 & 1 & 2 \\ 0 & 0 & 1 & -1 \end{array}\right]$$

represents the augmented matrix of a linear system. Then the solution is quickly found from the corresponding equations

$$\begin{aligned} x_1 + 2x_2 \qquad &= \ \ 3 \\ x_2 + x_3 &= \ \ 2 \\ x_3 &= -1. \end{aligned}$$

as

$$\begin{aligned} x_3 &= -1 \\ x_2 &= 2 - x_3 = 2 + 1 = 3 \\ x_1 &= 3 - 2x_2 = 3 - 6 = -3. \end{aligned}$$

The task of this section is to manipulate the augmented matrix representing a given linear system into a form from which the solution can be found more easily.

Definition 1.10

An $m \times n$ matrix A is said to be in **reduced row echelon form** if it satisfies the following properties:

(a) All zero rows, if there are any, appear at the bottom of the matrix.

(b) The first entry from the left of a nonzero row is a 1. This entry is called a **leading one** of its row.

(c) For each nonzero row, the leading one appears to the right and below any leading one's in preceding rows.

(d) If a column contains a leading one, then all other entries in that column are zero. ▲

A matrix in reduced row echelon form appears as a staircase ("echelon") pattern of leading ones descending from the upper left corner of the matrix.

An $m \times n$ matrix satisfying properties (a), (b), and (c) is said to be in **row echelon form**. In Definition 1.10, there may be no zero rows.

A similar definition can be formulated in the obvious manner for **reduced column echelon form** and **column echelon form**.

EXAMPLE 1

The following are matrices in reduced row echelon form since they satisfy properties (a), (b), (c), and (d):

$$A = \begin{bmatrix} 1 & 0 & 0 & 0 \\ 0 & 1 & 0 & 0 \\ 0 & 0 & 1 & 0 \\ 0 & 0 & 0 & 1 \end{bmatrix}, \quad B = \begin{bmatrix} 1 & 0 & 0 & 0 & -2 & 4 \\ 0 & 1 & 0 & 0 & 4 & 8 \\ 0 & 0 & 0 & 1 & 7 & -2 \\ 0 & 0 & 0 & 0 & 0 & 0 \\ 0 & 0 & 0 & 0 & 0 & 0 \end{bmatrix},$$

and

$$C = \begin{bmatrix} 1 & 2 & 0 & 0 & 1 \\ 0 & 0 & 1 & 2 & 3 \\ 0 & 0 & 0 & 0 & 0 \end{bmatrix}.$$

The following matrices are not in reduced row echelon form. (Why not?)

$$D = \begin{bmatrix} 1 & 2 & 0 & 4 \\ 0 & 0 & 0 & 0 \\ 0 & 0 & 1 & -3 \end{bmatrix}, \quad E = \begin{bmatrix} 1 & 0 & 3 & 4 \\ 0 & 2 & -2 & 5 \\ 0 & 0 & 1 & 2 \end{bmatrix},$$

$$F = \begin{bmatrix} 1 & 0 & 3 & 4 \\ 0 & 1 & -2 & 5 \\ 0 & 1 & 2 & 2 \\ 0 & 0 & 0 & 0 \end{bmatrix}, \quad G = \begin{bmatrix} 1 & 2 & 3 & 4 \\ 0 & 1 & -2 & 5 \\ 0 & 0 & 1 & 2 \\ 0 & 0 & 0 & 0 \end{bmatrix}.$$

■

EXAMPLE 2

The following are matrices in row echelon form:

$$H = \begin{bmatrix} 1 & 5 & 0 & 2 & -2 & 4 \\ 0 & 1 & 0 & 3 & 4 & 8 \\ 0 & 0 & 0 & 1 & 7 & -2 \\ 0 & 0 & 0 & 0 & 0 & 0 \\ 0 & 0 & 0 & 0 & 0 & 0 \end{bmatrix}, \quad I = \begin{bmatrix} 1 & 0 & 0 & 0 \\ 0 & 1 & 0 & 0 \\ 0 & 0 & 1 & 0 \\ 0 & 0 & 0 & 1 \end{bmatrix},$$

and

$$J = \begin{bmatrix} 0 & 0 & 1 & 3 & 5 & 7 & 9 \\ 0 & 0 & 0 & 0 & 1 & -2 & 3 \\ 0 & 0 & 0 & 0 & 0 & 1 & 2 \\ 0 & 0 & 0 & 0 & 0 & 0 & 1 \\ 0 & 0 & 0 & 0 & 0 & 0 & 0 \end{bmatrix}.$$

■

A useful property of matrices in reduced row echelon form (see Exercise 3) is that if A is an $n \times n$ matrix in reduced row echelon form $\neq I_n$, then A has a row consisting entirely of zeros.

We shall now show that every matrix can be put into row (column) echelon form, or into reduced row (column) echelon form, by means of certain row (column) operations.

Definition 1.11

An **elementary row (column) operation** on a matrix A is any one of the following operations:

(a) **Type I:** Interchange any two rows (columns).

(b) **Type II:** Multiply a row (column) by a nonzero number.

(c) **Type III:** Add a multiple of one row (column) to another. ▲

Observe that when a matrix is viewed as the augmented matrix of a linear system, the elementary row operations are equivalent, respectively, to interchanging two equations, multiplying an equation by a nonzero constant, and adding a multiple of one equation to another equation.

EXAMPLE 3

Let

$$A = \begin{bmatrix} 0 & 0 & 1 & 2 \\ 2 & 3 & 0 & -2 \\ 3 & 3 & 6 & -9 \end{bmatrix}.$$

Interchanging rows 1 and 3 of A, we obtain

$$B = \begin{bmatrix} 3 & 3 & 6 & -9 \\ 2 & 3 & 0 & -2 \\ 0 & 0 & 1 & 2 \end{bmatrix}.$$

Multiplying the third row of A by $\frac{1}{3}$, we obtain

$$C = \begin{bmatrix} 0 & 0 & 1 & 2 \\ 2 & 3 & 0 & -2 \\ 1 & 1 & 2 & -3 \end{bmatrix}.$$

Adding (-2) times row 2 of A to row 3 of A, we obtain

$$D = \begin{bmatrix} 0 & 0 & 1 & 2 \\ 2 & 3 & 0 & -2 \\ -1 & -3 & 6 & -5 \end{bmatrix}.$$

Observe that in obtaining D from A, row 2 of A *did not change*. ■

Definition 1.12

An $m \times n$ matrix B is said to be **row (column) equivalent** to an $m \times n$ matrix A if B can be obtained by applying a finite sequence of elementary row (column) operations to A. ▲

EXAMPLE 4

Let

$$A = \begin{bmatrix} 1 & 2 & 4 & 3 \\ 2 & 1 & 3 & 2 \\ 1 & -2 & 2 & 3 \end{bmatrix}.$$

If we add 2 times row 3 of A to its second row, we obtain

$$B = \begin{bmatrix} 1 & 2 & 4 & 3 \\ 4 & -3 & 7 & 8 \\ 1 & -2 & 2 & 3 \end{bmatrix},$$

so B is row equivalent to A.

Interchanging rows 2 and 3 of B, we obtain

$$C = \begin{bmatrix} 1 & 2 & 4 & 3 \\ 1 & -1 & 2 & 3 \\ 4 & -3 & 7 & 8 \end{bmatrix},$$

so C is row equivalent to B.

Multiplying row 1 of C by 2, we obtain

$$D = \begin{bmatrix} 2 & 4 & 8 & 6 \\ 1 & -1 & 2 & 3 \\ 4 & -3 & 7 & 8 \end{bmatrix},$$

so D is row equivalent to C. It then follows that D is row equivalent to A, since we obtained D by applying three successive elementary row operations to A. ■

We can readily show (see Exercise 4) that (a) every matrix is row equivalent to itself; (b) if B is row equivalent to A, then A is row equivalent to B; and (c) if C is row equivalent to B and B is row equivalent to A, then C is row equivalent to A. In view of (b), both statements "B is row equivalent to A" and "A is row equivalent to B" can be replaced by "A and B are row equivalent." A similar statement holds for column equivalence.

Theorem 1.9

Every nonzero $m \times n$ matrix $A = \left[a_{ij}\right]$ is row (column) equivalent to a matrix in row (column) echelon form.

Proof

We shall prove that A is row equivalent to a matrix in row echelon form. That is, by using only elementary row operations we can transform A into a matrix in row echelon form. A completely analogous proof using elementary column operations establishes the result for column equivalence.

We start by looking in matrix A for the first column with a nonzero entry. This column is called the **pivot column**, the first nonzero entry in the pivot column is called the **pivot**. Suppose the pivot column is column j and that the pivot occurs in row i. Now interchange, if necessary, rows 1 and i, getting matrix $B = \left[b_{ij}\right]$. Thus the pivot b_{1j} is $\neq 0$. Multiply the first row of B by the reciprocal of the pivot, that is, by $1/b_{1j}$, obtaining matrix $C = \left[c_{ij}\right]$. Note that $c_{1j} = 1$. Now if c_{hj}, $2 \leq h \leq m$, is not zero, then to row h of C we add $-c_{hj}$ times row 1; we do this for each value of h. It follows that the elements in column j, in rows $2, 3, \ldots, m$ of C are zero. Denote the resulting matrix by D.

Next, consider the $(m-1) \times n$ submatrix A_1 of D obtained by deleting the first row of D. We now repeat the procedure above with matrix A_1 instead of matrix A. Continuing this way, we obtain a matrix in row echelon form that is row equivalent to A. ●

EXAMPLE 5

Let

$$A = \begin{bmatrix} 0 & 2 & 3 & -4 & 1 \\ 0 & 0 & 2 & 3 & 4 \\ \textcircled{2} & 2 & -5 & 2 & 4 \\ 2 & 0 & -6 & 9 & 7 \end{bmatrix}.$$

Pivot column (column 1) — **Pivot** ($a_{31} = 2$)

Column 1 is the first (counting from left to right) column in A with a nonzero entry, so column 1 is the pivot column of A. The first (counting from top to bottom) nonzero entry in the pivot column occurs in the third row, so the pivot is $a_{31} = 2$. We interchange the first and third rows of A, obtaining

$$B = \begin{bmatrix} \textcircled{2} & 2 & -5 & 2 & 4 \\ 0 & 0 & 2 & 3 & 4 \\ 0 & 2 & 3 & -4 & 1 \\ 2 & 0 & -6 & 9 & 7 \end{bmatrix}.$$

Multiply the first row of B by the reciprocal of the pivot, that is, by $\dfrac{1}{b_{11}} = \dfrac{1}{2}$, to obtain

$$C = \begin{bmatrix} 1 & 1 & -\frac{5}{2} & 1 & 2 \\ 0 & 0 & 2 & 3 & 4 \\ 0 & 2 & 3 & -4 & 1 \\ 2 & 0 & -6 & 9 & 7 \end{bmatrix}.$$

Add (-2) times the first row of C to the fourth row of C to produce a matrix D in which the only nonzero entry in the pivot column is $d_{11} = 1$:

$$D = \begin{bmatrix} 1 & 1 & -\frac{5}{2} & 1 & 2 \\ 0 & 0 & 2 & 3 & 4 \\ 0 & 2 & 3 & -4 & 1 \\ 0 & -2 & -1 & 7 & 3 \end{bmatrix}.$$

Identify A_1 as the submatrix of D obtained by deleting the first row of D: do not erase the first row of D. Repeat the steps above with A_1 instead of A.

$$\begin{array}{ccccc} 1 & 1 & -\frac{5}{2} & 1 & 2 \end{array}$$

$$A_1 = \begin{bmatrix} 0 & 0 & 2 & 3 & 4 \\ 0 & \textcircled{2} & 3 & -4 & 1 \\ 0 & -2 & -1 & 7 & 3 \end{bmatrix}.$$

Pivot — Pivot column

Interchange the first and second rows of A_1 to obtain

$$\begin{array}{ccccc} 1 & 1 & -\frac{5}{2} & 1 & 2 \end{array}$$

$$B_1 = \begin{bmatrix} 0 & 2 & 3 & -4 & 1 \\ 0 & 0 & 2 & 3 & 4 \\ 0 & -2 & -1 & 7 & 3 \end{bmatrix}.$$

Multiply the first row of B_1 by $\frac{1}{2}$ to obtain

$$\begin{array}{ccccc} 1 & 1 & -\frac{5}{2} & 1 & 2 \end{array}$$

$$C_1 = \begin{bmatrix} 0 & 1 & \frac{3}{2} & -2 & \frac{1}{2} \\ 0 & 0 & 2 & 3 & 4 \\ 0 & -2 & -1 & 7 & 3 \end{bmatrix}.$$

Add 2 times the first row of C_1 to its third row to obtain

$$\begin{array}{ccccc} 1 & 1 & -\frac{5}{2} & 1 & 2 \end{array}$$

$$D_1 = \begin{bmatrix} 0 & 1 & \frac{3}{2} & -2 & \frac{1}{2} \\ 0 & 0 & 2 & 3 & 4 \\ 0 & 0 & 2 & 3 & 4 \end{bmatrix}.$$

Deleting the first row of D_1 yields the matrix A_2. We repeat the procedure above with A_2 instead of A. No rows of A_2 have to be interchanged.

$$\begin{array}{ccccc} 1 & 1 & -\frac{5}{2} & 1 & 2 \\ 0 & 1 & \frac{3}{2} & -2 & \frac{1}{2} \end{array}$$

$$A_2 = \begin{bmatrix} 0 & 0 & \textcircled{2} & 3 & 4 \\ 0 & 0 & 2 & 3 & 4 \end{bmatrix} = B_2.$$

Pivot — Pivot column of A_2

Multiply the first row of B_2 by $\frac{1}{2}$ to obtain

$$\begin{array}{ccccc} 1 & 1 & -\frac{5}{2} & 1 & 2 \\ 0 & 1 & \frac{3}{2} & -2 & \frac{1}{2} \end{array}$$

$$C_2 = \begin{bmatrix} 0 & 0 & 1 & \frac{3}{2} & 2 \\ 0 & 0 & 2 & 3 & 4 \end{bmatrix}.$$

Finally, add (-2) times the first row of C_2 to its second row to obtain

$$\begin{array}{ccccc} 1 & 1 & -\frac{5}{2} & 1 & 2 \\ 0 & 1 & \frac{3}{2} & -2 & \frac{1}{2} \end{array}$$

$$D_2 = \begin{bmatrix} 0 & 0 & 1 & \frac{3}{2} & 2 \\ 0 & 0 & 0 & 0 & 0 \end{bmatrix}.$$

The matrix

$$H = \begin{bmatrix} 1 & 1 & -\frac{5}{2} & 1 & 2 \\ 0 & 1 & \frac{3}{2} & -2 & \frac{1}{2} \\ 0 & 0 & 1 & \frac{3}{2} & 2 \\ 0 & 0 & 0 & 0 & 0 \end{bmatrix}$$

is in row echelon form and is row equivalent to A. ■

When doing hand computations, it is sometimes possible to avoid messy fractions by suitably modifying the steps in the procedure.

Theorem 1.10

Every nonzero $m \times n$ matrix $A = \left[a_{ij}\right]$ is row (column) equivalent to a unique matrix in reduced row (column) echelon form.

Proof

We proceed as in Theorem 1.9, obtaining matrix H in row echelon form that is row equivalent to A. Suppose that rows $1, 2, \ldots, r$ of H are nonzero and that the leading ones in these rows occur in columns $c_1, c_2, \ldots, c_r$. Then $c_1 < c_2 < \cdots < c_r$. Starting with the last nonzero row of H, we add suitable multiples of this row to all rows above it to make all entries in column c_r above the leading one in row r equal to zero. We repeat this process with rows $r-1, r-2, \ldots,$ and 2, making all entries above a leading one equal to zero. The result is a matrix K in reduced row echelon form which has been obtained from H by elementary row operations, and is thus row equivalent to H. Since A is row equivalent to H, and H is row equivalent to K, then A is row equivalent to K. An analogous proof can be given to show that A is column equivalent to a matrix in reduced column echelon form. It can be shown, with some difficulty, that there is only one matrix in reduced row echelon form that is row equivalent to a given matrix. For a proof, see K. Hoffman and R. Kunze, *Linear Algebra*, 2nd ed. (Englewood Cliffs, N.J.: Prentice Hall, 1971). ●

Remark It should be noted that a row echelon form of a matrix is not unique.

EXAMPLE 6

Find a matrix in reduced row echelon form that is row equivalent to the matrix A of Example 5.

Solution We start with the matrix H which is row equivalent to A that we obtained in the solution to Example 5. We now add $(-\frac{3}{2})$ times the third row of H to its second row:

$$J_1 = \begin{bmatrix} 1 & 1 & -\frac{5}{2} & 1 & 2 \\ 0 & 1 & 0 & -\frac{17}{4} & -\frac{5}{2} \\ 0 & 0 & 1 & \frac{3}{2} & 2 \\ 0 & 0 & 0 & 0 & 0 \end{bmatrix}.$$

Next, we add $\frac{5}{2}$ times the third row of J_1 to its first row:

$$J_2 = \begin{bmatrix} 1 & 1 & 0 & \frac{19}{4} & 7 \\ 0 & 1 & 0 & -\frac{17}{4} & -\frac{5}{2} \\ 0 & 0 & 1 & \frac{3}{2} & 2 \\ 0 & 0 & 0 & 0 & 0 \end{bmatrix}.$$

Finally, we add (-1) times the second row of J_2 to its first row:

$$K = \begin{bmatrix} 1 & 0 & 0 & 9 & \frac{19}{2} \\ 0 & 1 & 0 & -\frac{17}{4} & -\frac{5}{2} \\ 0 & 0 & 1 & \frac{3}{2} & 2 \\ 0 & 0 & 0 & 0 & 0 \end{bmatrix},$$

which is in reduced row echelon form and is row equivalent to A. ■

Remark The procedure given here for finding a matrix K in reduced row echelon form that is row equivalent to a given matrix A is not the only one possible. For example, instead of first obtaining a matrix H in row echelon form that is row equivalent to A and then transforming H to reduced row echelon form, we could proceed as follows. First, zero out the entries below a leading 1 and then immediately zero out the entries above the leading 1. This procedure is not as efficient as the procedure given above.

We now apply these results to the solution of linear systems.

Theorem 1.11

Let $AX = B$ and $CX = D$ be two linear systems, each of m equations in n unknowns. If the augmented matrices $\left[A \mid B\right]$ and $\left[C \mid D\right]$ are row equivalent, then the linear systems are equivalent; that is, they have exactly the same solutions.

Proof

This follows from the definition of row equivalence and from the fact that the three elementary row operations on the augmented matrix are the three manipulations on linear systems, discussed in Section 1.1, which yield equivalent linear systems. We also note that if one system has no solution, then the other system has no solution. ●

Corollary 1.2

If A and C are row equivalent $m \times n$ matrices, then the homogeneous systems $AX = O$ and $CX = O$ are equivalent.

Proof

Exercise. ●

Solving Linear Systems

We now pause to observe that we have developed the essential features of two very straightforward methods for solving linear systems. The idea consists of starting with the linear system $AX = B$, then obtaining a partitioned matrix $\left[C \mid D\right]$ in either row echelon form or reduced row echelon form that is row equivalent to the augmented matrix $\left[A \mid B\right]$. Now $\left[C \mid D\right]$ represents the linear system $CX = D$, which is quite simple to solve because of the structure of $\left[C \mid D\right]$, and the set of solutions to this system gives precisely the set of solutions to $AX = B$. The method where $\left[C \mid D\right]$ is in row echelon form is called **Gaussian elimination**; the method where $\left[C \mid D\right]$ is in reduced row echelon form is called **Gauss†–Jordan‡ reduc-**

†Carl Friedrich Gauss (1777–1855) was born into a poor working-class family in Brunswick, Germany, and died in Göttingen, Germany, the most famous mathematician in the world. He was a child prodigy with a genius that did not impress his father, who called him a "star-gazer." However, his teachers were impressed enough to arrange for the Duke of Brunswick to provide a scholarship for Gauss at the local secondary school. As a teenager there, he made original discoveries in number theory and began to speculate about non-Euclidean geometry. His scientific publications include important contributions in number theory, mathematical astronomy, mathematical geography, statistics, differential geometry, and magnetism. His diaries and private notes contain many other discoveries that he never published.

An austere, conservative man who had few friends and whose private life was generally unhappy, he was very concerned that proper credit be given for scientific discoveries. When he relied on the results of others, he was careful to acknowledge them; and when others independently discovered results in his private notes, he was quick to claim priority.

In his research Gauss used a method of calculation that later generations generalized to row reduction of matrices and named in his honor, although the method was used in China almost 2000 years earlier.

‡Wilhelm Jordan (1842–1899) was born in southern Germany. He attended college in Stuttgart and in 1868 became full professor of geodesy at the technical college in Karlsruhe, Germany. He participated in surveying several regions of Germany. Jordan was a prolific writer whose major work, *Handbuch der Vermessungskunde* (*Handbook of Geodesy*), was translated into French, Italian, and Russian. He was considered a superb writer and an excellent teacher. Unfortunately, the Gauss–Jordan reduction

tion. Strictly speaking, the original Gauss–Jordan reduction was more along the lines described in the Remark above. The version presented in this book is more efficient. In actual practice, neither Gaussian elimination nor Gauss–Jordan reduction are used as much as the method involving the LU-factorization of A that is discussed in Section 1.8. However, Gaussian elimination and Gauss–Jordan reduction are fine for small problems, and we use the latter heavily in this book.

Gaussian elimination consists of two steps:

STEP 1 The transformation of the augmented matrix $\left[A \mid B\right]$ to the matrix $\left[C \mid D\right]$ in row echelon form using elementary row operations.

STEP 2 Solution of the linear system corresponding to the augmented matrix $\left[C \mid D\right]$ using **back substitution**.

For the case in which A is $n \times n$, and the linear system $AX = B$ has a unique solution, the matrix $\left[C \mid D\right]$ has the following form:

$$\left[\begin{array}{cccccc|c} 1 & c_{12} & c_{13} & \cdots & & c_{1n} & d_1 \\ 0 & 1 & c_{23} & \cdots & & c_{2n} & d_2 \\ \vdots & \vdots & \vdots & & & \vdots & \vdots \\ 0 & 0 & 0 & \cdots & 1 & c_{n-1\,n} & d_{n-1} \\ 0 & 0 & 0 & \cdots & 0 & 1 & d_n \end{array}\right].$$

(The remaining cases are treated below.) This augmented matrix represents the linear system

$$\begin{aligned} x_1 + c_{12}x_2 + c_{13}x_3 + \cdots + \quad c_{1n}x_n &= d_1 \\ x_2 + c_{23}x_3 + \cdots + \quad c_{2n}x_n &= d_2 \\ \vdots \qquad\qquad \vdots \quad &\quad \vdots \\ x_{n-1} + c_{n-1\,n}x_n &= d_{n-1} \\ x_n &= d_n. \end{aligned}$$

Back substitution proceeds from the nth equation upward, solving for one variable from each equation:

$$\begin{aligned} x_n &= d_n \\ x_{n-1} &= d_{n-1} - c_{n-1\,n}x_n \\ &\vdots \\ x_2 &= d_2 - c_{23}x_3 - c_{24}x_4 - \cdots - c_{2n}x_n \\ x_1 &= d_1 - c_{12}x_2 - c_{13}x_3 - \cdots - c_{1n}x_n. \end{aligned}$$

method has been widely attributed to Camille Jordan (1838–1922), a well-known French mathematician. Moreover, it seems that the method was also discovered independently at the same time by B. I. Clasen, a priest who lived in Luxembourg. This biographical sketch is based on an excellent article: S. C. Althoen and R. McLaughlin, "Gauss–Jordan reduction: A brief history," *MAA Monthly*, 94 (1987), 130–142.

EXAMPLE 7

The linear system

$$\begin{aligned} x_1 + 2x_2 + 3x_3 &= 9 \\ 2x_1 - x_2 + x_3 &= 8 \\ 3x_1 \phantom{{}+2x_2} - x_3 &= 3 \end{aligned}$$

has the augmented matrix

$$\left[A \;\vdots\; B\right] = \left[\begin{array}{rrr:r} 1 & 2 & 3 & 9 \\ 2 & -1 & 1 & 8 \\ 3 & 0 & -1 & 3 \end{array}\right].$$

Transforming this matrix to row echelon form, we obtain (verify)

$$\left[C \;\vdots\; D\right] = \left[\begin{array}{rrr:r} 1 & 2 & 3 & 9 \\ 0 & 1 & 1 & 2 \\ 0 & 0 & 1 & 3 \end{array}\right].$$

Using back substitution, we now have

$$\begin{aligned} x_3 &= 3 \\ x_2 &= 2 - x_3 = 2 - 3 = -1 \\ x_1 &= 9 - 2x_2 - 3x_3 = 9 + 2 - 9 = 2; \end{aligned}$$

thus the solution is $x_1 = 2$, $x_2 = -1$, $x_3 = 3$. ■

The general case in which A is $m \times n$ is handled in a similar fashion, but we need to elaborate upon several situations than can occur. We thus consider $CX = D$, where C is $m \times n$, and $\left[C \;\vdots\; D\right]$ is in row echelon form. Then, for example, $\left[C \;\vdots\; D\right]$ might be of the following form:

$$\left[\begin{array}{ccccccc:c} 1 & c_{12} & c_{13} & \cdots & & & c_{1n} & d_1 \\ 0 & 0 & 1 & c_{24} & \cdots & & c_{2n} & d_2 \\ \vdots & \vdots & \vdots & \vdots & & & \vdots & \vdots \\ 0 & 0 & \cdots & & 0 & 1 & c_{k-1\,n} & d_{k-1} \\ 0 & \cdots & & & \vdots & 0 & 1 & d_k \\ 0 & \cdots & & & & \vdots & 0 & d_{k+1} \\ \vdots & & & & & & \vdots & \vdots \\ 0 & \cdots & & & & & 0 & d_m \end{array}\right].$$

This augmented matrix represents the linear system

$$\begin{aligned}
x_1 + c_{12}x_2 + c_{13}x_3 + \cdots \qquad\qquad + \quad c_{1n}x_n &= d_1 \\
x_3 + c_{24}x_4 + \cdots + \quad c_{2n}x_n &= d_2 \\
\vdots \qquad\qquad & \\
x_{n-1} + c_{k-1\,n}x_n &= d_{k-1} \\
x_n &= d_k \\
0x_1 + \qquad + \qquad \cdots \qquad + \quad 0x_n &= d_{k+1} \\
\vdots \qquad\qquad \vdots \qquad\qquad & \vdots \\
0x_1 + \qquad + \qquad \cdots \qquad + \quad 0x_n &= d_m.
\end{aligned}$$

First, if $d_{k+1} = 1$, then $CX = D$ has no solution, since at least one equation is not satisfied. If $d_{k+1} = 0$, which implies that $d_{k+2} = \cdots = d_m = 0$ (since $\left[C \mid D\right]$ was assumed to be in row echelon form), we then obtain $x_n = d_k$, $x_{n-1} = d_{k-1} - c_{k-1\,n}x_n = d_{k-1} - c_{k-1\,n}d_k$, and continue using back substitution to find the remaining unknowns corresponding to the leading entry in each row. Of course, in the solution some of the unknowns may be expressed in terms of others that can take on any values whatever. This merely indicates that $CX = D$ has infinitely many solutions. On the other hand, every unknown may have a determined value, indicating that the solution is unique.

EXAMPLE 8

Let

$$\left[C \mid D\right] = \left[\begin{array}{ccccc|c} 1 & 2 & 3 & 4 & 5 & 6 \\ 0 & 1 & 2 & 3 & -1 & 7 \\ 0 & 0 & 1 & 2 & 3 & 7 \\ 0 & 0 & 0 & 1 & 2 & 9 \end{array}\right].$$

Then

$$\begin{aligned}
x_4 &= 9 - 2x_5 \\
x_3 &= 7 - 2x_4 - 3x_5 = 7 - 2(9 - 2x_5) - 3x_5 = -11 + x_5 \\
x_2 &= 7 - 2x_3 - 3x_4 + x_5 = 2 + 5x_5 \\
x_1 &= 6 - 2x_2 - 3x_3 - 4x_4 - 5x_5 = -1 - 10x_5 \\
x_5 &= \text{any real number.}
\end{aligned}$$

Thus all solutions are of the form

$$\begin{aligned}
x_1 &= -1 - 10r \\
x_2 &= 2 + 5r \\
x_3 &= -11 + r \\
x_4 &= 9 - 2r \\
x_5 &= r, \text{ any real number.}
\end{aligned}$$

Since r can be assigned any real number, the given linear system has infinitely many solutions. ■

EXAMPLE 9

If

$$\left[C \mid D\right] = \left[\begin{array}{cccc|c} 1 & 2 & 3 & 4 & 5 \\ 0 & 1 & 2 & 3 & 6 \\ 0 & 0 & 0 & 0 & 1 \end{array}\right],$$

then $CX = D$ has no solution, since the last equation is

$$0x_1 + 0x_2 + 0x_3 + 0x_4 = 1,$$

which can never be satisfied. ■

When using the Gauss–Jordan reduction procedure, we transform the augmented matrix $\left[A \mid B\right]$ to $\left[C \mid D\right]$, which is in reduced row echelon form. This means that we can solve the linear system $CX = D$ without back substitution as the following examples show, but, of course, it takes more effort to put a matrix in reduced row echelon form than to put it in row echelon form. It turns out that the techniques of Gaussian elimination and Gauss-Jordan reduction, as described in this book, require the same number of operations.

EXAMPLE 10

If

$$\left[C \mid D\right] = \left[\begin{array}{cccc|c} 1 & 0 & 0 & 0 & 5 \\ 0 & 1 & 0 & 0 & 6 \\ 0 & 0 & 1 & 0 & 7 \\ 0 & 0 & 0 & 1 & 8 \end{array}\right],$$

then

$$\begin{aligned} x_1 &= 5 \\ x_2 &= 6 \\ x_3 &= 7 \\ x_4 &= 8. \end{aligned}$$

■

EXAMPLE 11

If

$$\left[C \mid D\right] = \left[\begin{array}{ccccc|c} 1 & 1 & 2 & 0 & -\frac{5}{2} & \frac{2}{3} \\ 0 & 0 & 0 & 1 & \frac{1}{2} & \frac{1}{2} \\ 0 & 0 & 0 & 0 & 0 & 0 \end{array}\right],$$

then

$$x_4 = \tfrac{1}{2} - \tfrac{1}{2}x_5$$

$$x_1 = \tfrac{2}{3} - x_2 - 2x_3 + \tfrac{5}{2}x_5,$$

where x_2, x_3, and x_5 can take on any real numbers. Thus a solution is of the form

$$\begin{aligned} x_1 &= \tfrac{2}{3} - r - 2s + \tfrac{5}{2}t \\ x_2 &= r \\ x_3 &= s \\ x_4 &= \tfrac{1}{2} - \tfrac{1}{2}t \\ x_5 &= t, \end{aligned}$$

where r, s, and t are any real numbers. ■

We now solve a linear system both by Gaussian elimination and by Gauss–Jordan reduction.

EXAMPLE 12

Consider the linear system

$$\begin{aligned} x_1 + 2x_2 + 3x_3 &= 6 \\ 2x_1 - 3x_2 + 2x_3 &= 14 \\ 3x_1 + x_2 - x_3 &= -2. \end{aligned}$$

We form the augmented matrix

$$\left[\begin{array}{ccc|c} 1 & 2 & 3 & 6 \\ 2 & -3 & 2 & 14 \\ 3 & 1 & -1 & -2 \end{array}\right].$$

Add (-2) times the first row to the second row:

$$\left[\begin{array}{ccc|c} 1 & 2 & 3 & 6 \\ 0 & -7 & -4 & 2 \\ 3 & 1 & -1 & -2 \end{array}\right].$$

Add (-3) times the first row to the third row:

$$\left[\begin{array}{ccc|c} 1 & 2 & 3 & 6 \\ 0 & -7 & -4 & 2 \\ 0 & -5 & -10 & -20 \end{array}\right].$$

Multiply the third row by $\left(-\frac{1}{5}\right)$ and interchange the second and third rows:

$$\left[\begin{array}{ccc|c} 1 & 2 & 3 & 6 \\ 0 & 1 & 2 & 4 \\ 0 & -7 & -4 & 2 \end{array}\right].$$

Add 7 times the second row to the third row:

$$\left[\begin{array}{ccc|c} 1 & 2 & 3 & 6 \\ 0 & 1 & 2 & 4 \\ 0 & 0 & 10 & 30 \end{array}\right].$$

Multiply the third row by $\frac{1}{10}$:

$$\left[\begin{array}{ccc|c} 1 & 2 & 3 & 6 \\ 0 & 1 & 2 & 4 \\ 0 & 0 & 1 & 3 \end{array}\right].$$

This matrix is in row echelon form. This means that $x_3 = 3$, and from the second row

$$x_2 + 2x_3 = 4$$

so that

$$x_2 = 4 - 2(3) = -2.$$

From the first row

$$x_1 + 2x_2 + 3x_3 = 6,$$

which implies that

$$x_1 = 6 - 2x_2 - 3x_3 = 6 - 2(-2) - 3(3) = 1.$$

Thus $x_1 = 1$, $x_2 = -2$, and $x_3 = 3$ is the solution. This gives the solution by Gaussian elimination.

To solve the given linear system by Gauss–Jordan reduction, we transform the last matrix to $\left[C \mid D\right]$, which is in reduced row echelon form, by the following steps:

Add (-2) times the third row to the second row:

$$\left[\begin{array}{ccc|c} 1 & 2 & 3 & 6 \\ 0 & 1 & 0 & -2 \\ 0 & 0 & 1 & 3 \end{array}\right].$$

Now add (-3) times the third row to the first row:

$$\left[\begin{array}{ccc|c} 1 & 2 & 0 & -3 \\ 0 & 1 & 0 & -2 \\ 0 & 0 & 1 & 3 \end{array}\right].$$

Finally, add (-2) times the second row to the first row:

$$\left[\begin{array}{ccc|c} 1 & 0 & 0 & 1 \\ 0 & 1 & 0 & -2 \\ 0 & 0 & 1 & 3 \end{array}\right].$$

The solution is $x_1 = 1$, $x_2 = -2$, and $x_3 = 3$, as before. ■

Remark In both Gaussian elimination and Gauss–Jordan reduction, we can only use row operations. Do not try to use any column operations.

Homogeneous Systems

Now we study a homogeneous system $AX = O$ of m linear equations in n unknowns.

EXAMPLE 13

Consider the homogeneous system whose augmented matrix is

$$\left[\begin{array}{ccccc|c} 1 & 0 & 0 & 0 & 2 & 0 \\ 0 & 0 & 1 & 0 & 3 & 0 \\ 0 & 0 & 0 & 1 & 4 & 0 \\ 0 & 0 & 0 & 0 & 0 & 0 \end{array}\right].$$

Since the augmented matrix is in reduced row echelon form, the solution is seen to be

$$\begin{aligned} x_1 &= -2r \\ x_2 &= s \\ x_3 &= -3r \\ x_4 &= -4r \\ x_5 &= r, \end{aligned}$$

where r and s are any real numbers. ■

In Example 13 we solved a homogeneous system of m (= 4) linear equations in n (= 5) unknowns, where $m < n$ and the augmented matrix A was in reduced row echelon form. We can ignore any row of the augmented matrix that consists entirely of zeros. Thus let rows $1, 2, \ldots, r$ of A be the nonzero rows, and let the 1 in row i occur in column c_i. We are then solving a homogeneous system of r equations in n unknowns, $r < n$, and in this special case (A is in reduced row echelon form) we can solve for $x_{c_1}, x_{c_2}, \ldots, x_{c_r}$ in terms of the remaining $n - r$ unknowns. Since the latter can take on any real values, there are infinitely many solutions to the system $AX = O$; in particular, there is a nontrivial solution. We now show that this situation holds whenever we have $m < n$; A does not have to be in reduced row echelon form.

Theorem 1.12

A homogeneous system of m linear equations in n unknowns always has a nontrivial solution if $m < n$, that is, if the number of unknowns exceeds the number of equations.

Proof

Let B be a matrix in reduced row echelon form that is row equivalent to A. Then the homogeneous systems $AX = O$ and $BX = O$ are equivalent. As we have just shown, the system $BX = O$ has a nontrivial solution, and therefore the same is true for the system $AX = O$. ●

We shall soon use this result in the following equivalent form: If A is $m \times n$ and $AX = O$ has only the trivial solution, then $m \geq n$.

EXAMPLE 14

Consider the homogeneous system

$$\begin{aligned} x_1 + x_2 + x_3 + x_4 &= 0 \\ x_1 \qquad\qquad\quad + x_4 &= 0 \\ x_1 + 2x_2 + x_3 \qquad &= 0. \end{aligned}$$

The augmented matrix

$$A = \left[\begin{array}{cccc|c} 1 & 1 & 1 & 1 & 0 \\ 1 & 0 & 0 & 1 & 0 \\ 1 & 2 & 1 & 0 & 0 \end{array}\right]$$

is row equivalent to (verify)

$$\left[\begin{array}{cccc|c} 1 & 0 & 0 & 1 & 0 \\ 0 & 1 & 0 & -1 & 0 \\ 0 & 0 & 1 & 1 & 0 \end{array}\right].$$

Hence the solution is

$$\begin{aligned} x_1 &= -r \\ x_2 &= r \\ x_3 &= -r \\ x_4 &= r, \text{ any real number.} \end{aligned}$$

■

Relationship Between Nonhomogeneous Linear Systems and Homogeneous Systems

Let $AX = B$, $B \neq O$, be a consistent linear system. If X_p is a particular solution to the given nonhomogeneous system and X_h is a solution to the associated homogeneous system $AX = O$, then $X_p + X_h$ is a solution to the given system $AX = B$. Moreover, every solution X to the nonhomogeneous linear system $AX = B$ can be written as $X_p + X_h$, where X_p is a particular solution to the given nonhomogeneous system and X_h is a solution to the associated homogeneous system $AX = O$. For a proof, see Exercise 28.

1.5 Exercises

1. Let

$$A = \begin{bmatrix} 1 & 2 & -3 & 1 \\ -1 & 0 & 3 & 4 \\ 0 & 1 & 2 & -1 \\ 2 & 3 & 0 & -3 \end{bmatrix}.$$

(a) Find matrices B and C in row echelon form that are row equivalent to A.

(b) Find a matrix D in reduced row echelon form that is row equivalent to A.

2. Let

$$A = \begin{bmatrix} 1 & -2 & 0 & 2 \\ 2 & -3 & -1 & 5 \\ 1 & 3 & 2 & 5 \\ 1 & 1 & 0 & 2 \end{bmatrix}.$$

(a) Find matrices B and C in row echelon form that are row equivalent to A.

(b) Find a matrix D in reduced row echelon form that is row equivalent to A.

3. Let A be an $n \times n$ matrix in reduced row echelon form. Prove that if $A \neq I_n$, then A has a row consisting entirely of zeros.

4. Prove:

(a) Every matrix is row equivalent to itself.

(b) If B is row equivalent to A, then A is row equivalent to B.

(c) If C is row equivalent to B and B is row equivalent to A, then C is row equivalent to A.

5. Consider the linear system

$$\begin{aligned} x_1 + x_2 + 2x_3 &= -1 \\ x_1 - 2x_2 + x_3 &= -5 \\ 3x_1 + x_2 + x_3 &= 3. \end{aligned}$$

(a) Find all solutions, if any exist, by using the Gaussian elimination method.

(b) Find all solutions, if any exist, by using the Gauss–Jordan reduction method.

6. Repeat Exercise 5 for each of the following linear systems.

(a) $$\begin{aligned} x_1 + x_2 + 2x_3 + 3x_4 &= 13 \\ x_1 - 2x_2 + x_3 + x_4 &= 8 \\ 3x_1 + x_2 + x_3 - x_4 &= 1. \end{aligned}$$

(b) $$\begin{aligned} x_1 + x_2 + x_3 &= 1 \\ x_1 + x_2 - 2x_3 &= 3 \\ 2x_1 + x_2 + x_3 &= 2. \end{aligned}$$

(c) $$\begin{aligned} 2x_1 + x_2 + x_3 - 2x_4 &= 1 \\ 3x_1 - 2x_2 + x_3 - 6x_4 &= -2 \\ x_1 + x_2 - x_3 - x_4 &= -1 \\ 6x_1 + x_3 - 9x_4 &= -2 \\ 5x_1 - x_2 + 2x_3 - 8x_4 &= 3. \end{aligned}$$

In Exercises 7 through 9, solve the linear system, if it is consistent, with the given augmented matrix.

7. (a) $\left[\begin{array}{ccc|c} 1 & 1 & 1 & 0 \\ 1 & 1 & 0 & 3 \\ 0 & 1 & 1 & 1 \end{array}\right]$. (b) $\left[\begin{array}{ccc|c} 1 & 2 & 3 & 0 \\ 1 & 1 & 1 & 0 \\ 1 & 1 & 2 & 0 \end{array}\right]$.

(c) $\left[\begin{array}{ccc|c} 1 & 2 & 3 & 0 \\ 1 & 1 & 1 & 0 \\ 5 & 7 & 9 & 0 \end{array}\right]$. (d) $\left[\begin{array}{ccc|c} 1 & 2 & 3 & 0 \\ 1 & 2 & 1 & 0 \end{array}\right]$.

8. (a) $\left[\begin{array}{cccc|c} 1 & 2 & 3 & 1 & 8 \\ 1 & 3 & 0 & 1 & 7 \\ 1 & 0 & 2 & 1 & 3 \end{array}\right]$.

(b) $\left[\begin{array}{cccc|c} 1 & 1 & 3 & -3 & 0 \\ 0 & 2 & 1 & -3 & 3 \\ 1 & 0 & 2 & -1 & -1 \end{array}\right]$.

9. (a) $\left[\begin{array}{ccc|c} 1 & 2 & 1 & 7 \\ 2 & 0 & 1 & 4 \\ 1 & 0 & 2 & 5 \\ 1 & 2 & 3 & 11 \\ 2 & 1 & 4 & 12 \end{array}\right]$.

(b) $\left[\begin{array}{ccc|c} 1 & 2 & 1 & 0 \\ 2 & 3 & 0 & 0 \\ 0 & 1 & 2 & 0 \\ 2 & 1 & 4 & 0 \end{array}\right]$.

10. Find a 2×1 matrix X with entries not all zero such that

$$AX = 4X, \quad \text{where } A = \begin{bmatrix} 4 & 1 \\ 0 & 2 \end{bmatrix}.$$

[*Hint*: Rewrite the matrix equation $AX = 4X$ as $4X - AX = (4I_2 - A)X = O$, and solve the homogeneous linear system.]

11. Find a 2×1 matrix X with entries not all zero such that

$$AX = 3X, \quad \text{where } A = \begin{bmatrix} 2 & 1 \\ 1 & 2 \end{bmatrix}.$$

12. Find a 3×1 matrix X with entries not all zero such that

$$AX = 3X, \quad \text{where } A = \begin{bmatrix} 1 & 2 & -1 \\ 1 & 0 & 1 \\ 4 & -4 & 5 \end{bmatrix}.$$

13. Find a 3×1 matrix X with entries not all zero such that

$$AX = 1X, \quad \text{where } A = \begin{bmatrix} 1 & 2 & -1 \\ 1 & 0 & 1 \\ 4 & -4 & 5 \end{bmatrix}.$$

14. In the following linear system, determine all values of a for which the resulting linear system has:

(a) No solution.

(b) A unique solution.

(c) Infinitely many solutions.

$$\begin{aligned} x_1 + x_2 - \quad\quad x_3 &= 2 \\ x_1 + 2x_2 + \quad\quad x_3 &= 3 \\ x_1 + x_2 + (a^2 - 5)x_3 &= a. \end{aligned}$$

15. Repeat Exercise 14 for the linear system

$$\begin{aligned} x_1 + x_2 + \quad\quad x_3 &= 2 \\ 2x_1 + 3x_2 + \quad\quad 2x_3 &= 5 \\ 2x_1 + 3x_2 + (a^2 - 1)x_3 &= a + 1. \end{aligned}$$

16. Repeat Exercise 14 for the linear system

$$\begin{aligned} x_1 + x_2 + \quad\quad x_3 &= 2 \\ x_1 + 2x_2 + \quad\quad x_3 &= 3 \\ x_1 + x_2 + (a^2 - 5)x_3 &= a. \end{aligned}$$

17. Repeat Exercise 14 for the linear system

$$\begin{aligned} x_1 + \quad\quad x_2 &= 3 \\ x_1 + (a^2 - 8)x_2 &= a. \end{aligned}$$

18. Let

$$A = \begin{bmatrix} a & b \\ c & d \end{bmatrix} \quad \text{and} \quad X = \begin{bmatrix} x_1 \\ x_2 \end{bmatrix}.$$

Show that the linear system $AX = O$ has only the trivial solution if and only if $ad - bc \neq 0$.

19. Show that $A = \begin{bmatrix} a & b \\ c & d \end{bmatrix}$ is row equivalent to I_2 if and only if $ad - bc \neq 0$.

Exercises 20 *through* 23 *are optional.*

20. (a) Formulate the definitions of column echelon form and reduced column echelon form of a matrix.

(b) Prove that every $m \times n$ matrix is column equivalent to a matrix in column echelon form.

21. Prove that every $m \times n$ matrix is column equivalent to a unique matrix in reduced column echelon form.

22. Let A be the matrix in Exercise 1.

(a) Find a matrix in column echelon form that is column equivalent to A.

(b) Find matrix in reduced column echelon form that is column equivalent to A.

23. Repeat Exercise 22 for the matrix

$$\begin{bmatrix} 1 & 2 & 3 & 4 & 5 \\ 2 & 1 & 3 & -1 & 2 \\ 3 & 1 & 2 & 4 & 1 \end{bmatrix}.$$

24. Determine the reduced row echelon form of

$$A = \begin{bmatrix} \cos\theta & \sin\theta \\ -\sin\theta & \cos\theta \end{bmatrix}.$$

25. Find an equation relating a, b, and c so that the linear system

$$\begin{aligned} x_1 + 2x_2 - 3x_3 &= a \\ 2x_1 + 3x_2 + 3x_3 &= b \\ 5x_1 + 9x_2 - 6x_3 &= c \end{aligned}$$

is consistent for any values of a, b, and c that satisfy that equation.

26. Find an equation relating a, b, and c so that the linear system

$$\begin{aligned} 2x_1 + 2x_2 + 3x_3 &= a \\ 3x_1 - x_2 + 5x_3 &= b \\ x_1 - 3x_2 + 2x_3 &= c \end{aligned}$$

is consistent for any values of a, b, and c that satisfy that equation.

27. Show that the homogeneous system

$$\begin{aligned} (a - r)x_1 + \quad\quad dx_2 &= 0 \\ cx_1 + (b - r)x_2 &= 0 \end{aligned}$$

has a nontrivial solution if and only if r satisfies the equation $(a - r)(b - r) - cd = 0$.

28. Let $AX = B$, $B \neq O$, be a consistent linear system.

(a) Show that if X_p is a particular solution to the given nonhomogeneous system and X_h is a solution to the associated homogeneous system $AX = O$, then $X_p + X_h$ is a solution to the given system $AX = B$.

(b) Show that every solution X to the non-homogeneous linear system $AX = B$ can be written as $X_p + X_h$, where X_p is a particular solution to the given non-homogeneous system and X_h is a solution to the associated homogeneous system $AX = O$.
[*Hint*: Let $X = X_p + (X - X_p)$.]

29. Determine if the software you are using has a command for computing the reduced row echelon form of a matrix. If it does, experiment with that command on some of the previous exercises.

1.6 ELEMENTARY MATRICES; FINDING A^{-1}

In this section we develop a method for finding the inverse of a matrix if it exists. To use this method we do not have to first find out whether A^{-1} exists. We start to find A^{-1}; if in the course of the computation we hit a certain situation, then we know that A^{-1} does not exist. Otherwise, we proceed to the end and obtain A^{-1}. This method requires that elementary row operations of types I, II, and III be performed on A. We clarify these notions by starting with the following definition.

Definition 1.13

An $n \times n$ **elementary matrix of type I, type II, or type III** is a matrix obtained from the identity matrix I_n by performing a single elementary row or elementary column operation of type I, type II, or type III, respectively. ▲

EXAMPLE 1

The following are elementary matrices:

$$E_1 = \begin{bmatrix} 0 & 0 & 1 \\ 0 & 1 & 0 \\ 1 & 0 & 0 \end{bmatrix}, \quad E_2 = \begin{bmatrix} 1 & 0 & 0 \\ 0 & -2 & 0 \\ 0 & 0 & 1 \end{bmatrix},$$

$$E_3 = \begin{bmatrix} 1 & 2 & 0 \\ 0 & 1 & 0 \\ 0 & 0 & 1 \end{bmatrix}, \quad \text{and} \quad E_4 = \begin{bmatrix} 1 & 0 & 3 \\ 0 & 1 & 0 \\ 0 & 0 & 1 \end{bmatrix}.$$

Matrix E_1 is of type I—we interchanged the first and third rows of I_3; E_2 is of type II—we multiplied the second row of I_3 by (-2); E_3 is of type III—we added twice the second row of I_3 to the first row of I_3; and E_4 is of type III—we added three times the first column of I_3 to the third column of I_3. ■

Theorem 1.13

Let A be an $m \times n$ matrix, and let an elementary row (column) operation of type I, *or type* II, *or type* III *be performed on A to yield matrix B. Let E be the elementary matrix obtained from I_m (I_n) by performing the same elementary row (column) operation as was performed on A. Then $B = EA$ ($B = AE$).*

Proof

Exercise. ●

Theorem 1.13 says that an elementary row operation on A can be achieved by premultiplying A (multiplying A on the left) by the corresponding elementary matrix E; an elementary column operation on A can be obtained by postmultiplying A (multiplying A on the right) by the corresponding elementary matrix.

EXAMPLE 2

Let

$$A = \begin{bmatrix} 1 & 3 & 2 & 1 \\ -1 & 2 & 3 & 4 \\ 3 & 0 & 1 & 2 \end{bmatrix}$$

and let B result from A by adding (-2) times the third row of A to the first row of A. Thus

$$B = \begin{bmatrix} -5 & 3 & 0 & -3 \\ -1 & 2 & 3 & 4 \\ 3 & 0 & 1 & 2 \end{bmatrix}.$$

Now let E be the matrix that is obtained from I_3 by adding (-2) times the third row of I_3 to the first row of I_3. Thus

$$E = \begin{bmatrix} 1 & 0 & -2 \\ 0 & 1 & 0 \\ 0 & 0 & 1 \end{bmatrix}.$$

We can readily verify that $B = EA$. ■

Theorem 1.14

If A and B are $m \times n$ matrices, then A is row (column) equivalent to B if and only if there exist elementary matrices $E_1, E_2, \ldots, E_k$ such that $B = E_k E_{k-1} \cdots E_2 E_1 A$ $(B = A E_1 E_2 \cdots E_{k-1} E_k)$.

Proof

We prove only the theorem for row equivalence. If A is row equivalent to B, then B results from A by a sequence of elementary row operations. This implies that there exist elementary matrices $E_1, E_2, \ldots, E_k$ such that $B = E_k E_{k-1} \cdots E_2 E_1 A$.

Conversely, if $B = E_k E_{k-1} \cdots E_2 E_1 A$, where the E_i are elementary matrices, then B results from A by a sequence of elementary row operations, which implies that A is row equivalent to B. ●

Theorem 1.15

An elementary matrix E is nonsingular and its inverse is an elementary matrix of the same type.

Proof

Exercise. ●

Thus an elementary row operation can be "undone" by another elementary row operation of the same type.

We now obtain an algorithm for finding A^{-1} if it exists; first, we prove the following lemma.

Lemma 1.1[†]

Let A be an $n \times n$ matrix and let the homogeneous system $AX = O$ have only the trivial solution $X = O$. Then A is row equivalent to I_n.

Proof

Let B be a matrix in reduced row echelon form that is row equivalent to A. Then the homogeneous systems $AX = O$ and $BX = O$ are equivalent, and thus $BX = O$ also has only the trivial solution. It is clear that if r is the number of nonzero rows of B, then the homogeneous system $BX = O$ is equivalent to the homogeneous system whose coefficient matrix consists of the nonzero rows of B and is therefore $r \times n$. Since this last homogeneous system only has the trivial solution, we conclude from Theorem 1.12 that $r \geq n$. Since B is $n \times n$, $r \leq n$. Hence $r = n$, which means that B has no zero rows. Thus $B = I_n$. ●

Theorem 1.16

A is nonsingular if and only if A is a product of elementary matrices.

Proof

If A is a product of elementary matrices $E_1, E_2, \ldots, E_k$, then $A = E_1 E_2 \cdots E_k$. Now each elementary matrix is nonsingular, and by Theorem 1.6, the product of nonsingular matrices is nonsingular; therefore, A is nonsingular.

Conversely, if A is nonsingular, then $AX = O$ implies that $A^{-1}(AX) = A^{-1}O = O$, so $I_n X = O$ or $X = O$. Thus $AX = O$ has only the trivial solution. Lemma 1.1 then implies that A is row equivalent to I_n. This means that there exist elementary matrices $E_1, E_2, \ldots, E_k$ such that

$$I_n = E_k E_{k-1} \cdots E_2 E_1 A.$$

It then follows that $A = (E_k E_{k-1} \cdots E_2 E_1)^{-1} = E_1^{-1} E_2^{-1} \cdots E_{k-1}^{-1} E_k^{-1}$. Since the inverse of an elementary matrix is an elementary matrix, we have established the result. ●

Corollary 1.3

A is nonsingular if and only if A is row equivalent to I_n.

Proof

If A is row equivalent to I_n, then $I_n = E_k E_{k-1} \cdots E_2 E_1 A$, where $E_1, E_2, \ldots, E_k$ are elementary matrices. Therefore, it follows that $A = E_1^{-1} E_2^{-1} \cdots E_k^{-1}$. Now the inverse of an elementary matrix is an elementary matrix, and so by Theorem 1.16, A is nonsingular.

Conversely, if A is nonsingular, then A is a product of elementary matrices, $A = E_k E_{k-1} \cdots E_2 E_1$. Now $A = AI_n = E_k E_{k-1} \cdots E_2 E_1 I_n$, which implies that A is row equivalent to I_n. ●

[†] A lemma is a theorem that is established for the purpose of proving another theorem

We can see that Lemma 1.1 and Corollary 1.3 imply that if the homogeneous system $AX = O$, where A is $n \times n$, has only the trivial solution $X = O$, then A is nonsingular. Conversely, consider $AX = O$, where A is $n \times n$, and let A be nonsingular. Then A^{-1} exists and we have $A^{-1}(AX) = A^{-1}O = O$. We also have $A^{-1}(AX) = (A^{-1}A)X = I_nX = X$, so $X = O$, which means that the homogeneous system has only the trivial solution. We have thus proved the following important theorem.

Theorem 1.17

The homogeneous system of n linear equations in n unknowns $AX = O$ has a nontrivial solution if and only if A is singular. ●

EXAMPLE 3

Let

$$A = \begin{bmatrix} 1 & 2 \\ 2 & 4 \end{bmatrix}$$

be the singular matrix defined in Example 11 of Section 1.4. Consider the homogeneous system $AX = O$; that is

$$\begin{bmatrix} 1 & 2 \\ 2 & 4 \end{bmatrix} \begin{bmatrix} x_1 \\ x_2 \end{bmatrix} = \begin{bmatrix} 0 \\ 0 \end{bmatrix}.$$

The reduced row echelon form of the augmented matrix is

$$\left[\begin{array}{cc|c} 1 & 2 & 0 \\ 0 & 0 & 0 \end{array}\right]$$

(verify), so a solution is

$$x_1 = -2r$$
$$x_2 = r,$$

where r is any real number. Thus the homogeneous system has a nontrivial solution. ■

In Section 1.4 we have shown that if the $n \times n$ matrix A is nonsingular, then the system $AX = B$ has a unique solution for every $n \times 1$ matrix B. The converse of this statement is also true (see Exercise 30).

Note that at this point we have shown that the following statements are equivalent for an $n \times n$ matrix A.

1. A is nonsingular.
2. $AX = O$ has only the trivial solution.
3. A is row (column) equivalent to I_n.
4. The linear system $AX = B$ has a unique solution for every $n \times 1$ matrix B.
5. A is a product of elementary matrices.

That is, any two of these five statements are pairwise equivalent. For example, statements 1 and 2 are equivalent by Theorem 1.17, while statements 1 and 3 are equivalent by Corollary 1.3. The importance of these five statements being equivalent is that we can always replace any one statement by any other one on the list. As you will see throughout this book, a given problem can often be solved in several alternative ways, and sometimes one procedure is easier to apply than another.

At the end of the proof of Theorem 1.16, A was nonsingular and

$$A = E_1^{-1} E_2^{-1} \cdots E_{k-1}^{-1} E_k^{-1},$$

from which it follows that

$$A^{-1} = (E_1^{-1} E_2^{-1} \cdots E_{k-1}^{-1} E_k^{-1})^{-1} = E_k E_{k-1} \cdots E_2 E_1.$$

This now provides an algorithm for finding A^{-1}. Thus we perform elementary row operations on A until we get I_n; the product of the elementary matrices $E_k E_{k-1} \cdots E_2 E_1$ then gives A^{-1}. A convenient way of organizing the computing process is to write down the partitioned matrix $\left[A \mid I_n \right]$. Then

$$\begin{aligned}(E_k E_{k-1} \cdots E_2 E_1)\left[A \mid I_n \right] &= \left[E_k E_{k-1} \cdots E_2 E_1 A \mid E_k E_{k-1} \cdots E_2 E_1 \right] \\ &= \left[I_n \mid A^{-1} \right].\end{aligned}$$

That is, we transform the partitioned matrix $\left[A \mid I_n \right]$ to reduced row echelon form, obtaining $\left[I_n \mid A^{-1} \right]$.

EXAMPLE 4

Let

$$A = \begin{bmatrix} 1 & 1 & 1 \\ 0 & 2 & 3 \\ 5 & 5 & 1 \end{bmatrix}.$$

Assuming that A is nonsingular, we form

$$\left[A \mid I_3 \right] = \left[\begin{array}{ccc|ccc} 1 & 1 & 1 & 1 & 0 & 0 \\ 0 & 2 & 3 & 0 & 1 & 0 \\ 5 & 5 & 1 & 0 & 0 & 1 \end{array}\right].$$

We now perform elementary row operations that transform $\left[A \mid I_3\right]$ to $\left[I_3 \mid A^{-1}\right]$; we consider $\left[A \mid I_3\right]$ as a 3×6 matrix, and whatever we do to a row of A we also do to the corresponding row of I_3. In place of using elementary matrices directly, we arrange our computations as follows:

	A			I_3		
1	1	1	1	0	0	Add (-5) times the first row to the third row to obtain
0	2	3	0	1	0	
5	5	1	0	0	1	
1	1	1	1	0	0	Multiply the second row by $\frac{1}{2}$ to obtain
0	2	3	0	1	0	
0	0	-4	-5	0	1	
1	1	1	1	0	0	Multiply the third row by $(-\frac{1}{4})$ to obtain
0	1	$\frac{3}{2}$	0	$\frac{1}{2}$	0	
0	0	-4	-5	0	1	
1	1	1	1	0	0	Add $(-\frac{3}{2})$ times the third row to the second row and add (-1) times the third row to the first row to obtain
0	1	$\frac{3}{2}$	0	$\frac{1}{2}$	0	
0	0	1	$\frac{5}{4}$	0	$-\frac{1}{4}$	
1	1	0	$-\frac{1}{4}$	0	$\frac{1}{4}$	Add (-1) times the second row to the first row to obtain
0	1	0	$-\frac{15}{8}$	$\frac{1}{2}$	$\frac{3}{8}$	
0	0	1	$\frac{5}{4}$	0	$-\frac{1}{4}$	
1	0	0	$\frac{13}{8}$	$-\frac{1}{2}$	$-\frac{1}{8}$	
0	1	0	$-\frac{15}{8}$	$\frac{1}{2}$	$\frac{3}{8}$	
0	0	1	$\frac{5}{4}$	0	$-\frac{1}{4}$	

Hence

$$A^{-1} = \begin{bmatrix} \frac{13}{8} & -\frac{1}{2} & -\frac{1}{8} \\ -\frac{15}{8} & \frac{1}{2} & \frac{3}{8} \\ \frac{5}{4} & 0 & -\frac{1}{4} \end{bmatrix}.$$

We can readily verify that $AA^{-1} = A^{-1}A = I_3$. ■

The question that arises at this point is how to tell when A is singular. The answer is that A is singular if and only if A is row equivalent to matrix B having at least one row that consists entirely of zeros. We now prove this result.

Theorem 1.18

An $n \times n$ matrix A is singular if and only if A is row equivalent to a matrix B that has a row of zeros.

Proof

First, let A be row equivalent to a matrix B that has a row consisting entirely of zeros. From Exercise 34 of Section 1.4 it follows that B is singular. Now $B = E_k E_{k-1} \cdots E_1 A$, where $E_1, E_2, \ldots, E_k$ are elementary matrices. If A is nonsingular, then B is nonsingular, a contradiction. Thus A is singular.

Conversely, if A is singular, then A is not row equivalent to I_n, by Corollary 1.3. Thus A is row equivalent to a matrix $B \neq I_n$, which is in reduced row echelon form. From Exercise 3 of Section 1.5 it follows that B must have a row of zeros. ●

This means that in order to find A^{-1}, we do not have to determine, in advance, whether it exists. We merely start to calculate A^{-1}; if at any point in the computation we find a matrix B that is row equivalent to A and has a row of zeros, then A^{-1} does not exist. That is, we transform the partitioned matrix $\left[A \mid I_n\right]$ to reduced row echelon form, obtaining $\left[C \mid D\right]$. If $C = I_n$, then $D = A^{-1}$. If $C \neq I_n$, then C has a row of zeros and we conclude that A is singular.

EXAMPLE 5

Let

$$A = \begin{bmatrix} 1 & 2 & -3 \\ 1 & -2 & 1 \\ 5 & -2 & -3 \end{bmatrix}.$$

To find A^{-1}, we proceed as follows:

A			I_3			
1	2	−3	1	0	0	Add (−1) times the first row to the second row to obtain
1	−2	1	0	1	0	
5	−2	−3	0	0	1	
1	2	−3	1	0	0	Add (−5) times the first row to the third row to obtain
0	−4	4	−1	1	0	
5	−2	−3	0	0	1	
1	2	−3	1	0	0	Add (−3) times the second row to the third row to obtain
0	−4	4	−1	1	0	
0	−12	12	−5	0	1	
1	2	−3	1	0	0	
0	−4	4	−1	1	0	
0	0	0	−2	−3	1	

At this point A is row equivalent to

$$B = \begin{bmatrix} 1 & 2 & -3 \\ 0 & -4 & 4 \\ 0 & 0 & 0 \end{bmatrix},$$

the last matrix under A. Since B has a row of zeros, we stop and conclude that A is a singular matrix. ■

In Section 1.4 we defined an $n \times n$ matrix B to be the inverse of the $n \times n$ matrix A if $AB = I_n$ and $BA = I_n$. We now show that one of these equations follows from the other.

Theorem 1.19

If A and B are $n \times n$ matrices such that $AB = I_n$, then $BA = I_n$. Thus $B = A^{-1}$.

Proof

We first show that if $AB = I_n$, then A is nonsingular. Suppose that A is singular. Then A is row equivalent to a matrix C with a row of zeros. Now $C = E_k E_{k-1} \cdots E_1 A$, where $E_1, E_2, \ldots, E_k$ are elementary matrices. Then $CB = E_k E_{k-1} \cdots E_1 AB$, so AB is row equivalent to CB. Since CB has a row of zeros, we conclude from Theorem 1.18 that AB is singular. Then $AB = I_n$ is impossible, because I_n is nonsingular. This contradiction shows that A is nonsingular, and so A^{-1} exists. Multiplying both sides of the equation $AB = I_n$ by A^{-1} on the left, we then obtain (verify) $B = A^{-1}$. ●

Remark Theorem 1.19 implies that if we want to check whether a given matrix B is A^{-1}, we need merely check whether $AB = I_n$ *or* $BA = I_n$. That is, we do not have to check both equalities.

1.6 Exercises

1. Prove Theorem 1.13.

2. Let A be a 4×3 matrix. Find the elementary matrix E, which as a premultiplier of A, that is, as EA, performs the following elementary row operations on A.

(a) Multiplies the second row of A by (-2).

(b) Adds 3 times the third row of A to the fourth row of A.

(c) Interchanges the first and third rows of A.

3. Let A be a 3×4 matrix. Find the elementary matrix F, which as a postmultiplier of A, that is, as AF, performs the following elementary column operations on A.

(a) Add (-4) times the first column of A to the second column of A.

(b) Interchanges the second and third columns of A.

(c) Multiplies the third column of A by 4.

4. Let

$$A = \begin{bmatrix} 1 & 0 & 0 \\ 0 & 1 & 0 \\ -2 & 0 & 1 \end{bmatrix}.$$

(a) Find a matrix C in reduced row echelon form that is row equivalent to A. Record the row operations used.

(b) Apply the same operations to I_3 that were used to obtain C. Denote the resulting matrix by B.

(c) How are A and B related? (*Hint*: Compute AB and BA.)

5. Let

$$A = \begin{bmatrix} 1 & 0 & 1 \\ 2 & 1 & 0 \\ 0 & -1 & 1 \end{bmatrix}.$$

(a) Find a matrix C in reduced row echelon form that is row equivalent to A. Record the row operations used.

(b) Apply the same operations to I_3 that were used to obtain C. Denote the resulting matrix by B.

(c) How are A and B related? (*Hint*: Compute AB and BA.)

6. Prove Theorem 1.15. (*Hint*: Find the inverse of the elementary matrix of type I, type II, and type III.)

7. Find the inverse of $A = \begin{bmatrix} 1 & 3 \\ 2 & 4 \end{bmatrix}$.

8. Find the inverse of $A = \begin{bmatrix} 1 & 2 & 3 \\ 0 & 2 & 3 \\ 1 & 2 & 4 \end{bmatrix}$.

9. Which of the following matrices are singular? For the nonsingular ones find the inverse.

(a) $\begin{bmatrix} 1 & 3 \\ 2 & 6 \end{bmatrix}$. (b) $\begin{bmatrix} 1 & 3 \\ -2 & 6 \end{bmatrix}$.

(c) $\begin{bmatrix} 1 & 2 & 3 \\ 1 & 1 & 2 \\ 0 & 1 & 2 \end{bmatrix}$. (d) $\begin{bmatrix} 1 & 2 & 3 \\ 1 & 1 & 2 \\ 0 & 1 & 1 \end{bmatrix}$.

10. Invert the following matrices, if possible.

(a) $\begin{bmatrix} 1 & 2 & -3 & 1 \\ -1 & 3 & -3 & -2 \\ 2 & 0 & 1 & 5 \\ 3 & 1 & -2 & 5 \end{bmatrix}$. (b) $\begin{bmatrix} 3 & 1 & 2 \\ 2 & 1 & 2 \\ 1 & 2 & 2 \end{bmatrix}$.

(c) $\begin{bmatrix} 1 & 2 & 3 \\ 1 & 1 & 2 \\ 1 & 1 & 0 \end{bmatrix}$. (d) $\begin{bmatrix} 2 & 1 & 3 \\ 0 & 1 & 2 \\ 1 & 0 & 3 \end{bmatrix}$.

11. Find the inverse, if it exists, of:

(a) $\begin{bmatrix} 1 & 1 & 1 \\ 1 & 2 & 3 \\ 0 & 1 & 1 \end{bmatrix}$. (b) $\begin{bmatrix} 1 & 1 & 1 & 1 \\ 1 & 2 & -1 & 2 \\ 1 & -1 & 2 & 1 \\ 1 & 3 & 3 & 2 \end{bmatrix}$.

(c) $\begin{bmatrix} 1 & 1 & 1 & 1 \\ 1 & 3 & 1 & 2 \\ 1 & 2 & -1 & 1 \\ 5 & 9 & 1 & 6 \end{bmatrix}$. (d) $\begin{bmatrix} 1 & 2 & 1 \\ 1 & 3 & 2 \\ 1 & 0 & 1 \end{bmatrix}$.

(e) $\begin{bmatrix} 1 & 2 & 2 \\ 1 & 3 & 1 \\ 1 & 1 & 3 \end{bmatrix}$.

12. Find the inverse, if it exists, of:

(a) $A = \begin{bmatrix} 1 & 1 & 2 & 1 \\ 0 & -2 & 0 & 0 \\ 1 & 2 & 1 & -2 \\ 0 & 3 & 2 & 1 \end{bmatrix}$.

(b) $A = \begin{bmatrix} 1 & 1 & 1 & 1 \\ 1 & 3 & 1 & 2 \\ 1 & 2 & -1 & 1 \\ 5 & 9 & 1 & 6 \end{bmatrix}$.

In Exercises 13 *and* 14, *prove that the given matrix A is nonsingular and write it as a product of elementary matrices.* (*Hint*: *First, write the inverse as a product of elementary matrices; then use Theorem* 1.15.)

13. $A = \begin{bmatrix} 1 & 2 \\ 3 & 4 \end{bmatrix}$.

14. $A = \begin{bmatrix} 1 & 2 & 3 \\ 0 & 1 & 2 \\ 1 & 0 & 3 \end{bmatrix}$.

15. If A is a nonsingular matrix whose inverse is $\begin{bmatrix} 4 & 2 \\ 1 & 1 \end{bmatrix}$, find A.

16. If $A^{-1} = \begin{bmatrix} 1 & 1 & 1 \\ 1 & 1 & 2 \\ 1 & -1 & 1 \end{bmatrix}$, find A.

17. Which of the following homogeneous systems have a nontrivial solution?

(a)
$$\begin{aligned} x_1 + 2x_2 + 3x_3 &= 0 \\ 2x_2 + 2x_3 &= 0 \\ x_1 + 2x_2 + 3x_3 &= 0. \end{aligned}$$

(b)
$$\begin{aligned} 2x_1 + x_2 - x_3 &= 0 \\ x_1 - 2x_2 - 3x_3 &= 0 \\ -3x_1 - x_2 + 2x_3 &= 0. \end{aligned}$$

(c)
$$\begin{aligned} 3x_1 + x_2 + 3x_3 &= 0 \\ -2x_1 + 2x_2 - 4x_3 &= 0 \\ 2x_1 - 3x_2 + 5x_3 &= 0. \end{aligned}$$

18. Which of the following homogeneous systems have a nontrivial solution?

(a)
$$\begin{aligned} x_1 + x_2 + 2x_3 &= 0 \\ 2x_1 + x_2 + x_3 &= 0 \\ 3x_1 - x_2 + x_3 &= 0. \end{aligned}$$

(b)
$$\begin{aligned} x_2 - x_2 + x_3 &= 0 \\ 2x_1 + x_2 \quad &= 0 \\ 2x_1 - 2x_2 + 2x_3 &= 0. \end{aligned}$$

(c)
$$\begin{aligned} 2x_1 - x_2 + 5x_3 &= 0 \\ 3x_1 + 2x_2 - 3x_3 &= 0 \\ x_1 - x_2 + 4x_3 &= 0. \end{aligned}$$

19. Find all value(s) of a for which the inverse of

$$A = \begin{bmatrix} 1 & 1 & 0 \\ 1 & 0 & 0 \\ 1 & 2 & a \end{bmatrix}$$

exists. What is A^{-1}?

20. For what values of a does the homogeneous system

$$\begin{aligned} (a-1)x_1 + \qquad 2x_2 &= 0 \\ 2x_1 + (a-1)x_2 &= 0 \end{aligned}$$

have a nontrivial solution?

21. Prove that

$$A = \begin{bmatrix} a & b \\ c & d \end{bmatrix}$$

is nonsingular if and only if $ad - bc \neq 0$. If this condition holds, show that

$$A^{-1} = \begin{bmatrix} \dfrac{d}{ad-bc} & \dfrac{-b}{ad-bc} \\ \dfrac{-c}{ad-bc} & \dfrac{a}{ad-bc} \end{bmatrix}.$$

22. Let

$$A = \begin{bmatrix} 2 & 3 & -1 \\ 1 & 0 & 3 \\ 0 & 2 & -3 \\ -2 & 1 & 3 \end{bmatrix}.$$

Find the elementary matrix that as a postmultiplier of A performs the following elementary column operations on A.

(a) Multiplies the third column of A by (-3).

(b) Interchanges the second and third columns of A.

(c) Adds (-5) times the first column of A to the third column of A.

23. Prove that two $m \times n$ matrices A and B are row equivalent if and only if there exists a nonsingular matrix P such that $B = PA$. (*Hint*: Use Theorems 1.14 and 1.16.)

24. Let A and B be row equivalent $n \times n$ matrices. Prove that A is nonsingular if and only if B is nonsingular.

25. Let A and B be $n \times n$ matrices. Show that if AB is nonsingular, then A and B must be nonsingular. (*Hint*: Use Theorem 1.17.)

26. Let A be an $m \times n$ matrix. Show that A is row equivalent to O_{mn} if and only if $A = O_{mn}$.

27. Let A and B be $m \times n$ matrices. Show that A is row equivalent to B if and only if A^T is column equivalent to B^T.

28. Show that a square matrix which has a row or a column consisting entirely of zeros must be singular.

29. (a) Is $(A+B)^{-1} = A^{-1} + B^{-1}$?

(b) Is $(cA)^{-1} = \dfrac{1}{c}A^{-1}$?

30. If A is an $n \times n$ matrix, prove that A is nonsingular if and only if the linear system $AX = B$ has a unique solution for every $n \times 1$ matrix B.

31. Prove that the inverse of a nonsingular upper (lower) triangular matrix is upper (lower) triangular.

32. If the software you use has a command for computing reduced row echelon form, use it to determine if the matrices A in Exercises 9, 10, and 11 have an inverse by operating on the matrix $\left[A \mid I_n\right]$ (see Example 4).

33. Repeat Exercise 32 on the matrices given in Exercise 47 of Section 1.4.

1.7 EQUIVALENT MATRICES

We have thus far considered A to be row (column) equivalent to B if B results from A by a finite sequence of elementary row (column) operations. A natural extension of this idea is that of considering B to arise from A by a finite sequence of elementary row *or* elementary column operations. This leads to the notion of equivalence of matrices. The material discussed in this section is used in Section 2.8.

Definition 1.14

If A and B are two $m \times n$ matrices, then A is **equivalent** to B if we obtain B from A by a finite sequence of elementary row or elementary column operations. ▲

As we have seen in the case of row equivalence, we can show (see Exercise 1) that (a) every matrix is equivalent to itself; (b) if B is equivalent to A, then A is equivalent to B; (c) if C is equivalent to B, and B is equivalent to A, then C is equivalent to A. In view of (b) both statements "A is equivalent to B" and "B is equivalent to A" can be replaced by "A and B are equivalent." We can also show that if two matrices are row equivalent, then they are equivalent (see Exercise 4).

Theorem 1.20

If A is any nonzero $m \times n$ matrix, then A is equivalent to a partitioned matrix of the form

$$\begin{bmatrix} I_r & O_{r\,n-r} \\ O_{m-r\,r} & O_{m-r\,n-r} \end{bmatrix}.$$

Proof

By Theorem 1.10, A is row equivalent to a matrix B that is in reduced row echelon form. Using elementary column operations of type I, we get B to be equivalent to a matrix C of the form

$$\begin{bmatrix} I_r & U_{r\,n-r} \\ O_{m-r\,r} & O_{m-r\,n-r} \end{bmatrix},$$

where r is the number of nonzero rows in B. By elementary column operations of type III, C is equivalent to a matrix D of the form

$$\begin{bmatrix} I_r & O_{r\,n-r} \\ O_{m-r\,r} & O_{m-r\,n-r} \end{bmatrix}.$$

From Exercise 1, it then follows that A is equivalent to D. ●

Of course, in Theorem 1.20, r may equal m, in which case there will not be any zero rows at the bottom of the matrix. (What happens if $r = n$? If $r = m = n$?)

EXAMPLE 1

Let

$$A = \begin{bmatrix} 1 & 1 & 2 & -1 \\ 1 & 2 & 1 & 0 \\ -1 & -4 & 1 & -2 \\ 1 & -2 & 5 & -4 \end{bmatrix}.$$

To find a matrix of the form described in Theorem 1.20, which is equivalent to A, we proceed as follows. Add (-1) times the first row of A to the second row of A to obtain

$$\begin{bmatrix} 1 & 1 & 2 & -1 \\ 0 & 1 & -1 & 1 \\ -1 & -4 & 1 & -2 \\ 1 & -2 & 5 & -4 \end{bmatrix}$$ Add the first row to the third row to obtain

$$\begin{bmatrix} 1 & 1 & 2 & -1 \\ 0 & 1 & -1 & 1 \\ 0 & -3 & 3 & -3 \\ 1 & -2 & 5 & -4 \end{bmatrix}$$ Add (-1) times the first row to the fourth row to obtain

$$\begin{bmatrix} 1 & 1 & 2 & -1 \\ 0 & 1 & -1 & 1 \\ 0 & -3 & 3 & -3 \\ 0 & -3 & 3 & -3 \end{bmatrix}$$ Add (-1) times the third row to the fourth row to obtain

$$\begin{bmatrix} 1 & 1 & 2 & -1 \\ 0 & 1 & -1 & 1 \\ 0 & -3 & 3 & -3 \\ 0 & 0 & 0 & 0 \end{bmatrix}$$ Multiply the third row by $(-\frac{1}{3})$ to obtain

$$\begin{bmatrix} 1 & 1 & 2 & -1 \\ 0 & 1 & -1 & 1 \\ 0 & 1 & -1 & 1 \\ 0 & 0 & 0 & 0 \end{bmatrix}$$ Add (-1) times the second row to the third row to obtain

$$\begin{bmatrix} 1 & 1 & 2 & -1 \\ 0 & 1 & -1 & 1 \\ 0 & 0 & 0 & 0 \\ 0 & 0 & 0 & 0 \end{bmatrix}$$ Add (-1) times the second row to the first row to obtain

$$\begin{bmatrix} 1 & 0 & 3 & -2 \\ 0 & 1 & -1 & 1 \\ 0 & 0 & 0 & 0 \\ 0 & 0 & 0 & 0 \end{bmatrix}$$ Add (-3) times the first column to the third column to obtain

$$\begin{bmatrix} 1 & 0 & 0 & -2 \\ 0 & 1 & -1 & 1 \\ 0 & 0 & 0 & 0 \\ 0 & 0 & 0 & 0 \end{bmatrix}$$ Add 2 times the first column to the fourth column to obtain

$$\begin{bmatrix} 1 & 0 & 0 & 0 \\ 0 & 1 & -1 & 1 \\ 0 & 0 & 0 & 0 \\ 0 & 0 & 0 & 0 \end{bmatrix}$$ Add the second column to the third column to obtain

$$\begin{bmatrix} 1 & 0 & 0 & 0 \\ 0 & 1 & 0 & 1 \\ 0 & 0 & 0 & 0 \\ 0 & 0 & 0 & 0 \end{bmatrix}$$ Add (-1) times the second column to the fourth column to obtain

$$\begin{bmatrix} 1 & 0 & 0 & 0 \\ 0 & 1 & 0 & 0 \\ 0 & 0 & 0 & 0 \\ 0 & 0 & 0 & 0 \end{bmatrix}.$$

This is the matrix desired. ■

The following theorem gives another useful way to look at the equivalence of matrices.

Theorem 1.21

Two $m \times n$ matrices A and B are equivalent if and only if $B = PAQ$ for some nonsingular matrices P and Q.

Proof

Exercise. ●

We next prove a theorem that is analogous to Corollary 1.3.

Theorem 1.22

An $n \times n$ matrix A is nonsingular if and only if A is equivalent to I_n.

Proof

If A is equivalent to I_n, then I_n arises from A by a sequence of elementary row or elementary column operations. Thus there exist elementary matrices $E_1, E_2, \ldots, E_r$, $F_1, F_2, \ldots, F_s$ such that

$$I_n = E_r E_{r-1} \cdots E_2 E_1 A F_1 F_2 \cdots F_s.$$

Let $E_r E_{r-1} \cdots E_2 E_1 = P$ and $F_1 F_2 \cdots F_s = Q$. Then $I_n = PAQ$, where P and Q are nonsingular. It then follows that $A = P^{-1} Q^{-1}$, and since P^{-1} and Q^{-1} are nonsingular, A is nonsingular.

Conversely, if A is nonsingular, then from Corollary 1.3 it follows that A is row equivalent to I_n. Hence A is equivalent to I_n. ●

1.7 Exercises

1. (a) Prove that every matrix A is equivalent to itself.

(b) Prove that if B is equivalent to A, then A is equivalent to B.

(c) Prove that if C is equivalent to B and B is equivalent to A, then C is equivalent to A.

2. For each of the following matrices, find a matrix of the form described in Theorem 1.20 that is equivalent to the given matrix.

(a) $\begin{bmatrix} 1 & 2 & -1 & 4 \\ 5 & 1 & 2 & -3 \\ 2 & 1 & 4 & 3 \\ 2 & 0 & 1 & 2 \\ 5 & 1 & 2 & 3 \end{bmatrix}$. (b) $\begin{bmatrix} 1 & 2 & 1 \\ 2 & 3 & 1 \\ 2 & 1 & 3 \end{bmatrix}$.

(c) $\begin{bmatrix} 1 & -2 & 1 \\ 2 & 3 & 2 \\ 3 & 1 & 3 \end{bmatrix}$. (d) $\begin{bmatrix} 1 & 3 & -1 & 2 \\ 2 & -4 & -2 & 1 \\ 3 & 1 & 2 & -3 \\ 7 & 3 & -2 & 5 \end{bmatrix}$.

3. Repeat Exercise 2 for the following matrices.

(a) $\begin{bmatrix} 1 & 2 & 3 & -1 \\ 1 & 0 & 2 & 3 \\ 3 & 4 & 8 & 1 \end{bmatrix}$. (b) $\begin{bmatrix} 3 & 4 & 1 \\ 1 & 2 & -2 \\ 5 & 6 & 4 \\ 5 & 8 & -1 \end{bmatrix}$.

(c) $\begin{bmatrix} 2 & 3 & 4 & -1 \\ 1 & 2 & 1 & -1 \\ 2 & -1 & 1 & 1 \\ 4 & 2 & 5 & 0 \\ 4 & 3 & 3 & -1 \end{bmatrix}$. (d) $\begin{bmatrix} 1 & 2 & 3 \\ 1 & -1 & 0 \\ 0 & 1 & 2 \end{bmatrix}$

4. Show that if A and B are row equivalent, then they are equivalent.

5. Prove Theorem 1.21.

6. Let

$$A = \begin{bmatrix} 1 & 2 & 3 \\ 1 & 1 & 2 \\ 0 & 1 & 1 \end{bmatrix}.$$

Find a matrix B of the form described in Theorem 1.20 which is equivalent to A. Also, find nonsingular matrices P and Q such that $B = PAQ$.

7. Repeat Exercise 6 for

$$A = \begin{bmatrix} 1 & -1 & 2 & 3 \\ 2 & -1 & 3 & 1 \\ 4 & -3 & 7 & 7 \\ 0 & -1 & 1 & 5 \end{bmatrix}.$$

8. Let A be an $m \times n$ matrix. Show that A is equivalent to O_{mn} if and only if $A = O_{mn}$.

9. Let A and B be $m \times n$ matrices. Show that A is equivalent to B if and only if A^T is equivalent to B^T.

10. For each of the following matrices A, find a matrix $B \neq A$ that is equivalent to A.

(a) $A = \begin{bmatrix} 1 & -2 & 3 & 1 \\ 0 & -1 & 4 & 3 \\ 1 & 0 & -2 & -1 \end{bmatrix}$.

(b) $A = \begin{bmatrix} 1 & 3 \\ 2 & 6 \end{bmatrix}$.

(c) $A = \begin{bmatrix} 1 & 2 & 3 & 4 & 3 \\ 0 & 1 & -2 & 0 & 2 \\ -1 & 3 & 2 & 0 & 1 \end{bmatrix}$.

11. Let A and B be equivalent square matrices. Prove that A is nonsingular if and only if B is nonsingular.

1.8 LU-FACTORIZATION (OPTIONAL)

In this section we discuss a variant of Gaussian elimination that decomposes a matrix as a product of a lower triangular matrix and an upper triangular matrix. This decomposition leads to an algorithm for solving a linear system $AX = B$ that is the most widely used method on computers for solving a linear system. A main reason for the popularity of this method is that it provides the cheapest way of solving a linear system for which we repeatedly have to change the right sides. This type of situation occurs often in applied problems. For example, an electric utility company must determine the inputs (the unknowns) needed to produce some required outputs (the right sides). The inputs and outputs might be related by a linear system, whose coefficient matrix is fixed, while the right side changes from day to day, or even hour to hour. The decomposition discussed in this section is also useful in solving other problems in linear algebra.

When U is an upper triangular matrix all of whose diagonal entries are different from zero, then the linear system $UX = B$ can be solved without transforming the augmented matrix $\left[U \mid B\right]$ to reduced row echelon form or to row echelon form. The augmented matrix of such a system is given by

$$\left[\begin{array}{ccccc|c} u_{11} & u_{12} & u_{13} & \cdots & u_{1n} & b_1 \\ 0 & u_{22} & u_{23} & \cdots & u_{2n} & b_2 \\ 0 & 0 & u_{33} & \cdots & u_{3n} & b_3 \\ \vdots & \vdots & \vdots & \cdots & \vdots & \vdots \\ 0 & 0 & 0 & \cdots & u_{nn} & b_n \end{array}\right].$$

The solution is obtained by the following algorithm:

$$x_n = \frac{b_n}{u_{nn}}$$

$$x_{n-1} = \frac{b_{n-1} - u_{n-1\,n}x_n}{u_{n-1\,n-1}}$$

$$\vdots$$

$$x_j = \frac{b_j - \displaystyle\sum_{k=n}^{j-1} u_{jk}x_k}{u_{jj}}, \qquad j = n, n-1, \ldots, 2, 1.$$

This procedure is merely **back substitution**, which we used in conjunction with Gaussian elimination in Section 1.5, where it was additionally required that the diagonal entries be 1. In a similar manner, if L is a lower triangular matrix all of whose diagonal entries are different from zero, then the linear system $LX = B$ can be solved by **forward substitution**, which consists of the following procedure: The augmented matrix has the form

$$\left[\begin{array}{ccccc|c} \ell_{11} & 0 & 0 & \cdots & 0 & b_1 \\ \ell_{21} & \ell_{22} & 0 & \cdots & 0 & b_2 \\ \ell_{31} & \ell_{32} & \ell_{33} & \cdots & 0 & b_3 \\ \vdots & \vdots & \vdots & \cdots & \vdots & \vdots \\ \ell_{n1} & \ell_{n2} & \ell_{n3} & \cdots & \ell_{nn} & b_n \end{array}\right]$$

and the solution is given by

$$x_1 = \frac{b_1}{\ell_{11}}$$

$$x_2 = \frac{b_2 - \ell_{21}x_1}{\ell_{22}}$$

$$\vdots$$

$$x_j = \frac{b_j - \sum_{k=1}^{j-1} \ell_{jk}x_k}{\ell_{jj}}, \qquad j = 2, \ldots, n.$$

That is, we proceed from the first equation downward, solving for one unknown from each equation.

We illustrate forward substitution in the following example.

EXAMPLE 1

To solve the linear system

$$\begin{aligned} 5x_1 \qquad\qquad\quad &= 10 \\ 4x_1 - 2x_2 \qquad\quad &= 28 \\ 2x_1 + 3x_2 + 4x_3 &= 26 \end{aligned}$$

we use forward substitution. Hence we obtain from the previous algorithm,

$$x_1 = \frac{10}{5} = 2$$

$$x_2 = \frac{28 - 4x_1}{-2} = -10$$

$$x_3 = \frac{26 - 2x_1 - 3x_2}{4} = 13,$$

which implies that the solution to the given lower triangular system of equations is

$$X = \begin{bmatrix} 2 \\ -10 \\ 13 \end{bmatrix}.$$

■

As illustrated above, the ease with which systems of equations with upper or lower triangular coefficient matrices can be solved is quite attractive. The forward substitution and back substitution algorithms are fast and simple to use. These are used in another important numerical procedure for solving linear systems of equations, which we develop next.

Suppose that an $n \times n$ matrix A can be written as a product of a matrix L in lower triangular form and a matrix U in upper triangular form; that is,

$$A = LU.$$

In this case we say that A has an **LU-factorization** or an **LU-decomposition**. The LU-factorization of a matrix A can be used to efficiently solve a linear system $AX = B$. Substituting LU for A, we have

$$(LU)X = B$$

or by (a) of Theorem 1.2,

$$L(UX) = B.$$

Letting $UX = Z$, this matrix equation becomes

$$LZ = B.$$

Since L is in lower triangular form, we solve directly for Z by forward substitution. Once we determine Z, since U is in upper triangular form, we solve $UX = Z$ by back substitution. In summary, if an $n \times n$ matrix A has an LU-factorization, then the solution of $AX = B$ can be determined by a forward substitution followed by a back substitution. We illustrate this procedure in the next example.

EXAMPLE 2

Consider the linear system

$$\begin{aligned} 6x_1 - 2x_2 - 4x_3 + 4x_4 &= 2 \\ 3x_1 - 3x_2 - 6x_3 + x_4 &= -4 \\ -12x_1 + 8x_2 + 21x_3 - 8x_4 &= 8 \\ -6x_1 \qquad\quad - 10x_3 + 7x_4 &= -43 \end{aligned}$$

whose coefficient matrix

$$A = \begin{bmatrix} 6 & -2 & -4 & 4 \\ 3 & -3 & -6 & 1 \\ -12 & 8 & 21 & -8 \\ -6 & 0 & -10 & 7 \end{bmatrix}$$

has an LU-factorization where

$$L = \begin{bmatrix} 1 & 0 & 0 & 0 \\ \frac{1}{2} & 1 & 0 & 0 \\ -2 & -2 & 1 & 0 \\ -1 & 1 & -2 & 1 \end{bmatrix} \quad \text{and} \quad U = \begin{bmatrix} 6 & -2 & -4 & 4 \\ 0 & -2 & -4 & -1 \\ 0 & 0 & 5 & -2 \\ 0 & 0 & 0 & 8 \end{bmatrix}$$

(verify). To solve the given system using this LU-factorization, we proceed as follows. Let

$$B = \begin{bmatrix} 2 \\ -4 \\ 8 \\ -43 \end{bmatrix}.$$

Then we solve $AX = B$ by writing it as $LUX = B$. First, let $UX = Z$ and solve $LZ = B$:

$$\begin{bmatrix} 1 & 0 & 0 & 0 \\ \frac{1}{2} & 1 & 0 & 0 \\ -2 & -2 & 1 & 0 \\ -1 & 1 & -2 & 1 \end{bmatrix} \begin{bmatrix} z_1 \\ z_2 \\ z_3 \\ z_4 \end{bmatrix} = \begin{bmatrix} 2 \\ -4 \\ 8 \\ -43 \end{bmatrix}$$

by forward substitution. We obtain

$$\begin{aligned} z_1 &= 2 \\ z_2 &= -4 - \tfrac{1}{2}z_1 = -5 \\ z_3 &= 8 + 2z_1 + 2z_2 = 2 \\ z_4 &= -43 + z_1 - z_2 + 2z_3 = -32. \end{aligned}$$

Next we solve $UX = Z$,

$$\begin{bmatrix} 6 & -2 & -4 & 4 \\ 0 & -2 & -4 & -1 \\ 0 & 0 & 5 & -2 \\ 0 & 0 & 0 & 8 \end{bmatrix} \begin{bmatrix} x_1 \\ x_2 \\ x_3 \\ x_4 \end{bmatrix} = \begin{bmatrix} 2 \\ -5 \\ 2 \\ -32 \end{bmatrix},$$

by back substitution. We obtain

$$\begin{aligned} x_4 &= \frac{-32}{8} = -4 \\ x_3 &= \frac{2 + 2x_4}{5} = -1.2 \\ x_2 &= \frac{-5 + 4x_3 + x_4}{-2} = 6.9 \\ x_1 &= \frac{2 + 2x_2 + 4x_3 - 4x_4}{6} = 4.5. \end{aligned}$$

Thus the solution to the given linear system is

$$X = \begin{bmatrix} 4.5 \\ 6.9 \\ -1.2 \\ -4 \end{bmatrix}.$$

■

Next, we show how to obtain an LU-factorization of a matrix by modifying the Gaussian elimination procedure from Section 1.5. No row interchanges will be permitted and we do not require that the diagonal entries have value 1. At the end of this section we provide a reference that indicates how to enhance the LU-factorization scheme presented to deal with matrices where row interchanges are necessary. We observe that the only elementary row operation permitted is the one that adds a multiple of one row to a different row.

To describe the LU-factorization, we present a step-by-step procedure in the next example.

EXAMPLE 3

Let A be the coefficient matrix of the linear system of Example 2.

$$A = \begin{bmatrix} 6 & -2 & -4 & 4 \\ 3 & -3 & -6 & 1 \\ -12 & 8 & 21 & -8 \\ -6 & 0 & -10 & 7 \end{bmatrix}.$$

We proceed to "zero out" entries below the diagonal entries, using only the row operation that adds a multiple of one row to a different row.

Procedure	***Matrices Used***
STEP 1 "Zero out" below the first diagonal entry of A. Add $\left(-\frac{1}{2}\right)$ times the first row of A to the second row of A. Add 2 times the first row of A to the third row of A. Add 1 times the first row of A to the fourth row of A. Call the new resulting matrix U_1.	$U_1 = \begin{bmatrix} 6 & -2 & -4 & 4 \\ 0 & -2 & -4 & -1 \\ 0 & 4 & 13 & 0 \\ 0 & -2 & -14 & 11 \end{bmatrix}$
We begin building a lower triangular matrix, L_1, with 1's on the main diagonal, to record the row operations. Enter the *negatives of the multipliers* used in the row operations in the first column of L_1, below the first diagonal entry of L_1.	$L_1 = \begin{bmatrix} 1 & 0 & 0 & 0 \\ \frac{1}{2} & 1 & 0 & 0 \\ -2 & * & 1 & 0 \\ -1 & * & * & 1 \end{bmatrix}$
STEP 2 "Zero out" below the second diagonal entry of U_1. Add 2 times the second row of U_1 to the third row of U_1. Add (-1) times the second row of U_1 to the fourth row of U_1. Call the new resulting matrix U_2.	$U_2 = \begin{bmatrix} 6 & -2 & -4 & 4 \\ 0 & -2 & -4 & -1 \\ 0 & 0 & 5 & -2 \\ 0 & 0 & -10 & 12 \end{bmatrix}$

Enter the negatives of the multipliers from the row operations below the second diagonal entry of L_1. Call the new matrix L_2.

$$L_2 = \begin{bmatrix} 1 & 0 & 0 & 0 \\ \frac{1}{2} & 1 & 0 & 0 \\ -2 & -2 & 1 & 0 \\ -1 & 1 & * & 1 \end{bmatrix}$$

STEP 3 "Zero out" below the third diagonal entry of U_2. Add 2 times the third row of U_2 to the fourth row of U_2. Call the new resulting matrix U_3.

$$U_3 = \begin{bmatrix} 6 & -2 & -4 & 4 \\ 0 & -2 & -4 & -1 \\ 0 & 0 & 5 & -2 \\ 0 & 0 & 0 & 8 \end{bmatrix}$$

Enter the negative of the multiplier below the third diagonal entry of L_2. Call the new matrix L_3.

$$L_3 = \begin{bmatrix} 1 & 0 & 0 & 0 \\ \frac{1}{2} & 1 & 0 & 0 \\ -2 & -2 & 1 & 0 \\ -1 & 1 & -2 & 1 \end{bmatrix}$$

Let $L = L_3$ and $U = U_3$. Then the product LU gives the original matrix A (verify). The linear system of equations was solved in Example 2 using the LU-factorization just obtained. ■

Remark In general, a given matrix may have more than one LU-factorization. For example, if A is the coefficient matrix considered in Example 2, then another LU-factorization is LU, where

$$L = \begin{bmatrix} 2 & 0 & 0 & 0 \\ 1 & -1 & 0 & 0 \\ -4 & 2 & 1 & 0 \\ -2 & -1 & -2 & 2 \end{bmatrix} \quad \text{and} \quad U = \begin{bmatrix} 3 & -1 & -2 & 2 \\ 0 & 2 & 4 & 1 \\ 0 & 0 & 5 & -2 \\ 0 & 0 & 0 & 4 \end{bmatrix}.$$

There are many methods for obtaining an LU-factorization of a matrix, besides the scheme for **storage of multipliers** described in Example 3. It is important to note that if $a_{11} = 0$, then the procedure used in Example 3 fails. Moreover, if the second diagonal entry of U_1 is zero or if the third diagonal entry of U_2 is zero, then the procedure also fails. In such cases we can try rearranging the equations of the system and beginning again or using one of the other methods for LU-factorization. Most computer programs for LU-factorization incorporate row interchanges into the storage of multipliers scheme and use additional strategies to help control roundoff error. If row interchanges are required, then the product of L and U is not necessarily A—it is a matrix that is a permutation of the rows of A. For example, if row interchanges occur when using the **lu** command in MATLAB in the form [L, U] = lu(A), then MATLAB responds as follows: the matrix that it yields as L is not lower triangular, U is upper triangular, and LU is A. The book *Experiments in Computational Matrix Algebra*, by David R. Hill (New York: Random House, 1988, distributed by McGraw-Hill) explores such a modification of the procedure for LU-factorization.

1.8 Exercises

In Exercises 1 through 4, solve the linear system $AX = B$ with the given LU-*factorization of the coefficient matrix A. Solve the linear system using a forward substitution followed by a back substitution.*

1. $A = \begin{bmatrix} 2 & 8 & 0 \\ 2 & 2 & -3 \\ 1 & 2 & 7 \end{bmatrix}$, $B = \begin{bmatrix} 18 \\ 3 \\ 12 \end{bmatrix}$,

$L = \begin{bmatrix} 2 & 0 & 0 \\ 2 & -3 & 0 \\ 1 & -1 & 4 \end{bmatrix}$, $U = \begin{bmatrix} 1 & 4 & 0 \\ 0 & 2 & 1 \\ 0 & 0 & 2 \end{bmatrix}$.

2. $A = \begin{bmatrix} 8 & 12 & -4 \\ 6 & 5 & 7 \\ 2 & 1 & 6 \end{bmatrix}$, $B = \begin{bmatrix} -36 \\ 11 \\ 16 \end{bmatrix}$,

$L = \begin{bmatrix} 4 & 0 & 0 \\ 3 & 2 & 0 \\ 1 & 1 & 1 \end{bmatrix}$, $U = \begin{bmatrix} 2 & 3 & -1 \\ 0 & -2 & 5 \\ 0 & 0 & 2 \end{bmatrix}$.

3. $A = \begin{bmatrix} 2 & 3 & 0 & 1 \\ 4 & 5 & 3 & 3 \\ -2 & -6 & 7 & 7 \\ 8 & 9 & 5 & 21 \end{bmatrix}$, $B = \begin{bmatrix} -2 \\ -2 \\ -16 \\ -66 \end{bmatrix}$,

$L = \begin{bmatrix} 1 & 0 & 0 & 0 \\ 2 & 1 & 0 & 0 \\ -1 & 3 & 1 & 0 \\ 4 & 3 & 2 & 1 \end{bmatrix}$,

$U = \begin{bmatrix} 2 & 3 & 0 & 1 \\ 0 & -1 & 3 & 1 \\ 0 & 0 & -2 & 5 \\ 0 & 0 & 0 & 4 \end{bmatrix}$.

4. $A = \begin{bmatrix} 4 & 2 & 1 & 0 \\ -4 & -6 & 1 & 3 \\ 8 & 16 & -3 & -4 \\ 20 & 10 & 4 & -3 \end{bmatrix}$, $B = \begin{bmatrix} 6 \\ 13 \\ -20 \\ 15 \end{bmatrix}$,

$L = \begin{bmatrix} 1 & 0 & 0 & 0 \\ -1 & 1 & 0 & 0 \\ 2 & -3 & 1 & 0 \\ 5 & 0 & -1 & 1 \end{bmatrix}$,

$U = \begin{bmatrix} 4 & 2 & 1 & 0 \\ 0 & -4 & 2 & 3 \\ 0 & 0 & 1 & 5 \\ 0 & 0 & 0 & 2 \end{bmatrix}$.

In Exercises 5 through 10, find an LU-*factorization of the coefficient matrix of the given linear system $AX = B$. Solve the linear system using a forward substitution followed by a back substitution.*

5. $A = \begin{bmatrix} 2 & 3 & 4 \\ 4 & 5 & 10 \\ 4 & 8 & 2 \end{bmatrix}$, $B = \begin{bmatrix} 6 \\ 16 \\ 2 \end{bmatrix}$.

6. $A = \begin{bmatrix} -3 & 1 & -2 \\ -12 & 10 & -6 \\ 15 & 13 & 12 \end{bmatrix}$, $B = \begin{bmatrix} 15 \\ 82 \\ -5 \end{bmatrix}$.

7. $A = \begin{bmatrix} 4 & 2 & 3 \\ 2 & 0 & 5 \\ 1 & 2 & 1 \end{bmatrix}$, $B = \begin{bmatrix} 1 \\ -1 \\ -3 \end{bmatrix}$.

8. $A = \begin{bmatrix} -5 & 4 & 0 & 1 \\ -30 & 27 & 2 & 7 \\ 5 & 2 & 0 & 2 \\ 10 & 1 & -2 & 1 \end{bmatrix}$, $B = \begin{bmatrix} -17 \\ -102 \\ -7 \\ -6 \end{bmatrix}$.

9. $A = \begin{bmatrix} 2 & 1 & 0 & -4 \\ 1 & 0 & 0.25 & -1 \\ -2 & -1.1 & 0.25 & 6.2 \\ 4 & 2.2 & 0.3 & -2.4 \end{bmatrix}$, $B = \begin{bmatrix} -3 \\ -1.5 \\ 5.6 \\ 2.2 \end{bmatrix}$.

10. $A = \begin{bmatrix} 4 & 1 & 0.25 & -0.5 \\ 0.8 & 0.6 & 1.25 & -2.6 \\ -1.6 & -0.08 & 0.01 & 0.2 \\ 8 & 1.52 & -0.6 & -1.3 \end{bmatrix}$,

$B = \begin{bmatrix} -0.15 \\ 9.77 \\ 1.69 \\ -4.576 \end{bmatrix}$.

11. In the software you are using investigate to see if there is a command for obtaining an LU-factorization of a matrix. If there is, use it to find the LU-factorization of matrix A in Example 2. The result obtained in your software need not be that given in Example 2 or 3, because there are many ways to compute such a factorization. Also, some software does not explicitly display L and U, but gives a matrix from which L and U can be "decoded." See the documentation on your software for more details.

12. In the software you are using investigate to see if there are commands for doing forward substitution or back substitution. Experiment with the use of such commands on the linear systems using L and U from Examples 2 and 3.

Supplementary Exercises

1. Determine the number of entries on or above the main diagonal of a $k \times k$ matrix when
   (a) $k = 2$; (b) $k = 3$; (c) $k = 4$; (d) $k = n$.

2. Let
$$A = \begin{bmatrix} 0 & 2 \\ 0 & 5 \end{bmatrix}.$$
   (a) Find a $2 \times k$ matrix $B \neq O$ such that $AB = O$ for $k = 1, 2, 3, 4$.
   (b) Are your answers to part (a) unique? Explain.

3. Find all 2×2 matrices with real entries of the form
$$A = \begin{bmatrix} a & b \\ 0 & c \end{bmatrix}$$
   such that $A^2 = I_2$.

4. An $n \times n$ matrix A (with real entries) is called a **square root** of the $n \times n$ matrix B (with real entries) if $A^2 = B$.
   (a) Find a square root of $B = \begin{bmatrix} 1 & 1 \\ 0 & 1 \end{bmatrix}$.
   (b) Find a square root of $B = \begin{bmatrix} 1 & 0 & 0 \\ 0 & 0 & 0 \\ 0 & 0 & 0 \end{bmatrix}$.
   (c) Find a square root of $B = I_4$.
   (d) Show that there is no square root of
$$B = \begin{bmatrix} 0 & 1 \\ 0 & 0 \end{bmatrix}.$$

5. Find all values of a for which the following linear systems have solutions.
   (a)
$$\begin{aligned} x_1 + 2x_2 + x_3 &= a^2 \\ x_1 + x_2 + 3x_3 &= a \\ 3x_1 + 4x_2 + 7x_3 &= 8. \end{aligned}$$
   (b)
$$\begin{aligned} x_1 + 2x_2 + x_3 &= a^2 \\ x_1 + x_2 + 3x_3 &= a \\ 3x_1 + 4x_2 + 8x_3 &= 8. \end{aligned}$$

6. Find all values of a for which the following homogeneous system has nontrivial solutions.
$$\begin{aligned} (1 - a)x \qquad\quad + z &= 0 \\ - ay + z &= 0 \\ y + z &= 0. \end{aligned}$$

7. Let A be an $m \times n$ matrix.
   (a) Describe the diagonal entries of $A^T A$ in terms of the columns of A.
   (b) Prove that the diagonal entries of $A^T A$ are nonnegative.
   (c) When is $A^T A = O_n$?

8. If A is an $n \times n$ matrix, show that $(A^k)^T = (A^T)^k$ for any positive integer k.

9. Prove that every symmetric upper (or lower) triangular matrix is diagonal.

10. Prove that if A is skew symmetric and nonsingular, then A^{-1} is skew symmetric.

11. Let A be an upper triangular matrix. Show that A is nonsingular if and only if all the entries on the main diagonal of A are nonzero.

12. Show that the product of two 2×2 skew symmetric matrices is diagonal. Is this true for $n \times n$ skew symmetric matrices with $n > 2$?

13. Prove that if $\operatorname{Tr}(A^T A) = 0$, then $A = O$.

14. For $n \times n$ matrices A and B, when does $(A+B)(A-B) = A^2 - B^2$?

15. Develop a simple expression for the entries of A^n, where n is a positive integer and
$$A = \begin{bmatrix} 1 & \frac{1}{2} \\ 0 & \frac{1}{2} \end{bmatrix}.$$

16. Let A be an $n \times n$ matrix. Prove that if $AX = O$ for all $n \times 1$ matrices X, then $A = O$.

17. Let A be an $n \times n$ matrix. Prove that if $AX = X$ for all $n \times 1$ matrices X, then $A = I_n$.

18. Let A and B be $n \times n$ matrices. Prove that if $AX = BX$ for all $n \times 1$ matrices X, then $A = B$.

19. If A is an $n \times n$ matrix, then A is called **idempotent** if $A^2 = A$.
   (a) Verify that I_n and O_n are idempotent.
   (b) Find an idempotent matrix that is not I_n or O_n.
   (c) Prove that the only $n \times n$ nonsingular idempotent matrix is I_n.

20. Let A and B be $n \times n$ idempotent matrices (see Exercise 19).

(a) Show that AB is idempotent if $AB = BA$.

(b) Show that if A is idempotent, then A^T is idempotent.

(c) Is $A + B$ idempotent? Justify your answer.

(d) Find all values of k for which kA is also idempotent.

21. Let A be an idempotent matrix.

(a) Show that $A^n = A$ for all integers $n \geq 1$.

(b) Show that $I_n - A$ is also idempotent.

22. If A is an $n \times n$ matrix, then A is called **nilpotent** if $A^k = O_n$ for some positive integer k.

(a) Prove that every nilpotent matrix is singular.

(b) Verify that $A = \begin{bmatrix} 0 & 1 & 1 \\ 0 & 0 & 1 \\ 0 & 0 & 0 \end{bmatrix}$ is nilpotent.

(c) If A is nilpotent, prove that $I_n - A$ is nonsingular. [*Hint*: Find $(I_n - A)^{-1}$ in the cases $A^k = O_n$, $k = 1, 2, \ldots$, and look for a pattern.]

23. For an $n \times n$ matrix A, the main counter diagonal elements are $a_{1n}, a_{2\,n-1}, \ldots a_{n1}$. (Note that a_{ij} is a main counter diagonal element provided $i + j = n + 1$.) The sum of the main counter diagonal elements is denoted $\text{Mcd}(A)$ and we have

$$\text{Mcd}(A) = \sum_{i+j=n+1} a_{ij},$$

meaning the sum of all the entries of A whose subscripts add to $n + 1$.

(a) Prove: $\text{Mcd}(cA) = c\,\text{Mcd}(A)$, where c is a real number.

(b) Prove: $\text{Mcd}(A + B) = \text{Mcd}(A) + \text{Mcd}(B)$.

(c) Prove: $\text{Mcd}(A^T) = \text{Mcd}(A)$.

(d) Show by example that $\text{Mcd}(AB)$ need not be equal to $\text{Mcd}(BA)$.

24. An $n \times n$ matrix A is called **block diagonal** if it can be partitioned in such a way that all the nonzero entries are contained in square blocks A_{ii}.

(a) Partition the following matrix into a block diagonal matrix:

$$\begin{bmatrix} 1 & 2 & 0 & 0 \\ 3 & 4 & 0 & 0 \\ 0 & 0 & 1 & 0 \\ 0 & 0 & 2 & 3 \end{bmatrix}.$$

(b) If A is block diagonal, then the linear system $AX = B$ is said to be **uncoupled**, because it can be solved by considering the linear systems with coefficient matrices A_{ii} and right sides an appropriate portion of B. Solve $AX = B$ by "uncoupling" the linear system when A is the 4×4 matrix of part (a) and

$$B = \begin{bmatrix} 1 \\ 1 \\ 0 \\ 3 \end{bmatrix}.$$

25. Let

$$A = \begin{bmatrix} A_{11} & O \\ O & A_{22} \end{bmatrix}$$

be a block diagonal $n \times n$ matrix. Prove that A is nonsingular if and only if A_{11} and A_{22} are nonsingular.

26. Let

$$A = \begin{bmatrix} A_{11} & A_{12} \\ O & A_{22} \end{bmatrix}$$

be a partitioned matrix. If A_{11} and A_{22} are nonsingular, show that A is nonsingular and find an expression for A^{-1}.

27. Let X and Y be column matrices with n elements. The **outer product** of X and Y is the matrix product XY^T which gives the $n \times n$ matrix

$$\begin{bmatrix} x_1y_1 & x_1y_2 & \cdots & x_1y_n \\ x_2y_1 & x_2y_2 & \cdots & x_2y_n \\ \vdots & \vdots & & \vdots \\ x_ny_1 & x_ny_2 & \cdots & x_ny_n \end{bmatrix}.$$

(a) Form the outer product of X and Y, where

$$X = \begin{bmatrix} 1 \\ 2 \\ 3 \end{bmatrix} \quad \text{and} \quad Y = \begin{bmatrix} 4 \\ 5 \\ 6 \end{bmatrix}.$$

(b) Form the outer product of X and Y, where

$$X = \begin{bmatrix} 1 \\ 2 \\ 1 \\ 2 \end{bmatrix} \quad \text{and} \quad Y = \begin{bmatrix} -1 \\ 0 \\ 3 \\ 5 \end{bmatrix}.$$

28. Prove or disprove: The outer product of X and Y equals the outer product of Y and X.

29. Prove that $\text{Tr}(XY^T) = X^TY$.

30. Show that the outer product of X and Y is row equivalent either to O_n or to a matrix with $n - 1$ rows of zeros.

31. Let

$$A = \begin{bmatrix} 1 & 7 \\ 3 & 9 \\ 5 & 11 \end{bmatrix} \quad \text{and} \quad B = \begin{bmatrix} 2 & 4 \\ 6 & 8 \end{bmatrix}.$$

Verify that

$$AB = \sum_{i=1}^{2} \text{outer product of } \operatorname{col}_i(A) \text{ with } \operatorname{row}_i(B).$$

32. Let W be an $n \times 1$ matrix such that $W^T W = 1$. The $n \times n$ matrix

$$H = I_n - 2WW^T$$

is called a **Householder matrix**. [Note that a Householder matrix is the identity matrix plus a scalar multiple of an outer product. (See Exercise 27.)]

(a) Show that H is symmetric.

(b) Show that $H^{-1} = H^T$.

33. A **circulant** of order n is the $n \times n$ matrix defined by

$$C = \operatorname{circ}(c_1, c_2, \ldots, c_n)$$
$$= \begin{bmatrix} c_1 & c_2 & c_3 & \cdots & c_n \\ c_n & c_1 & c_2 & \cdots & c_{n-1} \\ c_{n-1} & c_n & c_1 & \cdots & c_{n-2} \\ \vdots & \vdots & \vdots & \cdots & \vdots \\ c_2 & c_3 & c_4 & \cdots & c_1 \end{bmatrix}.$$

The elements of each row of C are the same as those in the previous rows, but shifted one position to the right and wrapped around.

(a) Form the circulant $C = \operatorname{circ}(1, 2, 3)$.

(b) Form the circulant $C = \operatorname{circ}(1, 2, 5, -1)$.

(c) Form the circulant $C = \operatorname{circ}(1, 0, 0, 0, 0)$.

(d) Form the circulant $C = \operatorname{circ}(1, 2, 1, 0, 0)$.

34. Let $C = \operatorname{circ}(c_1, c_2, c_3)$. Under what conditions is C symmetric?

35. Let $C = \operatorname{circ}(c_1, c_2, \ldots, c_n)$ and let X be the $n \times 1$ matrix of all ones. Determine a simple expression for CX.

36. Verify that for $C = \operatorname{circ}(c_1, c_2, c_3)$, $C^T C = CC^T$.

Chapter 2

Real Vector Spaces

2.1 VECTORS IN THE PLANE AND IN 3-SPACE

In many applications we deal with measurable quantities such as pressure, mass, and speed, which can be completely described by giving their magnitude. They are called **scalars** and will be denoted by lowercase *italic* letters such as c, d, r, s, and t. There are many other measurable quantities, such as velocity, force, and acceleration, which require for their description not only magnitude but, also, a sense of direction. These are called **vectors** and their study comprises this chapter. Vectors will be denoted by lowercase **boldface** letters, such as **u**, **v**, **x**, **y**, and **z**. The reader may already have encountered vectors in elementary physics and in calculus.

Vectors in the Plane

We draw a pair of perpendicular lines intersecting at a point O, called the **origin**. One of the lines, the ***x*-axis**, is usually taken in a horizontal position. The other line, the ***y*-axis**, is then taken in a vertical position. The x- and y-axes together are called **coordinate axes** (Figure 2.1) and they form a **rectangular coordinate system**, or a **Cartesian** (after René Descartes[†]) **coordinate system**. We now choose a point

[†]René Descartes (1596–1650) was one of the best-known scientists and philosophers of his day; he was considered by some to be the founder of modern philosophy. After completing a university degree in law, he turned to the private study of mathematics, simultaneously pursuing interests in Parisian night life and in the military, volunteering for brief periods in the Dutch, Bavarian, and French armies. The most productive period of his life was 1628–1648, when he lived in Holland. In 1649 he accepted an invitation from Queen Christina of Sweden to be her private tutor and to establish an Academy of Sciences there. Unfortunately, he did not carry out this project since he died of pneumonia in 1650.

In 1619 Descartes had a dream in which he realized that the method of mathematics is the best way for obtaining truth. However, his only mathematical publication was *La Géométrie*, which appeared as an appendix to his major philosophical work, *Discours de la méthode pour bien conduire sa raison, et chercher la vérité dans les sciences* (*Discourse on the Method of Reasoning Well and Seeking Truth in the Sciences*). In *La Géométrie* he proposes the radical idea of doing geometry algebraically. To express a curve algebraically, one chooses any convenient line of reference and, on the line, a point of reference. If y represents the distance from any point of the curve to the reference line and x represents the distance along the line to the reference point, there is an equation relating x and y that represents the curve.

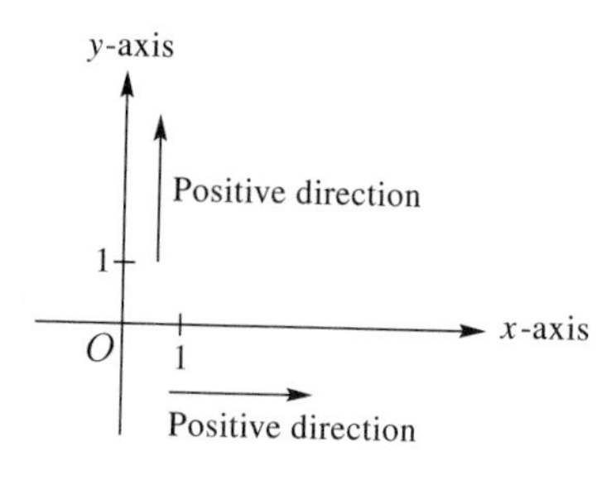

FIGURE 2.1

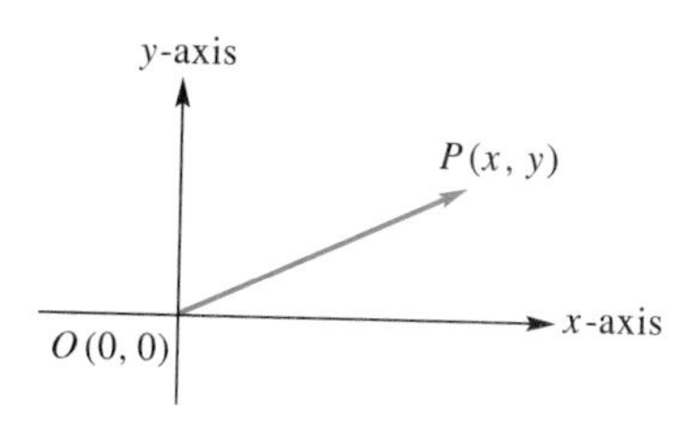

FIGURE 2.2

on the x-axis to the right of O and a point on the y-axis above O to fix the units of length and positive directions on the x- and y-axes. Frequently, but not always, these points are chosen so that they are both equidistant from O, that is, so that the same unit of length is used for both axes.

With each point P in the plane we associate an ordered pair (x, y) of real numbers, its **coordinates**. Conversely, we can associate a point in the plane with each ordered pair of real numbers. Point P with coordinates (x, y) is denoted by $P(x, y)$, or simply by (x, y). The set of all points in the plane is denoted by R^2; it is called **2-space**.

Consider the 2×1 matrix

$$X = \begin{bmatrix} x \\ y \end{bmatrix},$$

where x and y are real numbers. With X we associate the directed line segment with the initial point the origin and terminal point $P(x, y)$. The directed line segment from O to P is denoted by $\overrightarrow{OP}$; O is called its **tail** and P its **head**. We distinguish tail and head by placing an arrow at the head (Figure 2.2). A directed line segment has a **direction**, indicated by the arrow at its head. The **magnitude** of a directed line segment is its length. Thus a directed line segment can be used to describe force, velocity, or acceleration. Conversely, with the directed line segment $\overrightarrow{OP}$ with tail $O(0, 0)$ and head $P(x, y)$ we can associate the matrix

$$\begin{bmatrix} x \\ y \end{bmatrix}.$$

Definition 2.1

A **vector in the plane** is a 2×1 matrix

$$\mathbf{x} = \begin{bmatrix} x \\ y \end{bmatrix},$$

where x and y are real numbers, called the **components** of $\mathbf{x}$. We refer to a vector in the plane merely as a **vector**.

▲

Thus, with every vector, we can associate a directed line segment and, conversely, with every directed line segment we can associate a vector. Frequently, the notions of directed line segment and vector are used interchangeably, and a directed line segment is called a **vector**.

Since a vector is a matrix, the vectors

$$\mathbf{u} = \begin{bmatrix} x_1 \\ y_1 \end{bmatrix} \quad \text{and} \quad \mathbf{v} = \begin{bmatrix} x_2 \\ y_2 \end{bmatrix}$$

are said to be **equal** if $x_1 = x_2$ and $y_1 = y_2$. That is, two vectors are equal if their respective components are equal.

The systematic use of "Cartesian" coordinates described above was introduced later in the seventeenth century by mathematicians carrying on Descartes' work.

EXAMPLE 1

The vectors

$$\begin{bmatrix} a+b \\ 2 \end{bmatrix} \quad \text{and} \quad \begin{bmatrix} 3 \\ a-b \end{bmatrix}$$

are equal if

$$\begin{aligned} a + b &= 3 \\ a - b &= 2, \end{aligned}$$

which means (verify) that $a = \frac{5}{2}$ and $b = \frac{1}{2}$. ■

Frequently, in physical applications it is necessary to deal with a directed line segment $\overrightarrow{PQ}$ from the point $P(x, y)$ (not the origin) to the point $Q(x', y')$, as shown in Figure 2.3(a). Such a directed line segment will also be called a **vector in the plane**, or simply a **vector** with **tail** $P(x, y)$ and **head** $Q(x', y')$. The **components** of such a vector are $x' - x$ and $y' - y$. Thus the vector $\overrightarrow{PQ}$ in Figure 2.3(a) can also be represented by the vector

$$\begin{bmatrix} x' - x \\ y' - y \end{bmatrix}$$

with tail O and head $P''(x' - x, y' - y)$. Two such vectors in the plane will be called **equal** if their respective components are equal. Consider the vectors $\overrightarrow{P_1Q_1}$, $\overrightarrow{P_2Q_2}$, and $\overrightarrow{P_3Q_3}$ joining the points $P_1(3, 2)$ and $Q_1(5, 5)$, $P_2(0, 0)$ and $Q_2(2, 3)$, $P_3(-3, 1)$ and $Q_3(-1, 4)$, respectively, as shown in Figure 2.3(b). Since they all have the same components, they are equal.

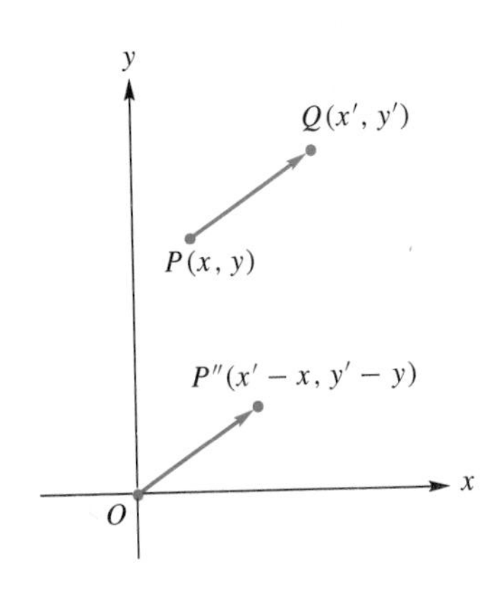

(a) Different directed line segments representing the same vector.

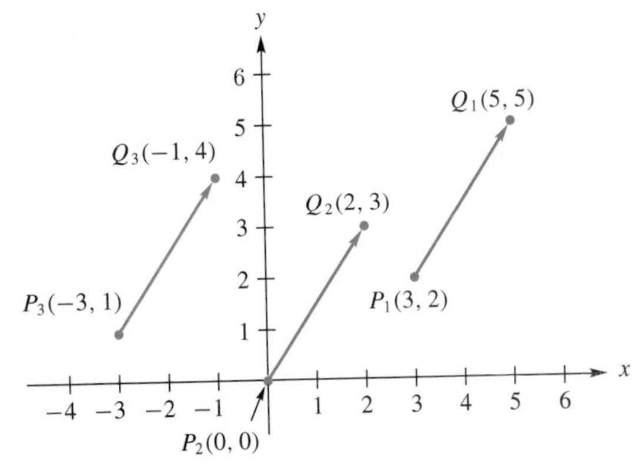

(b) Vectors in the plane.

FIGURE 2.3

Moreover, the head $Q_4(x_4', y_4')$ of the vector

$$\overrightarrow{P_4Q_4} = \begin{bmatrix} 2 \\ 3 \end{bmatrix} = \overrightarrow{P_2Q_2},$$

with tail $P_4(-5, 2)$, can be determined as follows. We must have $x_4' - (-5) = 2$ and $y_4' - 2 = 3$, so that $x_4' = 2 - 5 = -3$ and $y_4' = 3 + 2 = 5$. Similarly, the tail $P_5(x_5, y_5)$ of the vector

$$\overrightarrow{P_5Q_5} = \begin{bmatrix} 2 \\ 3 \end{bmatrix}$$

with head $Q_5(8, 6)$ is determined as follows. We must have $8 - x_5 = 2$ and $6 - y_5 = 3$, so that $x_5 = 8 - 2 = 6$ and $y_5 = 6 - 3 = 3$.

With each vector

$$\mathbf{x} = \begin{bmatrix} x \\ y \end{bmatrix}$$

we can also associate the unique point $P(x, y)$; conversely, with each point $P(x, y)$ we associate the unique vector

$$\begin{bmatrix} x \\ y \end{bmatrix}.$$

Hence we also write the vector $\mathbf{x}$ as (x, y). Of course, this association is carried out by means of the directed line segment $\overrightarrow{OP}$, where O is the origin and P is the point with coordinates (x, y) (Figure 2.2).

Thus the plane may be viewed both as the set of all points or as the set of all vectors. For this reason, and depending upon the context, we sometimes take R^2 as the set of all ordered pairs (x, y) and sometimes as the set of all 2×1 matrices

$$\begin{bmatrix} x \\ y \end{bmatrix}.$$

Definition 2.2

Let

$$\mathbf{u} = \begin{bmatrix} u_1 \\ u_2 \end{bmatrix} \quad \text{and} \quad \mathbf{v} = \begin{bmatrix} v_1 \\ v_2 \end{bmatrix}$$

be two vectors in the plane. The **sum** of the vectors $\mathbf{u}$ and $\mathbf{v}$ is the vector

$$\mathbf{u} + \mathbf{v} = \begin{bmatrix} u_1 + v_1 \\ u_2 + v_2 \end{bmatrix}.$$

▲

Remark Observe that vector addition is a special case of matrix addition.

EXAMPLE 2

Let $\mathbf{u} = \begin{bmatrix} 2 \\ 3 \end{bmatrix}$ and $\mathbf{v} = \begin{bmatrix} 3 \\ -4 \end{bmatrix}$. Then

$$\mathbf{u} + \mathbf{v} = \begin{bmatrix} 2+3 \\ 3+(-4) \end{bmatrix} = \begin{bmatrix} 5 \\ -1 \end{bmatrix}.$$

See Figure 2.4. ■

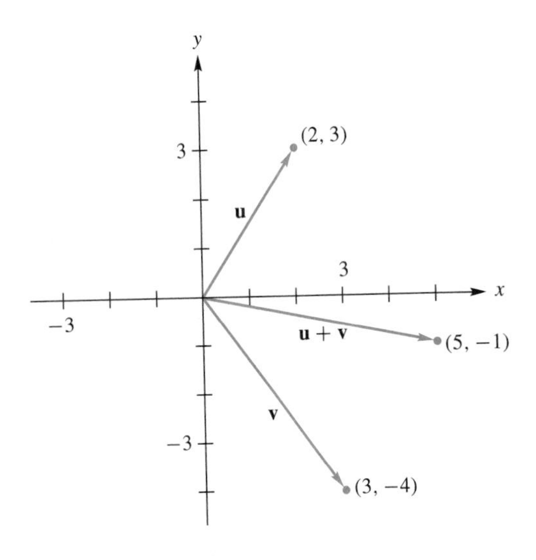

FIGURE 2.4

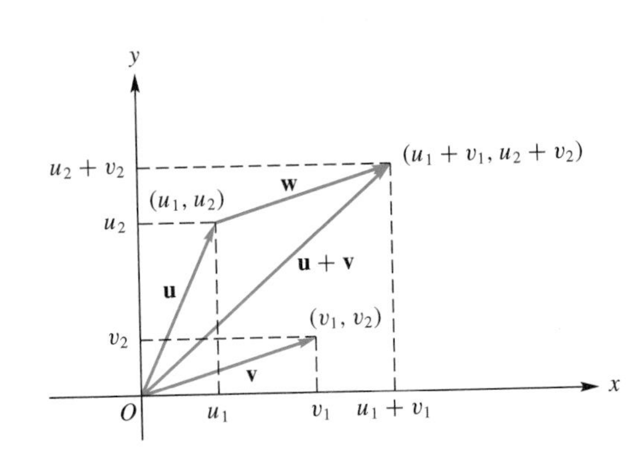

FIGURE 2.5
Vector addition.

We can interpret vector addition geometrically, as follows. In Figure 2.5 the directed line segment **w** is parallel to **v**, it has the same length as **v**, and its tail is the head (u_1, u_2) of **u**, so its head is $(u_1 + v_1, u_2 + v_2)$. Thus the vector with tail O and head $(u_1 + v_1, u_2 + v_2)$ is $\mathbf{u} + \mathbf{v}$. We can also describe $\mathbf{u} + \mathbf{v}$ as the diagonal of the parallelogram defined by **u** and **v**, as shown in Figure 2.6.

Definition 2.3

If $\mathbf{u} = \begin{bmatrix} u_1 \\ u_2 \end{bmatrix}$ is a vector and c is a scalar (a real number), then the **scalar multiple** $c\mathbf{u}$ of **u** by c is the vector $\begin{bmatrix} cu_1 \\ cu_2 \end{bmatrix}$. Thus the scalar multiple $c\mathbf{u}$ is obtained by multiplying each component of **u** by c. If $c > 0$, then $c\mathbf{u}$ is in the same direction as **u**, whereas if $d < 0$, then $d\mathbf{u}$ is in the opposite direction (Figure 2.7). ▲

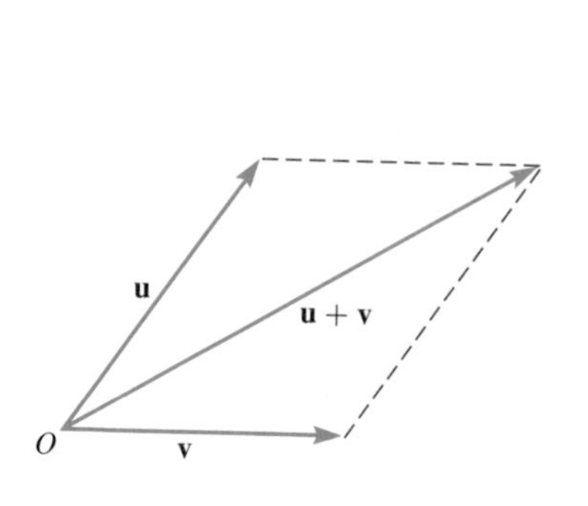

FIGURE 2.6

Vector addition.

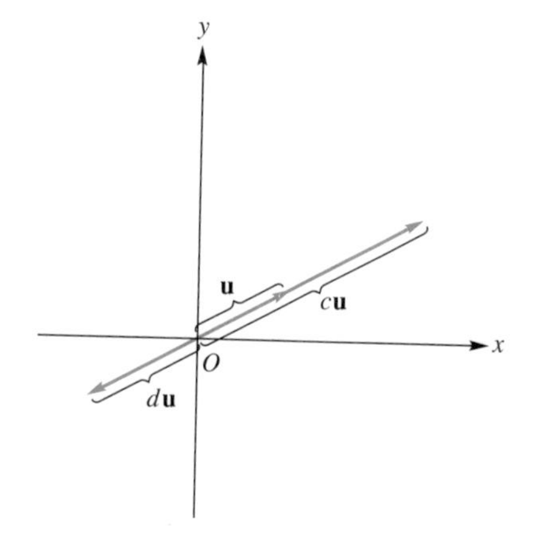

FIGURE 2.7

Scalar multiplication.

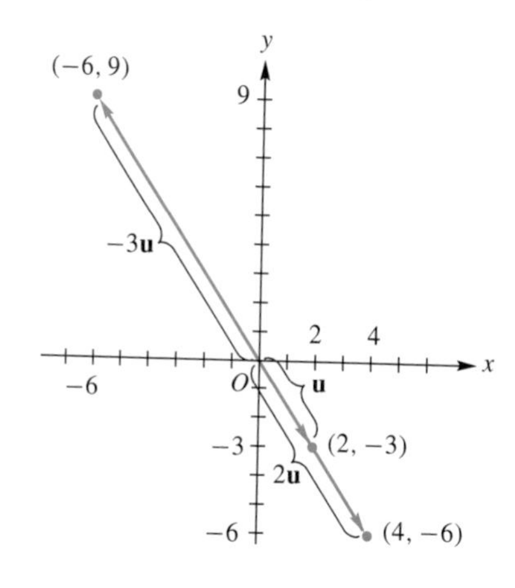

FIGURE 2.8

EXAMPLE 3

If $c = 2$, $d = -3$, and $\mathbf{u} = \begin{bmatrix} 2 \\ -3 \end{bmatrix}$, then

$$c\mathbf{u} = 2\begin{bmatrix} 2 \\ -3 \end{bmatrix} = \begin{bmatrix} 2(2) \\ 2(-3) \end{bmatrix} = \begin{bmatrix} 4 \\ -6 \end{bmatrix}$$

and

$$d\mathbf{u} = -3\begin{bmatrix} 2 \\ -3 \end{bmatrix} = \begin{bmatrix} (-3)(2) \\ (-3)(-3) \end{bmatrix} = \begin{bmatrix} -6 \\ 9 \end{bmatrix},$$

which are shown in Figure 2.8. ■

The vector

$$\begin{bmatrix} 0 \\ 0 \end{bmatrix}$$

is called the **zero vector** and is denoted by $\mathbf{0}$. If $\mathbf{u}$ is any vector, it follows that (Exercise 21)

$$\mathbf{u} + \mathbf{0} = \mathbf{u}.$$

We can also show (Exercise 22) that

$$\mathbf{u} + (-1)\mathbf{u} = \mathbf{0},$$

and we write $(-1)\mathbf{u}$ as $-\mathbf{u}$ and call it the **negative** of $\mathbf{u}$. Moreover, we write $\mathbf{u} + (-1)\mathbf{v}$ as $\mathbf{u} - \mathbf{v}$ and call it the **difference between u** and $\mathbf{v}$. It is shown in Figure 2.9(a). Observe that while vector addition gives one diagonal of a parallelogram, vector subtraction gives the other diagonal [see Figure 2.9(b)].

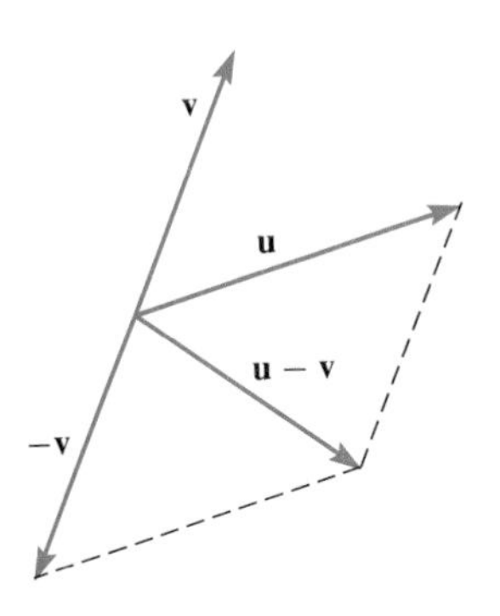

(a) Difference between vectors.

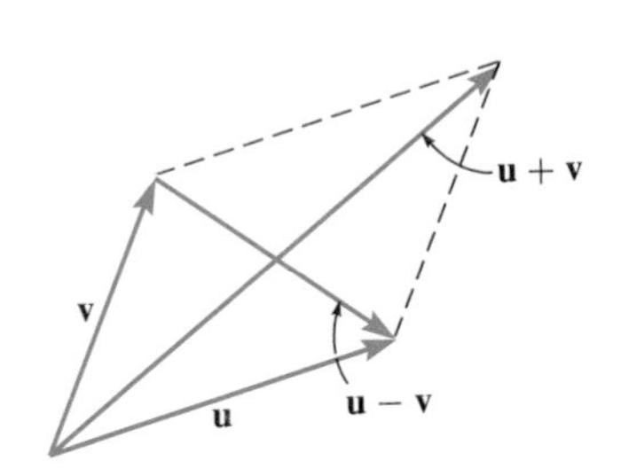

(b) Vector sum and vector difference.

FIGURE 2.9

Vectors in Space

The foregoing discussion of vectors in the plane can be generalized to vectors in space, as follows. We first fix a **coordinate system** by choosing a point, called the **origin**, and three lines, called the **coordinate axes**, each passing through the origin, so that each line is perpendicular to the other two. These lines are individually called the x-, y-, and z-axes. On each of these axes we choose a point fixing the units of length and positive directions on the coordinate axes. Frequently, but not always, the same unit of length is used for all the coordinate axes. In Figure 2.10 we show two of the many possible coordinate systems.

The coordinate system shown in Figure 2.10(a) is called a **right-handed coordinate system**; the one in Figure 2.10(b) is called **left-handed**. A right-handed system is characterized by the following property. If we curl the fingers of the right hand in the direction of a 90° rotation from the positive x-axis to the positive y-axis, then the thumb will point in the direction of the positive z-axis (see Figure 2.11).

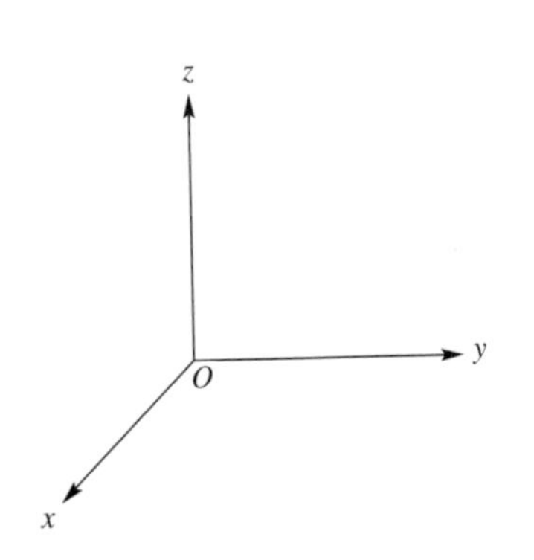

(a) Right-handed coordinate system.

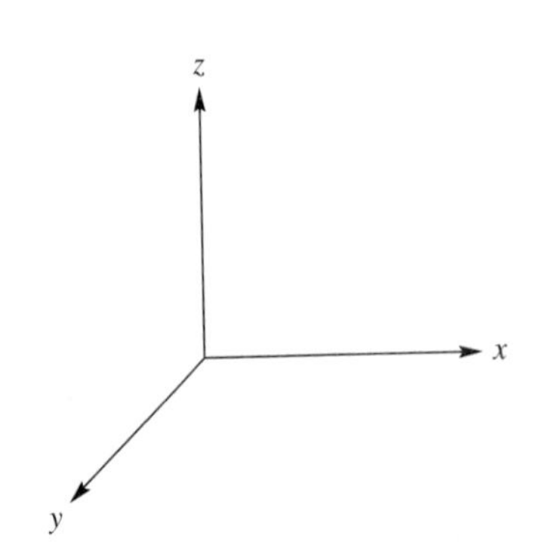

(b) Left-handed coordinate system.

FIGURE 2.10

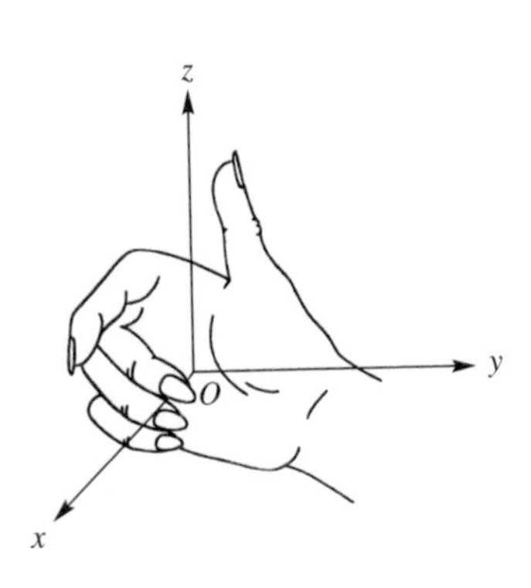

FIGURE 2.11

If we rotate the x-axis counterclockwise toward the y-axis, then a right-hand screw will move in the positive z-direction (see Figure 2.11).

With each point P in space we associate an ordered triple (x, y, z) of real numbers, its coordinates. Conversely, we can associate a point in space with each ordered triple of real numbers. The point P with coordinates x, y, and z is denoted by $P(x, y, z)$, or simply by (x, y, z). The set of all points in space is called **3-space** and it is denoted by R^3.

A **vector in space**, or simply a **vector**, is a 3×1 matrix

$$\mathbf{x} = \begin{bmatrix} x \\ y \\ z \end{bmatrix},$$

where x, y, and z are real numbers, called the **components** of vector $\mathbf{x}$. Two vectors in space are said to be **equal** if their respective components are equal.

As in the plane, with the vector

$$\mathbf{x} = \begin{bmatrix} x \\ y \\ z \end{bmatrix}$$

we associate the directed line segment $\overrightarrow{OP}$, whose tail is $O(0, 0, 0)$ and whose head is $P(x, y, z)$; conversely, with each such directed line segment we associate the vector $\mathbf{x}$ [see Figure 2.12(a)]. Thus we can also write the vector $\mathbf{x}$ as (x, y, z). Again, as in the plane, in physical applications we often deal with a directed line segment $\overrightarrow{PQ}$, from the point $P(x, y, z)$ (not the origin) to the point $Q(x', y', z')$, as shown in Figure 2.12(b). Such a directed line segment will also be called a **vector in R^3**, or simply a **vector** with tail $P(x, y, z)$ and head $Q(x', y', z')$. The components of such a vector are $x' - x$, $y' - y$, and $z' - z$. Two such vectors in R^3 will be called **equal** if their respective components are equal. Thus the vector $\overrightarrow{PQ}$

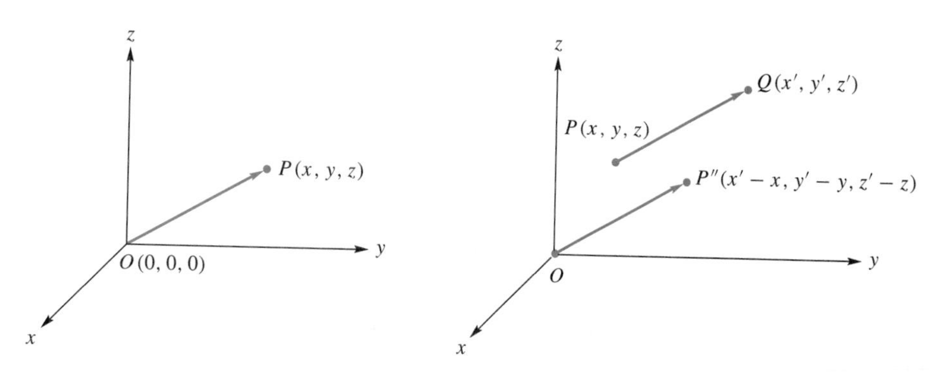

(a) A vector in R^3.

(b) Different directed line segments representing the same vector.

FIGURE 2.12

in Figure 2.12(b) can also be represented by the vector $\begin{bmatrix} x'-x \\ y'-y \\ z'-z \end{bmatrix}$ with tail O and head $P''(x'-x, y'-y, z'-z)$.

If $\mathbf{u} = \begin{bmatrix} u_1 \\ u_2 \\ u_3 \end{bmatrix}$ and $\mathbf{v} = \begin{bmatrix} v_1 \\ v_2 \\ v_3 \end{bmatrix}$ are vectors in R^3 and c is a scalar, then the **sum** $\mathbf{u}+\mathbf{v}$ and the **scalar multiple** $c\mathbf{u}$ are defined, respectively, as

$$\mathbf{u}+\mathbf{v} = \begin{bmatrix} u_1+v_1 \\ u_2+v_2 \\ u_3+v_3 \end{bmatrix} \quad \text{and} \quad c\mathbf{u} = \begin{bmatrix} cu_1 \\ cu_2 \\ cu_3 \end{bmatrix}.$$

The sum is shown in Figure 2.13, which resembles Figure 2.5, and the scalar multiple is shown in Figure 2.14, which resembles Figure 2.8.

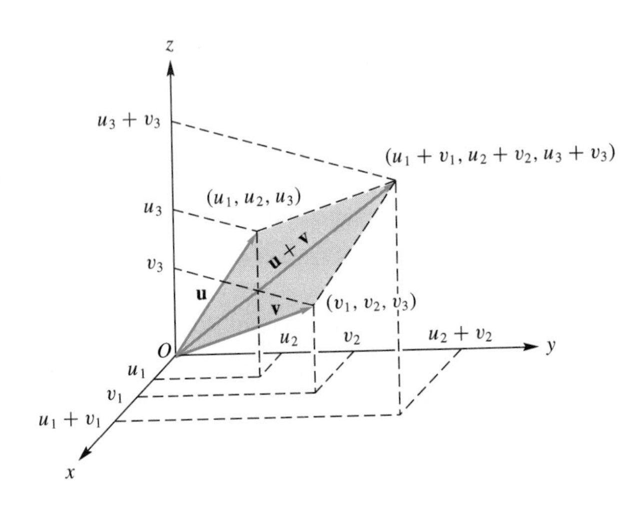

FIGURE 2.13

Vector addition.

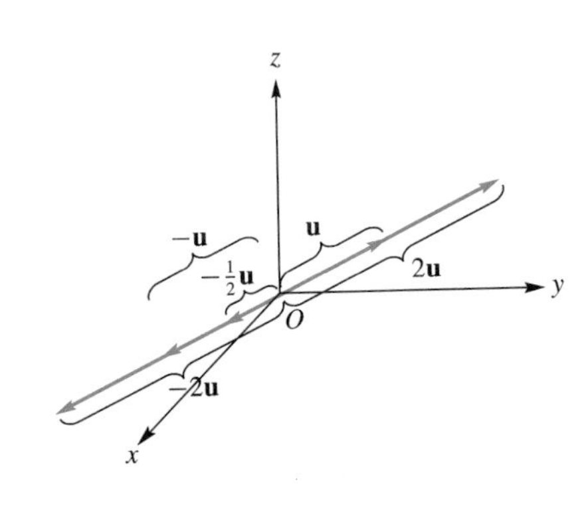

FIGURE 2.14

Scalar multiplication.

EXAMPLE 4

Let

$$\mathbf{u} = \begin{bmatrix} 2 \\ 3 \\ -1 \end{bmatrix} \quad \text{and} \quad \mathbf{v} = \begin{bmatrix} 3 \\ -4 \\ 2 \end{bmatrix}.$$

Compute: (a) $\mathbf{u}+\mathbf{v}$; (b) $-2\mathbf{u}$; (c) $3\mathbf{u}-2\mathbf{v}$.

Solution (a) $\mathbf{u}+\mathbf{v} = \begin{bmatrix} 2+3 \\ 3+(-4) \\ -1+2 \end{bmatrix} = \begin{bmatrix} 5 \\ -1 \\ 1 \end{bmatrix}.$

(b) $-2\mathbf{u} = \begin{bmatrix} -2(2) \\ -2(3) \\ -2(-1) \end{bmatrix} = \begin{bmatrix} -4 \\ -6 \\ 2 \end{bmatrix}.$

(c) $3\mathbf{u} - 2\mathbf{v} = \begin{bmatrix} 3(2) \\ 3(3) \\ 3(-1) \end{bmatrix} - \begin{bmatrix} 2(3) \\ 2(-4) \\ 2(2) \end{bmatrix} = \begin{bmatrix} 0 \\ 17 \\ -7 \end{bmatrix}.$ ■

The zero vector in R^3 is denoted by $\mathbf{0}$, where

$$\mathbf{0} = \begin{bmatrix} 0 \\ 0 \\ 0 \end{bmatrix}.$$

The vector $\mathbf{0}$ has the property that if $\mathbf{u}$ is any vector in R^3, then

$$\mathbf{u}+\mathbf{0} = \mathbf{u}.$$

The negative of the vector $\mathbf{u} = \begin{bmatrix} u_1 \\ u_2 \\ u_3 \end{bmatrix}$ is the vector $-\mathbf{u} = \begin{bmatrix} -u_1 \\ -u_2 \\ -u_3 \end{bmatrix}$ and

$$\mathbf{u}+(-\mathbf{u}) = \mathbf{0}.$$

Observe that we have defined a vector in the plane as an ordered pair of real numbers, or as a 2×1 matrix. Similarly, a vector in space is an ordered triple of real numbers, or a 3×1 matrix. However, in physics we often treat a vector as a directed line segment. Thus we have three very different representations of a vector, and one can then ask why all three are legitimately valid. That is, why are we justified in referring to an ordered pair of real numbers, a 2×1 matrix, and a directed line segment by the same name, "vector"?

To answer this question, we first observe that, mathematically speaking, the only thing that concerns us is the behavior of the object we call "vector." It turns out that all three objects behave, from an algebraic point of view, in exactly the same manner. Moreover, many other objects that arise naturally in applied problems behave, algebraically speaking, as do the above-mentioned objects. To a mathematician this is a perfect situation. For we can now abstract those features that all such objects have in common (i.e., those properties that make them all behave alike) and define a new structure. The great advantage of doing this is that we can now talk about properties of all such objects at the same time without having to refer to any one object in particular. This, of course, is much more efficient than studying the properties of each object separately. For example, the following theorem summarizes the properties of addition and scalar multiplication for vectors in the plane and in space. Moreover, this theorem will serve as the model for the generalization of the set of all vectors in the plane or in space to a more abstract setting.

Theorem 2.1

If **u**, **v**, *and* **w** *are vectors in* R^2 *or* R^3, *and* c *and* d *are real scalars, then*

(a) $\mathbf{u}+\mathbf{v}=\mathbf{v}+\mathbf{u}$.
(b) $\mathbf{u}+(\mathbf{v}+\mathbf{w})=(\mathbf{u}+\mathbf{v})+\mathbf{w}$.
(c) $\mathbf{u}+\mathbf{0}=\mathbf{0}+\mathbf{u}=\mathbf{u}$.
(d) $\mathbf{u}+(-\mathbf{u})=\mathbf{0}$.
(e) $c(\mathbf{u}+\mathbf{v})=c\mathbf{u}+c\mathbf{v}$.
(f) $(c+d)\mathbf{u}=c\mathbf{u}+d\mathbf{u}$.
(g) $c(d\mathbf{u})=(cd)\mathbf{u}$.
(h) $1\mathbf{u}=\mathbf{u}$.

Proof

(a) Suppose that **u** and **v** are vectors in R^2 so that

$$\mathbf{u}=\begin{bmatrix}u_1\\u_2\end{bmatrix}\quad\text{and}\quad\mathbf{v}=\begin{bmatrix}v_1\\v_2\end{bmatrix}.$$

Then

$$\mathbf{u}+\mathbf{v}=\begin{bmatrix}u_1+v_1\\u_2+v_2\end{bmatrix}\quad\text{and}\quad\mathbf{v}+\mathbf{u}=\begin{bmatrix}v_1+u_1\\v_2+u_2\end{bmatrix}.$$

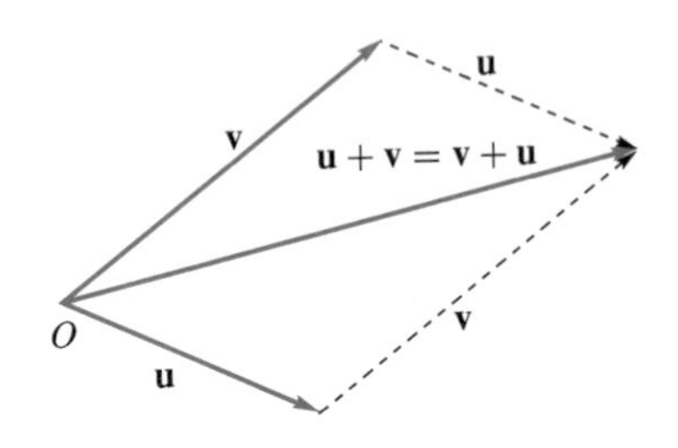

FIGURE 2.15
Vector addition.

Since the components of **u** and **v** are real numbers, $u_1+v_1=v_1+u_1$ and $u_2+v_2=v_2+u_2$. Therefore,

$$\mathbf{u}+\mathbf{v}=\mathbf{v}+\mathbf{u}.$$

A similar proof can be given if **u** and **v** are vectors in R^3.

Property (a) can also be established geometrically, as shown in Figure 2.15. The proofs of the remaining properties will be left as exercises. Remember, they can all be proved by either an algebraic or a geometric approach. ●

2.1 Exercises

1. Sketch a directed line segment in R^2, representing each of the following vectors.

   (a) $\mathbf{u}=\begin{bmatrix}-2\\3\end{bmatrix}$. (b) $\mathbf{v}=\begin{bmatrix}3\\4\end{bmatrix}$.

   (c) $\mathbf{w}=\begin{bmatrix}-3\\-3\end{bmatrix}$. (d) $\mathbf{z}=\begin{bmatrix}0\\-3\end{bmatrix}$.

2. Determine the head of the vector $\begin{bmatrix}-2\\5\end{bmatrix}$ whose tail is $(-3, 2)$. Make a sketch.

3. Determine the tail of the vector $\begin{bmatrix}2\\6\end{bmatrix}$ whose head is $(1, 2)$. Make a sketch.

4. Determine the tail of the vector $\begin{bmatrix}2\\4\\-1\end{bmatrix}$ whose head is $(3, -2, 2)$.

5. For what values of a and b are the vectors $\begin{bmatrix}a-b\\2\end{bmatrix}$ and $\begin{bmatrix}4\\a+b\end{bmatrix}$ equal?

6. For what values of a, b, and c are the vectors $\begin{bmatrix} 2a-b \\ a-2b \\ 6 \end{bmatrix}$ and $\begin{bmatrix} -2 \\ 2 \\ a+b-2c \end{bmatrix}$ equal?

In Exercises 7 and 8, determine the components of the vector $\overrightarrow{PQ}$.

7. (a) $P(1, 2)$, $Q(3, 5)$.
(b) $P(-2, 2, 3)$, $Q(-3, 5, 2)$.

8. (a) $P(-1, 0)$, $Q(-3, -4)$.
(b) $P(1, 1, 2)$, $Q(1, -2, -4)$.

In Exercises 9 and 10, find a vector whose tail is the origin representing the vector $\overrightarrow{PQ}$.

9. (a) $P(-1, 2)$, $Q(3, 5)$.
(b) $P(1, 1, -2)$, $Q(3, 4, 5)$.

10. (a) $P(2, -3)$, $Q(-2, 4)$.
(b) $P(-2, -3, 4)$, $Q(0, 0, 1)$.

11. Compute $\mathbf{u}+\mathbf{v}$, $\mathbf{u}-\mathbf{v}$, $2\mathbf{u}$, and $3\mathbf{u}-2\mathbf{v}$ if:

(a) $\mathbf{u} = \begin{bmatrix} 2 \\ 3 \end{bmatrix}$, $\mathbf{v} = \begin{bmatrix} -2 \\ 5 \end{bmatrix}$.

(b) $\mathbf{u} = \begin{bmatrix} 0 \\ 3 \end{bmatrix}$, $\mathbf{v} = \begin{bmatrix} 3 \\ 2 \end{bmatrix}$.

(c) $\mathbf{u} = \begin{bmatrix} 2 \\ 6 \end{bmatrix}$, $\mathbf{v} = \begin{bmatrix} 3 \\ 2 \end{bmatrix}$.

12. Compute $\mathbf{u}+\mathbf{v}$, $2\mathbf{u}-\mathbf{v}$, $3\mathbf{u}-2\mathbf{v}$, and $\mathbf{0}-3\mathbf{v}$ if:

(a) $\mathbf{u} = \begin{bmatrix} 1 \\ 2 \\ 3 \end{bmatrix}$, $\mathbf{v} = \begin{bmatrix} 2 \\ 0 \\ 1 \end{bmatrix}$.

(b) $\mathbf{u} = \begin{bmatrix} 2 \\ -1 \\ 4 \end{bmatrix}$, $\mathbf{v} = \begin{bmatrix} 1 \\ 2 \\ -3 \end{bmatrix}$.

(c) $\mathbf{u} = \begin{bmatrix} 1 \\ 0 \\ -1 \end{bmatrix}$, $\mathbf{v} = \begin{bmatrix} -1 \\ 1 \\ 4 \end{bmatrix}$.

13. Let

$$\mathbf{u} = \begin{bmatrix} 2 \\ 3 \\ -1 \end{bmatrix}, \quad \mathbf{v} = \begin{bmatrix} -1 \\ 2 \\ 4 \end{bmatrix}, \quad \mathbf{w} = \begin{bmatrix} 0 \\ 1 \\ -1 \end{bmatrix}.$$

$c = -2$, and $d = 3$. Compute:

(a) $\mathbf{u}+\mathbf{v}$.

(b) $c\mathbf{u}+d\mathbf{w}$.

(c) $\mathbf{u}+\mathbf{v}+\mathbf{w}$.

(d) $c\mathbf{u}+d\mathbf{v}+\mathbf{w}$.

14. Let

$$\mathbf{x} = \begin{bmatrix} 1 \\ 2 \end{bmatrix}, \quad \mathbf{y} = \begin{bmatrix} -3 \\ 4 \end{bmatrix},$$

$$\mathbf{z} = \begin{bmatrix} r \\ 4 \end{bmatrix}, \quad \text{and} \quad \mathbf{u} = \begin{bmatrix} -2 \\ s \end{bmatrix}.$$

Find r and s so that:

(a) $\mathbf{z} = 2\mathbf{x}$. (b) $\frac{3}{2}\mathbf{u} = \mathbf{y}$. (c) $\mathbf{z}+\mathbf{u} = \mathbf{x}$.

15. Let

$$\mathbf{x} = \begin{bmatrix} 1 \\ -2 \\ 3 \end{bmatrix}, \quad \mathbf{y} = \begin{bmatrix} -3 \\ 1 \\ 3 \end{bmatrix}, \quad \mathbf{z} = \begin{bmatrix} r \\ -1 \\ s \end{bmatrix},$$

$$\text{and} \quad \mathbf{u} = \begin{bmatrix} 3 \\ t \\ 2 \end{bmatrix}.$$

Find r, s, and t so that:

(a) $\mathbf{z} = \frac{1}{2}\mathbf{x}$. (b) $\mathbf{z}+\mathbf{u} = \mathbf{x}$. (c) $\mathbf{z}-\mathbf{x} = \mathbf{y}$.

16. If possible, find scalars c_1 and c_2 so that

$$c_1 \begin{bmatrix} 1 \\ -2 \end{bmatrix} + c_2 \begin{bmatrix} 3 \\ -4 \end{bmatrix} = \begin{bmatrix} -5 \\ 6 \end{bmatrix}.$$

17. If possible, find scalars c_1, c_2, and c_3 so that

$$c_1 \begin{bmatrix} 1 \\ 2 \\ -3 \end{bmatrix} + c_2 \begin{bmatrix} -1 \\ 1 \\ 1 \end{bmatrix} + c_3 \begin{bmatrix} -1 \\ 4 \\ -1 \end{bmatrix} = \begin{bmatrix} 2 \\ -2 \\ 3 \end{bmatrix}.$$

18. If possible, find scalars c_1 and c_2, not both zero, so that

$$c_1 \begin{bmatrix} 1 \\ 2 \end{bmatrix} + c_2 \begin{bmatrix} 3 \\ 4 \end{bmatrix} = \begin{bmatrix} 0 \\ 0 \end{bmatrix}.$$

19. If possible, find scalars c_1, c_2, and c_3, not all zero, so that

$$c_1 \begin{bmatrix} 1 \\ 2 \\ -1 \end{bmatrix} + c_2 \begin{bmatrix} 1 \\ 3 \\ -2 \end{bmatrix} + c_3 \begin{bmatrix} 3 \\ 7 \\ -4 \end{bmatrix} = \begin{bmatrix} 0 \\ 0 \\ 0 \end{bmatrix}.$$

20. Let

$$\mathbf{i} = \begin{bmatrix} 1 \\ 0 \\ 0 \end{bmatrix}, \quad \mathbf{j} = \begin{bmatrix} 0 \\ 1 \\ 0 \end{bmatrix}, \quad \text{and} \quad \mathbf{k} = \begin{bmatrix} 0 \\ 0 \\ 1 \end{bmatrix}.$$

Find scalars c_1, c_2, and c_3 so that any vector $\mathbf{u} = \begin{bmatrix} r \\ s \\ t \end{bmatrix}$ can be written as

$$\mathbf{u} = c_1\mathbf{i} + c_2\mathbf{j} + c_3\mathbf{k}.$$

21. Show that if $\mathbf{u}$ is a vector in R^2 or R^3, then $\mathbf{u} + \mathbf{0} = \mathbf{u}$.

22. Show that if $\mathbf{u}$ is a vector in R^2 or R^3, then $\mathbf{u} + (-1)\mathbf{u} = \mathbf{0}$.

23. Prove parts (b) and (d) through (h) of Theorem 2.1.

24. Determine if the software you use supports graphics. If it does, experiment with plotting vectors in R^2. Usually, you must supply coordinates for the head and tail of the vector and then tell the software to connect these points. The points in Exercises 7(a) and 8(a) can be used in this regard.

25. Assuming that the software you use supports graphics (see Exercise 24), plot the vector

$$\mathbf{v} = \begin{bmatrix} 3 \\ 4 \end{bmatrix}$$

on the same coordinate axes for each of the following.

(a) $\mathbf{v}$ is to have head $(1, 1)$.

(b) $\mathbf{v}$ is to have head $(2, 3)$.

26. Determine if the software you use supports three-dimensional graphics, that is, plots points in R^3. If it does, experiment with plotting points and connecting them to form vectors in R^3.

2.2 VECTOR SPACES

Definition 2.4

A **real vector space** is a set V of elements on which we have two operations $\oplus$ and $\odot$ defined with the following properties:

(a) If $\mathbf{u}$ and $\mathbf{v}$ are any elements in V, then $\mathbf{u} \oplus \mathbf{v}$ is in V. (We say that V is **closed** under the operation $\oplus$.)

(1) $\mathbf{u} \oplus \mathbf{v} = \mathbf{v} \oplus \mathbf{u}$ for all $\mathbf{u}$, $\mathbf{v}$ in V.

(2) $\mathbf{u} \oplus (\mathbf{v} \oplus \mathbf{w}) = (\mathbf{u} \oplus \mathbf{v}) \oplus \mathbf{w}$ for all $\mathbf{u}$, $\mathbf{v}$, $\mathbf{w}$ in V.

(3) There exists an element $\mathbf{0}$ in V such that $\mathbf{u} \oplus \mathbf{0} = \mathbf{0} \oplus \mathbf{u} = \mathbf{u}$ for any $\mathbf{u}$ in V.

(4) For each $\mathbf{u}$ in V there exists an element $-\mathbf{u}$ in V such that $\mathbf{u} \oplus -\mathbf{u} = -\mathbf{u} \oplus \mathbf{u} = \mathbf{0}$.

(b) If $\mathbf{u}$ is any element in V and c is any real number, then $c \odot \mathbf{u}$ is in V (i.e., V is closed under the operation $\odot$).

(5) $c \odot (\mathbf{u} \oplus \mathbf{v}) = c \odot \mathbf{u} \oplus c \odot \mathbf{v}$ for any $\mathbf{u}$, $\mathbf{v}$ in V and any real number c.

(6) $(c + d) \odot \mathbf{u} = c \odot \mathbf{u} \oplus d \odot \mathbf{u}$ for any $\mathbf{u}$ in V and any real numbers c and d.

(7) $c \odot (d \odot \mathbf{u}) = (cd) \odot \mathbf{u}$ for any $\mathbf{u}$ in V and any real numbers c and d.

(8) $1 \odot \mathbf{u} = \mathbf{u}$ for any $\mathbf{u}$ in V.

The elements of V are called **vectors**; the elements of the set of real numbers R are called **scalars**. The operation $\oplus$ is called **vector addition**; the operation $\odot$ is called

scalar multiplication. The vector **0** in property (3) is called a **zero vector**. The vector $-\mathbf{u}$ in property (4) is called a **negative of u**. It can be shown (see Exercises 15 and 16) that **0** and $-\mathbf{u}$ are unique. ▲

If we allow the scalars to be complex numbers, we obtain a **complex vector space**. More generally, the scalars can be members of a field F, and we obtain a vector space over F. Such spaces are important in many applications in mathematics and the physical sciences. We provide a brief introduction to complex vector spaces in Appendix B. However, in this book we limit our study to real vector spaces.

In order to specify a vector space, we must be given a set V and two operations $\oplus$ and $\odot$ satisfying all the properties of the definition. We shall often refer to a real vector space merely as a **vector space**. Thus a "vector" is now an element of a vector space and no longer needs to be interpreted as a directed line segment. In our examples we shall see, however, how this name came about in a natural manner. We now consider some examples of vector spaces, leaving it to the reader to verify that all the properties of Definition 2.4 hold.

EXAMPLE 1

Consider R^n, the set of all $n \times 1$ matrices

$$\begin{bmatrix} a_1 \\ a_2 \\ \vdots \\ a_n \end{bmatrix}$$

with real entries. Let the operation $\oplus$ be matrix addition and let the operation $\odot$ be multiplication of a matrix by a real number (scalar multiplication).

By the use of the properties of matrices established in Section 1.3, it is not difficult to show that R^n is a vector space by verifying that the properties of Definition 2.4 hold. Thus the matrix $\begin{bmatrix} a_1 \\ a_2 \\ \vdots \\ a_n \end{bmatrix}$, as an element of R^n, is now called a *vector*. We have already discussed R^2 and R^3 in Section 2.1. See Figure 2.16 for geometric representations of R^2 and R^3. Although we shall see later that many geometric notions such as length and the angle between vectors can be defined in R^n for $n > 3$, we cannot draw pictures in these cases. ■

EXAMPLE 2

The set of all $m \times n$ matrices with matrix addition as $\oplus$ and multiplication of a matrix by a real number as $\odot$ is a vector space (verify). We denote this vector space by M_{mn}. ■

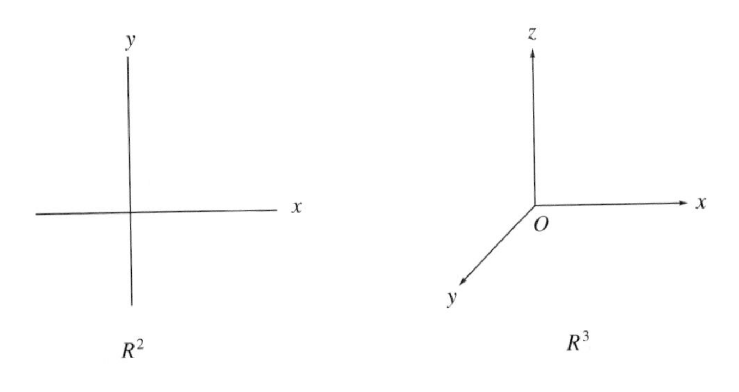

FIGURE 2.16

EXAMPLE 3

The set of all real numbers with $\oplus$ as the usual addition of real numbers and $\odot$ as the usual multiplication of real numbers is a vector space (verify). In this case the real numbers play the dual roles of both vectors and scalars. This vector space is essentially the case with $n = 1$ of Example 1. ■

EXAMPLE 4

Let R_n be the set of all $1 \times n$ matrices $\begin{bmatrix} a_1 & a_2 & \cdots & a_n \end{bmatrix}$, where we define $\oplus$ by

$$\begin{bmatrix} a_1 & a_2 & \cdots & a_n \end{bmatrix} \oplus \begin{bmatrix} b_1 & b_2 & \cdots & b_n \end{bmatrix} = \begin{bmatrix} a_1 + b_1 & a_2 + b_2 & \cdots & a_n + b_n \end{bmatrix}$$

and we define $\odot$ by

$$c \odot \begin{bmatrix} a_1 & a_2 & \cdots & a_n \end{bmatrix} = \begin{bmatrix} ca_1 & ca_2 & \cdots & ca_n \end{bmatrix}.$$

Then R_n is a vector space (verify). This is just a special case of Example 2. ■

EXAMPLE 5

Another source of examples are sets of polynomials; therefore, we recall some well-known facts about such functions. A **polynomial** (in t) is a function which is expressible as

$$p(t) = a_n t^n + a_{n-1} t^{n-1} + \cdots + a_1 t + a_0,$$

where a_0, $a_1 \ldots, a_n$ are real numbers and n is a nonnegative integer. If $a_n \neq 0$, then $p(t)$ is said to have **degree** $\boldsymbol{n}$. Thus the degree of a polynomial is the highest power having a nonzero coefficient; $p(t) = 2t + 1$ has degree 1 and the constant polynomial $p(t) = 3$ has degree 0. The **zero polynomial**, denoted by $\mathbf{0}$, has no degree. We now let P_n be the set of all polynomials of degree $\leq n$ together with the zero polynomial. If $p(t)$ and $q(t)$ are in P_n, we can write

$$p(t) = a_n t^n + a_{n-1} t^{n-1} + \cdots + a_1 t + a_0$$

and

$$q(t) = b_n t^n + b_{n-1}t^{n-1} + \cdots + b_1 t + b_0.$$

We define $p(t) \oplus q(t)$ as

$$p(t) \oplus q(t) = (a_n + b_n)t^n + (a_{n-1} + b_{n-1})t^{n-1} + \cdots + (a_1 + b_1)t + (a_0 + b_0).$$

If c is a scalar, we also define $c \odot p(t)$ as

$$c \odot p(t) = (ca_n)t^n + (ca_{n-1})t^{n-1} + \cdots + (ca_1)t + (ca_0).$$

We now show that P_n is a vector space.

Let $p(t)$ and $q(t)$ as above be elements of P_n; that is, they are polynomials of degree $\leq n$ or the zero polynomial. Then the definitions above of the operations $\oplus$ and $\odot$ show that $p(t) \oplus q(t)$ and $c \odot p(t)$, for any scalar c, are polynomials of degree $\leq n$ or the zero polynomial. That is, $p(t) \oplus q(t)$ and $c \odot p(t)$ are in P_n so that (a) and (b) in Definition 2.4 hold. To verify property (1), we observe that

$$q(t) \oplus p(t) = (b_n + a_n)t^n + (b_{n-1} + a_{n-1})t^{n-1} + \cdots + (b_1 + a_1)t + (a_0 + b_0),$$

and since $a_i + b_i = b_i + a_i$ holds for the real numbers, we conclude that $p(t) \oplus q(t) = q(t) \oplus p(t)$. Similarly, we verify property (2). The zero polynomial is the element $\mathbf{0}$ needed in property (3). If $p(t)$ is as given above, then its negative, $-p(t)$, is

$$-a_n t^n - a_{n-1}t^{n-1} - \cdots - a_1 t - a_0.$$

We shall now verify property (6) and will leave the verification of the remaining properties to the reader. Thus

$$\begin{aligned}
(c + d) \odot p(t) &= (c + d)a_n t^n + (c + d)a_{n-1}t^{n-1} + \cdots + (c + d)a_1 t \\
&\quad + (c + d)a_0 \\
&= ca_n t^n + da_n t^n + ca_{n-1}t^{n-1} + da_{n-1}t^{n-1} + \cdots + ca_1 t \\
&\quad + da_1 t + ca_0 + da_0 \\
&= c(a_n t^n + a_{n-1}t^{n-1} + \cdots + a_1 t + a_0) \\
&\quad + d(a_n t^n + a_{n-1}t^{n-1} + \cdots + a_1 t + a_0) \\
&= c \odot p(t) \oplus d \odot p(t).
\end{aligned}$$

■

Remark We shall show later that the vector space P_n behaves algebraically in exactly the same manner as R^{n+1}.

For each natural number n, we have just defined the vector space P_n of all polynomials of degree $\leq n$ together with the zero polynomial. We could also consider the space P of *all* polynomials (of any degree), together with the zero polynomial. Here P is the mathematical union of all the vector spaces P_n. Two polynomials $p(t)$ of degree n and $q(t)$ of degree m are added in P in the same way as they would be added in P_r, where r is the maximum of the two numbers m and n. Then P is a vector space (Exercise 2).

As in the case of ordinary real number arithmetic, in an expression containing both $\odot$ and $\oplus$, the $\odot$ operation is performed first. Moreover, the familiar arithmetic rules, when parentheses are encountered, apply in this case also.

EXAMPLE 6

Let V be the set of all real-valued continuous functions defined on R^1. If f and g are in V, we define $f \oplus g$ by $(f \oplus g)(t) = f(t) + g(t)$. If f is in V and c is a scalar, we define $c \odot f$ by $(c \odot f)(t) = cf(t)$. Then V is a vector space (verify), which is denoted by $C(-\infty, \infty)$. ■

EXAMPLE 7

Let V be the set of all real numbers with the operations $\mathbf{u} \oplus \mathbf{v} = \mathbf{u} - \mathbf{v}$ ($\oplus$ is ordinary subtraction) and $c \odot \mathbf{u} = c\mathbf{u}$ ($\odot$ is ordinary multiplication). Is V a vector space? If it is not, which properties in Definition 2.4 fail to hold?

Solution If $\mathbf{u}$ and $\mathbf{v}$ are in V, and c is a scalar, then $\mathbf{u} \oplus \mathbf{v}$ and $c \odot \mathbf{u}$ are in V, so that (a) and (b) in Definition 2.4 hold. However, property (1) fails to hold, since

$$\mathbf{u} \oplus \mathbf{v} = \mathbf{u} - \mathbf{v} \quad \text{and} \quad \mathbf{v} \oplus \mathbf{u} = \mathbf{v} - \mathbf{u},$$

and these are not the same, in general. (Find $\mathbf{u}$ and $\mathbf{v}$ such that $\mathbf{u} - \mathbf{v} \neq \mathbf{v} - \mathbf{u}$.) Also, we shall let the reader verify that properties (2), (3), and (4) fail to hold. Properties (5), (7), and (8) hold, but property (6) does not hold:

$$(c + d) \odot \mathbf{u} = (c + d)\mathbf{u} = c\mathbf{u} + d\mathbf{u}$$

whereas

$$c \odot \mathbf{u} \oplus d \odot \mathbf{u} = c\mathbf{u} \oplus d\mathbf{u} = c\mathbf{u} - d\mathbf{u}$$

and these are not equal, in general. Thus V is not a vector space. ■

EXAMPLE 8

Let V be the set of all ordered triples of real numbers (x, y, z) with the operations $(x, y, z) \oplus (x', y', z') = (x', y + y', z + z')$; $c \odot (x, y, z) = (cx, cy, cz)$. We can readily verify that properties (1), (3), (4), and (6) of Definition 2.4 fail to hold. For example, if $\mathbf{u} = (x, y, z)$ and $\mathbf{v} = (x', y', z')$, then

$$\mathbf{u} \oplus \mathbf{v} = (x, y, z) \oplus (x', y', z') = (x', y + y', z + z')$$

whereas

$$\mathbf{v} \oplus \mathbf{u} = (x', y', z') \oplus (x, y, z) = (x, y' + y, z' + z),$$

so property (1) fails to hold when $x \neq x'$. Also,

$$\begin{aligned} (c + d) \odot \mathbf{u} &= (c + d) \odot (x, y, z) \\ &= ((c + d)x, (c + d)y, (c + d)z) \\ &= (cx + dx, cy + dy, cz + dz) \end{aligned}$$

whereas

$$\begin{aligned} c \odot \mathbf{u} \oplus d \odot \mathbf{u} &= c \odot (x, y, z) \oplus d \odot (x, y, z) \\ &= (cx, cy, cz) \oplus (dx, dy, dz) \\ &= (dx, cy + dy, cz + dz), \end{aligned}$$

so property (6) fails to hold when $cx \neq 0$. Thus V is not a vector space. ■

EXAMPLE 9

Let V be the set of all integers; define $\oplus$ as ordinary addition and $\odot$ as ordinary multiplication. Here V is not a vector because if $\mathbf{u}$ is any nonzero vector in V and $c = \sqrt{3}$, then $c \odot \mathbf{u}$ is not in V. Thus (b) fails to hold. ■

To verify that a given set V with two operations $\oplus$ and $\odot$ is a real vector space, we must show that it satisfies all the properties of Definition 2.4. The first thing to check is whether (a) and (b) hold, for, if either of these fails, we do not have a vector space. If both (a) and (b) hold, it is recommended that (3), the existence of a zero element, be verified next. Naturally, if (3) fails to hold, we do not have a vector space and do not have to check the remaining properties.

The following theorem presents some useful properties common to all vector spaces.

Theorem 2.2

If V is a vector space, then

(a) $0 \odot \mathbf{u} = \mathbf{0}$ *for any vector* $\mathbf{u}$ *in* V.

(b) $c \odot \mathbf{0} = \mathbf{0}$ *for any scalar* c.

(c) *If* $c \odot \mathbf{u} = \mathbf{0}$, *then either* $c = 0$ *or* $\mathbf{u} = \mathbf{0}$.

(d) $(-1) \odot \mathbf{u} = -\mathbf{u}$ *for any vector* $\mathbf{u}$ *in* V.

Proof

(a) We have

$$0 \odot \mathbf{u} = (0 + 0) \odot \mathbf{u} = 0 \odot \mathbf{u} + 0 \odot \mathbf{u} \tag{1}$$

by (6) of Definition 2.4. Adding $-0 \odot \mathbf{u}$ to both sides of Equation (1), we obtain by (2), (3), and (4) of Definition 2.4,

$$0 \odot \mathbf{u} = \mathbf{0}.$$

(d) $(-1) \odot \mathbf{u} \oplus \mathbf{u} = (-1) \odot \mathbf{u} \oplus 1 \odot \mathbf{u} = (-1 + 1) \odot \mathbf{u} = 0 \odot \mathbf{u} = \mathbf{0}$. Since $-\mathbf{u}$ is unique, we conclude that

$$(-1) \odot \mathbf{u} = -\mathbf{u}.$$

Parts (b) and (c) are left as exercises. ●

2.2 Exercises

1. Prove in detail that R^n is a vector space.

2. Show that P, the set of all polynomials, is a vector space.

In Exercises 3 *through* 7, *the given set together with the given operations is not a vector space. List the properties of Definition* 2.4 *that fail to hold.*

3. The set of all positive real numbers with the operations of $\oplus$ as ordinary addition and $\odot$ as ordinary multiplication.

4. The set of all ordered pairs of real numbers with the operations
$$(x, y) \oplus (x', y') = (x + x', y + y')$$
and
$$r \odot (x, y) = (x, ry).$$

5. The set of all ordered triples of real numbers with the operations
$$(x, y, z) \oplus (x', y', z') = (x + x', y + y', z + z')$$
and
$$r \odot (x, y, z) = (x, 1, z).$$

6. The set of all 2×1 matrices $\begin{bmatrix} x \\ y \end{bmatrix}$, where $x \leq 0$, with the usual operations in R^2.

7. The set of all ordered pairs of real numbers with the operations $(x, y) \oplus (x', y') = (x + x', y + y')$ and $r \odot (x, y) = (0, 0)$.

8. Let V be the set of all positive real numbers; define $\oplus$ by $\mathbf{u} \oplus \mathbf{v} = \mathbf{uv}$ ($\oplus$ is ordinary multiplication) and define $\odot$ by $c \odot \mathbf{v} = \mathbf{v}^c$. Prove that V is a vector space.

9. Let V be the set of all real-valued continuous functions. If f and g are in V, define $f \oplus g$ by $(f \oplus g)(t) = f(t) + g(t)$. If f is in V, define $c \odot f$ by $(c \odot f)(t) = cf(t)$. Prove that V is a vector space (this is the vector space defined in Example 6).

10. Let V be the set consisting of a single element $\mathbf{0}$. Let $\mathbf{0} \oplus \mathbf{0} = \mathbf{0}$ and $c \odot \mathbf{0} = \mathbf{0}$. Prove that V is a vector space.

11. Consider the differential equation $y'' - y' + 2y = 0$. A solution is a real-valued function f satisfying the equation. Let V be the set of all solutions to the given differential equation; define $\oplus$ and $\odot$ as in Exercise 9. Prove that V is a vector space. (See also Chapter 7.)

12. Let V be the set of all positive real numbers; define $\oplus$ by $\mathbf{u} \oplus \mathbf{v} = \mathbf{uv} - 1$ and $\odot$ by $c \odot \mathbf{v} = \mathbf{v}$. Is V a vector space?

13. Let V be the set of all real numbers; define $\oplus$ by $\mathbf{u} \oplus \mathbf{v} = \mathbf{uv}$ and $\odot$ by $c \odot \mathbf{u} = c + \mathbf{u}$. Is V a vector space?

14. Let V be the set of all real numbers; define $\oplus$ by $\mathbf{u} \oplus \mathbf{v} = 2\mathbf{u} - \mathbf{v}$ and $\odot$ by $c \odot \mathbf{u} = c\mathbf{u}$. Is V a vector space?

15. Prove that a vector space has only one zero vector.

16. Prove that a vector $\mathbf{u}$ in a vector space has only one negative, $-\mathbf{u}$.

17. Prove parts (b) and (c) of Theorem 2.2.

18. Prove that the set V of all real-valued functions is a vector space under the operations defined as in Exercise 9.

19. Prove that $-(-\mathbf{v}) = \mathbf{v}$.

20. Prove that if $\mathbf{u} \oplus \mathbf{v} = \mathbf{u} \oplus \mathbf{w}$, then $\mathbf{v} = \mathbf{w}$.

21. Prove that if $\mathbf{u} \neq \mathbf{0}$ and $a \odot \mathbf{u} = b \odot \mathbf{u}$, then $a = b$.

22. Example 5 discusses the vector space P_n of polynomials of degree n or less. Operations on polynomials can be performed in linear algebra software by associating a row matrix of size $n + 1$ with polynomial $p(t)$ of P_n. The row matrix consists of the coefficients of $p(t)$, using the association
$$p(t) = a_n t^n + a_{n-1} t^{n-1} + \cdots + a_1 t + a_0$$
$$\rightarrow \begin{bmatrix} a_n & a_{n-1} & \cdots & a_1 & a_0 \end{bmatrix}.$$
If any term of $p(t)$ is explicitly missing, a zero is used for its coefficient. Then the addition of polynomials corresponds to matrix addition, and multiplication of a polynomial by a scalar corresponds to scalar multiplication of matrices. With your software perform the following operations on polynomials, using the matrix association as just described. Let $n = 3$ and
$$p(t) = 2t^3 + 5t^2 + t - 2, \quad q(t) = t^3 + 3t + 5.$$
(a) $p(t) + q(t)$. (b) $5p(t)$.
(c) $3p(t) - 4q(t)$.

2.3 SUBSPACES

In this section we begin to analyze the structure of a vector space. First, it is convenient to have a name for a subset of a given vector space that is itself a vector space with respect to the same operations as those in V. Thus we have a definition.

Definition 2.5

Let V be a vector space and W a nonempty subset of V. If W is a vector space with respect to the operations in V, then W is called a **subspace** of V. ▲

Examples of subspaces of a given vector space occur frequently. We shall list several of these, leaving the verifications that they are subspaces to the reader. More examples will be found in the exercises.

EXAMPLE 1

Every vector space has at least two subspaces, itself and the subspace $\{\mathbf{0}\}$ consisting only of the zero vector (recall that $\mathbf{0} \oplus \mathbf{0} = \mathbf{0}$ and $c \odot \mathbf{0} = \mathbf{0}$ in any vector space). The subspace $\{\mathbf{0}\}$ is called the **zero subspace**. ■

EXAMPLE 2

Let P_2 be the set consisting of all polynomials of degree ≤ 2 and the zero polynomial; P_2 is a subset of P, the vector space of all polynomials. It is easy to verify that P_2 is a *subspace* of P. In general, the set P_n consisting of all polynomials of degree $\leq n$ and the zero polynomial is a subspace of P. Also, P_n is a subspace of P_{n+1}. ■

EXAMPLE 3

Let V be the set of all polynomials of degree exactly $= 2$; V is a *subset* of P, the vector space of all polynomials, but not a *subspace* of P because the sum of the polynomials $2t^2 + 3t + 1$ and $-2t^2 + t + 2$ is not in V, since it is a polynomial of degree 1. ■

We now pause in our listing of subspaces to develop a labor-saving result. We just noted that to verify that a subset W of a vector space V is a subspace, one must check that (a), (b), and (1) through (8) of Definition 2.4 hold. However, the following theorem says that it is enough to merely check that (a) and (b) hold to verify that a subset W of a vector space V is a subspace. Property (a) is called the **closure** property for $\oplus$ and (b) is called the **closure** property for $\odot$.

Theorem 2.3

Let V be a vector space with operations $\oplus$ and $\odot$ and let W be a nonempty subset of V. Then W is a subspace of V if and only if the following conditions hold:

(a) *If* $\mathbf{u}$ *and* $\mathbf{v}$ *are any vectors in* W, *then* $\mathbf{u} \oplus \mathbf{v}$ *is in* W.

(b) *If* c *is any real number and* $\mathbf{u}$ *is any vector in* W, *then* $c \odot \mathbf{u}$ *is in* W.

Proof

If W is a subspace of V, then it is a vector space and so (a) and (b) of Definition 2.4 hold; these are precisely (a) and (b) of the theorem.

Conversely, suppose that (a) and (b) hold. We wish to show that W is a subspace of V. First, from (b) we have that $(-1) \odot \mathbf{u}$ is in W for any $\mathbf{u}$ in W. From (a) we have that $\mathbf{u} \oplus (-1) \odot \mathbf{u}$ is in W. But $\mathbf{u} \oplus (-1) \odot \mathbf{u} = \mathbf{0}$, so $\mathbf{0}$ is in W. Then $\mathbf{u} \oplus \mathbf{0} = \mathbf{u}$ for any $\mathbf{u}$ in W. Finally, properties (1), (2), (5), (6), (7), and (8) hold in W because they hold in V. Hence W is a subspace of V. ●

EXAMPLE 4

Let W be the set of all vectors in R^3 of the form $\begin{bmatrix} a \\ b \\ a+b \end{bmatrix}$, where a and b are any real numbers. To verify Theorem 2.3(a) and (b), we let

$$\mathbf{u} = \begin{bmatrix} a_1 \\ b_1 \\ a_1 + b_1 \end{bmatrix} \quad \text{and} \quad \mathbf{v} = \begin{bmatrix} a_2 \\ b_2 \\ a_2 + b_2 \end{bmatrix}$$

be two vectors in W. Then

$$\mathbf{u} \oplus \mathbf{v} = \begin{bmatrix} a_1 + a_2 \\ b_1 + b_2 \\ (a_1 + b_1) + (a_2 + b_2) \end{bmatrix} = \begin{bmatrix} a_1 + a_2 \\ b_1 + b_2 \\ (a_1 + a_2) + (b_1 + b_2) \end{bmatrix}$$

is in W, for W consists of all those vectors whose third entry is the sum of the first two entries. Similarly,

$$c \odot \begin{bmatrix} a_1 \\ b_1 \\ a_1 + b_1 \end{bmatrix} = \begin{bmatrix} ca_1 \\ cb_1 \\ c(a_1 + b_1) \end{bmatrix} = \begin{bmatrix} ca_1 \\ cb_1 \\ ca_1 + cb_1 \end{bmatrix}$$

is in W. Hence W is a subspace of R^3. ■

Henceforth, we shall usually denote $\mathbf{u} \oplus \mathbf{v}$ and $c \odot \mathbf{u}$ in a vector space V as $\mathbf{u} + \mathbf{v}$ and $c\mathbf{u}$, respectively.

We can also show that a nonempty subset W of a vector space V is a subspace of V if and only if $c\mathbf{u} + d\mathbf{v}$ is in W for any vectors $\mathbf{u}$ and $\mathbf{v}$ in W and any scalars c and d.

EXAMPLE 5

A simple way of constructing subspaces in a vector space is as follows. Let $\mathbf{v}_1$ and $\mathbf{v}_2$ be fixed vectors in a vector space V and let W be the set of all vectors in V of the form

$$a_1\mathbf{v}_1 + a_2\mathbf{v}_2,$$

where a_1 and a_2 are any real numbers. To show that W is a subspace of V, we verify properties (a) and (b) of Theorem 2.3. Thus let

$$\mathbf{w}_1 = a_1\mathbf{v}_1 + a_2\mathbf{v}_2 \quad \text{and} \quad \mathbf{w}_2 = b_1\mathbf{v}_1 + b_2\mathbf{v}_2$$

be vectors in W. Then

$$\mathbf{w}_1 + \mathbf{w}_2 = (a_1\mathbf{v}_1 + a_2\mathbf{v}_2) + (b_1\mathbf{v}_1 + b_2\mathbf{v}_2) = (a_1 + b_1)\mathbf{v}_1 + (a_2 + b_2)\mathbf{v}_2,$$

which is in W. Also, if c is a scalar, then

$$c\mathbf{w}_1 = c(a_1\mathbf{v}_1 + a_2\mathbf{v}_2) = (ca_1)\mathbf{v}_1 + (ca_2)\mathbf{v}_2$$

is in W. Hence W is a subspace of V. ■

The construction carried out in Example 5 for two vectors can be performed for more than two vectors. We now consider the following definition.

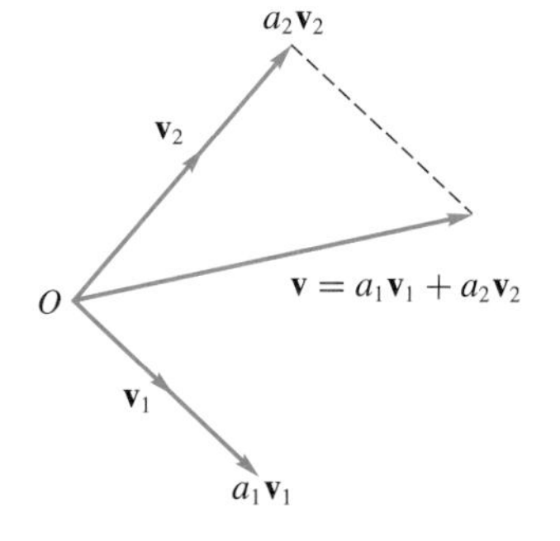

FIGURE 2.17

Linear combination of two vectors.

Definition 2.6

Let $\mathbf{v}_1, \mathbf{v}_2, \ldots, \mathbf{v}_k$ be vectors in a vector space V. A vector $\mathbf{v}$ in V is called a **linear combination** of $\mathbf{v}_1, \mathbf{v}_2, \ldots, \mathbf{v}_k$ if

$$\mathbf{v} = a_1\mathbf{v}_1 + a_2\mathbf{v}_2 + \cdots + a_k\mathbf{v}_k$$

for some real numbers $a_1, a_2, \ldots, a_k$. ▲

In Figure 2.17 we show the vector $\mathbf{v}$ in R^2 or R^3 as a linear combination of the vectors $\mathbf{v}_1$ and $\mathbf{v}_2$.

EXAMPLE 6

In R^3 let

$$\mathbf{v}_1 = \begin{bmatrix} 1 \\ 2 \\ 1 \end{bmatrix}, \quad \mathbf{v}_2 = \begin{bmatrix} 1 \\ 0 \\ 2 \end{bmatrix}, \quad \text{and} \quad \mathbf{v}_3 = \begin{bmatrix} 1 \\ 1 \\ 0 \end{bmatrix}.$$

The vector

$$\mathbf{v} = \begin{bmatrix} 2 \\ 1 \\ 5 \end{bmatrix}$$

is a linear combination of $\mathbf{v}_1$, $\mathbf{v}_2$, and $\mathbf{v}_3$ if we can find real numbers a_1, a_2, and a_3 so that

$$a_1\mathbf{v}_1 + a_2\mathbf{v}_2 + a_3\mathbf{v}_3 = \mathbf{v}.$$

Substituting for $\mathbf{v}$, $\mathbf{v}_1$, $\mathbf{v}_2$, and $\mathbf{v}_3$, we have

$$a_1 \begin{bmatrix} 1 \\ 2 \\ 1 \end{bmatrix} + a_2 \begin{bmatrix} 1 \\ 0 \\ 2 \end{bmatrix} + a_3 \begin{bmatrix} 1 \\ 1 \\ 0 \end{bmatrix} = \begin{bmatrix} 2 \\ 1 \\ 5 \end{bmatrix}.$$

Equating corresponding entries leads to the linear system (verify)

$$\begin{aligned} a_1 + a_2 + a_3 &= 2 \\ 2a_1 + a_3 &= 1 \\ a_1 + 2a_2 &= 5. \end{aligned}$$

Solving this linear system by the methods of Chapter 1 gives (verify) $a_1 = 1$, $a_2 = 2$, and $a_3 = -1$, which means that $\mathbf{v}$ is a linear combination of $\mathbf{v}_1$, $\mathbf{v}_2$, and $\mathbf{v}_3$. Thus

$$\mathbf{v} = \mathbf{v}_1 + 2\mathbf{v}_2 - \mathbf{v}_3.$$

Figure 2.18 shows $\mathbf{v}$ as a linear combination of $\mathbf{v}_1$, $\mathbf{v}_2$, and $\mathbf{v}_3$. ■

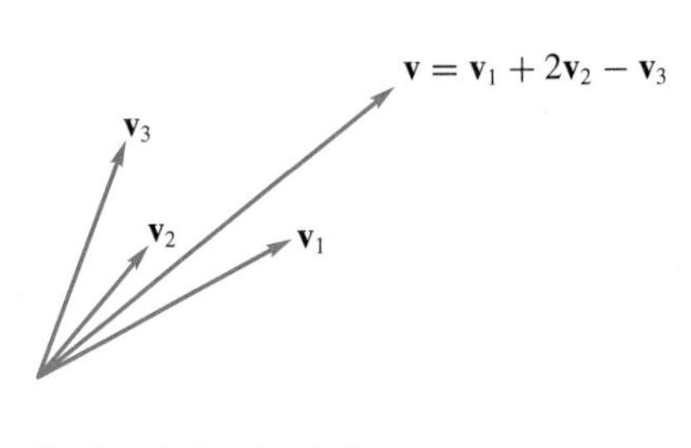

FIGURE 2.18

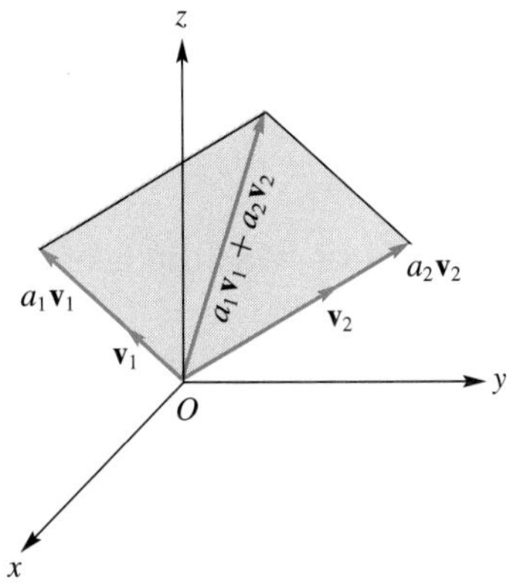

FIGURE 2.19

Definition 2.7

If $S = \{\mathbf{v}_1, \mathbf{v}_2, \ldots, \mathbf{v}_k\}$ is a set of vectors in a vector space V, then the set of all vectors in V that are linear combinations of the vectors in S is denoted by

$$\text{span } S \quad \text{or} \quad \text{span } \{\mathbf{v}_1, \mathbf{v}_2, \ldots, \mathbf{v}_k\}.$$ ▲

In Figure 2.19 we show span $\{\mathbf{v}_1, \mathbf{v}_2\}$, where $\mathbf{v}_1$ and $\mathbf{v}_2$ are noncollinear vectors in R^3. As you can see, it is a plane that passes through the origin and contains the vectors $\mathbf{v}_1$ and $\mathbf{v}_2$.

EXAMPLE 7

Consider the set S of 2×3 matrices given by

$$S = \left\{ \begin{bmatrix} 1 & 0 & 0 \\ 0 & 0 & 0 \end{bmatrix}, \begin{bmatrix} 0 & 1 & 0 \\ 0 & 0 & 0 \end{bmatrix}, \begin{bmatrix} 0 & 0 & 0 \\ 0 & 1 & 0 \end{bmatrix}, \begin{bmatrix} 0 & 0 & 0 \\ 0 & 0 & 1 \end{bmatrix} \right\}.$$

Then span S is the set in M_{23} consisting of all vectors of the form

$$a\begin{bmatrix}1 & 0 & 0\\ 0 & 0 & 0\end{bmatrix} + b\begin{bmatrix}0 & 1 & 0\\ 0 & 0 & 0\end{bmatrix} + c\begin{bmatrix}0 & 0 & 0\\ 0 & 1 & 0\end{bmatrix} + d\begin{bmatrix}0 & 0 & 0\\ 0 & 0 & 1\end{bmatrix}$$
$$= \begin{bmatrix}a & b & 0\\ 0 & c & d\end{bmatrix}, \text{ where } a, b, c, \text{ and } d \text{ are real numbers.}$$

That is, span S is the subset of M_{23} consisting of all matrices of the form

$$\begin{bmatrix}a & b & 0\\ 0 & c & d\end{bmatrix},$$

where a, b, c, and d are real numbers. ■

The following theorem is a generalization of Example 5.

Theorem 2.4

Let $S = \{\mathbf{v}_1, \mathbf{v}_2, \ldots, \mathbf{v}_k\}$ be a set of vectors in a vector space V. Then span S is a subspace of V.

Proof

See Exercise 17. ●

EXAMPLE 8

In P_2, let

$$\mathbf{v}_1 = 2t^2 + t + 2, \quad \mathbf{v}_2 = t^2 - 2t, \quad \mathbf{v}_3 = 5t^2 - 5t + 2, \quad \mathbf{v}_4 = -t^2 - 3t - 2.$$

Determine if the vector

$$\mathbf{v} = t^2 + t + 2$$

belongs to span $\{\mathbf{v}_1, \mathbf{v}_2, \mathbf{v}_3, \mathbf{v}_4\}$.

Solution If we can find scalars a_1, a_2, a_3, and a_4 so that

$$a_1\mathbf{v}_1 + a_2\mathbf{v}_2 + a_3\mathbf{v}_3 + a_4\mathbf{v}_4 = \mathbf{v},$$

then $\mathbf{v}$ belongs to span $\{\mathbf{v}_1, \mathbf{v}_2, \mathbf{v}_3, \mathbf{v}_4\}$. Substituting for $\mathbf{v}_1$, $\mathbf{v}_2$, $\mathbf{v}_3$, and $\mathbf{v}_4$, we have

$$a_1(2t^2 + t + 2) + a_2(t^2 - 2t) + a_3(5t^2 - 5t + 2) + a_4(-t^2 - 3t - 2)$$
$$= t^2 + t + 2$$

or

$$(2a_1 + a_2 + 5a_3 - a_4)t^2 + (a_1 - 2a_2 - 5a_3 - 3a_4)t + (2a_1 + 2a_3 - 2a_4)$$
$$= t^2 + t + 2.$$

Now two polynomials agree for all values of t only if the coefficients of respective powers of t agree. Thus we get the linear system

$$\begin{aligned} 2a_1 + a_2 + 5a_3 - a_4 &= 1 \\ a_1 - 2a_2 - 5a_3 - 3a_4 &= 1 \\ 2a_1 \qquad + 2a_3 - 2a_4 &= 2. \end{aligned}$$

To determine whether this system of linear equations is consistent, we form the augmented matrix and transform it to reduced row echelon form, obtaining (verify)

$$\left[\begin{array}{cccc|c} 1 & 0 & 1 & -1 & 0 \\ 0 & 1 & 3 & 1 & 0 \\ 0 & 0 & 0 & 0 & 1 \end{array}\right],$$

which indicates that the system is inconsistent; that is, it has no solution. Hence $\mathbf{v}$ does not belong to span $\{\mathbf{v}_1, \mathbf{v}_2, \mathbf{v}_3, \mathbf{v}_4\}$. ■

Remark In general, to determine if a specific vector $\mathbf{v}$ belongs to span S, we investigate the consistency of an appropriate linear system.

Another important example of a subspace is provided by Example 9. Before turning to this example, we discuss a slight change in the notation used to denote a linear system. In Chapter 1, we denoted a linear system consisting of m equations in n unknowns by $AX = B$, where X is an $n \times 1$ matrix and B is an $m \times 1$ matrix. We shall now take the point of view that X is a vector in R^n and B is a vector in R^m and will write these as $\mathbf{x}$ and $\mathbf{b}$, respectively. Thus the linear system $AX = B$ will now be written as $A\mathbf{x} = \mathbf{b}$.

EXAMPLE 9

If A is an $m \times n$ matrix, the homogeneous system of m equations in n unknowns with coefficient matrix A can be written as

$$A\mathbf{x} = \mathbf{0},$$

where $\mathbf{x}$ is a vector in R^n and $\mathbf{0}$ is the zero vector. Thus the set W of all solutions is a subset of R^n. We now show that W is a subspace of R^n (called the **solution space** of the homogeneous system or the **null space** of the matrix A) by verifying (a) and (b) of Theorem 2.3. Let $\mathbf{x}$ and $\mathbf{y}$ be solutions. Then

$$A\mathbf{x} = \mathbf{0} \quad \text{and} \quad A\mathbf{y} = \mathbf{0}.$$

Now

$$A(\mathbf{x} + \mathbf{y}) = A\mathbf{x} + A\mathbf{y} = \mathbf{0} + \mathbf{0} = \mathbf{0},$$

so $\mathbf{x} + \mathbf{y}$ is a solution. Also, if c is a scalar, then

$$A(c\mathbf{x}) = c(A\mathbf{x}) = c\mathbf{0} = \mathbf{0},$$

so $c\mathbf{x}$ is a solution. Thus W is closed under addition and scalar multiplication of vectors and is therefore a subspace of R^n. ■

It should be noted that the set of all solutions to the linear system $A\mathbf{x} = \mathbf{b}$, $\mathbf{b} \neq \mathbf{0}$, is not a subspace of R^n (see Exercise 15).

We leave it as an exercise to show that the subspaces of R^1 are $\{\mathbf{0}\}$ and R^1 itself (see Exercise 20). As for R^2, its subspaces are $\{\mathbf{0}\}$, R^2, and any set consisting of all scalar multiples of a nonzero vector (Exercise 21), that is, any line passing through the origin. Exercise 43 in Section 2.5 asks you to show that all the subspaces of R^3 are $\{\mathbf{0}\}$, R^3 itself, and any line or plane passing through the origin.

Lines in R^3

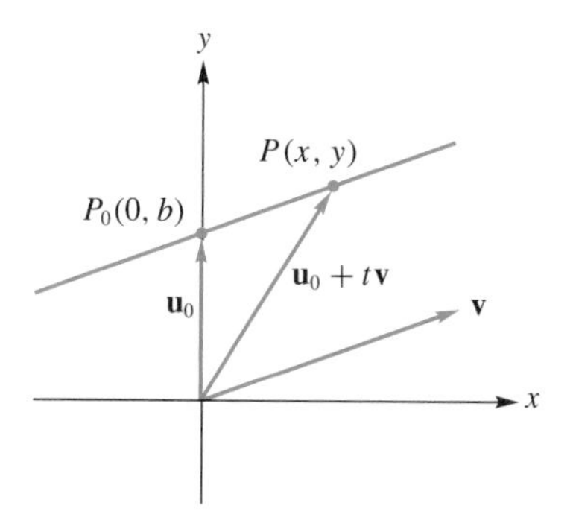

FIGURE 2.20

As you will recall, a line in the xy-plane, R^2, is often described by the equation $y = mx + b$, where m is the slope of the line and b is the y-intercept [i.e., the line intersects the y-axis at the point $P_0(0, b)$]. We may described a line in R^2 in terms of vectors by specifying its direction and a point on the line. Thus let $\mathbf{v}$ be the vector giving the direction of the line, and let $\mathbf{u}_0 = \begin{bmatrix} 0 \\ b \end{bmatrix}$ be the position vector of the point $P_0(0, b)$ at which the line intersects the y-axis. Then the line through P_0 and parallel to $\mathbf{v}$ consists of the points $P(x, y)$, whose position vector $\mathbf{u} = \begin{bmatrix} x \\ y \end{bmatrix}$ satisfies (see Figure 2.20)

$$\mathbf{u} = \mathbf{u}_0 + t\mathbf{v}, \qquad -\infty < t < +\infty.$$

We now turn to lines in R^3. In R^3 a line is determined by specifying its direction and one of its points. Let

$$\mathbf{v} = \begin{bmatrix} a \\ b \\ c \end{bmatrix}$$

be a nonzero vector in R^3. Then the line ℓ_0 through the origin and parallel to $\mathbf{v}$ consists of all the points $P(x, y, z)$ whose position vector $\mathbf{u} = \begin{bmatrix} x \\ y \\ z \end{bmatrix}$ is of the form $\mathbf{u} = t\mathbf{v}$, $-\infty < t < \infty$ [see Figure 2.21(a)]. It is easy to verify that the line ℓ_0 is a subspace of R^3. Now let $P_0(x_0, y_0, z_0)$ be a point in R^3, and let $\mathbf{u}_0 = \begin{bmatrix} x_0 \\ y_0 \\ z_0 \end{bmatrix}$ be the position vector of P_0. Then the line ℓ through P_0 and parallel to $\mathbf{v}$ consists of the points $P(x, y, z)$ whose position vector, $\mathbf{u} = \begin{bmatrix} x \\ y \\ z \end{bmatrix}$ satisfies [see Figure 2.21(b)]

$$\mathbf{u} = \mathbf{u}_0 + t\mathbf{v}, \qquad -\infty < t < \infty. \tag{1}$$

Equation (1) is called a **parametric equation** of ℓ, since it contains the parameter t, which can be assigned any real number. Equation (1) can also be written

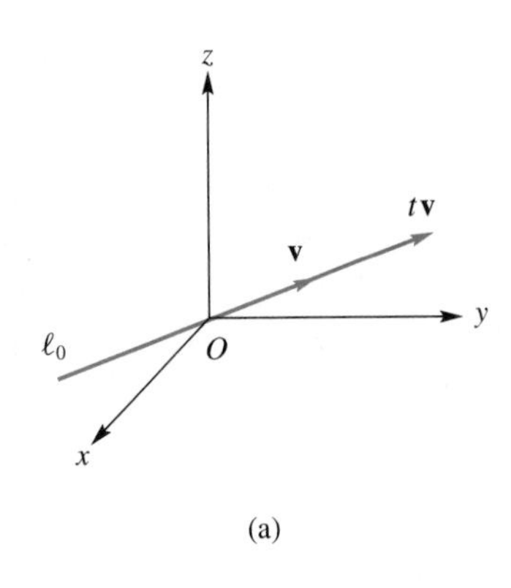

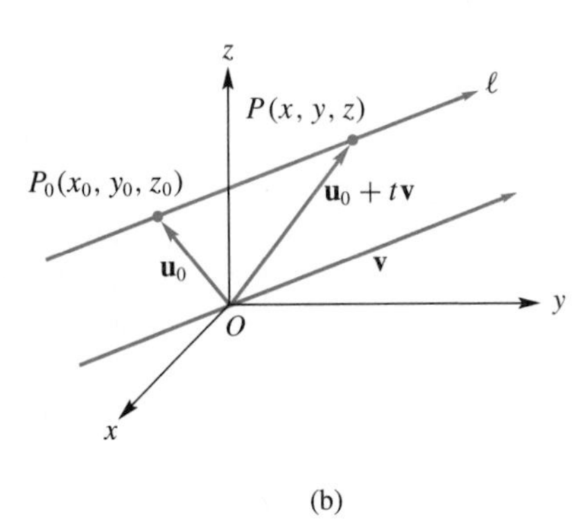

FIGURE 2.21

Line in R^3.

in terms of the components as

$$\begin{aligned} x &= x_0 + ta \\ y &= y_0 + tb \qquad -\infty < t < \infty \\ z &= z_0 + tc, \end{aligned}$$

which are called **parametric equations** of ℓ.

EXAMPLE 10

Parametric equations of the line passing through the point $P_0(-3, 2, 1)$ and parallel to the vector $\mathbf{v} = \begin{bmatrix} 2 \\ -3 \\ 4 \end{bmatrix}$ are

$$\begin{aligned} x &= -3 + 2t \\ y &= 2 - 3t \qquad -\infty < t < \infty. \\ z &= 1 + 4t, \end{aligned}$$

■

EXAMPLE 11

Find parametric equations of the line ℓ through the points $P_0(2, 3, -4)$ and $P_1(3, -2, 5)$.

Solution The desired line is parallel to the vector $\mathbf{v} = \overrightarrow{P_0P_1}$. Now

$$\mathbf{v} = \begin{bmatrix} 3 - 2 \\ -2 - 3 \\ 5 - (-4) \end{bmatrix} = \begin{bmatrix} 1 \\ -5 \\ 9 \end{bmatrix}.$$

Since P_0 is on the line, we can write the following parametric equations of ℓ:

$$\begin{aligned} x &= 2 + t \\ y &= 3 - 5t \qquad -\infty < t < \infty. \\ z &= -4 + 9t, \end{aligned}$$

Of course, we could have used the point P_1 instead of P_0. In fact, we could use any point on the line in a parametric equation for ℓ. Thus a line can be represented in infinitely many ways in parametric form. ■

2.3 Exercises

In Exercises 1 and 2, which of the given subsets of R^3 are subspaces?

1. The set of all vectors of the form:

(a) $\begin{bmatrix} a \\ b \\ 1 \end{bmatrix}$. (b) $\begin{bmatrix} a \\ b \\ a+2b \end{bmatrix}$. (c) $\begin{bmatrix} a \\ 0 \\ 0 \end{bmatrix}$.

(d) $\begin{bmatrix} a \\ b \\ c \end{bmatrix}$, where $a + 2b - c = 0$.

2. The set of all vectors of the form:

(a) $\begin{bmatrix} a \\ b \\ 0 \end{bmatrix}$. (b) $\begin{bmatrix} a \\ b \\ c \end{bmatrix}$, where $a > 0$.

(c) $\begin{bmatrix} a \\ a \\ c \end{bmatrix}$. (d) $\begin{bmatrix} a \\ b \\ c \end{bmatrix}$, where $2a - b + c = 1$.

In Exercises 3 and 4, which of the given subsets of R_4 are subspaces?

3. (a) $\begin{bmatrix} a & b & c & d \end{bmatrix}$, where $a - b = 2$.

(b) $\begin{bmatrix} a & b & c & d \end{bmatrix}$, where $c = a + 2b$ and $d = a - 3b$.

(c) $\begin{bmatrix} a & b & c & d \end{bmatrix}$, where $a = 0$ and $b = -d$.

4. (a) $\begin{bmatrix} a & b & c & d \end{bmatrix}$, where $a = b = 0$.

(b) $\begin{bmatrix} a & b & c & d \end{bmatrix}$, where $a = 1, b = 0$, and $a + d = 1$.

(c) $\begin{bmatrix} a & b & c & d \end{bmatrix}$, where $a > 0$ and $b < 0$.

In Exercises 5 and 6, which of the given subsets of the vector space, M_{23}, of all 2×3 matrices are subspaces?

5. The set of all matrices of the form:

(a) $\begin{bmatrix} a & b & c \\ d & 0 & 0 \end{bmatrix}$, where $b = a + c$.

(b) $\begin{bmatrix} a & b & c \\ d & 0 & 0 \end{bmatrix}$, where $c > 0$.

(c) $\begin{bmatrix} a & b & c \\ d & e & f \end{bmatrix}$, where $a = -2c$ and $f = 2e + d$.

6. (a) $\begin{bmatrix} a & b & c \\ d & e & f \end{bmatrix}$, where $a = 2c + 1$.

(b) $\begin{bmatrix} 0 & 1 & a \\ b & c & 0 \end{bmatrix}$.

(c) $\begin{bmatrix} a & b & c \\ d & e & f \end{bmatrix}$, where $a + c = 0$ and $b + d + f = 0$.

In Exercises 7 and 8, which of the given subsets of the vector space P_2 are subspaces?

7. The set of all polynomials of the form:

(a) $a_2t^2 + a_1t + a_0$, where $a_0 = 0$.

(b) $a_2t^2 + a_1t + a_0$, where $a_0 = 2$.

(c) $a_2t^2 + a_1t + a_0$, where $a_2 + a_1 = a_0$.

8. (a) $a_2t^2 + a_1t + a_0$, where $a_1 = 0$ and $a_0 = 0$.

(b) $a_2t^2 + a_1t + a_0$, where $a_1 = 2a_0$.

(c) $a_2t^2 + a_1t + a_0$, where $a_2 + a_1 + a_0 = 2$.

9. Which of the following subsets of the vector space M_{nn} are subspaces?

(a) The set of all $n \times n$ symmetric matrices.

(b) The set of all $n \times n$ diagonal matrices.

(c) The set of all $n \times n$ nonsingular matrices.

10. Which of the following subsets of the vector space M_{nn} are subspaces?

(a) The set of all $n \times n$ singular matrices.

(b) The set of all $n \times n$ upper triangular matrices.

(c) The set of all $n \times n$ skew symmetric matrices.

11. (**Calculus required**) Which of the following subsets are subspaces of the vector space $C(-\infty, \infty)$ defined in Example 6 of Section 2.2?

(a) All nonnegative functions.

(b) All constant functions.

(c) All functions f such that $f(0) = 0$.

(d) All functions f such that $f(0) = 5$.

(e) All differentiable functions.

12. (**Calculus required**) Which of the following subsets are subspaces of the vector space $C(-\infty, \infty)$ defined in Example 6 of Section 2.2?

(a) All integrable functions.

(b) All bounded functions.

(c) All functions that are integrable on $[a, b]$?

(d) All functions that are bounded on $[a, b]$?

13. Show that P is a subspace of the vector space $C(-\infty, \infty)$ defined in Example 6 of Section 2.2.

14. Prove that P_2 is a subspace of P_3.

15. Show that the set of all solutions to the linear system $A\mathbf{x} = \mathbf{b}$, $\mathbf{b} \neq \mathbf{0}$, is not a subspace of R^n.

16. If A is a nonsingular matrix, what is the null space of A?

17. Let V be a vector space and $S = \{\mathbf{v}_1, \mathbf{v}_2, \ldots, \mathbf{v}_k\}$ be a set of vectors in V. Prove that span S is a subspace of V.

18. Let $\mathbf{x}_0$ be a fixed vector in a vector space V. Show that the set W consisting of all scalar multiples $c\mathbf{x}_0$ of $\mathbf{x}_0$ is a subspace of V.

19. Let A be an $m \times n$ matrix. Is the set W of all vectors $\mathbf{x}$ in R^n such that $A\mathbf{x} \neq \mathbf{0}$ a subspace of R^n? Justify your answer.

20. Show that the only subspaces of R^1 are $\{\mathbf{0}\}$ and R^1 itself.

21. Show that the only subspaces of R^2 are $\{\mathbf{0}\}$, R^2, and any set consisting of all scalar multiples of a nonzero vector.

22. Determine which of the following subsets of R^2 are subspaces.

(a)

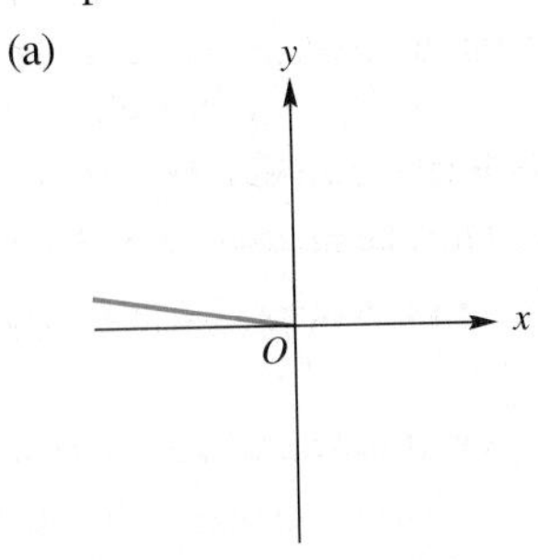

(b)

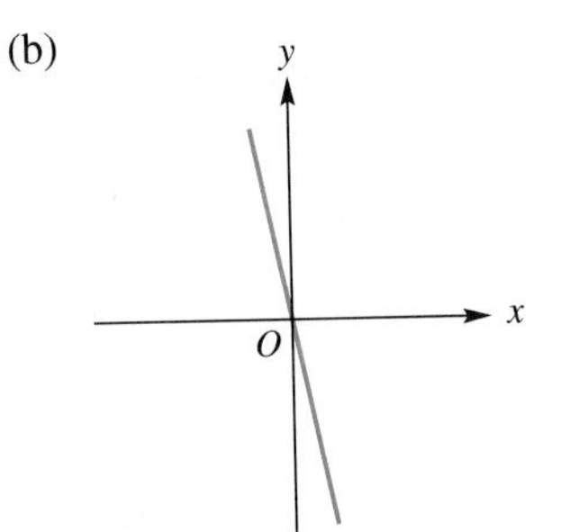

(c)

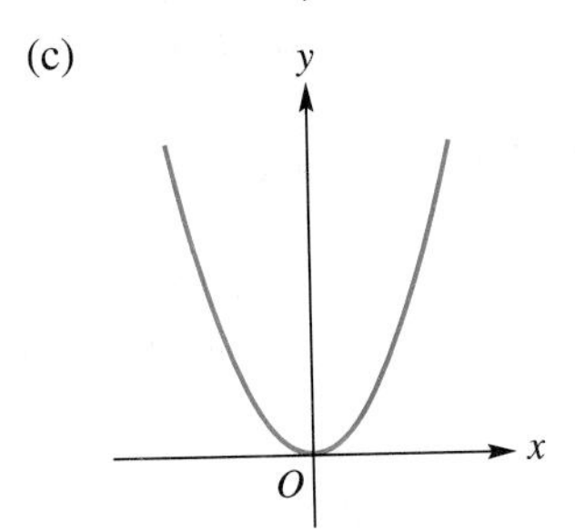

(d)

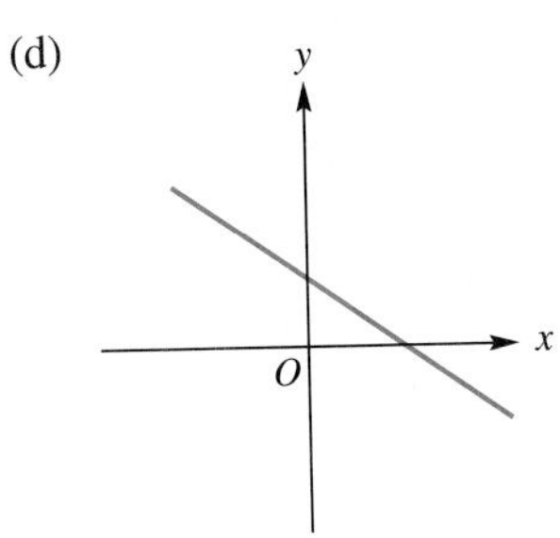

23. Which of the following vectors in R^3 are linear combinations of

$$\mathbf{v}_1 = \begin{bmatrix} 4 \\ 2 \\ -3 \end{bmatrix}, \quad \mathbf{v}_2 = \begin{bmatrix} 2 \\ 1 \\ -2 \end{bmatrix}, \quad \text{and} \quad \mathbf{v}_3 = \begin{bmatrix} -2 \\ -1 \\ 0 \end{bmatrix}?$$

(a) $\begin{bmatrix} 1 \\ 1 \\ 1 \end{bmatrix}$. (b) $\begin{bmatrix} 4 \\ 2 \\ -6 \end{bmatrix}$. (c) $\begin{bmatrix} -2 \\ -1 \\ 1 \end{bmatrix}$. (d) $\begin{bmatrix} -1 \\ 2 \\ 3 \end{bmatrix}$.

24. Which of the following vectors in R_4 are linear combinations of

$$\mathbf{v}_1 = \begin{bmatrix} 1 & 2 & 1 & 0 \end{bmatrix}, \quad \mathbf{v}_2 = \begin{bmatrix} 4 & 1 & -2 & 3 \end{bmatrix},$$
$$\mathbf{v}_3 = \begin{bmatrix} 1 & 2 & 6 & -5 \end{bmatrix}, \quad \mathbf{v}_4 = \begin{bmatrix} -2 & 3 & -1 & 2 \end{bmatrix}?$$

(a) $\begin{bmatrix} 3 & 6 & 3 & 0 \end{bmatrix}$. (b) $\begin{bmatrix} 1 & 0 & 0 & 0 \end{bmatrix}$.

(c) $\begin{bmatrix} 3 & 6 & -2 & 5 \end{bmatrix}$. (d) $\begin{bmatrix} 0 & 0 & 0 & 1 \end{bmatrix}$.

25. In each part, determine whether the given vector $p(t)$ in P_2 belongs to span $\{p_1(t), p_2(t), p_3(t)\}$, where

$$p_1(t) = t^2 + 2t + 1, \quad p_2(t) = t^2 + 3, \quad \text{and} \quad p_3(t) = t - 1.$$

(a) $p(t) = t^2 + t + 2$.
(b) $p(t) = 2t^2 + 2t + 3$.
(c) $p(t) = -t^2 + t - 4$.
(d) $p(t) = -2t^2 + 3t + 1$.

26. In each part, determine whether the given vector A in M_{22} belongs to span $\{A_1, A_2, A_3\}$, where

$$A_1 = \begin{bmatrix} 1 & -1 \\ 0 & 3 \end{bmatrix}, \quad A_2 = \begin{bmatrix} 1 & 1 \\ 0 & 2 \end{bmatrix}, \quad \text{and} \quad A_3 = \begin{bmatrix} 2 & 2 \\ -1 & 1 \end{bmatrix}.$$

(a) $A = \begin{bmatrix} 5 & 1 \\ -1 & 9 \end{bmatrix}$. (b) $A = \begin{bmatrix} -3 & -1 \\ 3 & 2 \end{bmatrix}$.
(c) $A = \begin{bmatrix} 3 & -2 \\ 3 & 2 \end{bmatrix}$. (d) $A = \begin{bmatrix} 1 & 0 \\ 2 & 1 \end{bmatrix}$.

27. (a) Show that a line ℓ_0 through the origin of R^3 is a subspace of R^3.
(b) Show that a line ℓ in R^3 not passing through the origin is not a subspace of R^3.

28. State which of the following points are on the line

$$\begin{aligned} x &= 3 + 2t \\ y &= -2 + 3t \qquad -\infty < t < \infty. \\ z &= 4 + 3t, \end{aligned}$$

(a) $(1, 1, 1)$. (b) $(1, -1, 0)$.
(c) $(1, 0, -2)$. (d) $\left(4, -\frac{1}{2}, \frac{5}{2}\right)$.

29. State which of the following points are on the line

$$\begin{aligned} x &= 4 - 2t \\ y &= -3 + 2t \qquad -\infty < t < \infty. \\ z &= 4 - 5t, \end{aligned}$$

(a) $(0, 1, -6)$. (b) $(1, 2, 3)$.
(c) $(4, -3, 4)$. (d) $(0, 1, -1)$.

30. Find parametric equations of the line through $P_0(x_0, y_0, z_0)$ parallel to $\mathbf{v}$.

(a) $P_0(3, 4, -2)$, $\mathbf{v} = \begin{bmatrix} 4 \\ -5 \\ 2 \end{bmatrix}$.

(b) $P_0(3, 2, 4)$, $\mathbf{v} = \begin{bmatrix} -2 \\ 5 \\ 1 \end{bmatrix}$.

31. Find parametric equations of the line through the given points.

(a) $(2, -3, 1)$, $(4, 2, 5)$.
(b) $(-3, -2, -2)$, $(5, 5, 4)$.

32. Numerical experiments in software *cannot* be used to verify that a set V with two operations $\oplus$ and $\odot$ is a vector space or a subspace. Such a verification must be done "abstractly" to take into account all possibilities for elements of V. However, numerical experiments can yield counterexamples which show that V is not a vector space or not a subspace. Use your software to verify that each of the following is *not* a subspace of M_{22}, with the usual operations of addition of matrices and scalar multiplication.

(a) The set of symmetric matrices with the $(1, 1)$ entry equal to 3.
(b) The set of matrices whose first column is $\begin{bmatrix} 0 & 1 \end{bmatrix}^T$.
(c) The set of matrices $\begin{bmatrix} a & b \\ c & d \end{bmatrix}$ such that $ad - bc \neq 0$.

33. A linear combination of vectors $\mathbf{v}_1, \mathbf{v}_2, \ldots, \mathbf{v}_k$ in R^n with coefficients $a_1, \ldots, a_k$ is given algebraically, as in Definition 2.6, by

$$\mathbf{v} = a_1\mathbf{v}_1 + a_2\mathbf{v}_2 + \cdots + a_k\mathbf{v}_k.$$

In software we can compute such a linear combination of columns by a matrix multiplication $\mathbf{v} = A\mathbf{c}$, where

$$A = \begin{bmatrix} \mathbf{v}_1 & \mathbf{v}_2 & \cdots & \mathbf{v}_k \end{bmatrix} \quad \text{and} \quad \mathbf{c} = \begin{bmatrix} a_1 \\ a_2 \\ \vdots \\ a_k \end{bmatrix}.$$

That is, matrix A has $\text{col}_j(A) = \mathbf{v}_j$ for $j = 1, 2, \ldots, k$. Experiment with your software with such linear combinations.

(a) Using $\mathbf{v}_1, \mathbf{v}_2, \mathbf{v}_3$ from Example 6, compute

$$5\mathbf{v}_1 - 2\mathbf{v}_2 + 4\mathbf{v}_3.$$

(b) Using $\mathbf{v}_j$, $j = 1, \ldots, 4$ in Example 8, compute

$$3\mathbf{v}_1 - \mathbf{v}_2 + 4\mathbf{v}_3 + 2\mathbf{v}_4.$$

(See also Exercise 32 in Section 1.2.)

34. In Exercise 33, suppose that the vectors were in R_n. Devise a procedure that uses matrix multiplication for forming linear combinations of vectors in R_n.

2.4 SPAN AND LINEAR INDEPENDENCE

Thus far we have defined a mathematical system called a real vector space and noted some of its properties. We further observe that the only real vector space having a finite number of vectors in it is the vector space whose only vector is $\mathbf{0}$, for if $\mathbf{v} \neq \mathbf{0}$ is in a vector space V, then by Exercise 21 in Section 2.2, $c \odot \mathbf{v} \neq c' \odot \mathbf{v}$, where c and c' are distinct real numbers, and so V has infinitely many vectors in it. However, in this section and the following one we show that most vector spaces V studied here have a set composed of a finite number of vectors that completely describe V. It should be noted that, in general, there is more than one such set describing V. We now turn to a formulation of these ideas. Remember that we will denote $\mathbf{u} \oplus \mathbf{v}$ and $c \odot \mathbf{u}$ in a vector space V as $\mathbf{u} + \mathbf{v}$ and $c\mathbf{u}$, respectively.

Definition 2.8

The vectors $\mathbf{v}_1, \mathbf{v}_2, \ldots, \mathbf{v}_k$ in a vector space V are said to **span** V if every vector in V is a linear combination of $\mathbf{v}_1, \mathbf{v}_2, \ldots, \mathbf{v}_k$. Moreover, if $S = \{\mathbf{v}_1, \mathbf{v}_2, \ldots, \mathbf{v}_k\}$, then we also say that the set S **spans** V, or that $\{\mathbf{v}_1, \mathbf{v}_2, \ldots, \mathbf{v}_k\}$ **spans** V, or that V is **spanned** by S or, in the language of Section 2.3, span $S = V$. ▲

Again, we investigate the consistency of a linear system, but this time for an arbitrary right side.

EXAMPLE 1

Let V be the vector space R^3. Let

$$\mathbf{v}_1 = \begin{bmatrix} 1 \\ 2 \\ 1 \end{bmatrix}, \quad \mathbf{v}_2 = \begin{bmatrix} 1 \\ 0 \\ 2 \end{bmatrix}, \quad \text{and} \quad \mathbf{v}_3 = \begin{bmatrix} 1 \\ 1 \\ 0 \end{bmatrix}.$$

To find out whether $\mathbf{v}_1, \mathbf{v}_2, \mathbf{v}_3$ span V, we pick any vector $\mathbf{v} = \begin{bmatrix} a \\ b \\ c \end{bmatrix}$ in V (a, b, and c are arbitrary real numbers) and must find out whether there are constants a_1, a_2, and a_3 such that

$$a_1\mathbf{v}_1 + a_2\mathbf{v}_2 + a_3\mathbf{v}_3 = \mathbf{v}.$$

This leads to the linear system (verify)

$$\begin{aligned} a_1 + a_2 + a_3 &= a \\ 2a_1 + a_3 &= b \\ a_1 + 2a_2 &= c. \end{aligned}$$

A solution is (verify)

$$a_1 = \frac{-2a + 2b + c}{3}, \quad a_2 = \frac{a - b + c}{3}, \quad a_3 = \frac{4a - b - 2c}{3}.$$

Thus $\mathbf{v}_1, \mathbf{v}_2, \mathbf{v}_3$ span V. This is equivalent to saying that

$$\operatorname{span}\{\mathbf{v}_1, \mathbf{v}_2, \mathbf{v}_3\} = R^3.$$

■

EXAMPLE 2

Let V be the vector space P_2. Let $\mathbf{v}_1 = t^2 + 2t + 1$ and $\mathbf{v}_2 = t^2 + 2$. Does $\{\mathbf{v}_1, \mathbf{v}_2\}$ span V?

Solution Let $\mathbf{v} = at^2 + bt + c$ be any vector in V, where a, b, and c are any real numbers. We must find out whether there are constants a_1 and a_2 such that

$$a_1\mathbf{v}_1 + a_2\mathbf{v}_2 = \mathbf{v}$$

or

$$a_1(t^2 + 2t + 1) + a_2(t^2 + 2) = at^2 + bt + c.$$

Thus

$$(a_1 + a_2)t^2 + (2a_1)t + (a_1 + 2a_2) = at^2 + bt + c.$$

Equating the coefficients of respective powers of t, we get the linear system

$$\begin{aligned} a_1 + a_2 &= a \\ 2a_1 &= b \\ a_1 + 2a_2 &= c. \end{aligned}$$

Transforming the augmented matrix of this linear system, we obtain (verify)

$$\left[\begin{array}{cc|c} 1 & 0 & 2a - c \\ 0 & 1 & c - a \\ 0 & 0 & b - 4a + 2c \end{array}\right].$$

If $b - 4a + 2c \neq 0$, then the system is inconsistent and there is no solution. Hence $\{\mathbf{v}_1, \mathbf{v}_2\}$ does not span V.

■

EXAMPLE 3

Consider the homogeneous linear system $A\mathbf{x} = \mathbf{0}$, where

$$A = \begin{bmatrix} 1 & 1 & 0 & 2 \\ -2 & -2 & 1 & -5 \\ 1 & 1 & -1 & 3 \\ 4 & 4 & -1 & 9 \end{bmatrix}.$$

From Example 9 in Section 2.3, the set of all solutions to $A\mathbf{x} = \mathbf{0}$ forms a subspace of R^4. To determine a spanning set for the solution space of this homogeneous system, we find that the reduced row echelon form of the augmented matrix is (verify)

$$\left[\begin{array}{cccc|c} 1 & 1 & 0 & 2 & 0 \\ 0 & 0 & 1 & -1 & 0 \\ 0 & 0 & 0 & 0 & 0 \\ 0 & 0 & 0 & 0 & 0 \end{array}\right].$$

The general solution is then given by

$$x_1 = -r - 2s, \quad x_2 = r, \quad x_3 = s, \quad x_4 = s,$$

where r and s are any real numbers. In matrix form we have that any member of the solution space is given by

$$\mathbf{x} = r \begin{bmatrix} -1 \\ 1 \\ 0 \\ 0 \end{bmatrix} + s \begin{bmatrix} -2 \\ 0 \\ 1 \\ 1 \end{bmatrix}.$$

Hence $\begin{bmatrix} -1 \\ 1 \\ 0 \\ 0 \end{bmatrix}$ and $\begin{bmatrix} -2 \\ 0 \\ 1 \\ 1 \end{bmatrix}$ span the solution space. ■

Definition 2.9

The vectors $\mathbf{v}_1, \mathbf{v}_2, \ldots, \mathbf{v}_k$, in a vector space V are said to be **linearly dependent** if there exist constants $a_1, a_2, \ldots, a_k$, not all zero, such that

$$a_1\mathbf{v}_2 + a_2\mathbf{v}_2 + \cdots + a_k\mathbf{v}_k = \mathbf{0}. \tag{1}$$

Otherwise, $\mathbf{v}_1, \mathbf{v}_2, \ldots, \mathbf{v}_k$ are called **linearly independent**. That is, $\mathbf{v}_1, \mathbf{v}_2, \ldots, \mathbf{v}_k$ are linearly independent if whenever $a_1\mathbf{v}_1 + a_2\mathbf{v}_2 + \cdots + a_k\mathbf{v}_k = \mathbf{0}$, then

$$a_1 = a_2 = \cdots = a_k = 0.$$

If $S = \{\mathbf{v}_1, \mathbf{v}_2, \ldots, \mathbf{v}_k\}$, then we also say that the set S is **linearly dependent** or **linearly independent** if the vectors have the corresponding property. ▲

It should be emphasized that for any vectors $\mathbf{v}_1, \mathbf{v}_2, \ldots, \mathbf{v}_k$, Equation (1) always holds if we choose all the scalars $a_1, a_2, \ldots, a_k$ equal to zero. The important point in this definition is whether it is possible to satisfy (1) with at least one of the scalars different from zero.

EXAMPLE 4

Determine whether the vectors

$$\begin{bmatrix} -1 \\ 1 \\ 0 \\ 0 \end{bmatrix} \quad \text{and} \quad \begin{bmatrix} -2 \\ 0 \\ 1 \\ 1 \end{bmatrix}$$

found in Example 3 as spanning the solution space of $A\mathbf{x} = \mathbf{0}$ are linearly dependent or linearly independent.

Solution Forming Equation (1),

$$a_1 \begin{bmatrix} -1 \\ 1 \\ 0 \\ 0 \end{bmatrix} + a_2 \begin{bmatrix} -2 \\ 0 \\ 1 \\ 1 \end{bmatrix} = \begin{bmatrix} 0 \\ 0 \\ 0 \\ 0 \end{bmatrix},$$

we obtain the homogeneous system

$$\begin{aligned} -a_1 - 2a_2 &= 0 \\ a_1 + 0a_2 &= 0 \\ 0a_1 + a_2 &= 0 \\ 0a_1 + a_2 &= 0, \end{aligned}$$

whose only solution is $a_1 = a_2 = 0$ (verify). Hence the given vectors are linearly independent. ■

EXAMPLE 5

Are the vectors $\mathbf{v}_1 = \begin{bmatrix} 1 & 0 & 1 & 2 \end{bmatrix}$, $\mathbf{v}_2 = \begin{bmatrix} 0 & 1 & 1 & 2 \end{bmatrix}$, and $\mathbf{v}_3 = \begin{bmatrix} 1 & 1 & 1 & 3 \end{bmatrix}$ in R_4 linearly dependent or linearly independent?

Solution We form Equation (1),

$$a_1\mathbf{v}_1 + a_2\mathbf{v}_2 + a_3\mathbf{v}_3 = \mathbf{0},$$

and solve for a_1, a_2, and a_3. The resulting homogeneous system is (verify)

$$\begin{aligned} a_1 + a_3 &= 0 \\ a_2 + a_3 &= 0 \\ a_1 + a_2 + a_3 &= 0 \\ 2a_1 + 2a_2 + 3a_3 &= 0, \end{aligned}$$

which has as its only solution $a_1 = a_2 = a_3 = 0$ (verify), showing that the given vectors are linearly independent. ■

EXAMPLE 6

Consider the vectors

$$\mathbf{v}_1 = \begin{bmatrix} 1 \\ 2 \\ -1 \end{bmatrix}, \quad \mathbf{v}_2 = \begin{bmatrix} 1 \\ -2 \\ 1 \end{bmatrix}, \quad \mathbf{v}_3 = \begin{bmatrix} -3 \\ 2 \\ -1 \end{bmatrix}, \quad \text{and} \quad \mathbf{v}_4 = \begin{bmatrix} 2 \\ 0 \\ 0 \end{bmatrix}$$

in R^3. Is $S = \{\mathbf{v}_1, \mathbf{v}_2, \mathbf{v}_3, \mathbf{v}_4\}$ linearly dependent or linearly independent?

Solution Setting up Equation (1), we are led to the homogeneous system

$$\begin{aligned} a_1 + a_2 - 3a_3 + 2a_4 &= 0 \\ 2a_1 - 2a_2 + 2a_3 &= 0 \\ -a_1 + a_2 - a_3 &= 0, \end{aligned}$$

of three equations in four unknowns. By Theorem 1.12, we are assured of the existence of a nontrivial solution. Hence S is linearly dependent. In fact, two of the infinitely many solutions are

$$\begin{aligned} a_1 &= 1, \quad a_2 = 2, \quad a_3 = 1, \quad a_4 = 0; \\ a_1 &= 1, \quad a_2 = 1, \quad a_3 = 0, \quad a_4 = -1. \end{aligned}$$

■

EXAMPLE 7

To find out whether the vectors $\mathbf{v}_1 = \begin{bmatrix} 1 & 0 & 0 \end{bmatrix}$, $\mathbf{v}_2 = \begin{bmatrix} 0 & 1 & 0 \end{bmatrix}$, and $\mathbf{v}_3 = \begin{bmatrix} 0 & 0 & 1 \end{bmatrix}$ in R_3 are linearly dependent or linearly independent, we form Equation (1),

$$a_1\mathbf{v}_1 + a_2\mathbf{v}_2 + a_3\mathbf{v}_3 = \mathbf{0},$$

and solve for a_1, a_2, and a_3. Since $a_1 = a_2 = a_3 = 0$ (verify), we conclude that the given vectors are linearly independent. ■

EXAMPLE 8

Are the vectors $\mathbf{v}_1 = t^2 + t + 2$, $\mathbf{v}_2 = 2t^2 + t$, and $\mathbf{v}_3 = 3t^2 + 2t + 2$ in P_2 linearly dependent or linearly independent?

Solution Forming Equation (1), we have (verify)

$$\begin{aligned} a_1 + 2a_2 + 3a_3 &= 0 \\ a_1 + a_2 + 2a_3 &= 0 \\ 2a_1 + 2a_3 &= 0, \end{aligned}$$

which has infinitely many solutions (verify). A particular solution is $a_1 = 1$, $a_2 = 1$, $a_3 = -1$, so

$$\mathbf{v}_1 + \mathbf{v}_2 - \mathbf{v}_3 = \mathbf{0}.$$

Hence the given vectors are linearly dependent. ■

Theorem 2.5

Let S_1 and S_2 be finite subsets of a vector space and let S_1 be a subset of S_2. Then:

(a) *If S_1 is linearly dependent, so is S_2.*

(b) *If S_2 is linearly independent, so is S_1.*

Proof

Let

$$S_1 = \{\mathbf{v}_1, \mathbf{v}_2, \ldots, \mathbf{v}_k\} \quad \text{and} \quad S_2 = \{\mathbf{v}_1, \mathbf{v}_2, \ldots, \mathbf{v}_k, \mathbf{v}_{k+1}, \ldots, \mathbf{v}_m\}.$$

We first prove (a). Since S_1 is linearly dependent, there exist $a_1, a_2, \ldots, a_k$, not all zero, such that

$$a_1\mathbf{v}_1 + a_2\mathbf{v}_2 + \cdots + a_k\mathbf{v}_k = \mathbf{0}.$$

Then

$$a_1\mathbf{v}_1 + a_2\mathbf{v}_2 + \cdots + a_k\mathbf{v}_k + 0\mathbf{v}_{k+1} + \cdots + 0\mathbf{v}_m = \mathbf{0}. \tag{2}$$

Since not all the coefficients in (2) are zero, we conclude that S_2 is linearly dependent.

Statement (b) is the contrapositive of statement (a), so it is logically equivalent to statement (a). If we insist on proving it, we may proceed as follows. Let S_2 be linearly independent. If S_1 is assumed as linearly dependent, then S_2 is linearly dependent, by (a), a contradiction. Hence, S_1 must be linearly independent. ●

We now note that the set $S = \{\mathbf{0}\}$ consisting only of $\mathbf{0}$ is linearly dependent, since, for example, $5\mathbf{0} = \mathbf{0}$ and $5 \neq 0$. From this it follows that if S is any set of vectors that contain $\mathbf{0}$, then S must be linearly dependent. Also, a set consisting of a single nonzero vector is linearly independent (verify). Moreover, if $\mathbf{v}_1, \mathbf{v}_2, \ldots, \mathbf{v}_k$ are vectors in a vector space V and any two of them are equal, then $\mathbf{v}_1, \mathbf{v}_2, \ldots, \mathbf{v}_k$ are linearly dependent (verify).

We consider next the meaning of linear independence in R^2 and R^3. Suppose that $\mathbf{v}_1$ and $\mathbf{v}_2$ are linearly dependent in R^2. Then there exist scalars a_1 and a_2, not both zero, such that

$$a_1\mathbf{v}_1 + a_2\mathbf{v}_2 = \mathbf{0}.$$

If $a_1 \neq 0$, then $\mathbf{v}_1 = \left(-\dfrac{a_2}{a_1}\right)\mathbf{v}_2$. If $a_2 \neq 0$, then $\mathbf{v}_2 = \left(-\dfrac{a_1}{a_2}\right)\mathbf{v}_1$. Thus one of the vectors is a multiple of the other. Conversely, suppose that $\mathbf{v}_1 = a\mathbf{v}_2$. Then

$$1\mathbf{v}_1 - a\mathbf{v}_2 = \mathbf{0},$$

and since the coefficients of $\mathbf{v}_1$ and $\mathbf{v}_2$ are not both zero, it follows that $\mathbf{v}_1$ and $\mathbf{v}_2$ are linearly dependent. Thus $\mathbf{v}_1$ and $\mathbf{v}_2$ are linearly dependent in R^2 if and only if one of the vectors is a multiple of the other [Figure 2.22(a)]. Hence two vectors in R^2 are linearly dependent if and only if they both lie on the same line passing through the origin [Figure 2.22(a)].

Suppose now that $\mathbf{v}_1$, $\mathbf{v}_2$, and $\mathbf{v}_3$ are linearly dependent in R^3. Then we can write

$$a_1\mathbf{v}_1 + a_2\mathbf{v}_2 + a_3\mathbf{v}_3 = \mathbf{0},$$

$\mathbf{v}_2$
$\mathbf{v}_1$
O

(a) Linearly dependent vectors in R^2.

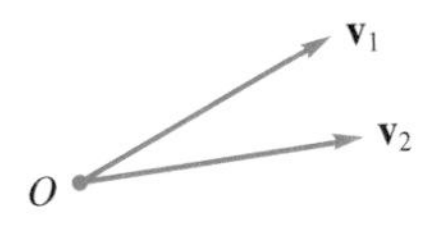

(b) Linearly independent vectors in R^2.

FIGURE 2.22

where a_1, a_2, and a_3 are not all zero, say $a_2 \neq 0$. Then

$$\mathbf{v}_2 = \left(-\frac{a_1}{a_2}\right)\mathbf{v}_1 - \left(\frac{a_3}{a_2}\right)\mathbf{v}_3,$$

which means that $\mathbf{v}_2$ is in the subspace W spanned by $\mathbf{v}_1$ and $\mathbf{v}_3$.

Now W is either a plane through the origin (when $\mathbf{v}_1$ and $\mathbf{v}_3$ are linearly independent), or a line through the origin (when $\mathbf{v}_1$ and $\mathbf{v}_3$ are linearly dependent), or $W = \{\mathbf{0}\}$. Since a line through the origin always lies in a plane through the origin, we conclude that $\mathbf{v}_1$, $\mathbf{v}_2$, and $\mathbf{v}_3$ all lie in the same plane through the origin. Conversely, suppose that $\mathbf{v}_1$, $\mathbf{v}_2$, and $\mathbf{v}_3$ all lie in the same plane through the origin. Then either all three vectors are the zero vector, or all three vectors lie on the same line through the origin, or all three vectors lie in a plane through the origin spanned by two vectors, say $\mathbf{v}_1$ and $\mathbf{v}_3$. Thus, in all these cases, $\mathbf{v}_2$ is a linear combination of $\mathbf{v}_1$ and $\mathbf{v}_3$:

$$\mathbf{v}_2 = c_1\mathbf{v}_1 + c_3\mathbf{v}_3.$$

Then

$$c_1\mathbf{v}_1 - 1\mathbf{v}_2 + c_3\mathbf{v}_3 = \mathbf{0},$$

which means that $\mathbf{v}_1$, $\mathbf{v}_2$, and $\mathbf{v}_3$ are linearly dependent. Hence three vectors in R^3 are linearly dependent if and only if they all lie in the same plane passing through the origin (Figure 2.23(a)).

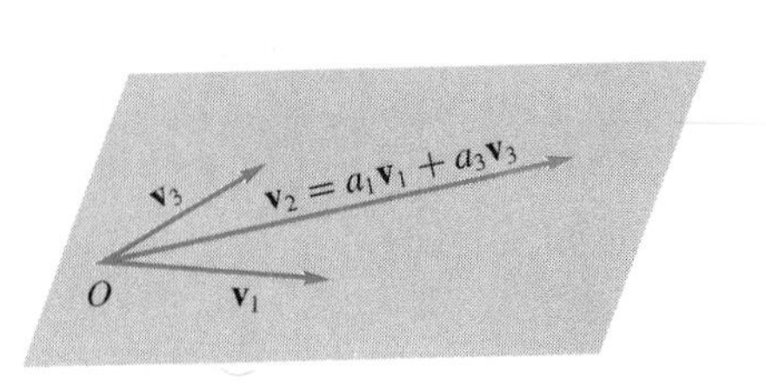

(a) Linearly dependent vectors in R^3.

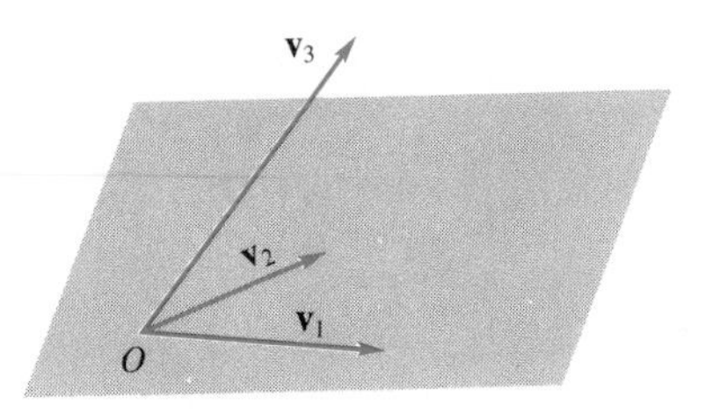

(b) Linearly independent vectors in R^3.

FIGURE 2.23

More generally, let $\mathbf{u}$ and $\mathbf{v}$ be nonzero vectors in a vector space V. We can show (Exercise 18) that $\mathbf{u}$ and $\mathbf{v}$ are linearly dependent if and only if there is a scalar k such that $\mathbf{v} = k\mathbf{u}$. Equivalently, $\mathbf{u}$ and $\mathbf{v}$ are linearly independent if and only if neither vector is a multiple of the other.

Theorem 2.6

The nonzero vectors $\mathbf{v}_1, \mathbf{v}_2, \ldots, \mathbf{v}_n$ *in a vector space* V *are linearly dependent if and only if one of the vectors* $\mathbf{v}_j$ $(j \geq 2)$ *is a linear combination of the preceding vectors* $\mathbf{v}_1, \mathbf{v}_2, \ldots, \mathbf{v}_{j-1}$.

Proof

If $\mathbf{v}_j$ is a linear combination of $\mathbf{v}_1, \mathbf{v}_2, \ldots, \mathbf{v}_{j-1}$, i.e.,

$$\mathbf{v}_j = a_1\mathbf{v}_1 + a_2\mathbf{v}_2 + \cdots + a_{j-1}\mathbf{v}_{j-1},$$

then

$$a_1\mathbf{v}_1 + a_2\mathbf{v}_2 + \cdots + a_{j-1}\mathbf{v}_{j-1} + (-1)\mathbf{v}_j + 0\mathbf{v}_{j+1} + \cdots + 0\mathbf{v}_n = \mathbf{0}.$$

Since at least one coefficient, -1, is nonzero, we conclude that $\mathbf{v}_1, \mathbf{v}_2, \ldots, \mathbf{v}_n$ are linearly dependent.

Conversely, suppose that $\mathbf{v}_1, \mathbf{v}_2, \ldots, \mathbf{v}_n$ are linearly dependent. Then there exist scalars, $a_1, a_2, \ldots, a_n$, not all zero, such that

$$a_1\mathbf{v}_1 + a_2\mathbf{v}_2 + \cdots + a_n\mathbf{v}_n = \mathbf{0}.$$

Now let j be the largest subscript for which $a_j \neq 0$. If $j \geq 2$, then

$$\mathbf{v}_j = -\left(\frac{a_1}{a_j}\right)\mathbf{v}_1 - \left(\frac{a_2}{a_j}\right)\mathbf{v}_2 - \cdots - \left(\frac{a_{j-1}}{a_j}\right)\mathbf{v}_{j-1}.$$

If $j = 1$, then $a_1\mathbf{v}_1 = \mathbf{0}$, which implies that $\mathbf{v}_1 = \mathbf{0}$, a contradiction of the hypothesis that none of the vectors is the zero vector. Thus one of the vectors $\mathbf{v}_j$ is a linear combination of the preceding vectors $\mathbf{v}_1, \mathbf{v}_2, \ldots, \mathbf{v}_{j-1}$. ●

EXAMPLE 9

Let $V = R_3$ and also $\mathbf{v}_1 = \begin{bmatrix} 1 & 2 & -1 \end{bmatrix}$, $\mathbf{v}_2 = \begin{bmatrix} 1 & -2 & 1 \end{bmatrix}$, $\mathbf{v}_3 = \begin{bmatrix} -3 & 2 & -1 \end{bmatrix}$, and $\mathbf{v}_4 = \begin{bmatrix} 2 & 0 & 0 \end{bmatrix}$. We find (verify) that

$$\mathbf{v}_1 + \mathbf{v}_2 + 0\mathbf{v}_3 - \mathbf{v}_4 = \mathbf{0}$$

so $\mathbf{v}_1$, $\mathbf{v}_2$, $\mathbf{v}_3$ and $\mathbf{v}_4$ are linearly dependent. We then have

$$\mathbf{v}_4 = \mathbf{v}_1 + \mathbf{v}_2 + 0\mathbf{v}_3.$$

■

Remarks

1. We observe that Theorem 2.6 does not say that *every* vector $\mathbf{v}$ is a linear combination of the preceding vectors. Thus, in Example 9, we also have $\mathbf{v}_1 + 2\mathbf{v}_2 + \mathbf{v}_3 + 0\mathbf{v}_4 = \mathbf{0}$. We cannot solve, in this equation, for $\mathbf{v}_4$ as a linear combination of $\mathbf{v}_1$, $\mathbf{v}_2$, and $\mathbf{v}_3$, since its coefficient is zero.

2. We can also prove that if $S = \{\mathbf{v}_1, \mathbf{v}_2, \ldots, \mathbf{v}_k\}$ is a set of vectors in a vector space V, then S is linearly dependent if and only if one of the vectors in S is a linear combination of all the other vectors in S (see Exercise 19). For instance, in Example 9,

$$\mathbf{v}_1 = -\mathbf{v}_2 - 0\mathbf{v}_3 + \mathbf{v}_4; \quad \mathbf{v}_2 = -\tfrac{1}{2}\mathbf{v}_1 - \tfrac{1}{2}\mathbf{v}_3 - 0\mathbf{v}_4.$$

3. Observe that if $\mathbf{v}_1, \mathbf{v}_2, \ldots, \mathbf{v}_k$ are linearly independent vectors in a vector space, then they must be distinct and nonzero.

The following result is used in Section 2.5 as well as in several other places. Suppose that $S = \{\mathbf{v}_1, \mathbf{v}_2, \ldots, \mathbf{v}_n\}$ spans a vector space V, and $\mathbf{v}_j$ is a linear combination of the preceding vectors in S. Then the set

$$S_1 = \{\mathbf{v}_1, \mathbf{v}_2, \ldots, \mathbf{v}_{j-1}, \mathbf{v}_{j+1}, \ldots, \mathbf{v}_n\},$$

consisting of S with $\mathbf{v}_j$ deleted, also spans V. To show this result, observe that if $\mathbf{v}$ is any vector in V, then, since S spans V, we can find scalars $a_1, a_2, \ldots, a_n$ such that

$$\mathbf{v} = a_1\mathbf{v}_1 + a_2\mathbf{v}_2 + \cdots + a_{j-1}\mathbf{v}_{j-1} + a_j\mathbf{v}_j + a_{j+1}\mathbf{v}_{j+1} + \cdots + a_n\mathbf{v}_n.$$

Now if

$$\mathbf{v}_j = b_1\mathbf{v}_1 + b_2\mathbf{v}_2 + \cdots + b_{j-1}\mathbf{v}_{j-1},$$

then

$$\begin{aligned}\mathbf{v} &= a_1\mathbf{v}_1 + a_2\mathbf{v}_2 + \cdots + a_{j-1}\mathbf{v}_{j-1} + a_j(b_1\mathbf{v}_1 + b_2\mathbf{v}_2 + \cdots + b_{j-1}\mathbf{v}_{j-1}) \\ &\quad + a_{j+1}\mathbf{v}_{j+1} + \cdots + a_n\mathbf{v}_n \\ &= c_1\mathbf{v}_1 + c_2\mathbf{v}_2 + \cdots + c_{j-1}\mathbf{v}_{j-1} + c_{j+1}\mathbf{v}_{j+1} + \cdots + c_n\mathbf{v}_n,\end{aligned}$$

which means that span $S_1 = V$.

EXAMPLE 10

Consider the set of vectors $S = \{\mathbf{v}_1, \mathbf{v}_2, \mathbf{v}_3, \mathbf{v}_4\}$ in R^4, where

$$\mathbf{v}_1 = \begin{bmatrix}1\\1\\0\\0\end{bmatrix}, \quad \mathbf{v}_2 = \begin{bmatrix}1\\0\\1\\0\end{bmatrix}, \quad \mathbf{v}_3 = \begin{bmatrix}0\\1\\1\\0\end{bmatrix}, \quad \text{and} \quad \mathbf{v}_4 = \begin{bmatrix}2\\1\\1\\0\end{bmatrix},$$

and let $W =$ span S. Since

$$\mathbf{v}_4 = \mathbf{v}_1 + \mathbf{v}_2,$$

we conclude that $W =$ span S_1, where $S_1 = \{\mathbf{v}_1, \mathbf{v}_2, \mathbf{v}_3\}$. ■

2.4 Exercises

1. Which of the following vectors span R_2?

(a) $\begin{bmatrix}1 & 2\end{bmatrix}, \begin{bmatrix}-1 & 1\end{bmatrix}$.

(b) $\begin{bmatrix}0 & 0\end{bmatrix}, \begin{bmatrix}1 & 1\end{bmatrix}, \begin{bmatrix}-2 & -2\end{bmatrix}$.

(c) $\begin{bmatrix}1 & 3\end{bmatrix}, \begin{bmatrix}2 & -3\end{bmatrix}, \begin{bmatrix}0 & 2\end{bmatrix}$.

(d) $\begin{bmatrix}2 & 4\end{bmatrix}, \begin{bmatrix}-1 & 2\end{bmatrix}$.

2. Which of the following sets of vectors span R^4?

(a) $\left\{\begin{bmatrix}1\\-1\\2\\0\end{bmatrix}, \begin{bmatrix}0\\1\\1\\1\end{bmatrix}\right\}$.

(b) $\left\{\begin{bmatrix}3\\2\\1\\0\end{bmatrix}, \begin{bmatrix}1\\2\\-1\\0\end{bmatrix}, \begin{bmatrix}0\\0\\0\\1\end{bmatrix}\right\}$.

(c) $\left\{\begin{bmatrix}3\\2\\-1\\2\end{bmatrix}, \begin{bmatrix}4\\0\\0\\2\end{bmatrix}, \begin{bmatrix}3\\2\\-1\\2\end{bmatrix}, \begin{bmatrix}5\\6\\-3\\2\end{bmatrix}, \begin{bmatrix}0\\4\\-2\\-1\end{bmatrix}\right\}$.

(d) $\left\{\begin{bmatrix}1\\1\\0\\0\end{bmatrix}, \begin{bmatrix}1\\2\\-1\\1\end{bmatrix}, \begin{bmatrix}0\\0\\1\\-1\end{bmatrix}, \begin{bmatrix}2\\1\\2\\-1\end{bmatrix}\right\}$.

3. Which of the following sets of vectors span R_4?

(a) $\begin{bmatrix}1 & 0 & 0 & 1\end{bmatrix}, \begin{bmatrix}0 & 1 & 0 & 0\end{bmatrix}, \begin{bmatrix}1 & 1 & 1 & 1\end{bmatrix}, \begin{bmatrix}1 & 1 & 1 & 0\end{bmatrix}$.

(b) $\begin{bmatrix}1 & 2 & 1 & 0\end{bmatrix}, \begin{bmatrix}1 & 1 & -1 & 0\end{bmatrix}, \begin{bmatrix}0 & 0 & 0 & 1\end{bmatrix}$.

(c) $\begin{bmatrix}6 & 4 & -2 & 4\end{bmatrix}, \begin{bmatrix}2 & 0 & 0 & 1\end{bmatrix}, \begin{bmatrix}3 & 2 & -1 & 2\end{bmatrix}, \begin{bmatrix}5 & 6 & -3 & 2\end{bmatrix}, \begin{bmatrix}0 & 4 & -2 & -1\end{bmatrix}$.

(d) $\begin{bmatrix}1 & 1 & 0 & 0\end{bmatrix}, \begin{bmatrix}1 & 2 & -1 & 1\end{bmatrix}, \begin{bmatrix}0 & 0 & 1 & 1\end{bmatrix}, \begin{bmatrix}2 & 1 & 2 & 1\end{bmatrix}$.

4. Which of the following sets of polynomials span P_2?

(a) $\{t^2+1, t^2+t, t+1\}$.

(b) $\{t^2+1, t-1, t^2+t\}$.

(c) $\{t^2+2, 2t^2-t+1, t+2, t^2+t+4\}$.

(d) $\{t^2+2t-1, t^2-1\}$.

5. Do the polynomials t^3+2t+1, t^2-t+2, t^3+2, $-t^3+t^2-5t+2$ span P_3?

6. Does the set

$$S = \left\{\begin{bmatrix}1 & 1\\0 & 0\end{bmatrix}, \begin{bmatrix}0 & 0\\1 & 1\end{bmatrix}, \begin{bmatrix}1 & 0\\0 & 1\end{bmatrix}, \begin{bmatrix}0 & 1\\1 & 1\end{bmatrix}\right\}$$

span M_{22}?

7. Find a set of vectors spanning the solution space of $A\mathbf{x} = \mathbf{0}$, where

$$A = \begin{bmatrix}1 & 0 & 1 & 0\\1 & 2 & 3 & 1\\2 & 1 & 3 & 1\\1 & 1 & 2 & 1\end{bmatrix}.$$

8. Find a set of vectors spanning the null space of

$$A = \begin{bmatrix}1 & 1 & 2 & -1\\2 & 3 & 6 & -2\\-2 & 1 & 2 & 2\\0 & -2 & -4 & 0\end{bmatrix}.$$

9. Let $\mathbf{x}_1 = \begin{bmatrix}2\\-1\\1\end{bmatrix}$, $\mathbf{x}_2 = \begin{bmatrix}4\\-7\\-1\end{bmatrix}$, $\mathbf{x}_3 = \begin{bmatrix}1\\2\\2\end{bmatrix}$ belong to the solution space of $A\mathbf{x} = \mathbf{0}$. Is $\{\mathbf{x}_1, \mathbf{x}_2, \mathbf{x}_3\}$ linearly independent?

10. Let $\mathbf{x}_1 = \begin{bmatrix}1\\2\\0\\1\end{bmatrix}$, $\mathbf{x}_2 = \begin{bmatrix}1\\0\\-1\\1\end{bmatrix}$, $\mathbf{x}_3 = \begin{bmatrix}1\\6\\2\\0\end{bmatrix}$ belong to the null space of A. Is $\{\mathbf{x}_1, \mathbf{x}_2, \mathbf{x}_3\}$ linearly independent?

11. Which of the following vectors in R_3 are linearly dependent? For those which are, express one vector as a linear combination of the rest.

(a) $\begin{bmatrix}1 & 1 & 0\end{bmatrix}, \begin{bmatrix}0 & 2 & 3\end{bmatrix}, \begin{bmatrix}1 & 2 & 3\end{bmatrix}, \begin{bmatrix}3 & 6 & 6\end{bmatrix}$.

(b) $\begin{bmatrix}1 & 1 & 0\end{bmatrix}, \begin{bmatrix}3 & 4 & 2\end{bmatrix}$.

(c) $\begin{bmatrix}1 & 1 & 0\end{bmatrix}, \begin{bmatrix}0 & 2 & 3\end{bmatrix}, \begin{bmatrix}1 & 2 & 3\end{bmatrix}, \begin{bmatrix}0 & 0 & 0\end{bmatrix}$.

12. Consider the vector space M_{22}. Follow the directions of Exercise 11.

(a) $\begin{bmatrix}1 & 1\\2 & 1\end{bmatrix}, \begin{bmatrix}1 & 0\\0 & 2\end{bmatrix}, \begin{bmatrix}0 & 3\\2 & 1\end{bmatrix}, \begin{bmatrix}4 & 6\\8 & 6\end{bmatrix}$.

(b) $\begin{bmatrix}1 & 1\\1 & 1\end{bmatrix}, \begin{bmatrix}1 & 0\\0 & 2\end{bmatrix}, \begin{bmatrix}0 & 1\\0 & 2\end{bmatrix}$.

(c) $\begin{bmatrix}1 & 1\\1 & 1\end{bmatrix}, \begin{bmatrix}2 & 3\\1 & 2\end{bmatrix}, \begin{bmatrix}3 & 1\\2 & 1\end{bmatrix}, \begin{bmatrix}2 & 2\\1 & 1\end{bmatrix}$.

13. Consider the vector space P_2. Follow the directions of Exercise 11.

(a) $t^2+1, t-2, t+3$.

(b) $2t^2+t, t^2+3, t$.

(c) $2t^2+t+1, 3t^2+t-5, t+13$.

14. Let V be the vector space of all real-valued continuous functions. Follow the directions of Exercise 11.

(a) $\cos t, \sin t, e^t$.

(b) $t, e^t, \sin t$.

(c) t^2, t, e^t.

(d) $\cos^2 t, \sin^2 t, \cos 2t$.

15. Consider the vector space R^3. Follow the directions of Exercise 11.

(a) $\begin{bmatrix}1\\0\\0\end{bmatrix}, \begin{bmatrix}0\\1\\1\end{bmatrix}, \begin{bmatrix}1\\2\\-1\end{bmatrix}$.

(b) $\begin{bmatrix} 1 \\ 1 \\ -1 \end{bmatrix}, \begin{bmatrix} 0 \\ 1 \\ 1 \end{bmatrix}, \begin{bmatrix} 1 \\ 1 \\ 1 \end{bmatrix}, \begin{bmatrix} 1 \\ 2 \\ -2 \end{bmatrix}.$

(c) $\begin{bmatrix} 1 \\ 0 \\ 0 \end{bmatrix}, \begin{bmatrix} 2 \\ 1 \\ 1 \end{bmatrix}, \begin{bmatrix} -1 \\ 2 \\ 1 \end{bmatrix}.$

16. For what values of c are the vectors $\begin{bmatrix} -1 & 0 & -1 \end{bmatrix}$, $\begin{bmatrix} 2 & 1 & 2 \end{bmatrix}$, and $\begin{bmatrix} 1 & 1 & c \end{bmatrix}$ in R_3 linearly dependent?

17. For what values of c are the vectors $t+3$ and $2t+c^2+2$ in P_1 linearly independent?

18. Let $\mathbf{u}$ and $\mathbf{v}$ be nonzero vectors in a vector space V. Show that $\mathbf{u}$ and $\mathbf{v}$ are linearly dependent if and only if there is a scalar k such that $\mathbf{v} = k\mathbf{u}$. Equivalently, $\mathbf{u}$ and $\mathbf{v}$ are linearly independent if and only if neither vector is a multiple of the other.

19. Let $S = \{\mathbf{v}_1, \mathbf{v}_2, \ldots, \mathbf{v}_k\}$ be a set of vectors in a vector space V. Prove that S is linearly dependent if and only if one of the vectors in S is a linear combination of all the other vectors in S.

20. Suppose that $S = \{\mathbf{v}_1, \mathbf{v}_2, \mathbf{v}_3\}$ is a linearly independent set of vectors in a vector space V. Prove that $T = \{\mathbf{w}_1, \mathbf{w}_2, \mathbf{w}_3\}$ is also linearly independent, where $\mathbf{w}_1 = \mathbf{v}_1 + \mathbf{v}_2 + \mathbf{v}_3$, $\mathbf{w}_2 = \mathbf{v}_2 + \mathbf{v}_3$, and $\mathbf{w}_3 = \mathbf{v}_3$.

21. Suppose that $S = \{\mathbf{v}_1, \mathbf{v}_2, \mathbf{v}_3\}$ is a linearly independent set of vectors in a vector space V. Is $T = \{\mathbf{w}_1, \mathbf{w}_2, \mathbf{w}_3\}$, where $\mathbf{w}_1 = \mathbf{v}_1 + \mathbf{v}_2$, $\mathbf{w}_2 = \mathbf{v}_1 + \mathbf{v}_3$, $\mathbf{w}_3 = \mathbf{v}_2 + \mathbf{v}_3$, linearly dependent or linearly independent? Justify your answer.

22. Suppose that $S = \{\mathbf{v}_1, \mathbf{v}_2, \mathbf{v}_3\}$ is a linearly dependent set of vectors in a vector space V. Is $T = \{\mathbf{w}_1, \mathbf{w}_2, \mathbf{w}_3\}$, where $\mathbf{w}_1 = \mathbf{v}_1$, $\mathbf{w}_2 = \mathbf{v}_1 + \mathbf{v}_3$, $\mathbf{w}_3 = \mathbf{v}_1 + \mathbf{v}_2 + \mathbf{v}_3$, linearly dependent or linearly independent? Justify your answer.

23. Show that if $\{\mathbf{v}_1, \mathbf{v}_2\}$ is linearly independent and $\mathbf{v}_3$ does not belong to span $\{\mathbf{v}_1, \mathbf{v}_2\}$, then $\{\mathbf{v}_1, \mathbf{v}_2, \mathbf{v}_3\}$ is linearly independent.

24. Suppose that $\{\mathbf{v}_1, \mathbf{v}_2, \ldots, \mathbf{v}_n\}$ is a linearly independent set of vectors in R^n. Show that if A is an $n \times n$ nonsingular matrix, then $\{A\mathbf{v}_1, A\mathbf{v}_2, \ldots, A\mathbf{v}_n\}$ is linearly independent.

25. Let A be an $m \times n$ matrix in reduced row echelon form. Prove that the nonzero rows of A, viewed as vectors in R_n, form a linearly independent set of vectors.

26. Let $S = \{\mathbf{u}_1, \mathbf{u}_2, \ldots, \mathbf{u}_k\}$ be a set of vectors in a vector space and let $T = \{\mathbf{v}_1, \mathbf{v}_2, \ldots, \mathbf{v}_m\}$, where each $\mathbf{v}_i$, $i = 1, 2, \ldots, m$, is a linear combination of the vectors in S. Prove that

$$\mathbf{w} = b_1\mathbf{v}_1 + b_2\mathbf{v}_2 + \cdots + b_m\mathbf{v}_m$$

is a linear combination of the vectors in S.

27. Let S_1 and S_2 be finite subsets of a vector space and let S_1 be a subset of S_2. If S_2 is linearly dependent, why or why not is S_1 linearly dependent? Give an example.

28. Let S_1 and S_2 be finite subsets of a vector space and let S_1 be a subset of S_2. If S_1 is linearly independent, why or why not is S_2 linearly independent? Give an example.

29. Determine if your software has a command for finding the null space (see Example 9 in Section 2.3) of a matrix A. If it does, use it on the matrix A in Example 3 and compare the command's output with the results in Example 3. To experiment further, use Exercises 7 and 8.

30. As noted in the Remark after Example 8 in Section 2.3, to determine if specific vector $\mathbf{v}$ belongs to span S, we investigate the consistency of an appropriate nonhomogeneous linear system $A\mathbf{x} = \mathbf{b}$. In addition, to determine if a set of vectors is linearly independent, we investigate the null space of an appropriate homogeneous system $A\mathbf{x} = \mathbf{0}$. These investigations can be performed computationally, using a command for reduced row echelon form if available. (See Exercise 29 in Section 1.5.) We summarize the use of a reduced row echelon form command in these cases as follows. Let RREF(C) represent the reduced row echelon form of matrix C.

(i) $\mathbf{v}$ belongs to span S provided RREF $\left(\left[A \mid \mathbf{b}\right]\right)$ contains no row of the form $\left[0 \ \cdots \ 0 \mid *\right]$, where $*$ represents a nonzero number.

(ii) The set of vectors is linearly independent if RREF $\left(\left[A \mid \mathbf{0}\right]\right)$ contains only rows from an identity matrix and possibly rows of all zeros.

Experiment in your software with this approach, using the data given in Example 6 in Section 2.3 and Examples 4, 5, 6, and 8.

31. **Warning:** The strategy given in Exercise 30 assumes the computations are performed by using exact arithmetic. Most software uses a model of exact arithmetic called floating-point arithmetic, hence the use of reduced row echelon form may yield incorrect results in these cases. Computationally, the "line between" linear independence and linear dependence may be blurred. Experiment in your software with the use of reduced row echelon form for the following vectors in R^2. Are they linearly independent or linearly dependent? Compare the theoretical answer with the computational answer from your software.

(a) $\begin{bmatrix} 1 \\ 0 \end{bmatrix}, \begin{bmatrix} 1 \\ 1 \times 10^{-5} \end{bmatrix}$.

(b) $\begin{bmatrix} 1 \\ 0 \end{bmatrix}, \begin{bmatrix} 1 \\ 1 \times 10^{-10} \end{bmatrix}$.

(c) $\begin{bmatrix} 1 \\ 0 \end{bmatrix}, \begin{bmatrix} 1 \\ 1 \times 10^{-16} \end{bmatrix}$.

2.5 BASIS AND DIMENSION

In this section we continue our study of the structure of a vector space V by determining a set of vectors in V that completely describes V.

Basis

Definition 2.10

The vectors $\mathbf{v}_1, \mathbf{v}_2, \ldots, \mathbf{v}_k$ in a vector space V are said to form a **basis** for V if (a) $\mathbf{v}_1, \mathbf{v}_2, \ldots, \mathbf{v}_k$ span V and (b) $\mathbf{v}_1, \mathbf{v}_2, \ldots, \mathbf{v}_k$ are linearly independent. ▲

Remark If $\mathbf{v}_1, \mathbf{v}_2, \ldots, \mathbf{v}_k$ form a basis for a vector space V, then they must be distinct and nonzero.

EXAMPLE 1

Let $V = R^3$. The vectors $\begin{bmatrix} 1 \\ 0 \\ 0 \end{bmatrix}, \begin{bmatrix} 0 \\ 1 \\ 0 \end{bmatrix}, \begin{bmatrix} 0 \\ 0 \\ 1 \end{bmatrix}$ form a basis for R^3, called the **natural basis** or **standard basis**, for R^3. One can readily see how to generalize this to obtain the natural basis for R^n. Similarly,

$$\begin{bmatrix} 1 & 0 & 0 \end{bmatrix}, \quad \begin{bmatrix} 0 & 1 & 0 \end{bmatrix}, \quad \begin{bmatrix} 0 & 0 & 1 \end{bmatrix}$$

is the natural basis for R_3. ■

The natural basis for R^n is denoted by $\{\mathbf{e}_1, \mathbf{e}_2, \ldots, \mathbf{e}_n\}$, where

$$\mathbf{e}_i = \begin{bmatrix} 0 \\ \vdots \\ 0 \\ 1 \\ 0 \\ \vdots \\ 0 \end{bmatrix} \leftarrow i\text{th row;}$$

that is, $\mathbf{e}_i$ is an $n \times 1$ matrix with a 1 in the ith row and zeros elsewhere.

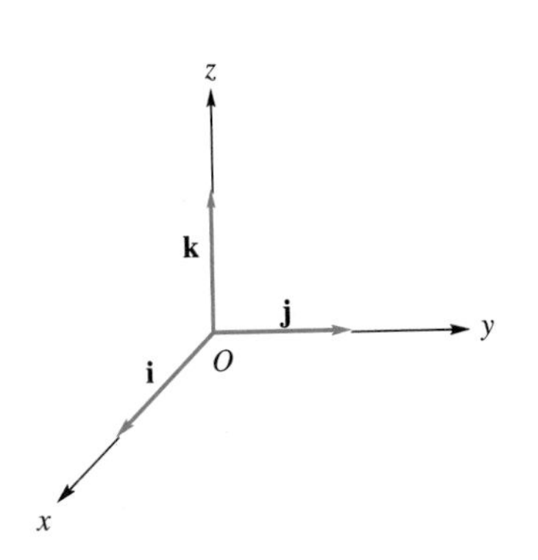

FIGURE 2.24

Natural basis for R^3.

The natural basis for R^3 is also often denoted by

$$\mathbf{i} = \begin{bmatrix} 1 \\ 0 \\ 0 \end{bmatrix}, \quad \mathbf{j} = \begin{bmatrix} 0 \\ 1 \\ 0 \end{bmatrix}, \quad \text{and} \quad \mathbf{k} = \begin{bmatrix} 0 \\ 0 \\ 1 \end{bmatrix}.$$

These vectors are shown in Figure 2.24. Thus any vector $\mathbf{v} = \begin{bmatrix} a_1 \\ a_2 \\ a_3 \end{bmatrix}$ in R^3 can be written as

$$\mathbf{v} = a_1\mathbf{i} + a_2\mathbf{j} + a_3\mathbf{k}.$$

EXAMPLE 2

Show that the set $S = \{t^2 + 1, t - 1, 2t + 2\}$ is a basis for the vector space P_2.

Solution To do this, we must show that S spans V and is linearly independent. To show that it spans V, we take any vector in V, that is, a polynomial $at^2 + bt + c$, and must find constants a_1, a_2, and a_3 such that

$$\begin{aligned} at^2 + bt + c &= a_1(t^2 + 1) + a_2(t - 1) + a_3(2t + 2) \\ &= a_1t^2 + (a_2 + 2a_3)t + (a_1 - a_2 + 2a_3). \end{aligned}$$

Since two polynomials agree for all values of t only if the coefficients of respective powers of t agree, we get the linear system

$$\begin{aligned} a_1 \qquad\qquad &= a \\ a_2 + 2a_3 &= b \\ a_1 - a_2 + 2a_3 &= c. \end{aligned}$$

Solving, we have

$$a_1 = a, \quad a_2 = \frac{a + b - c}{2}, \quad a_3 = \frac{c + b - a}{4}.$$

Hence S spans V.

To illustrate this result, suppose that we are given the vector $2t^2 + 6t + 13$. Here $a = 2$, $b = 6$, and $c = 13$. Substituting in the foregoing expressions for a_1, a_2, and a_3, we find that

$$a_1 = 2, \quad a_2 = -\tfrac{5}{2}, \quad a_3 = \tfrac{17}{4}.$$

Hence

$$2t^2 + 6t + 13 = 2(t^2 + 1) - \tfrac{5}{2}(t - 1) + \tfrac{17}{4}(2t + 2).$$

To show that S is linearly independent, we form

$$a_1(t^2 + 1) + a_2(t - 1) + a_3(2t + 2) = 0.$$

Then

$$a_1t^2 + (a_2 + 2a_3)t + (a_1 - a_2 + 2a_3) = 0.$$

Again, this can hold for all values of t only if

$$\begin{aligned} a_1 \qquad\qquad &= 0 \\ a_2 + 2a_3 &= 0 \\ a_1 - a_2 + 2a_3 &= 0. \end{aligned}$$

The only solution to this homogeneous system is $a_1 = a_2 = a_3 = 0$, which implies that S is linearly independent. Thus S is a basis for P_2. ■

The set of vectors $\{t^n, t^{n-1}, \dots, t, 1\}$ forms a basis for the vector space P_n called the **natural**, or **standard basis**, for P_n.

EXAMPLE 3

Show that the set $S = \{\mathbf{v}_1, \mathbf{v}_2, \mathbf{v}_3, \mathbf{v}_4\}$, where

$$\mathbf{v}_1 = \begin{bmatrix} 1 & 0 & 1 & 0 \end{bmatrix}, \quad \mathbf{v}_2 = \begin{bmatrix} 0 & 1 & -1 & 2 \end{bmatrix},$$
$$\mathbf{v}_3 = \begin{bmatrix} 0 & 2 & 2 & 1 \end{bmatrix}, \quad \text{and} \quad \mathbf{v}_4 = \begin{bmatrix} 1 & 0 & 0 & 1 \end{bmatrix},$$

is a basis for R_4.

Solution To show that S is linearly independent, we form the equation

$$a_1\mathbf{v}_1 + a_2\mathbf{v}_2 + a_3\mathbf{v}_3 + a_4\mathbf{v}_4 = \mathbf{0}$$

and solve for a_1, a_2, a_3, and a_4. Substituting for $\mathbf{v}_1$, $\mathbf{v}_2$, $\mathbf{v}_3$, and $\mathbf{v}_4$, we obtain the linear system

$$\begin{aligned} a_1 \qquad\qquad\qquad + a_4 &= 0 \\ a_2 + 2a_3 \qquad &= 0 \\ a_1 - a_2 + 2a_3 \qquad &= 0 \\ 2a_2 + a_3 + a_4 &= 0, \end{aligned}$$

which has as its only solution $a_1 = a_2 = a_3 = a_4 = 0$ (verify), showing that S is linearly independent.

To show that S spans R_4, we let $\mathbf{v} = \begin{bmatrix} a & b & c & d \end{bmatrix}$ be any vector in R_4. We now seek constants a_1, a_2, a_3, and a_4 such that

$$a_1\mathbf{v}_1 + a_2\mathbf{v}_2 + a_3\mathbf{v}_3 + a_4\mathbf{v}_4 = \mathbf{v}.$$

Substituting for $\mathbf{v}_1$, $\mathbf{v}_2$, $\mathbf{v}_3$, $\mathbf{v}_4$, and $\mathbf{v}$, we find a solution for a_1, a_2, a_3, and a_4 (verify) to the resulting linear system. Hence S spans R_4 and is a basis for R_4. ■

EXAMPLE 4

Find a basis for the subspace V of P_2, consisting of all vectors of the form $at^2 + bt + c$, where $c = a - b$.

Solution Every vector in V is of the form

$$at^2 + bt + a - b$$

which can be written as

$$a(t^2 + 1) + b(t - 1),$$

so the vectors $t^2 + 1$ and $t - 1$ span V. Moreover, these vectors are linearly independent because neither one is a multiple of the other. This conclusion could also be reached (with more work) by writing the equation

$$a_1(t^2 + 1) + a_2(t - 1) = 0$$

or

$$t^2 a_1 + t a_2 + (a_1 - a_2) = 0.$$

Since this equation is to hold for all values of t, we must have $a_1 = 0$ and $a_2 = 0$. ■

A vector space V is called **finite-dimensional** if there is a finite subset of V that is a basis for V. If there is no such finite subset of V, then V is called **infinite-dimensional**.

We now establish some results about finite-dimensional vector spaces that will tell about the number of vectors in a basis, compare two different bases, and give properties of bases. First, we observe that if $\{\mathbf{v}_1, \mathbf{v}_2, \ldots, \mathbf{v}_k\}$ is a basis for a vector space V, then $\{c\mathbf{v}_1, \mathbf{v}_2, \ldots, \mathbf{v}_k\}$ is also a basis when $c \neq 0$ (Exercise 35). Thus a basis for a nonzero vector space is never unique.

Theorem 2.7

If $S = \{\mathbf{v}_1, \mathbf{v}_2, \ldots, \mathbf{v}_n\}$ is a basis for a vector space V, then every vector in V can be written in one and only one way as a linear combination of the vectors in S.

Proof

First, every vector $\mathbf{v}$ in V can be written as a linear combination of the vectors in S because S spans V. Now let

$$\mathbf{v} = a_1\mathbf{v}_1 + a_2\mathbf{v}_2 + \cdots + a_n\mathbf{v}_n$$

and

$$\mathbf{v} = b_1\mathbf{v}_1 + b_2\mathbf{v}_2 + \cdots + b_n\mathbf{v}_n.$$

We must show that $a_i = b_i$ for $i = 1, 2, \ldots, n$. We have

$$\mathbf{0} = \mathbf{v} - \mathbf{v} = (a_1 - b_1)\mathbf{v}_1 + (a_2 - b_2)\mathbf{v}_2 + \cdots + (a_n - b_n)\mathbf{v}_n.$$

Since S is linearly independent, we conclude that

$$a_i - b_i = 0 \qquad \text{for } i = 1, 2, \ldots, n.$$ ●

We can also prove (Exercise 44) that if $S = \{\mathbf{v}_1, \mathbf{v}_2, \ldots, \mathbf{v}_n\}$ is a set of nonzero vectors in a vector space V such that every vector in V can be written in one and only one way as a linear combination of the vectors in S, then S is a basis for V.

Even though a nonzero vector space contains an infinite number of elements, a vector space with a finite basis is in a sense completely described by a finite number of vectors, namely, by those vectors in the basis.

Theorem 2.8

Let $S = \{\mathbf{v}_1, \mathbf{v}_2, \ldots, \mathbf{v}_n\}$ be a set of nonzero vectors in a vector space V and let $W = \text{span } S$. Then some subset of S is a basis for W.

Proof

Case I If S is linearly independent, then since S already spans W, we conclude that S is a basis for W.

Case II If S is linearly dependent, then

$$a_1\mathbf{v}_1 + a_2\mathbf{v}_2 + \cdots + a_n\mathbf{v}_n = \mathbf{0}, \tag{1}$$

where $a_1, a_2, \ldots, a_n$ are not all zero. Thus some $\mathbf{v}_j$ is a linear combination of the preceding vectors in S (Theorem 2.5). We now delete $\mathbf{v}_j$ from S, getting a subset S_1 of S. Then, by the remark made at the end of Section 2.4, we conclude that $S_1 = \{\mathbf{v}_1, \mathbf{v}_2, \ldots, \mathbf{v}_{j-1}\mathbf{v}_{j+1}, \ldots, \mathbf{v}_n\}$ also spans W.

If S_1 is linearly independent, then S_1 is a basis. If S_1 is linearly dependent, delete a vector of S_1 that is a linear combination of the preceding vectors of S_1 and get a new set S_2 which spans W. Continuing, since S is a finite set, we will eventually find a subset T of S that is linearly independent and spans W. The set T is a basis for W.

Alternative Constructive Proof When V is R^m or R_m, $n \geq m$. (By the results to be seen in Section 2.6, this proof will also be applicable when V is P_m or M_{pq}, where $n \geq pq$.) We take the vectors in S as $m \times 1$ matrices and form Equation (1) above. This equation leads to a homogeneous system in the n unknowns $a_1, a_2, \ldots, a_n$; the columns of its $m \times n$ coefficient matrix A are $\mathbf{v}_1, \mathbf{v}_2, \ldots, \mathbf{v}_n$. We now transform A to a matrix B in reduced row echelon form, having r nonzero rows, $1 \leq r \leq m$. Without loss of generality, we may assume that the r leading 1's in the r nonzero rows of B occur in the first r columns. Thus we have

$$B = \begin{bmatrix} 1 & 0 & 0 & \cdots & 0 & b_{1\,r+1} & \cdots & b_{1n} \\ 0 & 1 & 0 & \cdots & 0 & b_{2\,r+1} & \cdots & b_{2n} \\ 0 & 0 & 1 & \cdots & 0 & b_{3\,r+1} & \cdots & b_{3n} \\ & & & \ddots & & \vdots & & \vdots \\ 0 & 0 & 0 & \cdots & 1 & b_{r\,r+1} & \cdots & b_{rn} \\ 0 & 0 & 0 & \cdots & 0 & 0 & \cdots & 0 \\ \vdots & \vdots & \vdots & & \vdots & \vdots & & \vdots \\ 0 & 0 & 0 & \cdots & 0 & 0 & \cdots & 0 \end{bmatrix}.$$

Solving for the unknowns corresponding to the leading 1's, we see that $a_1, a_2, \ldots, a_r$ can be solved for in terms of the other unknowns $a_{r+1}, a_{r+2}, \ldots, a_n$. Thus

$$\begin{aligned} a_1 &= -b_{1\,r+1}a_{r+1} - b_{1\,r+2}a_{r+2} - \cdots - b_{1n}a_n \\ a_2 &= -b_{2\,r+1}a_{r+1} - b_{2\,r+2}a_{r+2} - \cdots - b_{2n}a_n \\ &\vdots \\ a_r &= -b_{r\,r+1}a_{r+1} - b_{r\,r+2}a_{r+2} - \cdots - b_{rn}a_n, \end{aligned} \tag{2}$$

where $a_{r+1}, a_{r+2}, \ldots, a_n$ can be assigned arbitrary real values. Letting

$$a_{r+1} = 1, \quad a_{r+2} = 0, \quad \ldots, \quad a_n = 0,$$

in Equation (2) and using these values in Equation (1), we have

$$-b_{1\,r+1}\mathbf{v}_1 - b_{2\,r+1}\mathbf{v}_2 - \cdots - b_{r\,r+1}\mathbf{v}_r + \mathbf{v}_{r+1} = \mathbf{0}$$

which implies that $\mathbf{v}_{r+1}$ is a linear combination of $\mathbf{v}_1, \mathbf{v}_2, \ldots, \mathbf{v}_r$. By the remark made at the end of Section 2.4, the set of vectors obtained from S by deleting $\mathbf{v}_{r+1}$ spans W. Similarly, letting $a_{r+1} = 0, a_{r+2} = 1, a_{r+3} = 0, \ldots, a_n = 0$, we find that $\mathbf{v}_{r+2}$ is a linear combination of $\mathbf{v}_1, \mathbf{v}_2, \ldots, \mathbf{v}_r$ and the set of vectors obtained from S by deleting $\mathbf{v}_{r+1}$ and $\mathbf{v}_{r+2}$ spans W. Continuing in this manner, $\mathbf{v}_{r+3}, \mathbf{v}_{r+4}, \ldots, \mathbf{v}_n$ are linear combinations of $\mathbf{v}_1, \mathbf{v}_2, \ldots, \mathbf{v}_r$, so it follows that $\{\mathbf{v}_1, \mathbf{v}_2, \ldots, \mathbf{v}_r\}$ spans W.

We next show that $\{\mathbf{v}_1, \mathbf{v}_2, \ldots, \mathbf{v}_r\}$ is linearly independent. Consider the matrix B_D obtained by deleting from B all columns not containing a leading 1. In this case, B_D consists of the first r columns of B. Thus

$$B_D = \begin{bmatrix} 1 & 0 & 0 & \cdots & 0 \\ 0 & 1 & 0 & \cdots & 0 \\ 0 & 0 & 1 & & \\ & & & \ddots & \vdots \\ 0 & 0 & & \cdots & 1 \\ 0 & 0 & & \cdots & 0 \\ \vdots & \vdots & & & \vdots \\ 0 & 0 & & \cdots & 0 \end{bmatrix}.$$

Let A_D be the matrix obtained from A by deleting the columns corresponding to the columns that were deleted in B to obtain B_D. In this case, the columns of A_D are $\mathbf{v}_1, \mathbf{v}_2, \ldots, \mathbf{v}_r$, the first r columns of A. Since A and B are row equivalent, so are A_D and B_D. Then the homogeneous systems

$$A_D\mathbf{x} = \mathbf{0} \quad \text{and} \quad B_D\mathbf{x} = \mathbf{0}$$

are equivalent. Recall now that the homogeneous system $B_D\mathbf{x} = \mathbf{0}$ can be written equivalently as

$$x_1\mathbf{y}_1 + x_2\mathbf{y}_2 + \cdots + x_r\mathbf{y}_r = \mathbf{0}, \tag{3}$$

where $\mathbf{x} = \begin{bmatrix} x_1 \\ x_2 \\ \vdots \\ x_r \end{bmatrix}$ and $\mathbf{y}_1, \mathbf{y}_2 \ldots, \mathbf{y}_r$ are the columns of B_D. Since the columns of B_D form a linearly independent set of vectors in R^m, Equation (3) has only the trivial solution. Hence $A_D\mathbf{x} = \mathbf{0}$ also has only the trivial solution. Thus the columns of A_D are linearly independent. That is, $\{\mathbf{v}_1, \mathbf{v}_2, \ldots, \mathbf{v}_r\}$ is linearly independent. ●

The first proof of Theorem 2.8 leads to a simple procedure for finding a subset T of a set S so that T is a basis for span S. Let $S = \{\mathbf{v}_1, \mathbf{v}_2, \ldots, \mathbf{v}_n\}$ be a set of nonzero vectors in a vector space V. The procedure for finding a subset T of S that is a basis for $W = \text{span } S$ is as follows:

STEP 1 Form Equation (1),

$$a_1\mathbf{v}_1 + a_2\mathbf{v}_2 + \cdots + a_n\mathbf{v}_n = \mathbf{0},$$

which we solve for $a_1, a_2, \ldots, a_n$. If these are all zero, then S is linearly independent and is then a basis for W.

STEP 2 If $a_1, a_2, \ldots, a_n$ are not all zero, then S is linearly dependent, so one of the vectors in S—say, $\mathbf{v}_j$—is a linear combination of the preceding vectors in S. Delete $\mathbf{v}_j$ from S, getting the subset S_1, which also spans W.

STEP 3 Repeat Step 1, using S_1 instead of S. By repeatedly deleting vectors of S we obtain a subset T of S that spans W and is linearly independent. Thus T is a basis for W.

This procedure can be rather tedious, since *every time* we delete a vector from S, we must solve a linear system. In Section 2.7 we present a much more efficient procedure for finding a basis for $W = \text{span } S$, but the basis is *not* guaranteed to be a subset of S. In many cases this is not a cause for concern, since one basis for $W = \text{span } S$ is as good as any other basis. However, there are cases when the vectors in S have some special properties and we want the basis for $W = \text{span } S$ to have the same properties, so we want the basis to be a subset of S. If $V = R^m$ or R_m, the alternative proof of Theorem 2.8 yields a very efficient procedure (see Example 5) for finding a basis for $W = \text{span } S$ consisting of vectors from S.

Let $V = R^m$ or R_m and let $S = \{\mathbf{v}_1, \mathbf{v}_2, \ldots, \mathbf{v}_n\}$ be a set of nonzero vectors in V. The procedure for finding a subset T of S that is a basis for $W = \text{span } S$ is as follows.

STEP 1 Form Equation (1),

$$a_1\mathbf{v}_1 + a_2\mathbf{v}_2 + \cdots + a_n\mathbf{v}_n = \mathbf{0}.$$

STEP 2 Construct the augmented matrix associated with the homogeneous system of Equation (1), and transform it to reduced row echelon form.

STEP 3 The vectors corresponding to the columns containing the leading 1's form a basis T for $W = \text{span } S$.

Recall that in the alternative proof of the theorem we assumed without loss of generality that the r leading 1's in the r nonzero rows of B occur in the first r columns. Thus, if $S = \{\mathbf{v}_1, \mathbf{v}_2, \ldots, \mathbf{v}_6\}$ and the leading 1's occur in columns 1, 3, and 4, then $\{\mathbf{v}_1, \mathbf{v}_3, \mathbf{v}_4\}$ is a basis for span S.

Remark In Step 2 of the procedure above, it is sufficient to transform the augmented matrix to row echelon form.

EXAMPLE 5

Let $V = R_3$ and $S = \{\mathbf{v}_1, \mathbf{v}_2, \mathbf{v}_3, \mathbf{v}_4, \mathbf{v}_5\}$, where $\mathbf{v}_1 = \begin{bmatrix} 1 & 0 & 1 \end{bmatrix}$, $\mathbf{v}_2 = \begin{bmatrix} 0 & 1 & 1 \end{bmatrix}$, $\mathbf{v}_3 = \begin{bmatrix} 1 & 1 & 2 \end{bmatrix}$, $\mathbf{v}_4 = \begin{bmatrix} 1 & 2 & 1 \end{bmatrix}$, and $\mathbf{v}_5 = \begin{bmatrix} -1 & 1 & -2 \end{bmatrix}$. We find that S spans R_3 (verify) and we now wish to find a subset of S that is a basis for R_3. Using the procedure just developed, we proceed as follows:

STEP 1 $a_1\begin{bmatrix} 1 & 0 & 1 \end{bmatrix} + a_2\begin{bmatrix} 0 & 1 & 1 \end{bmatrix} + a_3\begin{bmatrix} 1 & 1 & 2 \end{bmatrix} + a_4\begin{bmatrix} 1 & 2 & 1 \end{bmatrix} +$
$a_5\begin{bmatrix} -1 & 1 & -2 \end{bmatrix} = \begin{bmatrix} 0 & 0 & 0 \end{bmatrix}$.

STEP 2 Equating corresponding components, we obtain the homogeneous system

$$\begin{aligned} a_1 \qquad\quad + a_3 + a_4 - a_5 &= 0 \\ a_2 + a_3 + 2a_4 + a_5 &= 0 \\ a_1 + a_2 + 2a_3 + a_4 - 2a_5 &= 0. \end{aligned}$$

The reduced row echelon form of the associated augmented matrix is (verify)

$$\left[\begin{array}{ccccc|c} 1 & 0 & 1 & 0 & -2 & 0 \\ 0 & 1 & 1 & 0 & -1 & 0 \\ 0 & 0 & 0 & 1 & 1 & 0 \end{array}\right].$$

STEP 3 The leading 1's appear in columns 1, 2, and 4, so $\{\mathbf{v}_1, \mathbf{v}_2, \mathbf{v}_4\}$ is a basis for R_3. ■

Remark In the alternative proof of Theorem 2.8, the order of the vectors in the original spanning set S determines which basis for V is obtained. If, for example, we consider Example 5, where $S = \{\mathbf{w}_1, \mathbf{w}_2, \mathbf{w}_3, \mathbf{w}_4, \mathbf{w}_5\}$ with $\mathbf{w}_1 = \mathbf{v}_5$, $\mathbf{w}_2 = \mathbf{v}_4$, $\mathbf{w}_3 = \mathbf{v}_3$, $\mathbf{w}_4 = \mathbf{v}_2$, and $\mathbf{w}_5 = \mathbf{v}_1$, then the reduced row echelon form of the augmented matrix is (verify)

$$\left[\begin{array}{ccccc|c} 1 & 0 & 0 & 1 & -1 & 0 \\ 0 & 1 & 0 & -1 & 1 & 0 \\ 0 & 0 & 1 & 2 & -1 & 0 \end{array}\right].$$

It then follows that $\{\mathbf{w}_1, \mathbf{w}_2, \mathbf{w}_3\} = \{\mathbf{v}_5, \mathbf{v}_4, \mathbf{v}_3\}$ is a basis for R_3.

We are now about to establish a major result (Corollary 2.1) of this section, which will tell us about the number of vectors in two different bases.

Theorem 2.9

If $S = \{\mathbf{v}_1, \mathbf{v}_2, \ldots, \mathbf{v}_n\}$ is a basis for a vector space V and $T = \{\mathbf{w}_1, \mathbf{w}_2, \ldots, \mathbf{w}_r\}$ is a linearly independent set of vectors in V, then $r \leq n$.

Proof

Let $T_1 = \{\mathbf{w}_1, \mathbf{v}_1, \ldots, \mathbf{v}_n\}$. Since S spans V, so does T_1. Since $\mathbf{w}_1$ is a linear combination of the vectors in S, we find that T_1 is linearly dependent. Then, by Theorem 2.6, some $\mathbf{v}_j$ is a linear combination of the preceding vectors in T_1. Delete that particular vector $\mathbf{v}_j$.

Let $S_1 = \{\mathbf{w}_1, \mathbf{v}_1, \ldots, \mathbf{v}_{j-1}, \mathbf{v}_{j+1}, \ldots, \mathbf{v}_n\}$. Note that S_1 spans V. Next, let $T_2 = \{\mathbf{w}_2, \mathbf{w}_1, \mathbf{v}_1, \ldots, \mathbf{v}_{j-1}, \mathbf{v}_{j+1}, \ldots, \mathbf{v}_n\}$. Then T_2 is linearly dependent and some vector in T_2 is a linear combination of the preceding vectors in T_2. Since T is linearly independent, this vector cannot be $\mathbf{w}_1$, so it is $\mathbf{v}_i$, $i \neq j$. Repeat this process over and over. Each time there is a new $\mathbf{w}$ vector available from the set T, it is possible to discard one of the $\mathbf{v}$ vectors from the set S. Thus the number r of $\mathbf{w}$ vectors must be no greater than the number n of $\mathbf{v}$ vectors. That is, $r \leq n$. ●

Corollary 2.1

If $S = \{\mathbf{v}_1, \mathbf{v}_2, \ldots, \mathbf{v}_n\}$ and $T = \{\mathbf{w}_1, \mathbf{w}_2, \ldots, \mathbf{w}_n\}$ are bases for a vector space V, then $n = m$.

Proof

Since S is a basis and T is linearly independent, Theorem 2.9 implies that $m \leq n$. Similarly, we obtain $n \leq m$ because T is a basis and S is linearly independent. Hence $n = m$. ●

Dimension

Although a vector space may have many bases, we have just shown that, for a particular vector space V, all bases have the same number of vectors. We can then make the following definition.

Definition 2.11

The **dimension** of a nonzero vector space V is the number of vectors in a basis for V. We often write **dim** V for the dimension of V. We also define the dimension of the trivial vector space $\{\mathbf{0}\}$ to be zero. ▲

EXAMPLE 6

The set $S = \{t^2, t, 1\}$ is a basis for P_2, so dim $P_2 = 3$. ■

EXAMPLE 7

Let V be the subspace of R_3 spanned by $S = \{\mathbf{v}_1, \mathbf{v}_2, \mathbf{v}_3\}$, where $\mathbf{v}_1 = \begin{bmatrix} 0 & 1 & 1 \end{bmatrix}$, $\mathbf{v}_2 = \begin{bmatrix} 1 & 0 & 1 \end{bmatrix}$, and $\mathbf{v}_3 = \begin{bmatrix} 1 & 1 & 2 \end{bmatrix}$. Thus every vector in V is of the form

$$a_1\mathbf{v}_1 + a_2\mathbf{v}_2 + a_3\mathbf{v}_3,$$

where a_1, a_2, and a_3 are arbitrary real numbers. We find that S is linearly dependent, and $\mathbf{v}_3 = \mathbf{v}_1 + \mathbf{v}_2$ (verify). Thus $S_1 = \{\mathbf{v}_1, \mathbf{v}_2\}$ also spans V. Since S_1 is linearly independent (verify), we conclude that it is a basis for V. Hence $\dim V = 2$. ■

Definition 2.12

Let S be a set of vectors in a vector space V. A subset T of S is called a **maximal independent subset** of S if T is a linearly independent set of vectors which is not properly contained in any other linearly independent subset of S. ▲

EXAMPLE 8

Let V be R^3 and consider the set $S = \{\mathbf{v}_1, \mathbf{v}_2, \mathbf{v}_3, \mathbf{v}_4\}$, where

$$\mathbf{v}_1 = \begin{bmatrix} 1 \\ 0 \\ 0 \end{bmatrix}, \quad \mathbf{v}_2 = \begin{bmatrix} 0 \\ 1 \\ 0 \end{bmatrix}, \quad \mathbf{v}_3 = \begin{bmatrix} 0 \\ 0 \\ 1 \end{bmatrix}, \quad \text{and} \quad \mathbf{v}_4 = \begin{bmatrix} 1 \\ 1 \\ 1 \end{bmatrix}.$$

Maximal independent subsets of S are

$$\{\mathbf{v}_1, \mathbf{v}_2, \mathbf{v}_3\}, \quad \{\mathbf{v}_1, \mathbf{v}_2, \mathbf{v}_4\}, \quad \{\mathbf{v}_1, \mathbf{v}_3, \mathbf{v}_4\}, \quad \text{and} \quad \{\mathbf{v}_2, \mathbf{v}_3, \mathbf{v}_4\}.$$ ■

Corollary 2.2

If the vector space V has dimension n, then a maximal independent subset of vectors in V contains n vectors.

Proof

Let $S = \{\mathbf{v}_1, \mathbf{v}_2, \ldots, \mathbf{v}_k\}$ be a maximal independent subset of V. If span $S \neq V$, then there exists a vector $\mathbf{v}$ in V that cannot be written as a linear combination of $\mathbf{v}_1, \mathbf{v}_2, \ldots, \mathbf{v}_k$. It follows by Theorem 2.6 that $\{\mathbf{v}_1, \mathbf{v}_2, \ldots, \mathbf{v}_k, \mathbf{v}\}$ is a linearly independent set of vectors. However, this contradicts the assumption that S is a maximal independent subset of V. Hence span $S = V$, which implies that set S is a basis for V and $k = n$ by Corollary 2.1. ●

Corollary 2.3

If a vector space V has dimension n, then a minimal† *spanning set for V contains n vectors.*

Proof

Exercise 38. ●

Although Corollaries 2.2 and 2.3 are theoretically of considerable importance, they can be computationally awkward.

From the results above, we can make the following observations. If V has dimension n, then any set of $n + 1$ vectors in V is necessarily linearly dependent;

†If S is a set of vectors spanning a vector space V, then S is called a **minimal spanning set** for V if S does not properly contain any other set spanning V.

also, any set of $n - 1$ vectors in V cannot span V. More generally, we can establish the following results.

Corollary 2.4

If vector space V has dimension n, then any subset of $m > n$ vectors must be linearly dependent.

Proof

Exercise 39. ●

Corollary 2.5

If vector space V has dimension n, then any subset of $m < n$ vectors cannot span V.

Proof

Exercise 40. ●

In Section 2.4, we have already observed the set $\{\mathbf{0}\}$ is linearly dependent. This is why in Definition 2.11 we defined the dimension of the trivial vector space $\{\mathbf{0}\}$ to be zero.

Thus R^3 has dimension 3, R_2 has dimension 2, and R^n and R_n both have dimension n. Similarly, P_3 has dimension 4 because $\{t^3, t^2, t, 1\}$ is a basis for P_3. In general, P_n has dimension $n + 1$. Most vector spaces considered henceforth in this book are finite dimensional. Although infinite-dimensional vector spaces are very important in mathematics and physics, their study lies beyond the scope of this book. The vector space P of all polynomials is an infinite-dimensional vector space (Exercise 36).

In Section 2.3, it was an exercise to show that the subspaces of R^2 are $\{\mathbf{0}\}$, R^2 itself, and any line passing through the origin. We can now establish this result using the material developed in this section. First, we have $\{\mathbf{0}\}$ and R^2, the trivial subspaces of dimensions 0 and 2, respectively. The subspace V of R^2 spanned by a vector $\mathbf{v} \neq \mathbf{0}$ is a one-dimensional subspace of R^2; V is a line through the origin. Thus the subspaces of R^2 are $\{\mathbf{0}\}$, R^2, and all the lines through the origin. In a similar way, Exercise 43 asks you to show that the subspaces of R^3 are $\{\mathbf{0}\}$, R^3 itself, and all lines and planes passing through the origin. We now prove a theorem that we shall have occasion to use several times in constructing a basis containing a given set of linearly independent vectors.

Theorem 2.10

If S is a linearly independent set of vectors in a finite-dimensional vector space V, then there is a basis T for V that contains S.

Proof

Let $S = \{\mathbf{v}_1, \mathbf{v}_2, \ldots, \mathbf{v}_m\}$ be a linearly independent set of vectors in the n-dimensional vector space V, where $m < n$. Now let $\{\mathbf{w}_1, \mathbf{w}_2, \ldots, \mathbf{w}_n\}$ be a basis for V and let $S_1 = \{\mathbf{v}_1, \mathbf{v}_2, \ldots, \mathbf{v}_m, \mathbf{w}_1, \mathbf{w}_2, \ldots, \mathbf{w}_n\}$. Since S_1 spans V, it contains, by Theorem 2.8, a basis T for V. Recall that T is obtained by deleting from S_1 every vector that is a linear combination of the preceding vectors. Since S is linearly independent, none of the $\mathbf{v}_i$ can be linear combinations of other $\mathbf{v}_j$ and thus are not deleted. Hence T will contain S. ●

EXAMPLE 9

Suppose that we wish to find a basis for R_4 that contains the vectors

$$\mathbf{v}_1 = \begin{bmatrix} 1 & 0 & 1 & 0 \end{bmatrix} \quad \text{and} \quad \mathbf{v}_2 = \begin{bmatrix} -1 & 1 & -1 & 0 \end{bmatrix}.$$

We use Theorem 2.10 as follows. First, let $\{\mathbf{e}'_1, \mathbf{e}'_2, \mathbf{e}'_3, \mathbf{e}'_4\}$ be the natural basis for R_4, where

$$\mathbf{e}'_1 = \begin{bmatrix} 1 & 0 & 0 & 0 \end{bmatrix}, \quad \mathbf{e}'_2 = \begin{bmatrix} 0 & 1 & 0 & 0 \end{bmatrix}, \quad \mathbf{e}'_3 = \begin{bmatrix} 0 & 0 & 1 & 0 \end{bmatrix},$$

and

$$\mathbf{e}'_4 = \begin{bmatrix} 0 & 0 & 0 & 1 \end{bmatrix}.$$

Form the set $S = \{\mathbf{v}_1, \mathbf{v}_2, \mathbf{e}'_1, \mathbf{e}'_2, \mathbf{e}'_3, \mathbf{e}'_4\}$. Since $\{\mathbf{e}'_1, \mathbf{e}'_2, \mathbf{e}'_3, \mathbf{e}'_4\}$ spans R_4, so does S. We now use the alternative proof of Theorem 2.8 to find a subset of S that is a basis for R_4. Thus we form Equation (1),

$$a_1\mathbf{v}_1 + a_2\mathbf{v}_2 + a_3\mathbf{e}'_1 + a_4\mathbf{e}'_2 + a_5\mathbf{e}'_3 + a_6\mathbf{e}'_4 = \begin{bmatrix} 0 & 0 & 0 & 0 \end{bmatrix},$$

which leads to the homogeneous system

$$\begin{aligned} a_1 - a_2 + a_3 \qquad\qquad\qquad &= 0 \\ - a_2 \qquad + a_4 \qquad\qquad &= 0 \\ a_1 - a_2 \qquad\qquad + a_5 \quad &= 0 \\ a_6 &= 0. \end{aligned}$$

Transforming the augmented matrix to reduced row echelon form, we obtain (verify)

$$\left[\begin{array}{cccccc|c} 1 & 0 & 0 & 1 & 1 & 0 & 0 \\ 0 & 1 & 0 & 1 & 0 & 0 & 0 \\ 0 & 0 & 1 & 0 & -1 & 0 & 0 \\ 0 & 0 & 0 & 0 & 0 & 1 & 0 \end{array}\right].$$

Since the leading 1's appear in columns 1, 2, 3, and 6, we conclude that $\{\mathbf{v}_1, \mathbf{v}_2, \mathbf{e}'_1, \mathbf{e}'_4\}$ is a basis for R_4 containing $\mathbf{v}_1$ and $\mathbf{v}_2$. ■

It can be shown (Exercise 41) that if W is a subspace of a finite-dimensional vector space V, then W is finite-dimensional and $\dim W \leq \dim V$.

As defined earlier, a given set S of vectors in a vector space V is a basis for V if it spans V and is linearly independent. However, if we are given the *additional* information that the dimension of V is n, we need only verify one of the two conditions. This is the content of the following theorem.

Theorem 2.11

Let V be an n-dimensional vector space.

(a) *If $S = \{\mathbf{v}_1, \mathbf{v}_2, \ldots, \mathbf{v}_n\}$ is a linearly independent set of vectors in V, then S is a basis for V.*

(b) *If $S = \{\mathbf{v}_1, \mathbf{v}_2, \ldots, \mathbf{v}_n\}$ spans V, then S is a basis for V.*

Proof

Exercise. ●

As a particular application of Theorem 2.11, we have the following. To determine if a subset S of R^n (R_n) is a basis for R^n (R_n), first count the number of elements in S. If S has n elements, we can use either part (a) or part (b) of Theorem 2.11 to determine whether S is or is not a basis. If S does not have n elements, it is not a basis for R^n (R_n). (Why?) The same line of reasoning applies to any vector space or subspace whose *dimension is known.*

EXAMPLE 10

In Example 5, since $\dim R_3 = 3$ and the set S contains five vectors, we conclude by Theorem 2.11 that S is not a basis for R_3. In Example 3, since $\dim R_4 = 4$ and the set S contains four vectors, it is possible for S to be a basis for R_4. If S is linearly independent *or* spans R_4, it is a basis; otherwise, it is not a basis. Thus we need check only one of the conditions in Theorem 2.11, not both. ■

Theorem 2.12

Let S be a finite subset of the vector space V that spans V. A maximal independent subset T of S is a basis for V.

Proof

Exercise. ●

2.5 Exercises

1. Which of the following sets of vectors are bases for R^2?

(a) $\left\{\begin{bmatrix}1\\3\end{bmatrix}, \begin{bmatrix}1\\-1\end{bmatrix}\right\}$. (b) $\left\{\begin{bmatrix}0\\0\end{bmatrix}, \begin{bmatrix}1\\2\end{bmatrix}, \begin{bmatrix}2\\4\end{bmatrix}\right\}$.

(c) $\left\{\begin{bmatrix}1\\2\end{bmatrix}, \begin{bmatrix}2\\-3\end{bmatrix}, \begin{bmatrix}3\\2\end{bmatrix}\right\}$. (d) $\left\{\begin{bmatrix}1\\3\end{bmatrix}, \begin{bmatrix}-2\\6\end{bmatrix}\right\}$.

2. Which of the following sets of vectors are bases for R^3?

(a) $\left\{\begin{bmatrix}1\\2\\0\end{bmatrix}, \begin{bmatrix}0\\1\\-1\end{bmatrix}\right\}$.

(b) $\left\{\begin{bmatrix}1\\1\\-1\end{bmatrix}, \begin{bmatrix}2\\3\\4\end{bmatrix}, \begin{bmatrix}4\\1\\-1\end{bmatrix}, \begin{bmatrix}0\\1\\-1\end{bmatrix}\right\}$.

(c) $\left\{\begin{bmatrix}3\\2\\2\end{bmatrix}, \begin{bmatrix}-1\\2\\1\end{bmatrix}, \begin{bmatrix}0\\1\\0\end{bmatrix}\right\}$.

(d) $\left\{\begin{bmatrix}1\\0\\0\end{bmatrix}, \begin{bmatrix}0\\2\\-1\end{bmatrix}, \begin{bmatrix}3\\4\\1\end{bmatrix}, \begin{bmatrix}0\\1\\0\end{bmatrix}\right\}$.

3. Which of the following sets of vectors are bases for R_4?

(a) $\{[1\ \ 0\ \ 0\ \ 1], [0\ \ 1\ \ 0\ \ 0], [1\ \ 1\ \ 1\ \ 1], [0\ \ 1\ \ 1\ \ 1]\}$.

(b) $\{[1\ \ -1\ \ 0\ \ 2], [3\ \ -1\ \ 2\ \ 1], [1\ \ 0\ \ 0\ \ 1]\}$.

(c) $\{[-2\ \ 4\ \ 6\ \ 4], [0\ \ 1\ \ 2\ \ 0], [-1\ \ 2\ \ 3\ \ 2], [-3\ \ 2\ \ 5\ \ 6], [-2\ \ -1\ \ 0\ \ 4]\}$.

(d) $\{[0\ \ 0\ \ 1\ \ 1], [-1\ \ 1\ \ 1\ \ 2], [1\ \ 1\ \ 0\ \ 0], [2\ \ 1\ \ 2\ \ 1]\}$.

4. Which of the following sets of vectors are bases for P_2?

(a) $\{-t^2+t+2, 2t^2+2t+3, 4t^2-1\}$.

(b) $\{t^2+2t-1, 2t^2+3t-2\}$.

(c) $\{t^2+1, 3t^2+2t+1, 6t^2+6t+3\}$.

(d) $\{3t^2+2t+1, t^2+t+1, t^2+1\}$.

5. Which of the following sets of vectors are bases for P_3?

(a) $\{t^3+2t^2+3t, 2t^3+1, 6t^3+8t^2+6t+4, t^3+2t^2+t+1\}$.

(b) $\{t^3+t^2+1, t^3-1, t^3+t^2+t\}$.

(c) $\{t^3+t^2+t+1, t^3+2t^2+t+3, 2t^3+t^2+3t+2, t^3+t^2+2t+2\}$.

(d) $\{t^3-t, t^3+t^2+1, t-1\}$.

6. Show that the set of matrices

$$\left\{\begin{bmatrix}1 & 1\\ 0 & 0\end{bmatrix}, \begin{bmatrix}0 & 0\\ 1 & 1\end{bmatrix}, \begin{bmatrix}1 & 0\\ 0 & 1\end{bmatrix}, \begin{bmatrix}0 & 1\\ 1 & 1\end{bmatrix}\right\}$$

forms a basis for the vector space M_{22}.

In Exercises 7 and 8, determine which of the given subsets forms a basis for R^3. Express the vector $\begin{bmatrix}2\\1\\3\end{bmatrix}$ as a linear combination of the vectors in each subset that is a basis.

7. (a) $\left\{\begin{bmatrix}1\\1\\1\end{bmatrix}, \begin{bmatrix}1\\2\\3\end{bmatrix}, \begin{bmatrix}0\\1\\0\end{bmatrix}\right\}$.

(b) $\left\{\begin{bmatrix}1\\2\\3\end{bmatrix}, \begin{bmatrix}2\\1\\3\end{bmatrix}, \begin{bmatrix}0\\0\\0\end{bmatrix}\right\}$.

8. (a) $\left\{\begin{bmatrix}2\\1\\3\end{bmatrix}, \begin{bmatrix}1\\2\\1\end{bmatrix}, \begin{bmatrix}1\\1\\4\end{bmatrix}, \begin{bmatrix}1\\5\\1\end{bmatrix}\right\}$.

(b) $\left\{\begin{bmatrix}1\\1\\2\end{bmatrix}, \begin{bmatrix}2\\2\\0\end{bmatrix}, \begin{bmatrix}3\\4\\-1\end{bmatrix}\right\}$.

In Exercises 9 and 10, determine which of the given subsets form a basis for P_2. Express $5t^2-3t+8$ as a linear combination of the vectors in each subset that is a basis.

9. (a) $\{t^2+t, t-1, t+1\}$.

(b) $\{t^2+1, t-1\}$.

10. (a) $\{t^2+t, t^2, t^2+1\}$.

(b) $\{t^2+1, t^2-t+1\}$.

11. Find a basis for the subspace W of R^3 spanned by

$$\left\{\begin{bmatrix}1\\2\\2\end{bmatrix}, \begin{bmatrix}3\\2\\1\end{bmatrix}, \begin{bmatrix}11\\10\\7\end{bmatrix}, \begin{bmatrix}7\\6\\4\end{bmatrix}\right\}.$$

What is the dimension of W?

12. Find a basis for the subspace W of R_4 spanned by the set of vectors

$$\{[1\ \ 1\ \ 0\ \ -1], [0\ \ 1\ \ 2\ \ 1], [1\ \ 0\ \ 1\ \ -1], [1\ \ 1\ \ -6\ \ -3], [-1\ \ -5\ \ 1\ \ 0]\}$$

What is $\dim W$?

13. Let W be the subspace of P_3 spanned by

$$\{t^3+t^2-2t+1, t^2+1, t^3-2t, 2t^3+3t^2-4t+3\}.$$

Find a basis for W. What is the dimension of W?

14. Let $S =$

$$\left\{\begin{bmatrix}1 & 0\\ 0 & 1\end{bmatrix}, \begin{bmatrix}0 & 1\\ 1 & 0\end{bmatrix}, \begin{bmatrix}1 & 1\\ 1 & 1\end{bmatrix}, \begin{bmatrix}-1 & 1\\ 1 & -1\end{bmatrix}\right\}.$$

Find a basis for the subspace $W = \operatorname{span} S$ of M_{22}.

15. Find all values of a for which

$$\{[a^2\ \ 0\ \ 1], [0\ \ a\ \ 2], [1\ \ 0\ \ 1]\}$$

is a basis for R_3.

16. Find a basis for the subspace W of M_{33} consisting of all symmetric matrices.

17. Find a basis for the subspace of M_{33} consisting of all diagonal matrices.

18. Let W be the subspace of the space of all continuous real-valued functions spanned by $\{\cos^2 t, \sin^2 t, \cos 2t\}$. Find a basis for W. What is the dimension of W?

In Exercises 19 *and* 20, *find a basis for the given subspaces of* R^3 *and* R^4.

19. (a) All vectors of the form $\begin{bmatrix} a \\ b \\ c \end{bmatrix}$, where $b = a + c$.

(b) All vectors of the form $\begin{bmatrix} a \\ b \\ c \end{bmatrix}$, where $b = a$.

(c) All vectors of the form $\begin{bmatrix} a \\ b \\ c \end{bmatrix}$, where $2a + b - c = 0$.

20. (a) All vectors of the form $\begin{bmatrix} a \\ b \\ c \end{bmatrix}$, where $a = 0$.

(b) All vectors of the form $\begin{bmatrix} a+c \\ a-b \\ b+c \\ -a+b \end{bmatrix}$.

(c) All vectors of the form $\begin{bmatrix} a \\ b \\ c \end{bmatrix}$, where $a - b + 5c = 0$.

21. Find a basis for the subspace of P_2 consisting of all vectors of the form $at^2 + bt + c$, where $c = 2a - 3b$.

22. Find a basis for the subspace of P_3 consisting of all vectors of the form $at^3 + bt^2 + ct + d$, where $c = a - 2d$ and $b = 5a + 3d$.

In Exercises 23 *and* 24, *find the dimensions of the given subspaces of* R_4.

23. (a) All vectors of the form $\begin{bmatrix} a & b & c & d \end{bmatrix}$, where $d = a + b$.

(b) All vectors of the form $\begin{bmatrix} a & b & c & d \end{bmatrix}$, where $c = a - b$ and $d = a + b$.

24. (a) All vectors of the form $\begin{bmatrix} a & b & c & d \end{bmatrix}$, where $a = b$.

(b) All vectors of the form $\begin{bmatrix} a+c & a-b & b+c & -a+b \end{bmatrix}$.

25. Find the dimensions of the subspaces of R^2 spanned by the vectors in Exercise 1.

26. Find the dimensions of the subspaces of R^3 spanned by the vectors in Exercise 2.

27. Find the dimensions of the subspaces of R_4 spanned by the vectors in Exercise 3.

28. Find a basis for R^3 that includes:

(a) The vector $\begin{bmatrix} 1 \\ 0 \\ 2 \end{bmatrix}$.

(b) The vectors $\begin{bmatrix} 1 \\ 0 \\ 2 \end{bmatrix}$ and $\begin{bmatrix} 0 \\ 1 \\ 3 \end{bmatrix}$.

29. Find a basis for P_3 that includes the vectors $t^3 + t$ and $t^2 - t$.

30. Find a basis for M_{23}. What is the dimension of M_{23}? Generalize to M_{mn}.

31. Find the dimension of the subspace of P_2 consisting of all vectors of the form $at^2 + bt + c$, where $c = b - 2a$.

32. Find the dimension of the subspace of P_3 consisting of all vectors of the form $at^3 + bt^2 + ct + d$, where $b = 3a - 5d$ and $c = d + 4a$.

33. Give an example of a two-dimensional subspace of R^4.

34. Give an example of a two-dimensional subspace of P_3.

35. Prove that if $\{\mathbf{v}_1, \mathbf{v}_2, \ldots, \mathbf{v}_k\}$ is a basis for a vector space V, then $\{c\mathbf{v}_1, \mathbf{v}_2, \ldots, \mathbf{v}_k\}$ for $c \neq 0$, is also a basis for V.

36. Prove that the vector space P of all polynomials is not finite-dimensional. [*Hint*: Suppose that $\{p_1(t), p_2(t), \ldots, p_k(t)\}$ is a finite basis for P. Let d_j = degree $p_j(t)$. Establish a contradiction.]

37. Let V be an n-dimensional vector space. Show that any $n + 1$ vectors in V form a linearly dependent set.

38. Prove Corollary 2.3.

39. Prove Corollary 2.4.

40. Prove Corollary 2.5.

41. Show that if W is a subspace of a finite-dimensional vector space V, then W is finite-dimensional and $\dim W \leq \dim V$.

42. Show that if W is a subspace of a finite-dimensional vector space V and $\dim W = \dim V$, then $W = V$.

43. Prove that the subspaces of R^3 are $\{\mathbf{0}\}$, R^3 itself, and any line or plane passing through the origin.

44. Let $S = \{\mathbf{v}_1, \mathbf{v}_2, \dots, \mathbf{v}_n\}$ be a set of nonzero vectors in a vector space V such that every vector in V can be written in one and only one way as a linear combination of the vectors in S. Prove that S is a basis for V.

45. Prove Theorem 2.11.

46. Prove Theorem 2.12.

47. Suppose that $\{\mathbf{v}_1, \mathbf{v}_2, \dots, \mathbf{v}_n\}$ is a basis for R^n. Show that if A is an $n \times n$ nonsingular matrix, then $\{A\mathbf{v}_1, A\mathbf{v}_2, \dots, A\mathbf{v}_n\}$ is also a basis for R^n. (*Hint*: See Exercise 24 in Section 2.4.)

48. Suppose that $\{\mathbf{v}_1, \mathbf{v}_2, \dots, \mathbf{v}_n\}$ is a linearly independent set of vectors in R^n and let A be a singular matrix. Prove or disprove that $\{A\mathbf{v}_1, A\mathbf{v}_2, \dots, A\mathbf{v}_n\}$ is linearly independent.

2.6 HOMOGENEOUS SYSTEMS

Homogeneous systems play a central role in linear algebra. This will be seen in Chapter 6, where the foundations of the subject are all integrated to solve one of the major problems occurring in a wide variety of applications. In this section we deal with several problems involving homogeneous systems that will arise in Chapter 6. Here we are able to focus our attention on these problems without being distracted by the additional material in Chapter 6.

Consider the homogeneous system

$$A\mathbf{x} = \mathbf{0},$$

where A is an $m \times n$ matrix. As we have already observed in Example 9 of Section 2.3, the set of all solutions to this homogeneous system is a subspace of R^n. An extremely important problem, which will occur repeatedly in Chapter 6, is that of finding a basis for this solution space. To find such a basis, we use the method of Gauss–Jordan reduction presented in Section 1.5. Thus we transform the augmented matrix $\left[A \mid \mathbf{0}\right]$ of the system to a matrix $\left[B \mid \mathbf{0}\right]$ in reduced row echelon form, where B has r nonzero rows, $1 \leq r \leq m$. Without loss of generality we may assume that the leading 1's in the r nonzero rows occur in the first r columns. If $r = n$, then

$$\left[B \mid \mathbf{0}\right] = \overbrace{\left[\begin{array}{ccccc|c} 1 & 0 & \cdots & 0 & 0 & 0 \\ 0 & 1 & \cdots & 0 & 0 & 0 \\ \vdots & \vdots & \ddots & \vdots & \vdots & \vdots \\ 0 & 0 & \cdots & 0 & 1 & 0 \\ 0 & 0 & \cdots & & 0 & 0 \\ \vdots & \vdots & & & \vdots & \vdots \\ 0 & 0 & \cdots & & 0 & 0 \end{array}\right]}^{n} \begin{array}{l} \left.\begin{array}{l} \\ \\ \\ \\ \end{array}\right\} r = n \\ \\ \\ \\ \end{array} \Bigg\} m$$

and the only solution to $A\mathbf{x} = \mathbf{0}$ is the trivial one. The solution space has no basis and its dimension is zero.

If $r < n$, then

$$
[B \mid \mathbf{0}] = \left[\begin{array}{cccccccc|c}
\multicolumn{8}{c}{\overbrace{\hphantom{1\quad 0\quad 0\quad \cdots\quad 0\quad b_{1\,r+1}\quad \cdots\quad b_{1n}}}^{n}} & \\
1 & 0 & 0 & \cdots & 0 & b_{1\,r+1} & \cdots & b_{1n} & 0 \\
0 & 1 & 0 & \cdots & 0 & b_{2\,r+1} & \cdots & b_{2n} & 0 \\
0 & 0 & 1 & \cdots & 0 & \vdots & & \vdots & \vdots \\
\vdots & \vdots & \vdots & \ddots & \vdots & & & & \\
0 & 0 & 0 & \cdots & 1 & b_{r\,r+1} & \cdots & b_{rn} & 0 \\
0 & 0 & 0 & \cdots & 0 & 0 & \cdots & 0 & 0 \\
\vdots & \vdots & \vdots & & \vdots & \vdots & & \vdots & \vdots \\
0 & 0 & 0 & \cdots & 0 & \cdots & & 0 & 0
\end{array}\right]
\left.\begin{array}{l}
\left.\begin{array}{l} \\ \\ \\ \\ \\ \end{array}\right\} r \\
\\ \\ \\
\end{array}\right\} m
$$

Solving for the unknowns corresponding to the leadings 1's, we have

$$
\begin{aligned}
x_1 &= -b_{1\,r+1}x_{r+1} - b_{1\,r+2}x_{r+2} - \cdots - b_{1n}x_n \\
x_2 &= -b_{2\,r+1}x_{r+1} - b_{2\,r+2}x_{r+2} - \cdots - b_{2n}x_n \\
&\vdots \\
x_r &= -b_{r\,r+1}x_{r+1} - b_{r\,r+2}x_{r+2} - \cdots - b_{rn}x_n,
\end{aligned}
$$

where $x_{r+1}, x_{r+2}, \ldots, x_n$ can be assigned arbitrary real values s_j, $j = 1, 2, \ldots, p$, and $p = n - r$. Thus

$$
\mathbf{x} = \begin{bmatrix} x_1 \\ x_2 \\ \vdots \\ x_r \\ x_{r+1} \\ x_{r+2} \\ \vdots \\ x_n \end{bmatrix}
= \begin{bmatrix}
-b_{1\,r+1}s_1 - b_{1\,r+2}s_2 - \cdots - b_{1n}s_p \\
-b_{2\,r+1}s_1 - b_{2\,r+2}s_2 - \cdots - b_{2n}s_p \\
\vdots \\
-b_{r\,r+1}s_1 - b_{r\,r+2}s_2 - \cdots - b_{rn}s_p \\
s_1 \\
s_2 \\
\vdots \\
s_p
\end{bmatrix}
$$

$$= s_1 \begin{bmatrix} -b_{1\,r+1} \\ -b_{2\,r+1} \\ \vdots \\ -b_{r\,r+1} \\ 1 \\ 0 \\ 0 \\ \vdots \\ 0 \\ 0 \end{bmatrix} + s_2 \begin{bmatrix} -b_{1\,r+2} \\ -b_{2\,r+2} \\ \vdots \\ -b_{r\,r+2} \\ 0 \\ 1 \\ 0 \\ \vdots \\ 0 \\ 0 \end{bmatrix} + \cdots + s_p \begin{bmatrix} -b_{1n} \\ -b_{2n} \\ \vdots \\ -b_{rn} \\ 0 \\ 0 \\ 0 \\ \vdots \\ 0 \\ 1 \end{bmatrix}.$$

Since $s_1, s_2, \dots, s_p$ can be assigned arbitrary real values, we make the following choices for these values:

$$\begin{array}{lllll} s_1 = 1, & s_2 = 0, & \dots, & & s_p = 0, \\ s_1 = 0, & s_2 = 1, & \dots, & & s_p = 0, \\ & & \vdots & & \\ s_1 = 0, & s_2 = 0, & \dots, & s_{p-1} = 0, & s_p = 1, \end{array}$$

obtaining the solutions

$$\mathbf{x}_1 = \begin{bmatrix} -b_{1\,r+1} \\ -b_{2\,r+1} \\ \vdots \\ -b_{r\,r+1} \\ 1 \\ 0 \\ 0 \\ \vdots \\ 0 \\ 0 \end{bmatrix}, \quad \mathbf{x}_2 = \begin{bmatrix} -b_{1\,r+2} \\ -b_{2\,r+2} \\ \vdots \\ -b_{r\,r+2} \\ 0 \\ 1 \\ 0 \\ \vdots \\ 0 \\ 0 \end{bmatrix}, \dots, \quad \mathbf{x}_p = \begin{bmatrix} -b_{1n} \\ -b_{2n} \\ \vdots \\ -b_{rn} \\ 0 \\ 0 \\ 0 \\ \vdots \\ 0 \\ 1 \end{bmatrix}.$$

Since

$$\mathbf{x} = s_1\mathbf{x}_1 + s_2\mathbf{x}_2 + \cdots + s_p\mathbf{x}_p,$$

we see that $\{\mathbf{x}_1, \mathbf{x}_2, \dots, \mathbf{x}_p\}$ spans the solution space of $A\mathbf{x} = \mathbf{0}$. Moreover, if we form the equation

$$a_1\mathbf{x}_1 + a_2\mathbf{x}_2 + \cdots + a_p\mathbf{x}_p = \mathbf{0},$$

its coefficient matrix is the matrix whose columns are $\mathbf{x}_1, \mathbf{x}_2, \dots, \mathbf{x}_p$. If we look at rows $r+1, r+2, \dots, n$ of this matrix, we readily see that

$$a_1 = a_2 = \cdots = a_p = 0.$$

Hence $\{\mathbf{x}_1, \mathbf{x}_2, \dots, \mathbf{x}_p\}$ is linearly independent and forms a basis for the solution space of $A\mathbf{x} = \mathbf{0}$ (the null space of A).

The procedure for finding a basis for the solution space of a homogeneous system $A\mathbf{x} = \mathbf{0}$, or the null space of A, where A is $m \times n$, is as follows:

STEP 1 Solve the given homogeneous system by Gauss–Jordan reduction. If the solution contains no arbitrary constants, then the solution space is $\{\mathbf{0}\}$, which has no basis; the dimension of the solution space is zero.

STEP 2 If the solution $\mathbf{x}$ contains arbitrary constants, write $\mathbf{x}$ as a linear combination of vectors $\mathbf{x}_1, \mathbf{x}_2, \ldots, \mathbf{x}_p$ with $s_1, s_2, \ldots, s_p$ as coefficients:

$$\mathbf{x} = s_1\mathbf{x}_1 + s_2\mathbf{x}_2 + \cdots + s_p\mathbf{x}_p.$$

STEP 3 The set of vectors $\{\mathbf{x}_1, \mathbf{x}_2, \ldots, \mathbf{x}_p\}$ is a basis for the solution space of $A\mathbf{x} = \mathbf{0}$; the dimension of the solution space is p.

Remark In Step 1, suppose that the matrix in reduced row echelon form to which $\left[A \mid \mathbf{0}\right]$ has been transformed has r nonzero rows (also, r leading 1's). Then $p = n - r$. That is, the dimension of the solution space is $n - r$. Moreover, a solution $\mathbf{x}$ to $A\mathbf{x} = \mathbf{0}$ has $n - r$ arbitrary constants.

If A is an $m \times n$ matrix, we refer to the dimension of the null space of A as the **nullity** of A, denoted by nullity A.

EXAMPLE 1

Find a basis for and the dimension of the solution space W of the homogeneous system

$$\begin{bmatrix} 1 & 1 & 4 & 1 & 2 \\ 0 & 1 & 2 & 1 & 1 \\ 0 & 0 & 0 & 1 & 2 \\ 1 & -1 & 0 & 0 & 2 \\ 2 & 1 & 6 & 0 & 1 \end{bmatrix} \begin{bmatrix} x_1 \\ x_2 \\ x_3 \\ x_4 \\ x_5 \end{bmatrix} = \begin{bmatrix} 0 \\ 0 \\ 0 \\ 0 \\ 0 \end{bmatrix}.$$

Solution

STEP 1 To solve the given system by the Gauss–Jordan reduction method, we transform the augmented matrix to reduced row echelon form, obtaining (verify)

$$\left[\begin{array}{ccccc|c} 1 & 0 & 2 & 0 & 1 & 0 \\ 0 & 1 & 2 & 0 & -1 & 0 \\ 0 & 0 & 0 & 1 & 2 & 0 \\ 0 & 0 & 0 & 0 & 0 & 0 \\ 0 & 0 & 0 & 0 & 0 & 0 \end{array}\right].$$

Every solution is of the form (verify)

$$\mathbf{x} = \begin{bmatrix} -2s - t \\ -2s + t \\ s \\ -2t \\ t \end{bmatrix}, \tag{1}$$

where s and t are any real numbers.

STEP 2 Every vector in W is a solution and is then of the form given by Equation (1). We can then write every vector in W as

$$\mathbf{x} = s \begin{bmatrix} -2 \\ -2 \\ 1 \\ 0 \\ 0 \end{bmatrix} + t \begin{bmatrix} -1 \\ 1 \\ 0 \\ -2 \\ 1 \end{bmatrix}. \tag{2}$$

Since s and t can take on any values, we first let $s = 1$, $t = 0$, and then let $s = 0$, $t = 1$, in Equation (2), obtaining as solutions

$$\mathbf{x}_1 = \begin{bmatrix} -2 \\ -2 \\ 1 \\ 0 \\ 0 \end{bmatrix} \quad \text{and} \quad \mathbf{x}_2 = \begin{bmatrix} -1 \\ 1 \\ 0 \\ -2 \\ 1 \end{bmatrix}.$$

STEP 3 The set $\{\mathbf{x}_1, \mathbf{x}_2\}$ is a basis for W. Moreover, $\dim W = 2$. ■

The following example illustrates a type of problem that we will be solving often in Chapter 6.

EXAMPLE 2

Find a basis for the solution space of the homogeneous system $(\lambda I_3 - A)\mathbf{x} = \mathbf{0}$ for $\lambda = -2$ and

$$A = \begin{bmatrix} -3 & 0 & -1 \\ 2 & 1 & 0 \\ 0 & 0 & -2 \end{bmatrix}.$$

Solution We form $-2I_3 - A$:

$$-2 \begin{bmatrix} 1 & 0 & 0 \\ 0 & 1 & 0 \\ 0 & 0 & 1 \end{bmatrix} - \begin{bmatrix} -3 & 0 & -1 \\ 2 & 1 & 0 \\ 0 & 0 & -2 \end{bmatrix} = \begin{bmatrix} 1 & 0 & 1 \\ -2 & -3 & 0 \\ 0 & 0 & 0 \end{bmatrix}.$$

This last matrix is the coefficient matrix of the homogeneous system, so we transform the augmented matrix

$$\left[\begin{array}{rrr|r} 1 & 0 & 1 & 0 \\ -2 & -3 & 0 & 0 \\ 0 & 0 & 0 & 0 \end{array}\right]$$

to reduced row echelon form, obtaining (verify)

$$\left[\begin{array}{rrr|r} 1 & 0 & 1 & 0 \\ 0 & 1 & -\frac{2}{3} & 0 \\ 0 & 0 & 0 & 0 \end{array}\right].$$

Every solution is then of the form (verify)

$$\mathbf{x} = \begin{bmatrix} -s \\ \frac{2}{3}s \\ s \end{bmatrix},$$

where s is any real number. Then every vector in the solution can be written as

$$\mathbf{x} = s\begin{bmatrix} -1 \\ \frac{2}{3} \\ 1 \end{bmatrix},$$

so $\left\{\begin{bmatrix} -1 \\ \frac{2}{3} \\ 1 \end{bmatrix}\right\}$ is a basis for the solution space. ■

Another important problem that we have to solve often in Chapter 6 is illustrated in the following example.

EXAMPLE 3

Find all real numbers λ such that the homogeneous system $(\lambda I_2 - A)\mathbf{x} = \mathbf{0}$ has a nontrivial solution for

$$A = \begin{bmatrix} 1 & 5 \\ 3 & -1 \end{bmatrix}.$$

Solution We form $\lambda I_2 - A$:

$$\lambda\begin{bmatrix} 1 & 0 \\ 0 & 1 \end{bmatrix} - \begin{bmatrix} 1 & 5 \\ 3 & -1 \end{bmatrix} = \begin{bmatrix} \lambda - 1 & -5 \\ -3 & \lambda + 1 \end{bmatrix}.$$

The homogeneous system $(\lambda I_2 - A)\mathbf{x} = \mathbf{0}$ is then

$$\begin{bmatrix} \lambda - 1 & -5 \\ -3 & \lambda + 1 \end{bmatrix}\begin{bmatrix} x_1 \\ x_2 \end{bmatrix} = \begin{bmatrix} 0 \\ 0 \end{bmatrix}.$$

From Exercise 18 or 27 in Section 1.5, it follows that this homogeneous system has a nontrivial solution if and only if

$$(\lambda - 1)(\lambda + 1) - 15 = 0$$
$$\lambda^2 - 16 = 0$$
$$\lambda = 4 \quad \text{or} \quad \lambda = -4.$$

Thus, when $\lambda = 4$ or -4, the homogeneous system $(\lambda I_2 - A)\mathbf{x} = \mathbf{0}$ for the given matrix A has a nontrivial solution. ■

Relationship Between Nonhomogeneous Linear Systems and Homogeneous Systems

We have already noted in Section 1.5 that if A is $m \times n$, then the set of all solutions to the linear system $A\mathbf{x} = \mathbf{b}$, $\mathbf{b} \neq \mathbf{0}$, is not a subspace of R^n. The following example illustrates a geometric relationship between the set of all solutions to the nonhomogeneous system $A\mathbf{x} = \mathbf{b}$, $\mathbf{b} \neq \mathbf{0}$, and the associated homogeneous system $A\mathbf{x} = \mathbf{0}$.

EXAMPLE 4

Consider the linear system

$$\begin{bmatrix} 1 & 2 & -3 \\ 2 & 4 & -6 \\ 3 & 6 & -9 \end{bmatrix} \begin{bmatrix} x_1 \\ x_2 \\ x_3 \end{bmatrix} = \begin{bmatrix} 2 \\ 4 \\ 6 \end{bmatrix}.$$

The set of all solutions to this linear system consists of all vectors of the form

$$\mathbf{x} = \begin{bmatrix} 2 - 2r + 3s \\ r \\ s \end{bmatrix}$$

(verify), which can be written as

$$\mathbf{x} = \begin{bmatrix} 2 \\ 0 \\ 0 \end{bmatrix} + r \begin{bmatrix} -2 \\ 1 \\ 0 \end{bmatrix} + s \begin{bmatrix} 3 \\ 0 \\ 1 \end{bmatrix}.$$

The set of all solutions to the associated homogeneous system is the two-dimensional subspace of R^3 consisting of all vectors of the form

$$\mathbf{x} = r \begin{bmatrix} -2 \\ 1 \\ 0 \end{bmatrix} + s \begin{bmatrix} 3 \\ 0 \\ 1 \end{bmatrix}.$$

This subspace is a plane Π_1 passing through the origin; the set of all solutions to the given nonhomogeneous system is a plane Π_2 that does not pass through the origin and is obtained by shifting Π_1 parallel to itself. This situation is illustrated in Figure 2.25. ■

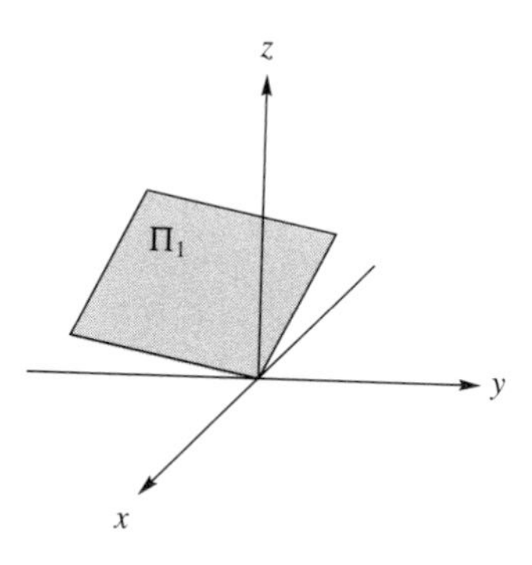

Π_1 is the solution space to $A\mathbf{x} = \mathbf{0}$.

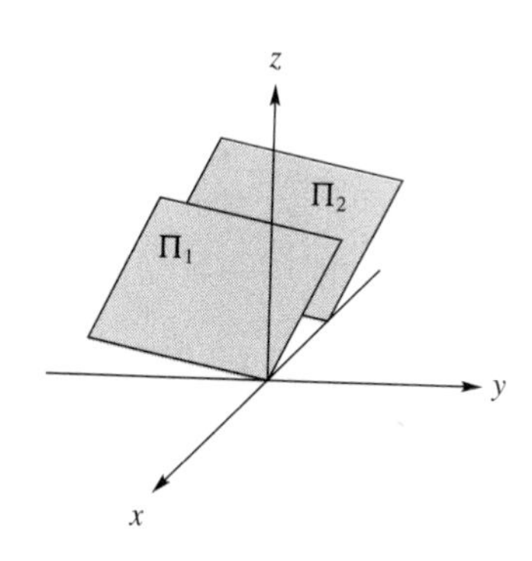

Π_2 is the set of all solutions to $A\mathbf{x} = \mathbf{b}$.

FIGURE 2.25

The following result, which is important in the study of differential equations, was presented in Section 1.5 and its proof left to Exercise 28(b) of that section.

If $\mathbf{x}_p$ is a particular solution to the nonhomogeneous system $A\mathbf{x} = \mathbf{b}$, $\mathbf{b} \neq \mathbf{0}$, and $\mathbf{x}_h$ is a solution to the associated homogeneous system $A\mathbf{x} = \mathbf{0}$, then $\mathbf{x}_p + \mathbf{x}_h$ is a solution to the given system $A\mathbf{x} = \mathbf{b}$. Moreover, every solution $\mathbf{x}$ to the nonhomogeneous linear system $A\mathbf{x} = \mathbf{b}$ can be written as $\mathbf{x}_p + \mathbf{x}_h$, where $\mathbf{x}_p$ is a particular solution to the given nonhomogeneous system and $\mathbf{x}_h$ is a solution to the associated homogeneous system $A\mathbf{x} = \mathbf{0}$.

2.6 Exercises

1. Let

$$A = \begin{bmatrix} 2 & -1 & -2 \\ -4 & 2 & 4 \\ -8 & 4 & 8 \end{bmatrix}.$$

(a) Find the set of all solutions to $A\mathbf{x} = \mathbf{0}$.

(b) Express each solution as a linear combination of two vectors in R^3.

(c) Sketch these vectors in a three-dimensional coordinate system to show that the solution space is a plane through the origin.

2. Let

$$A = \begin{bmatrix} 1 & 1 & -2 \\ -2 & -2 & 4 \\ -1 & -1 & 2 \end{bmatrix}.$$

(a) Find the set of all solutions to $A\mathbf{x} = \mathbf{0}$.

(b) Express each solution as a linear combination of two vectors in R^3.

(c) Sketch these vectors in a three-dimensional coordinate system to show that the solution space is a plane through the origin.

In Exercises 3 through 10, find a basis for and the dimension of the solution space of the given homogeneous system.

3.
$$\begin{aligned} x_1 + x_2 + x_3 + x_4 &= 0 \\ 2x_1 + x_2 - x_3 + x_4 &= 0. \end{aligned}$$

4.
$$\begin{bmatrix} 1 & -1 & 1 & -2 & 1 \\ 3 & -3 & 2 & 0 & 2 \end{bmatrix} \begin{bmatrix} x_1 \\ x_2 \\ x_3 \\ x_4 \\ x_5 \end{bmatrix} = \begin{bmatrix} 0 \\ 0 \end{bmatrix}.$$

5.
$$\begin{aligned} x_1 + 2x_2 - x_3 + 3x_4 &= 0 \\ 2x_1 + 2x_2 - x_3 + 2x_4 &= 0 \\ x_1 + 3x_3 + 3x_4 &= 0. \end{aligned}$$

6. $$\begin{aligned} x_1 - x_2 + 2x_3 + 3x_4 + 4x_5 &= 0 \\ -x_1 + 2x_2 + 3x_3 + 4x_4 + 5x_5 &= 0 \\ x_1 - x_2 + 3x_3 + 5x_4 + 6x_5 &= 0 \\ 3x_1 - 4x_2 + x_3 + 2x_4 + 3x_5 &= 0. \end{aligned}$$

7. $$\begin{bmatrix} 1 & 2 & 1 & 2 & 1 \\ 1 & 2 & 2 & 1 & 2 \\ 2 & 4 & 3 & 3 & 3 \\ 0 & 0 & 1 & -1 & -1 \end{bmatrix} \begin{bmatrix} x_1 \\ x_2 \\ x_3 \\ x_4 \\ x_5 \end{bmatrix} = \begin{bmatrix} 0 \\ 0 \\ 0 \\ 0 \end{bmatrix}.$$

8. $$\begin{bmatrix} 1 & 0 & 2 \\ 2 & 1 & 3 \\ 3 & 1 & 2 \end{bmatrix} \begin{bmatrix} x_1 \\ x_2 \\ x_3 \end{bmatrix} = \begin{bmatrix} 0 \\ 0 \\ 0 \end{bmatrix}.$$

9. $$\begin{bmatrix} 1 & 2 & 2 & -1 & 1 \\ 0 & 2 & 2 & -2 & -1 \\ 2 & 6 & 2 & -4 & 1 \\ 1 & 4 & 0 & -3 & 0 \end{bmatrix} \begin{bmatrix} x_1 \\ x_2 \\ x_3 \\ x_4 \\ x_5 \end{bmatrix} = \begin{bmatrix} 0 \\ 0 \\ 0 \\ 0 \end{bmatrix}.$$

10. $$\begin{bmatrix} 1 & 2 & -3 & -2 & 1 & 3 \\ 1 & 2 & -4 & 3 & 3 & 4 \\ -2 & -4 & 6 & 4 & -3 & 2 \\ 0 & 0 & -1 & 5 & 1 & 9 \\ 1 & 2 & -3 & -2 & 0 & 7 \end{bmatrix} \begin{bmatrix} x_1 \\ x_2 \\ x_3 \\ x_4 \\ x_5 \\ x_6 \end{bmatrix} = \begin{bmatrix} 0 \\ 0 \\ 0 \\ 0 \\ 0 \end{bmatrix}.$$

In Exercises 11 *and* 12, *find a basis for the null space of the given matrix A.*

11. $A = \begin{bmatrix} 1 & 2 & 3 & -1 \\ 2 & 3 & 2 & 0 \\ 3 & 4 & 1 & 1 \\ 1 & 1 & -1 & 1 \end{bmatrix}.$

12. $A = \begin{bmatrix} 1 & -1 & 2 & 1 & 0 \\ 2 & 0 & 1 & -1 & 3 \\ 5 & -1 & 3 & 0 & 3 \\ 4 & -2 & 5 & 1 & 3 \\ 1 & 3 & -4 & -5 & 6 \end{bmatrix}.$

In Exercises 13 *through* 16, *find a basis for the solution space of the homogeneous system $(\lambda I_n - A)\mathbf{x} = \mathbf{0}$ for the given scalar λ and given matrix A.*

13. $\lambda = 1$, $A = \begin{bmatrix} 3 & 2 \\ 1 & 2 \end{bmatrix}.$

14. $\lambda = -3$, $A = \begin{bmatrix} -4 & -3 \\ 2 & 3 \end{bmatrix}.$

15. $\lambda = 1$, $A = \begin{bmatrix} 0 & 0 & 1 \\ 1 & 0 & -3 \\ 0 & 1 & 3 \end{bmatrix}.$

16. $\lambda = 3$, $A = \begin{bmatrix} 1 & 1 & -2 \\ -1 & 2 & 1 \\ 0 & 1 & -1 \end{bmatrix}.$

In Exercises 17 *through* 20, *find all real numbers λ such that the homogeneous system $(\lambda I_n - A)\mathbf{x} = \mathbf{0}$ has a nontrivial solution.*

17. $A = \begin{bmatrix} 2 & 3 \\ 2 & -3 \end{bmatrix}.$

18. $A = \begin{bmatrix} 3 & 0 \\ 2 & -2 \end{bmatrix}.$

19. $A = \begin{bmatrix} 0 & 0 & 0 \\ 0 & 1 & -1 \\ 1 & 0 & 0 \end{bmatrix}.$

20. $A = \begin{bmatrix} -2 & 0 & 0 \\ 0 & -2 & -3 \\ 0 & 4 & 5 \end{bmatrix}.$

21. Let $S = \{\mathbf{x}_1, \mathbf{x}_2, \ldots, \mathbf{x}_k\}$ be a set of solutions to a homogeneous system $A\mathbf{x} = \mathbf{0}$. Show that every vector in span S is a solution to $A\mathbf{x} = \mathbf{0}$.

22. Show that if the $n \times n$ coefficient matrix A of the homogeneous system $A\mathbf{x} = \mathbf{0}$ has a row or column of zeros, then $A\mathbf{x} = \mathbf{0}$ has a nontrivial solution.

23. (a) Show that the zero matrix is the only 3×3 matrix whose null space has dimension 3.

(b) Let A be a nonzero 3×3 matrix and suppose that $A\mathbf{x} = \mathbf{0}$ has a nontrivial solution. Show that the dimension of the null space of A is either 1 or 2.

24. Matrices A and B are $m \times n$ and their reduced row echelon forms are the same. What is the relationship between the null space of A and the null space of B?

2.7 COORDINATES AND ISOMORPHISMS

Coordinates

If V is an n-dimensional vector space, we know that V has a basis S with n vectors in it; thus far we have not paid much attention to the order of the vectors in S. However, in the discussion of this section we speak of an **ordered basis** $S = \{\mathbf{v}_1, \mathbf{v}_2, \ldots, \mathbf{v}_n\}$ for V; thus $S_1 = \{\mathbf{v}_2, \mathbf{v}_1, \ldots, \mathbf{v}_n\}$ is a different ordered basis for V.

If $S = \{\mathbf{v}_1, \mathbf{v}_2, \ldots, \mathbf{v}_n\}$ is an ordered basis for the n-dimensional vector space V, then by Theorem 2.7 every vector $\mathbf{v}$ in V can be uniquely expressed in the form

$$\mathbf{v} = a_1\mathbf{v}_1 + a_2\mathbf{v}_2 + \cdots + a_n\mathbf{v}_n,$$

where $a_1, a_2, \ldots, a_n$ are real numbers. We shall refer to

$$[\mathbf{v}]_S = \begin{bmatrix} a_1 \\ a_2 \\ \vdots \\ a_n \end{bmatrix}$$

as the **coordinate vector of v with respect to the ordered basis** S. The entries of $[\mathbf{v}]_S$ are called the **coordinates of v with respect to** S.

EXAMPLE 1

Consider the vector space P_1 and let $S = \{\mathbf{v}_1, \mathbf{v}_2\}$ be an ordered basis for P_1, where $\mathbf{v}_1 = t$ and $\mathbf{v}_2 = 1$. If $\mathbf{v} = p(t) = 5t - 2$, then $[\mathbf{v}]_S = \begin{bmatrix} 5 \\ -2 \end{bmatrix}$ is the coordinate vector of $\mathbf{v}$ with respect to the ordered basis S. On the other hand, if $T = \{t+1, t-1\}$ is the ordered basis, we have $5t - 2 = \frac{3}{2}(t+1) + \frac{7}{2}(t-1)$, which implies that

$$[\mathbf{v}]_T = \begin{bmatrix} \frac{3}{2} \\ \frac{7}{2} \end{bmatrix}.$$

■

Notice that the coordinate vector $[\mathbf{v}]_S$ depends upon the order in which the vectors in S are listed; a change in the order of this listing may change the coordinates of $\mathbf{v}$ with respect to S.

EXAMPLE 2

Consider the vector space R^3 and let $S = \{\mathbf{v}_1, \mathbf{v}_2, \mathbf{v}_3\}$ be an ordered basis for R^3, where

$$\mathbf{v}_1 = \begin{bmatrix} 1 \\ 1 \\ 0 \end{bmatrix}, \quad \mathbf{v}_2 = \begin{bmatrix} 2 \\ 0 \\ 1 \end{bmatrix}, \quad \text{and} \quad \mathbf{v}_3 = \begin{bmatrix} 0 \\ 1 \\ 2 \end{bmatrix}.$$

If

$$\mathbf{v} = \begin{bmatrix} 1 \\ 1 \\ -5 \end{bmatrix},$$

compute $[\mathbf{v}]_S$.

Solution To find $\left[\mathbf{v}\right]_S$ we need to find the constants a_1, a_2, and a_3 such that

$$a_1\mathbf{v}_1 + a_2\mathbf{v}_2 + a_3\mathbf{v}_3 = \mathbf{v},$$

which leads to the linear system whose augmented matrix is (verify)

$$\left[\begin{array}{ccc|c} 1 & 2 & 0 & 1 \\ 1 & 0 & 1 & 1 \\ 0 & 1 & 2 & -5 \end{array}\right] \tag{1}$$

or, equivalently,

$$\left[\begin{array}{ccc|c} \mathbf{v}_1 & \mathbf{v}_2 & \mathbf{v}_3 & \mathbf{v} \end{array}\right].$$

Transforming the matrix in (1) to reduced row echelon form, we obtain the solution (verify)

$$a_1 = 3, \quad a_2 = -1, \quad a_3 = -2,$$

so

$$\left[\mathbf{v}\right]_S = \begin{bmatrix} 3 \\ -1 \\ -2 \end{bmatrix}.$$

■

Let $S = \{\mathbf{v}_1, \mathbf{v}_2, \ldots, \mathbf{v}_n\}$ be an ordered basis for a vector space V and let $\mathbf{v}$ and $\mathbf{w}$ be vectors in V such that

$$\left[\mathbf{v}\right]_S = \left[\mathbf{w}\right]_S.$$

We shall now show that $\mathbf{v} = \mathbf{w}$. Let

$$\left[\mathbf{v}\right]_S = \begin{bmatrix} a_1 \\ a_2 \\ \vdots \\ a_n \end{bmatrix}, \quad \text{and} \quad \left[\mathbf{w}\right]_S = \begin{bmatrix} b_1 \\ b_2 \\ \vdots \\ b_n \end{bmatrix}.$$

Since $\left[\mathbf{v}\right]_S = \left[\mathbf{w}\right]_S$, we have

$$a_i = b_i, \qquad i = 1, 2, \ldots, n.$$

Then

$$\mathbf{v} = a_1\mathbf{v}_1 + a_2\mathbf{v}_2 + \cdots + a_n\mathbf{v}_n$$

and

$$\begin{aligned} \mathbf{w} &= b_1\mathbf{v}_1 + b_2\mathbf{v}_2 + \cdots + b_n\mathbf{v}_n \\ &= a_1\mathbf{v}_1 + a_2\mathbf{v}_2 + \cdots + a_n\mathbf{v}_n \\ &= \mathbf{v}. \end{aligned}$$

The choice of an ordered basis and the consequent assignment of a coordinate vector for every $\mathbf{v}$ in V enables us to "picture" the vector space. We illustrate this notion by using Example 1. Choose a fixed point O in the plane R^2, and draw any two arrows $\mathbf{w}_1$ and $\mathbf{w}_2$ from O which depict the basis vectors t and 1 in the ordered

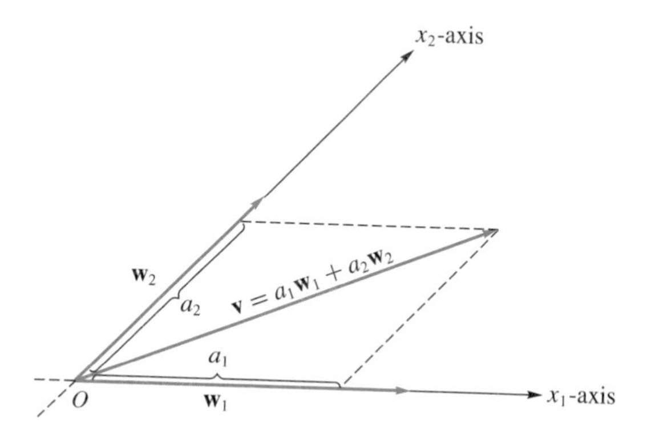

FIGURE 2.26

basis $S = \{t, 1\}$ for P_1 (see Figure 2.26). The directions of $\mathbf{w}_1$ and $\mathbf{w}_2$ determine two lines, which we call the $\boldsymbol{x_1}$**- and** $\boldsymbol{x_2}$**-axes**, respectively. The positive direction on the x_1-axis is in the direction of $\mathbf{w}_1$; the negative direction on the x_1-axis is along $-\mathbf{w}_1$. Similarly, the positive direction on the x_2-axis is in the direction of $\mathbf{w}_2$; the negative direction on the x_2-axis is along $-\mathbf{w}_2$. The lengths of $\mathbf{w}_1$ and $\mathbf{w}_2$ determine the scales on the x_1- and x_2-axes, respectively. If $\mathbf{v}$ is a vector in P_1, we can write $\mathbf{v}$, uniquely, as $\mathbf{v} = a_1\mathbf{w}_1 + a_2\mathbf{w}_2$. We now mark off a segment of length $|a_1|$ on the x_1-axis (in the positive direction if a_1 is positive and in the negative direction if a_1 is negative) and draw a line through the endpoint of this segment parallel to $\mathbf{w}_2$. Similarly, mark off a segment of length $|a_2|$ on the x_2-axis (in the positive direction if a_2 is positive and in the negative direction if a_2 is negative) and draw a line through the endpoint of this segment parallel to $\mathbf{w}_1$. We draw a directed line segment from O to the point of intersection of these two lines. This directed line segment represents $\mathbf{v}$.

Isomorphisms

If $\mathbf{v}$ and $\mathbf{w}$ are vectors in an n-dimensional vector space V with an ordered basis $S = \{\mathbf{v}_1, \mathbf{v}_2, \ldots, \mathbf{v}_n\}$, then we can write $\mathbf{v}$ and $\mathbf{w}$ uniquely, as

$$\mathbf{v} = a_1\mathbf{v}_1 + a_2\mathbf{v}_2 + \cdots + a_n\mathbf{v}_n, \quad \mathbf{w} = b_1\mathbf{v}_1 + b_2\mathbf{v}_2 + \cdots + b_n\mathbf{v}_n.$$

Thus with $\mathbf{v}$ and $\mathbf{w}$, we associate $\left[\mathbf{v}\right]_S$ and $\left[\mathbf{w}\right]_S$, respectively, elements in R^n:

$$\begin{aligned} \mathbf{v} &\to \left[\mathbf{v}\right]_S \\ \mathbf{w} &\to \left[\mathbf{w}\right]_S. \end{aligned}$$

The sum $\mathbf{v} + \mathbf{w} = (a_1 + b_1)\mathbf{v}_1 + (a_2 + b_2)\mathbf{v}_2 + \cdots + (a_n + b_n)\mathbf{v}_n$, which means that with $\mathbf{v} + \mathbf{w}$ we associate the vector

$$\left[\mathbf{v} + \mathbf{w}\right]_S = \begin{bmatrix} a_1 + b_1 \\ a_2 + b_2 \\ \vdots \\ a_n + b_n \end{bmatrix} = \left[\mathbf{v}\right]_S + \left[\mathbf{w}\right]_S.$$

Therefore

$$\mathbf{v} + \mathbf{w} \rightarrow \left[\mathbf{v} + \mathbf{w}\right]_S = \left[\mathbf{v}\right]_S + \left[\mathbf{w}\right]_S.$$

That is, when we add $\mathbf{v}$ and $\mathbf{w}$ in V, we add their associated coordinate vectors $\left[\mathbf{v}\right]_S$ and $\left[\mathbf{w}\right]_S$ to obtain the coordinate vector $\left[\mathbf{v} + \mathbf{w}\right]_S$ in R^n associated with $\mathbf{v} + \mathbf{w}$.

Similarly, if c is a real number, then

$$c\mathbf{v} = (ca_1)\mathbf{v}_1 + (ca_2)\mathbf{v}_2 + \cdots + (ca_n)\mathbf{v}_n,$$

which implies that

$$\left[c\mathbf{v}\right]_S = \begin{bmatrix} ca_1 \\ ca_2 \\ \vdots \\ ca_n \end{bmatrix} = c\left[\mathbf{v}\right]_S.$$

Therefore,

$$c\mathbf{v} \rightarrow \left[c\mathbf{v}\right]_S = c\left[\mathbf{v}\right]_S,$$

and thus when $\mathbf{v}$ is multiplied by a scalar c, we multiply $\left[\mathbf{v}\right]_S$ by c to obtain the coordinate vector in R^n associated with $c\mathbf{v}$.

This discussion suggests that, from an algebraic point of view, V and R^n behave "rather similarly." We now clarify this notion.

Let L be a function mapping a vector space V into a vector space W. Recall that L is **one-to-one** if $L(\mathbf{v}_1) = L(\mathbf{v}_2)$, for $\mathbf{v}_1$, $\mathbf{v}_2$ in V, implies that $\mathbf{v}_1 = \mathbf{v}_2$. Also, L is **onto** if for each $\mathbf{w}$ in W there is at least one $\mathbf{v}$ in V for which $L(\mathbf{v}) = \mathbf{w}$.[†] Thus the mapping $L\colon R^3 \rightarrow R^2$ defined by

$$L\left(\begin{bmatrix} a_1 \\ a_2 \\ a_3 \end{bmatrix}\right) = \begin{bmatrix} a_1 + a_2 \\ a_1 \end{bmatrix}$$

is onto. To see this, suppose that $\mathbf{w} = \begin{bmatrix} b_1 \\ b_2 \end{bmatrix}$; we seek $\mathbf{v} = \begin{bmatrix} a_1 \\ a_2 \\ a_3 \end{bmatrix}$ such that

$$L(\mathbf{v}) = \begin{bmatrix} a_1 + a_2 \\ a_1 \end{bmatrix} = \mathbf{w} = \begin{bmatrix} b_1 \\ b_2 \end{bmatrix}.$$

Thus we obtain the solution: $a_1 = b_2$, $a_2 = b_1 - b_2$, and a_3 is arbitrary. However, L is not one-to-one, for if $\mathbf{v}_1 = \begin{bmatrix} 1 \\ 2 \\ 3 \end{bmatrix}$ and $\mathbf{v}_2 = \begin{bmatrix} 1 \\ 2 \\ 4 \end{bmatrix}$, then

$$L(\mathbf{v}_1) = L(\mathbf{v}_2) = \begin{bmatrix} 3 \\ 1 \end{bmatrix} \quad \text{but} \quad \mathbf{v}_1 \neq \mathbf{v}_2.$$

[†] See Appendix A for further discussion of one-to-one and onto functions.

Definition 2.13

Let V be a real vector space with operations $\oplus$ and $\odot$, and let W be a real vector space with operations $\boxplus$ and $\boxdot$. A one-to-one function L mapping V onto W is called an **isomorphism** (from the Greek *isos*, meaning "the same," and *morphos*, meaning "structure") of V onto W if:

(a) $L(\mathbf{v} \oplus \mathbf{w}) = L(\mathbf{v}) \boxplus L(\mathbf{w})$ for $\mathbf{v}$, $\mathbf{w}$ in V.

(b) $L(c \odot \mathbf{v}) = c \boxdot L(\mathbf{v})$ for $\mathbf{v}$ in V, c a real number.

In this case we say that **V is isomorphic to W**. ▲

It also follows from Definition 2.13 that if L is an isomorphism of V onto W, then

$$L(a_1 \odot \mathbf{v}_1 \oplus a_2 \odot \mathbf{v}_2 \oplus \cdots \oplus a_k \odot \mathbf{v}_k) = a_1 \boxdot L(\mathbf{v}_1) \boxplus a_2 \boxdot L(\mathbf{v}_2) \boxplus \cdots \boxplus a_k \boxdot L(\mathbf{v}_k),$$

where $\mathbf{v}_1, \mathbf{v}_2, \ldots, \mathbf{v}_k$ are vectors in V and $a_1, a_2, \ldots, a_k$ are scalars [see Exercise 27(c)].

Remark A function L mapping a vector space V into a vector space W satisfying properties (a) and (b) of Definition 2.13 is called a linear transformation. These functions will be studied in depth in Chapter 4. Thus an isomorphism of a vector space V onto a vector space W is a linear transformation that is one-to-one and onto.

As a result of Theorem 2.14 below we can replace the expressions "V is isomorphic to W" and "W is isomorphic to V" by "V and W are isomorphic."

Isomorphic vector spaces differ only in the nature of their elements; their algebraic properties are identical. That is, if the vector spaces V and W are isomorphic, under the isomorphism L, then for each $\mathbf{v}$ in V there is a unique $\mathbf{w}$ in W so that $L(\mathbf{v}) = \mathbf{w}$ and, conversely, for each $\mathbf{w}$ in W there is a unique $\mathbf{v}$ in V so that $L(\mathbf{v}) = \mathbf{w}$. If we now replace each element of V by it image under L and replace the operations $\oplus$ and $\odot$ by $\boxplus$ and $\boxdot$, respectively, we get precisely W. The most important example of isomorphic vector spaces is given in the following theorem.

Theorem 2.13

If V is an n-dimensional real vector space, then V is isomorphic to R^n.

Proof

Let $S = \{\mathbf{v}_1, \mathbf{v}_2, \ldots, \mathbf{v}_n\}$ be an ordered basis for V, and let $L: V \to R^n$ be defined by

$$L(\mathbf{v}) = \left[\mathbf{v}\right]_S = \begin{bmatrix} a_1 \\ a_2 \\ \vdots \\ a_n \end{bmatrix},$$

where $\mathbf{v} = a_1\mathbf{v}_1 + a_2\mathbf{v}_2 + \cdots + a_n\mathbf{v}_n$.

We show that L is an isomorphism. First, L is one-to-one. Let

$$\left[\mathbf{v}\right]_S = \begin{bmatrix} a_1 \\ a_2 \\ \vdots \\ a_n \end{bmatrix} \quad \text{and} \quad \left[\mathbf{w}\right]_S = \begin{bmatrix} b_1 \\ b_2 \\ \vdots \\ b_n \end{bmatrix}$$

and suppose that $L(\mathbf{v}) = L(\mathbf{w})$. Then $\left[\mathbf{v}\right]_S = \left[\mathbf{w}\right]_S$, and from our earlier remarks it follows that $\mathbf{v} = \mathbf{w}$.

Next, L is onto, for if $\mathbf{w} = \begin{bmatrix} b_1 \\ b_2 \\ \vdots \\ b_n \end{bmatrix}$ is a given vector in R^n and we let

$$\mathbf{v} = b_1\mathbf{v}_1 + b_2\mathbf{v}_2 + \cdots + b_n\mathbf{v}_n,$$

then $L(\mathbf{v}) = \mathbf{w}$.

Finally, L satisfies Definition 2.13(a) and (b). Let $\mathbf{v}$ and $\mathbf{w}$ be vectors in V such that $\left[\mathbf{v}\right]_S = \begin{bmatrix} a_1 \\ a_2 \\ \vdots \\ a_n \end{bmatrix}$ and $\left[\mathbf{w}\right]_S = \begin{bmatrix} b_1 \\ b_2 \\ \vdots \\ b_n \end{bmatrix}$. Then

$$L(\mathbf{v} + \mathbf{w}) = \left[\mathbf{v} + \mathbf{w}\right]_S = \left[\mathbf{v}\right]_S + \left[\mathbf{w}\right]_S = L(\mathbf{v}) + L(\mathbf{w})$$

and

$$L(c\mathbf{v}) = \left[c\mathbf{v}\right]_S = c\left[\mathbf{v}\right]_S = cL(\mathbf{v}),$$

as we saw before. Hence V and R^n are isomorphic. ●

Another example of isomorphism is given by the vector spaces discussed in the review section at the beginning of this chapter: R^2, the vector space of directed line segments emanating from a point in the plane and the vector space of all ordered pairs of real numbers. There is a corresponding isomorphism for R^3.

Some important properties of isomorphisms are given in Theorem 2.14.

Theorem 2.14

(a) *Every vector space V is isomorphic to itself.*

(b) *If V is isomorphic to W, then W is isomorphic to V.*

(c) *If U is isomorphic to V and V is isomorphic to W, then U is isomorphic to W.*

Proof

Exercise. [Parts (a) and (c) are not difficult to show; (b) is slightly harder and will essentially be proved in Theorem 4.6.] ●

The following theorem shows that all vector spaces of the same dimension are, algebraically speaking, alike, and conversely, that isomorphic vector spaces have the same dimensions.

Theorem 2.15

Two finite-dimensional vector spaces are isomorphic if and only if their dimensions are equal.

Proof

Let V and W be n-dimensional vector spaces. Then V and R^n are isomorphic and W and R^n are isomorphic. From Theorem 2.14 it follows that V and W are isomorphic.

Conversely, let V and W be isomorphic finite-dimensional vector spaces; let $L: V \to W$ be an isomorphism. Assume that $\dim V = n$, and let $S = \{\mathbf{v}_1, \mathbf{v}_2, \ldots, \mathbf{v}_n\}$ be a basis for V.

We now prove that the set $T = \{L(\mathbf{v}_1), L(\mathbf{v}_2), \ldots, L(\mathbf{v}_n)\}$ is a basis for W. First, T spans W. If $\mathbf{w}$ is any vector in W, then $\mathbf{w} = L(\mathbf{v})$ for some $\mathbf{v}$ in V. Since S is a basis for V, $\mathbf{v} = a_1\mathbf{v}_1 + a_2\mathbf{v}_2 + \cdots + a_n\mathbf{v}_n$, where the a_i are uniquely determined real numbers, so

$$\begin{aligned} L(\mathbf{v}) &= L(a_1\mathbf{v}_1 + a_2\mathbf{v}_2 + \cdots + a_n\mathbf{v}_n) \\ &= L(a_1\mathbf{v}_1) + L(a_2\mathbf{v}_2) + \cdots + L(a_n\mathbf{v}_n) \\ &= a_1L(\mathbf{v}_1) + a_2L(\mathbf{v}_2) + \cdots + a_nL(\mathbf{v}_n). \end{aligned}$$

Thus T spans W.

Now suppose that

$$a_1L(\mathbf{v}_1) + a_2L(\mathbf{v}_2) + \cdots + a_nL(\mathbf{v}_n) = \mathbf{0}_W.$$

Then $L(a_1\mathbf{v}_1 + a_2\mathbf{v}_2 + \cdots + a_n\mathbf{v}_n) = \mathbf{0}_W$. From Exercise 27(a), $L(\mathbf{0}_V) = \mathbf{0}_W$. Since L is one-to-one, we get $a_1\mathbf{v}_1 + a_2\mathbf{v}_2 + \cdots + a_n\mathbf{v}_n = \mathbf{0}_V$. Since S is linearly independent, we conclude that $a_1 = a_2 = \cdots = a_n = 0$, which means that T is linearly independent. Hence T is a basis for W and $\dim W = n$. ●

As a consequence of Theorem 2.15, the spaces R^n and R^m are isomorphic if and only if $n = m$ (see Exercise 28). Moreover, the vector spaces P_n and R^{n+1} are isomorphic. (See Exercise 30.)

We can now establish the converse of Theorem 2.13, as follows.

Corollary 2.6

If V is a finite-dimensional vector space that is isomorphic to R^n, then $\dim V = n$.

Proof

This result follows from Theorem 2.15. ●

If $L: V \to W$ is an isomorphism, then since L is a one-to-one onto mapping, it has an inverse L^{-1}. (This will be shown in Theorem 4.6.) It is not difficult to show that $L^{-1}: W \to V$ is also an isomorphism. (This will also be essentially shown in Theorem 4.6.) Moreover, if $S = \{\mathbf{v}_1, \mathbf{v}_2, \ldots, \mathbf{v}_n\}$ is a basis for V, then $T = L(S) = \{L(\mathbf{v}_1), L(\mathbf{v}_2), \ldots, L(\mathbf{v}_n)\}$ is a basis for W, as we have seen in the proof of Theorem 2.15.

As an example of isomorphism, we note that the vector spaces P_3, R_4, and R^4 are all isomorphic, since each has dimension four.

We have shown in this section that the idea of a finite-dimensional vector space, which at first seemed fairly abstract, is not so mysterious. In fact, such a vector space does not differ much from R^n in its algebraic behavior.

Transition Matrices

We now look at the relationship between two coordinate vectors for the same vector $\mathbf{v}$ with respect to different bases. Thus let $S = \{\mathbf{v}_1, \mathbf{v}_2, \ldots, \mathbf{v}_n\}$ and $T = \{\mathbf{w}_1, \mathbf{w}_2, \ldots, \mathbf{w}_n\}$ be two ordered bases for the n-dimensional vector space V. If $\mathbf{v}$ is any vector in V, then

$$\mathbf{v} = c_1\mathbf{w}_1 + c_2\mathbf{w}_2 + \cdots + c_n\mathbf{w}_n \quad \text{and} \quad \left[\mathbf{v}\right]_T = \begin{bmatrix} c_1 \\ c_2 \\ \vdots \\ c_n \end{bmatrix}.$$

Then

$$\begin{aligned} \left[\mathbf{v}\right]_S &= \left[c_1\mathbf{w}_1 + c_2\mathbf{w}_2 + \cdots + c_n\mathbf{w}_n\right]_S \\ &= \left[c_1\mathbf{w}_1\right]_S + \left[c_2\mathbf{w}_2\right]_S + \cdots + \left[c_n\mathbf{w}_n\right]_S \\ &= c_1\left[\mathbf{w}_1\right]_S + c_2\left[\mathbf{w}_2\right]_S + \cdots + c_n\left[\mathbf{w}_n\right]_S. \end{aligned} \tag{2}$$

Let the coordinate vector of $\mathbf{w}_j$ with respect to S be denoted by

$$\left[\mathbf{w}_j\right]_S = \begin{bmatrix} a_{1j} \\ a_{2j} \\ \vdots \\ a_{nj} \end{bmatrix}.$$

The $n \times n$ matrix whose jth column is $\left[\mathbf{w}_j\right]_S$ is called the **transition matrix from the T-basis to the S-basis** and is denoted by $P_{S\leftarrow T}$. Then Equation (2) can be written in matrix form as

$$\left[\mathbf{v}\right]_S = P_{S\leftarrow T}\left[\mathbf{v}\right]_T. \tag{3}$$

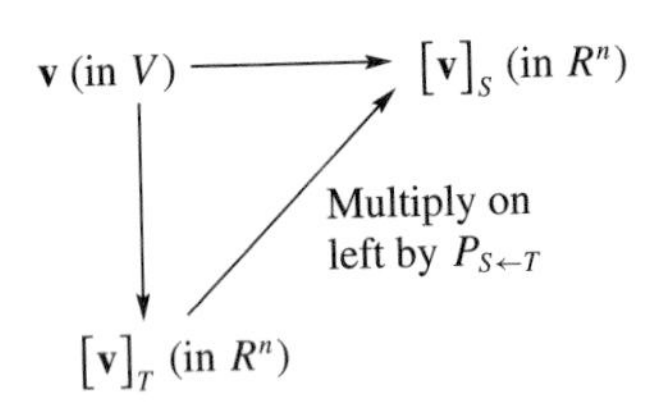

FIGURE 2.27

Thus, to find the transition matrix $P_{S\leftarrow T}$ from the T-basis to the S-basis, we first compute the coordinate vector of each member of the T-basis with respect to the S-basis. Forming a matrix with these vectors as columns arranged in their natural order, we obtain the transition matrix. Equation (3) says that the coordinate vector of $\mathbf{v}$ with respect to the basis S is the transition matrix $P_{S\leftarrow T}$ times the coordinate vector of $\mathbf{v}$ with respect to the basis T. Figure 2.27 illustrates Equation (3).

EXAMPLE 3

Let V be R^3 and let $S = \{\mathbf{v}_1, \mathbf{v}_2, \mathbf{v}_3\}$ and $T = \{\mathbf{w}_1, \mathbf{w}_2, \mathbf{w}_3\}$ be ordered bases for R^3, where

$$\mathbf{v}_1 = \begin{bmatrix} 2 \\ 0 \\ 1 \end{bmatrix}, \quad \mathbf{v}_2 = \begin{bmatrix} 1 \\ 2 \\ 0 \end{bmatrix}, \quad \mathbf{v}_3 = \begin{bmatrix} 1 \\ 1 \\ 1 \end{bmatrix}$$

and

$$\mathbf{w}_1 = \begin{bmatrix} 6 \\ 3 \\ 3 \end{bmatrix}, \quad \mathbf{w}_2 = \begin{bmatrix} 4 \\ -1 \\ 3 \end{bmatrix}, \quad \mathbf{w}_3 = \begin{bmatrix} 5 \\ 5 \\ 2 \end{bmatrix}.$$

(a) Compute the transition matrix $P_{S \leftarrow T}$ from the T-basis to the S-basis.

(b) Verify Equation (3) for $\mathbf{v} = \begin{bmatrix} 4 \\ -9 \\ 5 \end{bmatrix}$.

Solution (a) To compute $P_{S \leftarrow T}$, we need to find a_1, a_2, a_3 such that

$$a_1\mathbf{v}_1 + a_2\mathbf{v}_2 + a_3\mathbf{v}_3 = \mathbf{w}_1,$$

which leads to a linear system of three equations in three unknowns, whose augmented matrix is

$$\left[\begin{array}{ccc|c} \mathbf{v}_1 & \mathbf{v}_2 & \mathbf{v}_3 & \mathbf{w}_1 \end{array}\right].$$

That is, the augmented matrix is

$$\left[\begin{array}{ccc|c} 2 & 1 & 1 & 6 \\ 0 & 2 & 1 & 3 \\ 1 & 0 & 1 & 3 \end{array}\right].$$

Similarly, we need to find b_1, b_2, b_3 and c_1, c_2, c_3 such that

$$\begin{aligned} b_1\mathbf{v}_1 + b_2\mathbf{v}_2 + b_3\mathbf{v}_3 &= \mathbf{w}_2 \\ c_1\mathbf{v}_1 + c_2\mathbf{v}_2 + c_3\mathbf{v}_3 &= \mathbf{w}_3. \end{aligned}$$

These vector equations lead to two linear systems, each of three equations in three unknowns, whose augmented matrices are

$$\left[\begin{array}{ccc|c} \mathbf{v}_1 & \mathbf{v}_2 & \mathbf{v}_3 & \mathbf{w}_2 \end{array}\right] \quad \text{and} \quad \left[\begin{array}{ccc|c} \mathbf{v}_1 & \mathbf{v}_2 & \mathbf{v}_3 & \mathbf{w}_3 \end{array}\right],$$

or specifically,

$$\left[\begin{array}{ccc|c} 2 & 1 & 1 & 4 \\ 0 & 2 & 1 & -1 \\ 1 & 0 & 1 & 3 \end{array}\right] \quad \text{and} \quad \left[\begin{array}{ccc|c} 2 & 1 & 1 & 5 \\ 0 & 2 & 1 & 5 \\ 1 & 0 & 1 & 2 \end{array}\right].$$

Since the coefficient matrix of all three linear systems is $\begin{bmatrix} \mathbf{v}_1 & \mathbf{v}_2 & \mathbf{v}_3 \end{bmatrix}$, we can transform the three augmented matrices to reduced row echelon form simultaneously by transforming the partitioned matrix

$$\left[\begin{array}{ccc|c|c|c} \mathbf{v}_1 & \mathbf{v}_2 & \mathbf{v}_3 & \mathbf{w}_1 & \mathbf{w}_2 & \mathbf{w}_3 \end{array}\right]$$

to reduced row echelon form. Thus we transform

$$\left[\begin{array}{ccc|c|c|c} 2 & 1 & 1 & 6 & 4 & 5 \\ 0 & 2 & 1 & 3 & -1 & 5 \\ 1 & 0 & 1 & 3 & 3 & 2 \end{array}\right]$$

to reduced row echelon form, obtaining (verify)

$$\left[\begin{array}{ccc|c|c|c} 1 & 0 & 0 & 2 & 2 & 1 \\ 0 & 1 & 0 & 1 & -1 & 2 \\ 0 & 0 & 1 & 1 & 1 & 1 \end{array}\right],$$

which implies that the transition matrix from the T-basis to the S-basis is

$$P_{S \leftarrow T} = \begin{bmatrix} 2 & 2 & 1 \\ 1 & -1 & 2 \\ 1 & 1 & 1 \end{bmatrix}.$$

(b) If $\mathbf{v} = \begin{bmatrix} 4 \\ -9 \\ 5 \end{bmatrix}$, then expressing $\mathbf{v}$ in terms of the T-basis, we have (verify)

$$\mathbf{v} = \begin{bmatrix} 4 \\ -9 \\ 5 \end{bmatrix} = 1 \begin{bmatrix} 6 \\ 3 \\ 3 \end{bmatrix} + 2 \begin{bmatrix} 4 \\ -1 \\ 3 \end{bmatrix} - 2 \begin{bmatrix} 5 \\ 5 \\ 2 \end{bmatrix}.$$

So $\left[\mathbf{v}\right]_T = \begin{bmatrix} 1 \\ 2 \\ -2 \end{bmatrix}$. Then

$$\left[\mathbf{v}\right]_S = P_{S \leftarrow T} \left[\mathbf{v}\right]_T = \begin{bmatrix} 2 & 2 & 1 \\ 1 & -1 & 2 \\ 1 & 1 & 1 \end{bmatrix} \begin{bmatrix} 1 \\ 2 \\ -2 \end{bmatrix} = \begin{bmatrix} 4 \\ -5 \\ 1 \end{bmatrix}.$$

If we compute $\left[\mathbf{v}\right]_S$ directly, we find that

$$\mathbf{v} = \begin{bmatrix} 4 \\ -9 \\ 5 \end{bmatrix} = 4 \begin{bmatrix} 2 \\ 0 \\ 1 \end{bmatrix} - 5 \begin{bmatrix} 1 \\ 2 \\ 0 \end{bmatrix} + 1 \begin{bmatrix} 1 \\ 1 \\ 1 \end{bmatrix}, \quad \text{so } \left[\mathbf{v}\right]_S = \begin{bmatrix} 4 \\ -5 \\ 1 \end{bmatrix}.$$

Hence

$$\left[\mathbf{v}\right]_S = P_{S \leftarrow T} \left[\mathbf{v}\right]_T. \quad ■$$

We next want to show that the transition matrix $P_{S \leftarrow T}$ from the T-basis to the S-basis is nonsingular. Suppose that $P_{S \leftarrow T} \left[\mathbf{v}\right]_T = \mathbf{0}_{R^n}$ for some $\mathbf{v}$ in V. From Equation (3) we have

$$P_{S \leftarrow T} \left[\mathbf{v}\right]_T = \left[\mathbf{v}\right]_S = \mathbf{0}_{R^n}.$$

If $\mathbf{v} = b_1\mathbf{v}_1 + b_2\mathbf{v}_2 + \cdots + b_n\mathbf{v}_n$, then

$$\begin{bmatrix} b_1 \\ b_2 \\ \vdots \\ b_n \end{bmatrix} = \left[\mathbf{v}\right]_S = \mathbf{0}_{R^n} = \begin{bmatrix} 0 \\ 0 \\ \vdots \\ 0 \end{bmatrix},$$

so

$$\mathbf{v} = 0\mathbf{v}_1 + 0\mathbf{v}_2 + \cdots + 0\mathbf{v}_n = \mathbf{0}_V.$$

Hence $\left[\mathbf{v}\right]_T = \mathbf{0}_{R^n}$. Thus the homogeneous system $P_{S\leftarrow T}\mathbf{x} = \mathbf{0}$ has only the trivial solution; it then follows from Theorem 1.17 that $P_{S\leftarrow T}$ is nonsingular. Of course, we then also have

$$\left[\mathbf{v}\right]_T = P_{S\leftarrow T}^{-1}\left[\mathbf{v}\right]_S.$$

That is, $P_{S\leftarrow T}^{-1}$ is then the transition matrix from the S-basis to the T-basis; the jth column of $P_{S\leftarrow T}^{-1}$ is $\left[\mathbf{v}_j\right]_T$.

Remark In Exercises 37 through 39 we ask you to show that if S and T are ordered bases for the vector space R^n, then

$$P_{S\leftarrow T} = M_S^{-1}M_T,$$

where M_S is the $n \times n$ matrix whose jth column is $\mathbf{v}_j$ and M_T is the $n \times n$ matrix whose jth column is $\mathbf{w}_j$. This formula implies that $P_{S\leftarrow T}$ is nonsingular and it is helpful in solving some of the exercises in this section.

EXAMPLE 4

Let S and T be the ordered bases for R^3 defined in Example 3. Compute the transition matrix $Q_{T\leftarrow S}$ from the S-basis to the T-basis directly and show that $Q_{T\leftarrow S} = P_{S\leftarrow T}^{-1}$.

Solution $Q_{T\leftarrow S}$ is the matrix whose columns are the solution vectors to the linear systems obtained from the vector equations

$$\begin{aligned} a_1\mathbf{w}_1 + a_2\mathbf{w}_2 + a_3\mathbf{w}_3 &= \mathbf{v}_1 \\ b_1\mathbf{w}_1 + b_2\mathbf{w}_2 + b_3\mathbf{w}_3 &= \mathbf{v}_2 \\ c_1\mathbf{w}_1 + c_2\mathbf{w}_2 + c_3\mathbf{w}_3 &= \mathbf{v}_3. \end{aligned}$$

As in Example 3, we can solve these linear systems simultaneously by transforming the partitioned matrix

$$\left[\begin{array}{ccc|c|c|c} \mathbf{w}_1 & \mathbf{w}_2 & \mathbf{w}_3 & \mathbf{v}_1 & \mathbf{v}_2 & \mathbf{v}_3 \end{array}\right]$$

to reduced row echelon form. That is, we transform

$$\left[\begin{array}{rrr|r|r|r} 6 & 4 & 5 & 2 & 1 & 1 \\ 3 & -1 & 5 & 0 & 2 & 1 \\ 3 & 3 & 2 & 1 & 0 & 1 \end{array}\right]$$

to reduced row echelon form, obtaining (verify)

$$\left[\begin{array}{ccc|c|c|c} 1 & 0 & 0 & \frac{3}{2} & \frac{1}{2} & -\frac{5}{2} \\ 0 & 1 & 0 & -\frac{1}{2} & -\frac{1}{2} & \frac{3}{2} \\ 0 & 0 & 1 & -1 & 0 & 2 \end{array}\right],$$

so

$$Q_{T \leftarrow S} = \begin{bmatrix} \frac{3}{2} & \frac{1}{2} & -\frac{5}{2} \\ -\frac{1}{2} & -\frac{1}{2} & \frac{3}{2} \\ -1 & 0 & 2 \end{bmatrix}.$$

Multiplying $Q_{T \leftarrow S}$ by $P_{S \leftarrow T}$, we find (verify) that $Q_{T \leftarrow S} P_{S \leftarrow T} = I_3$, so we conclude that $Q_{T \leftarrow S} = P_{S \leftarrow T}^{-1}$. ■

EXAMPLE 5

Let V be P_1 and let $S = \{\mathbf{v}_1, \mathbf{v}_2\}$ and $T = \{\mathbf{w}_1, \mathbf{w}_2\}$ be ordered bases for P_1, where

$$\mathbf{v}_1 = t, \quad \mathbf{v}_2 = t - 3, \quad \mathbf{w}_1 = t - 1, \quad \mathbf{w}_2 = t + 1.$$

(a) Compute the transition matrix $P_{S \leftarrow T}$ from the T-basis to the S-basis.

(b) Verify Equation (3) for $\mathbf{v} = 5t + 1$.

(c) Compute the transition matrix $Q_{T \leftarrow S}$ from the S-basis to the T-basis and show that $Q_{T \leftarrow S} = P_{S \leftarrow T}^{-1}$.

Solution (a) To compute $P_{S \leftarrow T}$, we need to solve the vector equations

$$a_1\mathbf{v}_1 + a_2\mathbf{v}_2 = \mathbf{w}_1$$
$$b_1\mathbf{v}_1 + b_2\mathbf{v}_2 = \mathbf{w}_2$$

simultaneously by transforming the resulting partitioned matrix (verify)

$$\left[\begin{array}{cc|c|c} 1 & 1 & 1 & 1 \\ 0 & -3 & -1 & 1 \end{array}\right]$$

to reduced row echelon form. The result is (verify)

$$\left[\begin{array}{cc|c|c} 1 & 0 & \frac{2}{3} & \frac{4}{3} \\ 0 & 1 & \frac{1}{3} & -\frac{1}{3} \end{array}\right],$$

so

$$P_{S \leftarrow T} = \begin{bmatrix} \frac{2}{3} & \frac{4}{3} \\ \frac{1}{3} & -\frac{1}{3} \end{bmatrix}.$$

(b) If $\mathbf{v} = 5t + 1$, then expressing $\mathbf{v}$ in terms of the T-basis, we have (verify)

$$\mathbf{v} = 5t + 1 = 2(t - 1) + 3(t + 1),$$

so $\left[\mathbf{v}\right]_T = \begin{bmatrix} 2 \\ 3 \end{bmatrix}$. Then

$$\left[\mathbf{v}\right]_S = P_{S \leftarrow T}\left[\mathbf{v}\right]_T = \begin{bmatrix} \frac{2}{3} & \frac{4}{3} \\ \frac{1}{3} & -\frac{1}{3} \end{bmatrix}\begin{bmatrix} 2 \\ 3 \end{bmatrix} = \begin{bmatrix} \frac{16}{3} \\ -\frac{1}{3} \end{bmatrix}.$$

Computing $\left[\mathbf{v}\right]_S$ directly, we find that

$$\mathbf{v} = 5t + 1 = \tfrac{16}{3}t - \tfrac{1}{3}(t - 3), \quad \text{so } \left[\mathbf{v}\right]_S = \begin{bmatrix} \frac{16}{3} \\ -\frac{1}{3} \end{bmatrix}.$$

Hence

$$\left[\mathbf{v}\right]_S = P_{S \leftarrow T}\left[\mathbf{v}\right]_T.$$

(c) The transition matrix $Q_{T \leftarrow S}$ from the S-basis to the T-basis is obtained (verify) by transforming the partitioned matrix

$$\left[\begin{array}{rr|r|r} 1 & 1 & 1 & 1 \\ -1 & 1 & 0 & -3 \end{array}\right]$$

to reduced row echelon form, obtaining (verify)

$$\left[\begin{array}{rr|r|r} 1 & 0 & \frac{1}{2} & 2 \\ 0 & 1 & \frac{1}{2} & -1 \end{array}\right].$$

Hence

$$Q_{T \leftarrow S} = \begin{bmatrix} \frac{1}{2} & 2 \\ \frac{1}{2} & -1 \end{bmatrix}.$$

Multiplying $Q_{T \leftarrow S}$ by $P_{S \leftarrow T}$, we find (verify) that $Q_{T \leftarrow S}P_{S \leftarrow T} = I_2$, so we conclude that $Q_{T \leftarrow S} = P_{S \leftarrow T}^{-1}$. ■

2.7 Exercises

In Exercises 1 *through* 6, *compute the coordinate vector of* $\mathbf{v}$ *with respect to the given ordered basis* S *for* V.

1. V is R^2, $S = \left\{\begin{bmatrix} 1 \\ 0 \end{bmatrix}, \begin{bmatrix} 0 \\ 1 \end{bmatrix}\right\}$, $\mathbf{v} = \begin{bmatrix} 3 \\ -2 \end{bmatrix}$.

2. V is R_3, $S = \{\begin{bmatrix} 1 & -1 & 0 \end{bmatrix}, \begin{bmatrix} 0 & 1 & 0 \end{bmatrix}, \begin{bmatrix} 1 & 0 & 2 \end{bmatrix}\}$, $\mathbf{v} = \begin{bmatrix} 2 & -1 & -2 \end{bmatrix}$.

3. V is P_1, $S = \{t + 1, t - 2\}$, $\mathbf{v} = t + 4$.

4. V is P_2, $S = \{t^2 - t + 1, t + 1, t^2 + 1\}$, $\mathbf{v} = 4t^2 - 2t + 3$.

5. V is M_{22}, $S = \left\{\begin{bmatrix} 1 & 0 \\ 0 & 0 \end{bmatrix}, \begin{bmatrix} 0 & 0 \\ 1 & 0 \end{bmatrix}, \begin{bmatrix} 0 & 1 \\ 0 & 0 \end{bmatrix}, \begin{bmatrix} 0 & 0 \\ 0 & 1 \end{bmatrix}\right\}$, $\mathbf{v} = \begin{bmatrix} 1 & 0 \\ -1 & 2 \end{bmatrix}$.

6. V is M_{22}, $S = \left\{ \begin{bmatrix} 1 & -1 \\ 0 & 0 \end{bmatrix}, \begin{bmatrix} 0 & 1 \\ 1 & 0 \end{bmatrix}, \begin{bmatrix} 1 & 0 \\ 0 & -1 \end{bmatrix}, \begin{bmatrix} 1 & 0 \\ -1 & 0 \end{bmatrix} \right\}$, $\mathbf{v} = \begin{bmatrix} 1 & 3 \\ -2 & 2 \end{bmatrix}$.

In Exercises 7 through 12, compute the vector $\mathbf{v}$ *if the coordinate vector* $[\mathbf{v}]_S$ *is given with respect to the ordered basis* S *for* V.

7. V is R^2, $S = \left\{ \begin{bmatrix} 2 \\ 1 \end{bmatrix}, \begin{bmatrix} -1 \\ 1 \end{bmatrix} \right\}$, $[\mathbf{v}]_S = \begin{bmatrix} 1 \\ 2 \end{bmatrix}$.

8. V is R_3, $S = \{\begin{bmatrix} 0 & 1 & -1 \end{bmatrix}, \begin{bmatrix} 1 & 0 & 0 \end{bmatrix}, \begin{bmatrix} 1 & 1 & 1 \end{bmatrix}\}$, $[\mathbf{v}]_S = \begin{bmatrix} -1 \\ 1 \\ 2 \end{bmatrix}$.

9. V is P_1, $S = \{t, 2t - 1\}$, $[\mathbf{v}]_S = \begin{bmatrix} -2 \\ 3 \end{bmatrix}$.

10. V is P_2, $S = \{t^2 + 1, t + 1, t^2 + t\}$, $[\mathbf{v}]_S = \begin{bmatrix} 3 \\ -1 \\ -2 \end{bmatrix}$.

11. V is M_{22}, $S = \left\{ \begin{bmatrix} -1 & 0 \\ 1 & 0 \end{bmatrix}, \begin{bmatrix} 2 & 2 \\ 0 & 1 \end{bmatrix}, \begin{bmatrix} 1 & 2 \\ -1 & 3 \end{bmatrix}, \begin{bmatrix} 0 & 0 \\ 2 & 3 \end{bmatrix} \right\}$, $[\mathbf{v}]_S = \begin{bmatrix} 2 \\ 1 \\ -1 \\ 3 \end{bmatrix}$.

12. V is M_{22}, $S = \left\{ \begin{bmatrix} 1 & -2 \\ 0 & 0 \end{bmatrix}, \begin{bmatrix} -1 & 3 \\ 0 & 1 \end{bmatrix}, \begin{bmatrix} 1 & 0 \\ 0 & 0 \end{bmatrix}, \begin{bmatrix} 0 & -1 \\ 1 & 0 \end{bmatrix} \right\}$, $[\mathbf{v}]_S = \begin{bmatrix} 0 \\ 1 \\ 0 \\ 2 \end{bmatrix}$.

13. Let $S = \left\{ \begin{bmatrix} 1 \\ 2 \end{bmatrix}, \begin{bmatrix} 0 \\ 1 \end{bmatrix} \right\}$ and $T = \left\{ \begin{bmatrix} 1 \\ 1 \end{bmatrix}, \begin{bmatrix} 2 \\ 3 \end{bmatrix} \right\}$ be ordered bases for R^2. Let $\mathbf{v} = \begin{bmatrix} 1 \\ 5 \end{bmatrix}$ and $\mathbf{w} = \begin{bmatrix} 5 \\ 4 \end{bmatrix}$.

(a) Find the coordinate vectors of $\mathbf{v}$ and $\mathbf{w}$ with respect to the basis T.

(b) What is the transition matrix $P_{S \leftarrow T}$ from the T- to the S-basis?

(c) Find the coordinate vectors of $\mathbf{v}$ and $\mathbf{w}$ with respect to S using $P_{S \leftarrow T}$.

(d) Find the coordinate vectors of $\mathbf{v}$ and $\mathbf{w}$ with respect to S directly.

(e) Find the transition matrix $Q_{T \leftarrow S}$ from the S- to the T-basis.

(f) Find the coordinate vectors of $\mathbf{v}$ and $\mathbf{w}$ with respect to T using $Q_{T \leftarrow S}$. Compare the answers with those of (a).

14. Let $S = \left\{ \begin{bmatrix} 1 \\ 0 \\ 1 \end{bmatrix}, \begin{bmatrix} -1 \\ 0 \\ 0 \end{bmatrix}, \begin{bmatrix} 0 \\ 1 \\ 2 \end{bmatrix} \right\}$ and $T = \left\{ \begin{bmatrix} -1 \\ 1 \\ 0 \end{bmatrix}, \begin{bmatrix} 1 \\ 2 \\ -1 \end{bmatrix}, \begin{bmatrix} 0 \\ 1 \\ 0 \end{bmatrix} \right\}$ be ordered bases for R^3. Let $\mathbf{v} = \begin{bmatrix} 1 \\ 3 \\ 8 \end{bmatrix}$ and $\mathbf{w} = \begin{bmatrix} -1 \\ 8 \\ -2 \end{bmatrix}$.

Follow the directions of Exercise 13.

15. Let $S = \{t^2 + 1, t - 2, t + 3\}$ and $T = \{2t^2 + t, t^2 + 3, t\}$ be ordered bases for P_2. Let $\mathbf{v} = 8t^2 - 4t + 6$ and $\mathbf{w} = 7t^2 - t + 9$. Follow the directions of Exercise 13.

16. Let $S = \{\begin{bmatrix} 1 & 1 & 1 \end{bmatrix}, \begin{bmatrix} 1 & 2 & 3 \end{bmatrix}, \begin{bmatrix} 1 & 0 & 1 \end{bmatrix}\}$ and $T = \{\begin{bmatrix} 0 & 1 & 1 \end{bmatrix}, \begin{bmatrix} 1 & 0 & 0 \end{bmatrix}, \begin{bmatrix} 1 & 0 & 1 \end{bmatrix}\}$ be ordered bases for R_3. Let $\mathbf{v} = \begin{bmatrix} -1 & 4 & 5 \end{bmatrix}$ and $\mathbf{w} = \begin{bmatrix} 2 & 0 & -6 \end{bmatrix}$. Follow the directions of Exercise 13.

17. Let

$$S = \left\{ \begin{bmatrix} 1 & 0 \\ 0 & 0 \end{bmatrix}, \begin{bmatrix} 0 & 1 \\ 1 & 0 \end{bmatrix}, \begin{bmatrix} 0 & 2 \\ 0 & 1 \end{bmatrix}, \begin{bmatrix} 0 & 0 \\ 1 & 1 \end{bmatrix} \right\}$$

and

$$T = \left\{ \begin{bmatrix} 1 & 1 \\ 0 & 0 \end{bmatrix}, \begin{bmatrix} 0 & 0 \\ 1 & 0 \end{bmatrix}, \begin{bmatrix} 0 & 0 \\ 0 & 1 \end{bmatrix}, \begin{bmatrix} 1 & 0 \\ 0 & 0 \end{bmatrix} \right\}$$

be ordered bases for M_{22}. Let

$$\mathbf{v} = \begin{bmatrix} 1 & 1 \\ 1 & 1 \end{bmatrix} \quad \text{and} \quad \mathbf{w} = \begin{bmatrix} 1 & 2 \\ -2 & 1 \end{bmatrix}.$$

Follow the directions for Exercise 13.

18. Let $S = \{\begin{bmatrix} 1 & -1 \end{bmatrix}, \begin{bmatrix} 2 & 1 \end{bmatrix}\}$ and $T = \{\begin{bmatrix} 3 & 0 \end{bmatrix}, \begin{bmatrix} 4 & -1 \end{bmatrix}\}$ be ordered bases for R_2. If $\mathbf{v}$ is in R_2 and $[\mathbf{v}]_T = \begin{bmatrix} 1 \\ 2 \end{bmatrix}$, determine $[\mathbf{v}]_S$.

19. Let $S = \{t + 1, t - 2\}$ and $T = \{t - 5, t - 2\}$ be ordered bases for P_1. If $\mathbf{v}$ is in P_1 and $[\mathbf{v}]_T = \begin{bmatrix} -1 \\ 3 \end{bmatrix}$, determine $[\mathbf{v}]_S$.

20. Let $S = \{[-1 \quad 2 \quad 1], [0 \quad 1 \quad 1], [-2 \quad 2 \quad 1]\}$ and $T = \{[-1 \quad 1 \quad 0], [0 \quad 1 \quad 0], [0 \quad 1 \quad 1]\}$ be ordered bases for R_3. If $\mathbf{v}$ is in R_3 and $\left[\mathbf{v}\right]_S = \begin{bmatrix} 2 \\ 0 \\ 1 \end{bmatrix}$, determine $\left[\mathbf{v}\right]_T$.

21. If the vector $\mathbf{v}$ in P_2 has the coordinate vector $\begin{bmatrix} 1 \\ 2 \\ 3 \end{bmatrix}$ with respect to the ordered basis $T = \{t^2, t-1, 1\}$, what is $\left[\mathbf{v}\right]_S$ if $S = \{t^2+t+1, t+1, 1\}$?

22. Let $S = \{\mathbf{v}_1, \mathbf{v}_2, \mathbf{v}_3\}$ and $T = \{\mathbf{w}_1, \mathbf{w}_2, \mathbf{w}_3\}$ be ordered bases for R^3, where

$$\mathbf{v}_1 = \begin{bmatrix} 1 \\ 0 \\ 1 \end{bmatrix}, \quad \mathbf{v}_2 = \begin{bmatrix} 1 \\ 1 \\ 0 \end{bmatrix}, \quad \mathbf{v}_3 = \begin{bmatrix} 0 \\ 0 \\ 1 \end{bmatrix}.$$

Suppose that the transition matrix from T to S is

$$P_{S \leftarrow T} = \begin{bmatrix} 1 & 1 & 2 \\ 2 & 1 & 1 \\ -1 & -1 & 1 \end{bmatrix}.$$

Determine T.

23. Let $S = \{\mathbf{v}_1, \mathbf{v}_2\}$ and $T = \{\mathbf{w}_1, \mathbf{w}_2\}$ be ordered bases for P_1, where

$$\mathbf{w}_1 = t, \quad \mathbf{w}_2 = t - 1.$$

If the transition matrix from S to T is $\begin{bmatrix} 2 & 3 \\ -1 & 2 \end{bmatrix}$, determine S.

24. Let $S = \{\mathbf{v}_1, \mathbf{v}_2\}$ and $T = \{\mathbf{w}_1, \mathbf{w}_2\}$ be ordered bases for R^2, where

$$\mathbf{v}_1 = \begin{bmatrix} 1 \\ 2 \end{bmatrix}, \quad \mathbf{v}_2 = \begin{bmatrix} 0 \\ 1 \end{bmatrix}.$$

If the transition matrix from S to T is $\begin{bmatrix} 2 & 1 \\ 1 & 1 \end{bmatrix}$, determine T.

25. Let $S = \{\mathbf{v}_1, \mathbf{v}_2\}$ and $T = \{\mathbf{w}_1, \mathbf{w}_2\}$ be ordered bases for P_1, where

$$\mathbf{w}_1 = t - 1, \quad \mathbf{w}_2 = t + 1.$$

If the transition matrix from T to S is $\begin{bmatrix} 1 & 2 \\ 2 & 3 \end{bmatrix}$, determine S.

26. Prove parts (a) and (c) of Theorem 2.14.

27. Let $L: V \to W$ be an isomorphism of vector space V onto vector space W.

(a) Prove that $L(\mathbf{0}_V) = \mathbf{0}_W$.

(b) Show that $L(\mathbf{v} - \mathbf{w}) = L(\mathbf{v}) - L(\mathbf{w})$.

(c) Show that

$$L(a_1\mathbf{v}_1 + a_2\mathbf{v}_2 + \cdots + a_k\mathbf{v}_k) = a_1 L(\mathbf{v}_1) + a_2 L(\mathbf{v}_2) + \cdots + a_k L(\mathbf{v}_k).$$

28. Prove that R^n and R^m are isomorphic if and only if $n = m$.

29. Find an isomorphism $L: R_n \to R^n$.

30. Find an isomorphism $L: P_2 \to R^3$. More generally, show that P_n and R^{n+1} are isomorphic.

31. (a) Show that M_{22} is isomorphic to R^4.

(b) What is $\dim M_{22}$?

32. Let V be the subspace of the vector space of all real-valued continuous functions that has basis $S = \{e^t, e^{-t}\}$. Show that V and R^2 are isomorphic.

33. Let V be the subspace of the vector space of all real-valued functions that is *spanned* by the set $S = \{\cos^2 t, \sin^2 t, \cos 2t\}$. Show that V and R_2 are isomorphic.

34. Let V and W be isomorphic vector spaces. Prove that if V_1 is a subspace of V, then V_1 is isomorphic to a subspace W_1 of W.

35. Let $S = \{\mathbf{v}_1, \mathbf{v}_2, \ldots, \mathbf{v}_n\}$ be an ordered basis for the n-dimensional vector space V, and let $\mathbf{v}$ and $\mathbf{w}$ be two vectors in V. Show that $\mathbf{v} = \mathbf{w}$ if and only if $\left[\mathbf{v}\right]_S = \left[\mathbf{w}\right]_S$.

36. Show that if S is an ordered basis for an n-dimensional vector space V, $\mathbf{v}$ and $\mathbf{w}$ are vectors in V, and c is a scalar, then

$$\left[\mathbf{v} + \mathbf{w}\right]_S = \left[\mathbf{v}\right]_S + \left[\mathbf{w}\right]_S$$

and

$$\left[c\mathbf{v}\right]_S = c\left[\mathbf{v}\right]_S.$$

In Exercises 37 through 39, let $S = \{\mathbf{v}_1, \mathbf{v}_2, \ldots, \mathbf{v}_n\}$ *and* $T = \{\mathbf{w}_1, \mathbf{w}_2, \ldots, \mathbf{w}_n\}$ *be ordered bases for the vector space* R^n.

37. Let M_S be the $n \times n$ matrix whose jth column is $\mathbf{v}_j$ and let M_T be the $n \times n$ matrix whose jth column is $\mathbf{w}_j$. Prove that M_S and M_T are nonsingular. (*Hint*: Consider the homogeneous systems $M_S\mathbf{x} = \mathbf{0}$ and $M_T\mathbf{x} = \mathbf{0}$.)

38. If $\mathbf{v}$ is a vector in V, show that

$$\mathbf{v} = M_S\left[\mathbf{v}\right]_S \quad \text{and} \quad \mathbf{v} = M_T\left[\mathbf{v}\right]_T.$$

39. (a) Use Equation (3) and Exercises 37 and 38 to show that

$$P_{S\leftarrow T} = M_S^{-1} M_T.$$

(b) Show that $P_{S\leftarrow T}$ is nonsingular.

(c) Verify the result in part (a) for Example 3.

40. Let S be an ordered basis for n-dimensional vector space V. Show that if $\{\mathbf{w}_1, \mathbf{w}_2, \ldots, \mathbf{w}_k\}$ is a linearly independent set of vectors in V, then

$$\left\{\left[\mathbf{w}_1\right]_S, \left[\mathbf{w}_2\right]_S, \ldots, \left[\mathbf{w}_k\right]_S\right\}$$

is a linearly independent set of vectors in R^n.

41. Let $S = \{\mathbf{v}_1, \mathbf{v}_2, \ldots, \mathbf{v}_n\}$ be an ordered basis for an n-dimensional vector space V. Show that

$$\left\{\left[\mathbf{v}_1\right]_S, \left[\mathbf{v}_2\right]_S, \ldots, \left[\mathbf{v}_n\right]_S\right\}$$

is an ordered basis for R^n.

2.8 RANK OF A MATRIX

In this section we obtain another effective method for finding a basis for a vector space V spanned by a given set of vectors $S = \{\mathbf{v}_1, \mathbf{v}_2, \ldots, \mathbf{v}_k\}$. In Section 2.5 we developed a technique for choosing a basis for V that is a subset of S (Theorem 2.8). The method to be developed in this section produces a basis for V that is not guaranteed to be a subset of S. We shall also attach a unique number to a matrix A that we later show gives us information about the dimension of the solution space of a homogeneous system with coefficient matrix A.

Definition 2.14

Let

$$A = \begin{bmatrix} a_{11} & a_{12} & \cdots & a_{1n} \\ a_{21} & a_{22} & \cdots & a_{2n} \\ \vdots & \vdots & & \vdots \\ a_{m1} & a_{m2} & \cdots & a_{mn} \end{bmatrix}$$

be an $m \times n$ matrix. The rows of A, considered as vectors in R_n, span a subspace of R_n, called the **row space** of A. Similarly, the columns of A, considered as vectors in R^m, span a subspace of R^m called the **column space** of A. ▲

Theorem 2.16

If A and B are two $m \times n$ row (column) equivalent matrices, then the row (column) spaces of A and B are equal.

Proof

If A and B are row equivalent, then the rows of B are obtained from the rows of A by a finite number of the three elementary row operations. Thus each row of B is a linear combination of the rows of A. Hence the row space of B is contained in the row space of A. If we apply the inverse elementary row operations to B, we get A, so the row space of A is contained in the row space of B. Hence the row spaces of A and B are identical. The proof for the column spaces is similar. ●

We can use this theorem to find a basis for a subspace spanned by a given set of vectors. We illustrate this method with the following example.

EXAMPLE 1

Find a basis for the subspace V of R_5 that is spanned by $S = \{\mathbf{v}_1, \mathbf{v}_2, \mathbf{v}_3, \mathbf{v}_4\}$, where

$$\mathbf{v}_1 = \begin{bmatrix} 1 & -2 & 0 & 3 & -4 \end{bmatrix}, \quad \mathbf{v}_2 = \begin{bmatrix} 3 & 2 & 8 & 1 & 4 \end{bmatrix},$$
$$\mathbf{v}_3 = \begin{bmatrix} 2 & 3 & 7 & 2 & 3 \end{bmatrix}, \quad \text{and} \quad \mathbf{v}_4 = \begin{bmatrix} -1 & 2 & 0 & 4 & -3 \end{bmatrix}.$$

Solution Note that V is the row space of the matrix A whose rows are the given vectors.

$$A = \begin{bmatrix} 1 & -2 & 0 & 3 & -4 \\ 3 & 2 & 8 & 1 & 4 \\ 2 & 3 & 7 & 2 & 3 \\ -1 & 2 & 0 & 4 & -3 \end{bmatrix}.$$

Using elementary row operations, we find that A is row equivalent to the matrix (verify)

$$B = \begin{bmatrix} 1 & 0 & 2 & 0 & 1 \\ 0 & 1 & 1 & 0 & 1 \\ 0 & 0 & 0 & 1 & -1 \\ 0 & 0 & 0 & 0 & 0 \end{bmatrix},$$

which is in reduced row echelon form. The row spaces of A and B are identical and a basis for the row space of B consists of

$$\mathbf{w}_1 = \begin{bmatrix} 1 & 0 & 2 & 0 & 1 \end{bmatrix}, \quad \mathbf{w}_2 = \begin{bmatrix} 0 & 1 & 1 & 0 & 1 \end{bmatrix},$$
$$\text{and} \quad \mathbf{w}_3 = \begin{bmatrix} 0 & 0 & 0 & 1 & -1 \end{bmatrix}$$

(see Exercise 25 in Section 2.4). Hence $\{\mathbf{w}_1, \mathbf{w}_2, \mathbf{w}_3\}$ is also a basis for V. ■

It is not necessary to find a matrix B in reduced row echelon form that is row equivalent to A. All that is required is that we have a matrix B that is row equivalent to A and such that we can easily obtain a basis for the row space of B. Often one does not have to reduce A all the way to reduced row echelon form to get such a matrix B. We can show that if A is row equivalent to a matrix B that is in row echelon form, then the nonzero rows of B form a basis for the row space of A.

Of course, the basis that has been obtained by the procedure used in Example 1 produced a basis that may not be a subset of the given spanning set. The method used in Example 5 of Section 2.5 always gives a basis that is a subset of the spanning set. However, the basis for a subspace V of R^n that is obtained by the procedure used in Example 1 is analogous, in its simplicity, to the natural basis for R^n. Thus if

$$\mathbf{v} = \begin{bmatrix} a_1 \\ a_2 \\ \vdots \\ a_n \end{bmatrix}$$

is a vector in V and $\{\mathbf{v}_1, \mathbf{v}_2, \ldots, \mathbf{v}_k\}$ is a basis for V obtained by the method of Example 1 where the leading 1's occur in columns $j_1, j_2, \ldots, j_k$, then it can be shown (Exercise 36) that

$$\mathbf{v} = a_{j_1}\mathbf{v}_1 + a_{j_2}\mathbf{v}_2 + \cdots + a_{j_k}\mathbf{v}_k.$$

EXAMPLE 2

Let V be the subspace of Example 1. Given that the vector

$$\mathbf{v} = \begin{bmatrix} 5 & 4 & 14 & 6 & 3 \end{bmatrix}$$

is in V, write $\mathbf{v}$ as a linear combination of the basis determined in Example 1.

Solution We have $j_1 = 1$, $j_2 = 2$, and $j_3 = 4$, so $\mathbf{v} = 5\mathbf{w}_1 + 4\mathbf{w}_2 + 6\mathbf{w}_3$. ■

Remark The following example illustrates how to use the procedure given in Example 1 to find a basis for a subspace of a vector space that is not R^n or R_n.

EXAMPLE 3

Let V be the subspace of P_4 spanned by $S = \{\mathbf{v}_1, \mathbf{v}_2, \mathbf{v}_3, \mathbf{v}_4\}$, where $\mathbf{v}_1 = t^4 + t^2 + 2t + 1$, $\mathbf{v}_2 = t^4 + t^2 + 2t + 2$, $\mathbf{v}_3 = 2t^4 + t^3 + t + 2$, and $\mathbf{v}_4 = t^4 + t^3 - t^2 - t$. Find a basis for V.

Solution Since P_4 is isomorphic to R_5 under the isomorphism L defined by

$$L(at^4 + bt^3 + ct^2 + dt + e) = \begin{bmatrix} a & b & c & d & e \end{bmatrix},$$

then $L(V)$ is isomorphic to a subspace W of R_5 (see Exercise 34 of Section 2.7). The subspace W is spanned by $\{L(\mathbf{v}_1), L(\mathbf{v}_2), L(\mathbf{v}_3), L(\mathbf{v}_4)\}$, as we have seen in the proof of Theorem 2.15. We now find a basis for W by proceeding as in Example 1. Thus W is the row space of the matrix

$$A = \begin{bmatrix} 1 & 0 & 1 & 2 & 1 \\ 1 & 0 & 1 & 2 & 2 \\ 2 & 1 & 0 & 1 & 2 \\ 1 & 1 & -1 & -1 & 0 \end{bmatrix},$$

and A is row equivalent to (verify)

$$B = \begin{bmatrix} 1 & 0 & 1 & 2 & 0 \\ 0 & 1 & -2 & -3 & 0 \\ 0 & 0 & 0 & 0 & 1 \\ 0 & 0 & 0 & 0 & 0 \end{bmatrix}.$$

A basis for W is therefore $T = \{\mathbf{w}_1, \mathbf{w}_2, \mathbf{w}_3\}$, where $\mathbf{w}_1 = \begin{bmatrix} 1 & 0 & 1 & 2 & 0 \end{bmatrix}$, $\mathbf{w}_2 = \begin{bmatrix} 0 & 1 & -2 & -3 & 0 \end{bmatrix}$, and $\mathbf{w}_3 = \begin{bmatrix} 0 & 0 & 0 & 0 & 1 \end{bmatrix}$. A basis for V is then

$$\left\{L^{-1}(\mathbf{w}_1), L^{-1}(\mathbf{w}_2), L^{-1}(\mathbf{w}_3)\right\} = \left\{t^4 + t^2 + 2t, t^3 - 2t^2 - 3t, 1\right\}.$$ ■

Definition 2.15

The dimension of the row (column) space of A is called the **row (column) rank** of A. ▲

If A and B are row equivalent, then row rank A = row rank B and if A and B are column equivalent, then column rank A = column rank B. Therefore, if we start out with an $m \times n$ matrix A and find a matrix B in reduced row echelon form that is row equivalent to A, then A and B have equal row ranks. But the row rank of B is clearly the number of nonzero rows. Thus we have a good method for finding the row rank of a given matrix A.

EXAMPLE 4

Find a basis for the row space of the matrix A defined in the solution of Example 1 that contains only row vectors from A. Also compute the row rank of A.

Solution Using the procedure in the alternative proof of Theorem 2.7, we form the equation

$$a_1\begin{bmatrix}1 & -2 & 0 & 3 & -4\end{bmatrix} + a_2\begin{bmatrix}3 & 2 & 8 & 1 & 4\end{bmatrix} + a_3\begin{bmatrix}2 & 3 & 7 & 2 & 3\end{bmatrix} + a_4\begin{bmatrix}-1 & 2 & 0 & 4 & -3\end{bmatrix} = \begin{bmatrix}0 & 0 & 0 & 0 & 0\end{bmatrix}$$

$$\left[\begin{array}{rrrr|r} 1 & 3 & 2 & -1 & 0 \\ -2 & 2 & 3 & 2 & 0 \\ 0 & 8 & 7 & 0 & 0 \\ 3 & 1 & 2 & 4 & 0 \\ -4 & 4 & 3 & -3 & 0 \end{array}\right] = \left[\begin{array}{c|c} A^T & \mathbf{0} \end{array}\right]; \tag{1}$$

that is, the coefficient matrix is A^T. Transforming the augmented matrix $\left[\begin{array}{c|c} A^T & \mathbf{0} \end{array}\right]$ in (1) to reduced row echelon form, we obtain (verify)

$$\left[\begin{array}{cccc|c} 1 & 0 & \frac{11}{24} & 0 & 0 \\ 0 & 1 & 0 & -\frac{49}{24} & 0 \\ 0 & 0 & 1 & \frac{7}{3} & 0 \\ 0 & 0 & 0 & 0 & 0 \\ 0 & 0 & 0 & 0 & 0 \end{array}\right]. \tag{2}$$

Since the leading 1's in (2) occur in columns 1, 2, and 3, we conclude that the first three rows of A form a basis for the row space of A. That is,

$$\left\{\begin{bmatrix}1 & -2 & 0 & 3 & -4\end{bmatrix}, \begin{bmatrix}3 & 2 & 8 & 1 & 4\end{bmatrix}, \begin{bmatrix}2 & 3 & 7 & 2 & 3\end{bmatrix}\right\}$$

is a basis for the row space of A. The row rank of A is 3. ■

EXAMPLE 5

Find a basis for the column space of the matrix A defined in the solution of Example 1 and compute the column rank of A.

Solution 1 Writing the columns of A as row vectors, we obtain the matrix A^T, which when transformed to reduced row echelon form is (as we saw in Example 4)

$$\begin{bmatrix} 1 & 0 & 0 & \frac{11}{24} \\ 0 & 1 & 0 & -\frac{49}{24} \\ 0 & 0 & 1 & \frac{7}{3} \\ 0 & 0 & 0 & 0 \\ 0 & 0 & 0 & 0 \end{bmatrix}.$$

Thus the vectors $\begin{bmatrix} 1 & 0 & 0 & \frac{11}{24} \end{bmatrix}$, $\begin{bmatrix} 0 & 1 & 0 & -\frac{49}{24} \end{bmatrix}$, and $\begin{bmatrix} 0 & 0 & 1 & \frac{7}{3} \end{bmatrix}$ form a basis for the row space of A^T. Hence the vectors

$$\begin{bmatrix} 1 \\ 0 \\ 0 \\ \frac{11}{24} \end{bmatrix}, \quad \begin{bmatrix} 0 \\ 1 \\ 0 \\ -\frac{49}{24} \end{bmatrix}, \quad \text{and} \quad \begin{bmatrix} 0 \\ 0 \\ 1 \\ \frac{7}{3} \end{bmatrix}$$

form a basis for the column space of A and we conclude that the column rank of A is 3.

Solution 2 If we want to find a basis for the column space of A that contains only the column vectors from A, we follow the procedure developed in the proof of Theorem 2.8, forming the equation

$$a_1 \begin{bmatrix} 1 \\ 3 \\ 2 \\ -1 \end{bmatrix} + a_2 \begin{bmatrix} -2 \\ 2 \\ 3 \\ 2 \end{bmatrix} + a_3 \begin{bmatrix} 0 \\ 8 \\ 7 \\ 0 \end{bmatrix} + a_4 \begin{bmatrix} 3 \\ 1 \\ 2 \\ 4 \end{bmatrix} + a_5 \begin{bmatrix} -4 \\ 4 \\ 3 \\ -3 \end{bmatrix} = \begin{bmatrix} 0 \\ 0 \\ 0 \\ 0 \end{bmatrix}$$

whose augmented matrix is $\begin{bmatrix} A & \vdots & \mathbf{0} \end{bmatrix}$. Transforming this matrix to reduced row echelon form, we obtain (as in Example 1)

$$\left[\begin{array}{ccccc|c} 1 & 0 & 2 & 0 & 1 & 0 \\ 0 & 1 & 1 & 0 & 1 & 0 \\ 0 & 0 & 0 & 1 & -1 & 0 \\ 0 & 0 & 0 & 0 & 0 & 0 \end{array}\right].$$

Since the leading 1's occur in columns 1, 2, and 4, we conclude that the first, second, and fourth columns of A form a basis for the column space of A. That is,

$$\left\{ \begin{bmatrix} 1 \\ 3 \\ 2 \\ -1 \end{bmatrix}, \begin{bmatrix} -2 \\ 2 \\ 3 \\ 2 \end{bmatrix}, \begin{bmatrix} 3 \\ 1 \\ 2 \\ 4 \end{bmatrix} \right\}$$

is a basis for the column space of A. The column rank of A is 3. ■

We may also conclude that if A is an $m \times n$ matrix and P is a nonsingular $m \times m$ matrix, then row rank $(PA) =$ row rank A, for A and PA are row equivalent (Exercise 23 in Section 1.6). Similarly, if Q is a nonsingular $n \times n$ matrix, then column rank $(AQ) =$ column rank A. Moreover, since dimension $R_n = n$, we see that row rank $A \leq n$. Also, since the row space of A is spanned by m vectors, row rank $A \leq m$. Thus row rank $A \leq$ minimum $\{m, n\}$.

In Examples 4 and 5 we observe that the row and column ranks of A are equal. This is always true and is a very important result in linear algebra. We now turn to the proof of this theorem.

Theorem 2.17

The row rank and column rank of the $m \times n$ matrix $A = \left[a_{ij}\right]$ are equal.

Proof

Let $\mathbf{v}_1, \mathbf{v}_2, \ldots, \mathbf{v}_m$ be the row vectors of A, where

$$\mathbf{v}_i = \begin{bmatrix} a_{i1} & a_{i2} & \cdots & a_{in} \end{bmatrix} \qquad i = 1, 2, \ldots, m.$$

Let row rank $A = r$ and let the set of vectors $\{\mathbf{w}_1, \mathbf{w}_2, \ldots, \mathbf{w}_r\}$ form a basis for the row space of A, where $\mathbf{w}_i = \begin{bmatrix} b_{i1} & b_{i2} & \cdots & b_{in} \end{bmatrix}$ for $i = 1, 2, \ldots, r$. Now each of the row vectors is a linear combination of $\mathbf{w}_1, \mathbf{w}_2, \ldots, \mathbf{w}_r$:

$$\begin{aligned} \mathbf{v}_1 &= c_{11}\mathbf{w}_1 + c_{12}\mathbf{w}_2 + \cdots + c_{1r}\mathbf{w}_r \\ \mathbf{v}_2 &= c_{21}\mathbf{w}_1 + c_{22}\mathbf{w}_2 + \cdots + c_{2r}\mathbf{w}_r \\ &\vdots \\ \mathbf{v}_m &= c_{m1}\mathbf{w}_1 + c_{m2}\mathbf{w}_2 + \cdots + c_{mr}\mathbf{w}_r, \end{aligned}$$

where the c_{ij} are uniquely determined real numbers. Recalling that two matrices are equal if and only if the corresponding entries are equal, we equate the entries of these vector equations to get

$$\begin{aligned} a_{1j} &= c_{11}b_{1j} + c_{12}b_{2j} + \cdots + c_{1r}b_{rj} \\ a_{2j} &= c_{21}b_{1j} + c_{22}b_{2j} + \cdots + c_{2r}b_{rj} \\ &\vdots \\ a_{mj} &= c_{m1}b_{1j} + c_{m2}b_{2j} + \cdots + c_{mr}b_{rj} \end{aligned}$$

or

$$\begin{bmatrix} a_{1j} \\ a_{2j} \\ \vdots \\ a_{mj} \end{bmatrix} = b_{1j} \begin{bmatrix} c_{11} \\ c_{21} \\ \vdots \\ c_{m1} \end{bmatrix} + b_{2j} \begin{bmatrix} c_{12} \\ c_{22} \\ \vdots \\ c_{m2} \end{bmatrix} + \cdots + b_{rj} \begin{bmatrix} c_{1r} \\ c_{2r} \\ \vdots \\ c_{mr} \end{bmatrix}$$

for $j = 1, 2, \ldots, n$.

Since every column of A is a linear combination of r vectors, the dimension of the column space of A is at most r, or column rank $A \leq r =$ row rank A. Similarly, we get row rank $A \leq$ column rank A. Hence the row and column ranks of A are equal.

Alternative Proof: Let $\mathbf{x}_1, \mathbf{x}_2, \ldots, \mathbf{x}_n$ denote the columns of A. To determine the dimension of the column space of A, we use the procedure in the alternate proof of Theorem 2.8. Thus we consider the equation

$$c_1\mathbf{x}_1 + c_2\mathbf{x}_2 + \cdots + c_n\mathbf{x}_n = \mathbf{0}.$$

We now transform the augmented matrix, $\left[A \mid \mathbf{0}\right]$, of this homogeneous system to reduced row echelon form. The vectors corresponding to the columns containing the leading 1's form a basis for the column space of A. Thus the column rank of A is the number of leading 1's. But this number is also the number of nonzero rows in the reduced row echelon form matrix that is row equivalent to A, so it is the row rank of A. Thus row rank $A =$ column rank A. ●

Since the row and column ranks of a matrix are equal, we now merely refer to the **rank** of a matrix. Note that rank $I_n = n$. Theorem 1.21 states that A is equivalent to B if and only if there exist nonsingular matrices P and Q such that $B = PAQ$. If A is equivalent to B, then rank $A =$ rank B, for rank $B =$ rank$(PAQ) =$ rank$(PA) =$ rank A.

We also recall from Section 1.7 that if A is an $m \times n$ matrix, then A is equivalent to a matrix $C = \begin{bmatrix} I_r & O \\ O & O \end{bmatrix}$. Now rank $A =$ rank $C = r$. We use these facts to establish the result that if A and B are $m \times n$ matrices of equal rank, then A and B are equivalent. Thus let rank $A = r =$ rank B. Then there exist nonsingular matrices P_1, Q_1, P_2, and Q_2 such that

$$P_1AQ_1 = \begin{bmatrix} I_r & O \\ O & O \end{bmatrix} = P_2BQ_2.$$

Then $P_2^{-1}P_1AQ_1Q_2^{-1} = B$. Letting $P = P_2^{-1}P_1$ and $Q = Q_1Q_2^{-1}$, we find that P and Q are nonsingular and $B = PAQ$. Hence A and B are equivalent.

If A is an $m \times n$ matrix, we have defined (see Section 2.6) the nullity of A as the dimension of the null space of A, that is, the dimension of the solution space of $A\mathbf{x} = \mathbf{0}$. If A is transformed to a matrix B in reduced row echelon form having r nonzero rows, then we know that the dimension of the solution space of $A\mathbf{x} = \mathbf{0}$ is $n - r$. Since r is also the rank of A, we have obtained a fundamental relationship between the rank and nullity of A, which we state in the following theorem.

Theorem 2.18

If A is an $m \times n$ matrix, then rank A + nullity $A = n$.

EXAMPLE 6

Let

$$A = \begin{bmatrix} 1 & 1 & 4 & 1 & 2 \\ 0 & 1 & 2 & 1 & 1 \\ 0 & 0 & 0 & 1 & 2 \\ 1 & -1 & 0 & 0 & 2 \\ 2 & 1 & 6 & 0 & 1 \end{bmatrix},$$

which was defined in Example 1 of Section 2.6. When A is transformed to reduced row echelon form, we obtain

$$\begin{bmatrix} 1 & 0 & 2 & 0 & 1 \\ 0 & 1 & 2 & 0 & -1 \\ 0 & 0 & 0 & 1 & 2 \\ 0 & 0 & 0 & 0 & 0 \\ 0 & 0 & 0 & 0 & 0 \end{bmatrix}.$$

Then rank $A = 3$ and nullity $A = 2$. This agrees with the result obtained in solving Example 1 of Section 2.6, where we found that the dimension of the solution space of $A\mathbf{x} = \mathbf{0}$ is 2. ■

The following example will be used to illustrate geometrically some of the ideas discussed above.

EXAMPLE 7

Let

$$A = \begin{bmatrix} 3 & -1 & 2 \\ 2 & 1 & 3 \\ 7 & 1 & 8 \end{bmatrix}.$$

Transforming A to reduced row echelon form we obtain (verify)

$$\begin{bmatrix} 1 & 0 & 1 \\ 0 & 1 & 1 \\ 0 & 0 & 0 \end{bmatrix},$$

so we conclude the following:

- rank $A = 2$
- dimension of row space of $A = 2$, so the row space of A is a two-dimensional subspace of R^3, that is, a plane passing through the origin.

From the reduced row echelon form matrix that A has been transformed to, we see that every solution to the homogeneous system $A\mathbf{x} = \mathbf{0}$ is of the form

$$\mathbf{x} = \begin{bmatrix} -r \\ r \\ r \end{bmatrix},$$

where r is an arbitrary constant (verify), so the solution space of this homogeneous system, or the null space of A, is a line passing through the origin. Moreover, the dimension of the null space of A, or the nullity of A, is 1. Thus, Theorem 2.18 has been verified.

Of course, we already know that the dimension of the column space of A is also 2. We could also obtain this result by finding a basis consisting of two vectors for the column space of A. Thus, the column space of A is also a two-dimensional subspace of R^3, that is, a plane passing through the origin. These results are illustrated in Figure 2.28. ■

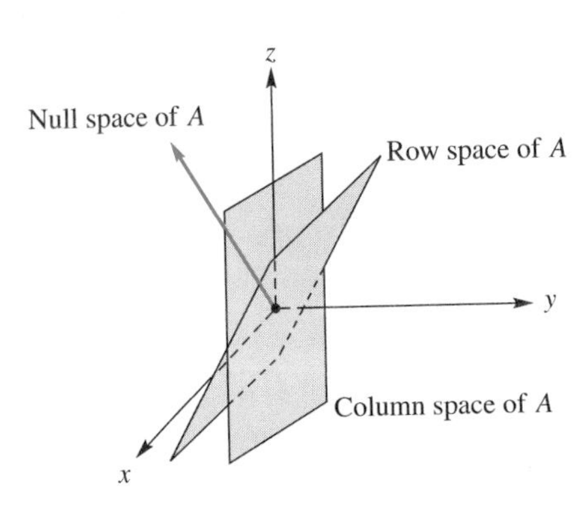

FIGURE 2.28

Rank and Singularity

The rank of a square matrix can be used to determine whether the matrix is singular or nonsingular. We first prove the following theorem.

Theorem 2.19

If A is an $n \times n$ matrix, then rank $A = n$ if and only if A is row equivalent to I_n.

Proof

If rank $A = n$, then A is row equivalent to a matrix B in reduced row echelon form, and rank $B = n$. Since rank $B = n$, we conclude that B has no zero rows, and this implies (by Exercise 3 of Section 1.5) that $B = I_n$. Hence A is row equivalent to I_n.

Conversely, if A is row equivalent to I_n, then rank A = rank $I_n = n$. ●

Corollary 2.7

A is nonsingular if and only if rank $A = n$.

Proof

This follows from Theorem 2.19 and Corollary 1.3. ●

From a practical point of view, this result is not too useful, since most of the time we want to know not only whether A is nonsingular but also its inverse. The method developed in Chapter 1 enables us to find A^{-1}, if it exists, and tells us if it does not exist. Thus we do not have to learn first if A^{-1} exists and then go through another procedure to obtain it.

Corollary 2.8

The homogeneous system $A\mathbf{x} = \mathbf{0}$, where A is $n \times n$, has a nontrivial solution if and only if rank $A < n$.

Proof

This follows from Corollary 2.7 and from the fact that $A\mathbf{x} = \mathbf{0}$ has a nontrivial solution if and only if A is singular (Theorem 1.17). ●

Corollary 2.9

Let A be an $n \times n$ matrix. The linear system $A\mathbf{x} = \mathbf{b}$ has a unique solution for every $n \times 1$ matrix $\mathbf{b}$ if and only if rank $A = n$.

Proof

Exercise. ●

Let $S = \{\mathbf{v}_1, \mathbf{v}_2, \ldots, \mathbf{v}_n\}$ be a set of n vectors in R_n, and let A be the matrix whose jth row is $\mathbf{v}_j$. It is not difficult to show (Exercise 31) that S is linearly independent if and only if rank $A = n$. Similarly, let $S = \{\mathbf{v}_1, \mathbf{v}_2, \ldots, \mathbf{v}_n\}$ be a set of n vectors in R^n, and let A be the matrix whose jth column is $\mathbf{v}_j$. It can then be shown (Exercise 32) that S is linearly independent if and only if rank $A = n$.

Applications of Rank to the Linear System $A\mathbf{x} = \mathbf{b}$

In Corollary 2.8 we have seen that the rank of A provides us with information about the existence of a nontrivial solution to the homogeneous system $A\mathbf{x} = \mathbf{0}$. We shall now obtain some results that use the rank of A to provide information about the solutions to the linear system $A\mathbf{x} = \mathbf{b}$, where $\mathbf{b}$ is an arbitrary $n \times 1$ matrix. When $\mathbf{b} \neq \mathbf{0}$, the linear system is said to be **nonhomogeneous**.

Theorem 2.20

The linear system $A\mathbf{x} = \mathbf{b}$ has a solution if and only if rank A = rank $\left[A \mid \mathbf{b} \right]$; that is, if and only if the ranks of the coefficient and augmented matrices are equal.

Proof

First, observe that if $A = \begin{bmatrix} a_{ij} \end{bmatrix}$ is $m \times n$, then the given linear system may be written as

$$x_1 \begin{bmatrix} a_{11} \\ a_{21} \\ \vdots \\ a_{m1} \end{bmatrix} + x_2 \begin{bmatrix} a_{12} \\ a_{22} \\ \vdots \\ a_{m2} \end{bmatrix} + \cdots + x_n \begin{bmatrix} a_{1n} \\ a_{2n} \\ \vdots \\ a_{mn} \end{bmatrix} = \begin{bmatrix} b_1 \\ b_2 \\ \vdots \\ b_m \end{bmatrix}. \tag{3}$$

Suppose now that $A\mathbf{x} = \mathbf{b}$ has a solution. Then there exist values of $x_1, x_2, \ldots, x_n$ that satisfy Equation (3). Thus $\mathbf{b}$ is a linear combination of the columns of A and so belongs to the column space of A. Hence rank A = rank $\begin{bmatrix} A \mid \mathbf{b} \end{bmatrix}$.

Conversely, suppose that rank A = rank $\begin{bmatrix} A \mid \mathbf{b} \end{bmatrix}$. Then $\mathbf{b}$ is in the column space of A, which means that we can find values of $x_1, x_2, \ldots, x_n$ that satisfy Equation (3). Hence $A\mathbf{x} = \mathbf{b}$ has a solution. ●

This result, although of interest, is not of great computational value, since we usually are interested in finding a solution rather than in merely knowing whether a solution exists.

EXAMPLE 8

Consider the linear system

$$\begin{bmatrix} 2 & 1 & 3 \\ 1 & -2 & 2 \\ 0 & 1 & 3 \end{bmatrix} \begin{bmatrix} x_1 \\ x_2 \\ x_3 \end{bmatrix} = \begin{bmatrix} 1 \\ 2 \\ 3 \end{bmatrix}.$$

Since rank A = rank $\begin{bmatrix} A \mid \mathbf{b} \end{bmatrix} = 3$ (verify), the linear system has a solution. ■

EXAMPLE 9

The linear system

$$\begin{bmatrix} 1 & 2 & 3 \\ 1 & -3 & 4 \\ 2 & -1 & 7 \end{bmatrix} \begin{bmatrix} x_1 \\ x_2 \\ x_3 \end{bmatrix} = \begin{bmatrix} 4 \\ 5 \\ 6 \end{bmatrix}$$

has no solution because rank $A = 2$ and rank $\begin{bmatrix} A \mid \mathbf{b} \end{bmatrix} = 3$ (verify). ■

has no solution because rank $A = 2$ and rank $\left[A \mid \mathbf{b} \right] = 3$ (verify). ■

The following statements are equivalent for an $n \times n$ matrix:

1. A is nonsingular.
2. $A\mathbf{x} = \mathbf{0}$ has only the trivial solution.
3. A is row (column) equivalent to I_n.
4. The system $A\mathbf{x} = \mathbf{b}$ has a unique solution for every vector $\mathbf{b}$ in R^n.
5. A is a product of elementary matrices.
6. A has rank n.
7. The nullity of A is zero.
8. The rows of A form a linearly independent set of vectors in R_n.
9. The columns of A form a linearly independent set of vectors in R^n.

2.8 Exercises

1. Find a basis for the subspace V of R^3 spanned by

$$S = \left\{ \begin{bmatrix} 1 \\ 2 \\ 3 \end{bmatrix}, \begin{bmatrix} 2 \\ 1 \\ 4 \end{bmatrix}, \begin{bmatrix} -1 \\ -1 \\ 2 \end{bmatrix}, \begin{bmatrix} 0 \\ 1 \\ 2 \end{bmatrix}, \begin{bmatrix} 1 \\ 1 \\ 1 \end{bmatrix} \right\}$$

and write each of the following vectors in terms of the basis vectors:

(a) $\begin{bmatrix} 3 \\ 4 \\ 12 \end{bmatrix}$. (b) $\begin{bmatrix} 3 \\ 2 \\ 2 \end{bmatrix}$. (c) $\begin{bmatrix} 1 \\ 2 \\ 6 \end{bmatrix}$.

2. Find a basis for the subspace of P_3 spanned by

$$S = \{t^3 + t^2 + 2t + 1, t^3 - 3t + 1, t^2 + t + 2, t + 1, t^3 + 1\}.$$

3. Find a basis for the subspace of M_{22} spanned by

$$S = \left\{ \begin{bmatrix} 1 & 2 \\ 1 & 1 \end{bmatrix}, \begin{bmatrix} 2 & 1 \\ 3 & 1 \end{bmatrix}, \begin{bmatrix} 0 & 2 \\ 1 & 2 \end{bmatrix}, \begin{bmatrix} 3 & 2 \\ 1 & 4 \end{bmatrix}, \begin{bmatrix} 5 & 0 \\ 0 & -1 \end{bmatrix} \right\}.$$

4. Find a basis for the subspace of R_2 spanned by

$$S = \{\begin{bmatrix} 1 & 2 \end{bmatrix}, \begin{bmatrix} 2 & 3 \end{bmatrix}, \begin{bmatrix} 3 & 1 \end{bmatrix}, \begin{bmatrix} -4 & 3 \end{bmatrix}\}.$$

In Exercises 5 and 6, find a basis for the row space of A (a) consisting of vectors that are not row vectors of A; (b) consisting of vectors that are row vectors of A.

5. $A = \begin{bmatrix} 1 & 2 & -1 \\ 1 & 9 & -1 \\ -3 & 8 & 3 \\ -2 & 3 & 2 \end{bmatrix}$.

6. $A = \begin{bmatrix} 1 & 2 & -1 & 3 \\ 3 & 5 & 2 & 0 \\ 0 & 1 & 2 & 1 \\ -1 & 0 & -2 & 7 \end{bmatrix}$.

In Exercises 7 and 8, find a basis for the column space of A (a) consisting of vectors that are not column vectors of A; (b) consisting of vectors that are column vectors of A.

7. $A = \begin{bmatrix} 1 & -2 & 7 & 0 \\ 1 & -1 & 4 & 0 \\ 3 & 2 & -3 & 5 \\ 2 & 1 & -1 & 3 \end{bmatrix}$.

8. $A = \begin{bmatrix} -2 & 2 & 3 & 7 & 1 \\ -2 & 2 & 4 & 8 & 0 \\ -3 & 3 & 2 & 8 & 4 \\ 4 & -2 & 1 & -5 & -7 \end{bmatrix}$.

In Exercises 9 and 10, find the row and column ranks of the given matrices.

9. (a) $\begin{bmatrix} 1 & 2 & 3 & 2 & 1 \\ 3 & 1 & -5 & -2 & 1 \\ 7 & 8 & -1 & 2 & 5 \end{bmatrix}$.

(b) $\begin{bmatrix} 1 & 3 & 2 & 0 & 0 & 1 \\ 2 & 1 & -5 & 1 & 2 & 0 \\ 3 & 2 & 5 & 1 & -2 & 1 \\ 5 & 8 & 9 & 1 & -2 & 2 \\ 9 & 9 & 4 & 2 & 0 & 2 \end{bmatrix}$.

10. (a) $\begin{bmatrix} 1 & 2 & 3 & 2 & 1 \\ 0 & 5 & 4 & 0 & -1 \\ 2 & -1 & 2 & 4 & 3 \end{bmatrix}$.

(b) $\begin{bmatrix} 1 & 1 & -1 & 2 & 0 \\ 2 & -4 & 0 & 1 & 1 \\ 5 & -1 & -3 & 7 & 1 \\ 3 & -9 & 1 & 0 & 2 \end{bmatrix}$.

11. Let A be an $m \times n$ matrix in row echelon form. Prove that rank $A =$ the number of nonzero rows of A.

12. For each of the following matrices, verify Theorem 2.17 by computing the row and column ranks.

(a) $\begin{bmatrix} 1 & 2 & 3 \\ -1 & 2 & 1 \\ 3 & 1 & 2 \end{bmatrix}$.

(b) $\begin{bmatrix} 1 & -2 & -1 \\ 2 & -1 & 3 \\ 7 & -8 & 3 \end{bmatrix}$.

(c) $\begin{bmatrix} 1 & -2 & -1 \\ 2 & -1 & 3 \\ 7 & -8 & 3 \\ 5 & -7 & 0 \end{bmatrix}$.

In Exercises 13 *and* 14, *compute the rank and nullity of the given matrix and verify Theorem* 2.17.

13. (a) $\begin{bmatrix} 1 & -1 & 2 & 3 \\ 2 & 6 & -8 & 1 \\ 5 & 3 & -2 & 10 \end{bmatrix}$.

(b) $\begin{bmatrix} 1 & 2 & 0 & 3 \\ 3 & 2 & -1 & 0 \\ 2 & -1 & 0 & 1 \end{bmatrix}$.

14. (a) $\begin{bmatrix} 1 & 3 & -2 & 4 \\ -1 & 4 & -5 & 10 \\ 3 & 2 & 1 & -2 \\ 3 & -5 & 8 & -16 \end{bmatrix}$.

(b) $\begin{bmatrix} 1 & 1 & 1 & 1 \\ 2 & -1 & 0 & 0 \\ 0 & 1 & -1 & 2 \\ 1 & 1 & -1 & 2 \end{bmatrix}$.

15. Which of the following matrices are equivalent?

$$A = \begin{bmatrix} 1 & 2 & 1 & 3 \\ 2 & 1 & -4 & -5 \\ 7 & 8 & -5 & -1 \\ 10 & 14 & -2 & 3 \end{bmatrix},$$

$$B = \begin{bmatrix} 1 & 2 & 1 & 3 \\ 2 & 1 & -4 & -5 \\ 1 & 1 & 0 & 0 \\ 0 & 0 & 1 & 1 \end{bmatrix},$$

$$C = \begin{bmatrix} 1 & 5 & 1 & 3 \\ 2 & 1 & 2 & 1 \\ -3 & 0 & 1 & 0 \\ 4 & 7 & -4 & 3 \end{bmatrix},$$

$$D = \begin{bmatrix} 1 & 2 & -4 & 3 \\ 4 & 7 & -4 & 1 \\ 7 & 12 & -4 & -1 \\ 2 & 3 & 4 & -5 \end{bmatrix},$$

$$E = \begin{bmatrix} 4 & 3 & -1 & -5 \\ -2 & -6 & -7 & 10 \\ -2 & -3 & -2 & 5 \\ 0 & -6 & -10 & 10 \end{bmatrix}.$$

In Exercises 16 *and* 17, *determine which of the given linear systems are consistent by comparing the ranks of the coefficient and augmented matrices.*

16. (a) $\begin{bmatrix} 1 & 2 & 5 & -2 \\ 2 & 3 & -2 & 4 \\ 5 & 1 & 0 & 2 \end{bmatrix} \begin{bmatrix} x_1 \\ x_2 \\ x_3 \\ x_4 \end{bmatrix} = \begin{bmatrix} 0 \\ 0 \\ 0 \end{bmatrix}$.

(b) $\begin{bmatrix} 1 & 2 & 5 & -2 \\ 2 & 3 & -2 & 4 \\ 5 & 1 & 0 & 2 \end{bmatrix} \begin{bmatrix} x_1 \\ x_2 \\ x_3 \\ x_4 \end{bmatrix} = \begin{bmatrix} -1 \\ -13 \\ 3 \end{bmatrix}$.

17. (a) $\begin{bmatrix} 1 & -2 & -3 & 4 \\ 4 & -1 & -5 & 6 \\ 2 & 3 & 1 & -2 \end{bmatrix} \begin{bmatrix} x_1 \\ x_2 \\ x_3 \\ x_4 \end{bmatrix} = \begin{bmatrix} 1 \\ 2 \\ 2 \end{bmatrix}$.

(b) $\begin{bmatrix} 1 & 1 & 1 \\ 1 & -1 & 1 \\ 5 & 1 & 5 \end{bmatrix} \begin{bmatrix} x_1 \\ x_2 \\ x_3 \end{bmatrix} = \begin{bmatrix} 6 \\ 2 \\ 5 \end{bmatrix}$.

In Exercises 18 and 19, use Corollary 2.7 to find which of the given matrices are nonsingular.

18. (a) $\begin{bmatrix} 1 & 2 & -3 \\ -1 & 2 & 3 \\ 0 & 8 & 0 \end{bmatrix}$.

(b) $\begin{bmatrix} 1 & 2 & -3 \\ -1 & 2 & 3 \\ 0 & 1 & 1 \end{bmatrix}$.

19. (a) $\begin{bmatrix} 1 & 1 & 2 \\ -1 & 3 & 4 \\ -5 & 7 & 8 \end{bmatrix}$.

(b) $\begin{bmatrix} 1 & 1 & 4 & -1 \\ 1 & 2 & 3 & 2 \\ -1 & 3 & 2 & 1 \\ -2 & 6 & 12 & -4 \end{bmatrix}$.

In Exercises 20 and 21, use Corollary 2.8 to find which of the given homogeneous systems have a nontrivial solution.

20. (a) $\begin{bmatrix} 1 & 2 & 3 \\ 0 & 1 & 0 \\ 1 & 0 & 3 \end{bmatrix} \begin{bmatrix} x_1 \\ x_2 \\ x_3 \end{bmatrix} = \begin{bmatrix} 0 \\ 0 \\ 0 \end{bmatrix}$.

(b) $\begin{bmatrix} 1 & 1 & 2 & -1 \\ 1 & 3 & -1 & 2 \\ 1 & 1 & 1 & 3 \\ 1 & 2 & 1 & 1 \end{bmatrix} \begin{bmatrix} x_1 \\ x_2 \\ x_3 \\ x_4 \end{bmatrix} = \begin{bmatrix} 0 \\ 0 \\ 0 \\ 0 \end{bmatrix}$.

21. (a) $\begin{bmatrix} 1 & 2 & -1 \\ 2 & -1 & 3 \\ 5 & -4 & 3 \end{bmatrix} \begin{bmatrix} x_1 \\ x_2 \\ x_3 \end{bmatrix} = \begin{bmatrix} 0 \\ 0 \\ 0 \end{bmatrix}$.

(b) $\begin{bmatrix} 1 & 2 & 3 \\ -1 & 2 & -1 \\ 1 & 6 & 5 \end{bmatrix} \begin{bmatrix} x_1 \\ x_2 \\ x_3 \end{bmatrix} = \begin{bmatrix} 0 \\ 0 \\ 0 \end{bmatrix}$.

In Exercises 22 and 23, find rank A by obtaining a matrix of the form $\begin{bmatrix} I_r & O \\ O & O \end{bmatrix}$ *that is equivalent to A.*

22. (a) $A = \begin{bmatrix} 1 & 1 & -2 \\ 1 & 2 & 3 \\ 0 & 1 & 3 \end{bmatrix}$.

(b) $A = \begin{bmatrix} 1 & 1 & -2 & 0 & 0 \\ 1 & 2 & 3 & 6 & 7 \\ 2 & 1 & 3 & 6 & 5 \end{bmatrix}$.

23. (a) $A = \begin{bmatrix} 1 & 1 & -2 \\ 1 & 2 & 3 \\ 3 & 4 & -1 \end{bmatrix}$.

(b) $A = \begin{bmatrix} 1 & -1 & 2 & 3 \\ 2 & 2 & 0 & 1 \\ 1 & -5 & 6 & 8 \\ 4 & 0 & 4 & 6 \end{bmatrix}$.

In Exercises 24 and 25, use Corollary 2.9 to determine whether the linear system $A\mathbf{x} = \mathbf{b}$ *has a unique solution for every* 3×1 *matrix* $\mathbf{b}$.

24. $A = \begin{bmatrix} 1 & 2 & -2 \\ 0 & 8 & -7 \\ 3 & -2 & 1 \end{bmatrix}$.

25. $A = \begin{bmatrix} 1 & -1 & 2 \\ 3 & 2 & 3 \\ 1 & -2 & 1 \end{bmatrix}$.

26. Is

$$S = \left\{ \begin{bmatrix} 2 \\ 2 \\ 3 \end{bmatrix}, \begin{bmatrix} 1 \\ 0 \\ 2 \end{bmatrix}, \begin{bmatrix} 0 \\ 1 \\ 3 \end{bmatrix} \right\}$$

a linearly independent set of vectors in R^3?

27. Is

$$S = \{[4 \quad 1 \quad 2], [2 \quad 5 \quad -5], [2 \quad -1 \quad 3]\}$$

a linearly independent set of vectors in R_3?

28. (a) If A is a 3×4 matrix, what is the largest possible value for rank A?

(b) If A is a 4×6 matrix, show that the columns of A are linearly dependent.

(c) If A is a 5×3 matrix, show that the rows of A are linearly dependent.

29. Let A be a 7×3 matrix whose rank is 3.

(a) Are the rows of A linearly dependent or linearly independent? Justify your answer.

(b) Are the columns of A linearly dependent or linearly independent? Justify your answer.

30. Let A be a 3×5 matrix.

(a) Give *all* possible values for the rank of A.

(b) If the rank of A is 3, what is the dimension of its column space?

(c) If the rank of A is 3, what is the dimension of the solution space of the homogeneous system $A\mathbf{x} = \mathbf{0}$?

31. Let $S = \{\mathbf{v}_1, \mathbf{v}_2, \ldots, \mathbf{v}_n\}$ be a set of n vectors in R_n, and let A be the matrix whose jth row is $\mathbf{v}_j$. Show that S is linearly independent if and only if rank $A = n$.

32. Let $S = \{\mathbf{v}_1, \mathbf{v}_2, \ldots, \mathbf{v}_n\}$ be a set of n vectors in R^n, and let A be the matrix whose jth column is $\mathbf{v}_j$. Show that S is linearly independent if and only if rank $A = n$.

33. Let A be an $n \times n$ matrix. Show that the homogeneous system $A\mathbf{x} = \mathbf{0}$ has a nontrivial solution if and only if the columns of A are linearly dependent.

34. Let A be an $n \times n$ matrix. Show that rank $A = n$ if and only if the columns of A are linearly independent.

35. Let A be an $n \times n$ matrix. Prove that the rows of A are linearly independent if and only if the columns of A span R^n.

36. Let $S = \{\mathbf{v}_1, \mathbf{v}_2, \ldots, \mathbf{v}_k\}$ be a basis for a subspace V of R_n that is obtained by the method of Example 1. If

$$\mathbf{v} = \begin{bmatrix} a_1 & a_2 & \cdots & a_n \end{bmatrix}$$

belongs to V and the leading 1's in the reduced row echelon form from the method in Example 1 occur in columns $j_1, j_2, \ldots, j_k$, then show that

$$\mathbf{v} = a_{j_1}\mathbf{v}_1 + a_{j_2}\mathbf{v}_2 + \cdots + a_{j_k}\mathbf{v}_k.$$

37. Prove Corollary 2.9.

38. Let A be an $m \times n$ matrix. Show that the linear system $A\mathbf{x} = \mathbf{b}$ has a solution for every $m \times 1$ matrix $\mathbf{b}$ if and only if rank $A = m$.

39. Let A be an $m \times n$ matrix with $m \neq n$. Show that either the rows or the columns of A are linearly dependent.

40. Suppose that the linear system $A\mathbf{x} = \mathbf{b}$, where A is $m \times n$, is consistent (i.e., has a solution). Prove that the solution is unique if and only if rank $A = n$.

41. What can you say about the dimension of the solution space of a homogeneous system of 8 equations in 10 unknowns?

42. Is it possible that all nontrivial solutions of a homogeneous system of 5 equations in 7 unknowns be multiples of each other? Explain.

43. Determine if your software has a command for computing the rank of a matrix. If it does, experiment with the command on matrices A in Examples 4 and 5 and Exercises 13 and 14.

44. Assuming that exact arithmetic is used, rank A is the number of nonzero rows in the reduced row echelon form of A. Compare the results using your rank command and the reduced row echelon form approach on the following matrices:

$$A = \begin{bmatrix} 1 & 1 \\ 0 & 1 \times 10^{-j} \end{bmatrix} \qquad j = 5, 10, 16.$$

(See Exercise 31 in Section 2.4.)

Supplementary Exercises

1. Let $C[a, b]$ denote the set of all real-valued continuous functions defined on $[a, b]$. If f and g are in $C[a, b]$, we define $f \oplus g$ by $(f \oplus g)(t) = f(t) + g(t)$, for t in $[a, b]$. If f is in $C[a, b]$ and c is a scalar, we define $c \odot f$ by $(c \odot f)(t) = cf(t)$, for t in $[a, b]$.

(a) Show that $C[a, b]$ is a real vector space.

(b) Let $W(k)$ be the set of all functions in $C[a, b]$ with $f(a) = k$. For what values of k will $W(k)$ be a subspace of $C[a, b]$?

(c) Let $t_1, t_2, \ldots, t_n$ be a fixed set of points in $[a, b]$. Show that the subset of all functions f in $C[a, b]$ that have roots at $t_1, t_2, \ldots, t_n$, that is, $f(t_i) = 0$ for $i = 1, 2, \ldots, n$, forms a subspace.

2. In R^4, let W be the subset of all vectors

$$\mathbf{v} = \begin{bmatrix} a_1 \\ a_2 \\ a_3 \\ a_4 \end{bmatrix}$$

that satisfy $a_4 - a_3 = a_2 - a_1$.

(a) Show that W is a subspace of R^4.

(b) Show that

$$S = \left\{ \begin{bmatrix} 1 \\ 0 \\ 0 \\ -1 \end{bmatrix}, \begin{bmatrix} 0 \\ 1 \\ 0 \\ 1 \end{bmatrix}, \begin{bmatrix} 1 \\ 1 \\ 1 \\ 1 \end{bmatrix}, \begin{bmatrix} 0 \\ 0 \\ 1 \\ 1 \end{bmatrix} \right\}$$

spans W.

(c) Find a subset of S that is a basis for W.

(d) Express $\mathbf{v} = \begin{bmatrix} 0 \\ 4 \\ 2 \\ 6 \end{bmatrix}$ as a linear combination of the basis obtained in part (c).

3. Consider

$$\mathbf{v}_1 = \begin{bmatrix} 1 \\ 1 \\ -2 \\ 1 \end{bmatrix}, \quad \mathbf{v}_2 = \begin{bmatrix} 1 \\ 5 \\ 2 \\ -1 \end{bmatrix}, \quad \text{and} \quad \mathbf{v}_3 = \begin{bmatrix} 3 \\ 0 \\ 2 \\ 1 \end{bmatrix}.$$

Determine whether the vector $\mathbf{v}$ belongs to span $\{\mathbf{v}_1, \mathbf{v}_2, \mathbf{v}_3\}$.

(a) $\mathbf{v} = \begin{bmatrix} 0 \\ 7 \\ 4 \\ 0 \end{bmatrix}$. (b) $\mathbf{v} = \begin{bmatrix} 5 \\ 6 \\ 2 \\ 1 \end{bmatrix}$. (c) $\mathbf{v} = \begin{bmatrix} 3 \\ -8 \\ -6 \\ 5 \end{bmatrix}$.

4. Let A be a fixed $n \times n$ matrix and let the set of all $n \times n$ matrices B such that $AB = BA$ be denoted by $C(A)$. Is $C(A)$ a subspace of M_{nn}?

5. Let W and U be subspaces of vector space V.

(a) Show that $W \cup U$, the set of all vectors $\mathbf{v}$ that are either in W or in U is not always a subspace of V.

(b) When is $W \cup U$ a subspace of V?

(c) Show that $W \cap U$, the set of all vectors $\mathbf{v}$ that are in both W and U, is a subspace of V.

6. Prove that a subspace W of R^3 coincides with R^3 if and only if it contains the vectors $\begin{bmatrix} 1 \\ 0 \\ 0 \end{bmatrix}$, $\begin{bmatrix} 0 \\ 1 \\ 0 \end{bmatrix}$, and $\begin{bmatrix} 0 \\ 0 \\ 1 \end{bmatrix}$.

7. Let A be a fixed $m \times n$ matrix and define W to be the subset of all $m \times 1$ matrices $\mathbf{b}$ in R^m for which the linear system $A\mathbf{x} = \mathbf{b}$ has a solution.

(a) Is W a subspace of R^m?

(b) What is the relationship between W and the column space of A?

8. Consider vector space R_2.

(a) For what values of m and b will all vectors of the form $\begin{bmatrix} x & mx + b \end{bmatrix}$ be a subspace of R_2?

(b) For what value of r will the set of all vectors of the form $\begin{bmatrix} x & rx^2 \end{bmatrix}$ be a subspce of R_2?

9. Let W be a nonempty subset of a vector space V. Prove that W is a subspace of V if and only if $r\mathbf{u} + s\mathbf{v}$ is in W for any vectors $\mathbf{u}$ and $\mathbf{v}$ in W and any scalars r and s.

10. Let A be an $n \times n$ matrix and λ a scalar. Show that the set W consisting of all vectors $\mathbf{x}$ in R^n such that $A\mathbf{x} = \lambda\mathbf{x}$ is a subspace of R^n.

11. For what values of a is the vector $\begin{bmatrix} a^2 \\ -3a \\ -2 \end{bmatrix}$ in

$$\text{span}\left\{\begin{bmatrix} 1 \\ 2 \\ 3 \end{bmatrix}, \begin{bmatrix} 0 \\ 1 \\ 1 \end{bmatrix}, \begin{bmatrix} 1 \\ 3 \\ 4 \end{bmatrix}\right\}?$$

12. For what values of a is the vector $\begin{bmatrix} a^2 \\ a \\ 1 \end{bmatrix}$ in

$$\text{span}\left\{\begin{bmatrix} 1 \\ 2 \\ 3 \end{bmatrix}, \begin{bmatrix} 1 \\ 1 \\ 1 \end{bmatrix}, \begin{bmatrix} 0 \\ 1 \\ 2 \end{bmatrix}\right\}?$$

13. For what values of k will the set S form a basis for R^6?

$$S = \left\{\begin{bmatrix} 1 \\ 2 \\ -2 \\ 1 \\ 1 \\ 1 \end{bmatrix}, \begin{bmatrix} 0 \\ 2 \\ 0 \\ 0 \\ 0 \\ 0 \end{bmatrix}, \begin{bmatrix} 0 \\ 5 \\ k \\ 1 \\ 0 \\ 0 \end{bmatrix}, \begin{bmatrix} 0 \\ 2 \\ 1 \\ k \\ 0 \\ 0 \end{bmatrix}, \begin{bmatrix} 0 \\ 1 \\ -2 \\ 4 \\ -3 \\ 1 \end{bmatrix}, \begin{bmatrix} 0 \\ 3 \\ 1 \\ 1 \\ 1 \\ 0 \end{bmatrix}\right\}.$$

14. Consider the subspace of R^4 given by

$$W = \text{span}\left\{\begin{bmatrix} 1 \\ 1 \\ 1 \\ 1 \end{bmatrix}, \begin{bmatrix} 3 \\ 2 \\ 1 \\ 1 \end{bmatrix}, \begin{bmatrix} 1 \\ 2 \\ 3 \\ 1 \end{bmatrix}, \begin{bmatrix} 0 \\ 2 \\ 4 \\ 1 \end{bmatrix}\right\}.$$

(a) Determine a subset S of the spanning set that is a basis for W.

(b) Find a basis T for W that is not a subset of the spanning set.

(c) Find the coordinate vector of $\mathbf{v} = \begin{bmatrix} 3 \\ -1 \\ -5 \\ 0 \end{bmatrix}$ with respect to each of the bases from parts (a) and (b).

15. Prove that if $S = \{\mathbf{v}_1, \mathbf{v}_2, \ldots, \mathbf{v}_k\}$ is a basis for a subspace W of vector space V, then there is a basis for V that includes the set S. (*Hint*: Use Theorem 2.10.)

16. Let $V = \text{span}\,\{\mathbf{v}_1, \mathbf{v}_2\}$, where

$$\mathbf{v}_1 = \begin{bmatrix} 1 \\ 0 \\ 2 \end{bmatrix} \quad \text{and} \quad \mathbf{v}_2 = \begin{bmatrix} 1 \\ 1 \\ 1 \end{bmatrix}.$$

Find a basis S for R^3 that includes $\mathbf{v}_1$ and $\mathbf{v}_2$. (*Hint*: Use the technique developed in the Alternative Constructive Proof of Theorem 2.8.)

17. Describe the set of all vectors $\mathbf{b}$ in R^3 for which the linear system $A\mathbf{x} = \mathbf{b}$ is consistent.

(a) $A = \begin{bmatrix} 1 & -2 & 1 & 0 \\ 2 & 1 & 1 & 2 \\ 1 & -7 & 2 & -2 \end{bmatrix}$.

(b) $A = \begin{bmatrix} 1 & 2 & 1 \\ 1 & 3 & 1 \\ 2 & 4 & 3 \end{bmatrix}$.

18. Find a basis for the solution space of the homogeneous system $(\lambda I_3 - A)\mathbf{x} = \mathbf{0}$ for the given scalar λ and given matrix A.

(a) $\lambda = 1$, $A = \begin{bmatrix} 0 & 0 & 1 \\ 1 & 0 & -3 \\ 0 & 1 & 3 \end{bmatrix}$.

(b) $\lambda = 3$, $A = \begin{bmatrix} 1 & 1 & -2 \\ -1 & 2 & 1 \\ 0 & 1 & -1 \end{bmatrix}$.

19. Show that $\text{rank}\,A = \text{rank}\,A^T$, for any $m \times n$ matrix A.

20. Let A and B be $m \times n$ matrices that are row equivalent.

(a) Prove that $\text{rank}\,A = \text{rank}\,B$.

(b) Prove that for $\mathbf{x}$ in R^n, $A\mathbf{x} = \mathbf{0}$ if and only if $B\mathbf{x} = \mathbf{0}$.

21. Let A be $m \times n$ and B be $n \times k$.

(a) Prove that $\text{rank}(AB) \leq \min\{\text{rank}\,A, \text{rank}\,B\}$.

(b) Find A and B such that $\text{rank}(AB) < \min\{\text{rank}\,A, \text{rank}\,B\}$.

(c) If $k = n$ and B is nonsingular, prove that $\text{rank}(AB) = \text{rank}\,A$.

(d) If $m = n$ and A is nonsingular, prove that $\text{rank}(AB) = \text{rank}\,B$.

(e) For nonsingular matrices P and Q, what is $\text{rank}(PAQ)$?

22. For an $m \times n$ matrix A, let the set of all vectors $\mathbf{x}$ in R^n such that $A\mathbf{x} = \mathbf{0}$ be denoted by $\text{NS}(A)$, which in Example 9 of Section 2.3 has been shown to be a subspace of R^n, called the null space of A.

(a) Prove that $\text{rank}\,A + \dim \text{NS}(A) = n$.

(b) For $m = n$, prove that A is nonsingular if and only if $\dim \text{NS}(A) = 0$.

23. Let A be an $m \times n$ matrix and B a nonsingular $m \times m$ matrix. Prove that $\text{NS}(BA) = \text{NS}(A)$ (see Exercise 22).

24. Find $\dim \text{NS}(A)$ (see Exercise 22) for each of the following matrices.

(a) $A = \begin{bmatrix} 1 & 2 & 1 \\ 2 & 0 & 3 \\ 0 & 4 & -1 \end{bmatrix}$.

(b) $A = \begin{bmatrix} 1 & 2 & 4 & 1 \\ 2 & 1 & 1 & 1 \\ -1 & 4 & 10 & 1 \end{bmatrix}$.

25. Any nonsingular 3×3 matrix P represents a transition matrix from some ordered basis $T = \{\mathbf{w}_1, \mathbf{w}_2, \mathbf{w}_3\}$ to some other ordered basis $S = \{\mathbf{v}_1, \mathbf{v}_2, \mathbf{v}_3\}$. Let

$$P = \begin{bmatrix} 1 & 1 & 1 \\ 0 & 2 & 1 \\ 1 & 0 & 3 \end{bmatrix}.$$

(a) If $\mathbf{v}_1 = \begin{bmatrix} 1 \\ 1 \\ 0 \end{bmatrix}$, $\mathbf{v}_2 = \begin{bmatrix} 1 \\ 0 \\ 1 \end{bmatrix}$, and $\mathbf{v}_3 = \begin{bmatrix} 1 \\ 1 \\ 1 \end{bmatrix}$, find T.

(b) If $\mathbf{w}_1 = \begin{bmatrix} 1 \\ 0 \\ 2 \end{bmatrix}$, $\mathbf{w}_2 = \begin{bmatrix} 2 \\ 0 \\ 1 \end{bmatrix}$, and $\mathbf{w}_3 = \begin{bmatrix} 1 \\ 1 \\ 0 \end{bmatrix}$, find S.

26. In Supplementary Exercise 32 for Chapter 1, we defined the outer product of two $n \times 1$ column matrices X and Y as XY^T. Determine the rank of an outer product.

27. Suppose that A is an $n \times n$ matrix and that there is no nonzero vector $\mathbf{x}$ in R^n such that $A\mathbf{x} = \mathbf{x}$. Show that $A - I_n$ is nonsingular.

28. Let A be an $m \times n$ matrix. Prove that if $A^T A$ is nonsingular, then $\text{rank}\,A = n$.

29. Prove or disprove by finding a counterexample.

(a) $\text{rank}(A + B) \leq \max\{\text{rank}\,A, \text{rank}\,B\}$.

(b) $\text{rank}(A + B) \geq \min\{\text{rank}\,A, \text{rank}\,B\}$.

(c) $\text{rank}(A + B) = \text{rank}\,A + \text{rank}\,B$.

30. Let A be an $n \times n$ matrix and $\{\mathbf{v}_1, \mathbf{v}_2, \dots, \mathbf{v}_k\}$ a linearly dependent set of vectors in R^n. Are $A\mathbf{v}_1, A\mathbf{v}_2, \dots, A\mathbf{v}_k$ linearly dependent or linearly independent vectors in R^n? Justify your answer.

31. Let A be an $m \times n$ matrix. Show that the linear system $A\mathbf{x} = \mathbf{b}$ has at most one solution for every $m \times 1$ matrix $\mathbf{b}$ if and only if the associated homogeneous system $A\mathbf{x} = \mathbf{0}$ has only the trivial solution.

32. Let A be an $m \times n$ matrix. Show that the linear system $A\mathbf{x} = \mathbf{b}$ has at most one solution for every $m \times 1$ matrix $\mathbf{b}$ if and only if the columns of A are linearly independent.

33. What can you say about the solutions to the consistent nonhomogeneous linear system $A\mathbf{x} = \mathbf{b}$ if the rank of A is less than the number of unknowns.

34. Let W_1 and W_2 be subspaces of a vector space V. Let $W_1 + W_2$ be the set of all vectors $\mathbf{v}$ in V such that $\mathbf{v} = \mathbf{w}_1 + \mathbf{w}_2$, where $\mathbf{w}_1$ is in W_1 and $\mathbf{w}_2$ is in W_2. Show that $W_1 + W_2$ is a subspace of V.

35. Let W_1 and W_2 be subspaces of a vector space V with $W_1 \cap W_2 = \{\mathbf{0}\}$. Let $W_1 + W_2$ be as defined in Exercise 34. Suppose that $V = W_1 + W_2$. Prove that every vector in V can be uniquely written as $\mathbf{w}_1 + \mathbf{w}_2$, where $\mathbf{w}_1$ is in W_1 and $\mathbf{w}_2$ is in W_2. In this case we write $V = W_1 \oplus W_2$ and say that V is the **direct sum** of the subspaces W_1 and W_2.

Chapter 3

Inner Product Spaces

As we noted in Chapter 2, when physicists talk about vectors in R^2 and R^3, they usually refer to objects that have magnitude and direction. However, thus far in our study of vectors spaces we have refrained from discussing these notions. In this chapter we deal with magnitude and direction in a vector space.

3.1 STANDARD INNER PRODUCT ON R^2 AND R^3

Length

In this section we discuss the notions of magnitude and direction in R^2 and R^3 and, in the next section, generalize these to R^n. We consider R^2 and R^3 with the usual Cartesian coordinate system. The **length**, or **magnitude**, of the vector $\mathbf{v} = \begin{bmatrix} v_1 \\ v_2 \end{bmatrix}$ in R^2, denoted by $\|\mathbf{v}\|$, is by the Pythagorean theorem (see Figure 3.1)

$$\|\mathbf{v}\| = \sqrt{v_1^2 + v_2^2}. \tag{1}$$

EXAMPLE 1

If $\mathbf{v} = \begin{bmatrix} 2 \\ -5 \end{bmatrix}$, then, by Equation (1),

$$\|\mathbf{v}\| = \sqrt{(2)^2 + (-5)^2} = \sqrt{4 + 25} = \sqrt{29}.$$

■

Consider now the points $P_1(u_1, u_2)$ and $P_2(v_1, v_2)$, as shown in Figure 3.2(a). Applying the Pythagorean Theorem to triangle $P_1 R P_2$, we find that the distance from P_1 to P_2, the length of the line segment from P_1 to P_2, is given by

$$\sqrt{(v_1 - u_1)^2 + (v_2 - u_2)^2}.$$

Note: This chapter may also be covered after Section 6.2 and before Section 6.3, which is where it is used.

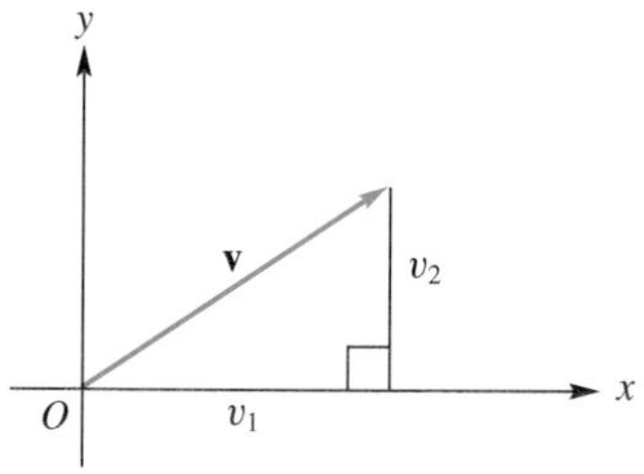

FIGURE 3.1

Length of **v**.

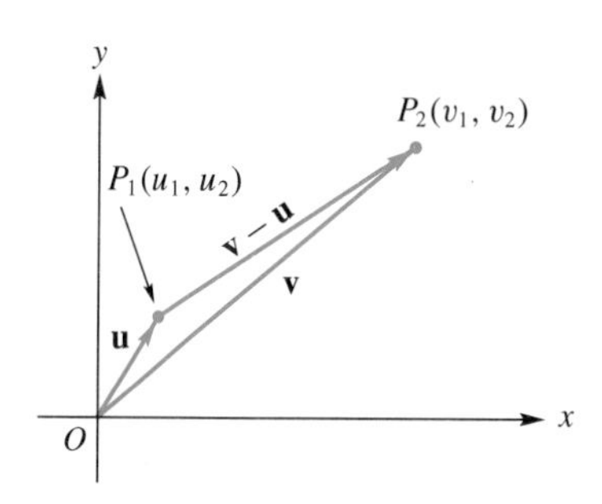

(a) Distance between the points $P_1(u_1, u_2)$ and $P_2(v_1, v_2)$.

(b) Distance between the vectors **u** and **v**

FIGURE 3.2

If $\mathbf{u} = \begin{bmatrix} u_1 \\ u_2 \end{bmatrix}$ and $\mathbf{v} = \begin{bmatrix} v_1 \\ v_2 \end{bmatrix}$ are vectors in R^2 as shown in Figure 3.2(b), then their heads are at the points $P_1(u_1, u_2)$ and $P_2(v_1, v_2)$, respectively. We then define the distance between the vectors **u** and **v** as the distance between the points P_1 and P_2. The distance between **u** and **v** is

$$\sqrt{(v_1 - u_1)^2 + (v_2 - u_2)^2} = \|\mathbf{v} - \mathbf{u}\|, \tag{2}$$

since $\mathbf{v} - \mathbf{u} = \begin{bmatrix} v_1 - u_1 \\ v_2 - u_2 \end{bmatrix}$.

EXAMPLE 2

Compute the distance between the vectors

$$\mathbf{u} = \begin{bmatrix} -1 \\ 5 \end{bmatrix} \quad \text{and} \quad \mathbf{v} = \begin{bmatrix} 3 \\ 2 \end{bmatrix}.$$

Solution By Equation (2), the distance between **u** and **v** is

$$\|\mathbf{v} - \mathbf{u}\| = \sqrt{(3+1)^2 + (2-5)^2} = \sqrt{4^2 + (-3)^2} = \sqrt{25} = 5.$$ ■

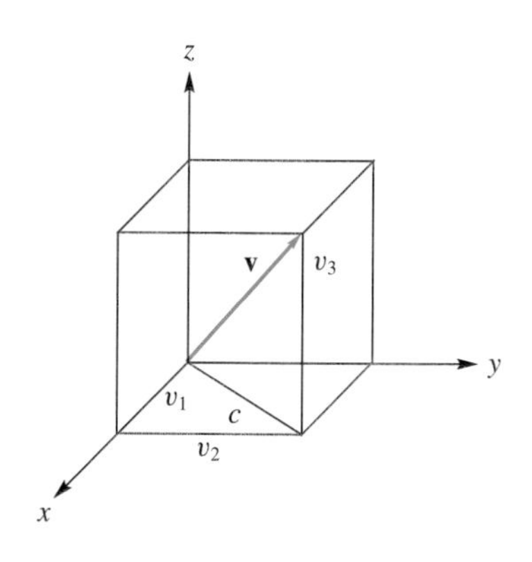

FIGURE 3.3

Length of **v**.

Now let $\mathbf{v} = \begin{bmatrix} v_1 \\ v_2 \\ v_3 \end{bmatrix}$ be a vector in R^3. Using the Pythagorean theorem twice (Figure 3.3), we can obtain the **length** of **v**, also denoted $\|\mathbf{v}\|$, as

$$\|\mathbf{v}\| = \sqrt{c^2 + v_3^2} = \sqrt{\left(\sqrt{v_1^2 + v_2^2}\right)^2 + v_3^2} = \sqrt{v_1^2 + v_2^2 + v_3^2}. \tag{3}$$

It follows from Equation (3) that the zero vector has length zero. It is easy to show that the zero vector is the only vector whose length is zero.

If $P_1(u_1, u_2, u_3)$ and $P_2(v_1, v_2, v_3)$ are points in R^3, then as in the case for R^2, the distance between P_1 and P_2 is given by

$$\sqrt{(v_1 - u_1)^2 + (v_2 - u_2)^2 + (v_3 - u_3)^2}.$$

Again, as in R^2, if $\mathbf{u} = \begin{bmatrix} u_1 \\ u_2 \\ u_3 \end{bmatrix}$ and $\mathbf{v} = \begin{bmatrix} v_1 \\ v_2 \\ v_3 \end{bmatrix}$ are vectors in R^3, then the distance between $\mathbf{u}$ and $\mathbf{v}$ is given by

$$\|\mathbf{v} - \mathbf{u}\| = \sqrt{(v_1 - u_1)^2 + (v_2 - u_2)^2 + (v_3 - u_3)^2}. \tag{4}$$

EXAMPLE 3

Compute the length of the vector

$$\mathbf{v} = \begin{bmatrix} 1 \\ 2 \\ 3 \end{bmatrix}.$$

Solution By Equation (3), the length of $\mathbf{v}$ is

$$\|\mathbf{v}\| = \sqrt{1^2 + 2^2 + 3^3} = \sqrt{14}.$$ ■

EXAMPLE 4

Compute the distance between the vectors

$$\mathbf{u} = \begin{bmatrix} 1 \\ 2 \\ 3 \end{bmatrix} \quad \text{and} \quad \mathbf{v} = \begin{bmatrix} -4 \\ 3 \\ 5 \end{bmatrix}.$$

Solution By Equation (4), the distance between $\mathbf{u}$ and $\mathbf{v}$ is

$$\|\mathbf{v} - \mathbf{u}\| = \sqrt{(-4 - 1)^2 + (3 - 2)^2 + (5 - 3)^2} = \sqrt{30}.$$ ■

The direction of a vector in R^2 is given by specifying its angle of inclination or slope. The direction of a vector $\mathbf{v}$ in R^3 is specified by giving the cosines of the angles that the vector $\mathbf{v}$ makes with the positive x-, y-, and z-axes (see Figure 3.4); these are called **direction cosines**.

Instead of dealing with the special problem of finding the cosines of these angles for a vector in R^3, or the angle of inclination for a vector in R^2, we consider the more general problem of determining the **angle** θ, $0 \leq \theta \leq \pi$, between two nonzero vectors in R^2 or R^3. As shown in Figure 3.5, let

$$\mathbf{u} = \begin{bmatrix} u_1 \\ u_2 \\ u_3 \end{bmatrix} \quad \text{and} \quad \mathbf{v} = \begin{bmatrix} v_1 \\ v_2 \\ v_3 \end{bmatrix}$$

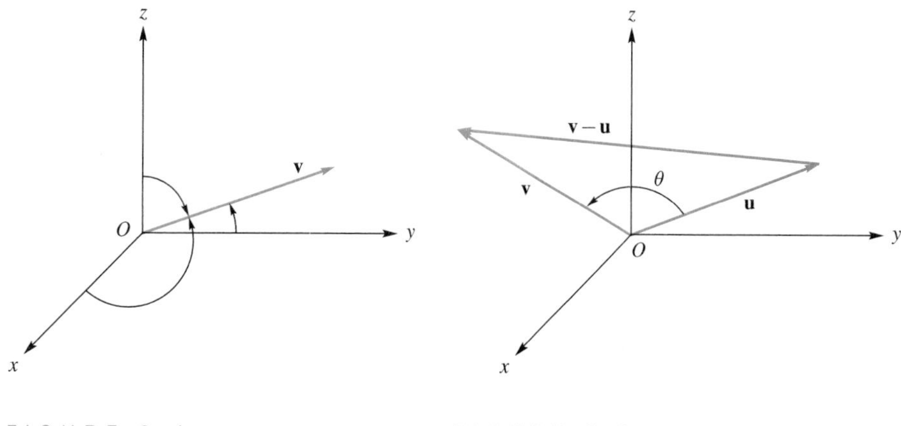

FIGURE 3.4

FIGURE 3.5

be two vectors in R^3. By the law of cosines, we have

$$\|\mathbf{v} - \mathbf{u}\|^2 = \|\mathbf{u}\|^2 + \|\mathbf{v}\|^2 - 2\|\mathbf{u}\|\,\|\mathbf{v}\|\cos\theta.$$

Hence

$$\cos\theta = \frac{\|\mathbf{u}\|^2 + \|\mathbf{v}\|^2 - \|\mathbf{v} - \mathbf{u}\|^2}{2\|\mathbf{u}\|\,\|\mathbf{v}\|} = \frac{(u_1^2 + u_2^2 + u_3^2) + (v_1^2 + v_2^2 + v_3^2)}{2\|\mathbf{u}\|\,\|\mathbf{v}\|}$$
$$- \frac{(v_1 - u_1)^2 + (v_2 - u_2)^2 + (v_3 - u_3)^2}{2\|\mathbf{u}\|\,\|\mathbf{v}\|}$$
$$= \frac{u_1v_1 + u_2v_2 + u_3v_3}{\|\mathbf{u}\|\,\|\mathbf{v}\|}.$$

Thus

$$\cos\theta = \frac{u_1v_1 + u_2v_2 + u_3v_3}{\|\mathbf{u}\|\,\|\mathbf{v}\|}. \tag{5}$$

In a similar way, if $\mathbf{u} = \begin{bmatrix} u_1 \\ u_2 \end{bmatrix}$ and $\mathbf{v} = \begin{bmatrix} v_1 \\ v_2 \end{bmatrix}$ are nonzero vectors in R^2 and θ is the angle between $\mathbf{u}$ and $\mathbf{v}$, then

$$\cos\theta = \frac{u_1v_1 + u_2v_2}{\|\mathbf{u}\|\,\|\mathbf{v}\|}. \tag{6}$$

The zero vector in R^2 or R^3 has no specific direction. The law of cosines expression above is true if $\mathbf{v} \neq \mathbf{0}$ and $\mathbf{u} = \mathbf{0}$ for any angle θ. Thus the zero vector can be assigned any direction.

EXAMPLE 5

Let $\mathbf{u} = \begin{bmatrix} 1 \\ 1 \\ 0 \end{bmatrix}$ and $\mathbf{v} = \begin{bmatrix} 0 \\ 1 \\ 1 \end{bmatrix}$. The angle θ between $\mathbf{u}$ and $\mathbf{v}$ is determined by

$$\cos\theta = \frac{(1)(0) + (1)(1) + (0)(1)}{\sqrt{1^2 + 1^2 + 0^2}\,\sqrt{0^2 + 1^2 + 1^2}} = \frac{1}{2}.$$

Since $0 \leq \theta \leq \pi$, $\theta = 60°$. ■

The cosine of an angle between two nonzero vectors in R^2 or R^3 can be expressed in terms of a very useful function, which we now define.

Standard Inner Product

Definition 3.1

Let $\mathbf{u} = \begin{bmatrix} u_1 \\ u_2 \end{bmatrix}$ and $\mathbf{v} = \begin{bmatrix} v_1 \\ v_2 \end{bmatrix}$ be vectors in R^2. The **standard inner product**, or **dot product**, on R^2 is defined as the number

$$u_1 v_1 + u_2 v_2$$

and is denoted by

$$\mathbf{u} \cdot \mathbf{v}.$$

Let $\mathbf{u} = \begin{bmatrix} u_1 \\ u_2 \\ u_3 \end{bmatrix}$ and $\mathbf{v} = \begin{bmatrix} v_1 \\ v_2 \\ v_3 \end{bmatrix}$ be vectors in R^3. The **standard inner product**, or **dot product**, on R^3 is defined as the number

$$u_1 v_1 + u_2 v_2 + u_3 v_3$$

and is denoted by $(\mathbf{u}, \mathbf{v})$ or by $\mathbf{u} \cdot \mathbf{v}$. ▲

EXAMPLE 6

If $\mathbf{u} = \begin{bmatrix} 2 \\ 3 \\ 2 \end{bmatrix}$ and $\mathbf{v} = \begin{bmatrix} 4 \\ 2 \\ -1 \end{bmatrix}$, then

$$\mathbf{u} \cdot \mathbf{v} = (2)(4) + (3)(2) + (2)(-1) = 12.$$ ■

Remark Observe that if we view the vectors $\mathbf{u}$ and $\mathbf{v}$ in R^2 or R^3 as matrices, then we can write the dot product of $\mathbf{u}$ and $\mathbf{v}$ in terms of matrix multiplication as $\mathbf{u}^T\mathbf{v}$, where we have ignored the brackets around the 1×1 matrix $\mathbf{u}^T\mathbf{v}$.

If we examine Equations (1) and (3), we see that if **v** is a vector in R^2 or R^3, then

$$\|\mathbf{v}\| = \sqrt{\mathbf{v} \cdot \mathbf{v}}. \tag{7}$$

Using Definition 3.1, we can also write Equations (5) and (6) for the cosine of the angle θ between two nonzero vectors **u** and **v** in R^2 and R^3 as

$$\cos\theta = \frac{\mathbf{u} \cdot \mathbf{v}}{\|\mathbf{u}\| \, \|\mathbf{v}\|} \qquad 0 \le \theta \le \pi. \tag{8}$$

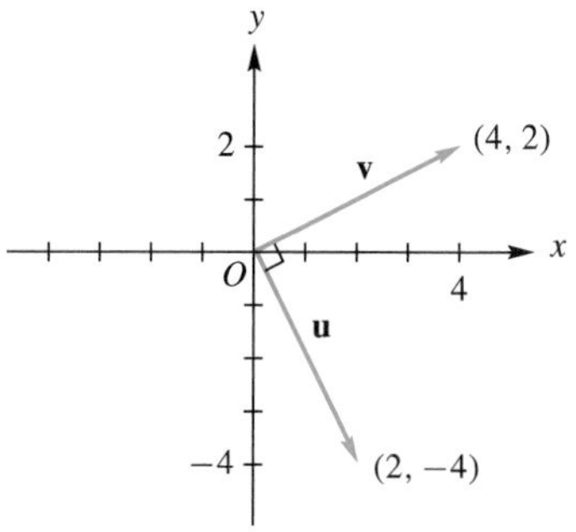

It then follows that two vectors **u** and **v** in R^2 or R^3 are **orthogonal**, or **perpendicular**, if and only if $\mathbf{u} \cdot \mathbf{v} = 0$.

FIGURE 3.6

EXAMPLE 7

The vectors $\mathbf{u} = \begin{bmatrix} 2 \\ -4 \end{bmatrix}$ and $\mathbf{v} = \begin{bmatrix} 4 \\ 2 \end{bmatrix}$ are orthogonal, since

$$\mathbf{u} \cdot \mathbf{v} = (2)(4) + (-4)(2) = 0.$$

See Figure 3.6. ■

We note the following properties of the standard inner product on R^2 and R^3 that will motivate our next section.

Theorem 3.1

Let **u**, **v**, *and* **w** *be vectors in* R^2 *or* R^3, *and let c be a scalar. The standard inner product on* R^2 *and* R^3 *has the following properties*:

(a) $\mathbf{u} \cdot \mathbf{u} \ge 0$; $\mathbf{u} \cdot \mathbf{u} = 0$ *if and only if* $\mathbf{u} = \mathbf{0}$.

(b) $\mathbf{u} \cdot \mathbf{v} = \mathbf{v} \cdot \mathbf{u}$.

(c) $(\mathbf{u} + \mathbf{v}) \cdot \mathbf{w} = \mathbf{u} \cdot \mathbf{w} + \mathbf{v} \cdot \mathbf{w}$.

(d) $c\mathbf{u} \cdot \mathbf{v} = c(\mathbf{u} \cdot \mathbf{v})$, *for any real scalar c*.

Proof

Exercise. ●

Unit Vectors

A **unit vector** in R^2 or R^3 is a vector whose length is 1. If **x** is any nonzero vector, then the vector

$$\mathbf{u} = \frac{1}{\|\mathbf{x}\|}\mathbf{x}$$

is a unit vector in the direction of **x** (Exercise 34.)

EXAMPLE 8

Let $\mathbf{x} = \begin{bmatrix} -3 \\ 4 \end{bmatrix}$. Then

$$\|\mathbf{x}\| = \sqrt{(-3)^2 + 4^2} = 5.$$

Hence the vector $\mathbf{u} = \frac{1}{5}\begin{bmatrix} -3 \\ 4 \end{bmatrix} = \begin{bmatrix} -\frac{3}{5} \\ \frac{4}{5} \end{bmatrix}$ is a unit vector, since

$$\|\mathbf{u}\| = \sqrt{\left(-\frac{3}{5}\right)^2 + \left(\frac{4}{5}\right)^2} = \sqrt{\frac{9+16}{25}} = 1.$$

FIGURE 3.7

Also, $\mathbf{u}$ points in the direction of $\mathbf{x}$ (Figure 3.7). ■

There are two unit vectors in R^2 that are of special importance. These are $\mathbf{i} = \begin{bmatrix} 1 \\ 0 \end{bmatrix}$ and $\mathbf{j} = \begin{bmatrix} 0 \\ 1 \end{bmatrix}$, the unit vectors along the positive x- and y-axes, respectively, shown in Figure 3.8. Observe that $\mathbf{i}$ and $\mathbf{j}$ are orthogonal. Since $\mathbf{i}$ and $\mathbf{j}$ form the natural basis for R^2, every vector in R^2 can be written uniquely as a linear combination of the orthogonal vectors $\mathbf{i}$ and $\mathbf{j}$. Thus, if $\mathbf{u} = \begin{bmatrix} u_1 \\ u_2 \end{bmatrix}$ is a vector in R^2, then

$$\mathbf{u} = u_1 \begin{bmatrix} 1 \\ 0 \end{bmatrix} + u_2 \begin{bmatrix} 0 \\ 1 \end{bmatrix} = u_1\mathbf{i} + u_2\mathbf{j}.$$

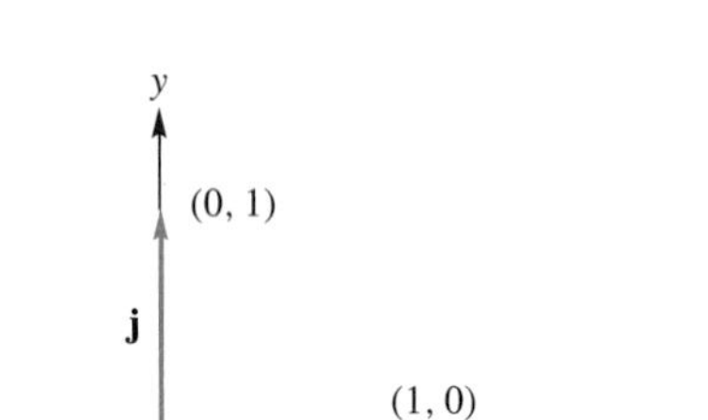

FIGURE 3.8

Similarly, the vectors in the natural basis for R^3,

$$\mathbf{i} = \begin{bmatrix} 1 \\ 0 \\ 0 \end{bmatrix}, \quad \mathbf{j} = \begin{bmatrix} 0 \\ 1 \\ 0 \end{bmatrix}, \quad \text{and} \quad \mathbf{k} = \begin{bmatrix} 0 \\ 0 \\ 1 \end{bmatrix},$$

are unit vectors that are mutually orthogonal. Thus, if $\mathbf{u} = \begin{bmatrix} u_1 \\ u_2 \\ u_3 \end{bmatrix}$ is a vector in R^3, then $\mathbf{u} = u_1\mathbf{i} + u_2\mathbf{j} + u_3\mathbf{k}$.

Resultant Force and Velocity

When several forces act on a body, we can find a single force, called the **resultant force**, having an equivalent effect. The resultant force can be determined using vectors. The following example illustrates the method.

EXAMPLE 9

Suppose that a force of 12 pounds is applied to an object along the negative x-axis and a force of 5 pounds is applied to the object along the positive y-axis. Find the magnitude and direction of the resultant force.

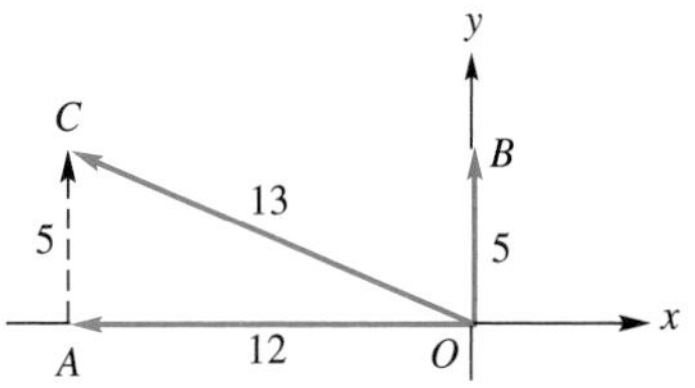

FIGURE 3.9

Solution In Figure 3.9 we have represented the force along the negative x-axis by the vector $\overrightarrow{OA}$ and the force along the positive y-axis by the vector $\overrightarrow{OB}$. The resultant force is the vector $\overrightarrow{OC} = \overrightarrow{OA} + \overrightarrow{OB}$. Thus the magnitude of the resultant force is 13 pounds and its direction is as indicated in the figure. ■

Vectors are also used in physics to deal with velocity problems, as the following example illustrates.

EXAMPLE 10

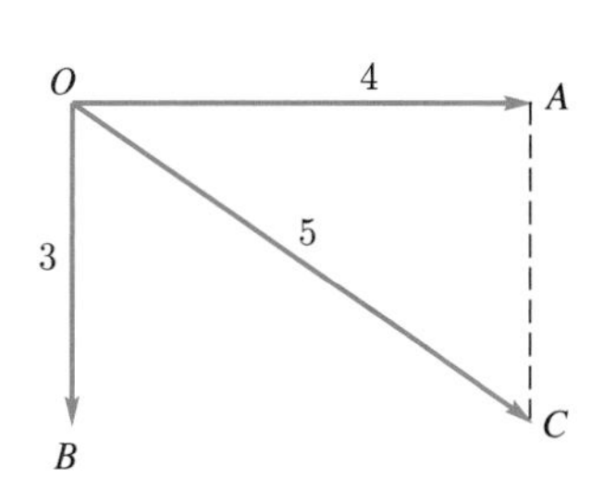

FIGURE 3.10

Suppose that a boat is traveling east across a river at a rate of 4 miles per hour while the river's current is flowing south at a rate of 3 miles per hour. Find the resultant velocity of the boat.

Solution In Figure 3.10 we have represented the velocity of the boat by the vector $\overrightarrow{OA}$ and the velocity of the river's current by the vector $\overrightarrow{OB}$. The resultant velocity is the vector $\overrightarrow{OC} = \overrightarrow{OA} + \overrightarrow{OB}$. Thus the magnitude of the resultant velocity is 5 miles per hour and its direction is as indicated in the figure. ■

3.1 Exercises

In Exercises 1 *and* 2, *find the length of each vector.*

1. (a) $\begin{bmatrix} 1 \\ 0 \end{bmatrix}$. (b) $\begin{bmatrix} 0 \\ 0 \end{bmatrix}$. (c) $\begin{bmatrix} 1 \\ 2 \end{bmatrix}$.

2. (a) $\begin{bmatrix} 0 \\ -2 \\ 0 \end{bmatrix}$. (b) $\begin{bmatrix} -1 \\ -3 \\ -4 \end{bmatrix}$. (c) $\begin{bmatrix} 1 \\ -2 \\ 4 \end{bmatrix}$.

In Exercises 3 *and* 4, *compute* $\|\mathbf{u} - \mathbf{v}\|$.

3. (a) $\mathbf{u} = \begin{bmatrix} 1 \\ 0 \end{bmatrix}, \mathbf{v} = \begin{bmatrix} 1 \\ 1 \end{bmatrix}$.

(b) $\mathbf{u} = \begin{bmatrix} 0 \\ 0 \end{bmatrix}, \mathbf{v} = \begin{bmatrix} 1 \\ -1 \end{bmatrix}$.

4. (a) $\mathbf{u} = \begin{bmatrix} 1 \\ 2 \\ 3 \end{bmatrix}, \mathbf{v} = \begin{bmatrix} 4 \\ 5 \\ 6 \end{bmatrix}$.

(b) $\mathbf{u} = \begin{bmatrix} -1 \\ -2 \\ -3 \end{bmatrix}, \mathbf{v} = \begin{bmatrix} -4 \\ -5 \\ -6 \end{bmatrix}$.

In Exercises 5 *and* 6, *find the distance between* **u** *and* **v**.

5. (a) $\mathbf{u} = \begin{bmatrix} 1 \\ 2 \end{bmatrix}, \mathbf{v} = \begin{bmatrix} -4 \\ -5 \end{bmatrix}$.

(b) $\mathbf{u} = \begin{bmatrix} 1 \\ 2 \end{bmatrix}, \mathbf{v} = \begin{bmatrix} 4 \\ -5 \end{bmatrix}$.

6. (a) $\mathbf{u} = \begin{bmatrix} -1 \\ -2 \\ -3 \end{bmatrix}, \mathbf{v} = \begin{bmatrix} 4 \\ 5 \\ 6 \end{bmatrix}$.

(b) $\mathbf{u} = \begin{bmatrix} 0 \\ 1 \\ -1 \end{bmatrix}, \mathbf{v} = \begin{bmatrix} 1 \\ 2 \\ 0 \end{bmatrix}$.

In Exercises 7 and 8, find $\mathbf{u} \cdot \mathbf{v}$.

7. (a) $\mathbf{u} = \begin{bmatrix} 0 \\ 0 \end{bmatrix}$, $\mathbf{v} = \begin{bmatrix} 0 \\ 0 \end{bmatrix}$.

(b) $\mathbf{u} = \begin{bmatrix} 3 \\ 0 \end{bmatrix}$, $\mathbf{v} = \begin{bmatrix} 1 \\ 2 \end{bmatrix}$.

(c) $\mathbf{u} = \begin{bmatrix} 2 \\ -3 \end{bmatrix}$, $\mathbf{v} = \begin{bmatrix} 1 \\ -2 \end{bmatrix}$.

8. (a) $\mathbf{u} = \begin{bmatrix} 0 \\ 0 \\ -3 \end{bmatrix}$, $\mathbf{v} = \begin{bmatrix} -4 \\ 5 \\ 6 \end{bmatrix}$.

(b) $\mathbf{u} = \begin{bmatrix} 1 \\ 3 \\ 2 \end{bmatrix}$, $\mathbf{v} = \begin{bmatrix} 2 \\ 3 \\ 4 \end{bmatrix}$.

(c) $\mathbf{u} = \begin{bmatrix} 1 \\ -1 \\ 1 \end{bmatrix}$, $\mathbf{v} = \begin{bmatrix} 1 \\ 2 \\ -1 \end{bmatrix}$.

9. For each pair of vectors $\mathbf{u}$ and $\mathbf{v}$ in Exercise 5, find the cosine of the angle θ between $\mathbf{u}$ and $\mathbf{v}$.

10. For each pair of vectors in Exercise 6, find the cosine of the angle θ between $\mathbf{u}$ and $\mathbf{v}$.

11. For each of the following vectors $\mathbf{v}$, find the direction cosines (the cosine of the angles between $\mathbf{v}$ and the positive x-, y-, and z-axes).

(a) $\mathbf{v} = \begin{bmatrix} 1 \\ 0 \\ 0 \end{bmatrix}$. (b) $\mathbf{v} = \begin{bmatrix} 1 \\ 3 \\ 2 \end{bmatrix}$.

(c) $\mathbf{u} = \begin{bmatrix} -1 \\ -2 \\ -3 \end{bmatrix}$. (d) $\mathbf{u} = \begin{bmatrix} 4 \\ -3 \\ 2 \end{bmatrix}$.

12. Let P and Q be the points in R^3 with respective coordinates $(3, -1, 2)$ and $(4, 2, -3)$. Find the length of the segment PQ.

13. Prove Theorem 3.1.

14. Verify Theorem 3.1 for

$$\mathbf{u} = \begin{bmatrix} 1 \\ 2 \\ 3 \end{bmatrix}, \quad \mathbf{v} = \begin{bmatrix} -2 \\ 4 \\ 3 \end{bmatrix}, \quad \mathbf{w} = \begin{bmatrix} 0 \\ 3 \\ -2 \end{bmatrix},$$

and $c = -3$.

15. Show that in R^2:

(a) $\mathbf{i} \cdot \mathbf{i} = \mathbf{j} \cdot \mathbf{j} = 1$. (b) $\mathbf{i} \cdot \mathbf{j} = 0$.

16. Show that in R^3:

(a) $\mathbf{i} \cdot \mathbf{i} = \mathbf{j} \cdot \mathbf{j} = \mathbf{k} \cdot \mathbf{k} = 1$.

(b) $\mathbf{i} \cdot \mathbf{j} = \mathbf{i} \cdot \mathbf{k} = \mathbf{j} \cdot \mathbf{k} = 0$.

17. Which of the vectors $\mathbf{v}_1 = \begin{bmatrix} 1 \\ 2 \end{bmatrix}$, $\mathbf{v}_2 = \begin{bmatrix} 0 \\ 1 \end{bmatrix}$, $\mathbf{v}_3 = \begin{bmatrix} -2 \\ -4 \end{bmatrix}$, $\mathbf{v}_4 = \begin{bmatrix} -2 \\ 1 \end{bmatrix}$, $\mathbf{v}_5 = \begin{bmatrix} 2 \\ 4 \end{bmatrix}$, and $\mathbf{v}_6 = \begin{bmatrix} -6 \\ 3 \end{bmatrix}$ are:

(a) Orthogonal? (b) In the same direction?

(c) In opposite directions?

18. Which of the vectors $\mathbf{v}_1 = \begin{bmatrix} 1 \\ -1 \\ -2 \end{bmatrix}$, $\mathbf{v}_2 = \begin{bmatrix} 3 \\ -1 \\ 2 \end{bmatrix}$, $\mathbf{v}_3 = \begin{bmatrix} 2 \\ 4 \\ -1 \end{bmatrix}$, $\mathbf{v}_4 = \begin{bmatrix} \frac{1}{2} \\ 0 \\ \frac{1}{4} \end{bmatrix}$, $\mathbf{v}_5 = \begin{bmatrix} \frac{1}{2} \\ -\frac{1}{2} \\ -1 \end{bmatrix}$, $\mathbf{v}_6 = \begin{bmatrix} -\frac{2}{3} \\ -\frac{4}{3} \\ \frac{1}{3} \end{bmatrix}$ are:

(a) Orthogonal? (b) In the same direction?

(c) In opposite directions?

19. **(Optional)** Which of the following pairs of lines are perpendicular?

(a) $x = 2 + 2t$, $y = -3 - 3t$, $z = 4 + 4t$ and $x = 2 + t$, $y = 4 - t$, $z = 5 - t$.

(b) $x = 3 - t$, $y = 4 + 4t$, $z = 2 + 2t$ and $x = 2t$, $y = 3 - 2t$, $z = 4 + 2t$.

20. **(Optional)** Find parametric equations of the line passing through $(3, -1, -3)$ and perpendicular to the line passing through $(3, -2, 4)$ and $(0, 3, 5)$.

21. A ship is being pushed by a tugboat with a force of 300 pounds along the negative y-axis while another tugboat is pushing along the negative x-axis with a force of 400 pounds. Find the magnitude and sketch the direction of the resultant force.

22. Suppose that an airplane is flying with an airspeed of 260 kilometers per hour while a wind is blowing to the west at 100 kilometers per hour. Indicate on a figure the appropriate direction that the plane must follow to result in a flight directly south. What will be the resultant speed?

23. Let points A, B, C, and D in R^3 have respective coordinates $(1, 2, 3)$, $(-2, 3, 5)$, $(0, 3, 6)$, and $(3, 2, 4)$. Prove that $ABCD$ is a parallelogram.

24. Find c so that the vector $\mathbf{v} = \begin{bmatrix} 1 \\ c \end{bmatrix}$ is orthogonal to $\mathbf{w} = \begin{bmatrix} 2 \\ -1 \end{bmatrix}$.

25. Find c so that the vector $\mathbf{v} = \begin{bmatrix} 2 \\ c \\ 3 \end{bmatrix}$ is orthogonal to $\mathbf{w} = \begin{bmatrix} 1 \\ -2 \\ 1 \end{bmatrix}$.

26. If possible, find a, b, and c so that $\mathbf{v} = \begin{bmatrix} a \\ b \\ c \end{bmatrix}$ is orthogonal to both $\mathbf{w} = \begin{bmatrix} 1 \\ 2 \\ 1 \end{bmatrix}$ and $\mathbf{x} = \begin{bmatrix} 1 \\ -1 \\ 1 \end{bmatrix}$.

27. If possible, find a and b so that $\mathbf{v} = \begin{bmatrix} a \\ b \\ 2 \end{bmatrix}$ is orthogonal to both $\mathbf{w} = \begin{bmatrix} 2 \\ 1 \\ 1 \end{bmatrix}$ and $\mathbf{x} = \begin{bmatrix} 1 \\ 0 \\ 1 \end{bmatrix}$.

28. Find c so that the vectors $\begin{bmatrix} c \\ 4 \end{bmatrix}$ and $\begin{bmatrix} 2 \\ 5 \end{bmatrix}$ are parallel.

29. Let θ be the angle between the nonzero vectors $\mathbf{u}$ and $\mathbf{v}$ in R^2 or R^3. Show that if $\mathbf{u}$ and $\mathbf{v}$ are parallel, then $\cos\theta = \pm 1$.

30. Show that the only vector $\mathbf{x}$ in R^2 or R^3 that is orthogonal to every other vector is the zero vector.

31. Prove that if $\mathbf{v}$, $\mathbf{w}$, and $\mathbf{x}$ are in R^2 or R^3 and $\mathbf{v}$ is orthogonal to both $\mathbf{w}$ and $\mathbf{x}$, then $\mathbf{v}$ is orthogonal to every vector in span $\{\mathbf{w}, \mathbf{x}\}$.

32. Let $\mathbf{u}$ be a fixed vector in R^2 (R^3). Prove that the set V of all vectors $\mathbf{v}$ in R^2 (R^3) such that $\mathbf{u}$ and $\mathbf{v}$ are orthogonal is a subspace of R^2 (R^3).

33. Prove that if c is a scalar and $\mathbf{v}$ is a vector in R^2 or R^3, then $\|c\mathbf{v}\| = |c|\,\|\mathbf{v}\|$.

34. Show that if $\mathbf{x}$ is a nonzero vector in R^2 or R^3, then $\mathbf{u} = \frac{1}{\|\mathbf{x}\|}\mathbf{x}$ is a unit vector in the direction of $\mathbf{x}$.

35. Let $S = \{\mathbf{v}_1, \mathbf{v}_2, \mathbf{v}_3\}$ be a set of nonzero vectors in R^3 such that any two vectors in S are orthogonal. Prove that S is linearly independent.

36. Prove that for any vectors $\mathbf{u}$, $\mathbf{v}$, and $\mathbf{w}$ in R^2 or R^3 we have

$$\mathbf{u} \cdot (\mathbf{v} + \mathbf{w}) = \mathbf{u} \cdot \mathbf{v} + \mathbf{u} \cdot \mathbf{w}.$$

37. Prove that for any vectors $\mathbf{u}$, $\mathbf{v}$, and $\mathbf{w}$ in R^2 or R^3 and any scalar c, we have:

(a) $(\mathbf{u} + c\mathbf{v}) \cdot \mathbf{w} = \mathbf{u} \cdot \mathbf{w} + c(\mathbf{v} \cdot \mathbf{w})$.

(b) $\mathbf{u} \cdot (c\mathbf{v}) = c(\mathbf{u} \cdot \mathbf{v})$.

(c) $(\mathbf{u} + \mathbf{v}) \cdot (c\mathbf{w}) = c(\mathbf{u} \cdot \mathbf{w}) + c(\mathbf{v} \cdot \mathbf{w})$.

38. Prove that the diagonals of a rectangle are of equal length. [*Hint*: Take the vertices of the rectangle as $(0, 0)$, $(0, b)$, $(a, 0)$, and (a, b).]

39. Prove that the angles at the base of an isosceles triangle are equal.

40. Prove that a parallelogram is a rhombus, a parallelogram with four equal sides, if and only if its diagonals are orthogonal.

41. To compute the dot product of a pair of vectors $\mathbf{u}$ and $\mathbf{v}$ in R^2 or R^3, use the matrix product operation in your software as follows. Let U and V be column matrices for vectors $\mathbf{u}$ and $\mathbf{v}$, respectively. Then $\mathbf{u} \cdot \mathbf{v}$ is the product of U^T and V (or V^T and U). Experiment with the vectors in Exercises 7 and 8. (Determine if your software has a particular command for computing a dot product.)

42. Determine if there is a command in your software to compute the length of a vector. If there is, use it on the vector in Example 3 and then compute the distance between the vectors in Example 4.

43. Assuming that your software has a command to compute the length of a vector (see Exercise 42), determine a unit vector in the direction of $\mathbf{v}$ for each of the following.

(a) $\mathbf{v} = \begin{bmatrix} 2 \\ 4 \end{bmatrix}$. (b) $\mathbf{v} = \begin{bmatrix} 7 \\ 1 \\ 0 \end{bmatrix}$. (c) $\mathbf{v} = \begin{bmatrix} 1 \\ 2 \\ -1 \end{bmatrix}$.

44. Referring to Exercise 41, how could your software check for orthogonal vectors?

3.2 CROSS PRODUCT IN R^3 (OPTIONAL)

In this section we discuss an operation that is meaningful only in R^3. Despite this limitation, it has a number of important applications, some of which we discuss in this section. Suppose that $\mathbf{u} = u_1\mathbf{i} + u_2\mathbf{j} + u_3\mathbf{k}$ and $\mathbf{v} = v_1\mathbf{i} + v_2\mathbf{j} + v_3\mathbf{k}$ and that we want to find a vector $\mathbf{w} = \begin{bmatrix} x \\ y \\ z \end{bmatrix}$ orthogonal (perpendicular) to both $\mathbf{u}$ and $\mathbf{v}$. Thus we want $\mathbf{u} \cdot \mathbf{w} = 0$ and $\mathbf{v} \cdot \mathbf{w} = 0$, which leads to the linear system

$$\begin{aligned} u_1x + u_2y + u_3z &= 0 \\ v_1x + v_2y + v_3z &= 0. \end{aligned} \tag{1}$$

It can be shown that

$$\mathbf{w} = \begin{bmatrix} u_2v_3 - u_3v_2 \\ u_3v_1 - u_1v_3 \\ u_1v_2 - u_2v_1 \end{bmatrix}$$

is a solution to (1) (verify). Of course, we can also write $\mathbf{w}$ as

$$\mathbf{w} = (u_2v_3 - u_3v_2)\mathbf{i} + (u_3v_1 - u_1v_3)\mathbf{j} + (u_1v_2 - u_2v_1)\mathbf{k}. \tag{2}$$

This vector is called the **cross product** of $\mathbf{u}$ and $\mathbf{v}$ and is denoted by $\mathbf{u} \times \mathbf{v}$. Note that the cross product, $\mathbf{u} \times \mathbf{v}$, is a vector, while the dot product, $\mathbf{u} \cdot \mathbf{v}$, is a scalar, or number. Although the cross product is not defined on R^n if $n \neq 3$, it has many applications; we shall use it when we study planes in R^3.

EXAMPLE 1

Let $\mathbf{u} = 2\mathbf{i} + \mathbf{j} + 2\mathbf{k}$ and $\mathbf{v} = 3\mathbf{i} - \mathbf{j} - 3\mathbf{k}$. From (2),

$$\mathbf{u} \times \mathbf{v} = -\mathbf{i} + 12\mathbf{j} - 5\mathbf{k}.$$

■

Let $\mathbf{u}$, $\mathbf{v}$, and $\mathbf{w}$ be vectors in R^3 and c a scalar. The cross product operation satisfies the following properties, whose verification we leave to the reader:

(a) $\mathbf{u} \times \mathbf{v} = -(\mathbf{v} \times \mathbf{u})$.
(b) $\mathbf{u} \times (\mathbf{v} + \mathbf{w}) = \mathbf{u} \times \mathbf{v} + \mathbf{u} \times \mathbf{w}$.
(c) $(\mathbf{u} + \mathbf{v}) \times \mathbf{w} = \mathbf{u} \times \mathbf{w} + \mathbf{v} \times \mathbf{w}$.
(d) $c(\mathbf{u} \times \mathbf{v}) = (c\mathbf{u}) \times \mathbf{v} = \mathbf{u} \times (c\mathbf{v})$.
(e) $\mathbf{u} \times \mathbf{u} = \mathbf{0}$.
(f) $\mathbf{0} \times \mathbf{u} = \mathbf{u} \times \mathbf{0} = \mathbf{0}$.
(g) $\mathbf{u} \times (\mathbf{v} \times \mathbf{w}) = (\mathbf{u} \cdot \mathbf{w})\mathbf{v} - (\mathbf{u} \cdot \mathbf{v})\mathbf{w}$.
(h) $(\mathbf{u} \times \mathbf{v}) \times \mathbf{w} = (\mathbf{w} \cdot \mathbf{u})\mathbf{v} - (\mathbf{w} \cdot \mathbf{v})\mathbf{u}$.

EXAMPLE 2

It follows from (2) that

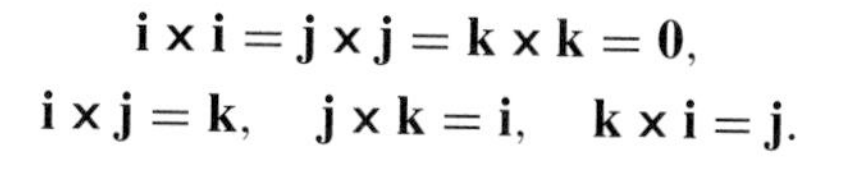

$$\mathbf{i} \times \mathbf{i} = \mathbf{j} \times \mathbf{j} = \mathbf{k} \times \mathbf{k} = \mathbf{0},$$
$$\mathbf{i} \times \mathbf{j} = \mathbf{k}, \quad \mathbf{j} \times \mathbf{k} = \mathbf{i}, \quad \mathbf{k} \times \mathbf{i} = \mathbf{j}.$$

Also,

$$\mathbf{j} \times \mathbf{i} = -\mathbf{k}, \quad \mathbf{k} \times \mathbf{j} = -\mathbf{i}, \quad \mathbf{i} \times \mathbf{k} = -\mathbf{j}.$$

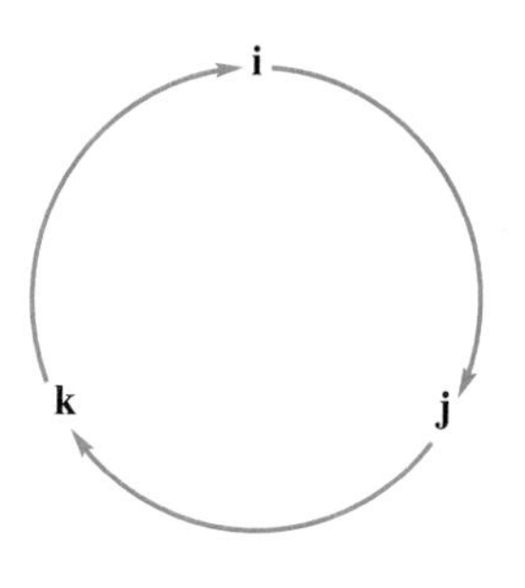

FIGURE 3.11

These rules can be remembered by the method illustrated by Figure 3.11. Moving around the circle in a clockwise direction, we see that the cross product of two vectors taken in the indicated order is the third vector; moving in a counterclockwise direction, we see that the cross product taken in the indicated order is the negative of the third vector. The cross product of a vector with itself is the zero vector. ■

Although many of the familiar properties of the real numbers hold for the cross product, it should be noted that two important properties do not hold. The commutative law does not hold, since $\mathbf{u} \times \mathbf{v} = -(\mathbf{v} \times \mathbf{u})$. Also, the associative law does not hold, since $\mathbf{i} \times (\mathbf{i} \times \mathbf{j}) = \mathbf{i} \times \mathbf{k} = -\mathbf{j}$ while $(\mathbf{i} \times \mathbf{i}) \times \mathbf{j} = \mathbf{0} \times \mathbf{j} = \mathbf{0}$.

We shall now take a closer look at the geometric properties of the cross product. First, we observe the following additional property of the cross product, whose proof we leave to the reader:

$$(\mathbf{u} \times \mathbf{v}) \cdot \mathbf{w} = \mathbf{u} \cdot (\mathbf{v} \times \mathbf{w}) \qquad \text{Exercise 7} \tag{3}$$

EXAMPLE 3

Let $\mathbf{u}$ and $\mathbf{v}$ be as in Example 1, and let $\mathbf{w} = \mathbf{i} + 2\mathbf{j} + 3\mathbf{k}$. Then

$$\mathbf{u} \times \mathbf{v} = -\mathbf{i} + 12\mathbf{j} - 5\mathbf{k} \quad \text{and} \quad (\mathbf{u} \times \mathbf{v}) \cdot \mathbf{w} = 8$$
$$\mathbf{v} \times \mathbf{w} = 3\mathbf{i} - 12\mathbf{j} + 7\mathbf{k} \quad \text{and} \quad \mathbf{u} \cdot (\mathbf{v} \times \mathbf{w}) = 8,$$

which illustrates Equation (3). ■

From the construction of $\mathbf{u} \times \mathbf{v}$, it follows that $\mathbf{u} \times \mathbf{v}$ is orthogonal to both $\mathbf{u}$ and $\mathbf{v}$; that is,

$$(\mathbf{u} \times \mathbf{v}) \cdot \mathbf{u} = 0, \tag{4}$$
$$(\mathbf{u} \times \mathbf{v}) \cdot \mathbf{v} = 0. \tag{5}$$

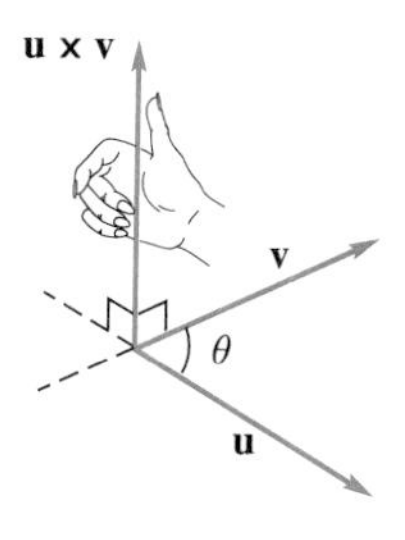

FIGURE 3.12

These equations can also be verified directly by using the definitions of $\mathbf{u} \times \mathbf{v}$ and dot product, or by using Equation (3) and properties (a) and (e) of the cross product operation. Then $\mathbf{u} \times \mathbf{v}$ is also orthogonal to the plane determined by $\mathbf{u}$ and $\mathbf{v}$. It can be shown that if θ is the angle between $\mathbf{u}$ and $\mathbf{v}$, then the direction of $\mathbf{u} \times \mathbf{v}$ is determined as follows. If we curl the fingers of the right hand in the direction of a rotation through the angle θ from $\mathbf{u}$ to $\mathbf{v}$, then the thumb will point in the direction of $\mathbf{u} \times \mathbf{v}$ (Figure 3.12).

The magnitude of $\mathbf{u} \times \mathbf{v}$ can be determined as follows. From Equation (7) of Section 3.1 it follows that

$$
\begin{aligned}
\|\mathbf{u} \times \mathbf{v}\|^2 &= (\mathbf{u} \times \mathbf{v}) \cdot (\mathbf{u} \times \mathbf{v}) & \\
&= \mathbf{u} \cdot [\mathbf{v} \times (\mathbf{u} \times \mathbf{v})] & \text{by (3)} \\
&= \mathbf{u} \cdot [(\mathbf{v} \cdot \mathbf{v})\mathbf{u} - (\mathbf{v} \cdot \mathbf{u})\mathbf{v}] & \text{by property (g) for cross product} \\
&= (\mathbf{u} \cdot \mathbf{u})(\mathbf{v} \cdot \mathbf{v}) - (\mathbf{v} \cdot \mathbf{u})(\mathbf{v} \cdot \mathbf{u}) & \text{by (d) and (b) of Theorem 3.1} \\
&= \|\mathbf{u}\|^2 \|\mathbf{v}\|^2 - (\mathbf{u} \cdot \mathbf{v})^2 & \text{by Equation (7) of Section 3.1 and (b) of Theorem 3.1.}
\end{aligned}
$$

From Equation (8) of Section 3.1 it follows that

$$\mathbf{u} \cdot \mathbf{v} = \|\mathbf{u}\| \|\mathbf{v}\| \cos\theta,$$

where θ is the angle between $\mathbf{u}$ and $\mathbf{v}$. Hence

$$
\begin{aligned}
\|\mathbf{u} \times \mathbf{v}\|^2 &= \|\mathbf{u}\|^2 \|\mathbf{v}\|^2 - \|\mathbf{u}\|^2 \|\mathbf{v}\|^2 \cos^2\theta \\
&= \|\mathbf{u}\|^2 \|\mathbf{v}\|^2 (1 - \cos^2\theta) \\
&= \|\mathbf{u}\|^2 \|\mathbf{v}\|^2 \sin^2\theta.
\end{aligned}
$$

Taking square roots, we obtain

$$\|\mathbf{u} \times \mathbf{v}\| = \|\mathbf{u}\| \|\mathbf{v}\| \sin\theta \qquad 0 \le \theta \le \pi. \tag{6}$$

Note that in (6) we do not have to write $|\sin\theta|$, since $\sin\theta$ is nonnegative for $0 \le \theta \le \pi$. It follows that vectors $\mathbf{u}$ and $\mathbf{v}$ are parallel if and only if $\mathbf{u} \times \mathbf{v} = \mathbf{0}$ (Exercise 9).

We now consider several applications of cross product.

Area of a Triangle

Consider the triangle with vertices P_1, P_2, and P_3 (Figure 3.13). The area of this triangle is $\frac{1}{2}bh$, where b is the base and h is the height. If we take the segment between P_1 and P_2 to be the base and denote $\overrightarrow{P_1P_2}$ by the vector $\mathbf{u}$, then

$$b = \|\mathbf{u}\|.$$

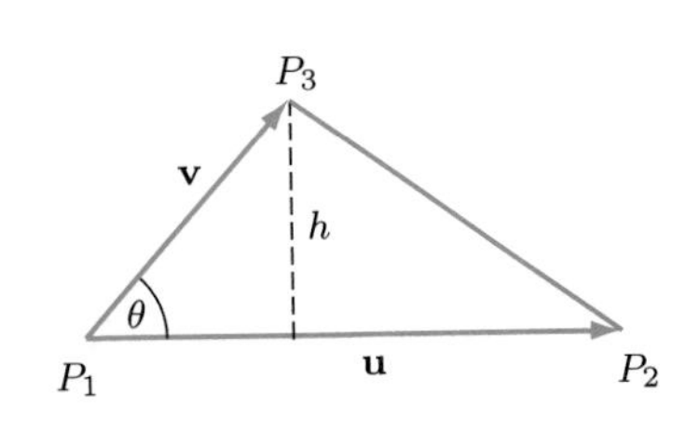

FIGURE 3.13

Letting $\overrightarrow{P_1P_3} = \mathbf{v}$, we find that the height h is given by

$$h = \|\mathbf{v}\| \sin\theta.$$

Hence, by (6), the area A_T of the triangle is

$$A_T = \tfrac{1}{2}\|\mathbf{u}\| \|\mathbf{v}\| \sin\theta = \tfrac{1}{2}\|\mathbf{u} \times \mathbf{v}\|.$$

EXAMPLE 4

Find the area of the triangle with vertices $P_1(2, 2, 4)$, $P_2(-1, 0, 5)$, and $P_3(3, 4, 3)$.

Solution We have

$$\mathbf{u} = \overrightarrow{P_1P_2} = -3\mathbf{i} - 2\mathbf{j} + \mathbf{k}$$
$$\mathbf{v} = \overrightarrow{P_1P_3} = \mathbf{i} + 2\mathbf{j} - \mathbf{k}.$$

Then

$$\begin{aligned} A_T &= \tfrac{1}{2}\|(-3\mathbf{i} - 2\mathbf{j} + \mathbf{k}) \times (\mathbf{i} + 2\mathbf{j} - \mathbf{k})\| \\ &= \tfrac{1}{2}\|-2\mathbf{j} - 4\mathbf{k}\| = \|-\mathbf{j} - 2\mathbf{k}\| = \sqrt{5}. \end{aligned}$$ ■

Area of a Parallelogram

The area A_P of the parallelogram with adjacent sides $\mathbf{u}$ and $\mathbf{v}$ (Figure 3.14) is $2A_T$, so

$$A_P = \|\mathbf{u} \times \mathbf{v}\|.$$

EXAMPLE 5

If P_1, P_2, and P_3 are as in Example 4, then the area of the parallelogram with adjacent sides $\overrightarrow{P_1P_2}$ and $\overrightarrow{P_1P_3}$ is $2\sqrt{5}$. (Verify.) ■

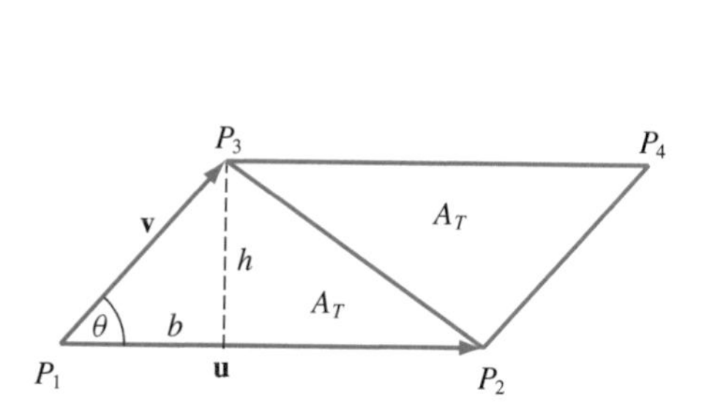

FIGURE 3.14

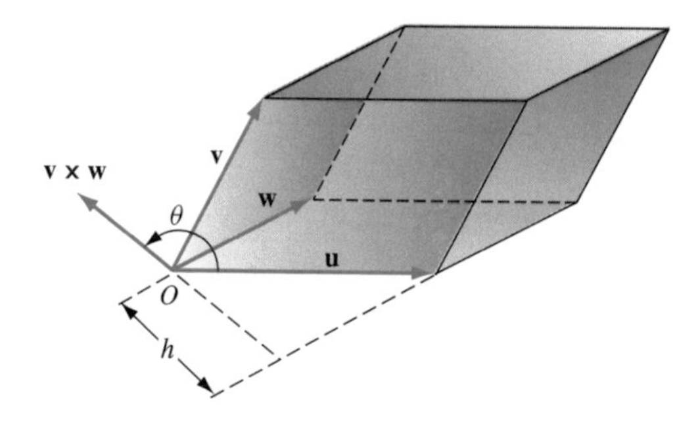

FIGURE 3.15

Volume of a Parallelepiped

Consider the parallelepiped with a vertex at the origin and edges $\mathbf{u}$, $\mathbf{v}$, and $\mathbf{w}$ (Figure 3.15). The volume V of the parallelepiped is the product of the area of the face containing $\mathbf{v}$ and $\mathbf{w}$ and the distance h from this face to the face parallel to it. Now

$$h = \|\mathbf{u}\|\,|\cos\theta|,$$

where θ is the angle between $\mathbf{u}$ and $\mathbf{v} \times \mathbf{w}$, and the area of the face determined by $\mathbf{v}$ and $\mathbf{w}$ is $\|\mathbf{v} \times \mathbf{w}\|$. Hence

$$V = \|\mathbf{v} \times \mathbf{w}\|\|\mathbf{u}\|\,|\cos\theta| = |\mathbf{u} \cdot (\mathbf{v} \times \mathbf{w})|.$$

EXAMPLE 6

Consider the parallelepiped with a vertex at the origin and edges $\mathbf{u} = \mathbf{i} - 2\mathbf{j} + 3\mathbf{k}$, $\mathbf{v} = \mathbf{i} + 3\mathbf{j} + \mathbf{k}$, and $\mathbf{w} = 2\mathbf{i} + \mathbf{j} + 2\mathbf{k}$. Then

$$\mathbf{v} \times \mathbf{w} = 5\mathbf{i} - 5\mathbf{k}.$$

Hence $\mathbf{u} \cdot (\mathbf{v} \times \mathbf{w}) = -10$. Thus the volume V is given by

$$V = |\mathbf{u} \cdot (\mathbf{v} \times \mathbf{w})| = |-10| = 10.$$ ■

Planes

A plane in R^3 can be determined by specifying a point in the plane and a vector perpendicular to the plane. This vector is called a **normal** to the plane.

To obtain an equation of the plane passing through the point $P_0(x_0, y_0, z_0)$ and having the nonzero vector $\mathbf{v} = a\mathbf{i} + b\mathbf{j} + c\mathbf{k}$ as a normal, we proceed as follows. A point $P(x, y, z)$ lies in the plane if and only if the vector $\overrightarrow{P_0P}$ is perpendicular to $\mathbf{v}$ (Figure 3.16). Thus $P(x, y, z)$ lies in the plane if and only if

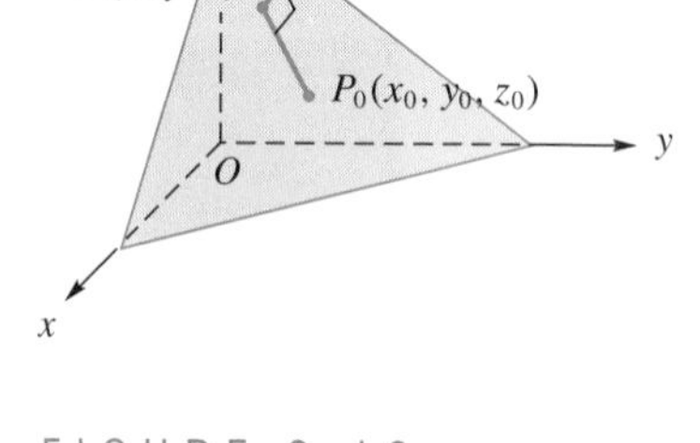

FIGURE 3.16

$$\mathbf{v} \cdot \overrightarrow{P_0P} = 0. \tag{7}$$

Since

$$\overrightarrow{P_0P} = (x - x_0)\mathbf{i} + (y - y_0)\mathbf{j} + (z - z_0)\mathbf{k},$$

we can write (7) as

$$a(x - x_0) + b(y - y_0) + c(z - z_0) = 0. \tag{8}$$

EXAMPLE 7

Find an equation of the plane passing through the point $(3, 4, -3)$ and perpendicular to the vector $\mathbf{v} = 5\mathbf{i} - 2\mathbf{j} + 4\mathbf{k}$.

Solution Substituting in (8), we obtain the equation of the plane as

$$5(x - 3) - 2(y - 4) + 4(z + 3) = 0.$$ ■

A plane is also determined by three noncollinear points in it, as we show in the following example.

EXAMPLE 8

Find an equation of the plane passing through the points $P_1(2, -2, 1)$, $P_2(-1, 0, 3)$, and $P_3(5, -3, 4)$.

Solution The nonparallel vectors $\overrightarrow{P_1P_2} = -3\mathbf{i} + 2\mathbf{j} + 2\mathbf{k}$ and $\overrightarrow{P_1P_3} = 3\mathbf{i} - \mathbf{j} + 3\mathbf{k}$ lie in the plane, since the points P_1, P_2, and P_3 lie in the plane. The vector

$$\mathbf{v} = \overrightarrow{P_1P_2} \times \overrightarrow{P_1P_3} = 8\mathbf{i} + 15\mathbf{j} - 3\mathbf{k}$$

is then perpendicular to both $\overrightarrow{P_1P_2}$ and $\overrightarrow{P_1P_3}$ and is thus a normal to the plane. Using the vector $\mathbf{v}$ and the point $P_1(2, -2, 1)$ in (8), we obtain

$$8(x - 2) + 15(y + 2) - 3(z - 1) = 0 \tag{9}$$

as an equation of the plane. ■

If we multiply out and simplify, (8) can be rewritten as

$$ax + by + cz + d = 0. \tag{10}$$

EXAMPLE 9

Equation (10) of the plane in Example 8 can be rewritten as

$$8x + 15y - 3z + 17 = 0. \tag{11}$$

■

It is not difficult to show (Exercise 24) that the graph of an equation of the form given in (10), where a, b, c, and d are constants (with a, b, and c not all zero), is a plane with normal $\mathbf{v} = a\mathbf{i} + b\mathbf{j} + c\mathbf{k}$; moreover, if $d = 0$, it is a two-dimensional subspace of R^3.

EXAMPLE 10

An alternative solution to Example 8 is as follows. Let the equation of the desired plane be

$$ax + by + cz + d = 0, \tag{12}$$

where a, b, c, and d are to be determined. Since P_1, P_2, and P_3 lie in the plane, their coordinates satisfy (12). Thus we obtain the linear system

$$\begin{aligned} 2a - 2b + c + d &= 0 \\ -a \phantom{{}- 2b} + 3c + d &= 0 \\ 5a - 3b + 4c + d &= 0. \end{aligned}$$

Solving this system, we have (verify)

$$a = \tfrac{8}{17}r, \quad b = \tfrac{15}{17}r, \quad c = -\tfrac{3}{17}r, \quad \text{and} \quad d = r,$$

where r is any real number. Letting $r = 17$, we obtain

$$a = 8, \quad b = 15, \quad c = -3, \quad \text{and} \quad d = 17,$$

which yields (11) as in the first solution. ■

FIGURE 3.17

EXAMPLE 11

Find parametric equations of the line of intersection of the planes

$$\Pi_1\colon 2x + 3y - 2z + 4 = 0 \quad \text{and} \quad \Pi_2\colon x - y + 2z + 3 = 0.$$

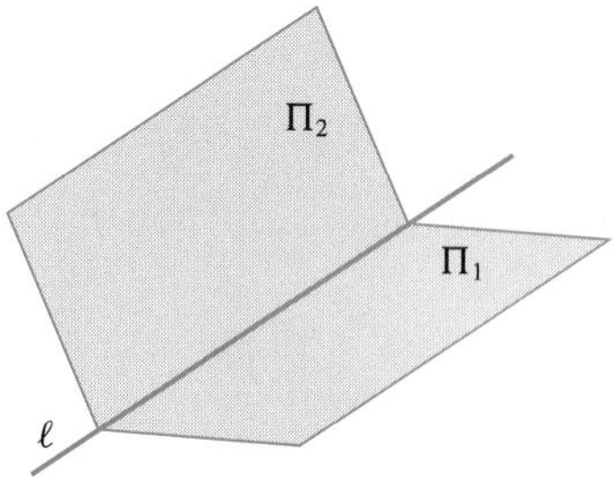

Solution Solving the linear system consisting of the equations of Π_1 and Π_2, we obtain (verify)

$$\begin{aligned} x &= -\tfrac{13}{5} - \tfrac{4}{5}t \\ y &= \tfrac{2}{5} + \tfrac{6}{5}t \qquad -\infty < t < \infty \\ z &= 0 + t \end{aligned}$$

as parametric equations (see Section 2.3) of the line ℓ of intersection of the planes (see Figure 3.17). ■

As we have indicated, the cross product cannot be generalized to R^n. However, we can generalize the notions of length, direction, and standard inner product to R^n in the natural manner, but there are some things to be checked. For example, if we define the cosine of the angle θ between two nonzero vectors $\mathbf{u}$ and $\mathbf{v}$ in R^n as

$$\cos\theta = \frac{\mathbf{u}\cdot\mathbf{v}}{\|\mathbf{u}\|\,\|\mathbf{v}\|},$$

we must check that $-1 \le \cos\theta \le 1$; otherwise, it would be misleading to call this fraction $\cos\theta$. Rather than verify this property for R^n now, we obtain this result in our next section, where we formulate the notion of inner product in any real vector space.

3.2 Exercises

1. Compute $\mathbf{u} \times \mathbf{v}$.

(a) $\mathbf{u} = 2\mathbf{i} + 3\mathbf{j} + 4\mathbf{k}$, $\mathbf{v} = -\mathbf{i} + 3\mathbf{j} - \mathbf{k}$.

(b) $\mathbf{u} = \mathbf{i} + \mathbf{k}$, $\mathbf{v} = 2\mathbf{i} + 3\mathbf{j} - \mathbf{k}$.

(c) $\mathbf{u} = \mathbf{i} - \mathbf{j} + 2\mathbf{k}$, $\mathbf{v} = 3\mathbf{i} - 4\mathbf{j} + \mathbf{k}$.

(d) $\mathbf{u} = \begin{bmatrix} 2 \\ -1 \\ 1 \end{bmatrix}$, $\mathbf{v} = -2\mathbf{u}$.

2. Compute $\mathbf{u} \times \mathbf{v}$.

(a) $\mathbf{u} = \mathbf{i} - \mathbf{j} + 2\mathbf{k}$, $\mathbf{v} = 3\mathbf{i} + \mathbf{j} + 2\mathbf{k}$.

(b) $\mathbf{u} = 2\mathbf{i} + \mathbf{j} - 2\mathbf{k}$, $\mathbf{v} = \mathbf{i} + 3\mathbf{k}$.

(c) $\mathbf{u} = 2\mathbf{j} + \mathbf{k}$, $\mathbf{v} = 3\mathbf{u}$.

(d) $\mathbf{u} = \begin{bmatrix} 4 \\ 0 \\ -2 \end{bmatrix}$, $\mathbf{v} = \begin{bmatrix} 0 \\ 2 \\ -1 \end{bmatrix}$.

3. Let $\mathbf{u} = \mathbf{i} + 2\mathbf{j} - 3\mathbf{k}$, $\mathbf{v} = 2\mathbf{i} + 3\mathbf{j} + \mathbf{k}$, $\mathbf{w} = 2\mathbf{i} - \mathbf{j} + 2\mathbf{k}$, and $c = -3$. Verify properties (a) through (h) for the cross product operation.

4. Prove properties (a) through (h) for the cross product operation.

5. Let $\mathbf{u} = 2\mathbf{i} - \mathbf{j} + 3\mathbf{k}$, $\mathbf{v} = 3\mathbf{i} + \mathbf{j} - \mathbf{k}$, and $\mathbf{w} = 3\mathbf{i} + \mathbf{j} + 2\mathbf{k}$. Verify Equation (3).

6. Verify that each of the cross products $\mathbf{u} \times \mathbf{v}$ in Exercise 1 is orthogonal to both $\mathbf{u}$ and $\mathbf{v}$.

7. Show that $(\mathbf{u} \times \mathbf{v}) \cdot \mathbf{w} = \mathbf{u} \cdot (\mathbf{v} \times \mathbf{w})$.

8. Verify Equation (6) for the pairs of vectors in Exercise 1.

9. Show that $\mathbf{u}$ and $\mathbf{v}$ are parallel if and only if $\mathbf{u} \times \mathbf{v} = \mathbf{0}$.

10. Show that $\|\mathbf{u} \times \mathbf{v}\|^2 + (\mathbf{u} \cdot \mathbf{v})^2 = \|\mathbf{u}\|^2\|\mathbf{v}\|^2$.

11. Prove the **Jacobi identity**

$$(\mathbf{u} \times \mathbf{v}) \times \mathbf{w} + (\mathbf{v} \times \mathbf{w}) \times \mathbf{u} + (\mathbf{w} \times \mathbf{u}) \times \mathbf{v} = \mathbf{0}.$$

12. Find the area of the triangle with vertices $P_1(1, -2, 3)$, $P_2(-3, 1, 4)$, and $P_3(0, 4, 3)$.

13. Find the area of the triangle with vertices P_1, P_2, and P_3, where $\overrightarrow{P_1P_2} = 2\mathbf{i} + 3\mathbf{j} - \mathbf{k}$ and $\overrightarrow{P_1P_3} = \mathbf{i} + 2\mathbf{j} + 2\mathbf{k}$.

14. Find the area of the parallelogram with adjacent sides $\mathbf{u} = \mathbf{i} + 3\mathbf{j} - 2\mathbf{k}$ and $\mathbf{v} = 3\mathbf{i} - \mathbf{j} - \mathbf{k}$.

15. Find the volume of the parallelepiped with a vertex at the origin and edges $\mathbf{u} = 2\mathbf{i} - \mathbf{j}$, $\mathbf{v} = \mathbf{i} - 2\mathbf{j} - 2\mathbf{k}$, and $\mathbf{w} = 3\mathbf{i} - \mathbf{j} + \mathbf{k}$.

16. Repeat Exercise 15 for $\mathbf{u} = \mathbf{i} - 2\mathbf{j} + 4\mathbf{k}$, $\mathbf{v} = 3\mathbf{i} + 4\mathbf{j} + \mathbf{k}$, and $\mathbf{w} = -\mathbf{i} + \mathbf{j} + \mathbf{k}$.

17. Determine which of the following points are in the plane

$$3(x - 2) + 2(y + 3) - 4(z - 4) = 0.$$

(a) $(0, -2, 3)$ (b) $(1, -2, 3)$.

18. Find an equation of the plane passing through the given point and perpendicular to the given vector.

(a) $(0, 2, -3)$, $3\mathbf{i} - 2\mathbf{j} + 4\mathbf{k}$.

(b) $(-1, 3, 2)$, $\mathbf{j} - 3\mathbf{k}$.

19. Find an equation of the plane passing through the given points.

(a) $(0, 1, 2)$, $(3, -2, 5)$, $(2, 3, 4)$.

(b) $(2, 3, 4)$, $(1, -2, 3)$, $(-5, -4, 2)$

20. Find parametric equations of the line of intersection of the given planes.

(a) $2x + 3y - 4z + 5 = 0$ and $-3x + 2y + 5z + 6 = 0$.

(b) $3x - 2y - 5z + 4 = 0$ and $2x + 3y + 4z + 8 = 0$.

21. Find an equation of the plane through $(-2, 3, 4)$ and perpendicular to the line through $(4, -2, 5)$ and $(0, 2, 4)$.

22. Find the point of intersection of the line

$$\begin{aligned} x &= 2 - 3t \\ y &= 4 + 2t \qquad -\infty < t < \infty \\ z &= 3 - 5t \end{aligned}$$

and the plane $2x + 3y + 4z + 8 = 0$.

23. Find a line passing through $(-2, 5, -3)$ and perpendicular to the plane $2x - 3y + 4z + 7 = 0$.

24. (a) Show that the graph of an equation of the form given in (10), with a, b, and c not all zero, is a plane with normal $\mathbf{v} = a\mathbf{i} + b\mathbf{j} + c\mathbf{k}$.

(b) Show that the set of all points on the plane $ax + by + cz = 0$ is a subspace of R^3.

(c) Find a basis for the subspace given by the plane $2x - 3y + 4z = 0$.

25. Find a basis for the subspace given by the plane $-3x + 2y + 5z = 0$.

26. Determine if your software has a command for computing cross products. If it does, check your results in Exercises 1 and 2.

3.3 INNER PRODUCT SPACES

In this section we use the properties of the standard inner product or dot product on R^3 listed in Theorem 3.1 as our foundation for generalizing the notion of the inner product to any real vector space. Here V is an arbitrary real vector space, not necessarily finite dimensional.

Definition 3.2

Let V be any real vector space. An **inner product** on V is a function that assigns to each ordered pair of vectors $\mathbf{u}$, $\mathbf{v}$ in V a real number $(\mathbf{u}, \mathbf{v})$ satisfying:

(a) $(\mathbf{u}, \mathbf{u}) \geq 0$; $(\mathbf{u}, \mathbf{u}) = 0$ if and only if $\mathbf{u} = \mathbf{0}_V$.
(b) $(\mathbf{v}, \mathbf{u}) = (\mathbf{u}, \mathbf{v})$ for any $\mathbf{u}$, $\mathbf{v}$ in V.
(c) $(\mathbf{u} + \mathbf{v}, \mathbf{w}) = (\mathbf{u}, \mathbf{w}) + (\mathbf{v}, \mathbf{w})$ for any $\mathbf{u}$, $\mathbf{v}$, $\mathbf{w}$ in V.
(d) $(c\mathbf{u}, \mathbf{v}) = c(\mathbf{u}, \mathbf{v})$ for $\mathbf{u}$, $\mathbf{v}$ in V and c a real scalar. ▲

From these properties it follows that $(\mathbf{u}, c\mathbf{v}) = c(\mathbf{u}, \mathbf{v})$ because $(\mathbf{u}, c\mathbf{v}) =$

$(c\mathbf{v}, \mathbf{u}) = c(\mathbf{v}, \mathbf{u}) = c(\mathbf{u}, \mathbf{v})$. Also, $(\mathbf{u}, \mathbf{v} + \mathbf{w}) = (\mathbf{u}, \mathbf{v}) + (\mathbf{u}, \mathbf{w})$.

EXAMPLE 1

In Section 3.1 we have defined the standard inner product or dot product on R^3 by defining $(\mathbf{u}, \mathbf{v})$ for $\mathbf{u} = \begin{bmatrix} u_1 \\ u_2 \\ u_3 \end{bmatrix}$ and $\mathbf{v} = \begin{bmatrix} v_1 \\ v_2 \\ v_3 \end{bmatrix}$ in R^3 as

$$(\mathbf{u}, \mathbf{v}) = \mathbf{u} \cdot \mathbf{v} = u_1v_1 + u_2v_2 + u_3v_3.$$

■

EXAMPLE 2

We can define the **standard inner product** on R^n by defining $(\mathbf{u}, \mathbf{v})$ for $\mathbf{u} = \begin{bmatrix} u_1 \\ u_2 \\ \vdots \\ u_n \end{bmatrix}$ and $\mathbf{v} = \begin{bmatrix} v_1 \\ v_2 \\ \vdots \\ v_n \end{bmatrix}$ in R^n, as

$$(\mathbf{u}, \mathbf{v}) = u_1v_1 + u_2v_2 + \cdots + u_nv_n.$$

Thus, if

$$\mathbf{u} = \begin{bmatrix} 1 \\ -2 \\ 3 \\ 4 \end{bmatrix} \quad \text{and} \quad \mathbf{v} = \begin{bmatrix} 3 \\ 2 \\ -2 \\ 1 \end{bmatrix}$$

are vectors in R^4, then

$$(\mathbf{u}, \mathbf{v}) = (1)(3) + (-2)(2) + (3)(-2) + (4)(1) = -3.$$

Of course, we must verify that this function satisfies the properties of Definition 3.2.

■

Remark If we view the vectors $\mathbf{u}$ and $\mathbf{v}$ in R^n as $n \times 1$ matrices, then we can write the standard inner product of $\mathbf{u}$ and $\mathbf{v}$ in terms of matrix multiplication as

$$(\mathbf{u}, \mathbf{v}) = \mathbf{u}^T\mathbf{v}, \tag{1}$$

where we have ignored the brackets around the 1×1 matrix $\mathbf{u}^T\mathbf{v}$ (Exercise 39). This observation can also be used in describing matrix multiplication as follows. If C is the matrix product of the $m \times n$ matrix A and the $n \times p$ matrix B, then the (i, j) entry c_{ij} in C is the matrix product of the ith row $\mathbf{a}_i$ of A and the jth column $\mathbf{b}_j$ of B. That is,

$$c_{ij} = \mathbf{a}_i\mathbf{b}_j.$$

Since $\mathbf{a}_i^T$ is a vector in R^n, we see that c_{ij} is the standard inner product of $\mathbf{a}_i^T$ and $\mathbf{b}_j$.

EXAMPLE 3

Let V be any finite-dimensional vector space and let $S = \{\mathbf{u}_1, \mathbf{u}_2, \ldots, \mathbf{u}_n\}$ be an ordered basis for V. If

$$\mathbf{v} = a_1\mathbf{u}_1 + a_2\mathbf{u}_2 + \cdots + a_n\mathbf{u}_n$$

and

$$\mathbf{w} = b_1\mathbf{u}_1 + b_2\mathbf{u}_2 + \cdots + b_n\mathbf{u}_n,$$

we define

$$(\mathbf{v}, \mathbf{w}) = \left(\left[\mathbf{v}\right]_S, \left[\mathbf{w}\right]_S\right) = a_1b_1 + a_2b_2 + \cdots + a_nb_n.$$

It is not difficult to verify that this defines an inner product on V. This definition of $(\mathbf{v}, \mathbf{w})$ as an inner product on V uses the standard inner product on R^n. ■

Example 3 shows that we can define an inner product on any finite-dimensional vector space. Of course, if we change the basis for V in Example 3, we obtain a different inner product.

EXAMPLE 4

Let $\mathbf{u} = \begin{bmatrix} u_1 \\ u_2 \end{bmatrix}$ and $\mathbf{v} = \begin{bmatrix} v_1 \\ v_2 \end{bmatrix}$ be vectors in R^2. We define

$$(\mathbf{u}, \mathbf{v}) = u_1v_1 - u_2v_1 - u_1v_2 + 3u_2v_2.$$

Show that this gives an inner product on R^2.

Solution We have

$$\begin{aligned}(\mathbf{u}, \mathbf{u}) &= u_1^2 - 2u_1u_2 + 3u_2^2 = u_1^2 - 2u_1u_2 + u_2^2 + 2u_2^2 \\ &= (u_1 - u_2)^2 + 2u_2^2 \geq 0.\end{aligned}$$

Moreover, if $(\mathbf{u}, \mathbf{u}) = 0$, then $u_1 = u_2$ and $u_2 = 0$, so $\mathbf{u} = \mathbf{0}$. Conversely, if $\mathbf{u} = \mathbf{0}$, then $(\mathbf{u}, \mathbf{u}) = 0$. We can also verify (see Exercise 2) the remaining three properties of Definition 3.2. This inner product is, of course, not the standard inner product on R^2. ■

Example 4 shows that on one vector space we may have more than one inner product, since we also have the standard inner product on R^2.

EXAMPLE 5

Let V be the vector space of all continuous real-valued functions on the unit interval $[0, 1]$. For f and g in V, we let $(f, g) = \int_0^1 f(t)g(t)\,dt$. We now verify that this is an inner product on V, that is, that the properties of Definition 3.2 are satisfied.

Using results from calculus, we have for $f \neq 0$, the zero function,

$$(f, f) = \int_0^1 (f(t))^2\,dt \geq 0.$$

Moreover, if $(f, f) = 0$, then $f = 0$. Conversely, if $f = 0$, then $(f, f) = 0$. Also,

$$(f, g) = \int_0^1 f(t)g(t)\,dt = \int_0^1 g(t)f(t)\,dt = (g, f).$$

Next,

$$\begin{aligned}(f + g, h) &= \int_0^1 (f(t) + g(t))h(t)\,dt = \int_0^1 f(t)h(t)\,dt + \int_0^1 g(t)h(t)\,dt \\ &= (f, h) + (g, h).\end{aligned}$$

Finally,

$$(cf, g) = \int_0^1 (cf(t))g(t)\,dt = c\int_0^1 f(t)g(t)\,dt = c(f, g).$$

Thus, if f and g are the functions defined by $f(t) = t + 1$, $g(t) = 2t + 3$, then

$$(f, g) = \int_0^1 (t + 1)(2t + 3)\,dt = \int_0^1 (2t^2 + 5t + 3)\,dt = \tfrac{37}{6}.$$ ■

EXAMPLE 6

Let $V = R_2$; if $\mathbf{u} = \begin{bmatrix} u_1 & u_2 \end{bmatrix}$ and $\mathbf{v} = \begin{bmatrix} v_1 & v_2 \end{bmatrix}$ are vectors in V, we define $(\mathbf{u}, \mathbf{v}) = u_1v_1 - u_2v_1 - u_1v_2 + 5u_2v_2$. The verification that this function is an inner product is entirely analogous to the verification required in Example 4. ■

EXAMPLE 7

Let $V = P$; if $p(t)$ and $q(t)$ are polynomials in P, we define

$$(p(t), q(t)) = \int_0^1 p(t)q(t)\,dt.$$

The verification that this function is an inner product is identical to the verification given for Example 5. ■

We now show that every inner product on a finite-dimensional vector space V is completely determined, in terms of a given basis, by a certain matrix.

Theorem 3.2

Let $S = \{\mathbf{u}_1, \mathbf{u}_2, \ldots, \mathbf{u}_n\}$ be an ordered basis for a finite-dimensional vector space V and assume that we are given an inner product on V. Let $c_{ij} = (\mathbf{u}_i, \mathbf{u}_j)$ and $C = \begin{bmatrix} c_{ij} \end{bmatrix}$. Then

(a) *C is a symmetric matrix.*

(b) *C determines $(\mathbf{v}, \mathbf{w})$ for every $\mathbf{v}$ and $\mathbf{w}$ in V.*

Proof

(a) Exercise.

(b) If $\mathbf{v}$ and $\mathbf{w}$ are in V, then

$$\mathbf{v} = a_1\mathbf{u}_1 + a_2\mathbf{u}_2 + \cdots + a_n\mathbf{u}_n$$
$$\mathbf{w} = b_1\mathbf{u}_1 + b_2\mathbf{u}_2 + \cdots + b_n\mathbf{u}_n,$$

which implies that

$$\left[\mathbf{v}\right]_S = \begin{bmatrix} a_1 \\ a_2 \\ \vdots \\ a_n \end{bmatrix} \quad \text{and} \quad \left[\mathbf{w}\right]_S = \begin{bmatrix} b_1 \\ b_2 \\ \vdots \\ b_n \end{bmatrix}.$$

The inner product $(\mathbf{v}, \mathbf{w})$ can then be expressed as

$$(\mathbf{v}, \mathbf{w}) = \left(\sum_{i=1}^{n} a_i\mathbf{u}_i, \mathbf{w}\right) = \sum_{i=1}^{n} a_i(\mathbf{u}_i, \mathbf{w})$$
$$= \sum_{i=1}^{n} a_i \left(\mathbf{u}_i, \sum_{j=1}^{n} b_j\mathbf{u}_j\right)$$
$$= \sum_{i=1}^{n} a_i \sum_{j=1}^{n} b_j(\mathbf{u}_i, \mathbf{u}_j) = \sum_{i=1}^{n}\sum_{j=1}^{n} a_i b_j(\mathbf{u}_i, \mathbf{u}_j)$$
$$= \sum_{i=1}^{n}\sum_{j=1}^{n} a_i c_{ij} b_j$$
$$= \left[\mathbf{v}\right]_S^T C \left[\mathbf{w}\right]_S,$$

so

$$(\mathbf{v}, \mathbf{w}) = \left[\mathbf{v}\right]_S^T C \left[\mathbf{w}\right]_S, \tag{2}$$

which means that C determines $(\mathbf{v}, \mathbf{w})$ for every $\mathbf{v}$ and $\mathbf{w}$ in V. ●

Thus the inner product in Equation (2) is the product of three matrices. We shall next show that the inner product in (2) can also be expressed in terms of a standard inner product on R^n. We first establish the following result for the standard inner product on R^n.

If $A = \left[a_{ij}\right]$ is an $n \times n$ matrix and $\mathbf{x}$ and $\mathbf{y}$ are vectors in R^n, then

$$(A\mathbf{x}, \mathbf{y}) = (\mathbf{x}, A^T\mathbf{y}). \tag{3}$$

Equation (1) together with associativity of matrix multiplication and Theorem 1.4(c) can now be used to prove (3):

$$(A\mathbf{x}, \mathbf{y}) = (A\mathbf{x})^T\mathbf{y} = (\mathbf{x}^T A^T)\mathbf{y} = \mathbf{x}^T(A^T\mathbf{y}) = (\mathbf{x}, A^T\mathbf{y}).$$

Using Equation (1), we can now write (2) as

$$(\mathbf{v}, \mathbf{w}) = \left(\left[\mathbf{v}\right]_S, C\left[\mathbf{w}\right]_S\right), \tag{4}$$

where the inner product on the left is in V and the inner product on the right is the standard inner product on R^n. Using (1) and the fact that C is symmetric, we have

$$(\mathbf{v}, \mathbf{w}) = \left(C\left[\mathbf{v}\right]_S, \left[\mathbf{w}\right]_S\right) \tag{5}$$

(verify). Thus C determines $(\mathbf{v}, \mathbf{w})$ for every $\mathbf{v}$ and $\mathbf{w}$ in V. In summary, we have shown than an inner product on a real finite-dimensional vector space V can be computed using the standard inner product on R^n, as in (4) or (5).

The matrix C in Theorem 3.2 is called the **matrix of the inner product with respect to the ordered basis** S. If the inner product is as defined in Example 3, then $C = I_n$ (verify).

There is another important property satisfied by the matrix of an inner product. If $\mathbf{u}$ is a nonzero vector in R^n, then $(\mathbf{u}, \mathbf{u}) > 0$, so letting $\mathbf{x} = \left[\mathbf{u}\right]_S$, Equation (2) says that

$$\mathbf{x}^T C\mathbf{x} > 0 \quad \text{for every nonzero } \mathbf{x} \text{ in } R^n.$$

This property of the matrix of an inner product is so important that we specifically identify such matrices. An $n \times n$ symmetric matrix C with the property that $\mathbf{x}^T C\mathbf{x} > 0$ for every nonzero vector $\mathbf{x}$ in R^n is called **positive definite**. A positive definite matrix C is nonsingular, for if C is singular, then the homogeneous system $C\mathbf{x} = \mathbf{0}$ has a nontrivial solution $\mathbf{x}_0$. Then $\mathbf{x}_0^T C\mathbf{x}_0 = 0$, contradicting the requirement that $\mathbf{x}^T C\mathbf{x} > 0$ for any nonzero vector $\mathbf{x}$.

If $C = \left[c_{ij}\right]$ is an $n \times n$ positive definite matrix, then we can use C to define an inner product on V. Using the same notation as above, we define

$$(\mathbf{v}, \mathbf{w}) = \left(\left[\mathbf{v}\right]_S, C\left[\mathbf{w}\right]_S\right) = \sum_{i=1}^{n}\sum_{j=1}^{n} a_i c_{ij} b_j.$$

It is not difficult to show that this defines an inner product on V (verify). The only gap in the discussion above is that we still do not know when a real symmetric matrix is positive definite, other than trying to verify the definition, which is usually not a fruitful approach. In Section 6.6 (see Theorem 6.14) we provide a characterization of positive definite matrices.

EXAMPLE 8

Let $C = \begin{bmatrix} 2 & 1 \\ 1 & 2 \end{bmatrix}$. In this case we may verify that C is positive definite as follows.

$$\begin{aligned} \mathbf{x}^T C\mathbf{x} &= \begin{bmatrix} x_1 & x_2 \end{bmatrix} \begin{bmatrix} 2 & 1 \\ 1 & 2 \end{bmatrix} \begin{bmatrix} x_1 \\ x_2 \end{bmatrix} \\ &= 2x_1^2 + 2x_1x_2 + 2x_2^2 \\ &= x_1^2 + x_2^2 + (x_1 + x_2)^2 > 0 \quad \text{if } \mathbf{x} \neq \mathbf{0}. \end{aligned}$$

■

We now define an inner product on P_1 whose matrix with respect to the ordered basis $S = \{t, 1\}$ is C. Thus let $p(t) = a_1 t + a_2$ and $q(t) = b_1 t + b_2$ be any two vectors in P_1. Let $(p(t), q(t)) = 2a_1b_1 + a_2b_1 + a_1b_2 + 2a_2b_2$. We must verify that $(p(t), p(t)) \geq 0$; that is, $2a_1^2 + 2a_1a_2 + 2a_2^2 \geq 0$. We now have

$$2a_1^2 + 2a_1a_2 + 2a_2^2 = a_1^2 + a_2^2 + (a_1 + a_2)^2 \geq 0.$$

Moreover, if $p(t) = 0$, so that $a_1 = 0$ and $a_2 = 0$, then $(p(t), p(t)) = 0$. Conversely, if $(p(t), p(t)) = 0$, then $a_1 = 0$ and $a_2 = 0$, so $p(t) = 0$. The remaining properties are not difficult to verify.

Definition 3.3

A real vector space that has an inner product defined on it is called an **inner product space**. If the space is finite dimensional, it is called a **Euclidean space**. ▲

If V is an inner product space, then by the **dimension** of V we mean the dimension of V as a real vector space, and a set S is a **basis** for V if S is a basis for the real vector space V. Examples 1 through 7 are inner product spaces. Examples 1, 2, 3, 4, and 6 are Euclidean spaces.

In an inner product space we define the **length** of a vector $\mathbf{u}$ by $\|\mathbf{u}\| = \sqrt{(\mathbf{u}, \mathbf{u})}$. This definition of length seems reasonable because at least we have $\|\mathbf{u}\| > 0$ if $\mathbf{u} \neq \mathbf{0}$. We can show (see Exercise 7) that $\|\mathbf{0}\| = 0$.

We shall now prove a result that will enable us to give a worthwhile definition for the cosine of an angle between two nonzero vectors $\mathbf{u}$ and $\mathbf{v}$ in an inner product space V. This result, called the **Cauchy–Schwarz inequality**, has many important applications in mathematics. The proof, although not difficult, is one that is not too natural and does call for a clever start.

Theorem 3.3 (Cauchy†–Schwarz‡ Inequality).

If $\mathbf{u}$ *and* $\mathbf{v}$ *are any two vectors in an inner product space* V*, then*

$$|(\mathbf{u}, \mathbf{v})| \leq \|\mathbf{u}\| \, \|\mathbf{v}\|.$$

†Augustin-Louis Cauchy (1789–1857) grew up, in a suburb of Paris, as a neighbor of several leading mathematicians of the day, attended the École Polytechnique and the École des Ponts et Chaussées, and was for a time a practicing engineer. He was a devout Roman Catholic, with an abiding interest in Catholic charities. He was also strongly devoted to royalty, especially to the Bourbon kings who ruled France after Napoleon's defeat. When Charles X was deposed in 1830, Cauchy voluntarily followed him into exile in Prague.

Cauchy wrote seven books and more than 700 papers of varying quality, touching all branches of mathematics. He made important contributions to the early theory of determinants, the theory of eigenvalues, the study of ordinary and partial differential equations, the theory of permutation groups, and the foundations of calculus, and he founded the theory of functions of a complex variable.

‡Karl Hermann Amandus Schwarz (1843–1921) was born in Poland but was educated and taught in Germany. He was a protégé of Karl Weierstrass and Ernst Eduard Kummer, whose daughter he married. His main contributions to mathematics were in the geometric aspects of analysis, such as conformal mappings and minimal surfaces. In connection with the latter he sought certain numbers associated with differential equations, numbers that have since come to be called eigenvalues. The inequality given above was used in the search for these numbers.

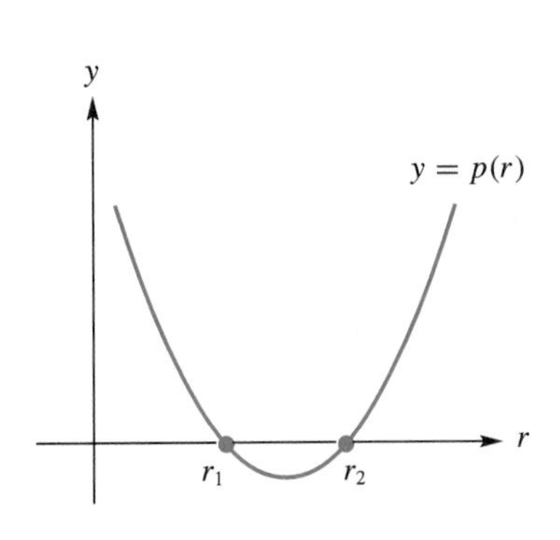

FIGURE 3.18

Proof

If $\mathbf{u} = 0$, then $\|\mathbf{u}\| = 0$ and by Exercise 7(b), $(\mathbf{u}, \mathbf{v}) = 0$, so the inequality holds. Now suppose that $\mathbf{u}$ is nonzero. Let r be a scalar and consider the vector $r\mathbf{u} + \mathbf{v}$. Since the inner product of a vector with itself is always nonnegative, we have

$$0 \leq (r\mathbf{u} + \mathbf{v}, r\mathbf{u} + \mathbf{v}) = (\mathbf{u}, \mathbf{u})r^2 + 2r(\mathbf{u}, \mathbf{v}) + (\mathbf{v}, \mathbf{v}) = ar^2 + 2br + c,$$

where $a = (\mathbf{u}, \mathbf{u})$, $b = (\mathbf{u}, \mathbf{v})$, and $c = (\mathbf{v}, \mathbf{v})$. If we fix $\mathbf{u}$ and $\mathbf{v}$, then $ar^2 + 2br + c = p(r)$ is a quadratic polynomial in r that is nonnegative for all values of r. This means that $p(r)$ has at most one real root, for if it had two distinct real roots, r_1 and r_2, it would be negative between r_1 and r_2 (Figure 3.18). From the quadratic formula, the roots of $p(r)$ are given by

$$\frac{-b + \sqrt{b^2 - ac}}{a} \quad \text{and} \quad \frac{-b - \sqrt{b^2 - ac}}{a}$$

($a \neq 0$ since $\mathbf{u} \neq \mathbf{0}$). Thus we must have $b^2 - ac \leq 0$, which means that $b^2 \leq ac$. Taking square roots, we have $|b| \leq \sqrt{a}\,\sqrt{c}$. Substituting for a, b, and c, we obtain the desired inequality. ●

Remark The result widely known as the Cauchy–Schwarz inequality (Theorem 3.3) provides a good example of how nationalistic feelings make their way into science. In Russia this result is generally known as Bunyakovsky's† inequality. In France it is often referred to as *Cauchy's inequality* and in Germany it is frequently called *Schwarz's inequality*. In an attempt to distribute credit for the result among all three contenders, a minority of authors refer to the result as the *CBS inequality*.

EXAMPLE 9

Let $\mathbf{u} = \begin{bmatrix} 1 \\ 2 \\ -3 \end{bmatrix}$ and $\mathbf{v} = \begin{bmatrix} -3 \\ 2 \\ 2 \end{bmatrix}$ be in the Euclidean space R^3 with the standard inner product. Then $(\mathbf{u}, \mathbf{v}) = -5$, $\|\mathbf{u}\| = \sqrt{14}$, and $\|\mathbf{v}\| = \sqrt{17}$. Therefore, $|(\mathbf{u}, \mathbf{v})| \leq \|\mathbf{u}\|\,\|\mathbf{v}\|$. ■

If $\mathbf{u}$ and $\mathbf{v}$ are any two nonzero vectors in an inner product space V, the Cauchy–Schwarz inequality can be written as

$$-1 \leq \frac{(\mathbf{u}, \mathbf{v})}{\|\mathbf{u}\|\,\|\mathbf{v}\|} \leq 1.$$

It then follows that there is one and only one angle θ such that

$$\cos\theta = \frac{(\mathbf{u}, \mathbf{v})}{\|\mathbf{u}\|\,\|\mathbf{v}\|} \qquad 0 \leq \theta \leq \pi.$$

†Viktor Yakovlevich Bunyakovsky (1804–1889) was born in Bar, Ukraine. He received a doctorate in Paris in 1825. He carried out additional studies in St. Petersburg and then had a long career there as a professor. Bunyakovsky made important contributions in number theory and also worked in geometry, applied mechanics, and hydrostatics. His proof of the Cauchy–Schwarz inequality appeared in one of his monographs in 1859, 25 years before Schwarz published his proof. He died in St. Petersburg.

We define this angle to be the **angle** between $\mathbf{u}$ and $\mathbf{v}$.

The triangle inequality is an easy consequence of the Cauchy–Schwarz inequality.

Corollary 3.1 (Triangle Inequality).

If $\mathbf{u}$ and $\mathbf{v}$ are any vectors in an inner product space V, then $\|\mathbf{u}+\mathbf{v}\| \leq \|\mathbf{u}\| + \|\mathbf{v}\|$.

Proof

We have

$$\begin{aligned}\|\mathbf{u}+\mathbf{v}\|^2 = (\mathbf{u}+\mathbf{v}, \mathbf{u}+\mathbf{v}) &= (\mathbf{u}, \mathbf{u}) + 2(\mathbf{u}, \mathbf{v}) + (\mathbf{v}, \mathbf{v}) \\ &= \|\mathbf{u}\|^2 + 2(\mathbf{u}, \mathbf{v}) + \|\mathbf{v}\|^2.\end{aligned}$$

The Cauchy–Schwarz inequality states that $(\mathbf{u}, \mathbf{v}) \leq |(\mathbf{u}, \mathbf{v})| \leq \|\mathbf{u}\| \, \|\mathbf{v}\|$, so

$$\|\mathbf{u}+\mathbf{v}\|^2 \leq \|\mathbf{u}\|^2 + 2\|\mathbf{u}\| \, \|\mathbf{v}\| + \|\mathbf{v}\|^2 = (\|\mathbf{u}\| + \|\mathbf{v}\|)^2.$$

Taking square roots, we obtain

$$\|\mathbf{u}+\mathbf{v}\| \leq \|\mathbf{u}\| + \|\mathbf{v}\|.$$

●

We now state the Cauchy–Schwarz inequality for the inner product spaces introduced in several of our examples. In Example 2, if

$$\mathbf{u} = \begin{bmatrix} u_1 \\ u_2 \\ \vdots \\ u_n \end{bmatrix} \quad \text{and} \quad \mathbf{v} = \begin{bmatrix} v_1 \\ v_2 \\ \vdots \\ v_n \end{bmatrix}$$

then

$$|(\mathbf{u}, \mathbf{v})| = \left| \sum_{i=1}^{n} u_i v_i \right| \leq \left(\sqrt{\sum_{i=1}^{n} u_i^2} \right) \left(\sqrt{\sum_{i=1}^{n} v_i^2} \right) = \|\mathbf{u}\| \, \|\mathbf{v}\|.$$

In Example 5, if f and g are continuous functions on $[0, 1]$, then

$$|(f, g)| = \left| \int_0^1 f(t)g(t)\,dt \right| \leq \left(\sqrt{\int_0^1 f^2(t)\,dt} \right) \left(\sqrt{\int_0^1 g^2(t)\,dt} \right).$$

EXAMPLE 10

Let V be the Euclidean space P_2 with inner product defined as in Example 7. If $p(t) = t + 2$, then the length of $p(t)$ is

$$\|p(t)\| = \sqrt{(p(t), p(t))} = \sqrt{\int_0^1 (t+2)^2\,dt} = \sqrt{\frac{19}{3}}.$$

If $q(t) = 2t - 3$, then to find the cosine of the angle θ between $p(t)$ and $q(t)$, we proceed as follows. First,

$$\|q(t)\| = \sqrt{\int_0^1 (2t-3)^2\,dt} = \sqrt{\frac{13}{3}}.$$

Next,

$$(p(t), q(t)) = \int_0^1 (t+2)(2t-3)\,dt = \int_0^1 (2t^2 + t - 6)\,dt = -\frac{29}{6}.$$

Then

$$\cos\theta = \frac{(p(t), q(t))}{\|p(t)\|\,\|q(t)\|} = \frac{-\frac{29}{6}}{\sqrt{\frac{19}{3}}\sqrt{\frac{13}{3}}} = \frac{-29}{2\sqrt{(19)(13)}}.$$ ■

Definition 3.4

If V is an inner product space, we define the **distance** between two vectors $\mathbf{u}$ and $\mathbf{v}$ in V as $d(\mathbf{u}, \mathbf{v}) = \|\mathbf{u} - \mathbf{v}\|$. ▲

Definition 3.5

Let V be an inner product space. Two vectors $\mathbf{u}$ and $\mathbf{v}$ in V are **orthogonal** if $(\mathbf{u}, \mathbf{v}) = 0$. ▲

EXAMPLE 11

Let V be the Euclidean space R^4 with the standard inner product. If

$$\mathbf{u} = \begin{bmatrix} 1 \\ 0 \\ 0 \\ 1 \end{bmatrix} \quad \text{and} \quad \mathbf{v} = \begin{bmatrix} 0 \\ 2 \\ 3 \\ 0 \end{bmatrix},$$

then $(\mathbf{u}, \mathbf{v}) = 0$, so $\mathbf{u}$ and $\mathbf{v}$ are orthogonal. ■

EXAMPLE 12

Let V be the inner product space P_2 considered in Example 10. The vectors t and $t - \frac{2}{3}$ are orthogonal, since

$$\left(t, t - \frac{2}{3}\right) = \int_0^1 t\left(t - \frac{2}{3}\right)dt = \int_0^1 \left(t^2 - \frac{2t}{3}\right)dt = 0.$$ ■

Of course, the vector $\mathbf{0}_V$ in an inner product space V is orthogonal to every vector in V [see Exercise 7(b)], and two nonzero vectors in V are orthogonal if the angle θ between them is $\pi/2$. Also, the subset of vectors in V orthogonal to a fixed vector in V is a subspace of V (see Exercise 23).

We know from calculus that we can work with any set of coordinate axes for R^3, but that the work becomes less burdensome, when we deal with Cartesian coordinates. The comparable notion in an inner product space is that of a basis whose vectors are mutually orthogonal. We now proceed to formulate this idea.

Definition 3.6

Let V be an inner product space. A set S of vectors in V is called **orthogonal** if any two distinct vectors in S are orthogonal. If, in addition, each vector in S is of unit length, then S is called **orthonormal**. ▲

We note here that if $\mathbf{x}$ is a nonzero vector in an inner product space, then we can always find a vector of unit length (called a **unit vector**) in the same direction as $\mathbf{x}$; we let $\mathbf{u} = \frac{1}{\|\mathbf{x}\|}\mathbf{x}$. Then

$$\|\mathbf{u}\| = \sqrt{(\mathbf{u}, \mathbf{u})} = \sqrt{\left(\frac{1}{\|\mathbf{x}\|}\mathbf{x}, \frac{1}{\|\mathbf{x}\|}\mathbf{x}\right)} = \sqrt{\frac{(\mathbf{x}, \mathbf{x})}{\|\mathbf{x}\|\,\|\mathbf{x}\|}} = \sqrt{\frac{\|\mathbf{x}\|^2}{\|\mathbf{x}\|\,\|\mathbf{x}\|}} = 1,$$

and the cosine of the angle between $\mathbf{x}$ and $\mathbf{u}$ is 1, so $\mathbf{x}$ and $\mathbf{u}$ have the same direction.

EXAMPLE 13

If $\mathbf{x}_1 = \begin{bmatrix} 1 \\ 0 \\ 2 \end{bmatrix}$, $\mathbf{x}_2 = \begin{bmatrix} -2 \\ 0 \\ 1 \end{bmatrix}$, and $\mathbf{x}_3 = \begin{bmatrix} 0 \\ 1 \\ 0 \end{bmatrix}$, then $\{\mathbf{x}_1, \mathbf{x}_2, \mathbf{x}_3\}$ is an orthogonal set (verify). The vectors

$$\mathbf{u}_1 = \begin{bmatrix} \frac{1}{\sqrt{5}} \\ 0 \\ \frac{2}{\sqrt{5}} \end{bmatrix} \quad \text{and} \quad \mathbf{u}_2 = \begin{bmatrix} -\frac{2}{\sqrt{5}} \\ 0 \\ \frac{1}{\sqrt{5}} \end{bmatrix}$$

are unit vectors in the directions of $\mathbf{x}_1$ and $\mathbf{x}_2$, respectively. Since $\mathbf{x}_3$ is also a unit vector, we conclude that $\{\mathbf{u}_1, \mathbf{u}_2, \mathbf{x}_3\}$ is an orthonormal set. ■

EXAMPLE 14

The natural bases for R^n and R_n are orthonormal sets with respect to the standard inner products on these vector spaces. ■

An important result about orthogonal sets of vectors in an inner product space is the following.

Theorem 3.4

Let $S = \{\mathbf{u}_1, \mathbf{u}_2, \ldots, \mathbf{u}_n\}$ be a finite orthogonal set of nonzero vectors in an inner product space V. Then S is linearly independent.

Proof

Suppose that

$$a_1\mathbf{u}_1 + a_2\mathbf{u}_2 + \cdots + a_n\mathbf{u}_n = \mathbf{0}.$$

Then taking the inner product of both sides with $\mathbf{u}_i$, we have

$$(a_1\mathbf{u}_1 + a_2\mathbf{u}_2 + \cdots + a_i\mathbf{u}_i + \cdots + a_n\mathbf{u}_n, \mathbf{u}_i) = (\mathbf{0}, \mathbf{u}_i) = 0.$$

The left side is

$$a_1(\mathbf{u}_1, \mathbf{u}_i) + a_2(\mathbf{u}_2, \mathbf{u}_i) + \cdots + a_i(\mathbf{u}_i, \mathbf{u}_i) + \cdots + a_n(\mathbf{u}_n, \mathbf{u}_i),$$

and since S is orthogonal, this is $a_i(\mathbf{u}_i, \mathbf{u}_i)$. Thus $a_i(\mathbf{u}_i, \mathbf{u}_i) = 0$. Since $\mathbf{u}_i \neq \mathbf{0}$, $(\mathbf{u}_i, \mathbf{u}_i) \neq 0$, so $a_i = 0$. Repeating this for $i = 1, 2, \ldots, n$, we find that $a_1 = a_2 = \cdots = a_n = 0$, so S is linearly independent. ●

EXAMPLE 15

Let V be the vector space of all continuous real-valued functions on $[-\pi, \pi]$. For f and g in V, we let

$$(f, g) = \int_{-\pi}^{\pi} f(t)g(t)\,dt,$$

which is shown to be an inner product on V (see Example 5). Consider the functions

$$1, \cos t, \sin t, \cos 2t, \sin 2t, \ldots, \cos nt, \sin nt, \ldots, \tag{6}$$

which are clearly in V. The relationships

$$\int_{-\pi}^{\pi} \cos nt\,dt = \int_{-\pi}^{\pi} \sin nt\,dt = \int_{-\pi}^{\pi} \sin nt\,\cos nt\,dt = 0,$$

$$\int_{-\pi}^{\pi} \cos mt\,\cos nt\,dt = \int_{-\pi}^{\pi} \sin mt\,\sin nt\,dt = 0 \quad \text{if } m \neq n$$

demonstrate that $(f, g) = 0$ whenever f and g are distinct functions from (6). Hence every finite subset of functions from (6) is an orthogonal set. Theorem 3.4 then implies that any finite subset of functions from (6) is linearly independent. The functions in (6) were studied by the French mathematician Jean Baptiste Joseph Fourier. We take a closer look at these functions in Section 3.5. ■

3.3 Exercises

1. Verify that the standard inner product on R^n satisfies the properties of Definition 3.2.

2. Verify that the function in Example 4 satisfies the remaining three properties of Definition 3.2.

3. Let $V = M_{nn}$ be the real vector space of all $n \times n$ matrices. If A and B are in V, we define $(A, B) = \mathrm{Tr}(B^T A)$, where Tr is the trace function defined in Exercise 29 of Section 1.2. Show that this function is an inner product on V.

4. Verify that the function defined on V in Example 3 is an inner product.

5. Verify that the function defined on R_2 in Example 6 is an inner product.

6. Verify that the function defined on P in Example 7 is an inner product.

7. Let V be an inner product space. Show the following.

(a) $\|\mathbf{0}\| = 0$.

(b) $(\mathbf{u}, \mathbf{0}) = (\mathbf{0}, \mathbf{u}) = 0$ for any $\mathbf{u}$ in V.

(c) If $(\mathbf{u}, \mathbf{v}) = 0$ for all $\mathbf{v}$ in V, then $\mathbf{u} = \mathbf{0}$.

(d) If $(\mathbf{u}, \mathbf{w}) = (\mathbf{v}, \mathbf{w})$ for all $\mathbf{w}$ in V, then $\mathbf{u} = \mathbf{v}$.

(e) If $(\mathbf{w}, \mathbf{u}) = (\mathbf{w}, \mathbf{v})$ for all $\mathbf{w}$ in V, then $\mathbf{u} = \mathbf{v}$.

In Exercises 8 and 9, let V be the Euclidean space R_4 with the standard inner product. Compute $(\mathbf{u}, \mathbf{v})$.

8. (a) $\mathbf{u} = \begin{bmatrix} 1 & 3 & -1 & 2 \end{bmatrix}, \mathbf{v} = \begin{bmatrix} -1 & 2 & 0 & 1 \end{bmatrix}$.

(b) $\mathbf{u} = \begin{bmatrix} 0 & 0 & 1 & 1 \end{bmatrix}, \mathbf{v} = \begin{bmatrix} 1 & 1 & 0 & 0 \end{bmatrix}$.

(c) $\mathbf{u} = \begin{bmatrix} -2 & 1 & 3 & 4 \end{bmatrix}, \mathbf{v} = \begin{bmatrix} 3 & 2 & 1 & -2 \end{bmatrix}$.

9. (a) $\mathbf{u} = \begin{bmatrix} 1 & 2 & 3 & 4 \end{bmatrix}, \mathbf{v} = \begin{bmatrix} -1 & 0 & -1 & -1 \end{bmatrix}$.

(b) $\mathbf{u} = \begin{bmatrix} 0 & -1 & 1 & 4 \end{bmatrix}, \mathbf{v} = \begin{bmatrix} 2 & 0 & -8 & 2 \end{bmatrix}$.

(c) $\mathbf{u} = \begin{bmatrix} 0 & 0 & -1 & 2 \end{bmatrix}, \mathbf{v} = \begin{bmatrix} 2 & 3 & -1 & 0 \end{bmatrix}$.

In Exercises 10 and 11, use the inner product space of continuous functions on $[0, 1]$ defined in Example 5. Find (f, g) for the following.

10. (a) $f(t) = 1 + t, g(t) = 2 - t$.

(b) $f(t) = 1, g(t) = 3$.

(c) $f(t) = 1, g(t) = 3 + 2t$.

11. (a) $f(t) = 3t, g(t) = 2t^2$.

(b) $f(t) = t, g(t) = e^t$.

(c) $f(t) = \sin t, g(t) = \cos t$.

In Exercises 12 and 13, let V be the Euclidean space of Example 5. Compute the length of the given vector.

12. (a) $\begin{bmatrix} 1 \\ 3 \end{bmatrix}$. (b) $\begin{bmatrix} 3 \\ -1 \end{bmatrix}$. (c) $\begin{bmatrix} 1 \\ 0 \end{bmatrix}$.

13. (a) $\begin{bmatrix} 0 \\ -2 \end{bmatrix}$. (b) $\begin{bmatrix} -2 \\ -4 \end{bmatrix}$. (c) $\begin{bmatrix} 2 \\ 2 \end{bmatrix}$.

In Exercises 14 and 15, let V be the inner product space of Example 7. Find the cosine of the angle between each pair of vectors in V.

14. (a) $p(t) = t, q(t) = t - 1$.

(b) $p(t) = t, q(t) = t$.

(c) $p(t) = 1, q(t) = 2t + 3$.

15. (a) $p(t) = 1, q(t) = 1$.

(b) $p(t) = t^2, q(t) = 2t^3 - \frac{4}{3}t$.

(c) $p(t) = \sin t, q(t) = \cos t$.

16. Prove the **parallelogram law** for any two vectors in an inner product space:

$$\|\mathbf{u} + \mathbf{v}\|^2 + \|\mathbf{u} - \mathbf{v}\|^2 = 2\|\mathbf{u}\|^2 + 2\|\mathbf{v}\|^2.$$

17. Let V be an inner product space. Show that $\|c\mathbf{u}\| = |c| \, \|\mathbf{u}\|$ for any vector $\mathbf{u}$ and any scalar c.

18. State the Cauchy–Schwarz inequality for the inner product spaces defined in Example 4, Example 6, and Exercise 3.

19. Let V be an inner product space. Prove that if $\mathbf{u}$ and $\mathbf{v}$ are any vectors in V, then $\|\mathbf{u} + \mathbf{v}\|^2 = \|\mathbf{u}\|^2 + \|\mathbf{v}\|^2$ if and only if $(\mathbf{u}, \mathbf{v}) = 0$, that is, if and only if $\mathbf{u}$ and $\mathbf{v}$ are orthogonal. This result is known as the **Pythagorean theorem**.

20. Let $\{\mathbf{u}, \mathbf{v}, \mathbf{w}\}$ be an orthonormal set of vectors in an inner product space V. Compute $\|\mathbf{u} + \mathbf{v} + \mathbf{w}\|^2$.

21. Let V be an inner product space. If $\mathbf{u}$ and $\mathbf{v}$ are vectors in V, show that

$$(\mathbf{u}, \mathbf{v}) = \tfrac{1}{4}\|\mathbf{u} + \mathbf{v}\|^2 - \tfrac{1}{4}\|\mathbf{u} - \mathbf{v}\|^2.$$

22. Let V be the Euclidean space R_4 considered in Exercise 8. Find which of the pairs of vectors listed there are orthogonal.

23. Let V be an inner product space and $\mathbf{u}$ a fixed vector in V. Prove that the set of all vectors in V that are orthogonal to $\mathbf{u}$ is a subspace of V.

24. For each of the inner products defined in Examples 1, 4, and 6, choose an ordered basis S for the vector space and find the matrix of the inner product with respect to S.

25. Let $C = \begin{bmatrix} 3 & -2 \\ -2 & 3 \end{bmatrix}$. Define an inner product on R_2 whose matrix with respect to the natural ordered basis is C.

26. If V is an inner product space, prove that the distance function of Definition 3.4 satisfies the following properties for all vectors $\mathbf{u}$, $\mathbf{v}$, and $\mathbf{w}$ in V.

(a) $d(\mathbf{u}, \mathbf{v}) \geq 0$.

(b) $d(\mathbf{u}, \mathbf{v}) = 0$ if and only if $\mathbf{u} = \mathbf{v}$.

(c) $d(\mathbf{u}, \mathbf{v}) = d(\mathbf{v}, \mathbf{u})$.

(d) $d(\mathbf{u}, \mathbf{v}) \leq d(\mathbf{u}, \mathbf{w}) + d(\mathbf{w}, \mathbf{v})$.

In Exercises 25 and 26, let V be the inner product space of Example 5. Compute the distance between the given vectors.

27. (a) $\sin t$, $\cos t$.

(b) t, t^2.

28. (a) $2t + 3$, $3t^2 - 1$.

(b) $3t + 1$, 1.

In Exercises 29 and 30, which of the given sets of vectors in R^3, with the standard inner product, are orthogonal, orthonormal, neither?

29. (a) $\left\{ \begin{bmatrix} \frac{1}{\sqrt{2}} \\ 0 \\ \frac{1}{\sqrt{2}} \end{bmatrix}, \begin{bmatrix} -\frac{1}{\sqrt{2}} \\ 0 \\ \frac{1}{\sqrt{2}} \end{bmatrix}, \begin{bmatrix} 0 \\ 1 \\ 0 \end{bmatrix} \right\}$.

(b) $\left\{ \begin{bmatrix} 1 \\ 1 \\ 0 \end{bmatrix}, \begin{bmatrix} 0 \\ 0 \\ 1 \end{bmatrix}, \begin{bmatrix} 0 \\ 1 \\ 0 \end{bmatrix} \right\}$.

(c) $\left\{ \begin{bmatrix} 1 \\ -1 \\ 0 \end{bmatrix}, \begin{bmatrix} 0 \\ 1 \\ 1 \end{bmatrix}, \begin{bmatrix} 0 \\ 0 \\ 1 \end{bmatrix} \right\}$.

30. (a) $\left\{ \begin{bmatrix} 0 \\ 1 \\ -1 \end{bmatrix}, \begin{bmatrix} 0 \\ 1 \\ 1 \end{bmatrix}, \begin{bmatrix} 2 \\ 0 \\ 0 \end{bmatrix} \right\}$.

(b) $\left\{ \begin{bmatrix} -1 \\ 1 \\ 0 \end{bmatrix}, \begin{bmatrix} \frac{1}{\sqrt{3}} \\ \frac{1}{\sqrt{3}} \\ \frac{1}{\sqrt{3}} \end{bmatrix}, \begin{bmatrix} 0 \\ \frac{1}{\sqrt{2}} \\ -\frac{1}{\sqrt{2}} \end{bmatrix} \right\}$.

(c) $\left\{ \begin{bmatrix} \frac{2}{3} \\ \frac{2}{3} \\ \frac{1}{3} \end{bmatrix}, \begin{bmatrix} -\frac{2}{3} \\ \frac{1}{3} \\ \frac{2}{3} \end{bmatrix}, \begin{bmatrix} \frac{1}{3} \\ -\frac{2}{3} \\ \frac{2}{3} \end{bmatrix} \right\}$

In Exercises 31 and 32, let V be the inner product space of Example 7.

31. Let $p(t) = 3t + 1$ and $q(t) = at$. For what values of a are $p(t)$ and $q(t)$ orthogonal?

32. Let $p(t) = 3t + 1$ and $q(t) = at + b$. For what values of a and b are $p(t)$ and $q(t)$ orthogonal?

In Exercises 33 and 34, let V be the Euclidean space R^3 with the standard inner product.

33. Let $\mathbf{u} = \begin{bmatrix} 1 \\ 1 \\ -2 \end{bmatrix}$ and $\mathbf{v} = \begin{bmatrix} a \\ -1 \\ 2 \end{bmatrix}$. For what values of a are $\mathbf{u}$ and $\mathbf{v}$ orthogonal?

34. Let $\mathbf{u} = \begin{bmatrix} \frac{1}{\sqrt{2}} \\ 0 \\ \frac{1}{\sqrt{2}} \end{bmatrix}$ and $\mathbf{v} = \begin{bmatrix} a \\ -1 \\ -b \end{bmatrix}$. For what values of a and b is $\{\mathbf{u}, \mathbf{v}\}$ an orthonormal set?

35. Let $A = \begin{bmatrix} 1 & 2 \\ 3 & 4 \end{bmatrix}$. Find a 2×2 matrix $B \neq O_2$ such that A and B are orthogonal in the inner product space defined in Exercise 3. Can there be more than one matrix B that is orthogonal to A?

36. Let V be the inner product space in Example 5.

(a) If $p(t) = \sqrt{t}$, find $q(t) = a + bt \neq 0$ such that $p(t)$ and $q(t)$ are orthogonal.

(b) If $p(t) = \sin t$, find $q(t) = a + be^t \neq 0$ such that $p(t)$ and $q(t)$ are orthogonal.

37. Let $C = \begin{bmatrix} c_{ij} \end{bmatrix}$ be an $n \times n$ positive definite symmetric matrix and let V be an n-dimensional vector space with ordered basis $S = \{\mathbf{u}_1, \mathbf{u}_2, \dots, \mathbf{u}_n\}$. For $\mathbf{v} = a_1\mathbf{u}_1 + a_2\mathbf{u}_2 + \cdots + a_n\mathbf{u}_n$ and $\mathbf{w} = b_1\mathbf{u}_1 + b_2\mathbf{u}_2 + \cdots + b_n\mathbf{u}_n$ in V define $(\mathbf{v}, \mathbf{w}) = \sum_{i=1}^{n}\sum_{j=1}^{n} a_i c_{ij} b_j$. Prove that this defines an inner product on V.

38. If A and B are $n \times n$ matrices, show that $(A\mathbf{u}, B\mathbf{v}) = (\mathbf{u}, A^T B\mathbf{v})$ for any vectors $\mathbf{u}$ and $\mathbf{v}$ in Euclidean space R^n with the standard inner product.

39. In the Euclidean space R^n with the standard inner product, prove that $(\mathbf{u}, \mathbf{v}) = \mathbf{u}^T\mathbf{v}$.

40. Consider Euclidean space R^4 with the standard inner product and let

$$\mathbf{u}_1 = \begin{bmatrix} 1 \\ 0 \\ 0 \\ 1 \end{bmatrix} \quad \text{and} \quad \mathbf{u}_2 = \begin{bmatrix} 0 \\ 1 \\ 0 \\ 1 \end{bmatrix}.$$

(a) Prove that the set W consisting of all vectors in R^4 that are orthogonal to both $\mathbf{u}_1$ and $\mathbf{u}_2$ is a subspace of R^4.

(b) Find a basis for W.

41. Let V be an inner product space. Show that if $\mathbf{v}$ is orthogonal to $\mathbf{w}_1, \mathbf{w}_2, \dots, \mathbf{w}_k$, then $\mathbf{v}$ is orthogonal to every vector in

$$\operatorname{span}\{\mathbf{w}_1, \mathbf{w}_2, \dots, \mathbf{w}_k\}.$$

42. Suppose that $\{\mathbf{v}_1, \mathbf{v}_2, \dots, \mathbf{v}_n\}$ is an orthonormal set in R^n with the standard inner product. Let the matrix A be given by $A = \begin{bmatrix} \mathbf{v}_1 & \mathbf{v}_2 & \cdots & \mathbf{v}_n \end{bmatrix}$. Show that A is nonsingular and compute its inverse. Give three different examples of such a matrix in R^2 or R^3.

43. Suppose that $\{\mathbf{v}_1, \mathbf{v}_2, \dots, \mathbf{v}_n\}$ is an orthogonal set in R^n with the standard inner product. Let A be the matrix whose jth column is $\mathbf{v}_j$, $j = 1, 2, \dots, n$. Prove or disprove: A is nonsingular.

44. If A is nonsingular, prove that $A^T A$ is positive definite.

45. If C is positive definite, and $\mathbf{x} \neq \mathbf{0}$ is such that $C\mathbf{x} = k\mathbf{x}$ for some scalar k, show that $k > 0$.

46. If C is positive definite, show that its diagonal entries are positive.

47. Let C be positive definite and r any scalar. Prove or disprove: rC is positive definite.

48. If B and C are $n \times n$ positive definite matrices, show that $B + C$ is positive definite.

49. Let S be the set of $n \times n$ positive definite matrices. Is S a subspace of M_{nn}?

50. To compute the standard inner product of a pair of vectors $\mathbf{u}$ and $\mathbf{v}$ in R^n, use the matrix product operation in your software as follows. Let U and V be column matrices for vectors $\mathbf{u}$ and $\mathbf{v}$, respectively. Then $(\mathbf{u}, \mathbf{v}) =$ the product of U^T and V (or V^T and U). Experiment with the vectors in Example 2 and Exercises 8 and 9 (see Exercise 39).

51. Exercise 41 in Section 3.1 can be generalized to R^n or even R_n in some software. Determine if this is the case for the software that you use.

52. Exercise 43 in Section 3.1 can be generalized to R^n or even R_n in some software. Determine if this is the case for the software that you use.

53. If your software incorporates a computer algebra system that computes definite integrals, then you can compute inner products of functions as in Examples 10 and 15. Use your software to check your results in Exercises 10 and 11.

3.4 GRAM–SCHMIDT PROCESS

In this section we prove that for every Euclidean space V we can obtain a basis S for V such that S is an orthonormal set; such a basis is called an **orthonormal basis**, and the method we use to obtain it is the **Gram–Schmidt process**. From our work with the natural bases for R^2, R^3, and, in general, for R^n, we know that when these bases are present, the computations are kept to a minimum. The reduction in the computational effort is due to the fact that we are dealing with an orthonormal basis. For example, if $S = \{\mathbf{u}_1, \mathbf{u}_2, \dots, \mathbf{u}_n\}$ is a basis for an n-dimensional Euclidean space V, then if $\mathbf{v}$ is any vector in V, we can write $\mathbf{v}$ as

$$\mathbf{v} = c_1\mathbf{u}_1 + c_2\mathbf{u}_2 + \cdots + c_n\mathbf{u}_n.$$

The coefficients $c_1, c_2, \ldots, c_n$ are obtained by solving a linear system of n equations in n unknowns.

However, if S is orthonormal, we can obtain the same result with much less work. This is the content of the following theorem.

Theorem 3.5

Let $S = \{\mathbf{u}_1, \mathbf{u}_2, \ldots, \mathbf{u}_n\}$ be an orthonormal basis for a Euclidean space V and let $\mathbf{v}$ be any vector in V. Then

$$\mathbf{v} = c_1\mathbf{u}_1 + c_2\mathbf{u}_2 + \cdots + c_n\mathbf{u}_n,$$

where

$$c_i = (\mathbf{v}, \mathbf{u}_i) \qquad i = 1, 2, \ldots, n.$$

Proof

Exercise 19. ●

EXAMPLE 1

Let $S = \{\mathbf{u}_1, \mathbf{u}_2, \mathbf{u}_3\}$ be an orthonormal basis for R^3, where

$$\mathbf{u}_1 = \begin{bmatrix} \frac{2}{3} \\ -\frac{2}{3} \\ \frac{1}{3} \end{bmatrix}, \quad \mathbf{u}_2 = \begin{bmatrix} \frac{2}{3} \\ \frac{1}{3} \\ -\frac{2}{3} \end{bmatrix}, \quad \text{and} \quad \mathbf{u}_3 = \begin{bmatrix} \frac{1}{3} \\ \frac{2}{3} \\ \frac{2}{3} \end{bmatrix}.$$

Write the vector $\mathbf{v} = \begin{bmatrix} 3 \\ 4 \\ 5 \end{bmatrix}$ as a linear combination of the vectors in S.

Solution We have

$$\mathbf{v} = c_1\mathbf{u}_1 + c_2\mathbf{u}_2 + c_3\mathbf{u}_3.$$

Theorem 3.5 shows that c_1, c_2, and c_3 can be obtained without having to solve a linear system of three equations in three unknowns. Thus

$$c_1 = (\mathbf{v}, \mathbf{u}_1) = 1, \quad c_2 = (\mathbf{v}, \mathbf{u}_2) = 0, \quad c_3 = (\mathbf{v}, \mathbf{u}_3) = 7$$

and $\mathbf{v} = \mathbf{u}_1 + 7\mathbf{u}_3$. ■

Theorem 3.6 (Gram[†]–Schmidt[‡] Process).

Let V be an inner product space and $W \neq \{\mathbf{0}\}$ an m-dimensional subspace of V. Then there exists an orthonormal basis $T = \{\mathbf{w}_1, \mathbf{w}_2, \ldots, \mathbf{w}_m\}$ for W.

Proof

The proof is constructive; that is, we exhibit the desired basis T. However, we first find an orthogonal basis $T^* = \{\mathbf{v}_1, \mathbf{v}_2, \ldots, \mathbf{v}_m\}$ for W.

Let $S = \{\mathbf{u}_1, \mathbf{u}_2, \ldots, \mathbf{u}_m\}$ be any basis for W. We start by selecting any one of the vectors in S, say $\mathbf{u}_1$, and call it $\mathbf{v}_1$. Thus $\mathbf{v}_1 = \mathbf{u}_1$. We now look for a vector $\mathbf{v}_2$ in the subspace W_1 of W spanned by $\{\mathbf{u}_1, \mathbf{u}_2\}$ that is orthogonal to $\mathbf{v}_1$. Since $\mathbf{v}_1 = \mathbf{u}_1$, W_1 is also the subspace spanned by $\{\mathbf{v}_1, \mathbf{u}_2\}$. Thus $\mathbf{v}_2 = a_1\mathbf{v}_1 + a_2\mathbf{u}_2$. We determine a_1 and a_2 so that $(\mathbf{v}_2, \mathbf{v}_1) = 0$. Now $0 = (\mathbf{v}_2, \mathbf{v}_1) = (a_1\mathbf{v}_1 + a_2\mathbf{u}_2, \mathbf{v}_1) = a_1(\mathbf{v}_1, \mathbf{v}_1) + a_2(\mathbf{u}_2, \mathbf{v}_1)$. Note that $\mathbf{v}_1 \neq \mathbf{0}$ (why?), so $(\mathbf{v}_1, \mathbf{v}_1) \neq 0$. Thus

$$a_1 = -a_2\frac{(\mathbf{u}_2, \mathbf{v}_1)}{(\mathbf{v}_1, \mathbf{v}_1)}.$$

We may assign an arbitrary nonzero value to a_2. Thus, letting $a_2 = 1$, we obtain

$$a_1 = -\frac{(\mathbf{u}_2, \mathbf{v}_1)}{(\mathbf{v}_1, \mathbf{v}_1)}.$$

Hence

$$\mathbf{v}_2 = a_1\mathbf{v}_1 + \mathbf{u}_2 = \mathbf{u}_2 - \frac{(\mathbf{u}_2, \mathbf{v}_1)}{(\mathbf{v}_1, \mathbf{v}_1)}\mathbf{v}_1.$$

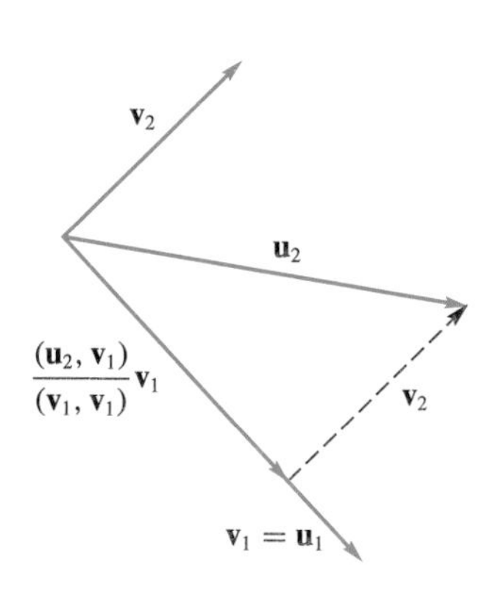

FIGURE 3.19

At this point we have an orthogonal subset $\{\mathbf{v}_1, \mathbf{v}_2\}$ of W (Figure 3.19).

Next, we look for a vector $\mathbf{v}_3$ in the subspace W_2 of W spanned by $\{\mathbf{u}_1, \mathbf{u}_2, \mathbf{u}_3\}$ which is orthogonal to both $\mathbf{v}_1$ and $\mathbf{v}_2$. Of course, W_2 is also the subspace spanned by $\{\mathbf{v}_1, \mathbf{v}_2, \mathbf{u}_3\}$ (why?). Thus $\mathbf{v}_3 = b_1\mathbf{v}_1 + b_2\mathbf{v}_2 + b_3\mathbf{u}_3$. We try to find b_1, b_2, and b_3 so that $(\mathbf{v}_3, \mathbf{v}_1) = 0$ and $(\mathbf{v}_3, \mathbf{v}_2) = 0$. Now

$$0 = (\mathbf{v}_3, \mathbf{v}_1) = (b_1\mathbf{v}_1 + b_2\mathbf{v}_2 + b_3\mathbf{u}_3, \mathbf{v}_1) = b_1(\mathbf{v}_1, \mathbf{v}_1) + b_3(\mathbf{u}_3, \mathbf{v}_1)$$
$$0 = (\mathbf{v}_3, \mathbf{v}_2) = (b_1\mathbf{v}_1 + b_2\mathbf{v}_2 + b_3\mathbf{u}_3, \mathbf{v}_2) = b_2(\mathbf{v}_2, \mathbf{v}_2) + b_3(\mathbf{u}_3, \mathbf{v}_2).$$

Observe that $\mathbf{v}_2 \neq \mathbf{0}$ (why?). Solving for b_1 and b_2, we have

$$b_1 = -b_3\frac{(\mathbf{u}_3, \mathbf{v}_1)}{(\mathbf{v}_1, \mathbf{v}_1)} \quad \text{and} \quad b_2 = -b_3\frac{(\mathbf{u}_3, \mathbf{v}_2)}{(\mathbf{v}_2, \mathbf{v}_2)}.$$

We may assign an arbitrary nonzero value to b_3. Thus, letting $b_3 = 1$, we have

$$\mathbf{v}_3 = \mathbf{u}_3 - \frac{(\mathbf{u}_3, \mathbf{v}_1)}{(\mathbf{v}_1, \mathbf{v}_1)}\mathbf{v}_1 - \frac{(\mathbf{u}_3, \mathbf{v}_2)}{(\mathbf{v}_2, \mathbf{v}_2)}\mathbf{v}_2.$$

[†]Jörgen Pederson Gram (1850–1916) was a Danish actuary.

[‡]Erhard Schmidt (1876–1959) taught at several leading German Universities and was a student of both Hermann Amandus Schwarz and David Hilbert. He made important contributions to the study of integral equations and partial differential equations and, as part of this study, he introduced the method for finding an orthonormal basis in 1907. In 1908 he wrote a paper on infinitely many linear equations in infinitely many unknowns, in which he founded the theory of Hilbert spaces and in which he again used his method.

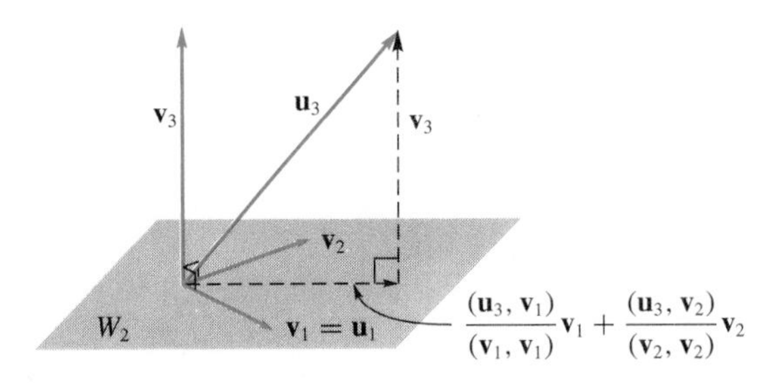

FIGURE 3.20

At this point we have an orthogonal subset $\{\mathbf{v}_1, \mathbf{v}_2, \mathbf{v}_3\}$ of W (Figure 3.20).

We next seek a vector $\mathbf{v}_4$ in the subspace W_3 spanned by $\{\mathbf{u}_1, \mathbf{u}_2, \mathbf{u}_3, \mathbf{u}_4\}$, and also by $\{\mathbf{v}_1, \mathbf{v}_2, \mathbf{v}_3, \mathbf{u}_4\}$, which is orthogonal to $\mathbf{v}_1, \mathbf{v}_2, \mathbf{v}_3$. We obtain

$$\mathbf{v}_4 = \mathbf{u}_4 - \frac{(\mathbf{u}_4, \mathbf{v}_1)}{(\mathbf{v}_1, \mathbf{v}_1)}\mathbf{v}_1 - \frac{(\mathbf{u}_4, \mathbf{v}_2)}{(\mathbf{v}_2, \mathbf{v}_2)}\mathbf{v}_2 - \frac{(\mathbf{u}_4, \mathbf{v}_3)}{(\mathbf{v}_3, \mathbf{v}_3)}\mathbf{v}_3.$$

Continue in this manner until we have an orthogonal set

$$T^* = \{\mathbf{v}_1, \mathbf{v}_2, \ldots, \mathbf{v}_m\}$$

of m vectors. By Theorem 3.4 we conclude that T^* is a basis for W. If we now let $\mathbf{w}_i = \frac{1}{\|\mathbf{v}_i\|}\mathbf{v}_i$ for $i = 1, 2, \ldots, m$, then $T = \{\mathbf{w}_1, \mathbf{w}_2, \ldots, \mathbf{w}_m\}$ is an orthonormal basis for W. ●

Remark It is not difficult to show that if $\mathbf{u}$ and $\mathbf{v}$ are vectors in an inner product space such that $(\mathbf{u}, \mathbf{v}) = 0$, then $(\mathbf{u}, c\mathbf{v}) = 0$ for any scalar c (Exercise 31). This result can often be used to simplify hand computations in the Gram–Schmidt process. As soon as a vector $\mathbf{v}_i$ is computed in Step 2, multiply it by a proper scalar to clear any fractions that may be present. We shall use this approach in our computational work with the Gram–Schmidt process.

EXAMPLE 2

Let W be the subspace of the Euclidean space R^4 with the standard inner product with basis $S = \{\mathbf{u}_1, \mathbf{u}_2, \mathbf{u}_3\}$, where

$$\mathbf{u}_1 = \begin{bmatrix} 1 \\ 1 \\ 1 \\ 0 \end{bmatrix}, \quad \mathbf{u}_2 = \begin{bmatrix} -1 \\ 0 \\ -1 \\ 1 \end{bmatrix}, \quad \mathbf{u}_3 = \begin{bmatrix} -1 \\ 0 \\ 0 \\ -1 \end{bmatrix}.$$

Transform S to an orthonormal basis $T = \{\mathbf{w}_1, \mathbf{w}_2, \mathbf{w}_3\}$.

Solution First, let $\mathbf{v}_1 = \mathbf{u}_1$. Then we find that

$$\mathbf{v}_2 = \mathbf{u}_2 - \frac{(\mathbf{u}_2, \mathbf{v}_1)}{(\mathbf{v}_1, \mathbf{v}_1)}\mathbf{v}_1 = \begin{bmatrix} -1 \\ 0 \\ -1 \\ 1 \end{bmatrix} - \left(-\frac{2}{3}\right)\begin{bmatrix} 1 \\ 1 \\ 1 \\ 0 \end{bmatrix} = \begin{bmatrix} -\frac{1}{3} \\ \frac{2}{3} \\ -\frac{1}{3} \\ 1 \end{bmatrix}.$$

Multiplying $\mathbf{v}_2$ by 3 to clear fractions, we obtain

$$\begin{bmatrix} -1 \\ 2 \\ -1 \\ 3 \end{bmatrix},$$

which we now use as $\mathbf{v}_2$. Next,

$$\mathbf{v}_3 = \mathbf{u}_3 - \frac{(\mathbf{u}_3, \mathbf{v}_1)}{(\mathbf{v}_1, \mathbf{v}_1)}\mathbf{v}_1 - \frac{(\mathbf{u}_3, \mathbf{v}_2)}{(\mathbf{v}_2, \mathbf{v}_2)}\mathbf{v}_2$$

$$= \begin{bmatrix} -1 \\ 0 \\ 0 \\ -1 \end{bmatrix} - \left(-\frac{1}{3}\right)\begin{bmatrix} 1 \\ 1 \\ 1 \\ 0 \end{bmatrix} - \left(-\frac{2}{15}\right)\begin{bmatrix} -1 \\ 2 \\ -1 \\ 3 \end{bmatrix} = \begin{bmatrix} -\frac{4}{5} \\ \frac{3}{5} \\ \frac{1}{5} \\ -\frac{3}{5} \end{bmatrix}.$$

Multiplying $\mathbf{v}_3$ by 5 to clear fractions, we obtain

$$\begin{bmatrix} -4 \\ 3 \\ 1 \\ -3 \end{bmatrix},$$

which we now take as $\mathbf{v}_3$. Thus

$$S = \{\mathbf{v}_1, \mathbf{v}_2, \mathbf{v}_3\} = \left\{ \begin{bmatrix} 1 \\ 1 \\ 1 \\ 0 \end{bmatrix}, \begin{bmatrix} -1 \\ 2 \\ -1 \\ 3 \end{bmatrix}, \begin{bmatrix} -4 \\ 3 \\ 1 \\ -3 \end{bmatrix} \right\}$$

is an orthogonal basis for W. Multiplying each vector in S by the reciprocal of its length yields

$$T = \{\mathbf{w}_1, \mathbf{w}_2, \mathbf{w}_3\} = \left\{ \begin{bmatrix} \frac{1}{\sqrt{3}} \\ \frac{1}{\sqrt{3}} \\ \frac{1}{\sqrt{3}} \\ 0 \end{bmatrix}, \begin{bmatrix} -\frac{1}{\sqrt{15}} \\ \frac{2}{\sqrt{15}} \\ -\frac{1}{\sqrt{15}} \\ \frac{3}{\sqrt{15}} \end{bmatrix}, \begin{bmatrix} -\frac{4}{\sqrt{35}} \\ \frac{3}{\sqrt{35}} \\ \frac{1}{\sqrt{35}} \\ -\frac{3}{\sqrt{35}} \end{bmatrix} \right\},$$

which is an orthonormal basis for W. ■

Remark In solving Example 2, as soon as a vector is computed we multiplied it by an appropriate scalar to eliminate any fractions that may be present. This optional step results in simpler computations when working by hand. If this approach is taken, the resulting basis, while orthonormal, may differ from the orthonormal basis obtained by not clearing fractions. Most computer implementations of the Gram–Schmidt process, including those developed with MATLAB, do not clear fractions.

EXAMPLE 3

Let V be the Euclidean space P_3 with the inner product defined in Example 7 of Section 3.3. Let W be the subspace of P_3 having $S = \{t^2, t\}$ as a basis. Find an orthonormal basis for W.

Solution First, let $\mathbf{u}_1 = t^2$ and $\mathbf{u}_2 = t$. Now let $\mathbf{v}_1 = \mathbf{u}_1 = t^2$. Then

$$\mathbf{v}_2 = \mathbf{u}_2 - \frac{(\mathbf{u}_2, \mathbf{v}_1)}{(\mathbf{v}_1, \mathbf{v}_1)}\mathbf{v}_1 = t - \frac{\frac{1}{4}}{\frac{1}{5}}t^2 = t - \frac{5}{4}t^2,$$

where

$$(\mathbf{v}_1, \mathbf{v}_1) = \int_0^1 t^2 t^2\, dt = \int_0^1 t^4\, dt = \frac{1}{5}$$

and

$$(\mathbf{u}_2, \mathbf{v}_1) = \int_0^1 t t^2\, dt = \int_0^1 t^3\, dt = \frac{1}{4}.$$

Since

$$(\mathbf{v}_2, \mathbf{v}_2) = \int_0^1 \left(t - \frac{5}{4}t^2\right)^2 dt = \frac{1}{48},$$

$\left\{\sqrt{5}\,t^2, \sqrt{48}\left(t - \frac{5}{4}t^2\right)\right\}$ is an orthonormal basis for W. If we choose $\mathbf{u}_1 = t$ and $\mathbf{u}_2 = t^2$, then we obtain (verify) the orthonormal basis $\left\{\sqrt{3}\,t, \sqrt{30}\left(t^2 - \frac{1}{2}t\right)\right\}$ for W. ■

In the proof of Theorem 3.6 we have also established the following result. At each stage of the Gram–Schmidt process, the ordered set $\{\mathbf{w}_1, \mathbf{w}_2, \ldots, \mathbf{w}_k\}$ is an orthonormal basis for the subspace spanned by

$$\{\mathbf{u}_1, \mathbf{u}_2, \ldots, \mathbf{u}_k\} \qquad 1 \leq k \leq n.$$

Also, the final orthonormal basis T depends upon the order of the vectors in the given basis S. Thus, if we change the order of the vectors in S, we might obtain a different orthonormal basis T for W.

Remark We make one final observation with regard to the Gram–Schmidt process. In our proof of Theorem 3.6 we first obtained an orthogonal basis T^* and then normalized all the vectors in T^* to find the orthonormal basis T. Of course, an alternative course of action is to normalize each vector as it is produced. However, normalizing at the end is simpler for hand computation.

One of the useful consequences of having an orthonormal basis in a Euclidean space V is that an arbitrary inner product on V, when it is expressed in terms of coordinates with respect to the orthonormal basis, behaves like the standard inner product on R^n.

Theorem 3.7

Let V be an n-dimensional Euclidean space, and let $S = \{\mathbf{u}_1, \mathbf{u}_2, \ldots, \mathbf{u}_n\}$ be an orthonormal basis for V. If $\mathbf{v} = a_1\mathbf{u}_1 + a_2\mathbf{u}_2 + \cdots + a_n\mathbf{u}_n$ and $\mathbf{w} = b_1\mathbf{u}_1 + b_2\mathbf{u}_2 + \cdots + b_n\mathbf{u}_n$, then

$$(\mathbf{v}, \mathbf{w}) = a_1b_1 + a_2b_2 + \cdots + a_nb_n.$$

Proof

We first compute the matrix $C = \begin{bmatrix} c_{ij} \end{bmatrix}$ of the given inner product with respect to the ordered basis S. We have

$$c_{ij} = (\mathbf{u}_i, \mathbf{u}_j) = \begin{cases} 1 & \text{if } i = j \\ 0 & \text{if } i \neq j. \end{cases}$$

Hence $C = I_n$, the identity matrix. Now we also know from Equation (2) of Section 3.3 that

$$(\mathbf{v}, \mathbf{w}) = \begin{bmatrix} \mathbf{v} \end{bmatrix}_S^T C \begin{bmatrix} \mathbf{w} \end{bmatrix}_S = \begin{bmatrix} \mathbf{v} \end{bmatrix}_S^T I_n \begin{bmatrix} \mathbf{w} \end{bmatrix}_S = \begin{bmatrix} \mathbf{v} \end{bmatrix}_S^T \begin{bmatrix} \mathbf{w} \end{bmatrix}_S$$

$$= \begin{bmatrix} a_1 & a_2 & \cdots & a_n \end{bmatrix} \begin{bmatrix} b_1 \\ b_2 \\ \vdots \\ b_n \end{bmatrix} = a_1b_1 + a_2b_2 + \cdots + a_nb_n,$$

which establishes the result. ●

The theorem that we just proved has some additional implications. Consider the Euclidean space R_3 with the standard inner product, and let W be the subspace with ordered basis $S = \{[2 \;\; 1 \;\; 1], [1 \;\; 1 \;\; 2]\}$. Let $\mathbf{u} = [5 \;\; 3 \;\; 4]$ be a vector in W. Then

$$[5 \;\; 3 \;\; 4] = 2[2 \;\; 1 \;\; 1] + 1[1 \;\; 1 \;\; 2],$$

so $[5 \;\; 3 \;\; 4]_S = \begin{bmatrix} 2 \\ 1 \end{bmatrix}$. Now the length of $\mathbf{u}$ is

$$\|\mathbf{u}\| = \sqrt{5^2 + 3^2 + 4^2} = \sqrt{25 + 9 + 16} = \sqrt{50}.$$

We might expect to compute the length of $\mathbf{u}$ by using the coordinate vector with respect to S, that is, $\|\mathbf{u}\| = \sqrt{2^2 + 1^2} = \sqrt{5}$. Obviously, we have the wrong answer. However, let us transform the given basis S for W into an orthonormal basis T for W. Using the Gram–Schmidt process, we find that (verify)

$$\{[2 \;\; 1 \;\; 1], [-\tfrac{4}{6} \;\; \tfrac{1}{6} \;\; \tfrac{7}{6}]\}$$

is an orthogonal basis for W. It then follows from Exercise 31 that

$$\{[2 \;\; 1 \;\; 1], [-4 \;\; 1 \;\; 7]\}$$

is also an orthogonal basis, so

$$T = \left\{ \begin{bmatrix} \frac{2}{\sqrt{6}} & \frac{1}{\sqrt{6}} & \frac{1}{\sqrt{6}} \end{bmatrix}, \begin{bmatrix} -\frac{4}{\sqrt{66}} & \frac{1}{\sqrt{66}} & \frac{7}{\sqrt{66}} \end{bmatrix} \right\}$$

is an orthonormal basis for W. Then the coordinate vector of $\mathbf{u}$ with respect to T is (verify)

$$[\mathbf{u}]_T = \begin{bmatrix} \frac{17}{6}\sqrt{6} \\ \frac{1}{6}\sqrt{66} \end{bmatrix}.$$

Computing the length of $\mathbf{u}$, using these coordinates, we find that

$$\|\mathbf{u}\|_T = \sqrt{\left(\frac{17}{6}\sqrt{6}\right)^2 + \left(\frac{1}{6}\sqrt{66}\right)^2} = \sqrt{\frac{1800}{36}} = \sqrt{50}.$$

It is not difficult to show (Exercise 21) that if T is an orthonormal basis for an inner product space and $[\mathbf{v}]_T = \begin{bmatrix} a_1 \\ a_2 \\ \vdots \\ a_n \end{bmatrix}$, then $\|\mathbf{v}\| = \sqrt{a_1^2 + a_2^2 + \cdots + a_n^2}$.

QR-Factorization

In Section 1.8 we discussed the LU-factorization of a matrix and showed how it leads to a very efficient method for solving a linear system. We now discuss another factorization of a matrix A, called the **QR-factorization** of A. This type of factorization is widely used in computer codes to find the eigenvalues of a matrix, to solve linear systems, and to find least squares approximations.

Theorem 3.8

If A is an $m \times n$ matrix with linearly independent columns, then A can be factored as $A = QR$, where Q is an $m \times n$ matrix whose columns form an orthonormal basis for the column space of A and R is an $n \times n$ nonsingular upper triangular matrix.

Proof

Let $\mathbf{u}_1, \mathbf{u}_2, \ldots, \mathbf{u}_n$ denote the linearly independent columns of A, which form a basis for the column space of A. By using the Gram–Schmidt process, we can obtain an orthonormal basis $\mathbf{w}_1, \mathbf{w}_2, \ldots, \mathbf{w}_n$ for the column space of A. Recall how this orthonormal basis was obtained. We first constructed an orthogonal basis $\mathbf{v}_1, \mathbf{v}_2, \ldots, \mathbf{v}_n$ as follows: $\mathbf{v}_1 = \mathbf{u}_1$ and then for $i = 2, 3, \ldots, n$ we have

$$\mathbf{v}_i = \mathbf{u}_i - \frac{(\mathbf{u}_i, \mathbf{v}_1)}{(\mathbf{v}_1, \mathbf{v}_1)}\mathbf{v}_1 - \frac{(\mathbf{u}_i, \mathbf{v}_2)}{(\mathbf{v}_2, \mathbf{v}_2)}\mathbf{v}_2 - \cdots - \frac{(\mathbf{u}_i, \mathbf{v}_{i-1})}{(\mathbf{v}_{i-1}, \mathbf{v}_{i-1})}\mathbf{v}_{i-1}. \tag{1}$$

Finally, $\mathbf{w}_i = \dfrac{1}{\|\mathbf{v}_i\|}\mathbf{v}_i$ for $i = 1, 2, 3, \ldots, n$. Now every $\mathbf{u}$-vector can be written as a linear combination of $\mathbf{w}_1, \mathbf{w}_2, \ldots, \mathbf{w}_n$:

$$\begin{aligned}
\mathbf{u}_1 &= r_{11}\mathbf{w}_1 + r_{21}\mathbf{w}_2 + \cdots + r_{n1}\mathbf{w}_n \\
\mathbf{u}_2 &= r_{12}\mathbf{w}_1 + r_{22}\mathbf{w}_2 + \cdots + r_{n2}\mathbf{w}_n \\
&\ \ \vdots \\
\mathbf{u}_n &= r_{1n}\mathbf{w}_1 + r_{2n}\mathbf{w}_2 + \cdots + r_{nn}\mathbf{w}_n.
\end{aligned} \tag{2}$$

From Theorem 3.5 we have

$$r_{ji} = (\mathbf{u}_i, \mathbf{w}_j).$$

Moreover, from Equation (1), we see that $\mathbf{u}_i$ lies in

$$\text{span}\,\{\mathbf{v}_1, \mathbf{v}_2, \ldots, \mathbf{v}_i\} = \text{span}\,\{\mathbf{w}_1, \mathbf{w}_2, \ldots, \mathbf{w}_i\}.$$

Since $\mathbf{w}_j$ is orthogonal to span $\{\mathbf{w}_1, \mathbf{w}_2, \ldots, \mathbf{w}_i\}$ for $j > i$, it is orthogonal to $\mathbf{u}_i$. Hence $r_{ji} = 0$ for $j > i$. Let Q be the matrix whose columns are $\mathbf{w}_1, \mathbf{w}_2, \ldots, \mathbf{w}_j$. Let

$$\mathbf{r}_j = \begin{bmatrix} r_{1j} \\ r_{2j} \\ \vdots \\ r_{nj} \end{bmatrix}.$$

Then the equations in (2) can be written in matrix form as

$$A = \begin{bmatrix} \mathbf{u}_1 & \mathbf{u}_2 & \cdots & \mathbf{u}_n \end{bmatrix} = \begin{bmatrix} Q\mathbf{r}_1 & Q\mathbf{r}_2 & \cdots & Q\mathbf{r}_n \end{bmatrix} = QR,$$

where R is the matrix whose columns are $\mathbf{r}_1, \mathbf{r}_2, \ldots, \mathbf{r}_n$. Thus

$$R = \begin{bmatrix} r_{11} & r_{12} & \cdots & r_{1n} \\ 0 & r_{22} & \cdots & r_{2n} \\ 0 & 0 & \cdots & \\ \vdots & \vdots & & \vdots \\ 0 & 0 & \cdots & r_{nn} \end{bmatrix}.$$

We now show that R is nonsingular. Let $\mathbf{x}$ be a solution to the linear system $R\mathbf{x} = \mathbf{0}$. Multiplying this equation by Q on the left, we have

$$Q(R\mathbf{x}) = (QR)\mathbf{x} = A\mathbf{x} = Q\mathbf{0} = \mathbf{0}.$$

As we know from Chapters 1 and 2, the homogeneous system $A\mathbf{x} = \mathbf{0}$ can be written as

$$x_1\mathbf{u}_1 + x_2\mathbf{u}_2 + \cdots + x_n\mathbf{u}_n = \mathbf{0},$$

where $x_1, x_2, \ldots, x_n$ are the components of the vector $\mathbf{x}$. Since the columns of A are linearly independent,

$$x_1 = x_2 = \cdots = x_n = 0,$$

so $\mathbf{x}$ must be the zero vector. Then Theorem 1.17 implies that R is nonsingular. In Exercise 36 we ask you to show that the diagonal entries r_{ii} of R are nonzero by first expressing $\mathbf{u}_i$ as a linear combination of $\mathbf{v}_1, \mathbf{v}_2, \ldots, \mathbf{v}_i$ and then computing $r_{ii} = (\mathbf{u}_i, \mathbf{w}_i)$. This provides another proof of the nonsingularity of R. ●

EXAMPLE 4

Find the QR-factorization of

$$A = \begin{bmatrix} 1 & -1 & -1 \\ 1 & 0 & 0 \\ 1 & -1 & 0 \\ 0 & 1 & -1 \end{bmatrix}.$$

Solution The columns of A are the vectors $\mathbf{u}_1$, $\mathbf{u}_2$, and $\mathbf{u}_3$, respectively, defined in Example 2. In that example we obtained the following orthonormal basis for the column space of A:

$$\mathbf{w}_1 = \begin{bmatrix} \frac{1}{\sqrt{3}} \\ \frac{1}{\sqrt{3}} \\ \frac{1}{\sqrt{3}} \\ 0 \end{bmatrix}, \quad \mathbf{w}_2 = \begin{bmatrix} -\frac{1}{\sqrt{15}} \\ \frac{2}{\sqrt{15}} \\ -\frac{1}{\sqrt{15}} \\ \frac{3}{\sqrt{15}} \end{bmatrix}, \quad \mathbf{w}_3 = \begin{bmatrix} -\frac{4}{\sqrt{35}} \\ \frac{3}{\sqrt{35}} \\ \frac{1}{\sqrt{35}} \\ -\frac{3}{\sqrt{35}} \end{bmatrix}.$$

Then

$$Q = \begin{bmatrix} \frac{1}{\sqrt{3}} & -\frac{1}{\sqrt{15}} & -\frac{4}{\sqrt{35}} \\ \frac{1}{\sqrt{3}} & \frac{2}{\sqrt{15}} & \frac{3}{\sqrt{35}} \\ \frac{1}{\sqrt{3}} & -\frac{1}{\sqrt{15}} & \frac{1}{\sqrt{35}} \\ 0 & \frac{3}{\sqrt{15}} & -\frac{3}{\sqrt{35}} \end{bmatrix} \approx \begin{bmatrix} 0.5774 & -0.2582 & -0.6761 \\ 0.5774 & 0.5164 & 0.5071 \\ 0.5774 & -0.2582 & 0.1690 \\ 0 & 0.7746 & -0.5071 \end{bmatrix}$$

and

$$R = \begin{bmatrix} r_{11} & r_{12} & r_{13} \\ 0 & r_{22} & r_{23} \\ 0 & 0 & r_{33} \end{bmatrix},$$

where $r_{ji} = (\mathbf{u}_i, \mathbf{w}_j)$. Thus

$$R = \begin{bmatrix} \frac{3}{\sqrt{3}} & -\frac{2}{\sqrt{3}} & -\frac{1}{\sqrt{3}} \\ 0 & \frac{5}{\sqrt{15}} & -\frac{2}{\sqrt{15}} \\ 0 & 0 & \frac{7}{\sqrt{35}} \end{bmatrix} \approx \begin{bmatrix} 1.7321 & -1.1547 & -0.5774 \\ 0 & 1.2910 & -0.5164 \\ 0 & 0 & 1.1832 \end{bmatrix}.$$

As you can verify, $A = QR$. ■

Remark State-of-the-art computer implementations (such as in MATLAB) yield an alternate QR-factorization of an $m \times n$ matrix A as the product of an $m \times m$

matrix Q and an $m \times n$ matrix $R = [r_{ij}]$, where $r_{ij} = 0$ if $i > j$. Thus, if A is 5×3, then

$$R = \begin{bmatrix} * & * & * \\ 0 & * & * \\ 0 & 0 & * \\ 0 & 0 & 0 \\ 0 & 0 & 0 \end{bmatrix}.$$

3.4 Exercises

In this set of exercises, the Euclidean spaces R_n and R^n have the standard inner products on them. Euclidean space P_n has on it the inner product defined in Example 7 of Section 3.3.

1. Use the Gram–Schmidt process to transform the basis $\left\{\begin{bmatrix} 1 \\ 2 \end{bmatrix}, \begin{bmatrix} -3 \\ 4 \end{bmatrix}\right\}$ for the Euclidean space R^2 into:

(a) An orthogonal basis.

(b) An orthonormal basis.

2. Use the Gram–Schmidt process to transform the basis $\left\{\begin{bmatrix} 1 \\ 0 \\ 1 \end{bmatrix}, \begin{bmatrix} -2 \\ 1 \\ 3 \end{bmatrix}\right\}$ for the subspace W of Euclidean space R^3 into:

(a) An orthogonal basis.

(b) An orthonormal basis.

3. Consider the Euclidean space R_4 and let W be the subspace that has

$$S = \{[1 \;\; 1 \;\; -1 \;\; 0], [0 \;\; 2 \;\; 0 \;\; 1]\}$$

as a basis. Use the Gram–Schmidt process to obtain an orthonormal basis for W.

4. Consider Euclidean space R^3 and let W be the subspace that has basis $S = \left\{\begin{bmatrix} 1 \\ 1 \\ 1 \end{bmatrix}, \begin{bmatrix} 1 \\ 0 \\ 2 \end{bmatrix}\right\}$. Use the Gram–Schmidt process to obtain an orthogonal basis for W.

5. Let $S = \{t, 1\}$ be a basis for a subspace W of the Euclidean space P_2. Find an orthonormal basis for W.

6. Repeat Exercise 5 with $S = \{t + 1, t - 1\}$.

7. Let $S = \{t, \sin 2\pi t\}$ be a basis for a subspace W of the inner product space of Example 5 in Section 3.3. Find an orthonormal basis for W.

8. Let $S = \{t, e^t\}$ be a basis for a subspace W of the inner product space of Example 5 in Section 3.3. Find an orthonormal basis for W.

9. Find an orthonormal basis for the Euclidean space R^3 that contains the vectors

$$\begin{bmatrix} \frac{2}{3} \\ -\frac{2}{3} \\ \frac{1}{3} \end{bmatrix} \quad \text{and} \quad \begin{bmatrix} \frac{2}{3} \\ \frac{1}{3} \\ -\frac{2}{3} \end{bmatrix}.$$

10. Use the Gram–Schmidt process to transform the basis

$$\left\{\begin{bmatrix} 1 \\ 1 \\ 1 \end{bmatrix}, \begin{bmatrix} 0 \\ 1 \\ 1 \end{bmatrix}, \begin{bmatrix} 1 \\ 2 \\ 3 \end{bmatrix}\right\}$$

for the Euclidean space R^3 into an orthonormal basis for R^3.

11. Use the Gram–Schmidt process to construct an orthonormal basis for the subspace W of the Euclidean space R^3 *spanned* by

$$\left\{\begin{bmatrix} 1 \\ 1 \\ 1 \end{bmatrix}, \begin{bmatrix} 2 \\ 2 \\ 2 \end{bmatrix}, \begin{bmatrix} 0 \\ 0 \\ 1 \end{bmatrix}, \begin{bmatrix} 1 \\ 2 \\ 3 \end{bmatrix}\right\}.$$

12. Use the Gram–Schmidt process to construct an orthonormal basis for the subspace W of the Euclidean space R_3 *spanned* by

$$\{[1 \;\; -1 \;\; 1], [-2 \;\; 2 \;\; -2], [2 \;\; -1 \;\; 2], [0 \;\; 0 \;\; 0]\}.$$

13. Find an orthonormal basis for the subspace of R^3 consisting of all vectors of the form

$$\begin{bmatrix} a \\ a+b \\ b \end{bmatrix}.$$

14. Find an orthonormal basis for the subspace of R_4 consisting of all vectors of the form

$$\begin{bmatrix} a & a+b & c & b+c \end{bmatrix}.$$

15. Find an orthonormal basis for the subspace of R^3 consisting of all vectors $\begin{bmatrix} a \\ b \\ c \end{bmatrix}$ such that $a+b+c=0$.

16. Find an orthonormal basis for the subspace of R_4 consisting of all vectors $\begin{bmatrix} a & b & c & d \end{bmatrix}$ such that

$$a-b-2c+d=0.$$

17. Find an orthonormal basis for the solution space of the homogeneous system

$$\begin{aligned} x_1 + x_2 - x_3 &= 0 \\ 2x_1 + x_2 + 2x_3 &= 0. \end{aligned}$$

18. Find an orthonormal basis for the solution space of the homogeneous system

$$\begin{bmatrix} 1 & 1 & -1 \\ 2 & 1 & 3 \\ 1 & 2 & -6 \end{bmatrix} \begin{bmatrix} x_1 \\ x_2 \\ x_3 \end{bmatrix} = \begin{bmatrix} 0 \\ 0 \\ 0 \end{bmatrix}.$$

19. Prove Theorem 3.5.

20. Let $S = \{\begin{bmatrix} 1 & -1 & 0 \end{bmatrix}, \begin{bmatrix} 1 & 0 & -1 \end{bmatrix}\}$ be a basis for a subspace W of the Euclidean space R_3.

(a) Use the Gram–Schmidt process to obtain an orthonormal basis for W.

(b) Using Theorem 3.5 write $\mathbf{u} = \begin{bmatrix} 5 & -2 & -3 \end{bmatrix}$ as a linear combination of the vectors obtained in part (a).

21. Prove that if T is an orthonormal basis for a Euclidean space and

$$\left[\mathbf{v}\right]_T = \begin{bmatrix} a_1 \\ a_2 \\ \vdots \\ a_n \end{bmatrix},$$

then $\|\mathbf{v}\| = \sqrt{a_1^2 + a_2^2 + \cdots + a_n^2}$.

22. Let W be the subspace of the Euclidean space R^3 with basis $S = \left\{ \begin{bmatrix} 1 \\ 0 \\ -2 \end{bmatrix}, \begin{bmatrix} -3 \\ 2 \\ 1 \end{bmatrix} \right\}$. Let $\mathbf{v} = \begin{bmatrix} -1 \\ 2 \\ -3 \end{bmatrix}$ be in W.

(a) Find the length of $\mathbf{v}$ directly.

(b) Using the Gram–Schmidt process, transform S into an orthonormal basis T for W.

(c) Find the length of $\mathbf{v}$ using the coordinate vector of $\mathbf{v}$ with respect to T.

23. (a) Verify that

$$S = \left\{ \begin{bmatrix} \frac{1}{3} \\ \frac{2}{3} \\ \frac{2}{3} \end{bmatrix}, \begin{bmatrix} \frac{2}{3} \\ \frac{1}{3} \\ -\frac{2}{3} \end{bmatrix}, \begin{bmatrix} \frac{2}{3} \\ -\frac{2}{3} \\ \frac{1}{3} \end{bmatrix} \right\}$$

is an orthonormal basis for the Euclidean space R^3.

(b) Use Theorem 3.5 to find the coordinate vector of $\mathbf{v} = \begin{bmatrix} 15 \\ 3 \\ 3 \end{bmatrix}$ with respect to S.

(c) Find the length of $\mathbf{v}$ directly and also using the coordinate vector found in part (b).

24. **(Calculus required)** Apply the Gram–Schmidt process to the basis $\{1, t, t^2\}$ for the Euclidean space P_2 and obtain an orthonormal basis for P_2.

25. Let V be the Euclidean space of all 2×2 matrices with inner product defined by $(A, B) = \mathrm{Tr}(B^T A)$.

(a) Prove that

$$S = \left\{ \begin{bmatrix} 1 & 0 \\ 0 & 0 \end{bmatrix}, \begin{bmatrix} 0 & 1 \\ 0 & 0 \end{bmatrix}, \begin{bmatrix} 0 & 0 \\ 1 & 0 \end{bmatrix}, \begin{bmatrix} 0 & 0 \\ 0 & 1 \end{bmatrix} \right\}$$

is an orthonormal basis for V.

(b) Use Theorem 3.5 to find the coordinate vector of $\mathbf{v} = \begin{bmatrix} 1 & 2 \\ 3 & 4 \end{bmatrix}$ with respect to S.

26. Let

$$S = \left\{ \begin{bmatrix} 0 & 0 \\ 0 & 1 \end{bmatrix}, \begin{bmatrix} 1 & 1 \\ 0 & 0 \end{bmatrix}, \begin{bmatrix} 1 & 0 \\ 0 & 1 \end{bmatrix} \right\}$$

be a basis for a subspace W of the Euclidean space defined in Exercise 25. Use the Gram–Schmidt process to find an orthonormal basis for W.

27. Repeat Exercise 26 if

$$S = \left\{ \begin{bmatrix} 1 & 0 \\ 0 & 0 \end{bmatrix}, \begin{bmatrix} 0 & 1 \\ 1 & 0 \end{bmatrix}, \begin{bmatrix} 1 & -1 \\ 0 & 1 \end{bmatrix} \right\}.$$

28. Consider the orthonormal basis

$$S = \left\{ \begin{bmatrix} \frac{1}{\sqrt{5}} \\ 0 \\ \frac{2}{\sqrt{5}} \end{bmatrix}, \begin{bmatrix} -\frac{2}{\sqrt{5}} \\ 0 \\ \frac{1}{\sqrt{5}} \end{bmatrix}, \begin{bmatrix} 0 \\ 1 \\ 0 \end{bmatrix} \right\}$$

for R^3. Using Theorem 3.5 write the vector $\begin{bmatrix} 2 \\ -3 \\ 1 \end{bmatrix}$ as a linear combination of the vectors in S.

In Exercises 29 *and* 30, *compute the* QR-*factorization of* A.

29. (a) $A = \begin{bmatrix} 1 & 2 \\ -1 & 3 \end{bmatrix}$. (b) $A = \begin{bmatrix} 1 & 2 \\ -1 & -2 \\ 1 & 1 \end{bmatrix}$.

(c) $A = \begin{bmatrix} 1 & 0 & -1 \\ 2 & -3 & 3 \\ -1 & 2 & 4 \end{bmatrix}$.

30. (a) $A = \begin{bmatrix} 2 & -1 \\ -1 & 3 \\ 0 & 1 \end{bmatrix}$.

(b) $A = \begin{bmatrix} 1 & 0 & 2 \\ -1 & 2 & 0 \\ -1 & -2 & 2 \end{bmatrix}$.

(c) $A = \begin{bmatrix} 2 & -1 & 1 \\ 1 & 2 & -2 \\ 0 & 1 & -2 \end{bmatrix}$.

31. Show that if $\mathbf{u}$ and $\mathbf{v}$ are orthogonal vectors in an inner product space, then $(\mathbf{u}, c\mathbf{v}) = 0$ for any scalar c.

32. Let $\mathbf{u}_1, \mathbf{u}_2, \dots, \mathbf{u}_n$ be vectors in R^n. Show that if $\mathbf{u}$ is orthogonal to $\mathbf{u}_1, \mathbf{u}_2, \dots, \mathbf{u}_n$, then $\mathbf{u}$ is orthogonal to every vector in

$$\text{span } \{\mathbf{u}_1, \mathbf{u}_2, \dots, \mathbf{u}_n\}.$$

33. Let $\mathbf{u}$ be a fixed vector in R^n. Prove that the set of all vectors in R^n that are orthogonal to $\mathbf{u}$ is a subspace of R^n.

34. Let $S = \{\mathbf{u}_1, \mathbf{u}_2, \dots, \mathbf{u}_k\}$ be an orthonormal basis for a subspace W of Euclidean space V which has dimension $n > k$. Discuss how to construct an orthonormal basis for V that includes S.

35. Let $S = \{\mathbf{v}_1, \mathbf{v}_2, \dots, \mathbf{v}_k\}$ be an orthonormal basis for the Euclidean space V and $\{a_1, a_2, \dots, a_k\}$ be any set of scalars none of which is zero. Prove that $T = \{a_1\mathbf{v}_1, a_2\mathbf{v}_2, \dots, a_k\mathbf{v}_k\}$ is an orthogonal basis for V. How should the scalars $a_1, a_2, \dots, a_k$ be chosen so that T is an orthonormal basis for V?

36. In the proof of Theorem 3.8, show that r_{ii} is nonzero by first expressing $\mathbf{u}_i$ as a linear combination of $\mathbf{v}_1, \mathbf{v}_2, \dots, \mathbf{v}_i$ and then computing $r_{ii} = (\mathbf{u}_i, \mathbf{w}_i)$.

37. Show that every nonsingular matrix has a QR-factorization.

38. Determine if the software that you use has a command to obtain an orthonormal set of vectors from a linearly independent set of vectors in R^n. (Assume that the standard inner product is used.) If it does, compare the output from your command with the results in Example 2. To experiment further, use Exercises 2, 7, 8, 11, and 12.

39. Determine if the software you use has a command to obtain the QR-factorization of a given matrix. If it does, compare the output produced by your command with the results obtained in Example 4. Experiment further with Exercises 29 and 30. Remember the remark following Example 4, pointing out that most software in use today will yield a different type of QR-factorization of a given matrix.

3.5 ORTHOGONAL COMPLEMENTS

In Supplementary Exercises 34 and 35 in Chapter 2, we asked you to show that if W_1 and W_2 are subspaces of a vector space V, then $W_1 + W_2$ (the set of all vectors $\mathbf{v}$ in W such that $\mathbf{v} = \mathbf{w}_1 + \mathbf{w}_2$, where $\mathbf{w}_1$ is in W_1 and $\mathbf{w}_2$ is in W_2) is a subspace of V. Moreover, if $V = W_1 + W_2$ and $W_1 \cap W_2 = \{\mathbf{0}\}$, then $V = W_1 \oplus W_2$; that is, V is the direct sum of W_1 and W_2, which means that every vector in V can be written uniquely as $\mathbf{w}_1 + \mathbf{w}_2$, where $\mathbf{w}_1$ is in W_1 and $\mathbf{w}_2$ is in W_2. In this section we show that if V is an inner product space and W is a finite-dimensional subspace of V, then V can be written as a direct sum of W and another subspace of V. This subspace will be used to examine a basic relationship between four vector spaces associated with a matrix.

Definition 3.7

Let W be a subspace of an inner product space V. A vector $\mathbf{u}$ in V is said to be **orthogonal** to W if it is orthogonal to every vector in W. The set of all vectors in V that are orthogonal to all the vectors in W is called the **orthogonal complement** of W in V and is denoted by $W^{\perp}$ (read "W perp"). ▲

EXAMPLE 1

Let W be the subspace of R^3 consisting of all multiples of the vector

$$\mathbf{w} = \begin{bmatrix} 2 \\ -3 \\ 4 \end{bmatrix}.$$

Thus $W = \text{span}\,\{\mathbf{w}\}$, so W is a one-dimensional subspace of W. Then a vector $\mathbf{u}$ in R^3 belongs to $W^{\perp}$ if and only if $\mathbf{u}$ is orthogonal to $c\mathbf{w}$, for any scalar c. Thus, geometrically, $W^{\perp}$ is the plane with normal $\mathbf{w}$. Using Equations (7) and (8) in optional Section 3.2, $W^{\perp}$ can also be described as the set of all points $P(x, y, z)$ in R^3 such that

$$2x - 3y + 4z = 0.$$ ■

Observe that if W is a subspace of an inner product space V, then the zero vector of V always belongs to $W^{\perp}$ (Exercise 24). Moreover, the orthogonal complement of V is the zero subspace and the orthogonal complement of the zero subspace is V itself (Exercise 25).

Theorem 3.9

Let W be a subspace of an inner product space V. Then

(a) $W^{\perp}$ *is a subspace of* V.

(b) $W \cap W^{\perp} = \{\mathbf{0}\}$.

Proof

(a) Let $\mathbf{u}_1$ and $\mathbf{u}_2$ be in $W^{\perp}$. Then $\mathbf{u}_1$ and $\mathbf{u}_2$ are orthogonal to each vector $\mathbf{w}$ in W. We now have

$$(\mathbf{u}_1 + \mathbf{u}_2, \mathbf{w}) = (\mathbf{u}_1, \mathbf{w}) + (\mathbf{u}_2, \mathbf{w}) = 0 + 0 = 0,$$

so $\mathbf{u}_1 + \mathbf{u}_2$ is in $W^{\perp}$. Also, let $\mathbf{u}$ be in $W^{\perp}$ and c be a real scalar. Then for any vector $\mathbf{w}$ in W we have

$$(c\mathbf{u}, \mathbf{w}) = c(\mathbf{u}, \mathbf{w}) = c\,0 = 0,$$

so $c\mathbf{u}$ is in W, which implies that $W^\perp$ is closed under vector addition and scalar multiplication and hence is a subspace of V.

(b) Let $\mathbf{u}$ be a vector in $W \cap W^\perp$. Then $\mathbf{u}$ is in both W and $W^\perp$, so $(\mathbf{u}, \mathbf{u}) = 0$. From Definition 3.2 it follows that $\mathbf{u} = \mathbf{0}$. ●

In Exercise 26 we ask you to show that if W is a subspace of an inner product space V that is spanned by a set of vectors S, then a vector $\mathbf{u}$ in V belongs to $W^\perp$ if and only if $\mathbf{u}$ is orthogonal to every vector in S. This result can be helpful in finding $W^\perp$, as shown by the next example.

EXAMPLE 2

Let V be the Euclidean space P_3 with the inner product defined in Example 7 of Section 3.3:

$$(p(t), q(t)) = \int_0^1 p(t)q(t)\,dt.$$

Let W be the subspace of P_3 with basis $\{1, t^2\}$. Find a basis for $W^\perp$.

Solution Let $p(t) = at^3 + bt^2 + ct + d$ be an element of $W^\perp$. Since $p(t)$ must be orthogonal to each of the vectors in the given basis for W, we have

$$(p(t), 1) = \int_0^1 (at^3 + bt^2 + ct + d)\,dt = \frac{a}{4} + \frac{b}{3} + \frac{c}{2} + d = 0,$$

$$(p(t), t^2) = \int_0^1 (at^5 + bt^4 + ct^3 + dt^2)\,dt = \frac{a}{6} + \frac{b}{5} + \frac{c}{4} + \frac{d}{3} = 0.$$

Solving the homogeneous system

$$\begin{aligned} \frac{a}{4} + \frac{b}{3} + \frac{c}{2} + d &= 0 \\ \frac{a}{6} + \frac{b}{5} + \frac{c}{4} + \frac{d}{3} &= 0, \end{aligned}$$

we obtain (verify)

$$a = 3r + 16s, \quad b = -\frac{15}{4}r - 15s, \quad c = r, \quad d = s.$$

Then

$$\begin{aligned} p(t) &= (3r + 16s)t^3 + \left(-\frac{15}{4}r - 15s\right)t^2 + rt + s \\ &= r\left(3t^3 - \frac{15}{4}t^2 + t\right) + s(16t^3 - 15t^2 + 1). \end{aligned}$$

Hence the vectors $3t^3 - \frac{15}{4}t^2 + t$ and $16t^3 - 15t^2 + 1$ span $W^\perp$. Since they are not multiples of each other, they are linearly independent and thus form a basis for $W^\perp$. ■

Theorem 3.10

Let W be a finite-dimensional subspace of an inner product space V. Then

$$V = W \oplus W^{\perp}.$$

Proof

Let $\dim W = m$. Then W has a basis consisting of m vectors. By the Gram–Schmidt process we can transform this basis to an orthonormal basis. Thus let $S = \{\mathbf{w}_1, \mathbf{w}_2, \ldots, \mathbf{w}_m\}$ be an orthonormal basis for W. If $\mathbf{v}$ is a vector in V, let

$$\mathbf{w} = (\mathbf{v}, \mathbf{w}_1)\mathbf{w}_1 + (\mathbf{v}, \mathbf{w}_2)\mathbf{w}_2 + \cdots + (\mathbf{v}, \mathbf{w}_m)\mathbf{w}_m \tag{1}$$

and

$$\mathbf{u} = \mathbf{v} - \mathbf{w}. \tag{2}$$

Since $\mathbf{w}$ is a linear combination of vectors in S, $\mathbf{w}$ belongs to W. We next show that $\mathbf{u}$ lies in $W^{\perp}$ by showing that $\mathbf{u}$ is orthogonal to every vector in S, a basis for W. For each $\mathbf{w}_i$ in S, we have

$$\begin{aligned}
(\mathbf{u}, \mathbf{w}_i) &= (\mathbf{v} - \mathbf{w}, \mathbf{w}_i) = (\mathbf{v}, \mathbf{w}_i) - (\mathbf{w}, \mathbf{w}_i) \\
&= (\mathbf{v}, \mathbf{w}_i) - ((\mathbf{v}, \mathbf{w}_1)\mathbf{w}_1 + (\mathbf{v}, \mathbf{w}_2)\mathbf{w}_2 + \cdots + (\mathbf{v}, \mathbf{w}_m)\mathbf{w}_m, \mathbf{w}_i) \\
&= (\mathbf{v}, \mathbf{w}_i) - (\mathbf{v}, \mathbf{w}_i)(\mathbf{w}_i, \mathbf{w}_i) \\
&= 0
\end{aligned}$$

since $(\mathbf{w}_i, \mathbf{w}_j) = 0$ for $i \neq j$ and $(\mathbf{w}_i, \mathbf{w}_i) = 1$, $1 \leq i \leq m$. Thus $\mathbf{u}$ is orthogonal to every vector in W and so lies in $W^{\perp}$. Hence

$$\mathbf{v} = \mathbf{w} + \mathbf{u},$$

which means that $V = W + W^{\perp}$. From part (b) of Theorem 3.9 it follows that

$$V = W \oplus W^{\perp}.$$

Remark As pointed out at the beginning of this section, we also conclude that the vectors $\mathbf{w}$ and $\mathbf{u}$ defined by Equations (1) and (2) are unique.

Theorem 3.11

If W is a finite-dimensional subspace of an inner product space V, then

$$(W^{\perp})^{\perp} = W.$$

Proof

First, if $\mathbf{w}$ is any vector in W, then $\mathbf{w}$ is orthogonal to every vector $\mathbf{u}$ in $W^{\perp}$, so $\mathbf{w}$ is in $(W^{\perp})^{\perp}$. Hence W is a subspace of $(W^{\perp})^{\perp}$. Conversely, let $\mathbf{v}$ be an arbitrary vector in $(W^{\perp})^{\perp}$. Then, by Theorem 3.10, $\mathbf{v}$ can be written as

$$\mathbf{v} = \mathbf{w} + \mathbf{u},$$

where $\mathbf{w}$ is in W and $\mathbf{u}$ is in $W^\perp$. Since $\mathbf{u}$ is in $W^\perp$, it is orthogonal to $\mathbf{v}$ and $\mathbf{w}$. Thus

$$0 = (\mathbf{u}, \mathbf{v}) = (\mathbf{u}, \mathbf{w} + \mathbf{u}) = (\mathbf{u}, \mathbf{w}) + (\mathbf{u}, \mathbf{u}) = (\mathbf{u}, \mathbf{u})$$

or

$$(\mathbf{u}, \mathbf{u}) = 0,$$

which implies that $\mathbf{u} = \mathbf{0}$. Then $\mathbf{v} = \mathbf{w}$, so $\mathbf{v}$ belongs to W. Hence $(W^\perp)^\perp = W$. ●

Remark Since W is the orthogonal complement of $W^\perp$ and $W^\perp$ is also the orthogonal complement of W, we say that W and $W^\perp$ are orthogonal complements.

Relations Among the Fundamental Vector Spaces Associated with a Matrix

If A is a given $m \times n$ matrix, we associate the following four fundamental vector spaces with A: the null space of A, the row space of A, the null space of A^T, and the column space of A. The following theorem shows that pairs of these four vector spaces are orthogonal complements.

Theorem 3.12

If A is a given $m \times n$ matrix, then

(a) *The null space of A is the orthogonal complement of the row space of A.*

(b) *The null space of A^T is the orthogonal complement of the column space of A.*

Proof

(a) Before proving the result, let us verify that the two vector spaces that we wish to show are the same have equal dimensions. If r is the rank of A, then the dimension of the null space of A is $n - r$ (Theorem 2.18). Since the dimension of the row space of A is r, then by Theorem 3.10, the dimension of its orthogonal complement is also $n - r$. Let the vector $\mathbf{x}$ in R^n be in the null space of A. Then $A\mathbf{x} = \mathbf{0}$. Let the vectors $\mathbf{v}_1, \mathbf{v}_2, \ldots, \mathbf{v}_m$ in R^n denote the rows of A. Then the entries in the $m \times 1$ matrix $A\mathbf{x}$ are $\mathbf{v}_1\mathbf{x}, \mathbf{v}_2\mathbf{x}, \ldots, \mathbf{v}_m\mathbf{x}$. Thus we have

$$\mathbf{v}_1\mathbf{x} = 0, \quad \mathbf{v}_2\mathbf{x} = 0, \quad \ldots, \quad \mathbf{v}_m\mathbf{x} = 0. \tag{3}$$

Hence $\mathbf{x}$ is orthogonal to the vectors $\mathbf{v}_1, \mathbf{v}_2, \ldots, \mathbf{v}_m$, which span the row space of A. It then follows that $\mathbf{x}$ is orthogonal to every vector in the row space of A, so $\mathbf{x}$ lies in the orthogonal complement of the row space of A. Hence the null space of A is contained in the orthogonal complement of the row space of A.

Conversely, if $\mathbf{x}$ is in the orthogonal complement of the row space of A, then $\mathbf{x}$ is orthogonal to the vectors $\mathbf{v}_1, \mathbf{v}_2, \ldots, \mathbf{v}_m$, so we have the relations given by (3), which imply that $A\mathbf{x} = \mathbf{0}$. Thus $\mathbf{x}$ belongs to the null space of A. Hence the orthogonal complement of the row space of A is contained in the null space of A. It then follows that the null space of A equals the orthogonal complement of the row space of A.

(b) To establish the result, replace A by A^T in part (a) to conclude that the null space of A^T is the orthogonal complement of the row space of A^T. Since the row space of A^T is the column space of A, we have established part (b). ●

Remark An alternative proof of the converse in part (a) can be given using Exercise 42 in Section 2.5. Explain.

EXAMPLE 3

Let

$$A = \begin{bmatrix} 1 & -2 & 1 & 0 & 2 \\ 1 & -1 & 4 & 1 & 3 \\ -1 & 3 & 2 & 1 & -1 \\ 2 & -3 & 5 & 1 & 5 \end{bmatrix}.$$

Compute the four fundamental vector spaces associated with A and verify Theorem 3.12.

Solution We first transform A to reduced row echelon form, obtaining (verify)

$$B = \begin{bmatrix} 1 & 0 & 7 & 2 & 4 \\ 0 & 1 & 3 & 1 & 1 \\ 0 & 0 & 0 & 0 & 0 \\ 0 & 0 & 0 & 0 & 0 \end{bmatrix}.$$

Solving the linear system $B\mathbf{x} = \mathbf{0}$, we find (verify) that

$$S = \left\{ \begin{bmatrix} -7 \\ -3 \\ 1 \\ 0 \\ 0 \end{bmatrix}, \begin{bmatrix} -2 \\ -1 \\ 0 \\ 1 \\ 0 \end{bmatrix}, \begin{bmatrix} -4 \\ -1 \\ 0 \\ 0 \\ 1 \end{bmatrix} \right\}$$

is a basis for the null space of A. Moreover,

$$T = \{\begin{bmatrix} 1 & 0 & 7 & 2 & 4 \end{bmatrix}, \begin{bmatrix} 0 & 1 & 3 & 1 & 1 \end{bmatrix}\}$$

is a basis for the row space of A. Since the vectors in S and T are orthogonal, it follows that S is a basis for the orthogonal complement of the row space of A, where we take the vectors in S in horizontal form. Next,

$$A^T = \begin{bmatrix} 1 & 1 & -1 & 2 \\ -2 & -1 & 3 & -3 \\ 1 & 4 & 2 & 5 \\ 0 & 1 & 1 & 1 \\ 2 & 3 & -1 & 5 \end{bmatrix}.$$

Solving the linear system $A^T\mathbf{x} = \mathbf{0}$, we find (verify) that

$$S' = \left\{ \begin{bmatrix} 2 \\ -1 \\ 1 \\ 0 \end{bmatrix}, \begin{bmatrix} -1 \\ -1 \\ 0 \\ 1 \end{bmatrix} \right\}$$

is a basis for the null space of A^T. Transforming A^T to reduced row echelon form, we obtain (verify)

$$C = \begin{bmatrix} 1 & 0 & -2 & 1 \\ 0 & 1 & 1 & 1 \\ 0 & 0 & 0 & 0 \\ 0 & 0 & 0 & 0 \\ 0 & 0 & 0 & 0 \end{bmatrix}.$$

Then the nonzero rows of C read vertically yield the following basis for the column space of A:

$$T' = \left\{ \begin{bmatrix} 1 \\ 0 \\ -2 \\ 1 \end{bmatrix}, \begin{bmatrix} 0 \\ 1 \\ 1 \\ 1 \end{bmatrix} \right\}.$$

Since the vectors in S' and T' are orthogonal, it follows that S' is a basis for the orthogonal complement of the column space of A. ■

EXAMPLE 4

Find a basis for the orthogonal complement of the subspace W of R_5 spanned by the vectors

$$\mathbf{w}_1 = \begin{bmatrix} 2 & -1 & 0 & 1 & 2 \end{bmatrix}, \quad \mathbf{w}_2 = \begin{bmatrix} 1 & 3 & 1 & -2 & -4 \end{bmatrix},$$
$$\mathbf{w}_3 = \begin{bmatrix} 3 & 2 & 1 & -1 & -2 \end{bmatrix}, \quad \mathbf{w}_4 = \begin{bmatrix} 7 & 7 & 3 & -4 & -8 \end{bmatrix},$$
$$\mathbf{w}_5 = \begin{bmatrix} 1 & -4 & -1 & -1 & -2 \end{bmatrix}.$$

Solution 1 Let $\mathbf{u} = \begin{bmatrix} a & b & c & d & e \end{bmatrix}$ be an arbitrary vector in $W^\perp$. Since $\mathbf{u}$ is orthogonal to each of the given vectors spanning W, we have a linear system of five equations in five unknowns, whose coefficient matrix is (verify)

$$A = \begin{bmatrix} 2 & -1 & 0 & 1 & 2 \\ 1 & 3 & 1 & -2 & -4 \\ 3 & 2 & 1 & -1 & -2 \\ 7 & 7 & 3 & -4 & -8 \\ 1 & -4 & -1 & -1 & -2 \end{bmatrix}.$$

Solving the homogeneous system $A\mathbf{x} = \mathbf{0}$, we obtain the following basis for the solution space (verify):

$$S = \left\{ \begin{bmatrix} -\frac{1}{7} \\ -\frac{2}{7} \\ 1 \\ 0 \\ 0 \end{bmatrix}, \begin{bmatrix} 0 \\ 0 \\ 0 \\ -2 \\ 1 \end{bmatrix} \right\}.$$

These vectors taken in horizontal form provide a basis for $W^{\perp}$.

Solution 2 Form the matrix whose rows are the given vectors. This matrix is A as shown in Solution 1 above, so the row space of A is W. By Theorem 3.12, $W^{\perp}$ is the null space of A. Thus we obtain the same basis for $W^{\perp}$ as in Solution 1. ■

The following example will be used to geometrically illustrate Theorem 3.12.

EXAMPLE 5

Let

$$A = \begin{bmatrix} 3 & -1 & 2 \\ 2 & 1 & 3 \\ 7 & 1 & 8 \end{bmatrix}.$$

Transforming A to reduced row echelon form we obtain

$$\begin{bmatrix} 1 & 0 & 1 \\ 0 & 1 & 1 \\ 0 & 0 & 0 \end{bmatrix},$$

so the row space of A is a two-dimensional subspace of R_3, that is, a plane passing through the origin, with basis $\{(1, 0, 1), (0, 1, 1)\}$. The null space of A is a one-dimensional subspace of R^3 with basis

$$\left\{ \begin{bmatrix} -1 \\ -1 \\ 1 \end{bmatrix} \right\}$$

(verify). Since this basis vector is orthogonal to the two basis vectors for the row space of A given above, the null space of A is orthogonal to the row space of A; that is, the null space of A is the orthogonal complement of the row space of A.

Next, transforming A^T to reduced row echelon form, we have

$$\begin{bmatrix} 1 & 0 & 1 \\ 0 & 1 & 2 \\ 0 & 0 & 0 \end{bmatrix},$$

(verify). It follows that

$$\left\{ \begin{bmatrix} -1 \\ -2 \\ 1 \end{bmatrix} \right\}$$

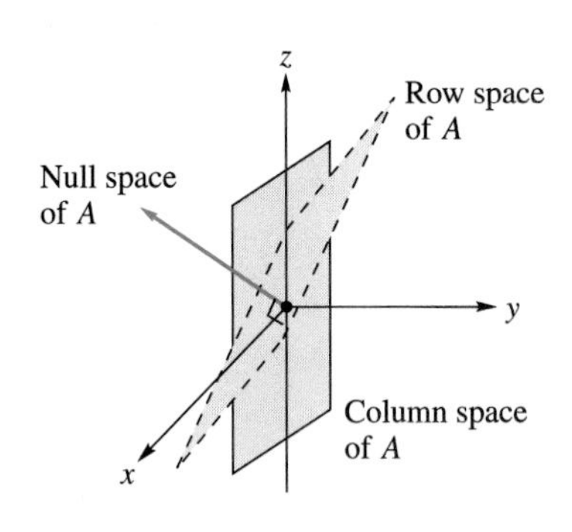

FIGURE 3.21

is a basis for the null space of A^T (verify). Hence the null space of A^T is a line through the origin. Moreover,

$$\left\{ \begin{bmatrix} 1 \\ 0 \\ 1 \end{bmatrix}, \begin{bmatrix} 0 \\ 1 \\ 2 \end{bmatrix} \right\}$$

is a basis for the column space of A^T (verify), so the column space of A^T is a plane through the origin. Since every basis vector for the null space of A^T is orthogonal to every basis vector for the column space of A^T, we conclude that the null space of A^T is the orthogonal complement of the column space of A^T. These results are illustrated in Figure 3.21. ■

Projections and Applications

In Theorem 3.10 and in the Remark following the theorem, we have shown that if W is a finite-dimensional subspace of an inner product space V with orthonormal basis $\{\mathbf{w}_1, \mathbf{w}_2, \ldots, \mathbf{w}_m\}$ and $\mathbf{v}$ is any vector in V, then there exist unique vectors $\mathbf{w}$ in W and $\mathbf{u}$ in $W^\perp$ such that

$$\mathbf{v} = \mathbf{w} + \mathbf{u}.$$

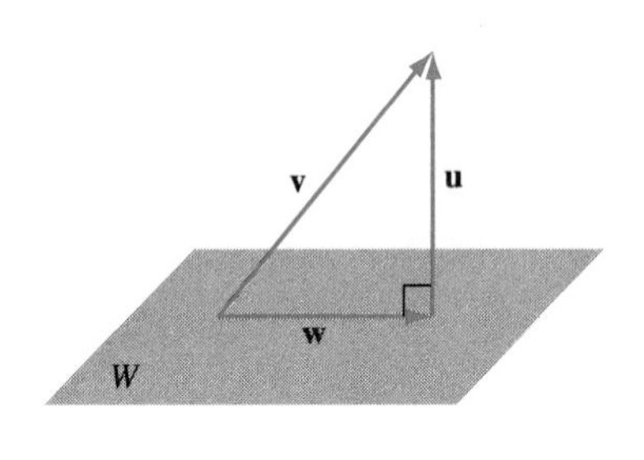

FIGURE 3.22

Moreover, as we saw in Equation (1),

$$\mathbf{w} = (\mathbf{v}, \mathbf{w}_1)\mathbf{w}_1 + (\mathbf{v}, \mathbf{w}_2)\mathbf{w}_2 + \cdots + (\mathbf{v}, \mathbf{w}_m)\mathbf{w}_m,$$

which is called the **orthogonal projection** of $\mathbf{v}$ on W and is denoted by $\text{proj}_W \mathbf{v}$. In Figure 3.22, we illustrate Theorem 3.10 when W is a two-dimensional subspace of R^3 (a plane through the origin).

Often an orthonormal basis has many fractions, so it is helpful to also have a formula giving $\text{proj}_W \mathbf{v}$ when W has an *orthogonal* basis. In Exercise 29, we ask you to show that if $\{\mathbf{w}_1, \mathbf{w}_2, \ldots, \mathbf{w}_m\}$ is an orthogonal basis for W, then

$$\text{proj}_W \mathbf{v} = \frac{(\mathbf{v}, \mathbf{w}_1)}{(\mathbf{w}_1, \mathbf{w}_1)}\mathbf{w}_1 + \frac{(\mathbf{v}, \mathbf{w}_2)}{(\mathbf{w}_2, \mathbf{w}_2)}\mathbf{w}_2 + \cdots + \frac{(\mathbf{v}, \mathbf{w}_m)}{(\mathbf{w}_m, \mathbf{w}_m)}\mathbf{w}_m.$$

EXAMPLE 6

Let W be the two-dimensional subspace of R^3 with orthonormal basis $\{\mathbf{w}_1, \mathbf{w}_2\}$, where

$$\mathbf{w}_1 = \begin{bmatrix} \frac{2}{3} \\ -\frac{1}{3} \\ -\frac{2}{3} \end{bmatrix} \quad \text{and} \quad \mathbf{w}_2 = \begin{bmatrix} \frac{1}{\sqrt{2}} \\ 0 \\ \frac{1}{\sqrt{2}} \end{bmatrix}.$$

Using the standard inner product on R^3, find the orthogonal projection of

$$\mathbf{v} = \begin{bmatrix} 2 \\ 1 \\ 3 \end{bmatrix}$$

on W and the vector $\mathbf{u}$ that is orthogonal to every vector in W.

Solution From Equation (1) we have

$$\mathbf{w} = \text{proj}_W \mathbf{v} = (\mathbf{v}, \mathbf{w}_1)\mathbf{w}_1 + (\mathbf{v}, \mathbf{w}_2)\mathbf{w}_2 = -1\mathbf{w}_1 + \frac{5}{\sqrt{2}}\mathbf{w}_2 = \begin{bmatrix} \frac{11}{6} \\ \frac{1}{3} \\ \frac{19}{6} \end{bmatrix}$$

and

$$\mathbf{u} = \mathbf{v} - \mathbf{w} = \begin{bmatrix} \frac{1}{6} \\ \frac{2}{3} \\ -\frac{1}{6} \end{bmatrix}.$$

■

It is clear from Figure 3.22 that the distance from $\mathbf{v}$ to the plane W is given by the length of the vector $\mathbf{u} = \mathbf{v} - \mathbf{w}$, that is, by

$$\left\| \mathbf{v} - \text{proj}_W \mathbf{v} \right\|.$$

We prove this result in general later in Theorem 3.13.

EXAMPLE 7

Let W be the subspace of R^3 defined in Example 6 and let $\mathbf{v} = \begin{bmatrix} 1 \\ 1 \\ 0 \end{bmatrix}$. Find the distance from $\mathbf{v}$ to W.

Solution We first compute

$$\text{proj}_W \mathbf{v} = (\mathbf{v}, \mathbf{w}_1)\mathbf{w}_1 + (\mathbf{v}, \mathbf{w}_2)\mathbf{w}_2 = \frac{1}{3}\mathbf{w}_1 + \frac{1}{\sqrt{2}}\mathbf{w}_2 = \begin{bmatrix} \frac{13}{18} \\ -\frac{1}{9} \\ \frac{5}{18} \end{bmatrix}.$$

Then

$$\mathbf{v} - \text{proj}_W \mathbf{v} = \begin{bmatrix} 1 \\ 1 \\ 0 \end{bmatrix} - \begin{bmatrix} \frac{13}{18} \\ -\frac{1}{9} \\ \frac{5}{18} \end{bmatrix} = \begin{bmatrix} \frac{5}{18} \\ \frac{10}{9} \\ -\frac{5}{18} \end{bmatrix}$$

and

$$\|\mathbf{v} - \text{proj}_W \mathbf{v}\| = \sqrt{\frac{25}{18^2} + \frac{400}{18^2} + \frac{25}{18^2}} = \frac{15}{18}\sqrt{2} = \frac{5}{6}\sqrt{2},$$

so the distance from $\mathbf{v}$ to W is $\frac{5}{6}\sqrt{2}$. ■

In Example 7, $\|\mathbf{v} - \text{proj}_W \mathbf{v}\|$ represented the distance in 3-space from $\mathbf{v}$ to the plane W. We can generalize this notion of distance from a vector in V to a subspace W of V. We can show that the vector in W that is closest to $\mathbf{v}$ is in fact $\text{proj}_W \mathbf{v}$, so $\|\mathbf{v} - \text{proj}_W \mathbf{v}\|$ represents the distance from $\mathbf{v}$ to W.

Theorem 3.13

Let W be a finite-dimensional subspace of the inner product space V. Then for vector $\mathbf{v}$ belonging to V, the vector in W closest to $\mathbf{v}$ is $\text{proj}_W \mathbf{v}$. That is, $\|\mathbf{v} - \mathbf{w}\|$, for $\mathbf{w}$ belonging to W, is minimized when $\mathbf{w} = \text{proj}_W \mathbf{v}$.

Proof

Let $\mathbf{w}$ be any vector in W. Then

$$\mathbf{v} - \mathbf{w} = (\mathbf{v} - \text{proj}_W \mathbf{v}) + (\text{proj}_W \mathbf{v} - \mathbf{w}).$$

Since $\mathbf{w}$ and $\text{proj}_W \mathbf{v}$ are both in W, $\text{proj}_W \mathbf{v} - \mathbf{w}$ is in W. By Theorem 3.10, $\mathbf{v} - \text{proj}_W \mathbf{v}$ is orthogonal to every vector in W, so

$$\begin{aligned}
\|\mathbf{v} - \mathbf{w}\|^2 &= (\mathbf{v} - \mathbf{w}, \mathbf{v} - \mathbf{w}) \\
&= ((\mathbf{v} - \text{proj}_W \mathbf{v}) + (\text{proj}_W \mathbf{v} - \mathbf{w}), (\mathbf{v} - \text{proj}_W \mathbf{v}) + (\text{proj}_W \mathbf{v} - \mathbf{w})) \\
&= \|\mathbf{v} - \text{proj}_W \mathbf{v}\|^2 + \|\text{proj}_W \mathbf{v} - \mathbf{w}\|^2.
\end{aligned}$$

If $\mathbf{w} \neq \text{proj}_W \mathbf{v}$, then $\|\text{proj}_W \mathbf{v} - \mathbf{w}\|^2$ is positive and

$$\|\mathbf{v} - \mathbf{w}\|^2 > \|\mathbf{v} - \text{proj}_W \mathbf{v}\|^2.$$

Thus it follows that $\text{proj}_W \mathbf{v}$ is the vector in W that minimizes $\|\mathbf{v} - \mathbf{w}\|^2$ and hence minimizes $\|\mathbf{v} - \mathbf{w}\|$. ●

In Example 6, $\mathbf{w} = \text{proj}_W \mathbf{v} = \begin{bmatrix} \frac{11}{6} \\ \frac{1}{3} \\ \frac{19}{6} \end{bmatrix}$ is the vector in $W = \text{span}\,\{\mathbf{w}_1, \mathbf{w}_2\}$ that is closest to $\mathbf{v} = \begin{bmatrix} 2 \\ 1 \\ 3 \end{bmatrix}$.

Fourier Series (calculus required)

In the study of calculus you most likely encountered functions $f(t)$, which had derivatives of all orders at a point $t = t_0$. Associated with $f(t)$ is its Taylor series, defined by

$$\sum_{k=0}^{\infty} \frac{f^{(k)}(t_0)}{k!}(t - t_0)^k. \tag{4}$$

The expression in (4) is called the **Taylor series of f at t_0** (or **about t_0** or **centered at t_0**). When $t_0 = 0$, the Taylor series is called a **Maclaurin series**. The coefficients of

Taylor and Maclaurin series expansions involve successive derivatives of the given function evaluated at the center of the expansion. If we take the first $n + 1$ terms of the series in (4), we obtain a Taylor or Maclaurin polynomial of degree n that approximates the given function.

The function $f(t) = |t|$ does not have a Taylor series expansion at $t_0 = 0$ (a Maclaurin series), because f does not have a derivative at $t = 0$. Thus there is no way to compute the coefficients in such an expansion. The expression in (4) is in terms of the functions $1, t, t^2, \ldots$. However, it is possible to obtain a series expansion for such a function by using a different type of expansion. One such important expansion involves the set of functions

$$1, \quad \cos t, \quad \sin t, \quad \cos 2t, \quad \sin 2t, \quad \ldots, \quad \cos nt, \quad \sin nt, \ldots,$$

which we discussed briefly in Example 15 of Section 3.3. The French mathematician Jean Baptiste Joseph Fourier[†] showed that every function f (continuous or not) that is defined on $[-\pi, \pi]$, can be represented by a series of the form

$$\begin{aligned} &\tfrac{1}{2}a_0 + a_1 \cos t + a_2 \cos 2t + \cdots + a_n \cos nt \\ &\quad + b_1 \sin t + b_2 \sin 2t + \cdots + b_n \sin nt + \cdots . \end{aligned}$$

It then follows that every function f (continuous or not) that is defined on $[-\pi, \pi]$, can be approximated as closely as we wish by a function of the form

$$\begin{aligned} &\tfrac{1}{2}a_0 + a_1 \cos t + a_2 \cos 2t + \cdots + a_n \cos nt \\ &\quad + b_1 \sin t + b_2 \sin 2t + \cdots + b_n \sin nt \end{aligned} \tag{5}$$

for n sufficiently large. The function in (5) is called a **trigonometric polynomial**, and if a_n and b_n are both nonzero we say that its **degree** is n. The topic of Fourier series is beyond the scope of this book. We limit ourselves to a brief discussion on how to obtain the best approximation of a function by trigonometric polynomials.

[†]Jean Baptiste Joseph Fourier (1768–1830) was born in Auxere, France. His father was a tailor. Fourier received much of his early education in the local military school, which was run by the Benedictine order, and at the age of 19 he decided to study for the priesthood. His strong interest in mathematics, which started developing at the age of 13, continued while studying for the priesthood. Two years later, he decided not to take his religious vows and became a teacher at the military school where he had studied.

Fourier was active in French politics throughout the French Revolution and the turbulent period that followed. In 1795, he was appointed to a chair at the prestigious École Polytechnique. In 1798, Fourier accompanied Napoleon as a scientific advisor in his invasion of Egypt. Upon returning to France, he served for 12 years as prefect of the department of Isére and lived in Grenoble. During this period he did his pioneering work on the theory of heat. In this work he showed that every function can be represented by a series of trigonometric polynomials. Such a series is now called a Fourier series. He died in Paris in 1830.

It is not difficult to show that

$$\int_{-\pi}^{\pi} 1\,dt = 2\pi, \qquad \int_{-\pi}^{\pi} \sin nt\,dt = 0, \qquad \int_{-\pi}^{\pi} \cos nt\,dt = 0,$$

$$\int_{-\pi}^{\pi} \sin nt\,\sin mt\,dt = 0 \quad (n \neq m), \qquad \int_{-\pi}^{\pi} \cos nt\,\cos mt\,dt = 0 \quad (n \neq m),$$

$$\int_{-\pi}^{\pi} \sin nt\,\cos mt\,dt = 0 \quad (n \neq m), \qquad \int_{-\pi}^{\pi} \sin nt\,\sin nt\,dt = \pi,$$

$$\int_{-\pi}^{\pi} \cos nt\,\cos nt\,dt = \pi.$$

Let V be the vector space of real-valued continuous functions on $[-\pi, \pi]$. If f and g belong to V, then $(f, g) = \int_{-\pi}^{\pi} f(t)g(t)\,dt$ defines an inner product on V as in Example 15 of Section 3.3. The relations above show that the following set of vectors is an orthonormal set in V

$$\frac{1}{\sqrt{2\pi}}, \quad \frac{1}{\sqrt{\pi}}\cos t, \quad \frac{1}{\sqrt{\pi}}\sin t, \quad \frac{1}{\sqrt{\pi}}\cos 2t, \quad \frac{1}{\sqrt{\pi}}\sin 2t, \quad \ldots,$$
$$\frac{1}{\sqrt{\pi}}\cos nt, \quad \frac{1}{\sqrt{\pi}}\sin nt, \quad \ldots.$$

Now

$$W = \operatorname{span}\left\{\frac{1}{\sqrt{2\pi}}, \frac{1}{\sqrt{\pi}}\cos t, \frac{1}{\sqrt{\pi}}\sin t, \frac{1}{\sqrt{\pi}}\cos 2t, \frac{1}{\sqrt{\pi}}\sin 2t, \quad \ldots, \right.$$
$$\left. \frac{1}{\sqrt{\pi}}\cos nt, \frac{1}{\sqrt{\pi}}\sin nt\right\}$$

is a finite-dimensional subspace of V. Theorem 3.13 implies that the best approximation to a given function f in V by a trigonometric polynomial of degree n is given by $\operatorname{proj}_W f$, the projection of f onto W. This polynomial is called the **Fourier polynomial of degree *n* for *f*.**

EXAMPLE 8

Find Fourier polynomials of degrees one and three for the function $f(t) = |t|$.

Solution First, we compute the Fourier polynomial of degree one. Using Theorem 3.10, we can compute $\operatorname{proj}_W \mathbf{v}$ for $\mathbf{v} = |t|$, as

$$\operatorname{proj}_W |t| = \left(|t|, \frac{1}{\sqrt{2\pi}}\right)\frac{1}{\sqrt{2\pi}} + \left(|t|, \frac{1}{\sqrt{\pi}}\cos t\right)\frac{1}{\sqrt{\pi}}\cos t$$
$$+ \left(|t|, \frac{1}{\sqrt{\pi}}\sin t\right)\frac{1}{\sqrt{\pi}}\sin t.$$

We have

$$\left(|t|, \frac{1}{\sqrt{2\pi}}\right) = \int_{-\pi}^{\pi} |t| \frac{1}{\sqrt{2\pi}}\, dt$$

$$= \frac{1}{\sqrt{2\pi}} \int_{-\pi}^{0} -t\, dt + \frac{1}{\sqrt{2\pi}} \int_{0}^{\pi} t\, dt = \frac{\pi^2}{\sqrt{2\pi}},$$

$$\left(|t|, \frac{1}{\sqrt{\pi}} \cos t\right) = \int_{-\pi}^{\pi} |t| \frac{1}{\sqrt{\pi}} \cos t\, dt$$

$$= \frac{1}{\sqrt{\pi}} \int_{-\pi}^{0} -t \cos t\, dt + \frac{1}{\sqrt{\pi}} \int_{0}^{\pi} t \cos t\, dt$$

$$= -\frac{2}{\sqrt{\pi}} - \frac{2}{\sqrt{\pi}} = -\frac{4}{\sqrt{\pi}},$$

and

$$\left(|t|, \frac{1}{\sqrt{\pi}} \sin t\right) = \int_{-\pi}^{\pi} |t| \frac{1}{\sqrt{\pi}} \sin t\, dt$$

$$= \frac{1}{\sqrt{\pi}} \int_{-\pi}^{0} -t \sin t\, dt + \frac{1}{\sqrt{\pi}} \int_{0}^{\pi} t \sin t\, dt$$

$$= -\sqrt{\pi} + \sqrt{\pi} = 0.$$

Then

$$\text{proj}_W |t| = \frac{\pi^2}{\sqrt{2\pi}} \frac{1}{\sqrt{2\pi}} - \frac{4}{\sqrt{\pi}} \frac{1}{\sqrt{\pi}} \cos t = \frac{\pi}{2} - \frac{4}{\pi} \cos t.$$

Next, we compute the Fourier polynomial of degree three. By Theorem 3.10,

$$\begin{aligned}\text{proj}_W |t| = {} & \left(|t|, \frac{1}{\sqrt{2\pi}}\right) \frac{1}{\sqrt{2\pi}} \\ & + \left(|t|, \frac{1}{\sqrt{\pi}} \cos t\right) \frac{1}{\sqrt{\pi}} \cos t + \left(|t|, \frac{1}{\sqrt{\pi}} \sin t\right) \frac{1}{\sqrt{\pi}} \sin t \\ & + \left(|t|, \frac{1}{\sqrt{\pi}} \cos 2t\right) \frac{1}{\sqrt{\pi}} \cos 2t + \left(|t|, \frac{1}{\sqrt{\pi}} \sin 2t\right) \frac{1}{\sqrt{\pi}} \sin 2t \\ & + \left(|t|, \frac{1}{\sqrt{\pi}} \cos 3t\right) \frac{1}{\sqrt{\pi}} \cos 3t + \left(|t|, \frac{1}{\sqrt{\pi}} \sin 3t\right) \frac{1}{\sqrt{\pi}} \sin 3t.\end{aligned}$$

We have

$$\int_{-\pi}^{\pi} |t| \frac{1}{\sqrt{\pi}} \cos 2t\, dt = 0, \qquad \int_{-\pi}^{\pi} |t| \frac{1}{\sqrt{\pi}} \sin 2t\, dt = 0,$$

$$\int_{-\pi}^{\pi} |t| \frac{1}{\sqrt{\pi}} \cos 3t\, dt = -\frac{4}{9\sqrt{\pi}}, \qquad \int_{-\pi}^{\pi} |t| \frac{1}{\sqrt{\pi}} \sin 3t\, dt = 0.$$

Hence

$$\text{proj}_W \mathbf{v} = \frac{\pi}{2} - \frac{4}{\pi}\cos t - \frac{4}{9\pi}\cos 3t.$$

Figure 3.23 shows the graphs of f and the Fourier polynomial of degree one. Figure 3.24 shows the graphs of f and the Fourier polynomial of degree three. Figure 3.25 shows the graphs of

$$\left||t| - \left(\tfrac{\pi}{2} - \tfrac{4}{\pi}\cos t\right)\right| \quad \text{and} \quad \left||t| - \left(\tfrac{\pi}{2} - \tfrac{4}{\pi}\cos t - \tfrac{4}{9\pi}\cos 3t\right)\right|.$$

Observe how much better the approximation by a Fourier polynomial of degree three is. ■

Fourier series play an important role in the study of heat distribution and in the analysis of sound waves. The study of projections is important in a number of areas in applied mathematics. We illustrate this in Section 3.6 by considering the topic of least squares, which provides a technique for dealing with inconsistent systems.

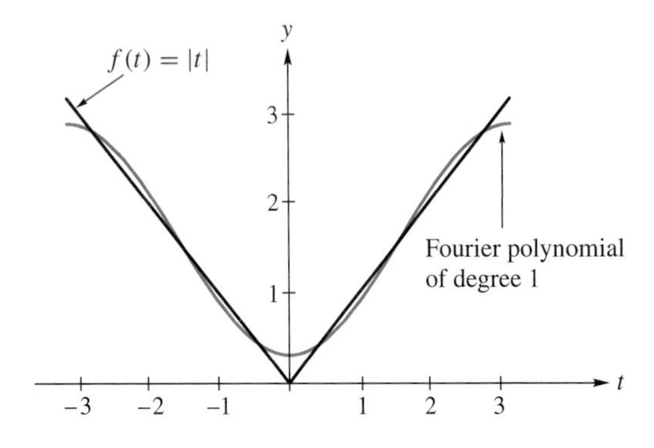

FIGURE 3.23

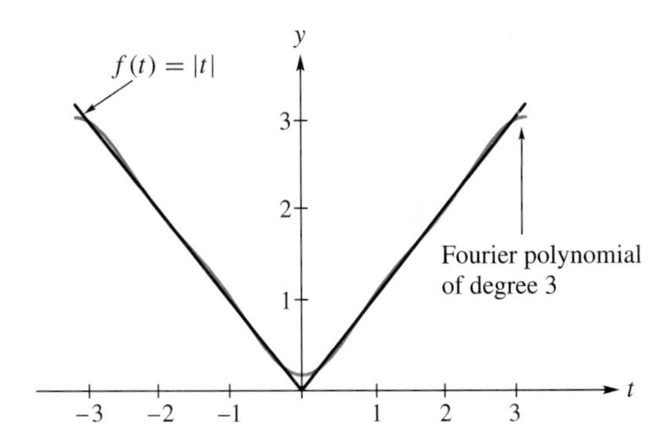

FIGURE 3.24

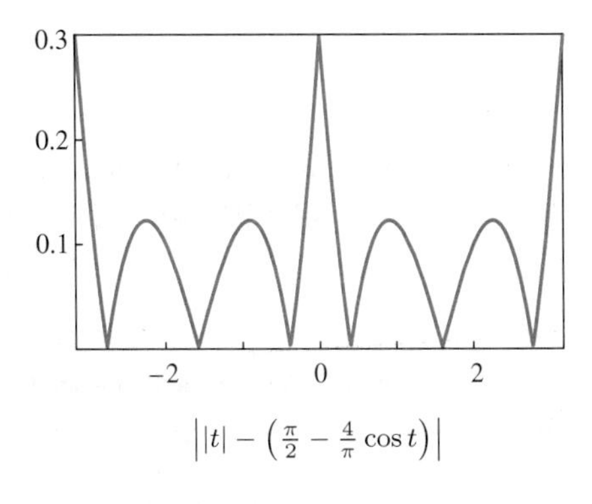

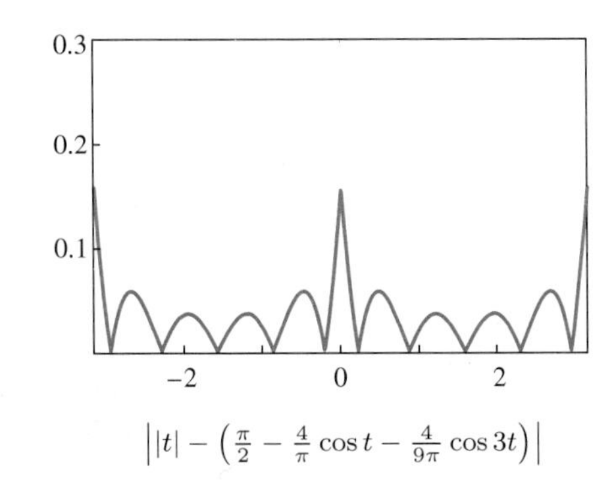

FIGURE 3.25

3.5 Exercises

1. Let W be the subspace of R^3 spanned by the vector

$$\mathbf{w} = \begin{bmatrix} 2 \\ -3 \\ 1 \end{bmatrix}.$$

(a) Find a basis for $W^\perp$.

(b) Describe $W^\perp$ geometrically. (You may use a verbal or pictorial description.)

2. Let

$$W = \text{span}\left\{ \begin{bmatrix} 1 \\ 2 \\ -1 \end{bmatrix}, \begin{bmatrix} -1 \\ 3 \\ 2 \end{bmatrix} \right\}.$$

(a) Find a basis for $W^\perp$.

(b) Describe $W^\perp$ geometrically. (You may use a verbal or pictorial description.)

3. Let W be the subspace of R_5 spanned by the vectors $\mathbf{w}_1$, $\mathbf{w}_2$, $\mathbf{w}_3$, $\mathbf{w}_4$, $\mathbf{w}_5$, where

$$\mathbf{w}_1 = \begin{bmatrix} 2 & -1 & 1 & 3 & 0 \end{bmatrix},$$
$$\mathbf{w}_2 = \begin{bmatrix} 1 & 2 & 0 & 1 & -2 \end{bmatrix},$$
$$\mathbf{w}_3 = \begin{bmatrix} 4 & 3 & 1 & 5 & -4 \end{bmatrix},$$
$$\mathbf{w}_4 = \begin{bmatrix} 3 & 1 & 2 & -1 & 1 \end{bmatrix},$$
$$\mathbf{w}_5 = \begin{bmatrix} 2 & -1 & 2 & -2 & 3 \end{bmatrix}.$$

Find a basis for $W^\perp$.

4. Let W be the subspace of R^4 spanned by the vectors $\mathbf{w}_1$, $\mathbf{w}_2$, $\mathbf{w}_3$, $\mathbf{w}_4$, where

$$\mathbf{w}_1 = \begin{bmatrix} 2 \\ 0 \\ -1 \\ 3 \end{bmatrix}, \quad \mathbf{w}_2 = \begin{bmatrix} 1 \\ 2 \\ 2 \\ -5 \end{bmatrix},$$

$$\mathbf{w}_3 = \begin{bmatrix} 3 \\ 2 \\ 1 \\ -2 \end{bmatrix}, \quad \mathbf{w}_4 = \begin{bmatrix} 7 \\ 2 \\ -1 \\ 4 \end{bmatrix}.$$

Find a basis for $W^\perp$.

5. (**Calculus required**). Let V be the Euclidean space P_3 with the inner product defined in Example 2. Let W be the subspace of P_3 spanned by $\{t-1, t^2\}$. Find a basis for $W^\perp$.

6. Let V be the Euclidean space P_4 with the inner product defined in Example 2. Let W be the subspace of P_4 spanned by $\{1, t\}$. Find a basis for $W^\perp$.

7. Let W be the plane $3x + 2y - z = 0$ in R^3. Find a basis for $W^\perp$.

8. Let V be the Euclidean space of all 2×2 matrices with the inner product defined by $(A, B) = \text{Tr}(B^T A)$, where Tr is the trace function defined in Exercise 29 of Section 1.2. Let

$$W = \text{span}\left\{ \begin{bmatrix} 1 & -1 \\ 2 & 3 \end{bmatrix}, \begin{bmatrix} 1 & 2 \\ 0 & -1 \end{bmatrix} \right\}.$$

Find a basis for $W^\perp$.

In Exercises 9 *and* 10, *compute the four fundamental vector spaces associated with A and verify Theorem* 3.12.

9. $A = \begin{bmatrix} 1 & 5 & 3 & 7 \\ 2 & 0 & -4 & -6 \\ 4 & 7 & -1 & 2 \end{bmatrix}.$

10. $A = \begin{bmatrix} 2 & -1 & 3 & 4 \\ 0 & -3 & 7 & -2 \\ 1 & 1 & -2 & 3 \\ 1 & 4 & -9 & 5 \end{bmatrix}.$

In Exercises 11 *through* 14, *find* $\text{proj}_W \mathbf{v}$ *for the given vector* $\mathbf{v}$ *and subspace* W.

11. Let V be the Euclidean space R^3, W the subspace with basis

$$\begin{bmatrix} \frac{1}{\sqrt{5}} \\ 0 \\ \frac{2}{\sqrt{5}} \end{bmatrix}, \quad \begin{bmatrix} -\frac{2}{\sqrt{5}} \\ 0 \\ \frac{1}{\sqrt{5}} \end{bmatrix}.$$

(a) $\mathbf{v} = \begin{bmatrix} 3 \\ 4 \\ -1 \end{bmatrix}$. (b) $\mathbf{v} = \begin{bmatrix} 2 \\ 1 \\ 3 \end{bmatrix}$. (c) $\mathbf{v} = \begin{bmatrix} -5 \\ 0 \\ 1 \end{bmatrix}$.

12. Let V be the Euclidean space R^4, W the subspace with basis

$$\begin{bmatrix} 1 & 1 & 0 & 1 \end{bmatrix}, \begin{bmatrix} 0 & 1 & 1 & 0 \end{bmatrix}, \begin{bmatrix} -1 & 0 & 0 & 1 \end{bmatrix}.$$

(a) $\mathbf{v} = \begin{bmatrix} 2 & 1 & 3 & 0 \end{bmatrix}$.
(b) $\mathbf{v} = \begin{bmatrix} 0 & -1 & 1 & 0 \end{bmatrix}$.
(c) $\mathbf{v} = \begin{bmatrix} 0 & 2 & 0 & 3 \end{bmatrix}$.

13. Let V be the vector space of real-valued continuous functions on $[-\pi, \pi]$, and let $W = \text{span}\{1, \cos t, \sin t\}$.

(a) $\mathbf{v} = t$. (b) $\mathbf{v} = t^2$. (c) $\mathbf{v} = e^t$.

14. Let W be the plane in R^3 given by the equation $x + y - 2z = 0$.

(a) $\mathbf{v} = \begin{bmatrix} 0 \\ 2 \\ -1 \end{bmatrix}$. (b) $\mathbf{v} = \begin{bmatrix} 3 \\ 1 \\ 4 \end{bmatrix}$. (c) $\mathbf{v} = \begin{bmatrix} 1 \\ 0 \\ -1 \end{bmatrix}$.

15. Let W be the subspace of R^3 with orthonormal basis $\{\mathbf{w}_1, \mathbf{w}_2\}$, where

$$\mathbf{w}_1 = \begin{bmatrix} 0 \\ 1 \\ 0 \end{bmatrix} \quad \text{and} \quad \mathbf{w}_2 = \begin{bmatrix} \frac{1}{\sqrt{5}} \\ 0 \\ \frac{2}{\sqrt{5}} \end{bmatrix}.$$

Write the vector $\mathbf{v} = \begin{bmatrix} 1 \\ 2 \\ -1 \end{bmatrix}$ as $\mathbf{w} + \mathbf{u}$, with $\mathbf{w}$ in W and $\mathbf{u}$ in $W^{\perp}$.

16. Let W be the subspace of R^4 with orthonormal basis $\{\mathbf{w}_1, \mathbf{w}_2, \mathbf{w}_3\}$, where

$$\mathbf{w}_1 = \begin{bmatrix} \frac{1}{\sqrt{2}} \\ 0 \\ 0 \\ -\frac{1}{\sqrt{2}} \end{bmatrix}, \quad \mathbf{w}_2 = \begin{bmatrix} 0 \\ 0 \\ 1 \\ 0 \end{bmatrix}, \quad \text{and} \quad \mathbf{w}_3 = \begin{bmatrix} \frac{1}{\sqrt{2}} \\ 0 \\ 0 \\ \frac{1}{\sqrt{2}} \end{bmatrix}.$$

Write the vector $\mathbf{v} = \begin{bmatrix} 1 \\ 0 \\ 2 \\ 3 \end{bmatrix}$ as $\mathbf{w} + \mathbf{u}$ with $\mathbf{w}$ in W and $\mathbf{u}$ in $W^{\perp}$.

17. Let W be the subspace of continuous functions on $[-\pi, \pi]$ defined in Exercise 13. Write the vector $\mathbf{v} = t - 1$ as $\mathbf{w} + \mathbf{u}$, with $\mathbf{w}$ in W and $\mathbf{u}$ in $W^{\perp}$.

18. Let W be the plane in R^3 given by the equation $x - y - z = 0$. Write the vector $\mathbf{v} = \begin{bmatrix} 2 \\ -1 \\ 0 \end{bmatrix}$ as $\mathbf{w} + \mathbf{u}$, with $\mathbf{w}$ in W and $\mathbf{u}$ in $W^{\perp}$.

19. Let W be the subspace of R^3 defined in Exercise 15, and let $\mathbf{v} = \begin{bmatrix} -1 \\ 0 \\ 1 \end{bmatrix}$. Find the distance from $\mathbf{v}$ to W.

20. Let W be the subspace of R^4 defined in Exercise 16, and let $\mathbf{v} = \begin{bmatrix} 1 \\ 2 \\ -1 \\ 0 \end{bmatrix}$. Find the distance from $\mathbf{v}$ to W.

21. Let W be the subspace of continuous functions on $[-\pi, \pi]$ defined in Exercise 13 and let $\mathbf{v} = t$. Find the distance from $\mathbf{v}$ to W.

In Exercises 22 and 23, find the Fourier polynomial of degree two for f.

22. (**Calculus required**) $f(t) = t^2$.

23. (**Calculus required**) $f(t) = e^t$.

24. Show that if V is an inner product space and W is a subspace of V, then the zero vector of V belongs to $W^{\perp}$

25. Let V be an inner product space. Show that the orthogonal complement of V is the zero subspace and the orthogonal complement of the zero subspace is V itself.

26. Show that if W is a subspace of an inner product space V that is spanned by a set of vectors S, then a vector $\mathbf{u}$ in V belongs to $W^{\perp}$ if and only if $\mathbf{u}$ is orthogonal to every vector in S.

27. Let A be an $m \times n$ matrix. Show that every vector $\mathbf{v}$ in R^n can be written uniquely as $\mathbf{w} + \mathbf{u}$, where $\mathbf{w}$ is in the null space of A and $\mathbf{u}$ is in the column space of A^T.

28. Let V be a Euclidean space and W a subspace of V. Show that if $\mathbf{w}_1, \mathbf{w}_2, \ldots, \mathbf{w}_r$ is a basis for W and $\mathbf{u}_1, \mathbf{u}_2, \ldots, \mathbf{u}_s$ is a basis for $W^{\perp}$, then $\mathbf{w}_1, \mathbf{w}_2, \ldots, \mathbf{w}_r, \mathbf{u}_1, \mathbf{u}_2, \ldots, \mathbf{u}_s$ is a basis for V, and that $\dim V = \dim W + \dim W^{\perp}$.

29. Let W be a subspace of an inner product space V and let $\{\mathbf{w}_1, \mathbf{w}_2, \ldots, \mathbf{w}_m\}$ be an orthogonal basis for W. Show that if $\mathbf{v}$ is any vector in V, then

$$\text{proj}_W \mathbf{v} = \frac{(\mathbf{v}, \mathbf{w}_1)}{(\mathbf{w}_1, \mathbf{w}_1)} \mathbf{w}_1 + \frac{(\mathbf{v}, \mathbf{w}_2)}{(\mathbf{w}_2, \mathbf{w}_2)} \mathbf{w}_2 + \cdots + \frac{(\mathbf{v}, \mathbf{w}_m)}{(\mathbf{w}_m, \mathbf{w}_m)} \mathbf{w}_m.$$

3.6 LEAST SQUARES (OPTIONAL)

From Chapter 1 we recall that an $m \times n$ linear system $A\mathbf{x} = \mathbf{b}$ is inconsistent if it has no solution. In the proof of Theorem 2.20 in Section 2.8 we show that $A\mathbf{x} = \mathbf{b}$ is consistent if and only if $\mathbf{b}$ belongs to the column space of A. Equivalently, $A\mathbf{x} = \mathbf{b}$ is inconsistent if and only if $\mathbf{b}$ is *not* in the column space of A. Inconsistent systems do indeed arise in many situations and we must determine how to deal with them. Our approach is to change the problem so that we do not require that the matrix equation $A\mathbf{x} = \mathbf{b}$ be satisfied. Instead, we seek a vector $\widehat{\mathbf{x}}$ in R^n such that $A\widehat{\mathbf{x}}$ is as close to $\mathbf{b}$ as possible. That is, find a vector in the column space of A that is closest to $\mathbf{b}$. From Theorem 3.13, if W is the column space of A, we determine $\widehat{\mathbf{x}}$ such that $A\widehat{\mathbf{x}} = \text{proj}_W\mathbf{b}$. As shown in the proof of Theorem 3.13, $\mathbf{b} - \text{proj}_W\mathbf{b} = \mathbf{b} - A\widehat{\mathbf{x}}$ is orthogonal to every vector in W. It follows that $\mathbf{b} - A\widehat{\mathbf{x}}$ is orthogonal to each column of A. In terms of a matrix equation, we have

$$A^T(A\widehat{\mathbf{x}} - \mathbf{b}) = \mathbf{0}$$

or, equivalently,

$$A^TA\widehat{\mathbf{x}} = A^T\mathbf{b}. \tag{1}$$

Any solution to (1) is called a **least squares solution** to the linear system $A\mathbf{x} = \mathbf{b}$. (**Warning:** In general, $A\widehat{\mathbf{x}} \neq \mathbf{b}$.) Equation (1) is called the **normal system** of equations associated with $A\mathbf{x} = \mathbf{b}$, or just the normal system. Observe that if A is nonsingular, a least squares solution to $A\mathbf{x} = \mathbf{b}$ is just the usual solution $\mathbf{x} = A^{-1}\mathbf{b}$ (see Exercise 1).

To compute a least squares solution $\widehat{\mathbf{x}}$ to the linear system $A\mathbf{x} = \mathbf{b}$, we can proceed as follows. Compute $\text{proj}_W\mathbf{b}$ using Equation (1) in Section 3.5 and then solve $A\widehat{\mathbf{x}} = \text{proj}_W\mathbf{b}$. To compute $\text{proj}_W\mathbf{b}$ requires that we have an orthonormal basis for W, the column space of A. We could first find a basis for W by determining the reduced row echelon form of A^T and taking the transposes of the nonzero rows. Next, apply the Gram–Schmidt process to the basis to find an orthonormal basis for W. The procedure just outlined is theoretically valid when we assume that exact arithmetic is used. However, even small numerical errors, due to, say, roundoff, may adversely affect the results. Thus more sophisticated algorithms are required for numerical applications. [See D. Hill, *Experiments in Computational Matrix Algebra* (New York: Random House, 1988), distributed by McGraw-Hill.] We shall not pursue the general case here, but turn our attention to an important special case.

Remark An alternative method for finding $\text{proj}_W\mathbf{b}$ is as follows. Solve Equation (1) for $\widehat{\mathbf{x}}$, the least squares solution to the linear system $A\mathbf{x} = \mathbf{b}$. Then $A\widehat{\mathbf{x}}$ will be $\text{proj}_W\mathbf{b}$.

Theorem 3.14

If A is an $m \times n$ matrix with rank $A = n$, then A^TA is nonsingular and the linear system $A\mathbf{x} = \mathbf{b}$ has a unique least squares solution given by $\widehat{\mathbf{x}} = (A^TA)^{-1}A^T\mathbf{b}$.

Proof

If A has rank n, then the columns of A are linearly independent. The matrix $A^T A$ is nonsingular provided the linear system $A^T A\mathbf{x} = \mathbf{0}$ has only the zero solution. Multiplying both sides of $A^T A\mathbf{x} = \mathbf{0}$ by $\mathbf{x}^T$ on the left gives

$$\mathbf{0} = \mathbf{x}^T A^T A\mathbf{x} = (A\mathbf{x})^T (A\mathbf{x}) = (A\mathbf{x}, A\mathbf{x}),$$

using the standard inner product on R^n. It follows from Definition 3.2(a) that $A\mathbf{x} = \mathbf{0}$. But this implies that we have a linear combination of the linearly independent columns of A that is zero; hence $\mathbf{x} = \mathbf{0}$. Thus $A^T A$ is nonsingular and Equation (1) has the unique solution $\widehat{\mathbf{x}} = (A^T A)^{-1} A^T\mathbf{b}$. ●

EXAMPLE 1

Determine a least squares solution to $A\mathbf{x} = \mathbf{b}$, where

$$A = \begin{bmatrix} 1 & 2 & -1 & 3 \\ 2 & 1 & 1 & 2 \\ -2 & 3 & 4 & 1 \\ 4 & 2 & 1 & 0 \\ 0 & 2 & 1 & 3 \\ 1 & -1 & 2 & 0 \end{bmatrix}, \qquad \mathbf{b} = \begin{bmatrix} 1 \\ 5 \\ -2 \\ 1 \\ 3 \\ 5 \end{bmatrix}.$$

Solution Using row reduction, we can show that rank $A = 4$ (verify). Then using Theorem 3.14, we form the normal system $A^T A\widehat{\mathbf{x}} = A^T\mathbf{b}$ (verify),

$$\begin{bmatrix} 26 & 5 & -1 & 5 \\ 5 & 23 & 13 & 17 \\ -1 & 13 & 24 & 6 \\ 5 & 17 & 6 & 23 \end{bmatrix} \widehat{\mathbf{x}} = \begin{bmatrix} 24 \\ 4 \\ 10 \\ 20 \end{bmatrix}.$$

Applying Gaussian elimination, we have the unique least squares solution (verify)

$$\widehat{\mathbf{x}} \approx \begin{bmatrix} 0.9990 \\ -2.0643 \\ 1.1039 \\ 1.8902 \end{bmatrix}.$$

If W is the column space of A, then (verify)

$$\text{proj}_W \mathbf{b} = A\widehat{\mathbf{x}} \approx \begin{bmatrix} 1.4371 \\ 4.8181 \\ -1.8852 \\ 0.9713 \\ 2.6459 \\ 5.2712 \end{bmatrix},$$

which is the vector in W such that $\|\mathbf{b} - \mathbf{y}\|$, $\mathbf{y}$ in W, is minimized. That is,

$$\min_{\mathbf{y} \text{ in } W} \|\mathbf{b} - \mathbf{w}\| = \|\mathbf{b} - A\widehat{\mathbf{x}}\|. \qquad \blacksquare$$

When A is an $m \times n$ matrix whose rank is n, it is computationally more efficient to solve Equation (1) by Gaussian elimination than to determine $(A^T A)^{-1}$ and then form the product $(A^T A)^{-1} A^T \mathbf{b}$. An even better approach is to use the QR-factorization of A as follows.

Suppose that $A = QR$ is a QR-factorization of A. Substituting this expression for A into Equation (1), we obtain

$$(QR)^T (QR)\widehat{\mathbf{x}} = (QR)^T \mathbf{b}$$

or

$$R^T (Q^T Q) R\widehat{\mathbf{x}} = R^T Q^T \mathbf{b}.$$

Since the columns of Q form an orthonormal set, we have $Q^T Q = I_m$, so

$$R^T R\widehat{\mathbf{x}} = R^T Q^T \mathbf{b}.$$

Since R^T is a nonsingular matrix, we obtain

$$R\widehat{\mathbf{x}} = Q^T \mathbf{b}.$$

Using the fact that R is upper triangular, we readily solve this linear system by back substitution.

EXAMPLE 2

Solve Example 1 using the QR-factorization of A.

Solution We use the Gram–Schmidt process, carrying out all computations in MATLAB. We find that Q is given by (verify)

$$Q = \begin{bmatrix} -0.1961 & -0.3851 & 0.5099 & 0.3409 \\ -0.3922 & -0.1311 & -0.1768 & 0.4244 \\ 0.3922 & -0.7210 & -0.4733 & -0.2177 \\ -0.7845 & -0.2622 & -0.1041 & -0.5076 \\ 0 & -0.4260 & 0.0492 & 0.4839 \\ -0.1961 & 0.2540 & -0.6867 & 0.4055 \end{bmatrix}$$

and R is given by (verify)

$$R = \begin{bmatrix} -5.0990 & -0.9806 & 0.1961 & -0.9806 \\ 0 & -4.6945 & -2.8102 & -3.4164 \\ 0 & 0 & -4.0081 & 0.8504 \\ 0 & 0 & 0 & 3.1054 \end{bmatrix}.$$

Then

$$Q^T\mathbf{b} = \begin{bmatrix} 4.7068 \\ -0.1311 \\ 2.8172 \\ 5.8699 \end{bmatrix}.$$

Finally, solving

$$R\widehat{\mathbf{x}} = Q^T\mathbf{b}$$

we find (verify) exactly the same $\widehat{\mathbf{x}}$ as in the solution to Example 1. ■

Remark As we have already observed in our discussion of the QR-factorization of a matrix, if you use a computer program such as MATLAB to find a QR-factorization of A in Example 2, you would find that the program yields Q as a 6×6 matrix, whose first four columns agree with the Q found in our solution, and R as a 6×4 matrix, whose first four rows agree with the R found in our solution, and whose last two rows consist entirely of zeros.

Least squares problems often arise when one tries to construct a mathematical model of the form

$$y(t) = x_1 f_1(t) + x_2 f_2(t) + \cdots + x_n f_n(t) \tag{2}$$

to a data set $D = \{(t_i, y_i),\ i = 1, 2, \ldots, m\}$, where $m > n$. Ideally, we would like to determine $x_1, x_2, \ldots, x_n$ such that

$$y_i = x_1 f_1(t_i) + x_2 f_2(t_i) + \cdots + x_n f_n(t_i)$$

for each data point t_i, $i = 1, 2, \ldots, m$. In matrix form we have the linear system $A\mathbf{x} = \mathbf{b}$, where

$$A = \begin{bmatrix} f_1(t_1) & f_2(t_1) & \cdots & f_n(t_1) \\ f_1(t_2) & f_2(t_2) & \cdots & f_n(t_2) \\ \vdots & \vdots & \vdots & \vdots \\ f_1(t_m) & f_2(t_m) & \cdots & f_n(t_m) \end{bmatrix}, \tag{3}$$

$$\mathbf{x} = \begin{bmatrix} x_1 & x_2 & \cdots & x_n \end{bmatrix}^T, \quad \text{and} \quad \mathbf{b} = \begin{bmatrix} y_1 & y_2 & \cdots & y_m \end{bmatrix}^T.$$

As is often the case, the system $A\mathbf{x} = \mathbf{b}$ is inconsistent, so we determine a least squares solution $\widehat{\mathbf{x}}$ to $A\mathbf{x} = \mathbf{b}$. If we set $x_i = \hat{x}_i$ in the model equation (2), we say that

$$\hat{y}(t) = \hat{x}_1 f_1(t) + \hat{x}_2 f_2(t) + \cdots + \hat{x}_n f_n(t)$$

gives a least squares model for data set D. In general, $\hat{y}(t_i) \neq y_i$, $i = 1, 2, \ldots, m$ (they may be equal, but there is no guarantee). Let $e_i = y_i - \hat{y}(t_i)$, $i = 1, 2, \ldots, m$, which represents the error incurred at t_i when $\hat{y}(t)$ is used as a mathematical model for data set D. If

$$\mathbf{e} = \begin{bmatrix} e_1 & e_2 & \cdots & e_m \end{bmatrix}^T,$$

then

$$\mathbf{e} = \mathbf{b} - A\widehat{\mathbf{x}},$$

and Theorem 3.13 in Section 3.5 guarantees that $\|\mathbf{e}\| = \|\mathbf{b} - A\widehat{\mathbf{x}}\|$ is as small as possible. That is,

$$\|\mathbf{e}\|^2 = (\mathbf{e}, \mathbf{e}) = (\mathbf{b} - A\widehat{\mathbf{x}}, \mathbf{b} - A\widehat{\mathbf{x}}) = \sum_{i=1}^{m} \left[y_i - \sum_{j=1}^{n} \hat{x}_j f_j(t_i) \right]^2$$

is minimized. We say that $\widehat{\mathbf{x}}$, the least squares solution, minimizes the sum of the squares of the deviations between the observations y_i and the values $\hat{y}(t_i)$ predicted by the model equation.

EXAMPLE 3

The following data show atmospheric pollutants y_i (relative to an EPA standard) at half-hour intervals t_i.

t_i	1	1.5	2	2.5	3	3.5	4	4.5	5
y_i	−0.15	0.24	0.68	1.04	1.21	1.15	0.86	0.41	−0.08

A plot of these data points, as shown in Figure 3.26, suggests that a quadratic polynomial

$$y(t) = x_1 + x_2 t + x_3 t^2$$

may produce a good model for these data. With $f_1(t) = 1$, $f_2(t) = t$, and $f_3(t) = t^2$, (3) gives

$$A = \begin{bmatrix} 1 & 1 & 1 \\ 1 & 1.5 & 2.25 \\ 1 & 2 & 4 \\ 1 & 2.5 & 6.25 \\ 1 & 3 & 9 \\ 1 & 3.5 & 12.25 \\ 1 & 4 & 16 \\ 1 & 4.5 & 20.25 \\ 1 & 5 & 25 \end{bmatrix}, \quad \widehat{\mathbf{x}} = \begin{bmatrix} \hat{x}_1 \\ \hat{x}_2 \\ \hat{x}_3 \end{bmatrix}, \quad \mathbf{b} = \begin{bmatrix} -0.15 \\ 0.24 \\ 0.68 \\ 1.04 \\ 1.21 \\ 1.15 \\ 0.86 \\ 0.41 \\ -0.08 \end{bmatrix}.$$

The rank of A is 3 (verify), and the normal system is

$$\begin{bmatrix} 9 & 27 & 96 \\ 27 & 96 & 378 \\ 96 & 378 & 1583.25 \end{bmatrix} \begin{bmatrix} \hat{x}_1 \\ \hat{x}_2 \\ \hat{x}_3 \end{bmatrix} = \begin{bmatrix} 5.36 \\ 16.71 \\ 54.65 \end{bmatrix}.$$

Applying Gaussian elimination gives (verify)

$$\widehat{\mathbf{x}} \approx \begin{bmatrix} -1.9317 \\ 2.0067 \\ -0.3274 \end{bmatrix},$$

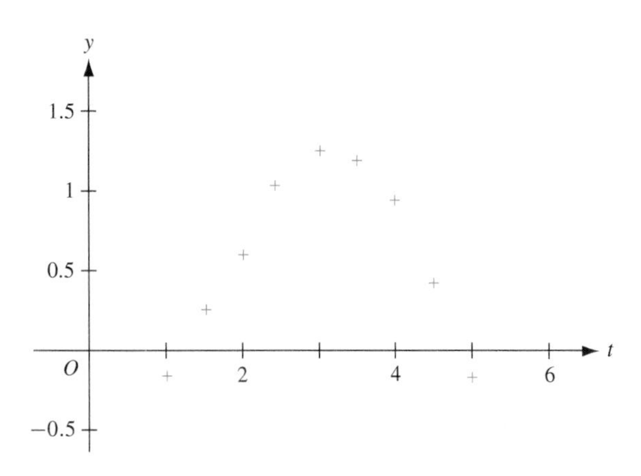

FIGURE 3.26

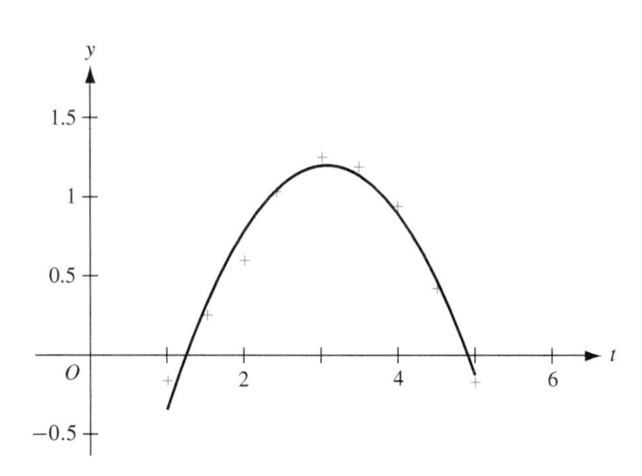

FIGURE 3.27

so we obtain the quadratic polynomial model

$$y(t) = -1.9317 + 2.0067t - 0.3274t^2.$$

Figure 3.27 shows the data set indicated with + and a graph of $y(t)$. We see that $y(t)$ is close to each data point but is not required to go through the data. ■

3.6 Exercises

1. Let A be $n \times n$ and nonsingular. From the normal system of equations in (1), show that the least squares solution to $A\mathbf{x} = \mathbf{b}$ is $\widehat{\mathbf{x}} = A^{-1}\mathbf{b}$.

2. Determine the least squares solution to $A\mathbf{x} = \mathbf{b}$, where

$$A = \begin{bmatrix} 2 & 1 \\ 1 & 0 \\ 0 & -1 \\ -1 & 1 \end{bmatrix} \quad \text{and} \quad \mathbf{b} = \begin{bmatrix} 3 \\ 1 \\ 2 \\ -1 \end{bmatrix}.$$

3. Determine the least squares solution to $A\mathbf{x} = \mathbf{b}$, where

$$A = \begin{bmatrix} 1 & 2 & 1 \\ 1 & 3 & 2 \\ 2 & 5 & 3 \\ 2 & 0 & 1 \\ 3 & 1 & 1 \end{bmatrix} \quad \text{and} \quad \mathbf{b} = \begin{bmatrix} -1 \\ 2 \\ 0 \\ 1 \\ -2 \end{bmatrix}.$$

4. Solve Exercise 2 using QR-factorization of A.

5. Solve Exercise 3 using QR-factorization of A.

6. In the manufacture of product Z, the amount of compound A present in the product is controlled by the amount of ingredient B used in the refining process. In manufacturing a liter of Z, the amount of B used and the amount of A present are recorded. The following data were obtained:

B used (grams/liter)	2	4	6	8	10
A present (grams/liter)	3.5	8.2	10.5	12.9	14.6

Determine the least squares line to the data. [In Equation (2), use $f_1(t) = 1$, $f_2(t) = t$.] Also compute $\|\mathbf{e}\|$.

7. In Exercise 6, the least squares line to the data set $D = \{(t_i, y_i),\ i = 1, 2, \ldots, m\}$ is the line $y = \hat{x}_1 + \hat{x}_2 t$, which minimizes

$$E_1 = \sum_{i=1}^{m} [y_i - (x_1 + x_2 t_i)]^2.$$

Similarly, the least squares quadratic (see Example 2) to the data set D is the parabola $y = \hat{x}_1 + \hat{x}_2 t + \hat{x}_3 t^2$, which minimizes

$$E_2 = \sum_{i=1}^{m} \left[y_i - (x_1 + x_2 t_i + x_3 t_i^2) \right]^2 .$$

Give a vector space argument to show that $E_2 \leq E_1$.

8. The following table is a sample set of seasonal farm employment data (t_i, y_i) over about a two-year period, where t_i represents months and y_i represents millions of people. A plot of the data is given in the accompanying figure. It is decided to develop a least squares mathematical model of the form

$$y(t) = x_1 + x_2 t + x_3 \cos t.$$

t_i	y_i	t_i	y_i
3.1	3.7	11.8	5.0
4.3	4.4	13.1	5.0
5.6	5.3	14.3	3.8
6.8	5.2	15.6	2.8
8.1	4.0	16.8	3.3
9.3	3.6	18.1	4.5
10.6	3.6		

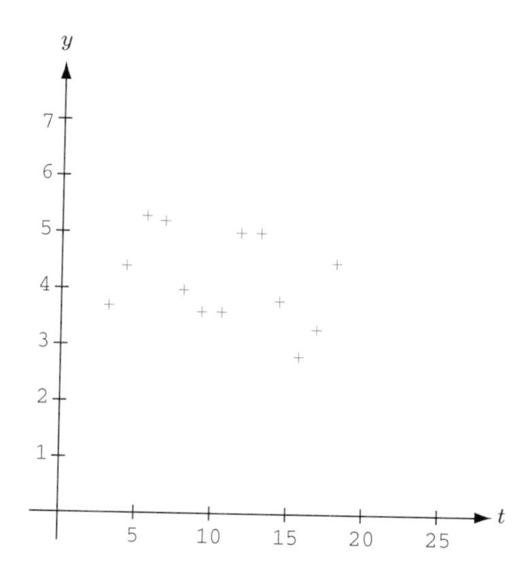

Determine the least squares model. Plot the resulting function $y(t)$ and the data set on the same coordinate system.

9. Given $A\mathbf{x} = \mathbf{b}$, where

$$A = \begin{bmatrix} 1 & 3 & -3 \\ 2 & 4 & -2 \\ 0 & -1 & 2 \\ 1 & 2 & -1 \end{bmatrix} \quad \text{and} \quad \mathbf{b} = \begin{bmatrix} 1 \\ 0 \\ 0 \\ 1 \end{bmatrix}.$$

(a) Show that rank $A = 2$.

(b) Since rank $A \neq$ number of columns, Theorem 3.14 cannot be used to determine a least squares solution $\widehat{\mathbf{x}}$. Follow the general procedure as discussed prior to Theorem 3.14 to find a least squares solution. Is the solution unique?

10. The following data showing U.S. Health Expenditures (in billions of dollars) appears in the World Almanac and Book of Facts 1997, edited by Robert Famighettie, K-III Reference Corporation, Mahwah, N.J. 1996.

Year	Personal health care
1985	376.4
1990	614.7
1991	676.2
1992	739.8
1993	786.5
1994	831.7

(a) Determine the line of best fit to the given data.

(b) Predict the expenditures for Personal Health Care in the year 1999.

11. The following data showing the size of the U.S. Debt per capita (in dollars) appears in the World Almanac and Book of Facts 1997, edited by Robert Famighettie, K-III Reference Corporation, Mahwah, N.J. 1996.

Year	Debt per capita
1970	1,814
1975	2,475
1980	3,985
1985	7,598
1990	13,000
1995	18,930

(a) Determine the line of best fit to the given data.

(b) Predict the debt per capita in the year 2000.

12. For the data in Exercise 11, find the least squares quadratic polynomial approximation. Compare this model with that obtained in Exercise 11 by computing the error in each case.

13. Gauge is a measure of shotgun bore. Gauge numbers originally referred to the number of lead balls with the diameter equal to that of the gun barrel that could be made from a pound of lead. Thus, a 16-gauge shotgun's bore was smaller than a 12-gauge shotgun's. (Ref., The World Almanac and Book of Facts, 1993, Pharos Books, N.Y., 1992, page 290.) Today, an international agreement assigns millimeter measures to each gauge. The following table gives such information for popular gauges of shotguns.

x	y
Gauge	Bore diameter (in mm)
6	23.34
10	19.67
12	18.52
14	17.60
16	16.81
20	15.90

We would like to develop a model of the form $y = re^{sx}$ for the data in this table. By taking the natural logarithm of both sides in this expression, we obtain

$$\ln y = \ln r + sx. \qquad (*)$$

Let $c_1 = \ln r$ and $c_2 = s$. Substituting the data from the table in $(*)$ we obtain a linear system of six equations in two unknowns.

(a) Show that this system is inconsistent.

(b) Find its least squares solution.

(c) Determine r and s.

(d) Estimate the bore diameter for an 18-gauge shotgun.

14. In some software, the command for solving a linear system produces a least squares solution when the coefficient matrix is not square or nonsingular. Determine if this is the case in your software. If it is, compare your software's output with the solution given in Example 1. To experiment further, use Exercise 9.

Supplementary Exercises

1. Exercise 33 of Section 3.4 proves that the set of all vectors in R^n that are orthogonal to a fixed vector $\mathbf{u}$ forms a subspace of R^n. For $\mathbf{u} = \begin{bmatrix} 1 \\ -2 \\ 1 \end{bmatrix}$, find an orthogonal basis for the subspace of vectors in R^3 that are orthogonal to $\mathbf{u}$. [*Hint*: Solve the linear system $(\mathbf{u}, \mathbf{v}) = 0$, when $\mathbf{v} = \begin{bmatrix} x_1 \\ x_2 \\ x_3 \end{bmatrix}$.]

2. Use the Gram–Schmidt process to find an orthonormal basis for the subspace of R^4 with basis

$$\left\{ \begin{bmatrix} 1 \\ 0 \\ 0 \\ -1 \end{bmatrix}, \begin{bmatrix} 1 \\ -1 \\ 0 \\ 0 \end{bmatrix}, \begin{bmatrix} 0 \\ 1 \\ 0 \\ 1 \end{bmatrix} \right\}.$$

3. Given the orthonormal basis

$$S = \left\{ \begin{bmatrix} \frac{1}{\sqrt{2}} \\ 0 \\ -\frac{1}{\sqrt{2}} \end{bmatrix}, \begin{bmatrix} 0 \\ 1 \\ 0 \end{bmatrix}, \begin{bmatrix} \frac{1}{\sqrt{2}} \\ 0 \\ \frac{1}{\sqrt{2}} \end{bmatrix} \right\}$$

for R^3, write the vector

$$\mathbf{v} = \begin{bmatrix} 1 \\ 2 \\ 3 \end{bmatrix}$$

as a linear combination of the vectors in S.

4. Use the Gram–Schmidt process to find an orthonormal basis for the subspace of R^4 with basis

$$\left\{ \begin{bmatrix} 1 \\ 0 \\ -1 \\ 0 \end{bmatrix}, \begin{bmatrix} 1 \\ -1 \\ 0 \\ 0 \end{bmatrix}, \begin{bmatrix} 2 \\ 1 \\ 0 \\ 0 \end{bmatrix} \right\}.$$

5. Given vector $\mathbf{v} = \mathbf{i} + 2\mathbf{j} + \mathbf{k}$ and the plane P determined by the vectors

$$\mathbf{w}_1 = \mathbf{i} - 3\mathbf{j} - 2\mathbf{k} \quad \text{and} \quad \mathbf{w}_2 = 3\mathbf{i} - \mathbf{j} - 3\mathbf{k},$$

find the vector in P closest to $\mathbf{v}$ and the distance from $\mathbf{v}$ to P.

6. Find the distance from the point $(2, 3, -1)$ to the plane $3x - 2y + z = 0$. (*Hint*: First find an orthonormal basis for the plane.)

7. Consider the vector space of continuous real-valued functions on $[0, \pi]$ with an inner product defined by $(f, g) = \int_0^{\pi} f(t)g(t)\,dt$. Show that the collection of functions $\sin nt$, for $n = 1, 2, \ldots,$ is an orthogonal set.

8. Let V be the inner product space defined in Exercise 13 of Section 3.5. In each of the following let W be the subspace spanned by the given orthonormal vectors $\mathbf{w}_1, \mathbf{w}_2, \ldots, \mathbf{w}_n$. Find $\text{proj}_W \mathbf{v}$, for the vector $\mathbf{v}$ in V.

(a) $\mathbf{v} = t + t^2$, $\mathbf{w}_1 = \frac{1}{\sqrt{\pi}} \cos t$, $\mathbf{w}_2 = \frac{1}{\sqrt{\pi}} \sin 2t$.

(b) $\mathbf{v} = \sin \frac{1}{2}t$, $\mathbf{w}_1 = \frac{1}{\sqrt{2\pi}}$, $\mathbf{w}_2 = \frac{1}{\sqrt{\pi}} \sin t$.

(c) $\mathbf{v} = \cos^2 t$, $\mathbf{w}_1 = \frac{1}{\sqrt{2\pi}}$, $\mathbf{w}_2 = \frac{1}{\sqrt{\pi}} \cos t$,

$\mathbf{w}_3 = \frac{1}{\sqrt{\pi}} \cos 2t$.

9. Find an orthonormal basis for the null space of A.

(a) $A = \begin{bmatrix} 1 & 0 & 5 & -2 \\ 0 & 1 & -2 & -5 \end{bmatrix}$.

(b) $A = \begin{bmatrix} 1 & 0 & 5 & -2 \\ 0 & 1 & -2 & 4 \end{bmatrix}$.

10. Find the QR-factorization for the given matrix A.

(a) $\begin{bmatrix} 1 & 0 & 1 \\ -1 & 1 & 2 \\ 2 & 2 & -1 \end{bmatrix}$. (b) $\begin{bmatrix} 2 & 1 \\ -1 & -1 \\ -2 & 3 \end{bmatrix}$.

11. Let $W = \text{span}\left\{ \begin{bmatrix} 1 \\ 0 \\ 1 \end{bmatrix}, \begin{bmatrix} 0 \\ 1 \\ 0 \end{bmatrix} \right\}$ in R^3.

(a) Find a basis for the orthogonal complement of W.

(b) Show that vectors $\begin{bmatrix} 1 \\ 0 \\ 1 \end{bmatrix}, \begin{bmatrix} 0 \\ 1 \\ 0 \end{bmatrix}$, and the basis for the orthogonal complement of W from part (a) form a basis for R^3.

(c) Express each of the given vectors $\mathbf{v}$ as $\mathbf{w} + \mathbf{u}$, where $\mathbf{w}$ is in W and $\mathbf{u}$ is in $W^{\perp}$.

(i) $\mathbf{v} = \begin{bmatrix} 1 \\ 0 \\ 0 \end{bmatrix}$. (ii) $\mathbf{v} = \begin{bmatrix} 1 \\ 2 \\ 3 \end{bmatrix}$.

12. Find the orthogonal complement of the null space of A.

(a) $A = \begin{bmatrix} 1 & -2 & 2 \\ 2 & 3 & 2 \\ 4 & -1 & 6 \end{bmatrix}$.

(b) $A = \begin{bmatrix} 1 & -1 & 3 & 2 \\ 1 & -4 & 7 & 8 \\ 2 & 1 & 2 & -2 \end{bmatrix}$.

13. Use the Gram–Schmidt process to find an orthonormal basis for P_2 with respect to the inner product $(f, g) = \int_{-1}^{1} f(t)g(t)\,dt$, starting with the standard basis $\{1, t, t^2\}$ for P_2. The polynomials thus obtained are called the **Legendre polynomials**.

14. Using the Legendre polynomials from Exercise 13 as an orthonormal basis for P_2 considered as a subspace of the vector space of real-valued continuous functions on $[-1, 1]$, find $\text{proj}_{P_2} \mathbf{v}$ for each of the following.

(a) $\mathbf{v} = t^3$. (b) $\mathbf{v} = \sin \pi t$. (c) $\mathbf{v} = \cos \pi t$.

15. Using the Legendre polynomials from Exercise 13 as an orthonormal basis for P_2 considered as a subspace of the vector space of real-valued continuous functions on $[-1, 1]$, find the distance from $\mathbf{v} = t^3 + 1$ to P_2.

16. Let V be the inner product space of real-valued continuous functions on $[-\pi, \pi]$ with inner product defined by $(f, g) = \int_{-\pi}^{\pi} f(t)g(t)\,dt$ (see Exercise 13 in Section 3.5). Find the distance between $\sin nx$ and $\cos mx$.

17. Let A be an $n \times n$ symmetric matrix, and suppose that R^n has the standard inner product. Prove that if $(\mathbf{u}, A\mathbf{u}) = (\mathbf{u}, \mathbf{u})$ for all $\mathbf{u}$ in R^n, then $A = I_n$.

18. An $n \times n$ symmetric matrix A is **positive semidefinite** if $\mathbf{x}^T A\mathbf{x} \geq 0$ for all $\mathbf{x}$ in R^n. Prove:

(a) Every positive definite matrix is positive semidefinite.

(b) If A is singular and positive semidefinite, then A is not positive definite.

(c) A diagonal matrix A is positive semidefinite if and only if $a_{ii} \geq 0$ for $i = 1, 2, \ldots, n$.

19. In Chapter 6 the notion of an orthogonal matrix is discussed. It is shown that an $n \times n$ matrix P is orthogonal if and only if the columns of P, denoted $\mathbf{p}_1, \mathbf{p}_2, \ldots, \mathbf{p}_n$, form an orthonormal set in R^n, using the standard inner product. Let P be an orthogonal matrix.

(a) For $\mathbf{x}$ in R^n, prove that $\|P\mathbf{x}\| = \|\mathbf{x}\|$, using Theorem 3.7.

(b) For $\mathbf{x}$ and $\mathbf{y}$ in R^n, prove that the angle between $P\mathbf{x}$ and $P\mathbf{y}$ is the same as that between $\mathbf{x}$ and $\mathbf{y}$.

20. Let A be an $n \times n$ skew symmetric matrix. Prove that $\mathbf{x}^T A\mathbf{x} = 0$ for all $\mathbf{x}$ in R^n.

21. Let B be an $m \times n$ matrix with orthonormal columns $\mathbf{b}_1, \mathbf{b}_2, \ldots, \mathbf{b}_n$.

(a) Prove that $m \geq n$.

(b) Prove that $B^T B = I_n$.

22. Let $\{\mathbf{u}_1, \ldots, \mathbf{u}_k, \mathbf{u}_{k+1}, \ldots, \mathbf{u}_n\}$ be an orthonormal basis for Euclidean space V, $S = \text{span}\,\{\mathbf{u}_1, \ldots, \mathbf{u}_k\}$, and $T = \text{span}\,\{\mathbf{u}_{k+1}, \ldots, \mathbf{u}_n\}$. For any $\mathbf{x}$ in S and any $\mathbf{y}$ in T, show that $(\mathbf{x}, \mathbf{y}) = 0$.

23. Let V be a Euclidean space and W a subspace of V. Use Exercise 28 in Section 3.5 to show that $(W^\perp)^\perp = W$.

24. Let V be a Euclidean space with basis $S = \{\mathbf{v}_1, \mathbf{v}_2, \ldots, \mathbf{v}_n\}$. Show that if $\mathbf{u}$ is a vector in V that is orthogonal to every vector in S, then $\mathbf{u} = \mathbf{0}$.

25. Show that if A is an $m \times n$ matrix such that AA^T is nonsingular, then rank $A = m$.

A **vector norm** on R^n is a function that assigns to each vector $\mathbf{v}$ in R^n a nonnegative real number, called the **norm** of $\mathbf{v}$ and denoted by $\|\mathbf{v}\|$, satisfying:

(a) $\|\mathbf{v}\| \geq 0$, and $\|\mathbf{v}\| = 0$ if and only if $\mathbf{v} = \mathbf{0}$.

(b) $\|c\mathbf{v}\| = |c|\,\|\mathbf{v}\|$ for any real scalar c and vector $\mathbf{v}$.

(c) $\|\mathbf{u} + \mathbf{v}\| \leq \|\mathbf{u}\| + \|\mathbf{v}\|$ for all vectors $\mathbf{u}$ and $\mathbf{v}$. (The triangle inequality)

There are three widely used norms in applications of linear algebra, called the **1-norm**, the **2-norm**, and the **∞-norm**, and denoted by $\|\ \|_1$, $\|\ \|_2$, $\|\ \|_\infty$, respectively, and which are defined as follows: Let

$$\mathbf{x} = \begin{bmatrix} x_1 \\ x_2 \\ \vdots \\ x_n \end{bmatrix}$$

be a vector in R^n.

$$\|\mathbf{x}\|_1 = |x_1| + |x_2| + \cdots + |x_n|$$
$$\|\mathbf{x}\|_2 = \sqrt{x_1^2 + x_2^2 + \cdots + x_n^2}$$
$$\|\mathbf{x}\|_\infty = \max\{|x_1|, |x_2|, \ldots, |x_n|\}.$$

Observe that $\|\mathbf{x}\|_2$ is the length of the vector $\mathbf{x}$ as defined in Section 3.1.

26. For each given vector in R^2, compute $\|\mathbf{x}\|_1$, $\|\mathbf{x}\|_2$, and $\|\mathbf{x}\|_\infty$.

(a) $\begin{bmatrix} 2 \\ 3 \end{bmatrix}$. (b) $\begin{bmatrix} 0 \\ -2 \end{bmatrix}$. (c) $\begin{bmatrix} -4 \\ -1 \end{bmatrix}$.

27. For each given vector in R_3, compute $\|\mathbf{x}\|_1$, $\|\mathbf{x}\|_2$, and $\|\mathbf{x}\|_\infty$.

(a) $\begin{bmatrix} 2 & -2 & 3 \end{bmatrix}$. (b) $\begin{bmatrix} 0 & 3 & -2 \end{bmatrix}$.

(c) $\begin{bmatrix} 2 & 0 & 0 \end{bmatrix}$.

28. Verify that $\|\ \|_1$ is a norm.

29. Verify that $\|\ \|_\infty$ is a norm.

30. Show the following properties:

(a) $\|\mathbf{x}\|_2^2 \leq \|\mathbf{x}\|_1^2$

(b) $\dfrac{\|\mathbf{x}\|_1}{n} \leq \|\mathbf{x}\|_\infty \leq \|\mathbf{x}\|_1$.

31. Sketch the set of points in R^2 such that

(a) $\|\mathbf{x}\|_1 = 1$. (b) $\|\mathbf{x}\|_2 = 1$. (c) $\|\mathbf{x}\|_\infty = 1$.

Chapter 4

Linear Transformations and Matrices

4.1 DEFINITION AND EXAMPLES

As we have noted earlier, much of calculus deals with the study of properties of functions. Indeed, properties of functions are of great importance in every branch of mathematics, and linear algebra is no exception. We have already encountered functions mapping one vector space into another vector space; these were the isomorphisms. If we drop some of the conditions that need to be satisfied by a function on a vector space to be an isomorphism, we get another very useful type of function, called a linear transformation. Linear transformations play an important role in many areas of mathematics, the physical and social sciences, and economics.

Definition 4.1

Let V and W be vector spaces. A function $L: V \to W$ is called a **linear transformation** of V into W if

(a) $L(\mathbf{u}+\mathbf{v}) = L(\mathbf{u}) + L(\mathbf{v})$ for $\mathbf{u}$ and $\mathbf{v}$ in V.

(b) $L(c\mathbf{u}) = cL(\mathbf{u})$ for $\mathbf{u}$ in V, and c a real number.

If $V = W$, the linear transformation $L: V \to W$ is also called a **linear operator** on V. ▲

Most of the vector spaces considered henceforth, but not all, are finite dimensional.

In Definition 4.1, observe that in (a) the $+$ in $\mathbf{u}+\mathbf{v}$ refers to the addition operation in V, whereas the $+$ in $L(\mathbf{u}) + L(\mathbf{v})$ refers to the addition operation in W. Similarly, in (b) the scalar product $c\mathbf{u}$ is in V, while the scalar product $cL(\mathbf{u})$ is in W.

We have already pointed out in Section 2.7 that an isomorphism is a linear transformation that is one-to-one and onto. Linear transformations occur very frequently, and we now look at some examples. (At this point it might be profitable to review the material of Section A.2.) It can be shown that $L: V \to W$ is a linear transformation if and only if $L(a\mathbf{u}+b\mathbf{v}) = aL(\mathbf{u}) + bL(\mathbf{v})$ for any real numbers a, b and any vectors $\mathbf{u}$, $\mathbf{v}$ in V (see Exercise 4.)

EXAMPLE 1

Let $L: R^3 \to R^2$ be defined by

$$L\left(\begin{bmatrix} a_1 \\ a_2 \\ a_3 \end{bmatrix}\right) = \begin{bmatrix} a_1 \\ a_2 \end{bmatrix}.$$

To show that L is a linear transformation, we let

$$\mathbf{u} = \begin{bmatrix} a_1 \\ a_2 \\ a_3 \end{bmatrix} \quad \text{and} \quad \mathbf{v} = \begin{bmatrix} b_1 \\ b_2 \\ b_3 \end{bmatrix}.$$

Then

$$L(\mathbf{u}+\mathbf{v}) = L\left(\begin{bmatrix} a_1+b_1 \\ a_2+b_2 \\ a_3+b_3 \end{bmatrix}\right) = \begin{bmatrix} a_1+b_1 \\ a_2+b_2 \end{bmatrix} = \begin{bmatrix} a_1 \\ a_2 \end{bmatrix} + \begin{bmatrix} b_1 \\ b_2 \end{bmatrix} = L(\mathbf{u}) + L(\mathbf{v}).$$

Also, if c is a real number, then

$$L(c\mathbf{u}) = L\left(\begin{bmatrix} ca_1 \\ ca_2 \\ ca_3 \end{bmatrix}\right) = \begin{bmatrix} ca_1 \\ ca_2 \end{bmatrix} = c\begin{bmatrix} a_1 \\ a_2 \end{bmatrix} = cL(\mathbf{u}).$$

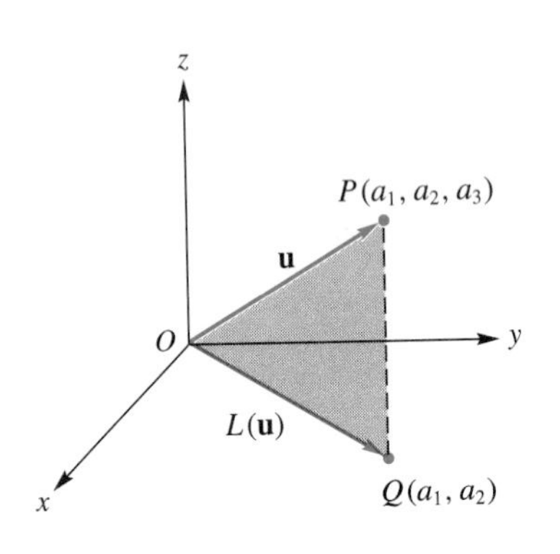

FIGURE 4.1

Projection

Hence L is a linear transformation, which is called **projection**. It is simple and helpful to describe geometrically the effect of L. The image under L of a vector in R^3 with head $P(a_1, a_2, a_3)$ is found by drawing a line through P perpendicular to R^2, the (x, y)-plane. We obtain the point $Q(a_1, a_2)$ of intersection of this line with the (x, y)-plane. The vector in R^2 with head Q is the image of $\mathbf{u}$ under L (Figure 4.1). Referring to the material in Section 3.5 on projections, we see that $L(\mathbf{u})$ is the orthogonal projection of the vector $\mathbf{u}$ in R^3 onto the (x, y)-plane. ■

EXAMPLE 2

Let $L: P_1 \to P_2$ be defined by

$$L[p(t)] = tp(t).$$

Show that L is a linear transformation.

Solution Let $p(t)$ and $q(t)$ be vectors in P_1 and let c be a scalar. Then

$$\begin{aligned} L[p(t)+q(t)] &= t[p(t)+q(t)] \\ &= tp(t) + tq(t) \\ &= L[p(t)] + L[q(t)], \end{aligned}$$

and

$$\begin{aligned} L[cp(t)] &= t[cp(t)] \\ &= c[tp(t)] \\ &= cL[p(t)]. \end{aligned}$$

Hence L is a linear transformation. ■

EXAMPLE 3

Let $L\colon R^3 \to R^3$ be defined by

$$L\left(\begin{bmatrix} a_1 \\ a_2 \\ a_3 \end{bmatrix}\right) = r\begin{bmatrix} a_1 \\ a_2 \\ a_3 \end{bmatrix},$$

where r is a real number. Then L is a linear operator on R^3 (verify). If $r > 1$, L is called a **dilation**; if $0 < r < 1$, L is called a **contraction**. Thus dilation stretches a vector, while contraction shrinks it (Figure 4.2). ■

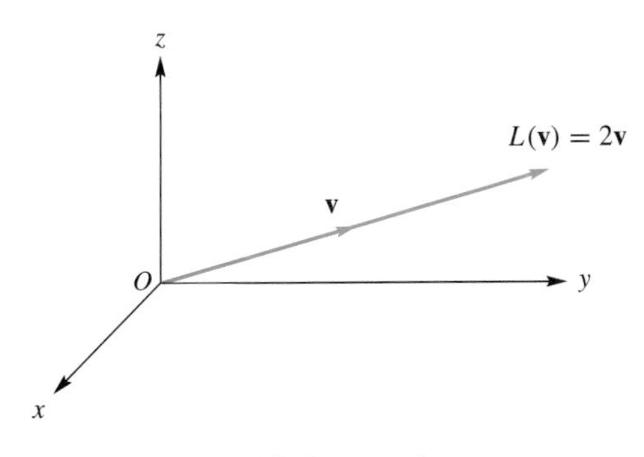

(a) Dilation: $r > 1$.

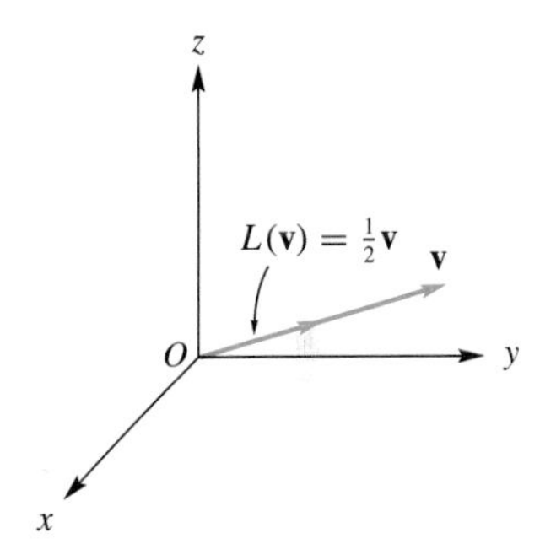

(b) Contraction: $0 < r < 1$.

FIGURE 4.2

Projection

EXAMPLE 4

Let W be the vector space of all real-valued functions and let V be the subspace of all differentiable functions. Let $L\colon V \to W$ be defined by

$$L(f) = f',$$

where f' is the derivative of f. We can show (Exercise 20), using the properties of differentiation, that L is a linear transformation. ■

EXAMPLE 5

Let $V = C[a, b]$ be the vector space of all real-valued functions that are integrable over the interval $[a, b]$ Let $W = R^2$. Define $L: V \to W$ by

$$L(f) = \int_a^b f(x)\,dx.$$

We can show (Exercise 23), using the properties of integration, that L is a linear transformation. ■

EXAMPLE 6

Let $L: R^3 \to R^2$ be defined by

$$L\left(\begin{bmatrix} a_1 \\ a_2 \\ a_3 \end{bmatrix}\right) = \begin{bmatrix} 1 & 0 & 1 \\ 0 & 1 & -1 \end{bmatrix} \begin{bmatrix} a_1 \\ a_2 \\ a_3 \end{bmatrix}.$$

Then L is a linear transformation (verify). ■

EXAMPLE 7

Let A be an $m \times n$ matrix. Define $L: R^n \to R^m$ by $L(\mathbf{x}) = A\mathbf{x}$ for $\mathbf{x}$ in R^n. Then L is a linear transformation (Exercise 21). ■

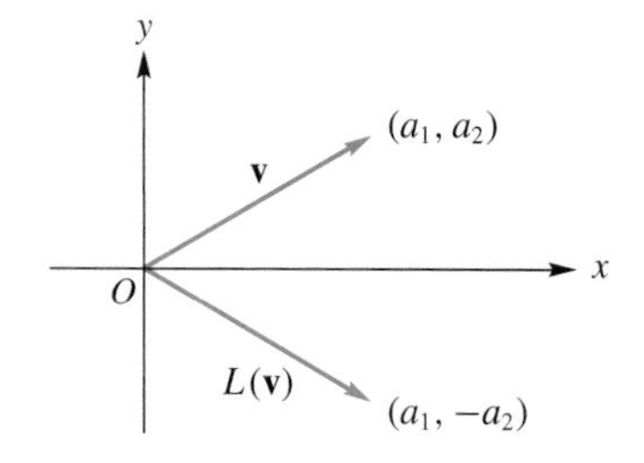

FIGURE 4.3 Reflection

EXAMPLE 8

Let $L: R^2 \to R^2$ be defined by

$$L\left(\begin{bmatrix} a_1 \\ a_2 \end{bmatrix}\right) = \begin{bmatrix} a_1 \\ -a_2 \end{bmatrix}.$$

Then L is a linear operator on R^2 (verify). Geometrically, the action of L is shown in Figure 4.3. Thus L is a **reflection** with respect to the x-axis. In a similar way we can also consider reflection with respect to the y-axis. ■

EXAMPLE 9

Suppose that we rotate every point in the (x, y)-plane counterclockwise through an angle ϕ about the origin of a Cartesian coordinate system. Thus, if P has coordinates (x, y), then after rotating we get P' with coordinates (x', y'). To obtain a relationship between the coordinates of P' and those of P, we let $\mathbf{v}$ be the vector from the origin to $P(x, y)$. See Figure 4.4(a). Also, let θ be the angle made by $\mathbf{v}$ with the positive x-axis.

Letting r denote the length of vector $\mathbf{v}$, we see from Figure 4.4(a) that

$$x = r\cos\theta, \qquad y = r\sin\theta \tag{1}$$

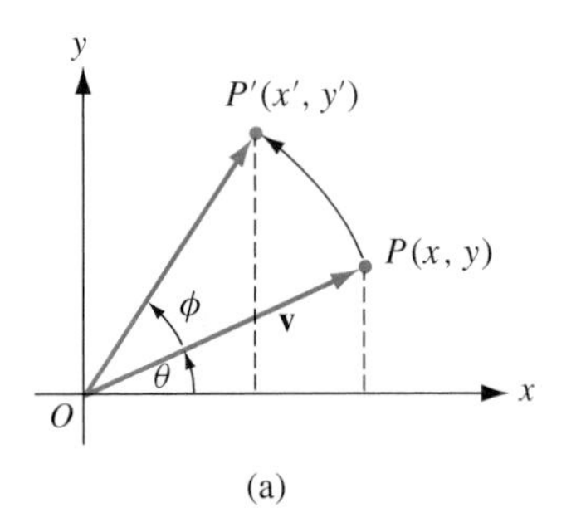

(a)

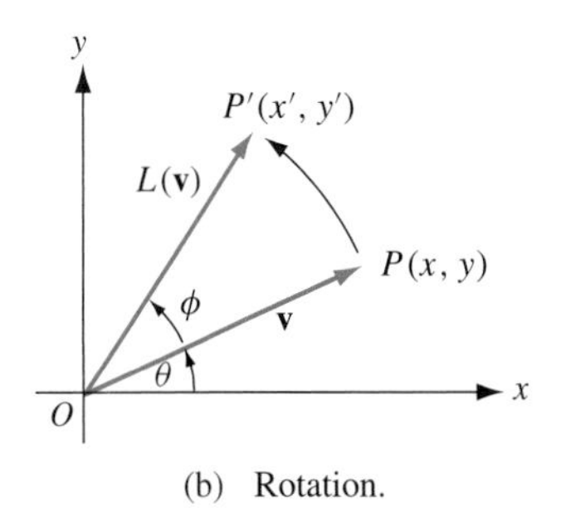

(b) Rotation.

FIGURE 4.4

and

$$x' = r\cos(\theta + \phi), \qquad y' = r\sin(\theta + \phi). \tag{2}$$

Using the formulas for the sine and cosine of a sum of angles (2) becomes

$$\begin{aligned} x' &= r\cos\theta\cos\phi - r\sin\theta\sin\phi \\ y' &= r\sin\theta\cos\phi + r\cos\theta\sin\phi. \end{aligned}$$

Substituting the expression in (1) into the last pair of equations, we obtain

$$x' = x\cos\phi - y\sin\phi, \qquad y' = x\sin\phi + y\cos\phi. \tag{3}$$

Solving (3) for x and y, we have

$$x = x'\cos\phi + y'\sin\phi \quad \text{and} \quad y = -x'\sin\phi + y'\cos\phi. \tag{4}$$

Equation (3) gives the coordinates of P' in terms of those of P, and (4) expresses the coordinates of P in terms of those of P'. This type of rotation is used in calculus to simplify the general equation of second degree $ax^2 + bxy + cy^2 + dx + ey + f = 0$. Substituting for x and y in terms of x' and y', we obtain

$$a'x'^2 + b'x'y' + c'y'^2 + d'x' + e'y' + f' = 0.$$

The key point is to choose ϕ so that $b' = 0$. Once this is done (we might now have to perform a translation of coordinates), we identify the general equation of second degree as a circle, ellipse, hyperbola, parabola, or a degenerate form of these. This topic will be treated from a linear algebra point of view in Section 6.6.

We may also perform this change of coordinates by considering the function $L\colon R^2 \to R^2$ defined by

$$L\left(\begin{bmatrix} x \\ y \end{bmatrix}\right) = \begin{bmatrix} \cos\phi & -\sin\phi \\ \sin\phi & \cos\phi \end{bmatrix} \begin{bmatrix} x \\ y \end{bmatrix}.$$

If $\mathbf{v} = \begin{bmatrix} x \\ y \end{bmatrix}$, then

$$L(\mathbf{v}) = \begin{bmatrix} x\cos\phi - y\sin\phi \\ x\sin\phi + y\cos\phi \end{bmatrix},$$

so $L(\mathbf{v})$ is the vector from the origin to $P'(x', y')$. See Figure 4.4(b). Then L is a linear transformation (verify) called a **rotation**. ■

EXAMPLE 10

Let $L\colon R^3 \to R^3$ be defined by

$$L\left(\begin{bmatrix} a_1 \\ a_2 \\ a_3 \end{bmatrix}\right) = \begin{bmatrix} a_1 + 1 \\ 2a_2 \\ a_3 \end{bmatrix}.$$

To determine whether L is a linear transformation, let

$$\mathbf{u} = \begin{bmatrix} a_1 \\ a_2 \\ a_3 \end{bmatrix} \quad \text{and} \quad \mathbf{v} = \begin{bmatrix} b_1 \\ b_2 \\ b_3 \end{bmatrix}.$$

Then

$$\begin{aligned} L(\mathbf{u} + \mathbf{v}) &= L\left(\begin{bmatrix} a_1 \\ a_2 \\ a_3 \end{bmatrix} + \begin{bmatrix} b_1 \\ b_2 \\ b_3 \end{bmatrix}\right) = L\left(\begin{bmatrix} a_1 + b_1 \\ a_2 + b_2 \\ a_3 + b_3 \end{bmatrix}\right) \\ &= \begin{bmatrix} (a_1 + b_1) + 1 \\ 2(a_2 + b_2) \\ a_3 + b_3 \end{bmatrix}. \end{aligned}$$

On the other hand,

$$L(\mathbf{u}) + L(\mathbf{v}) = \begin{bmatrix} a_1 + 1 \\ 2a_2 \\ a_3 \end{bmatrix} + \begin{bmatrix} b_1 + 1 \\ 2b_2 \\ b_3 \end{bmatrix} = \begin{bmatrix} (a_1 + b_1) + 2 \\ 2(a_2 + b_2) \\ a_3 + b_3 \end{bmatrix}.$$

Letting $a_1 = 1$, $a_2 = 3$, $b_1 = 2$, and $b_2 = 4$, we see that $L(\mathbf{u} + \mathbf{v}) \neq L(\mathbf{u}) + L(\mathbf{v})$. Hence, we conclude that the function L is not a linear transformation. ■

EXAMPLE 11

Let $L\colon R_2 \to R_2$ be defined by

$$L\left(\begin{bmatrix} a_1 & a_2 \end{bmatrix}\right) = \begin{bmatrix} a_1^2 & 2a_2 \end{bmatrix}.$$

Is L a linear transformation?

Solution Let

$$\mathbf{u} = \begin{bmatrix} a_1 & a_2 \end{bmatrix} \quad \text{and} \quad \mathbf{v} = \begin{bmatrix} b_1 & b_2 \end{bmatrix}.$$

Then

$$\begin{aligned} L(\mathbf{u} + \mathbf{v}) &= L(\begin{bmatrix} a_1 & a_2 \end{bmatrix} + \begin{bmatrix} b_1 & b_2 \end{bmatrix}) \\ &= L(\begin{bmatrix} a_1 + b_1 & a_2 + b_2 \end{bmatrix}) \\ &= \begin{bmatrix} (a_1 + b_1)^2 & 2(a_2 + b_2) \end{bmatrix}. \end{aligned}$$

On the other hand,

$$\begin{aligned} L(\mathbf{u}) + L(\mathbf{v}) &= \begin{bmatrix} a_1^2 & 2a_2 \end{bmatrix} + \begin{bmatrix} b_1^2 & 2b_2 \end{bmatrix} \\ &= \begin{bmatrix} a_1^2 + b_1^2 & 2(a_2 + b_2) \end{bmatrix}. \end{aligned}$$

Since there are some choices of a and b such that $L(\mathbf{u} + \mathbf{v}) \neq L(\mathbf{u}) + L(\mathbf{v})$, we conclude that L is not a linear transformation. ■

Theorem 4.1

Let $L: V \to W$ be a linear transformation. Then

(a) $L(\mathbf{0}_V) = \mathbf{0}_W$.

(b) $L(\mathbf{u} - \mathbf{v}) = L(\mathbf{u}) - L(\mathbf{v})$, *for* $\mathbf{u}, \mathbf{v}$ *in* V.

Proof

(a) We have

$$\mathbf{0}_V = \mathbf{0}_V + \mathbf{0}_V,$$

so

$$\begin{aligned} L(\mathbf{0}_V) &= L(\mathbf{0}_V + \mathbf{0}_V) \\ L(\mathbf{0}_V) &= L(\mathbf{0}_V) + L(\mathbf{0}_V). \end{aligned}$$

Adding $-L(\mathbf{0}_V)$ to both sides, we obtain

$$L(\mathbf{0}_V) = \mathbf{0}_W.$$

(b) $$\begin{aligned} L(\mathbf{u} - \mathbf{v}) &= L(\mathbf{u} + (-1)\mathbf{v}) = L(\mathbf{u}) + L((-1)\mathbf{v}) \\ &= L(\mathbf{u}) + (-1)L(\mathbf{v}) = L(\mathbf{u}) - L(\mathbf{v}). \end{aligned}$$ ●

Remark Example 10 can be solved more easily by observing that

$$L\left(\begin{bmatrix} 0 \\ 0 \\ 0 \end{bmatrix}\right) = \begin{bmatrix} 1 \\ 0 \\ 0 \end{bmatrix},$$

so, by part (a) of Theorem 4.1, L is not a linear transformation.

EXAMPLE 12

Let V be an n-dimensional vector space and $S = \{\mathbf{v}_1, \mathbf{v}_2, \ldots, \mathbf{v}_n\}$ an ordered basis for V. If $\mathbf{v}$ is a vector in V, then $\mathbf{v}$ can be written in one and only one way as

$$\mathbf{v} = a_1\mathbf{v}_1 + a_2\mathbf{v}_2 + \cdots + a_n\mathbf{v}_n,$$

where $a_1, a_2, \ldots, a_n$ are real numbers, which were called in Section 2.7 the coordinates of $\mathbf{v}$ with respect to S. Recall that in Section 2.7, we defined the coordinate vector of $\mathbf{v}$ with respect to S as

$$\left[\mathbf{v}\right]_S = \begin{bmatrix} a_1 \\ a_2 \\ \vdots \\ a_n \end{bmatrix}.$$

We define $L: V \to R^n$ by

$$L(\mathbf{v}) = \left[\mathbf{v}\right]_S.$$

It is not difficult to show (Exercise 25) that L is a linear transformation. ■

EXAMPLE 13

Let V be an n-dimensional vector space and let $S = \{\mathbf{v}_1, \mathbf{v}_2, \ldots, \mathbf{v}_n\}$ and $T = \{\mathbf{w}_1, \mathbf{w}_2, \ldots, \mathbf{w}_n\}$ be ordered bases for V. If $\mathbf{v}$ is any vector in V, then Equation (3) in Section 2.7 gives the relationship between the coordinate vector of $\mathbf{v}$ with respect to S and the coordinate vector of $\mathbf{v}$ with respect to T as

$$\left[\mathbf{v}\right]_S = P_{S \leftarrow T} \left[\mathbf{v}\right]_T,$$

where $P_{S \leftarrow T}$ is the transition matrix from T to S. Let $L: R^n \to R^n$ be defined by

$$L(\mathbf{v}) = P_{S \leftarrow T} \mathbf{v}$$

for $\mathbf{v}$ in R^n. By Example 7, it follows that L is a linear transformation. ■

We know from calculus that a function can be specified by a formula that assigns to every member of the domain a unique element of the range. On the other hand, we can also specify a function by listing next to each member of the domain its assigned element of the range. An example of this would be provided by listing the names of all charge account customers of a department store along with their charge account number. At first glance it appears impossible to describe a linear transformation $L: V \to W$ of a vector space $V \neq \{\mathbf{0}\}$ into a vector space W in this latter manner, since V has infinitely many members in it. However, the following very useful theorem tells us that *once we say what a linear transformation L does to a basis for V, then we have completely specified L*. Thus, since in this book we deal mostly with finite-dimensional vector spaces, it is possible to describe L by giving only the images of a finite number of vectors in the domain V.

Theorem 4.2

Let $L: V \to W$ be a linear transformation of an n-dimensional vector space V into a vector space W. Let $S = \{\mathbf{v}_1, \mathbf{v}_2, \ldots, \mathbf{v}_n\}$ be a basis for V. If $\mathbf{v}$ is any vector in V, then $L(\mathbf{v})$ is completely determined by $\{L(\mathbf{v}_1), L(\mathbf{v}_2), \ldots, L(\mathbf{v}_n)\}$.

Proof

Since $\mathbf{v}$ is in V, we can write $\mathbf{v} = a_1\mathbf{v}_1 + a_2\mathbf{v}_2 + \cdots + a_n\mathbf{v}_n$, where $a_1, a_2, \ldots, a_n$ are uniquely determined real numbers. Then

$$\begin{aligned} L(\mathbf{v}) &= L(a_1\mathbf{v}_1 + a_2\mathbf{v}_2 + \cdots + a_n\mathbf{v}_n) \\ &= L(a_1\mathbf{v}_1) + L(a_2\mathbf{v}_2) + \cdots + L(a_n\mathbf{v}_n) \\ &= a_1L(\mathbf{v}_1) + a_2L(\mathbf{v}_2) + \cdots + a_nL(\mathbf{v}_n). \end{aligned}$$

Thus $L(\mathbf{v})$ has been completely determined by the vectors $L(\mathbf{v}_1), L(\mathbf{v}_2), \ldots, L(\mathbf{v}_n)$. ●

Theorem 4.2 can also be stated in the following useful form: Let $L\colon V \to W$ and $L'\colon V \to W$ be linear transformations of the n-dimensional vector space V into a vector space W. Let $S = \{\mathbf{v}_1, \mathbf{v}_2, \ldots, \mathbf{v}_n\}$ be a basis for V. If $L'(\mathbf{v}_i) = L(\mathbf{v}_i)$ for $i = 1, 2, \ldots, n$, then $L'(\mathbf{v}) = L(\mathbf{v})$ for every $\mathbf{v}$ in V; that is, *if L and L' agree on a basis for V, then L and L' are identical linear transformations.*

EXAMPLE 14

Let $L\colon R_4 \to R_2$ be a linear transformation and let $S = \{\mathbf{v}_1, \mathbf{v}_2, \mathbf{v}_3, \mathbf{v}_4\}$ be a basis for R_4, where $\mathbf{v}_1 = \begin{bmatrix} 1 & 0 & 1 & 0 \end{bmatrix}$, $\mathbf{v}_2 = \begin{bmatrix} 0 & 1 & -1 & 2 \end{bmatrix}$, $\mathbf{v}_3 = \begin{bmatrix} 0 & 2 & 2 & 1 \end{bmatrix}$, and $\mathbf{v}_4 = \begin{bmatrix} 1 & 0 & 0 & 1 \end{bmatrix}$. Suppose that

$$L(\mathbf{v}_1) = \begin{bmatrix} 1 & 2 \end{bmatrix}, \quad L(\mathbf{v}_2) = \begin{bmatrix} 0 & 3 \end{bmatrix},$$
$$L(\mathbf{v}_3) = \begin{bmatrix} 0 & 0 \end{bmatrix}, \quad \text{and} \quad L(\mathbf{v}_4) = \begin{bmatrix} 2 & 0 \end{bmatrix}.$$

Let

$$\mathbf{v} = \begin{bmatrix} 3 & -5 & -5 & 0 \end{bmatrix}.$$

Find $L(\mathbf{v})$.

Solution We first write $\mathbf{v}$ as a linear combination of the vectors in S, obtaining (verify)

$$\mathbf{v} = \begin{bmatrix} 3 & -5 & -5 & 0 \end{bmatrix} = 2\mathbf{v}_1 + \mathbf{v}_2 - 3\mathbf{v}_3 + \mathbf{v}_4.$$

It then follows by Theorem 4.2 that

$$\begin{aligned} L(\mathbf{v}) &= L(2\mathbf{v}_1 + \mathbf{v}_2 - 3\mathbf{v}_3 + \mathbf{v}_4) \\ &= 2L(\mathbf{v}_1) + L(\mathbf{v}_2) - 3L(\mathbf{v}_3) + L(\mathbf{v}_4) = \begin{bmatrix} 4 & 7 \end{bmatrix}. \end{aligned}$$ ■

In Example 7, we have seen that if A is an $m \times n$ matrix, then the function $L\colon R^n \to R^m$ defined by $L(\mathbf{x}) = A\mathbf{x}$ for $\mathbf{x}$ in R^n is a linear transformation. In the next example, we show that if $L\colon R^n \to R^m$ is a linear transformation, then L must be of this form.

Theorem 4.3

Let $L: R^n \to R^m$ be a linear transformation and consider the natural basis $\{\mathbf{e}_1, \mathbf{e}_2, \dots, \mathbf{e}_n\}$ for R^n. Let A be the $m \times n$ matrix whose jth column is $L(\mathbf{e}_j)$. The matrix A has the following property: If $\mathbf{x} = \begin{bmatrix} x_1 \\ x_2 \\ \vdots \\ x_n \end{bmatrix}$ is any vector in R^n, then

$$L(\mathbf{x}) = A\mathbf{x}. \tag{5}$$

Moreover, A is the only matrix satisfying Equation (5). It is called the ***standard matrix representing*** *L.*

Proof

Writing $\mathbf{x}$ as a linear combination of the natural basis for R^n, we have

$$\mathbf{x} = x_1\mathbf{e}_1 + x_2\mathbf{e}_2 + \cdots + x_n\mathbf{e}_n;$$

so by Theorem 4.2,

$$\begin{aligned} L(\mathbf{x}) &= L(x_1\mathbf{e}_1 + x_2\mathbf{e}_2 + \cdots + x_n\mathbf{e}_n) \\ &= x_1L(\mathbf{e}_1) + x_2L(\mathbf{e}_2) + \cdots + x_nL(\mathbf{e}_n). \end{aligned} \tag{6}$$

Since A is the $m \times n$ matrix whose jth column is $L(\mathbf{e}_j)$, we can write Equation (6) in matrix form as

$$L(\mathbf{x}) = A\mathbf{x}.$$

We leave it as an exercise (Exercise 33) to show that A is unique. ●

EXAMPLE 15

Let $L: R^3 \to R^2$ be the linear transformation defined by

$$L\left(\begin{bmatrix} x_1 \\ x_2 \\ x_3 \end{bmatrix}\right) = \begin{bmatrix} x_1 + 2x_2 \\ 3x_2 - 2x_3 \end{bmatrix}.$$

Find the standard matrix representing L.

Solution Let $\{\mathbf{e}_1, \mathbf{e}_2, \mathbf{e}_3\}$ be the natural basis for R^3. We now compute $L(\mathbf{e}_j)$ for

$j = 1, 2, 3$ as follows:

$$L(\mathbf{e}_1) = L\left(\begin{bmatrix} 1 \\ 0 \\ 0 \end{bmatrix}\right) = \begin{bmatrix} 1 \\ 0 \end{bmatrix}$$

$$L(\mathbf{e}_2) = L\left(\begin{bmatrix} 0 \\ 1 \\ 0 \end{bmatrix}\right) = \begin{bmatrix} 2 \\ 3 \end{bmatrix}$$

$$L(\mathbf{e}_3) = L\left(\begin{bmatrix} 0 \\ 0 \\ 1 \end{bmatrix}\right) = \begin{bmatrix} 0 \\ -2 \end{bmatrix}$$

Hence

$$A = \begin{bmatrix} L(\mathbf{e}_1) & L(\mathbf{e}_2) & L(\mathbf{e}_3) \end{bmatrix} = \begin{bmatrix} 1 & 2 & 0 \\ 0 & 3 & -2 \end{bmatrix}.$$

■

4.1 Exercises

1. Which of the following functions are linear transformations?

(a) $L\colon R_2 \to R_3$ defined by $L\left(\begin{bmatrix} a_1 & a_2 \end{bmatrix}\right) = \begin{bmatrix} a_1 + 1 & a_2 & a_1 + a_2 \end{bmatrix}$.

(b) $L\colon R_2 \to R_3$ defined by $L\left(\begin{bmatrix} a_1 & a_2 \end{bmatrix}\right) = \begin{bmatrix} a_1 + a_2 & a_2 & a_1 - a_2 \end{bmatrix}$.

2. Which of the following functions are linear transformations?

(a) $L\colon R_3 \to R_3$ defined by $L\left(\begin{bmatrix} a_1 & a_2 & a_3 \end{bmatrix}\right) = \begin{bmatrix} a_1 & a_2^2 + a_3^2 & a_3^2 \end{bmatrix}$.

(b) $L\colon R_3 \to R_3$ defined by $L\left(\begin{bmatrix} a_1 & a_2 & a_3 \end{bmatrix}\right) = \begin{bmatrix} 1 & a_3 & a_2 \end{bmatrix}$.

(c) $L\colon R_3 \to R_3$ defined by $L\left(\begin{bmatrix} a_1 & a_2 & a_3 \end{bmatrix}\right) = \begin{bmatrix} 0 & a_3 & a_2 \end{bmatrix}$.

3. Which of the following functions are linear transformations? [Here $p'(t)$ denotes the derivative of $p(t)$ with respect to t.]

(a) $L\colon P_2 \to P_3$ defined by $L(p(t)) = t^3 p'(0) + t^2 p(0)$.

(b) $L\colon P_1 \to P_2$ defined by $L(p(t)) = tp(t) + p(0)$.

(c) $L\colon P_1 \to P_2$ defined by $L(p(t)) = tp(t) + 1$.

4. Let $L\colon V \to W$ be a mapping of a vector space V into a vector space W. Prove that L is a linear transformation if and only if $L(a\mathbf{u} + b\mathbf{v}) = aL(\mathbf{u}) + bL(\mathbf{v})$ for any real numbers a, b and any vectors $\mathbf{u}$, $\mathbf{v}$ in V.

5. Prove that the functions in Example 3 are linear transformations.

6. Prove that the functions in Examples 6 and 9 are linear transformations.

7. Consider the function $L\colon M_{34} \to M_{24}$ defined by $L(A) = \begin{bmatrix} 2 & 3 & 1 \\ 1 & 2 & -3 \end{bmatrix} A$ for A in M_{34}.

(a) Find $L\left(\begin{bmatrix} 1 & 2 & 0 & -1 \\ 3 & 0 & 2 & 3 \\ 4 & 1 & -2 & 1 \end{bmatrix}\right)$.

(b) Show that L is a linear transformation.

8. Find the standard matrix representing the given linear transformation.

(a) The projection from R^3 to R^2 in Example 1.

(b) The dilation from R^3 to R^3 in Example 3.

(c) The reflection from R^2 to R^2 in Example 8.

9. Find the standard matrix representing the given linear transformation.

(a) $L\colon R^2 \to R^2$ defined by $L\left(\begin{bmatrix} a_1 \\ a_2 \end{bmatrix}\right) = \begin{bmatrix} a_2 \\ a_1 \end{bmatrix}$.

(b) $L: R^2 \to R^3$ defined by

$$L\left(\begin{bmatrix} a_1 \\ a_2 \end{bmatrix}\right) = \begin{bmatrix} a_1 - 3a_2 \\ 2a_1 - a_2 \\ 2a_2 \end{bmatrix}.$$

(c) $L: R^3 \to R^3$ defined by

$$L\left(\begin{bmatrix} a_1 \\ a_2 \\ a_3 \end{bmatrix}\right) = \begin{bmatrix} a_1 + 4a_2 \\ -a_3 \\ a_2 + a_3 \end{bmatrix}.$$

10. Let $A = \begin{bmatrix} 0 & -1 & 2 \\ -2 & 1 & 3 \\ 1 & 2 & -3 \end{bmatrix}$ be the standard matrix representing the linear transformation $L: R^3 \to R^3$.

(a) Find $L\left(\begin{bmatrix} 2 \\ -3 \\ 1 \end{bmatrix}\right)$. (b) Find $L\left(\begin{bmatrix} a_1 \\ a_2 \\ a_3 \end{bmatrix}\right)$.

11. Let $L: R^3 \to R^2$ be a linear transformation for which we know that

$$L\left(\begin{bmatrix} 1 \\ 0 \\ 0 \end{bmatrix}\right) = \begin{bmatrix} 2 \\ -4 \end{bmatrix},$$

$$L\left(\begin{bmatrix} 0 \\ 1 \\ 0 \end{bmatrix}\right) = \begin{bmatrix} 3 \\ -5 \end{bmatrix}, \quad L\left(\begin{bmatrix} 0 \\ 0 \\ 1 \end{bmatrix}\right) = \begin{bmatrix} 2 \\ 3 \end{bmatrix}.$$

(a) What is $L\left(\begin{bmatrix} 1 \\ -2 \\ 3 \end{bmatrix}\right)$?

(b) What is $L\left(\begin{bmatrix} a_1 \\ a_2 \\ a_2 \end{bmatrix}\right)$?

12. Let $L: R_2 \to R_2$ be a linear transformation for which we know that

$$L\left(\begin{bmatrix} 1 & 1 \end{bmatrix}\right) = \begin{bmatrix} 1 & -2 \end{bmatrix},$$
$$L\left(\begin{bmatrix} -1 & 1 \end{bmatrix}\right) = \begin{bmatrix} 2 & 3 \end{bmatrix}.$$

(a) What is $L\left(\begin{bmatrix} -1 & 5 \end{bmatrix}\right)$?

(b) What is $L\left(\begin{bmatrix} a_1 & a_2 \end{bmatrix}\right)$?

13. Let $L: P_2 \to P_3$ be a linear transformation for which we know that $L(1) = 1$, $L(t) = t^2$, $L(t^2) = t^3 + t$.

(a) Find $L(2t^2 - 5t + 3)$. (b) Find $L(at^2 + bt + c)$.

14. Let A be a fixed 3×3 matrix; also let $L: M_{33} \to M_{33}$ be defined by $L(X) = AX - XA$, for X in M_{33}. Show that L is a linear transformation.

15. Let $L: R \to R$ be defined by $L(\mathbf{v}) = a\mathbf{v} + b$, where a and b are real numbers (of course, $\mathbf{v}$ is a vector in R, which in this case means that $\mathbf{v}$ is also a real number). Find all values of a and b such that L is a linear transformation.

16. Let V be an inner product space, and let $\mathbf{w}$ be a fixed vector in V. Let $L: V \to R$ be defined by $L(\mathbf{v}) = (\mathbf{v}, \mathbf{w})$ for $\mathbf{v}$ in V. Show that L is a linear transformation.

17. Describe the following linear transformations geometrically.

(a) $L\left(\begin{bmatrix} a_1 \\ a_2 \end{bmatrix}\right) = \begin{bmatrix} -a_1 \\ a_2 \end{bmatrix}$.

(b) $L\left(\begin{bmatrix} a_1 \\ a_2 \end{bmatrix}\right) = \begin{bmatrix} -a_1 \\ -a_2 \end{bmatrix}$.

(c) $L\left(\begin{bmatrix} a_1 \\ a_2 \end{bmatrix}\right) = \begin{bmatrix} -a_2 \\ a_1 \end{bmatrix}$.

18. Let V be a vector space and r a fixed scalar. Prove that the function $L: V \to V$ defined by $L(\mathbf{v}) = r\mathbf{v}$ is a linear operator on V.

19. Let V and W be vector spaces. Prove that the function $O: V \to W$ defined by $O(\mathbf{v}) = \mathbf{0}_W$ is a linear transformation, which is called the **zero linear transformation**.

20. Let $I: V \to V$ be defined by $I(\mathbf{v}) = \mathbf{v}$, for $\mathbf{v}$ in V. Show that I is a linear transformation, which is called the **identity operator** on V.

21. Let A be an $m \times n$ matrix, and suppose that $L: R^n \to R^m$ is defined by $L(\mathbf{x}) = A\mathbf{x}$ for $\mathbf{x}$ in R^n. Show that L is a linear transformation.

22. Let W be the vector space of all real-valued functions and let V be the subspace of all differentiable functions. Define $L: V \to W$ by $L(f) = f'$, where f' is the derivative of f. Prove that L is a linear transformation.

23. Let $V = C[a, b]$ be the vector space of all real-valued functions that are integrable over the interval $[a, b]$. Let $W = R^1$. Define $L: V \to W$ by $L(f) = \int_a^b f(x)\,dx$. Prove that L is a linear transformation.

24. Let A be an $n \times n$ matrix and suppose that $L: M_{nn} \to M_{nn}$ is defined by $L(X) = AX$, for X in M_{nn}. Show that L is a linear transformation.

25. Show that the function L defined in Example 12 is a linear transformation.

26. For the linear transformation defined in Example 14, find $L\left(\begin{bmatrix} a & b & c & d \end{bmatrix}\right)$.

27. Let V be an n-dimensional vector space with ordered basis $S = \{\mathbf{v}_1, \mathbf{v}_2, \ldots, \mathbf{v}_n\}$ and let $T = \{\mathbf{w}_1, \mathbf{w}_2, \ldots, \mathbf{w}_n\}$ be an ordered set of vectors in V. Prove that there is a unique linear transformation $L: V \to V$ such that $L(\mathbf{v}_i) = \mathbf{w}_i$ for $i = 1, 2, \ldots, n$. [*Hint*: Let L be a mapping from V into V such that $L(\mathbf{v}_i) = \mathbf{w}_i$; then show how to extend L to be a linear transformation defined on all of V.]

28. Let $L: V \to W$ be a linear transformation from a vector space V into a vector space W. The **image** of a subspace V_1 of V is defined as

$$L(V_1) = \{\mathbf{w} \text{ in } W \mid \mathbf{w} = L(\mathbf{v}) \text{ for some } \mathbf{v} \text{ in } V\}.$$

Show that $L(V_1)$ is a subspace of V.

29. Let L_1 and L_2 be linear transformations from a vector space V into a vector space W. Let $\{\mathbf{v}_1, \mathbf{v}_2, \ldots, \mathbf{v}_n\}$ be a basis for V. Show that if $L_1(\mathbf{v}_i) = L_2(\mathbf{v}_i)$ for $i = 1, 2, \ldots, n$, then $L_1(\mathbf{v}) = L_2(\mathbf{v})$ for any $\mathbf{v}$ in V.

30. Let $L: V \to W$ be a linear transformation from a vector space V into a vector space W. The **preimage** of a subspace W_1 of W is defined as

$$L^{-1}(W_1) = \{\mathbf{v} \text{ in } V \mid L(\mathbf{v}) \text{ is in } W_1\}.$$

Show that $L^{-1}(W_1)$ is a subspace of V.

31. Let $O: R^n \to R^n$ be the zero linear transformation defined by $O(\mathbf{v}) = \mathbf{0}$ for $\mathbf{v}$ in R^n (see Exercise 19). Find the standard matrix representing O.

32. Let $I: R^n \to R^n$ be the identity linear transformation defined by $I(\mathbf{v}) = \mathbf{v}$ for $\mathbf{v}$ in R^n (see Exercise 20). Find the standard matrix representing I.

33. Complete the proof of Theorem 4.3 by showing that the matrix A is unique (*Hint*: Suppose that there is another matrix B such that $L(\mathbf{x}) = B\mathbf{x}$ for $\mathbf{x}$ in R^n. Consider $L(\mathbf{e}_j)$ for $j = 1, 2, \ldots, n$. Show that $A = B$.)

4.2 KERNEL AND RANGE OF A LINEAR TRANSFORMATION

In this section we study special types of linear transformations; we formulate the notions of one-to-one linear transformations and onto linear transformations. We also develop methods for determining when a linear transformation is one-to-one or onto, and examine some applications of these notions.

Definition 4.2

A linear transformation $L: V \to W$ is called **one-to-one** if it is a one-to-one function; that is, if $\mathbf{v}_1 \neq \mathbf{v}_2$ implies that $L(\mathbf{v}_1) \neq L(\mathbf{v}_2)$. An equivalent statement is that L is one-to-one if $L(\mathbf{v}_1) = L(\mathbf{v}_2)$ implies that $\mathbf{v}_1 = \mathbf{v}_2$ (see Figure A.2 in Appendix A). ▲

EXAMPLE 1

Let $L: R^2 \to R^2$ be defined by

$$L\left(\begin{bmatrix} a_1 \\ a_2 \end{bmatrix}\right) = \begin{bmatrix} a_1 + a_2 \\ a_1 - a_2 \end{bmatrix}.$$

To determine whether L is one-to-one, we let

$$\mathbf{v}_1 = \begin{bmatrix} a_1 \\ a_2 \end{bmatrix} \quad \text{and} \quad \mathbf{v}_2 = \begin{bmatrix} b_1 \\ b_2 \end{bmatrix}.$$

Then if $L(\mathbf{v}_1) = L(\mathbf{v}_2)$, we have

$$\begin{aligned} a_1 + a_2 &= b_1 + b_2 \\ a_1 - a_2 &= b_1 - b_2. \end{aligned}$$

Adding these equations, we obtain $2a_1 = 2b_1$, or $a_1 = b_1$, which implies that $a_2 = b_2$. Hence $\mathbf{v}_1 = \mathbf{v}_2$ and L is one-to-one. ■

EXAMPLE 2

Let $L: R^3 \to R^2$ be the linear transformation defined in Example 1 of Section 4.1 (a projection) by

$$L\left(\begin{bmatrix} a_1 \\ a_2 \\ a_3 \end{bmatrix}\right) = \begin{bmatrix} a_1 \\ a_2 \end{bmatrix}.$$

Since $L\left(\begin{bmatrix} 1 \\ 3 \\ 3 \end{bmatrix}\right) = L\left(\begin{bmatrix} 1 \\ 3 \\ -2 \end{bmatrix}\right)$ yet $\begin{bmatrix} 1 \\ 3 \\ 3 \end{bmatrix} \neq \begin{bmatrix} 1 \\ 3 \\ -2 \end{bmatrix}$, we conclude that L is not one-to-one. ■

We shall now develop some more efficient ways of determining whether a linear transformation is one-to-one.

Definition 4.3

Let $L: V \to W$ be a linear transformation of a vector space V into a vector space W. The **kernel** of L, $\ker L$, is the subset of V consisting of all elements $\mathbf{v}$ of V such that $L(\mathbf{v}) = \mathbf{0}_W$. ▲

We observe that Theorem 4.1 assures us that $\ker L$ is never an empty set, because if $L: V \to W$ is a linear transformation, then $\mathbf{0}_V$ is in $\ker L$.

EXAMPLE 3

Let $L: R^3 \to R^2$ be as defined in Example 2. The vector $\begin{bmatrix} 0 \\ 0 \\ 2 \end{bmatrix}$ is in $\ker L$, since $L\left(\begin{bmatrix} 0 \\ 0 \\ 2 \end{bmatrix}\right) = \begin{bmatrix} 0 \\ 0 \end{bmatrix}$. However, the vector $\begin{bmatrix} 2 \\ -3 \\ 4 \end{bmatrix}$ is not in $\ker L$, since $L\left(\begin{bmatrix} 2 \\ -3 \\ 4 \end{bmatrix}\right) = \begin{bmatrix} 2 \\ -3 \end{bmatrix}$. To find $\ker L$, we must determine all $\mathbf{v}$ in R^3 so that $L(\mathbf{v}) = \mathbf{0}_{R^2}$. That is, we seek $\mathbf{v} = \begin{bmatrix} a_1 \\ a_2 \\ a_3 \end{bmatrix}$ so that

$$L(\mathbf{v}) = L\left(\begin{bmatrix} a_1 \\ a_2 \\ a_3 \end{bmatrix}\right) = \begin{bmatrix} 0 \\ 0 \end{bmatrix} = \mathbf{0}_{R^2}.$$

However, $L(\mathbf{v}) = \begin{bmatrix} a_1 \\ a_2 \end{bmatrix}$. Thus $\begin{bmatrix} a_1 \\ a_2 \end{bmatrix} = \begin{bmatrix} 0 \\ 0 \end{bmatrix}$, so $a_1 = 0$, $a_2 = 0$, and a_3 can be any real number. Hence $\ker L$ consists of all vectors in R^3 of the form $\begin{bmatrix} 0 \\ 0 \\ a \end{bmatrix}$, where a is any real number. It is clear that $\ker L$ consists of the z-axis in (x, y, z) three-dimensional space R^3. ■

An examination of the elements in $\ker L$ allows us to decide whether L is or is not one-to-one.

Theorem 4.4

Let $L: V \to W$ be a linear transformation of a vector space V into a vector space W. Then

(a) $\ker L$ *is a subspace of* V.

(b) L *is one-to-one if and only if* $\ker L = \{\mathbf{0}_V\}$.

Proof

(a) We show that if $\mathbf{v}$ and $\mathbf{w}$ are in $\ker L$, then so are $\mathbf{v} + \mathbf{w}$ and $c\mathbf{v}$ for any real number c. If $\mathbf{v}$ and $\mathbf{w}$ are in $\ker L$, then $L(\mathbf{v}) = \mathbf{0}_W$, and $L(\mathbf{w}) = \mathbf{0}_W$.

Then since L is a linear transformation,

$$L(\mathbf{v} + \mathbf{w}) = L(\mathbf{v}) + L(\mathbf{w}) = \mathbf{0}_W + \mathbf{0}_W = \mathbf{0}_W.$$

Thus $\mathbf{v} + \mathbf{w}$ is in $\ker L$. Also,

$$L(c\mathbf{v}) = cL(\mathbf{v}) = c\,\mathbf{0}_W = \mathbf{0},$$

so $c\mathbf{v}$ is in $\ker L$. Hence $\ker L$ is a subspace of V.

(b) Let L be one-to-one. We show that $\ker L = \{\mathbf{0}_V\}$. Let $\mathbf{v}$ be in $\ker L$. Then $L(\mathbf{v}) = \mathbf{0}_W$. Also, we already know that $L(\mathbf{0}_V) = \mathbf{0}_W$. Then $L(\mathbf{v}) = L(\mathbf{0}_V)$. Since L is one-to-one, we conclude that $\mathbf{v} = \mathbf{0}_V$. Hence $\ker L = \{\mathbf{0}_V\}$.

Conversely, suppose that $\ker L = \{\mathbf{0}_V\}$. We wish to show that L is one-to-one. Let $L(\mathbf{v}_1) = L(\mathbf{v}_2)$ for $\mathbf{v}_1$ and $\mathbf{v}_2$ in V. Then

$$L(\mathbf{v}_1) - L(\mathbf{v}_2) = \mathbf{0}_W,$$

so that $L(\mathbf{v}_1 - \mathbf{v}_2) = \mathbf{0}_W$. This means that $\mathbf{v}_1 - \mathbf{v}_2$ is in $\ker L$, so $\mathbf{v}_1 - \mathbf{v}_2 = \mathbf{0}_V$. Hence $\mathbf{v}_1 = \mathbf{v}_2$, and L is one-to-one. ●

Note that we can also state Theorem 4.4(b) as: *L is one-to-one if and only if* $\dim \ker L = 0$.

The proof of Theorem 4.4 has also established the following result, which we state as Corollary 4.1.

Corollary 4.1

If $L(\mathbf{x}) = \mathbf{b}$ *and* $L(\mathbf{y}) = \mathbf{b}$, *then* $\mathbf{x} - \mathbf{y}$ *belongs to* $\ker L$. *In other words, any two solutions to* $L(\mathbf{x}) = \mathbf{b}$ *differ by an element of the kernel of* L.

Proof

Exercise. ●

EXAMPLE 4

Let $L\colon P_2 \to R$ be the linear transformation defined by

$$L(at^2 + bt + c) = \int_0^1 (at^2 + bt + c)\, dt.$$

(a) Find $\ker L$.

(b) Find $\dim \ker L$.

(c) Is L one-to-one?

Solution (a) To find $\ker L$, we seek an element $\mathbf{v} = at^2 + bt + c$ in P_2 such that $L(\mathbf{v}) = L(at^2 + bt + c) = \mathbf{0}_R = 0$. Now

$$L(\mathbf{v}) = \left.\frac{at^3}{3} + \frac{bt^2}{2} + ct\right|_0^1 = \frac{a}{3} + \frac{b}{2} + c.$$

Thus $c = -a/3 - b/2$. Then $\ker L$ consists of all polynomials in P_2 of the form $at^2 + bt + (-a/3 - b/2)$, for a and b any real numbers.

(b) To find the dimension of $\ker L$, we obtain a basis for $\ker L$. Any vector in $\ker L$ can be written as

$$at^2 + bt + \left(-\frac{a}{3} - \frac{b}{2}\right) = a\left(t^2 - \frac{1}{3}\right) + b\left(t - \frac{1}{2}\right).$$

Thus the elements $\left(t^2 - \frac{1}{3}\right)$ and $\left(t - \frac{1}{2}\right)$ in P_2 span $\ker L$. Now, these elements are also linearly independent, since they are not constant multiples of each other. Thus $\left\{t^2 - \frac{1}{3}, t - \frac{1}{2}\right\}$ is a basis for $\ker L$, and $\dim \ker L = 2$.

(c) Since $\dim \ker L = 2$, L is not one-to-one. ■

Definition 4.4

If $L\colon V \to W$ is a linear transformation of a vector space V into a vector space W, then the **range** of L or **image** of V under L, denoted by range L, consists of all those vectors in W that are images under L of vectors in V. Thus $\mathbf{w}$ is in range L if there exists some vector $\mathbf{v}$ in V such that $L(\mathbf{v}) = \mathbf{w}$. The linear transformation L is called **onto** if range $L = W$. ▲

Theorem 4.5

If $L: V \to W$ is a linear transformation of a vector space V into a vector space W, then range L is a subspace of W.

Proof

Let $\mathbf{w}_1$ and $\mathbf{w}_2$ be in range L. Then $\mathbf{w}_1 = L(\mathbf{v}_1)$ and $\mathbf{w}_2 = L(\mathbf{v}_2)$ for some $\mathbf{v}_1$ and $\mathbf{v}_2$ in V. Now

$$\mathbf{w}_1 + \mathbf{w}_2 = L(\mathbf{v}_1) + L(\mathbf{v}_2) = L(\mathbf{v}_1 + \mathbf{v}_2),$$

which implies that $\mathbf{w}_1 + \mathbf{w}_2$ is in range L. Also, if $\mathbf{w}$ is in range L, then $\mathbf{w} = L(\mathbf{v})$ for some $\mathbf{v}$ in V. Then $c\mathbf{w} = cL(\mathbf{v}) = L(c\mathbf{v})$ where c is a scalar, so that $c\mathbf{w}$ is in range L. Hence range L is a subspace of W. ●

EXAMPLE 5

Consider Example 2 of this section again. Is the projection L onto?

Solution We choose any vector $\mathbf{w} = \begin{bmatrix} c \\ d \end{bmatrix}$ in R^2 and seek a vector $\mathbf{v} = \begin{bmatrix} a_1 \\ a_2 \\ a_3 \end{bmatrix}$ in V such that $L(\mathbf{v}) = \mathbf{w}$. Now $L(\mathbf{v}) = \begin{bmatrix} a_1 \\ a_2 \end{bmatrix}$, so if $a_1 = c$ and $a_2 = d$, then $L(\mathbf{v}) = \mathbf{w}$. Therefore, L is onto and $\dim \operatorname{range} L = 2$. ■

EXAMPLE 6

Consider Example 4 of this section; is L onto?

Solution Given a vector $\mathbf{w}$ in R, $\mathbf{w} = r$, a real number, can we find a vector $\mathbf{v} = at^2 + bt + c$ in P_2 so that $L(\mathbf{v}) = \mathbf{w} = r$?

Now

$$L(\mathbf{v}) = \int_0^1 (at^2 + bt + c)\,dt = \frac{a}{3} + \frac{b}{2} + c.$$

We can let $a = b = 0$ and $c = r$. Hence L is onto. Moreover, $\dim \operatorname{range} L = 1$. ■

EXAMPLE 7

Let $L: R^3 \to R^3$ be defined by

$$L\left(\begin{bmatrix} a_1 \\ a_2 \\ a_3 \end{bmatrix}\right) = \begin{bmatrix} 1 & 0 & 1 \\ 1 & 1 & 2 \\ 2 & 1 & 3 \end{bmatrix} \begin{bmatrix} a_1 \\ a_2 \\ a_3 \end{bmatrix}.$$

(a) Is L onto?

(b) Find a basis for range L.

(c) Find $\ker L$.

(d) Is L one-to-one?

Solution (a) Given any $\mathbf{w} = \begin{bmatrix} a \\ b \\ c \end{bmatrix}$ in R^3, where a, b, and c are any real numbers, can we find $\mathbf{v} = \begin{bmatrix} a_1 \\ a_2 \\ a_3 \end{bmatrix}$ so that $L(\mathbf{v}) = \mathbf{w}$? We seek a solution to the linear system

$$\begin{bmatrix} 1 & 0 & 1 \\ 1 & 1 & 2 \\ 2 & 1 & 3 \end{bmatrix} \begin{bmatrix} a_1 \\ a_2 \\ a_3 \end{bmatrix} = \begin{bmatrix} a \\ b \\ c \end{bmatrix},$$

and we find the reduced row echelon form of the augmented matrix to be (verify)

$$\left[\begin{array}{ccc|c} 1 & 0 & 1 & a \\ 0 & 1 & 1 & b-a \\ 0 & 0 & 0 & c-b-a \end{array}\right].$$

Thus a solution exists only for $c - b - a = 0$, so L is not onto.

(b) To find a basis for range L, we note that

$$L\left(\begin{bmatrix} a_1 \\ a_2 \\ a_3 \end{bmatrix}\right) = \begin{bmatrix} 1 & 0 & 1 \\ 1 & 1 & 2 \\ 2 & 1 & 3 \end{bmatrix} \begin{bmatrix} a_1 \\ a_2 \\ a_3 \end{bmatrix} = \begin{bmatrix} a_1 + a_3 \\ a_1 + a_2 + 2a_3 \\ 2a_1 + a_2 + 3a_3 \end{bmatrix}$$

$$= a_1 \begin{bmatrix} 1 \\ 1 \\ 2 \end{bmatrix} + a_2 \begin{bmatrix} 0 \\ 1 \\ 1 \end{bmatrix} + a_3 \begin{bmatrix} 1 \\ 2 \\ 3 \end{bmatrix}.$$

This means that

$$\left\{ \begin{bmatrix} 1 \\ 1 \\ 2 \end{bmatrix}, \begin{bmatrix} 0 \\ 1 \\ 1 \end{bmatrix}, \begin{bmatrix} 1 \\ 2 \\ 3 \end{bmatrix} \right\}$$

spans range L. That is, range L is the subspace of R^3 spanned by the columns of the matrix defining L.

The first two vectors in this set are linearly independent, since they are not constant multiples of each other. The third vector is the sum of the first two. Therefore, the first two vectors form a basis for range L, and $\dim \text{range } L = 2$.

(c) To find $\ker L$, we wish to find all $\mathbf{v}$ in R^3 so that $L(\mathbf{v}) = \mathbf{0}_{R^3}$. Solving the resulting homogeneous system, we find (verify) that $a_1 = -a_3$ and $a_2 = -a_3$. Thus $\ker L$ consists of all vectors of the form

$$\begin{bmatrix} -a \\ -a \\ a \end{bmatrix} = a \begin{bmatrix} -1 \\ -1 \\ 1 \end{bmatrix},$$

where a is any real number. Moreover, $\dim \ker L = 1$.
(d) Since $\ker L \neq \{\mathbf{0}_{R^3}\}$, it follows from Theorem 4.4(b) that L is not one-to-one. ■

The problem of finding a basis for $\ker L$ always reduces to the problem of finding a basis for the solution space of a homogeneous system; this later problem has been solved in Section 2.6.

If range L is a subspace of R^m or R_m, then a basis for range L can be obtained by the method discussed in Theorem 2.8 or by the procedure given in Section 2.8. Both approaches are illustrated in the next example.

EXAMPLE 8

Let $L\colon R_4 \to R_3$ be defined by

$$L\left(\begin{bmatrix} a_1 & a_2 & a_3 & a_4 \end{bmatrix}\right) = \begin{bmatrix} a_1 + a_2 & a_3 + a_4 & a_1 + a_3 \end{bmatrix}.$$

Find a basis for range L.

Solution We have

$$\begin{aligned} L\left(\begin{bmatrix} a_1 & a_2 & a_3 & a_4 \end{bmatrix}\right) &= a_1\begin{bmatrix} 1 & 0 & 1 \end{bmatrix} + a_2\begin{bmatrix} 1 & 0 & 0 \end{bmatrix} \\ &\quad + a_3\begin{bmatrix} 0 & 1 & 1 \end{bmatrix} + a_4\begin{bmatrix} 0 & 1 & 0 \end{bmatrix}. \end{aligned}$$

Thus

$$S = \left\{\begin{bmatrix} 1 & 0 & 1 \end{bmatrix}, \begin{bmatrix} 1 & 0 & 0 \end{bmatrix}, \begin{bmatrix} 0 & 1 & 1 \end{bmatrix}, \begin{bmatrix} 0 & 1 & 0 \end{bmatrix}\right\}$$

spans range L. To find a subset of S that is a basis for range L, we proceed as in Theorem 2.8 by first writing

$$a_1\begin{bmatrix} 1 & 0 & 1 \end{bmatrix} + a_2\begin{bmatrix} 1 & 0 & 0 \end{bmatrix} + a_3\begin{bmatrix} 0 & 1 & 1 \end{bmatrix} + a_4\begin{bmatrix} 0 & 1 & 0 \end{bmatrix} = \begin{bmatrix} 0 & 0 & 0 \end{bmatrix}.$$

The reduced row echelon form of the augmented matrix of this homogeneous system is (verify)

$$\left[\begin{array}{cccc|c} 1 & 0 & 0 & -1 & 0 \\ 0 & 1 & 0 & 1 & 0 \\ 0 & 0 & 1 & 1 & 0 \end{array}\right].$$

Since the leading 1's appear in columns 1, 2, and 3, we conclude that the first three vectors in S form a basis for range L. Thus

$$\left\{\begin{bmatrix} 1 & 0 & 1 \end{bmatrix}, \begin{bmatrix} 1 & 0 & 0 \end{bmatrix}, \begin{bmatrix} 0 & 1 & 1 \end{bmatrix}\right\}$$

is a basis for range L.

Alternatively, we may proceed as in Section 2.8 to form the matrix whose rows are the given vectors

$$\begin{bmatrix} 1 & 0 & 1 \\ 1 & 0 & 0 \\ 0 & 1 & 1 \\ 0 & 1 & 0 \end{bmatrix}.$$

Transforming this matrix to reduced row echelon form, we obtain (verify)

$$\begin{bmatrix} 1 & 0 & 0 \\ 0 & 1 & 0 \\ 0 & 0 & 1 \\ 0 & 0 & 0 \end{bmatrix}.$$

Hence $\{\begin{bmatrix} 1 & 0 & 0 \end{bmatrix}, \begin{bmatrix} 0 & 1 & 0 \end{bmatrix}, \begin{bmatrix} 0 & 0 & 1 \end{bmatrix}\}$ is a basis for range L. ■

To determine if a linear transformation is one-to-one or onto, we must solve a linear system. This is one further demonstration of the frequency with which linear systems must be solved to answer many questions in linear algebra. Finally, from Example 7, where $\dim \ker L = 1$, $\dim \operatorname{range} L = 2$, and $\dim \operatorname{domain} L = 3$ we saw that

$$\dim \ker L + \dim \operatorname{range} L = \dim \operatorname{domain} L.$$

This very important result is always true, and we now prove it in the following theorem.

Theorem 4.6

If $L: V \to W$ is a linear transformation of an n-dimensional vector space V into a vector space W, then

$$\dim \ker L + \dim \operatorname{range} L = \dim V.$$

Proof

Let $k = \dim \ker L$. If $k = n$, then $\ker L = V$ (Exercise 42, Section 2.5), which implies that $L(\mathbf{v}) = \mathbf{0}_W$ for every $\mathbf{v}$ in V. Hence range $L = \{\mathbf{0}_W\}$, $\dim \operatorname{range} L = 0$, and the conclusion holds. Next, suppose that $1 \leq k < n$. We shall prove that $\dim \operatorname{range} L = n - k$. Let $\{\mathbf{v}_1, \mathbf{v}_2, \ldots, \mathbf{v}_k\}$ be a basis for $\ker L$. By Theorem 2.10 we can extend this basis to a basis

$$S = \{\mathbf{v}_1, \mathbf{v}_2, \ldots, \mathbf{v}_k \mathbf{v}_{k+1}, \ldots, \mathbf{v}_n\}$$

for V. We prove that the set $T = \{L(\mathbf{v}_{k+1}), L(\mathbf{v}_{k+2}), \ldots, L(\mathbf{v}_n)\}$ is a basis for range L.

First, we show that T spans range L. Let $\mathbf{w}$ be any vector in range L. Then $\mathbf{w} = L(\mathbf{v})$ for some $\mathbf{v}$ in V. Since S is a basis for V, we can find a unique set of real numbers $a_1, a_2, \ldots, a_n$ such that

$$\mathbf{v} = a_1\mathbf{v}_1 + a_2\mathbf{v}_2 + \ldots + a_n\mathbf{v}_n.$$

Then

$$\begin{aligned}\mathbf{w} = L(\mathbf{v}) &= L(a_1\mathbf{v}_1 + a_2\mathbf{v}_2 + \cdots + a_k\mathbf{v}_k + a_{k+1}\mathbf{v}_{k+1} + \cdots + a_n\mathbf{v}_n)\\ &= a_1L(\mathbf{v}_1) + a_2L(\mathbf{v}_2) + \cdots + a_kL(\mathbf{v}_k) + a_{k+1}L(\mathbf{v}_{k+1}) + \cdots + a_nL(\mathbf{v}_n)\\ &= a_{k+1}L(\mathbf{v}_{k+1}) + \cdots + a_nL(\mathbf{v}_n)\end{aligned}$$

because $\mathbf{v}_1, \mathbf{v}_2, \ldots, \mathbf{v}_k$ are in ker L. Hence T spans range L.

Now we show that T is linearly independent. Suppose that

$$a_{k+1}L(\mathbf{v}_{k+1}) + a_{k+2}L(\mathbf{v}_{k+2}) + \cdots + a_nL(\mathbf{v}_n) = \mathbf{0}_W.$$

Then

$$L(a_{k+1}\mathbf{v}_{k+1} + a_{k+2}\mathbf{v}_{k+2} + \cdots + a_n\mathbf{v}_n) = \mathbf{0}_W.$$

Hence the vector $a_{k+1}\mathbf{v}_{k+1} + a_{k+2}\mathbf{v}_{k+2} + \cdots + a_n\mathbf{v}_n$ is in ker L, and we can write

$$a_{k+1}\mathbf{v}_{k+1} + a_{k+2}\mathbf{v}_{k+2} + \cdots + a_n\mathbf{v}_n = b_1\mathbf{v}_1 + b_2\mathbf{v}_2 + \cdots + b_k\mathbf{v}_k,$$

where $b_1, b_2, \ldots, b_k$ are uniquely determined real numbers. We then have

$$b_1\mathbf{v}_1 + b_2\mathbf{v}_2 + \cdots + b_k\mathbf{v}_k - a_{k+1}\mathbf{v}_{k+1} - a_{k+2}\mathbf{v}_{k+2} - \cdots - a_n\mathbf{v}_n = \mathbf{0}_V.$$

Since S is linearly independent, we find that

$$b_1 = b_2 = \cdots = b_k = a_{k+1} = \cdots = a_n = 0.$$

Hence T is linearly independent and forms a basis for range L.

If $k = 0$, then ker L has no basis; we let $\{\mathbf{v}_1, \mathbf{v}_2, \ldots, \mathbf{v}_n\}$ be a basis for V. The proof now proceeds as above. ●

The dimension of ker L is also called the **nullity** of L. In Section 4.5, we define the rank of L and show that it is equal to dim range L. With this terminology the conclusion of Theorem 4.6 is very similar to that of Theorem 2.18. This is not a coincidence, since in the next section we show how to attach a unique $m \times n$ matrix to L, whose properties reflect those of L.

The following example illustrates Theorem 4.6 graphically.

EXAMPLE 9

Let $L: R^3 \to R^3$ be the linear transformation defined by

$$L\left(\begin{bmatrix} a_1 \\ a_2 \\ a_3 \end{bmatrix}\right) = \begin{bmatrix} a_1 + a_3 \\ a_1 + a_2 \\ a_2 - a_3 \end{bmatrix}.$$

A vector $\begin{bmatrix} a_1 \\ a_2 \\ a_3 \end{bmatrix}$ is in ker L if

$$L\left(\begin{bmatrix} a_1 \\ a_2 \\ a_3 \end{bmatrix}\right) = \begin{bmatrix} 0 \\ 0 \\ 0 \end{bmatrix}.$$

We must then find a basis for the solution space of the homogeneous system

$$\begin{aligned} a_1 + \quad\quad a_3 &= 0 \\ a_1 + a_2 \quad\quad &= 0 \\ a_2 - a_3 &= 0. \end{aligned}$$

We find (verify) that a basis for ker L is $\left\{ \begin{bmatrix} -1 \\ 1 \\ 1 \end{bmatrix} \right\}$, so dim ker $L = 1$, and ker L is a line through the origin.

Next, every vector in range L is of the form $\begin{bmatrix} a_1 + a_3 \\ a_1 + a_2 \\ a_2 - a_3 \end{bmatrix}$, which can be written as

$$a_1 \begin{bmatrix} -1 \\ 1 \\ 0 \end{bmatrix} + a_2 \begin{bmatrix} 0 \\ 1 \\ 1 \end{bmatrix} + a_3 \begin{bmatrix} -1 \\ 1 \\ 0 \end{bmatrix}.$$

Then a basis for range L is

$$\left\{ \begin{bmatrix} -1 \\ 1 \\ 0 \end{bmatrix}, \begin{bmatrix} 0 \\ 1 \\ 1 \end{bmatrix} \right\}$$

(explain), so dim range $L = 2$ and range L is a plane passing through the origin. These results are illustrated in Figure 4.5. Moreover,

$$\dim R^3 = 3 = \dim \ker L + \dim \operatorname{range} L = 1 + 2,$$

verifying Theorem 4.6. ■

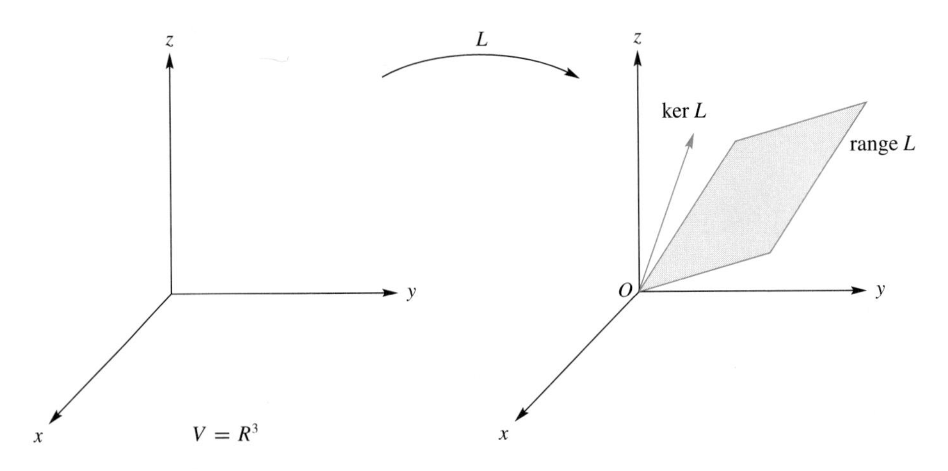

FIGURE 4.5

EXAMPLE 10

Let $L: P_2 \to P_2$ be the linear transformation defined by

$$L(at^2 + bt + c) = (a + 2b)t + (b + c).$$

(a) Find a basis for $\ker L$.
(b) Find a basis for range L.
(c) Verify Theorem 4.6

Solution (a) The vector $at^2 + bt + c$ is in $\ker L$ if

$$L(at^2 + bt + c) = \mathbf{0},$$

that is, if

$$(a + 2b)t + (b + c) = 0.$$

Then

$$\begin{aligned} a + 2b \quad\quad &= 0 \\ b + c &= 0. \end{aligned}$$

Transforming the augmented matrix of this linear system to reduced row echelon form, we find that a basis for the solution space, and then a basis for $\ker L$, is (verify)

$$\{2t^2 - t + 1\}.$$

(b) Every vector in range L has the form

$$(a + 2b)t + (b + c),$$

so the vectors t and 1 span range L. Since these vectors are also linearly independent, they form a basis for range L.
(c) From (a), $\dim \ker L = 1$, and from (b), $\dim \operatorname{range} L = 2$, so

$$\dim \ker L + \dim \operatorname{range} L = \dim P_2 = 3.$$ ■

We have seen that a linear transformation may be one-to-one and not onto or onto and not one-to-one. However, the following corollary shows that each of these properties implies the other if the vector spaces V and W have the same dimensions.

Corollary 4.2

If $L: V \to W$ is a linear transformation of a vector space V into a vector space W and $\dim V = \dim W$, *then:*

(a) *If L is one-to-one, then it is onto.*
(b) *If L is onto, then it is one-to-one.*

Proof
Exercise. ●

A linear transformation $L: V \to W$ of a vector space V into a vector space W is called **invertible** if it is an invertible function, that is, if there exists a unique function $L^{-1}: W \to V$ such that $L \circ L^{-1} = I_W$ and $L^{-1} \circ L = I_V$, where $I_V =$ identity linear transformation on V and $I_W =$ identity linear transformation on W. We now prove the following theorem.

Theorem 4.7

A linear transformation $L: V \to W$ is invertible if and only if L is one-to-one and onto. Moreover, L^{-1} is a linear transformation and $(L^{-1})^{-1} = L$.

Proof

Let L be one-to-one and onto. We define a function $H: W \to V$ as follows. If $\mathbf{w}$ is in W, then since L is onto, $\mathbf{w} = L(\mathbf{v})$ for some $\mathbf{v}$ in V, and since L is one-to-one, $\mathbf{v}$ is unique. Let $H(\mathbf{w}) = \mathbf{v}$; H is a function and $L(H(\mathbf{w})) = L(\mathbf{v}) = \mathbf{w})$, so that $L \circ H = I_W$. Also, $H(L(\mathbf{v})) = H(\mathbf{w}) = \mathbf{v}$, so $H \circ L = I_V$. Thus H is an inverse of L. Now H is unique, for if $H_1: W \to V$ is a function such that $L \circ H_1 = I_W$ and $H_1 \circ L = I_V$, then $L(H(\mathbf{w})) = \mathbf{w} = L(H_1(\mathbf{w}))$ for any $\mathbf{w}$ in W. Since L is one-to-one, we conclude that $H(\mathbf{w}) = H_1(\mathbf{w})$. Hence $H = H_1$. Thus $H = L^{-1}$ and L is invertible.

Conversely, let L be invertible; that is, $L \circ L^{-1} = I_W$ and $L^{-1} \circ L = I_V$. We show that L is one-to-one and onto. Suppose that $L(\mathbf{v}_1) = L(\mathbf{v}_2)$ for $\mathbf{v}_1, \mathbf{v}_2$ in V. Then $L^{-1}(L(\mathbf{v}_1)) = L^{-1}(L(\mathbf{v}_2))$, so $\mathbf{v}_1 = \mathbf{v}_2$, which means that L is one-to-one. Also, if $\mathbf{w}$ is a vector in W, then $L(L^{-1}(\mathbf{w})) = \mathbf{w}$, so if we let $L^{-1}(\mathbf{w}) = \mathbf{v}$, then $L(\mathbf{v}) = \mathbf{w}$. Thus L is onto.

We now show that L^{-1} is a linear transformation. Let $\mathbf{w}_1, \mathbf{w}_2$ be in W, where $L(\mathbf{v}_1) = \mathbf{w}_1$ and $L(\mathbf{v}_2) = \mathbf{w}_2$ for $\mathbf{v}_1, \mathbf{v}_2$ in V. Then since

$$L(a\mathbf{v}_1 + b\mathbf{v}_2) = aL(\mathbf{v}_1) + bL(\mathbf{v}_2) = a\mathbf{w}_1 + b\mathbf{w}_2 \quad \text{for } a, b \text{ real numbers},$$

we have

$$L^{-1}(a\mathbf{w}_1 + b\mathbf{w}_2) = a\mathbf{v}_1 + b\mathbf{v}_2 = aL^{-1}(\mathbf{w}_1) + bL^{-1}(\mathbf{w}_2),$$

which implies that L^{-1} is a linear transformation.

Finally, since $L \circ L^{-1} = I_W$, $L^{-1} \circ L = I_V$, and inverses are unique, we conclude that $(L^{-1})^{-1} = L$. ●

Remark If $L: V \to V$ is a linear operator that is one-to-one and onto, then L is an isomorphism. See Definition 2.13 in Section 2.7.

EXAMPLE 11

Consider the linear operator $L: R^3 \to R^3$ defined by

$$L\left(\begin{bmatrix} a_1 \\ a_2 \\ a_3 \end{bmatrix}\right) = \begin{bmatrix} 1 & 1 & 1 \\ 2 & 2 & 1 \\ 0 & 1 & 1 \end{bmatrix} \begin{bmatrix} a_1 \\ a_2 \\ a_3 \end{bmatrix}.$$

Since $\ker L = \{\mathbf{0}\}$ (verify), L is one-to-one and by Corollary 4.2 it is also onto, so it is invertible. To obtain L^{-1}, we proceed as follows. Since $L^{-1}(\mathbf{w}) = \mathbf{v}$, we must solve $L(\mathbf{v}) = \mathbf{w}$ for $\mathbf{v}$. We have

$$L(\mathbf{v}) = L\left(\begin{bmatrix} a_1 \\ a_2 \\ a_3 \end{bmatrix}\right) = \begin{bmatrix} a_1 + a_2 + a_3 \\ 2a_1 + 2a_2 + a_3 \\ a_2 + a_3 \end{bmatrix} = \mathbf{w} = \begin{bmatrix} b_1 \\ b_2 \\ b_3 \end{bmatrix}.$$

We are then solving the linear system

$$\begin{aligned} a_1 + a_2 + a_3 &= b_1 \\ 2a_1 + 2a_2 + a_3 &= b_2 \\ a_2 + a_3 &= b_3 \end{aligned}$$

for a_1, a_2, and a_3. We find that (verify)

$$\begin{bmatrix} a_1 \\ a_2 \\ a_3 \end{bmatrix} = \mathbf{v} = L^{-1}(\mathbf{w}) = L^{-1}\left(\begin{bmatrix} b_1 \\ b_2 \\ b_3 \end{bmatrix}\right) = \begin{bmatrix} b_1 - b_3 \\ -2b_1 + b_2 + b_3 \\ 2b_1 - b_2 \end{bmatrix}.$$

■

The following useful theorem shows that one-to-one linear transformations preserve linear independence of a set of vectors. Moreover, if this property holds, then L is one-to-one.

Theorem 4.8

A linear transformation $L: V \to W$ is one-to-one if and only if the image of every linearly independent set of vectors in V is a linearly independent set of vectors in W.

Proof

Let $S = \{\mathbf{v}_1, \mathbf{v}_2, \ldots, \mathbf{v}_k\}$ be a linearly independent set of vectors in V and let $T = \{L(\mathbf{v}_1), L(\mathbf{v}_2), \ldots, L(\mathbf{v}_k)\}$. Suppose that L is one-to-one; we show that T is linearly independent. Let

$$a_1 L(\mathbf{v}_1) + a_2 L(\mathbf{v}_2) + \cdots + a_k L(\mathbf{v}_k) = \mathbf{0}_W,$$

where $a_1, a_2, \ldots, a_k$ are real numbers. Then

$$L(a_1\mathbf{v}_1 + a_2\mathbf{v}_2 + \cdots + a_k\mathbf{v}_k) = \mathbf{0}_W = L(\mathbf{0}_V).$$

Since L is one-to-one, we conclude that

$$a_1\mathbf{v}_1 + a_2\mathbf{v}_2 + \cdots + a_k\mathbf{v}_k = \mathbf{0}_V.$$

Now S is linearly independent, so $a_1 = a_2 = \cdots = a_k = 0$. Hence T is linearly independent.

Conversely, suppose that the image of any linearly independent set of vectors in V is a linearly independent set of vectors in W. Now $\{\mathbf{v}\}$, where $\mathbf{v} \neq \mathbf{0}_V$, is a linearly independent set in V. Since the set $\{L(\mathbf{v})\}$ is linearly independent, $L(\mathbf{v}) \neq \mathbf{0}_W$, so $\ker L = \{\mathbf{0}_V\}$, which means that L is one-to-one. ●

It follows from this theorem that *if* $L: V \to W$ *is a linear transformation and* $\dim V = \dim W$, *then* L *is one-to-one, and thus invertible, if and only if the image of a basis for* V *under* L *is a basis for* W (see Exercise 18).

We now make one final remark for a linear system $A\mathbf{x} = \mathbf{b}$, where A is $n \times n$. We again consider the linear transformation $L: R^n \to R^n$ defined by $L(\mathbf{x}) = A\mathbf{x}$, for $\mathbf{x}$ in R^n. If A is a nonsingular matrix, then $\dim \operatorname{range} L = \operatorname{rank} A = n$, so $\dim \ker L = 0$. Thus L is one-to-one and hence onto. This means that the given linear system has a unique solution (of course, we already knew this result from other considerations). Now assume that A is singular. Then $\operatorname{rank} A < n$. This means that $\dim \ker L = n - \operatorname{rank} A > 0$, so L is not one-to-one and not onto. Therefore, there exists a vector $\mathbf{b}$ in R^n, for which the system $A\mathbf{x} = \mathbf{b}$ has no solution. Moreover, since A is singular, $A\mathbf{x} = \mathbf{0}$ has a nontrivial solution $\mathbf{x}_0$. If $A\mathbf{x} = \mathbf{b}$ has a solution $\mathbf{y}$, then $\mathbf{x}_0 + \mathbf{y}$ is a solution to $A\mathbf{x} = \mathbf{b}$ (verify). Thus, for A singular, if a solution to $A\mathbf{x} = \mathbf{b}$ exists, then it is not unique.

The following statements are then equivalent:

1. A is nonsingular.
2. $A\mathbf{x} = \mathbf{0}$ has only the trivial solution.
3. A is row (column) equivalent to I_n.
4. The linear system $A\mathbf{x} = \mathbf{b}$ has a unique solution for every vector $\mathbf{b}$ in R^n.
5. A is a product of elementary matrices.
6. A has rank n.
7. The rows (columns) of A form a linearly independent set of n vectors in R_n (R^n).
8. The dimension of the solution space of $A\mathbf{x} = \mathbf{0}$ is zero.
9. The linear transformation $L: R^n \to R^n$ defined by $L(\mathbf{x}) = A\mathbf{x}$, for $\mathbf{x}$ in R^n, is one-to-one and onto.

We can summarize the conditions under which a linear transformation L of an n-dimensional vector space V into itself (or more generally to an n-dimensional vector space W) is invertible by the following equivalent statements:

1. L is invertible.
2. L is one-to-one.
3. L is onto.

4.2 Exercises

1. Let $L: R^2 \to R^2$ be the linear transformation defined by $L\left(\begin{bmatrix} a_1 \\ a_2 \end{bmatrix}\right) = \begin{bmatrix} a_1 \\ 0 \end{bmatrix}$.

(a) Is $\begin{bmatrix} 0 \\ 2 \end{bmatrix}$ in ker L?

(b) Is $\begin{bmatrix} 2 \\ 2 \end{bmatrix}$ in ker L?

(c) Is $\begin{bmatrix} 3 \\ 0 \end{bmatrix}$ in range L?

(d) Is $\begin{bmatrix} 3 \\ 2 \end{bmatrix}$ in range L?

(e) Find ker L.

(f) Find range L.

2. Let $L: R^2 \to R^2$ be the linear operator defined by $L\left(\begin{bmatrix} a_1 \\ a_2 \end{bmatrix}\right) = \begin{bmatrix} 1 & 2 \\ 2 & 4 \end{bmatrix}\begin{bmatrix} a_1 \\ a_2 \end{bmatrix}$.

(a) Is $\begin{bmatrix} 1 \\ 2 \end{bmatrix}$ in ker L?

(b) Is $\begin{bmatrix} 2 \\ -1 \end{bmatrix}$ in ker L?

(c) Is $\begin{bmatrix} 3 \\ 6 \end{bmatrix}$ in range L?

(d) Is $\begin{bmatrix} 2 \\ 3 \end{bmatrix}$ in range L?

(e) Find ker L.

(f) Find a set of vectors spanning range L.

3. Let $L: R_4 \to R_2$ be the linear transformation defined by

$$L\left(\begin{bmatrix} a_1 & a_2 & a_3 & a_4 \end{bmatrix}\right) = \begin{bmatrix} a_1 + a_3 & a_2 + a_4 \end{bmatrix}.$$

(a) Is $\begin{bmatrix} 2 & 3 & -2 & 3 \end{bmatrix}$ in ker L?

(b) Is $\begin{bmatrix} 4 & -2 & -4 & 2 \end{bmatrix}$ in ker L?

(c) Is $\begin{bmatrix} 1 & 2 \end{bmatrix}$ in range L?

(d) Is $\begin{bmatrix} 0 & 0 \end{bmatrix}$ in range L?

(e) Find ker L.

(f) Find a set of vectors spanning range L.

4. Let $L: R_2 \to R_3$ be the linear transformation defined by $L\left(\begin{bmatrix} a_1 & a_2 \end{bmatrix}\right) = \begin{bmatrix} a_1 & a_1 + a_2 & a_2 \end{bmatrix}$.

(a) Find ker L.

(b) Is L one-to-one?

(c) Is L onto?

5. Let $L: R_4 \to R_3$ be the linear transformation defined by

$$L\left(\begin{bmatrix} a_1 & a_2 & a_3 & a_4 \end{bmatrix}\right) = \begin{bmatrix} a_1 + a_2 & a_3 + a_4 & a_1 + a_3 \end{bmatrix}.$$

(a) Find a basis for ker L.

(b) What is dim ker L?

(c) Find a basis for range L.

(d) What is dim range L?

6. Let $L: P_2 \to P_3$ be the linear transformation defined by $L(p(t)) = t^2 p'(t)$.

(a) Find a basis for and the dimension of ker L.

(b) Find a basis for and the dimension of range L.

7. Let $L: M_{23} \to M_{33}$ be the linear transformation defined by

$$L(A) = \begin{bmatrix} 2 & -1 \\ 1 & 2 \\ 3 & 1 \end{bmatrix} A \quad \text{for } A \text{ in } M_{23}.$$

(a) Find the dimension of ker L.

(b) Find the dimension of range L.

8. Let $L: P_2 \to P_1$ be the linear transformation defined by

$$L(at^2 + bt + c) = (a + b)t + (b - c).$$

(a) Find a basis for ker L.

(b) Find a basis for range L.

9. Let $L: P_2 \to R_2$ be the linear transformation defined by $L(at^2 + bt + c) = \begin{bmatrix} a & b \end{bmatrix}$.

(a) Find a basis for ker L.

(b) Find a basis for range L.

10. Let $L: M_{22} \to M_{22}$ be the linear transformation defined by $L(A) = \begin{bmatrix} 1 & 2 \\ 1 & 1 \end{bmatrix} A - A \begin{bmatrix} 1 & 2 \\ 1 & 1 \end{bmatrix}$.

(a) Find a basis for ker L.

(b) Find a basis for range L.

11. Let $L: M_{22} \to M_{22}$ be the linear operator defined by $L\left(\begin{bmatrix} a & b \\ c & d \end{bmatrix}\right) = \begin{bmatrix} a + b & b + c \\ a + d & b + d \end{bmatrix}$.

(a) Find a basis for ker L.

(b) Find a basis for range L.

12. Let $L: V \to W$ be a linear transformation.

(a) Show that dim range $L \leq \dim V$.

(b) Prove that if L is onto, then $\dim W \leq \dim V$.

13. Verify Theorem 4.6 for the following linear transformations.

(a) $L: P_2 \to P_2$ defined by $L(p(t)) = tp'(t)$.

(b) $L: R_3 \to R_2$ defined by $L\left(\begin{bmatrix} a_1 & a_2 & a_3 \end{bmatrix}\right) = \begin{bmatrix} a_1 + a_2 & a_1 + a_3 \end{bmatrix}$.

(c) $L: M_{22} \to M_{23}$ defined by

$$L(A) = A\begin{bmatrix} 1 & 2 & 3 \\ 2 & 1 & 3 \end{bmatrix}, \quad \text{for } A \text{ in } M_{22}.$$

14. Verify Theorem 4.5 for the linear transformation given in Exercise 11.

15. Let A be an $m \times n$ matrix, and consider the linear transformation $L: R^n \to R^m$ defined by $L(\mathbf{x}) = A\mathbf{x}$, for $\mathbf{x}$ in R^n. Show that

$$\text{range } L = \text{column space of } A.$$

16. Let $L: R^5 \to R^4$ be the linear transformation defined by

$$L\left(\begin{bmatrix} a_1 \\ a_2 \\ a_3 \\ a_4 \\ a_5 \end{bmatrix}\right) = \begin{bmatrix} 1 & 0 & -1 & 3 & -1 \\ 1 & 0 & 0 & 2 & -1 \\ 2 & 0 & -1 & 5 & -1 \\ 0 & 0 & -1 & 1 & 0 \end{bmatrix}\begin{bmatrix} a_1 \\ a_2 \\ a_3 \\ a_4 \\ a_5 \end{bmatrix}.$$

(a) Find a basis for and the dimension of ker L.

(b) Find a basis for and the dimension of range L.

17. Let $L: R_3 \to R_3$ be the linear transformation defined by
$L(\mathbf{e}_1^T) = L\left(\begin{bmatrix} 1 & 0 & 0 \end{bmatrix}\right) = \begin{bmatrix} 3 & 0 & 0 \end{bmatrix}$,
$L(\mathbf{e}_2^T) = L\left(\begin{bmatrix} 0 & 1 & 0 \end{bmatrix}\right) = \begin{bmatrix} 1 & 1 & 1 \end{bmatrix}$, and
$L(\mathbf{e}_3^T) = L\left(\begin{bmatrix} 0 & 0 & 1 \end{bmatrix}\right) = \begin{bmatrix} 2 & 1 & 1 \end{bmatrix}$.
Is the set $\{L(\mathbf{e}_1^T), L(\mathbf{e}_2^T), L(\mathbf{e}_3^T)\} =$
$\{\begin{bmatrix} 3 & 0 & 0 \end{bmatrix}, \begin{bmatrix} 1 & 1 & 1 \end{bmatrix}, \begin{bmatrix} 2 & 1 & 1 \end{bmatrix}\}$ a basis for R_3?

18. Let $L: V \to W$ be a linear transformation, and let $\dim V = \dim W$. Prove that L is invertible if and only if the image of a basis for V under L is a basis for W.

19. Let $L: R^3 \to R^3$ be defined by

$$L\left(\begin{bmatrix} 1 \\ 0 \\ 0 \end{bmatrix}\right) = \begin{bmatrix} 1 \\ 2 \\ 3 \end{bmatrix}, \quad L\left(\begin{bmatrix} 0 \\ 1 \\ 0 \end{bmatrix}\right) = \begin{bmatrix} 0 \\ 1 \\ 1 \end{bmatrix},$$

$$L\left(\begin{bmatrix} 0 \\ 0 \\ 1 \end{bmatrix}\right) = \begin{bmatrix} 1 \\ 1 \\ 0 \end{bmatrix}.$$

(a) Prove that L is invertible.

(b) Find $L^{-1}\left(\begin{bmatrix} 2 \\ 3 \\ 4 \end{bmatrix}\right)$.

20. Let $L: V \to W$ be a linear transformation, and let $S = \{\mathbf{v}_1, \mathbf{v}_2, \ldots, \mathbf{v}_n\}$ be a set of vectors in V. Prove that if $T = \{L(\mathbf{v}_1), L(\mathbf{v}_2), \ldots, L(\mathbf{v}_n)\}$ is linearly independent, then so is S. What can we say about the converse?

21. Find the dimension of the solution space for the following homogeneous system:

$$\begin{bmatrix} 1 & 2 & 1 & 3 \\ 2 & 1 & -1 & 2 \\ 1 & 0 & 0 & -1 \\ 4 & 1 & -1 & 0 \end{bmatrix}\begin{bmatrix} x_1 \\ x_2 \\ x_3 \\ x_4 \end{bmatrix} = \begin{bmatrix} 0 \\ 0 \\ 0 \\ 0 \end{bmatrix}.$$

22. Find a linear transformation $L: R_2 \to R_3$ such that $S = \{\begin{bmatrix} 1 & -1 & 2 \end{bmatrix}, \begin{bmatrix} 3 & 1 & -1 \end{bmatrix}\}$ is a basis for range L.

23. Let $L: R^3 \to R^3$ be the linear transformation defined by

$$L\left(\begin{bmatrix} a_1 \\ a_2 \\ a_3 \end{bmatrix}\right) = \begin{bmatrix} 1 & 1 & 1 \\ 0 & 1 & 2 \\ 1 & 2 & 2 \end{bmatrix}\begin{bmatrix} a_1 \\ a_2 \\ a_3 \end{bmatrix}.$$

(a) Prove that L is invertible.

(b) Find $L^{-1}\left(\begin{bmatrix} a_1 \\ a_2 \\ a_3 \end{bmatrix}\right)$.

24. Let $L: V \to W$ be a linear transformation. Prove that L is one-to-one if and only if $\dim \text{range } L = \dim V$.

25. Let $L: R^4 \to R^6$ be a linear transformation.

(a) If $\dim \ker L = 2$, what is $\dim \text{range } L$?

(b) If $\dim \text{range } L = 3$, what is $\dim \ker L$?

26. Let $L: V \to R^5$ be a linear transformation.

(a) If L is onto and $\dim \ker L = 2$, what is $\dim V$?

(b) If L is one-to-one and onto, what is $\dim V$?

27. Let L be the linear transformation defined in Exercise 22, Section 4.1. Prove or disprove:

(a) L is one-to-one.

(b) L is onto.

28. Let L be the linear transformation defined in Exercise 23, Section 4.1. Prove or disprove:

(a) L is one-to-one.

(b) L is onto.

29. Prove Corollary 4.1.

30. Let $L: R^n \to R^m$ be a linear transformation defined by $L(\mathbf{x}) = A\mathbf{x}$, for $\mathbf{x}$ in R^n. Prove that L is onto if and only if rank $A = m$.

31. Prove Corollary 4.2.

4.3 MATRIX OF A LINEAR TRANSFORMATION

In Section 4.2 we saw that if A is an $m \times n$ matrix, then we can define a linear transformation $L: R^n \to R^m$ by $L(\mathbf{x}) = A\mathbf{x}$ for $\mathbf{x}$ in R^n. We shall now develop the following notion; if $L: V \to W$ is a linear transformation of an n-dimensional vector space V into an m-dimensional vector space W, and if we choose ordered bases for V and W, then we can associate a unique $m \times n$ matrix A with L that will enable us to find $L(\mathbf{x})$ for $\mathbf{x}$ in V by merely performing matrix multiplication.

Theorem 4.9

Let $L: V \to W$ be a linear transformation of an n-dimensional vector space V into an m-dimensional vector space W ($n \neq 0$, $m \neq 0$) and let $S = \{\mathbf{v}_1, \mathbf{v}_2, \dots, \mathbf{v}_n\}$ and $T = \{\mathbf{w}_1, \mathbf{w}_2, \dots, \mathbf{w}_m\}$ be ordered bases for V and W, respectively. Then the $m \times n$ matrix A whose jth column is the coordinate vector $\left[L(\mathbf{v}_j)\right]_T$ of $L(\mathbf{v}_j)$ with respect to T has the following property:

$$\left[L(\mathbf{x})\right]_T = A\left[\mathbf{x}\right]_S \qquad \textit{for every } \mathbf{x} \textit{ in } V. \tag{1}$$

Moreover, A is the only matrix with this property.

Proof

We show how to construct the matrix A. Consider the vector $\mathbf{v}_j$ in V for $j = 1, 2, \dots, n$. Then $L(\mathbf{v}_j)$ is a vector in W, and since T is an ordered basis for W, we can express this vector as a linear combination of the vectors in T in a unique manner. Thus

$$L(\mathbf{v}_j) = c_{1j}\mathbf{w}_1 + c_{2j}\mathbf{w}_2 + \cdots + c_{mj}\mathbf{w}_m. \tag{2}$$

This means that the coordinate vector of $L(\mathbf{v}_j)$ with respect to T is

$$\left[L(\mathbf{v}_j)\right]_T = \begin{bmatrix} c_{1j} \\ c_{2j} \\ \vdots \\ c_{mj} \end{bmatrix}.$$

Recall from Section 2.7 that to find the coordinate vector $\left[L(\mathbf{v}_j)\right]_T$ we must solve a linear system. We now define an $m \times n$ matrix A by choosing $\left[L(\mathbf{v}_j)\right]_T$ as the jth column of A and show that this matrix satisfies the properties stated in the theorem.

Let $\mathbf{x}$ be any vector in V. Then $L(\mathbf{x})$ is in W. Now let

$$\left[\mathbf{x}\right]_S = \begin{bmatrix} a_1 \\ a_2 \\ \vdots \\ a_n \end{bmatrix} \quad \text{and} \quad \left[L(\mathbf{x})\right]_T = \begin{bmatrix} b_1 \\ b_2 \\ \vdots \\ b_m \end{bmatrix}.$$

This means that $\mathbf{x} = a_1\mathbf{v}_1 + a_2\mathbf{v}_2 + \cdots + a_n\mathbf{v}_n$. Then

$$\begin{aligned} L(\mathbf{x}) &= a_1 L(\mathbf{v}_1) + a_2 L(\mathbf{v}_2) + \cdots + a_n L(\mathbf{v}_n) \\ &= a_1(c_{11}\mathbf{w}_1 + c_{21}\mathbf{w}_2 + \cdots + c_{m1}\mathbf{w}_m) \\ &\quad + a_2(c_{12}\mathbf{w}_1 + c_{22}\mathbf{w}_2 + \cdots + c_{m2}\mathbf{w}_m) \\ &\quad + \cdots + a_n(c_{1n}\mathbf{w}_1 + c_{2n}\mathbf{w}_2 + \cdots + c_{mn}\mathbf{w}_m) \\ &= (c_{11}a_1 + c_{12}a_2 + \cdots + c_{1n}a_n)\mathbf{w}_1 + (c_{21}a_1 + c_{22}a_2 + \cdots + c_{2n}a_n)\mathbf{w}_2 \\ &\quad + \cdots + (c_{m1}a_1 + c_{m2}a_2 + \cdots + c_{mn}a_n)\mathbf{w}_m. \end{aligned}$$

Now $L(\mathbf{x}) = b_1\mathbf{w}_1 + b_2\mathbf{w}_2 + \cdots + b_m\mathbf{w}_m$. Hence

$$b_i = c_{i1}a_1 + c_{i2}a_2 + \cdots + c_{in}a_n \quad \text{for } i = 1, 2, \ldots, m.$$

Next, we verify Equation (1). We have

$$\begin{aligned} A\left[\mathbf{x}\right]_S &= \begin{bmatrix} c_{11} & c_{12} & \cdots & c_{1n} \\ c_{21} & c_{22} & \cdots & c_{2n} \\ \vdots & \vdots & & \vdots \\ c_{m1} & c_{m2} & \cdots & c_{mn} \end{bmatrix} \begin{bmatrix} a_1 \\ a_2 \\ \vdots \\ a_n \end{bmatrix} \\ &= \begin{bmatrix} c_{11}a_1 + c_{12}a_2 + \cdots + c_{1n}a_n \\ c_{21}a_1 + c_{22}a_2 + \cdots + c_{2n}a_n \\ \vdots \\ c_{m1}a_1 + c_{m2}a_2 + \cdots + c_{mn}a_n \end{bmatrix} = \begin{bmatrix} b_1 \\ b_2 \\ \vdots \\ b_m \end{bmatrix} = \left[L(\mathbf{x})\right]_T. \end{aligned}$$

Finally, we show that $A = \left[c_{ij}\right]$ is the only matrix with this property. Suppose that we have another matrix $A^* = \left[c^*_{ij}\right]$ with the same properties as A, and that $A^* \neq A$. All the elements of A and A^* cannot be equal, so say that the kth columns of these matrices are unequal. Now the coordinate vector of $\mathbf{v}_k$ with respect to the basis S is

$$\left[\mathbf{v}_k\right]_S = \begin{bmatrix} 0 \\ 0 \\ \vdots \\ 1 \\ 0 \\ \vdots \\ 0 \end{bmatrix} \quad \leftarrow k\text{th row}.$$

Then

$$\left[L(\mathbf{v}_k)\right]_T = A \begin{bmatrix} 0 \\ 0 \\ \vdots \\ 1 \\ 0 \\ \vdots \\ 0 \end{bmatrix} = \begin{bmatrix} a_{1k} \\ a_{2k} \\ \vdots \\ a_{mk} \end{bmatrix} = k\text{th column of } A$$

and

$$\left[L(\mathbf{v}_k)\right]_T = A^* \begin{bmatrix} 0 \\ 0 \\ \vdots \\ 1 \\ 0 \\ \vdots \\ 0 \end{bmatrix} = \begin{bmatrix} a^*_{1k} \\ a^*_{2k} \\ \vdots \\ a^*_{mk} \end{bmatrix} = k\text{th column of } A^*.$$

This means that $L(\mathbf{v}_k)$ has two different coordinate vectors with respect to the same ordered basis, which is impossible. Hence the matrix A is unique. ●

We now summarize the procedure given in Theorem 4.9 for computing the matrix of a linear transformation $L: V \to W$ with respect to the ordered bases $S = \{\mathbf{v}_1, \mathbf{v}_2, \ldots, \mathbf{v}_n\}$ and $T = \{\mathbf{w}_1, \mathbf{w}_2, \ldots, \mathbf{w}_m\}$ for V and W, respectively.

STEP 1 Compute $L(\mathbf{v}_j)$ for $j = 1, 2, \ldots, n$.

STEP 2 Find the coordinate vector $\left[L(\mathbf{v}_j)\right]_T$ of $L(\mathbf{v}_j)$ with respect to T. This means that we have to express $L(\mathbf{v}_j)$ as a linear combination of the vectors in T [see Equation (2)] and this requires the solution of a linear system.

STEP 3 The matrix A of the linear transformation L with respect to the ordered bases S and T is formed by choosing for each j from 1 to n, $\left[L(\mathbf{v}_j)\right]_T$ as the jth column of A.

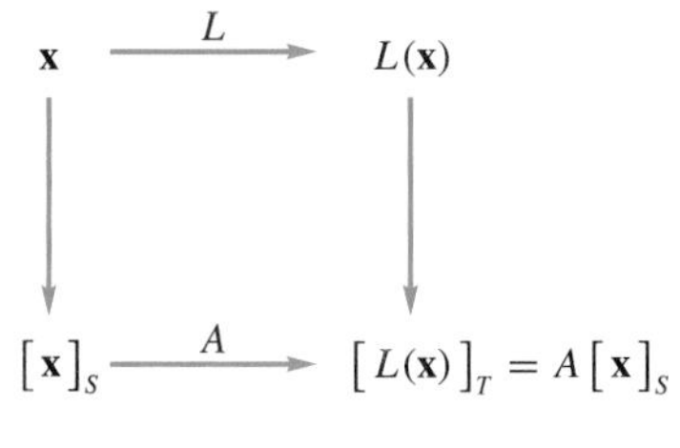

$$\begin{array}{ccc} \mathbf{x} & \xrightarrow{L} & L(\mathbf{x}) \\ \downarrow & & \downarrow \\ [\mathbf{x}]_S & \xrightarrow{A} & [L(\mathbf{x})]_T = A[\mathbf{x}]_S \end{array}$$

FIGURE 4.6

Figure 4.6 gives a graphical interpretation of Equation (1), that is, of Theorem 4.9. The top horizontal arrow represents the linear transformation L from the n-dimensional vector space V into the m-dimensional vector space W and takes the vector $\mathbf{x}$ in V to the vector $L(\mathbf{x})$ in W. The bottom horizontal line represents the matrix A. Then $\left[L(\mathbf{x})\right]_T$, a coordinate vector in R^m, is obtained simply by multiplying $\left[\mathbf{x}\right]_S$, a coordinate vector in R^n, by the matrix A on the left. We can thus work with matrices rather than with linear transformations.

EXAMPLE 1

Let $L: P_2 \to P_1$ be defined by $L(p(t)) = p'(t)$, and consider the ordered bases $S = \{t^2, t, 1\}$ and $T = \{t, 1\}$ for P_2 and P_1, respectively.

(a) Find the matrix A associated with L.
(b) If $p(t) = 5t^2 - 3t + 2$, compute $L(p(t))$ directly and using A.

Solution (a) We have

$$L(t^2) = 2t = 2t + 0(1), \qquad \text{so } \left[L(t^2)\right]_T = \begin{bmatrix} 2 \\ 0 \end{bmatrix}.$$

$$L(t) = 1 = 0(t) + 1(1), \qquad \text{so } \left[L(t)\right]_T = \begin{bmatrix} 0 \\ 1 \end{bmatrix}.$$

$$L(1) = 0 = 0(t) + 0(1), \qquad \text{so } \left[L(1)\right]_T = \begin{bmatrix} 0 \\ 0 \end{bmatrix}.$$

In this case, the coordinates of $L(t^2)$, $L(t)$, and $L(1)$ with respect to the T-basis are obtained by observation since the T-basis is quite simple. Thus

$$A = \begin{bmatrix} 2 & 0 & 0 \\ 0 & 1 & 0 \end{bmatrix}.$$

(b) Since $p(t) = 5t^2 - 3t + 2$, then $L(p(t)) = 10t - 3$. However, we can find $L(p(t))$ using the matrix A as follows. Since

$$\left[p(t)\right]_S = \begin{bmatrix} 5 \\ -3 \\ 2 \end{bmatrix},$$

then

$$\left[L(p(t))\right]_T = \begin{bmatrix} 2 & 0 & 0 \\ 0 & 1 & 0 \end{bmatrix} \begin{bmatrix} 5 \\ -3 \\ 2 \end{bmatrix} = \begin{bmatrix} 10 \\ -3 \end{bmatrix},$$

which means that $L(p(t)) = 10t - 3$. ■

EXAMPLE 2

Let $L\colon P_2 \to P_1$ be defined as in Example 1 and consider the ordered bases $S = \{1, t, t^2\}$ and $T = \{t, 1\}$ for P_2 and P_1, respectively. We then find that the matrix A associated with L is $\begin{bmatrix} 0 & 0 & 2 \\ 0 & 1 & 0 \end{bmatrix}$ (verify). Notice that if we change the order of the vectors in S or T, the matrix may change. ■

EXAMPLE 3

Let $L\colon P_2 \to P_1$ be defined as in Example 1, and consider the ordered bases $S = \{t^2, t, 1\}$ and $T = \{t + 1, t - 1\}$ for P_2 and P_1, respectively.

(a) Find the matrix A associated with L.
(b) If $p(t) = 5t^2 - 3t + 2$, compute $L(p(t))$.

Solution (a) We have

$$L(t^2) = 2t.$$

To find the coordinates of $L(t^2)$ with respect to the T-basis, we form

$$L(t^2) = 2t = a_1(t+1) + a_2(t-1),$$

which leads to the linear system

$$\begin{aligned} a_1 + a_2 &= 2 \\ a_1 - a_2 &= 0, \end{aligned}$$

whose solution is $a_1 = 1$, $a_2 = 1$ (verify). Hence

$$\left[L(t^2)\right]_T = \begin{bmatrix} 1 \\ 1 \end{bmatrix}.$$

Similarly,

$$L(t) = 1 = \tfrac{1}{2}(t+1) - \tfrac{1}{2}(t-1), \qquad \text{so } \left[L(t)\right]_T = \begin{bmatrix} \frac{1}{2} \\ -\frac{1}{2} \end{bmatrix}.$$

$$L(1) = 0 = 0(t+1) + 0(t-1), \qquad \text{so } \left[L(1)\right]_T = \begin{bmatrix} 0 \\ 0 \end{bmatrix}.$$

Hence $A = \begin{bmatrix} 1 & \frac{1}{2} & 0 \\ 1 & -\frac{1}{2} & 0 \end{bmatrix}$.

(b) We have

$$\left[L(p(t))\right]_T = \begin{bmatrix} 1 & \frac{1}{2} & 0 \\ 1 & -\frac{1}{2} & 0 \end{bmatrix} \begin{bmatrix} 5 \\ -3 \\ 2 \end{bmatrix} = \begin{bmatrix} \frac{7}{2} \\ \frac{13}{2} \end{bmatrix},$$

so $L(p(t)) = \frac{7}{2}(t+1) + \frac{13}{2}(t-1) = 10t - 3$, which agrees with the result obtained in Example 1. ■

Notice that the matrices obtained in Examples 1, 2, and 3 are different even though L is the same in all three examples. In Section 4.5 we discuss the relationship between any two of these three matrices.

The matrix A is called the **representation of L with respect to the ordered bases S and T**. We also say that **A represents L with respect to S and T**. Having A enables us to replace L by A and $\mathbf{x}$ by $\left[\mathbf{x}\right]_S$, to get $A\left[\mathbf{x}\right]_S = \left[L(\mathbf{x})\right]_T$. Thus the result of applying L to $\mathbf{x}$ in V to obtain $L(\mathbf{x})$ in W can be obtained by multiplying the matrix A by the matrix $\left[\mathbf{x}\right]_S$. That is, we can work with matrices rather than with linear transformations. Physicists and others who deal at great length with linear transformations perform most of their computations with the matrix representations of the linear transformations. Of course, it is easier to work with matrices on a computer than with our abstract definition of a linear transformation. The relationship between linear transformations and matrices is a much stronger one than mere computational convenience. In the next section we shall show that the set of all linear transformations from an n-dimensional vector space V to an m-dimensional vector space W is a vector space that is isomorphic to the vector space M_{mn} of all $m \times n$ matrices.

We might also mention that if $L: R^n \to R^m$ is a linear transformation, then one often uses the natural bases for R^n and R^m, which simplifies the task of obtaining a representation of L.

EXAMPLE 4

Let $L: R^3 \to R^2$ be defined by

$$L\left(\begin{bmatrix} x_1 \\ x_2 \\ x_3 \end{bmatrix}\right) = \begin{bmatrix} 1 & 1 & 1 \\ 1 & 2 & 3 \end{bmatrix} \begin{bmatrix} x_1 \\ x_2 \\ x_3 \end{bmatrix}.$$

Let

$$\mathbf{e}_1 = \begin{bmatrix} 1 \\ 0 \\ 0 \end{bmatrix}, \quad \mathbf{e}_2 = \begin{bmatrix} 0 \\ 1 \\ 0 \end{bmatrix}, \quad \mathbf{e}_3 = \begin{bmatrix} 0 \\ 0 \\ 1 \end{bmatrix},$$

$$\overline{\mathbf{e}}_1 = \begin{bmatrix} 1 \\ 0 \end{bmatrix}, \quad \text{and} \quad \overline{\mathbf{e}}_2 = \begin{bmatrix} 0 \\ 1 \end{bmatrix}.$$

Then $S = \{\mathbf{e}_1, \mathbf{e}_2, \mathbf{e}_3\}$ and $T = \{\overline{\mathbf{e}}_1, \overline{\mathbf{e}}_2\}$ are the natural bases for R^3 and R^2, respectively.

Now

$$L(\mathbf{e}_1) = \begin{bmatrix} 1 & 1 & 1 \\ 1 & 2 & 3 \end{bmatrix} \begin{bmatrix} 1 \\ 0 \\ 0 \end{bmatrix} = \begin{bmatrix} 1 \\ 1 \end{bmatrix} = 1\overline{\mathbf{e}}_1 + 1\overline{\mathbf{e}}_2, \qquad \text{so } \left[L(\mathbf{e}_1)\right]_T = \begin{bmatrix} 1 \\ 1 \end{bmatrix},$$

$$L(\mathbf{e}_2) = \begin{bmatrix} 1 & 1 & 1 \\ 1 & 2 & 3 \end{bmatrix} \begin{bmatrix} 0 \\ 1 \\ 0 \end{bmatrix} = \begin{bmatrix} 1 \\ 2 \end{bmatrix} = 1\overline{\mathbf{e}}_1 + 2\overline{\mathbf{e}}_2, \qquad \text{so } \left[L(\mathbf{e}_2)\right]_T = \begin{bmatrix} 1 \\ 2 \end{bmatrix},$$

$$L(\mathbf{e}_3) = \begin{bmatrix} 1 & 1 & 1 \\ 1 & 2 & 3 \end{bmatrix} \begin{bmatrix} 0 \\ 0 \\ 1 \end{bmatrix} = \begin{bmatrix} 1 \\ 3 \end{bmatrix} = 1\overline{\mathbf{e}}_1 + 3\overline{\mathbf{e}}_2, \qquad \text{so } \left[L(\mathbf{e}_3)\right]_T = \begin{bmatrix} 1 \\ 3 \end{bmatrix}.$$

In this case, the coordinate vectors of $L(\mathbf{e}_1)$, $L(\mathbf{e}_2)$, and $L(\mathbf{e}_3)$ with respect to the T-basis are readily obtained because T is the natural basis for R^2. Then the representation of L with respect to S and T is

$$A = \begin{bmatrix} 1 & 1 & 1 \\ 1 & 2 & 3 \end{bmatrix}.$$

The reason that A is the same matrix as the one involved in the definition of L is that the natural bases are being used for R^3 and R^2. ■

EXAMPLE 5

Let $L: R^3 \to R^2$ be defined as in Example 4, and consider the ordered bases

$$S = \left\{ \begin{bmatrix} 1 \\ 1 \\ 0 \end{bmatrix}, \begin{bmatrix} 0 \\ 1 \\ 1 \end{bmatrix}, \begin{bmatrix} 0 \\ 0 \\ 1 \end{bmatrix} \right\} \quad \text{and} \quad T = \left\{ \begin{bmatrix} 1 \\ 2 \end{bmatrix}, \begin{bmatrix} 1 \\ 3 \end{bmatrix} \right\}$$

for R^3 and R^2, respectively. Then

$$L\left(\begin{bmatrix} 1 \\ 1 \\ 0 \end{bmatrix} \right) = \begin{bmatrix} 1 & 1 & 1 \\ 1 & 2 & 3 \end{bmatrix} \begin{bmatrix} 1 \\ 1 \\ 0 \end{bmatrix} = \begin{bmatrix} 2 \\ 3 \end{bmatrix}.$$

Similarly,

$$L\left(\begin{bmatrix} 0 \\ 1 \\ 1 \end{bmatrix} \right) = \begin{bmatrix} 2 \\ 5 \end{bmatrix} \quad \text{and} \quad L\left(\begin{bmatrix} 0 \\ 0 \\ 1 \end{bmatrix} \right) = \begin{bmatrix} 1 \\ 3 \end{bmatrix}.$$

To determine the coordinates of the images of the S-basis, we must solve the three linear systems

$$a_1 \begin{bmatrix} 1 \\ 2 \end{bmatrix} + a_2 \begin{bmatrix} 1 \\ 3 \end{bmatrix} = \mathbf{b},$$

where $\mathbf{b} = \begin{bmatrix} 2 \\ 3 \end{bmatrix}, \begin{bmatrix} 2 \\ 5 \end{bmatrix}$, and $\begin{bmatrix} 1 \\ 3 \end{bmatrix}$. This can be done simultaneously as in Section 2.7, by transforming the partitioned matrix

$$\left[\begin{array}{cc|c|c|c} 1 & 1 & 2 & 2 & 1 \\ 2 & 3 & 3 & 5 & 3 \end{array}\right]$$

to reduced row echelon form, obtaining (verify)

$$\left[\begin{array}{cc|c|c|c} 1 & 0 & 3 & 1 & 0 \\ 0 & 1 & -1 & 1 & 1 \end{array}\right].$$

The last three columns of this matrix are the desired coordinate vectors of the S-basis with respect to the T-basis. That is, the last three columns form the matrix A representing L with respect to S and T. Thus

$$A = \begin{bmatrix} 3 & 1 & 0 \\ -1 & 1 & 1 \end{bmatrix}.$$

This matrix, of course, differs from the one that defined L. Thus, although a matrix A may be involved in the definition of a linear transformation L, we cannot conclude that it is necessarily the representation of L that we seek. ■

From Example 5 we see that if $L: R^n \to R^m$ is a linear transformation, then a computationally efficient way to obtain a matrix representation A of L with respect to the ordered bases $S = \{\mathbf{v}_1, \mathbf{v}_2, \ldots, \mathbf{v}_n\}$ for R^n and $T = \{\mathbf{w}_1, \mathbf{w}_2, \ldots, \mathbf{w}_m\}$ for R^m is to proceed as follows: Transform the partitioned matrix

$$\left[\mathbf{w}_1 \quad \mathbf{w}_2 \quad \cdots \quad \mathbf{w}_m \;\vdots\; L(\mathbf{v}_1) \;\vdots\; L(\mathbf{v}_2) \;\vdots\; \cdots \;\vdots\; L(\mathbf{v}_n)\right]$$

to reduced row echelon form. The matrix A consists of the last n columns of this last matrix.

If $L: V \to V$ is a linear operator on an n-dimensional space V, then to obtain a representation of L we fix ordered bases S and T for V, and obtain a matrix A representing L with respect to S and T. However, it is often convenient in this case to choose $S = T$. To avoid verbosity in this case, we refer to A as the **representation of L with respect to S**. If $L: R^n \to R^n$ is a linear operator, then the matrix representing L with respect to the natural basis for R^n has already been discussed in Theorem 4.3 in Section 4.1, where it was called the standard matrix respresenting L.

Also, we can show readily that the matrix of the identity operator (see Exercise 20 in Section 4.1) on an n-dimensional space, with respect to any basis, is I_n.

Let $I: V \to V$ be the identity operator on an n-dimensional vector space V and let $S = \{\mathbf{v}_1, \mathbf{v}_2, \ldots, \mathbf{v}_n\}$ and $T = \{\mathbf{w}_1, \mathbf{w}_2, \ldots, \mathbf{w}_m\}$ be ordered bases for V. It is not difficult to show (Exercise 23) that the matrix of the identity operator with respect to S and T is the transition matrix from the S-basis to the T-basis (see Section 2.7).

If $L: V \to V$ is an invertible linear operator and if A is the representation of L with respect to an ordered basis S for V, then A^{-1} is the representation of L^{-1} with respect to S. This fact, which can be proved directly at this point, will follow almost trivially in Section 4.4.

Suppose that $L: V \to W$ is a linear transformation and that A is the matrix representing L with respect to ordered bases for V and W. Then the problem of finding $\ker L$ reduces to the problem of finding the solution space of $A\mathbf{x} = \mathbf{0}$. Moreover, the problem of finding range L reduces to the problem of finding the column space of A.

We can summarize the conditions under which a linear transformation L of an n-dimensional vector space V into itself (or more generally into a n-dimensional vector space W) is invertible by the following equivalent statement:

1. L is invertible.
2. L is one-to-one.
3. L is onto.
4. The matrix A representing L with respect to ordered bases S and T for V and W is nonsingular.

4.3 Exercises

1. Let $L\colon R^2 \to R^2$ be defined by

$$L\left(\begin{bmatrix} x_1 \\ x_2 \end{bmatrix}\right) = \begin{bmatrix} x_1 + 2x_2 \\ 2x_1 - x_2 \end{bmatrix}.$$

Let S be the natural basis for R^2 and let

$$T = \left\{\begin{bmatrix} -1 \\ 2 \end{bmatrix}, \begin{bmatrix} 2 \\ 0 \end{bmatrix}\right\}.$$

Find the representation of L with respect to:

(a) S. (b) S and T. (c) T and S. (d) T.

(e) Find $L\left(\begin{bmatrix} 1 \\ 2 \end{bmatrix}\right)$ using the definition of L and also using the matrices obtained in parts (a) through (d).

2. Let $L\colon R_4 \to R_3$ be defined by

$$L\left(\begin{bmatrix} x_1 & x_2 & x_3 & x_4 \end{bmatrix}\right) = \begin{bmatrix} x_1 & x_2 + x_3 & x_3 + x_4 \end{bmatrix}.$$

Let S and T be the natural bases for R_4 and R_3, respectively. Let $S' = \{\begin{bmatrix} 1 & 0 & 0 & 1 \end{bmatrix}, \begin{bmatrix} 0 & 0 & 0 & 1 \end{bmatrix}, \begin{bmatrix} 1 & 1 & 0 & 0 \end{bmatrix}, \begin{bmatrix} 0 & 1 & 1 & 0 \end{bmatrix}\}$ and $T' = \{\begin{bmatrix} 1 & 1 & 0 \end{bmatrix}, \begin{bmatrix} 0 & 1 & 0 \end{bmatrix}, \begin{bmatrix} 1 & 0 & 1 \end{bmatrix}\}$.

(a) Find the representation of L with respect to S and T.

(b) Find the representation of L with respect to S' and T'.

(c) Find $L\left(\begin{bmatrix} 2 & 1 & -1 & 3 \end{bmatrix}\right)$ using the matrices obtained in parts (a) and (b) and compare this answer with that obtained from the definition for L.

3. Let $L\colon R^4 \to R^3$ be defined by

$$L\left(\begin{bmatrix} x_1 \\ x_2 \\ x_3 \\ x_4 \end{bmatrix}\right) = \begin{bmatrix} 1 & 0 & 1 & 1 \\ 0 & 1 & 2 & 1 \\ -1 & -2 & 1 & 0 \end{bmatrix}\begin{bmatrix} x_1 \\ x_2 \\ x_3 \\ x_4 \end{bmatrix}.$$

Let S and T be the natural bases for R^4 and R^3, respectively, and consider the ordered bases

$$S' = \left\{\begin{bmatrix} 1 \\ 1 \\ 0 \\ 0 \end{bmatrix}, \begin{bmatrix} 0 \\ 1 \\ 0 \\ 0 \end{bmatrix}, \begin{bmatrix} 0 \\ 0 \\ 1 \\ 1 \end{bmatrix}, \begin{bmatrix} 0 \\ 1 \\ 1 \\ 0 \end{bmatrix}\right\} \quad \text{and}$$

$$T' = \left\{\begin{bmatrix} 1 \\ 0 \\ 1 \end{bmatrix}, \begin{bmatrix} 0 \\ 1 \\ 1 \end{bmatrix}, \begin{bmatrix} 0 \\ 0 \\ 1 \end{bmatrix}\right\}$$

for R^4 and R^3, respectively. Find the representation of L with respect to (a) S and T; (b) S' and T'.

4. Let $L\colon R^2 \to R^2$ be the linear transformation rotating R^2 counterclockwise through an angle ϕ. Find the representation of L with respect to the natural basis for R^2.

5. Let $L\colon R^3 \to R^3$ be defined by

$$L\left(\begin{bmatrix} 1 \\ 0 \\ 0 \end{bmatrix}\right) = \begin{bmatrix} 1 \\ 1 \\ 0 \end{bmatrix}, \quad L\left(\begin{bmatrix} 0 \\ 1 \\ 0 \end{bmatrix}\right) = \begin{bmatrix} 2 \\ 0 \\ 1 \end{bmatrix},$$

$$L\left(\begin{bmatrix} 0 \\ 0 \\ 1 \end{bmatrix}\right) = \begin{bmatrix} 1 \\ 0 \\ 1 \end{bmatrix}.$$

(a) Find the representation of L with respect to the natural basis S for R^3.

(b) Find $L\left(\begin{bmatrix} 1 \\ 2 \\ 3 \end{bmatrix}\right)$ using the definition of L and also using the matrix obtained in part (a).

6. Let $L\colon R^3 \to R^3$ be defined as in Exercise 5. Let $T = \{L(\mathbf{e}_1), L(\mathbf{e}_2), L(\mathbf{e}_3)\}$ be an ordered basis for R^3, and let S be the natural basis for R^3.

(a) Find the representation of L with respect to S and T.

(b) Find $L\left(\begin{bmatrix} 1 \\ 2 \\ 3 \end{bmatrix}\right)$ using the matrix obtained in part (a).

7. Let $L\colon R^3 \to R^3$ be the linear transformation represented by the matrix

$$\begin{bmatrix} 1 & 3 & 1 \\ 1 & 2 & 0 \\ 0 & 1 & 1 \end{bmatrix}$$

with respect to the natural basis for R^3. Find:

(a) $L\left(\begin{bmatrix} 1 \\ 2 \\ 3 \end{bmatrix}\right)$. (b) $L\left(\begin{bmatrix} 0 \\ 1 \\ 1 \end{bmatrix}\right)$.

8. Let $L: M_{22} \to M_{22}$ be defined by

$$L(A) = \begin{bmatrix} 1 & 2 \\ 3 & 4 \end{bmatrix} A$$

for A in M_{22}. Consider the ordered bases

$$S = \left\{ \begin{bmatrix} 1 & 0 \\ 0 & 0 \end{bmatrix}, \begin{bmatrix} 0 & 1 \\ 0 & 0 \end{bmatrix}, \begin{bmatrix} 0 & 0 \\ 1 & 0 \end{bmatrix}, \begin{bmatrix} 0 & 0 \\ 0 & 1 \end{bmatrix} \right\}$$

and

$$T = \left\{ \begin{bmatrix} 1 & 0 \\ 0 & 1 \end{bmatrix}, \begin{bmatrix} 1 & 1 \\ 0 & 0 \end{bmatrix}, \begin{bmatrix} 1 & 0 \\ 1 & 0 \end{bmatrix}, \begin{bmatrix} 0 & 1 \\ 0 & 0 \end{bmatrix} \right\}$$

for M_{22}. Find the representation of L with respect to:

(a) S. (b) T. (c) S and T. (d) T and S.

9. Let V be the vector space with basis $S = \{1, t, e^t, te^t\}$, and let $L: V \to V$ be a linear operator defined by $L(f) = f' = df/dt$. Find the representation of L with respect to S.

10. Let $L: P_1 \to P_2$ be defined by $L(p(t)) = tp(t) + p(0)$. Consider the ordered bases $S = \{t, 1\}$ and $S' = \{t + 1, t - 1\}$ for P_1, and $T = \{t^2, t, 1\}$ and $T' = \{t^2 + 1, t - 1, t + 1\}$ for P_2. Find the representation of L with respect to:

(a) S and T. (b) S' and T'.

(c) Find $L(-3t - 3)$ using the definition of L and the matrices obtained in parts (a) and (b).

11. Let $A = \begin{bmatrix} 1 & 2 \\ 3 & 4 \end{bmatrix}$, and let $L: M_{22} \to M_{22}$ be the linear transformation defined by $L(X) = AX - XA$ for X in M_{22}. Let S and T be the ordered bases for M_{22} defined in Exercise 8. Find the representation of L with respect to:

(a) S. (b) T. (c) S and T. (d) T and S.

12. Let $L: V \to V$ be a linear operator. A nonempty subspace U of V is called **invariant** under L if $L(U)$ is contained in U. Let L be a linear operator with invariant subspace U. Show that if $\dim U = m$, and $\dim V = n$, then L has a representation with respect to a basis S for V of the form $\begin{bmatrix} A & B \\ O & C \end{bmatrix}$, where A is $m \times m$, B is $m \times (n - m)$, O is the zero $(n - m) \times m$ matrix, and C is $(n - m) \times (n - m)$.

13. Let $L: R^2 \to R^2$ be defined by

$$L\left(\begin{bmatrix} x \\ y \end{bmatrix}\right) = \begin{bmatrix} x \\ -y \end{bmatrix},$$

a reflection about the x-axis. Consider the natural basis S and the ordered basis $T = \left\{ \begin{bmatrix} 1 \\ 1 \end{bmatrix}, \begin{bmatrix} -1 \\ 1 \end{bmatrix} \right\}$ for R^2. Find the representation of L with respect to:

(a) S. (b) T. (c) S and T. (d) T and S.

14. If $L: R_3 \to R_2$ is the linear transformation whose representation with respect to the natural bases for R_3 and R_2 is $\begin{bmatrix} 1 & -1 & 2 \\ 2 & 1 & 3 \end{bmatrix}$, find:

(a) $L\left(\begin{bmatrix} 1 & 2 & 3 \end{bmatrix}\right)$. (b) $L\left(\begin{bmatrix} -1 & 2 & -1 \end{bmatrix}\right)$.

(c) $L\left(\begin{bmatrix} 0 & 1 & 2 \end{bmatrix}\right)$. (d) $L\left(\begin{bmatrix} 0 & 1 & 0 \end{bmatrix}\right)$.

(e) $L\left(\begin{bmatrix} 0 & 0 & 1 \end{bmatrix}\right)$.

15. If $O: V \to W$ is the zero linear transformation, show that the matrix representation of O with respect to any ordered bases for V and W is the $m \times n$ zero matrix, where $n = \dim V$ and $m = \dim W$.

16. If $I: V \to V$ is the identity linear operator on V defined by $I(\mathbf{v}) = \mathbf{v}$ for $\mathbf{v}$ in V, prove that the matrix representation of I with respect to any ordered basis S for V is I_n, where $\dim V = n$.

17. Let $I: R_2 \to R_2$ be the identity linear operator on R_2. Let $S = \{\begin{bmatrix} 1 & 0 \end{bmatrix}, \begin{bmatrix} 0 & 1 \end{bmatrix}\}$ and $T = \{\begin{bmatrix} 1 & -1 \end{bmatrix}, \begin{bmatrix} 2 & 3 \end{bmatrix}\}$ be ordered bases for R_2. Find the representation of I with respect to:

(a) S. (b) T. (c) S and T. (d) T and S.

18. Let V be the vector space of real-valued continuous functions with basis $S = \{e^t, e^{-t}\}$. Find the representation of the linear operator $L: V \to V$ defined by $L(f) = f'$ with respect to S.

19. Let V be the vector space of real-valued continuous functions with ordered basis $S = \{\sin t, \cos t\}$. Find the representation of the linear operator $L: V \to V$ defined by $L(f) = f'$ with respect to S.

20. Let V be the vector space of real-valued continuous functions with ordered basis $S = \{\sin t, \cos t\}$ and consider $T = \{\sin t - \cos t, \sin t + \cos t\}$, another ordered basis for V. Find the representation of the linear operator $L: V \to V$ defined by $L(f) = f'$ with respect to:

(a) S. (b) T. (c) S and T. (d) T and S.

21. Let $L: V \to V$ be a linear operator defined by $L(\mathbf{v}) = c\mathbf{v}$, where c is a fixed constant. Prove that the representation of L with respect to any ordered basis for V is a scalar matrix (see Section 1.4).

22. Let the representation of $L: R^3 \to R^2$ with respect to the ordered bases $S = \{\mathbf{v}_1, \mathbf{v}_2, \mathbf{v}_3\}$ and $T = \{\mathbf{w}_1, \mathbf{w}_2\}$ be

$$A = \begin{bmatrix} 1 & 2 & 1 \\ -1 & 1 & 0 \end{bmatrix},$$

where

$$\mathbf{v}_1 = \begin{bmatrix} -1 \\ 1 \\ 0 \end{bmatrix}, \quad \mathbf{v}_2 = \begin{bmatrix} 0 \\ 1 \\ 1 \end{bmatrix}, \quad \mathbf{v}_3 = \begin{bmatrix} 1 \\ 0 \\ 0 \end{bmatrix},$$

$$\mathbf{w}_1 = \begin{bmatrix} 1 \\ 2 \end{bmatrix}, \quad \text{and} \quad \mathbf{w}_2 = \begin{bmatrix} 1 \\ -1 \end{bmatrix}.$$

(a) Compute $\left[L(\mathbf{v}_1)\right]_T$, $\left[L(\mathbf{v}_2)\right]_T$, and $\left[L(\mathbf{v}_3)\right]_T$.

(b) Compute $L(\mathbf{v}_1)$, $L(\mathbf{v}_2)$, and $L(\mathbf{v}_3)$.

(c) Compute $L\left(\begin{bmatrix} 2 \\ 1 \\ -1 \end{bmatrix}\right)$.

23. Let $I: V \to V$ be the identity operator on an n-dimensional vector space V and let $S = \{\mathbf{v}_1, \mathbf{v}_2, \dots, \mathbf{v}_n\}$ and $T = \{\mathbf{w}_1, \mathbf{w}_2, \dots, \mathbf{w}_n\}$ be ordered bases for V. Show that the matrix of the identity operator with respect to S and T is the transition matrix from the S-basis to the T-basis (see Section 2.7).

4.4 VECTOR SPACE OF MATRICES AND VECTOR SPACE OF LINEAR TRANSFORMATIONS (OPTIONAL)

We have already seen in Section 2.2 that the set M_{mn} of all $m \times n$ matrices is a vector space under the operations of matrix addition and scalar multiplication. We now show in this section that the set of all linear transformations of an n-dimensional vector space V into an m-dimensional vector space W is also a vector space U under two suitably defined operations, and we shall examine the relation between U and M_{mn}.

Definition 4.5

Let V and W be two vector spaces of dimensions n and m, respectively. Also, let $L_1: V \to W$ and $L_2: V \to W$ be linear transformations. We define a mapping $L: V \to W$ by $L(\mathbf{x}) = L_1(\mathbf{x}) + L_2(\mathbf{x})$, for $\mathbf{x}$ in V. Of course, the $+$ here is vector addition in W. We shall denote L by $L_1 \boxplus L_2$ and call it the **sum** of L_1 and L_2. Also, if $L: V \to W$ is a linear transformation and c is a real number, we define a mapping $H: V \to W$ by $H(\mathbf{x}) = cL(\mathbf{x})$ for $\mathbf{x}$ in V. Of course, the operation on the right side is scalar multiplication in W. We denote H by $c \boxdot L$ and call it the **scalar multiple** of L by c. ▲

EXAMPLE 1

Let $V = R_3$ and $W = R_2$. Let $L_1: R_3 \to R_2$ and $L_2: R_3 \to R_2$ be defined by

$$L_1(\mathbf{x}) = L_1\left(\begin{bmatrix} a_1 & a_2 & a_3 \end{bmatrix}\right) = \begin{bmatrix} a_1 + a_2 & a_2 + a_3 \end{bmatrix}$$

and

$$L_2(\mathbf{x}) = L_2\left(\begin{bmatrix} a_1 & a_2 & a_3 \end{bmatrix}\right) = \begin{bmatrix} a_1 + a_3 & a_2 \end{bmatrix}.$$

Then

$$(L_1 \boxplus L_2)(\mathbf{x}) = \begin{bmatrix} 2a_1 + a_2 + a_3 & 2a_2 + a_3 \end{bmatrix}$$

and

$$(3 \boxdot L_1)(\mathbf{x}) = \begin{bmatrix} 3a_1 + 3a_2 & 3a_2 + 3a_3 \end{bmatrix}.$$ ■

We leave it to the reader (see the exercises in this section) to verify that if L, L_1, and L_2 are linear transformations of V into W and if c is a real number, then $L_1 \boxplus L_2$ and $c \boxdot L$ are linear transformations. We also let the reader show that the set U of all linear transformations of V into W is a vector space under the operations $\boxplus$ and $\boxdot$. The linear transformation $O\colon V \to W$ defined by $O(\mathbf{x}) = \mathbf{0}_W$ for $\mathbf{x}$ in V is the zero vector in U. That is, $L \boxplus O = O \boxplus L = L$ for any L in U. Also, if L is in U, then $L \boxplus (-1 \boxdot L) = O$, so we may write $(-1) \boxdot L$ as $-L$. Of course, to say that $S = \{L_1, L_2, \ldots, L_k\}$ is a linearly dependent set in U merely means that there exist k scalars $a_1, a_2, \ldots, a_k$, not all zero, such that

$$(a_1 \boxdot L_1) \boxplus (a_2 \boxdot L_2) \boxplus \cdots \boxplus (a_k \boxdot L_k) = O,$$

where O is the zero linear transformation.

EXAMPLE 2

Let $L_1\colon R_2 \to R_3$, $L_2\colon R_2 \to R_3$, and $L_3\colon R_2 \to R_3$ be defined by

$$\begin{aligned} L_1\left(\begin{bmatrix} a_1 & a_2 \end{bmatrix}\right) &= \begin{bmatrix} a_1 + a_2 & 2a_2 & a_2 \end{bmatrix}, \\ L_2\left(\begin{bmatrix} a_1 & a_2 \end{bmatrix}\right) &= \begin{bmatrix} a_2 - a_1 & 2a_1 + a_2 & a_1 \end{bmatrix}, \\ L_3\left(\begin{bmatrix} a_1 & a_2 \end{bmatrix}\right) &= \begin{bmatrix} 3a_1 & -2a_2 & a_1 + 2a_2 \end{bmatrix}. \end{aligned}$$

Determine whether $S = \{L_1, L_2, L_3\}$ is linearly independent.

Solution Suppose that

$$(a_1 \boxdot L_1) \boxplus (a_2 \boxdot L_2) \boxplus (a_3 \boxdot L_3) = O,$$

where a_1, a_2, and a_3 are real numbers. Then for $\mathbf{e}_1^T = \begin{bmatrix} 1 & 0 \end{bmatrix}$, we have

$$((a_1 \boxdot L_1) \boxplus (a_2 \boxdot L_2) \boxplus (a_3 \boxdot L_3))(\mathbf{e}_1^T) = O(\mathbf{e}_1^T) = \begin{bmatrix} 0 & 0 & 0 \end{bmatrix},$$

so

$$\begin{aligned} &a_1 L_1(\mathbf{e}_1^T) + a_2 L_2(\mathbf{e}_1^T) + a_3 L_3(\mathbf{e}_1^T) \\ &\qquad = a_1 \begin{bmatrix} 1 & 2 & 0 \end{bmatrix} + a_2 \begin{bmatrix} -1 & 2 & 1 \end{bmatrix} + a_3 \begin{bmatrix} 3 & 0 & 1 \end{bmatrix} = \begin{bmatrix} 0 & 0 & 0 \end{bmatrix}. \end{aligned}$$

Thus we must solve the homogeneous system

$$\begin{bmatrix} 1 & -1 & 3 \\ 2 & 2 & 0 \\ 0 & 1 & 1 \end{bmatrix} \begin{bmatrix} a_1 \\ a_2 \\ a_3 \end{bmatrix} = \begin{bmatrix} 0 \\ 0 \\ 0 \end{bmatrix},$$

obtaining $a_1 = a_2 = a_3 = 0$ (verify). Hence S is linearly independent. ■

Theorem 4.10

Let U be the vector space of all linear transformations of an n-dimensional vector space V into an m-dimensional vector space W, $n \neq 0$ and $m \neq 0$, under the operations $\boxplus$ and $\boxdot$. Then U is isomorphic to the vector space M_{mn} of all $m \times n$ matrices.

Proof

Let $S = \{\mathbf{v}_1, \mathbf{v}_2, \ldots, \mathbf{v}_n\}$ and $T = \{\mathbf{w}_1, \mathbf{w}_2, \ldots, \mathbf{w}_m\}$ be ordered bases for V and W, respectively. We define a function $M: U \to M_{mn}$ by letting $M(L) =$ the matrix representing L with respect to the bases S and T. We now show that M is an isomorphism.

First, M is one-to-one, for if L_1 and L_2 are two different elements in U, then $L_1(\mathbf{v}_j) \neq L_2(\mathbf{v}_j)$ for some $j = 1, 2, \ldots, n$. This means that the jth columns of $M(L_1)$ and $M(L_2)$, which are the coordinates vectors of $L_1(\mathbf{v}_j)$ and $L_2(\mathbf{v}_j)$, respectively, with respect to T, are different, so $M(L_1) \neq M(L_2)$. Hence M is one-to-one.

Next, M is onto. Let $A = \begin{bmatrix} a_{ij} \end{bmatrix}$ be a given $m \times n$ matrix; that is, A is an element of M_{mn}. Then we define a function $L: V \to W$ by

$$L(\mathbf{v}_i) = \sum_{k=1}^{m} a_{ki}\mathbf{w}_k \qquad i = 1, 2, \ldots, n,$$

and if $\mathbf{x} = c_1\mathbf{v}_1 + c_2\mathbf{v}_2 + \cdots + c_n\mathbf{v}_n$, we define $L(\mathbf{x})$ by

$$L(\mathbf{x}) = \sum_{i=1}^{n} c_i L(\mathbf{v}_i).$$

It is not difficult to show that L is a linear transformation; moreover, the matrix representing L with respect to S and T is $A = \begin{bmatrix} a_{ij} \end{bmatrix}$ (verify). Thus $M(L) = A$, so M is onto.

Now let $M(L_1) = A = \begin{bmatrix} a_{ij} \end{bmatrix}$ and $M(L_2) = B = \begin{bmatrix} b_{ij} \end{bmatrix}$. We show that $M(L_1 \boxplus L_2) = A + B$. First, note that the jth column of $M(L_1 \boxplus L_2)$ is

$$\left[(L_1 \boxplus L_2)(\mathbf{v}_j)\right]_T = \left[L_1(\mathbf{v}_j) + L_2(\mathbf{v}_j)\right]_T = \left[L_1(\mathbf{v}_j)\right]_T + \left[L_2(\mathbf{v}_j)\right]_T.$$

Thus the jth column of $M(L_1 \boxplus L_2)$ is the sum of the jth columns of $M(L_1) = A$ and $M(L_2) = B$. Hence $M(L_1 \boxplus L_2) = A + B$.

Finally, let $M(L) = A$ and $c =$ a real number. Following the idea in the paragraph above, we can show that $M(c \boxdot L) = cA$ (verify). Hence U and M_{mn} are isomorphic. ●

This theorem implies that the dimension of U is mn, for $\dim M_{mn} = mn$. Also, it means that when dealing with finite-dimensional vector spaces, we can always replace all linear transformations by their matrix representations and work only with the matrices. Moreover, it should be noted again that matrices lend themselves much more readily than linear transformations to computer implementations.

EXAMPLE 3

Let $A = \begin{bmatrix} 1 & 2 & -1 \\ 2 & -1 & 3 \end{bmatrix}$ and $S = \{\mathbf{e}_1, \mathbf{e}_2, \mathbf{e}_3\}$ and $T = \{\overline{\mathbf{e}}_1, \overline{\mathbf{e}}_2\}$ be the natural bases for R^3 and R^2, respectively.

(a) Find the unique linear transformation $L\colon R^3 \to R^2$ whose representation with respect to S and T is A.

(b) Let

$$S' = \left\{ \begin{bmatrix} 1 \\ 0 \\ 1 \end{bmatrix}, \begin{bmatrix} 1 \\ 1 \\ 0 \end{bmatrix}, \begin{bmatrix} 0 \\ 1 \\ 1 \end{bmatrix} \right\} \quad \text{and} \quad T' = \left\{ \begin{bmatrix} 1 \\ 3 \end{bmatrix}, \begin{bmatrix} 2 \\ -1 \end{bmatrix} \right\}$$

be ordered bases for R^3 and R^2, respectively. Determine the linear transformation $L\colon R^3 \to R^2$, whose representation with respect to S' and T' is A.

(c) Compute $L\left(\begin{bmatrix} 1 \\ 2 \\ 3 \end{bmatrix}\right)$ using L as determined in part (b).

Solution (a) Let

$$L(\mathbf{e}_1) = 1\overline{\mathbf{e}}_1 + 2\overline{\mathbf{e}}_2 = \begin{bmatrix} 1 \\ 2 \end{bmatrix},$$

$$L(\mathbf{e}_2) = 2\overline{\mathbf{e}}_1 - 1\overline{\mathbf{e}}_2 = \begin{bmatrix} 2 \\ -1 \end{bmatrix},$$

$$L(\mathbf{e}_3) = -\overline{\mathbf{e}}_1 + 3\overline{\mathbf{e}}_2 = \begin{bmatrix} -1 \\ 3 \end{bmatrix}.$$

Now if $\mathbf{x} = \begin{bmatrix} a_1 \\ a_2 \\ a_3 \end{bmatrix}$ is in R^3, we define $L(\mathbf{x})$ by

$$\begin{aligned} L(\mathbf{x}) &= L(a_1\mathbf{e}_1 + a_2\mathbf{e}_2 + a_3\mathbf{e}_3) \\ &= a_1L(\mathbf{e}_1) + a_2L(\mathbf{e}_2) + a_3L(\mathbf{e}_3), \end{aligned}$$

$$\begin{aligned} L(\mathbf{x}) &= a_1\begin{bmatrix} 1 \\ 2 \end{bmatrix} + a_2\begin{bmatrix} 2 \\ -1 \end{bmatrix} + a_3\begin{bmatrix} -1 \\ 3 \end{bmatrix} \\ &= \begin{bmatrix} a_1 + 2a_2 - a_3 \\ 2a_1 - a_2 + 3a_3 \end{bmatrix}. \end{aligned}$$

Note that

$$L(\mathbf{x}) = \begin{bmatrix} 1 & 2 & -1 \\ 2 & -1 & 3 \end{bmatrix} \begin{bmatrix} a_1 \\ a_2 \\ a_3 \end{bmatrix},$$

so we could have defined L by $L(\mathbf{x}) = A\mathbf{x}$ for $\mathbf{x}$ in R^3. We can do this when the bases S and T are the natural bases.

(b) Let

$$L\left(\begin{bmatrix}1\\0\\1\end{bmatrix}\right) = 1\begin{bmatrix}1\\3\end{bmatrix} + 2\begin{bmatrix}2\\-1\end{bmatrix} = \begin{bmatrix}5\\1\end{bmatrix},$$

$$L\left(\begin{bmatrix}1\\1\\0\end{bmatrix}\right) = 2\begin{bmatrix}1\\3\end{bmatrix} - 1\begin{bmatrix}2\\-1\end{bmatrix} = \begin{bmatrix}0\\7\end{bmatrix},$$

$$L\left(\begin{bmatrix}0\\1\\1\end{bmatrix}\right) = -1\begin{bmatrix}1\\3\end{bmatrix} + 3\begin{bmatrix}2\\-1\end{bmatrix} = \begin{bmatrix}5\\-6\end{bmatrix}.$$

Then if $\mathbf{x} = \begin{bmatrix}a_1\\a_2\\a_3\end{bmatrix}$, we express $\mathbf{x}$ in terms of the basis S' as

$$\mathbf{x} = b_1\begin{bmatrix}1\\0\\1\end{bmatrix} + b_2\begin{bmatrix}1\\1\\0\end{bmatrix} + b_3\begin{bmatrix}0\\1\\1\end{bmatrix}; \quad \text{hence } \left[\mathbf{x}\right]_{S'} = \begin{bmatrix}b_1\\b_2\\b_3\end{bmatrix}.$$

Define $L(\mathbf{x})$ by

$$\begin{aligned} L(\mathbf{x}) &= b_1 L\left(\begin{bmatrix}1\\0\\1\end{bmatrix}\right) + b_2 L\left(\begin{bmatrix}1\\1\\0\end{bmatrix}\right) + b_3 L\left(\begin{bmatrix}0\\1\\1\end{bmatrix}\right) \\ &= b_1\begin{bmatrix}5\\1\end{bmatrix} + b_2\begin{bmatrix}0\\7\end{bmatrix} + b_3\begin{bmatrix}5\\-6\end{bmatrix}, \end{aligned}$$

so

$$L(\mathbf{x}) = \begin{bmatrix}5b_1 + 5b_3\\ b_1 + 7b_2 - 6b_3\end{bmatrix}. \tag{1}$$

(c) To find $L\left(\begin{bmatrix}1\\2\\3\end{bmatrix}\right)$, we first have (verify)

$$\begin{bmatrix}1\\2\\3\end{bmatrix} = 1\begin{bmatrix}1\\0\\1\end{bmatrix} + 0\begin{bmatrix}1\\1\\0\end{bmatrix} + 2\begin{bmatrix}0\\1\\1\end{bmatrix}; \quad \text{hence } \left[\begin{bmatrix}1\\2\\3\end{bmatrix}\right]_{S'} = \begin{bmatrix}1\\0\\2\end{bmatrix}.$$

Then using $b_1 = 1$, $b_2 = 0$, and $b_3 = 2$ in Equation (1), we obtain

$$L\left(\begin{bmatrix}1\\2\\3\end{bmatrix}\right) = \begin{bmatrix}15\\-11\end{bmatrix}.$$

■

The linear transformations obtained in Example 3 depend on the ordered bases for R^2 and R^3. Thus if L is as in part (a), then

$$L\left(\begin{bmatrix}1\\2\\3\end{bmatrix}\right)=\begin{bmatrix}2\\9\end{bmatrix},$$

which differs from the answer obtained in part (c), since the linear transformation in part (b) differs from that in part (a).

Now let V_1 be an n-dimensional vector space, V_2 an m-dimensional vector space, and V_3 a p-dimensional vector space. Let $L_1: V_1 \to V_2$ and $L_2: V_2 \to V_3$ be linear transformations. We can define the composite function

$$L_2 \circ L_1: V_1 \to V_3 \quad \text{by} \quad (L_2 \circ L_1)(\mathbf{x}) = L_2(L_1(\mathbf{x}))$$

for $\mathbf{x}$ in V_1. It follows that $L_2 \circ L_1$ is a linear transformation. If $L: V \to V$, then $L \circ L$ is written as L^2.

EXAMPLE 4

Let $L_1: R^2 \to R^2$ and $L_2: R^2 \to R^2$ be defined by

$$L_1\left(\begin{bmatrix}a_1\\a_2\end{bmatrix}\right)=\begin{bmatrix}a_1\\-a_2\end{bmatrix} \quad \text{and} \quad L_2\left(\begin{bmatrix}a_1\\a_2\end{bmatrix}\right)=\begin{bmatrix}a_2\\a_1\end{bmatrix}.$$

Then

$$(L_1 \circ L_2)\left(\begin{bmatrix}a_1\\a_2\end{bmatrix}\right)=L_1\left(\begin{bmatrix}a_2\\a_1\end{bmatrix}\right)=\begin{bmatrix}a_2\\-a_1\end{bmatrix},$$

while

$$(L_2 \circ L_1)\left(\begin{bmatrix}a_1\\a_2\end{bmatrix}\right)=L_2\left(\begin{bmatrix}a_1\\-a_2\end{bmatrix}\right)=\begin{bmatrix}-a_2\\a_1\end{bmatrix}.$$

Thus $L_1 \circ L_2 \neq L_2 \circ L_1$. ■

Theorem 4.11

Let V_1 be an n-dimensional vector space, V_2 an m-dimensional vector space, and V_3 a p-dimensional vector space with linear transformations L_1 and L_2 such that $L_1: V_1 \to V_2$ and $L_2: V_2 \to V_3$. If the ordered bases P, S, and T are chosen for V_1, V_2, and V_3, respectively, then $M(L_2 \circ L_1) = M(L_2)M(L_1)$.

Proof

Let $M(L_1) = A$, with respect to the P and S ordered bases for V_1 and V_2, respectively, and let $M(L_2) = B$, with respect to the S and T ordered bases for V_2 and V_3, respectively. For any vector $\mathbf{x}$ in V_1, $\left[L_1(\mathbf{x})\right]_S = A\left[\mathbf{x}\right]_P$ and, for any vector $\mathbf{y}$ in V_2, $\left[L_2(\mathbf{y})\right]_T = B\left[\mathbf{y}\right]_S$. Then it follows that

$$\begin{aligned}\left[(L_2 \circ L_1)(\mathbf{x})\right]_T &= \left[L_2(L_1(\mathbf{x}))\right]_T\\ &= B\left[L_1(\mathbf{x})\right]_S = B\left(A\left[\mathbf{x}\right]_P\right) = (BA)\left[\mathbf{x}\right]_P.\end{aligned}$$

Since a linear transformation has a unique representation with respect to ordered bases (see Theorem 4.9), we have $M(L_2 \circ L_1) = BA$ and it follows that $M(L_2 \circ L_1) = M(L_2)M(L_1)$. ●

Since AB need not equal BA for matrices A and B, it is thus not surprising that $L_1 \circ L_2$ need not be the same linear transformation as $L_2 \circ L_1$.

If $L: V \to V$ is an invertible linear operator and if A is a representation of L with respect to an ordered basis S for V, then the representation of the identity operator $L \circ L^{-1}$ is the identity matrix I_n. Thus $M(L)M(L^{-1}) = I_n$, which means that A^{-1} is the representation of L^{-1} with respect to S.

4.4 Exercises

1. Let L_1, L_2, and L be linear transformations of V into W. Prove:

(a) $L_1 \boxplus L_2$ is a linear transformation of V into W.

(b) If c is a real number, then $c \boxdot L$ is a linear transformation of V into W.

(c) If A represents L with respect to the ordered bases S and T for V and W, respectively, then cA represents $c \boxdot L$ with respect to S and T, where c = a real number.

2. Let U be the set of all linear transformations of V into W, and let $O: V \to W$ be the zero linear transformation defined by $O(\mathbf{x}) = \mathbf{0}_W$ for all $\mathbf{x}$ in V.

(a) Prove that $O \boxplus L = L \boxplus O = L$ for any L in U.

(b) Show that if L is in U, then

$$L \boxplus ((-1) \boxdot L) = O.$$

3. Let $L_1: R_3 \to R_3$ and $L_2: R_3 \to R_3$ be linear transformations such that $L(\mathbf{e}_1^T) = \begin{bmatrix} -1 & 2 & 1 \end{bmatrix}$, $L_1(\mathbf{e}_2^T) = \begin{bmatrix} 0 & 1 & 2 \end{bmatrix}$, $L_1(\mathbf{e}_3^T) = \begin{bmatrix} -1 & 1 & 3 \end{bmatrix}$, and $L_2(\mathbf{e}_1^T) = \begin{bmatrix} 0 & 1 & 3 \end{bmatrix}$, $L_2(\mathbf{e}_2^T) = \begin{bmatrix} 4 & -2 & 1 \end{bmatrix}$, $L_2(\mathbf{e}_3^T) = \begin{bmatrix} 0 & 2 & 2 \end{bmatrix}$, where $S = \{\mathbf{e}_1^T, \mathbf{e}_2^T, \mathbf{e}_3^T\}$ is the natural basis for R_3. Find:

(a) $(L_1 \boxplus L_2)\left(\begin{bmatrix} a_1 & a_2 & a_3 \end{bmatrix}\right)$.

(b) $(L_1 \boxplus L_2)\left(\begin{bmatrix} 2 & 1 & -3 \end{bmatrix}\right)$.

(c) The representation of $L_1 \boxplus L_2$ with respect to S.

(d) $(-2 \boxdot L_1)\left(\begin{bmatrix} a_1 & a_2 & a_3 \end{bmatrix}\right)$.

(e) $(-2 \boxdot L_1) \boxplus (4 \boxdot L_2)\left(\begin{bmatrix} 2 & 1 & -3 \end{bmatrix}\right)$.

(f) The representation of $(-2 \boxdot L_1) \boxplus (4 \boxdot L_2)$ with respect to S.

4. Verify that the set U of all linear transformations of V into W is a vector space under the operations $\boxplus$ and $\boxdot$.

5. Let $L_i: R_3 \to R_3$ be a linear transformation defined by

$$L_i(\mathbf{x}) = A_i\mathbf{x}, \qquad i = 1, 2,$$

where

$$A_1 = \begin{bmatrix} 1 & 1 & 1 \\ 2 & 2 & 2 \\ 3 & 3 & 3 \end{bmatrix} \quad \text{and} \quad A_2 = \begin{bmatrix} 1 & 1 & 2 \\ 2 & 2 & 4 \\ -2 & -2 & -4 \end{bmatrix}.$$

(a) For $\mathbf{x} = \begin{bmatrix} 3 & -2 & 1 \end{bmatrix}^T$, find $(L_1 \boxplus L_2)(\mathbf{x})$.

(b) Determine $\ker L_1$, $\ker L_2$, and $\ker L_1 \cap \ker L_2$ (intersection of kernels).

(c) Determine $\ker(L_1 \boxplus L_2)$.

(d) What is the relationship between $\ker L_1 \cap \ker L_2$ and $\ker(L_1 \boxplus L_2)$?

6. Let V_1, V_2, and V_3 be vector spaces of dimensions n, m, and p, respectively. Also let $L_1: V_1 \to V_2$ and $L_2: V_2 \to V_3$ be linear transformations. Prove that $L_2 \circ L_1: V_1 \to V_3$ is a linear transformation.

7. Let $L_1: P_1 \to P_2$ be the linear transformation defined by $L_1(p(t)) = tp(t)$ and let $L_2: P_2 \to P_3$ be the linear transformation defined by $L_2(p(t)) = t^2p'(t)$. Let $R = \{t+1, t-1\}$, $S = \{t^2, t-1, t+2\}$, and $T = \{t^3, t^2-1, t, t+1\}$ be ordered bases for P_1, P_2, and P_3, respectively.

(a) Find the representation C of $L_2 \circ L_1$ with respect to R and T.

(b) Compute the representation A of L_1 with respect to R and S and the representation B of L_2 with respect to S and T. Verify that BA is the matrix C obtained in part (a).

8. Let L_1, L_2, and S be as in Exercise 3. Find:

(a) $(L_1 \circ L_2)\left(\begin{bmatrix} a_1 & a_2 & a_3 \end{bmatrix}\right)$.

(b) $(L_2 \circ L_1)\left(\begin{bmatrix} a_1 & a_2 & a_3 \end{bmatrix}\right)$.

(c) The representation of $L_1 \circ L_2$ with respect to S.

(d) The representation of $L_2 \circ L_1$ with respect to S.

9. If $\begin{bmatrix} 1 & 4 & -1 \\ 2 & 1 & 3 \\ 1 & -1 & 2 \end{bmatrix}$ is the representation of a linear operator $L\colon R^3 \to R^3$ with respect to ordered bases S and T for R^3, find the representation with respect to S and T of:

(a) $2 \boxdot L$. (b) $2 \boxdot L \boxplus L \circ L$.

10. Let L_1, L_2, and L_3 be linear transformations of R_3 into R_2 defined by

$$L_1\left(\begin{bmatrix} a_1 & a_2 & a_3 \end{bmatrix}\right) = \begin{bmatrix} a_1 + a_2 & a_1 - a_3 \end{bmatrix},$$
$$L_2\left(\begin{bmatrix} a_1 & a_2 & a_3 \end{bmatrix}\right) = \begin{bmatrix} a_1 - a_2 & a_3 \end{bmatrix}, \quad \text{and}$$
$$L_3\left(\begin{bmatrix} a_1 & a_2 & a_3 \end{bmatrix}\right) = \begin{bmatrix} a_1 & a_2 + a_3 \end{bmatrix}.$$

Prove that $S = \{L_1, L_2, L_3\}$ is a linearly independent set in the vector space U of all linear transformations of R_3 into R_2.

11. Find the dimension of the vector space U of all linear transformations of V into W for each of the following.

(a) $V = R^2$, $W = R^3$. (b) $V = P_2$, $W = P_1$.

(c) $V = M_{21}$, $W = M_{32}$. (d) $V = R_3$, $W = R_4$.

12. Repeat Exercise 11 for each of the following.

(a) $V = W$ is the vector space with basis $\{\sin t, \cos t\}$.

(b) $V = W$ is the vector space with basis $\{1, t, e^t, te^t\}$.

(c) V is the vector space *spanned* by $\{1, t, 2t\}$, and W is the vector space with basis $\{t^2, t, 1\}$.

13. Let $A = \begin{bmatrix} a_{ij} \end{bmatrix}$ be a given $m \times n$ matrix, and let V and W be given vector spaces of dimensions n and m, respectively. Let $S = \{\mathbf{v}_1, \mathbf{v}_2, \ldots, \mathbf{v}_n\}$ be an ordered basis for V and $T = \{\mathbf{w}_1, \mathbf{w}_2, \ldots, \mathbf{w}_m\}$ an ordered basis for W. Define a function $L\colon V \to W$ by

$$L(\mathbf{v}_i) = \sum_{k=1}^{m} a_{ki}\mathbf{w}_k, \qquad i = 1, 2, \ldots, n,$$

and if $\mathbf{x} = c_1\mathbf{v}_1 + c_2\mathbf{v}_2 + \cdots + c_n\mathbf{v}_n$, we define $L(\mathbf{x})$ by

$$L(\mathbf{x}) = \sum_{i=1}^{n} c_i L(\mathbf{v}_i).$$

(a) Show that L is a linear transformation.

(b) Show that A represents L with respect to S and T.

14. Let $A = \begin{bmatrix} 1 & 2 & -2 \\ 3 & 4 & -1 \end{bmatrix}$. Let S be the natural basis for R^3 and T the natural basis for R^2.

(a) Find the linear transformation $L\colon R^3 \to R^2$ determined by A.

(b) Find $L\left(\begin{bmatrix} a_1 \\ a_2 \\ a_3 \end{bmatrix}\right)$. (c) Find $L\left(\begin{bmatrix} 1 \\ 2 \\ 3 \end{bmatrix}\right)$.

15. Let A be as in Exercise 14. Consider the ordered bases $S = \{t^2, t, 1\}$ and $T = \{t, 1\}$ for P_2 and P_1, respectively.

(a) Find the linear transformation $L\colon P_2 \to P_1$ determined by A.

(b) Find $L(at^2 + bt + c)$.

(c) Find $L(2t^2 - 5t + 4)$.

16. Find two linear transformations $L_1\colon R^2 \to R^2$ and $L_2\colon R^2 \to R^2$ such that $L_2 \circ L_1 \neq L_1 \circ L_2$.

17. Find a linear transformation $L\colon R^2 \to R^2$, $L \neq I$, the identity operator, such that $L^2 = L \circ L = I$.

18. Find a linear transformation $L\colon R^2 \to R^2$, $L \neq O$, the zero transformation, such that $L^2 = O$.

19. Find a linear transformation $L\colon R^2 \to R^2$, $L \neq I$, $L \neq O$, such that $L^2 = L$.

20. Let $L\colon R^3 \to R^3$ be the linear transformation defined in Exercise 19 of Section 4.2. Find the matrix representing L^{-1} with respect to the natural basis for R^3.

21. Let $L\colon R^3 \to R^3$ be the linear transformation defined in Exercise 23 of Section 4.2. Find the matrix representing L^{-1} with respect to the natural basis for R^3.

22. Let $L\colon R^3 \to R^3$ be the invertible linear transformation represented by $A = \begin{bmatrix} 2 & 0 & 4 \\ -1 & 1 & -2 \\ 2 & 3 & 3 \end{bmatrix}$ with respect to an ordered basis S for R^3. Find the representation of L^{-1} with respect to S.

23. Let $L\colon V \to V$ be a linear transformation represented by a matrix A with respect to an ordered basis S for V. Show that A^2 represents $L^2 = L \circ L$ with respect to S. Moreover, show that if k is a positive integer, then A^k represents $L^k = L \circ L \circ \cdots \circ L$ (k times) with respect to S.

24. Let $L\colon P_1 \to P_1$ be the invertible linear transformation represented by $A = \begin{bmatrix} 4 & 2 \\ 3 & -1 \end{bmatrix}$ with respect to an ordered basis S for P_1. Find the representation of L^{-1} with respect to S.

4.5 SIMILARITY

In Section 4.3 we saw how the matrix representing a linear transformation of an n-dimensional vector space V into an m-dimensional vector space W depends upon the ordered bases we choose for V and W. We now see how this matrix changes when the bases for V and W are changed. For simplicity in this section we represent the transition matrices $P_{S \leftarrow S'}$ and $Q_{T \leftarrow T'}$ as P and Q, respectively.

Theorem 4.12

Let $L: V \to W$ be a linear transformation of an n-dimensional vector space V into an m-dimensional vector space W. Let $S = \{\mathbf{v}_1, \mathbf{v}_2, \ldots, \mathbf{v}_n\}$ and $S' = \{\mathbf{v}'_1, \mathbf{v}'_2, \ldots, \mathbf{v}'_n\}$ be ordered bases for V, with transition matrix P from S' to S; let $T = \{\mathbf{w}_1, \mathbf{w}_2, \ldots, \mathbf{w}_m\}$ and $T' = \{\mathbf{w}'_1, \mathbf{w}'_2, \ldots, \mathbf{w}'_m\}$ be ordered bases for W with transition matrix Q from T' to T. If A is the representation of L with respect to S and T, then $Q^{-1}AP$ is the representation of L with respect to S' and T'.

Proof

Recall Section 2.7, where the transition matrix was first introduced. If P is the transition matrix from S' to S, and $\mathbf{x}$ is a vector in V, then

$$\left[\mathbf{x}\right]_S = P\left[\mathbf{x}\right]_{S'}, \tag{1}$$

where the jth column of P is the coordinate vector $\left[\mathbf{v}'_j\right]_S$ of $\mathbf{v}'_j$ with respect to S. Similarly, if Q is the transition matrix from T' to T and $\mathbf{y}$ is a vector in W, then

$$\left[\mathbf{y}\right]_T = Q\left[\mathbf{y}\right]_{T'} \tag{2}$$

where the jth column of Q is the coordinate vector $\left[\mathbf{w}'_j\right]_T$ of $\mathbf{w}'_j$ with respect to T. If A is the representation of L with respect to S and T, then

$$\left[L(\mathbf{x})\right]_T = A\left[\mathbf{x}\right]_S \tag{3}$$

for $\mathbf{x}$ in V. Substituting $\mathbf{y} = L(\mathbf{x})$ in (2), we have $\left[L(\mathbf{x})\right]_T = Q\left[L(\mathbf{x})\right]_{T'}$. Now, using first (3) and then (1) in this last equation, we obtain

$$Q\left[L(\mathbf{x})\right]_{T'} = AP\left[\mathbf{x}\right]_{S'},$$

so

$$\left[L(\mathbf{x})\right]_{T'} = Q^{-1}AP\left[\mathbf{x}\right]_{S'}.$$

This means that $Q^{-1}AP$ is the representation of L with respect to S' and T'. ●

Theorem 4.12 can be illustrated by the diagram shown in Figure 4.7 where P is the transition matrix from the S'-basis to the S-basis and Q^{-1} is the transition matrix from the T-basis to the T'-basis. Recall from Section 4.3 that P is the matrix representation of the identity operator on V and Q is the matrix representation of the identity operator on W. This figure shows that there are two ways of going from $\mathbf{x}$ in V to $L(\mathbf{x})$ in W: directly using matrix B, or indirectly using the matrices P, A, and Q^{-1}.

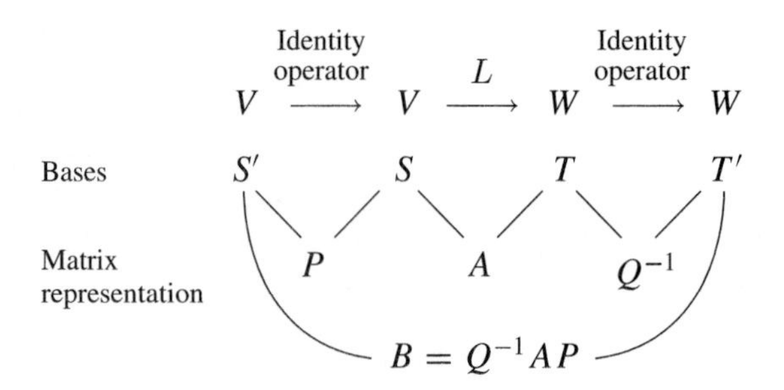

FIGURE 4.7

From Section 1.7 we see that *two representations of a linear transformation with respect to different pairs of bases are equivalent.*

EXAMPLE 1

Let $L: R^3 \to R^2$ be defined by $L\left(\begin{bmatrix} a_1 \\ a_2 \\ a_3 \end{bmatrix}\right) = \begin{bmatrix} a_1 + a_3 \\ a_2 - a_3 \end{bmatrix}$. Consider the ordered bases

$$S = \left\{ \begin{bmatrix} 1 \\ 0 \\ 0 \end{bmatrix}, \begin{bmatrix} 0 \\ 1 \\ 0 \end{bmatrix}, \begin{bmatrix} 0 \\ 0 \\ 1 \end{bmatrix} \right\} \quad \text{and} \quad S' = \left\{ \begin{bmatrix} 1 \\ 1 \\ 0 \end{bmatrix}, \begin{bmatrix} 0 \\ 1 \\ 1 \end{bmatrix}, \begin{bmatrix} 0 \\ 0 \\ 1 \end{bmatrix} \right\}$$

for R^3, and

$$T = \left\{ \begin{bmatrix} 1 \\ 0 \end{bmatrix}, \begin{bmatrix} 0 \\ 1 \end{bmatrix} \right\} \quad \text{and} \quad T' = \left\{ \begin{bmatrix} 1 \\ 1 \end{bmatrix}, \begin{bmatrix} 1 \\ 3 \end{bmatrix} \right\}$$

for R^2. We can establish (verify) that $A = \begin{bmatrix} 1 & 0 & 1 \\ 0 & 1 & -1 \end{bmatrix}$ is the representation of L with respect to S and T.

The transition matrix P from S' to S is the matrix whose jth column is the coordinate vector of the jth vector in the basis S' with respect to S. Thus $P = \begin{bmatrix} 1 & 0 & 0 \\ 1 & 1 & 0 \\ 0 & 1 & 1 \end{bmatrix}$, and the transition matrix Q from T' to T is $Q = \begin{bmatrix} 1 & 1 \\ 1 & 3 \end{bmatrix}$.

Now $Q^{-1} = \begin{bmatrix} \frac{3}{2} & -\frac{1}{2} \\ -\frac{1}{2} & \frac{1}{2} \end{bmatrix}$. (We could also obtain Q^{-1} as the transition matrix from T to T'.) Then the representation of L with respect to S' and T' is

$$B = Q^{-1}AP = \begin{bmatrix} 1 & \frac{3}{2} & 2 \\ 0 & -\frac{1}{2} & -1 \end{bmatrix}.$$

On the other hand, we can compute the representation of L with respect to S' and T' directly. We have

$$L\left(\begin{bmatrix}1\\1\\0\end{bmatrix}\right)=\begin{bmatrix}1\\1\end{bmatrix}=1\begin{bmatrix}1\\1\end{bmatrix}+0\begin{bmatrix}1\\3\end{bmatrix},\quad\text{so}\quad\left[L\left(\begin{bmatrix}1\\1\\0\end{bmatrix}\right)\right]_{T'}=\begin{bmatrix}1\\0\end{bmatrix},$$

$$L\left(\begin{bmatrix}0\\1\\1\end{bmatrix}\right)=\begin{bmatrix}1\\0\end{bmatrix}=\frac{3}{2}\begin{bmatrix}1\\1\end{bmatrix}-\frac{1}{2}\begin{bmatrix}1\\3\end{bmatrix},\quad\text{so}\quad\left[L\left(\begin{bmatrix}0\\1\\1\end{bmatrix}\right)\right]_{T'}=\begin{bmatrix}\frac{3}{2}\\-\frac{1}{2}\end{bmatrix},$$

$$L\left(\begin{bmatrix}0\\0\\1\end{bmatrix}\right)=\begin{bmatrix}1\\-1\end{bmatrix}=2\begin{bmatrix}1\\1\end{bmatrix}-1\begin{bmatrix}1\\3\end{bmatrix},\quad\text{so}\quad\left[L\left(\begin{bmatrix}0\\0\\1\end{bmatrix}\right)\right]_{T'}=\begin{bmatrix}2\\-1\end{bmatrix}.$$

Then the representation of L with respect to S' and T' is

$$\begin{bmatrix}1 & \frac{3}{2} & 2\\0 & -\frac{1}{2} & -1\end{bmatrix},$$

which agrees with our earlier result. ■

Taking $V = W$ in Theorem 4.11, we obtain an important result, which we state as Corollary 4.3.

Corollary 4.3

Let $L: V \to V$ be a linear operator on an n-dimensional vector space. Let $S = \{\mathbf{v}_1, \mathbf{v}_2, \ldots, \mathbf{v}_n\}$ and $S' = \{\mathbf{v}'_1, \mathbf{v}'_2, \ldots, \mathbf{v}'_n\}$ be ordered bases for V with transition matrix P from S' to S. If A is the representation of L with respect to S, then $P^{-1}AP$ is the representation of L with respect to S'. ●

We may define the **rank** of a linear transformation $L: V \to W$, rank L, as the rank of any matrix representing L. This definition makes sense, since if A and B represent L, then A and B are equivalent (see Section 1.7); by Section 2.8 we know that equivalent matrices have the same rank.

We can now restate Theorem 4.6 as follows. If $L: V \to W$ is a linear transformation, then

$$\text{nullity } L + \text{rank } L = \dim V.$$

If $L: R^n \to R^m$ is a linear transformation defined by $L(\mathbf{x}) = A\mathbf{x}$, for $\mathbf{x}$ in R^n, where A is an $m \times n$ matrix, then it follows that nullity L = nullity A (see Section 2.6.)

Theorem 4.13

Let $L: V \to W$ be a linear transformation. Then rank $L = \dim \text{range } L$.

Proof

Let $n = \dim V$, $m = \dim W$, and $r = \dim \text{range } L$. Then, from Theorem 4.6, $\dim \ker L = n - r$. Let $\mathbf{v}_{r+1}, \mathbf{v}_{r+2}, \dots, \mathbf{v}_n$ be a basis for $\ker L$. By Theorem 2.10, there exist vectors $\mathbf{v}_1, \mathbf{v}_2, \dots, \mathbf{v}_r$ in V such that $S = \{\mathbf{v}_1, \mathbf{v}_2, \dots, \mathbf{v}_r, \mathbf{v}_{r+1}, \dots, \mathbf{v}_n\}$ is a basis for V. The vectors $\mathbf{w}_1 = L(\mathbf{v}_1)$, $\mathbf{w}_2 = L(\mathbf{v}_2)$, ..., $\mathbf{w}_r = L(\mathbf{v}_r)$ form a basis for range L (they clearly span range L and there are r of them, so Theorem 2.11 applies). Again by Theorem 2.10, there exist vectors $\mathbf{w}_{r+1}, \mathbf{w}_{r+2}, \dots, \mathbf{w}_m$ in W such that $T = \{\mathbf{w}_1, \mathbf{w}_2, \dots, \mathbf{w}_r, \mathbf{w}_{r+1}, \mathbf{w}_{r+2}, \dots, \mathbf{w}_m\}$ is a basis for W. Let A denote the $m \times n$ matrix which represents L with respect to S and T. The columns of A are (Theorem 4.9)

$$\left[L(\mathbf{v}_i)\right]_T = \left[\mathbf{w}_i\right]_T = \mathbf{e}_i \qquad i = 1, 2, \dots, r$$

and

$$\left[L(\mathbf{v}_j)\right]_T = \left[\mathbf{0}_W\right]_T = \mathbf{0}_{R^m} \qquad j = r+1, r+2, \dots, n.$$

Hence

$$A = \begin{bmatrix} I_r & O \\ O & O \end{bmatrix}.$$

Therefore

$$\text{rank } L = \text{rank } A = r = \dim \text{range } L.$$ ●

Definition 4.6

If A and B are $n \times n$ matrices, we say that B is **similar** to A if there is a nonsingular matrix P such that $B = P^{-1}AP$. ▲

We can show readily (Exercise 1) that B is similar to A if and only if A is similar to B. Thus we replace the statements "A is similar to B" and "B is similar to A" by "A and B are similar."

By Corollary 4.3 we then see that any two representations of a linear operator $L: V \to V$ are similar. Conversely, let $A = \left[a_{ij}\right]$ and $B = \left[b_{ij}\right]$ be similar $n \times n$ matrices and let V be an n-dimensional vector space (we may take V as R^n). We wish to show that A and B represent the same linear transformation $L: V \to V$ with respect to different bases. Since A and B are similar, $B = P^{-1}AP$ for some nonsingular matrix $P = \left[p_{ij}\right]$. Let $S = \{\mathbf{v}_1, \mathbf{v}_2, \dots, \mathbf{v}_n\}$ be an ordered basis for V; from the proof of Theorem 4.10, we know that there exists a linear transformation $L: V \to V$, which is represented by A with respect to S. Then

$$\left[L(\mathbf{x})\right]_S = A\left[\mathbf{x}\right]_S. \tag{4}$$

We wish to prove that B also represents L with respect to some basis for V. Let

$$\mathbf{w}_j = \sum_{i=1}^{n} p_{ij}\mathbf{v}_i. \tag{5}$$

We first show that $T = \{\mathbf{w}_1, \mathbf{w}_2, \dots, \mathbf{w}_n\}$ is also a basis for V. Suppose that

$$a_1\mathbf{w}_1 + a_2\mathbf{w}_2 + \cdots + a_n\mathbf{w}_n = \mathbf{0}.$$

Then, from (5), we have

$$a_1\left(\sum_{i=1}^{n} p_{i1}\mathbf{v}_i\right) + a_2\left(\sum_{i=1}^{n} p_{i2}\mathbf{v}_i\right) + \cdots + a_n\left(\sum_{i=1}^{n} p_{in}\mathbf{v}_i\right) = \mathbf{0},$$

which can be rewritten as

$$\left(\sum_{j=1}^{n} p_{1j}a_j\right)\mathbf{v}_1 + \left(\sum_{j=1}^{n} p_{2j}a_j\right)\mathbf{v}_2 + \cdots + \left(\sum_{j=1}^{n} p_{nj}a_j\right)\mathbf{v}_n = \mathbf{0}. \tag{6}$$

Since S is linearly independent, each of the coefficients in (6) is zero. Thus

$$\sum_{j=1}^{n} p_{ij}a_j = 0 \qquad i = 1, 2, \ldots, n,$$

or equivalently,

$$P\mathbf{a} = \mathbf{0},$$

where $\mathbf{a} = \begin{bmatrix} a_1 & a_2 & \cdots & a_n \end{bmatrix}^T$. This is a homogeneous system of n equations in the n unknowns $a_1, a_2, \ldots, a_n$, whose coefficient matrix is P. Since P is nonsingular, the only solution is the trivial one. Hence $a_1 = a_2 = \cdots = a_n = 0$, and T is linearly independent. Moreover, Equation (5) implies that P is the transition matrix from T to S (see Section 2.7). Thus

$$\left[\mathbf{x}\right]_S = P\left[\mathbf{x}\right]_T. \tag{7}$$

In (7) replace $\mathbf{x}$ by $L(\mathbf{x})$ giving

$$\left[L(\mathbf{x})\right]_S = P\left[L(\mathbf{x})\right]_T.$$

Using (4) we have

$$A\left[\mathbf{x}\right]_S = P\left[L(\mathbf{x})\right]_T$$

and by (7)

$$AP\left[\mathbf{x}\right]_T = P\left[L(\mathbf{x})\right]_T.$$

Hence,

$$\left[L(\mathbf{x})\right]_T = P^{-1}AP\left[x\right]_T,$$

which means that the representation of L with respect to T is $B = P^{-1}AP$. We can summarize these results in the following theorem.

Theorem 4.14

Let V be any n-dimensional vector space and let A and B be any $n \times n$ matrices. Then A and B are similar if and only if A and B represent the same linear transformation $L\colon V \to V$ with respect to two ordered bases for V. ●

EXAMPLE 2

Let $L\colon R_3 \to R_3$ be defined by

$$L\left(\begin{bmatrix} a_1 & a_2 & a_3 \end{bmatrix}\right) = \begin{bmatrix} 2a_1 - a_3 & a_1 + a_2 - a_3 & a_3 \end{bmatrix}.$$

Let $S = \{\begin{bmatrix} 1 & 0 & 0 \end{bmatrix}, \begin{bmatrix} 0 & 1 & 0 \end{bmatrix}, \begin{bmatrix} 0 & 0 & 1 \end{bmatrix}\}$ be the natural basis for R_3. The representation of L with respect to S is

$$A = \begin{bmatrix} 2 & 0 & -1 \\ 1 & 1 & -1 \\ 0 & 0 & 1 \end{bmatrix}.$$

Now consider the ordered basis

$$S' = \{\begin{bmatrix} 1 & 0 & 1 \end{bmatrix}, \begin{bmatrix} 0 & 1 & 0 \end{bmatrix}, \begin{bmatrix} 1 & 1 & 0 \end{bmatrix}\}$$

for R_3. The transition matrix P from S' to S is

$$P = \begin{bmatrix} 1 & 0 & 1 \\ 0 & 1 & 1 \\ 1 & 0 & 0 \end{bmatrix}; \quad \text{moreover, } P^{-1} = \begin{bmatrix} 0 & 0 & 1 \\ -1 & 1 & 1 \\ 1 & 0 & -1 \end{bmatrix}.$$

Then the representation of L with respect to S' is

$$B = P^{-1}AP = \begin{bmatrix} 1 & 0 & 0 \\ 0 & 1 & 0 \\ 0 & 0 & 2 \end{bmatrix}.$$

The same result can be obtained directly (verify). The matrices A and B are similar. ■

Observe that the matrix B obtained in Example 2 is diagonal. We can now ask a number of related questions. First, given $L\colon V \to V$, when can we choose a basis S for V such that the representation of L with respect to S is diagonal? How do we choose such a basis? In Example 2 we apparently pulled our basis S' "out of the air." If we cannot choose a basis giving a representation of L that is diagonal, can we choose a basis giving a matrix that is close in appearance to a diagonal matrix? What do we gain from having such simple representations? First, we already know from Section 4.4 that if A represents $L\colon V \to V$ with respect to some ordered basis S for V, then A^k represents $L \circ L \circ \cdots \circ L = L^k$ with respect to S; now, if A is similar to B, then B^k also represents L^k. Of course, if B is diagonal, then it is a trivial matter to compute B^k: the diagonal elements of B^k are those of B raised to the kth power. We shall also find that if A is similar to a diagonal matrix, then we can easily solve a homogeneous linear system of differential equations with constant coefficients. The answers to these questions will be taken up in detail in Chapter 6.

Similar matrices enjoy some other nice properties. For example, if A and B are similar, then $\mathrm{Tr}(A) = \mathrm{Tr}(B)$ (see Exercise 29 in Section 1.2 for a definition of trace). Also, if A and B are similar, then A^k and B^k are similar for any positive integer k. Proofs of these results are not difficult and are left as exercises.

We obtain one final result on similar matrices.

Theorem 4.15

If A and B are similar $n \times n$ matrices, then rank A = rank B.

Proof

We know from Theorem 4.14 that A and B represent the same linear transformation $L: R^n \to R^n$ with respect to different bases. Hence rank A = rank L = rank B. ●

4.5 Exercises

1. Let A, B, and C be square matrices. Show that:
   (a) A is similar to A.
   (b) If B is similar to A, then A is similar to B.
   (c) If C is similar to B and B is similar to A, then C is similar to A.

2. Let L be the linear transformation defined in Exercise 2, Section 4.3
   (a) Find the transition matrix P from S' to S.
   (b) Find the transition matrix from S to S' and verify that it is P^{-1}.
   (c) Find the transition matrix Q from T' to T.
   (d) Find the representation of L with respect to S' and T'.
   (e) What is the dimension of range L?

3. Do Exercise 1(d) of Section 4.3 using transition matrices.

4. Do Exercise 8(b) of Section 4.3 using transition matrices.

5. Let $L: R^2 \to R^2$ be defined by
$$L\left(\begin{bmatrix} a_1 \\ a_2 \end{bmatrix}\right) = \begin{bmatrix} a_1 \\ -a_2 \end{bmatrix}.$$
   (a) Find the representation of L with respect to the natural basis S for R^2.
   (b) Find the representation of L with respect to the ordered basis
$$T = \left\{\begin{bmatrix} 1 \\ -1 \end{bmatrix}, \begin{bmatrix} 1 \\ 2 \end{bmatrix}\right\}.$$
   (c) Verify that the matrices obtained in parts (a) and (b) are similar.
   (d) Verify that the ranks of the matrices obtained in parts (a) and (b) are equal.

6. Show that if A and B are similar matrices, then A^k and B^k are similar for any positive integer k. (*Hint*: If $B = P^{-1}AP$, find $B^2 = BB$, and so on.)

7. Show that if A and B are similar, then A^T and B^T are similar.

8. Prove that if A and B are similar, then $\text{Tr}(A) = \text{Tr}(B)$. (*Hint*: See Exercise 29 in Section 1.2 for a definition of trace.)

9. Let $L: R_3 \to R_2$ be the linear transformation whose representation is $A = \begin{bmatrix} 2 & -1 & 3 \\ 3 & 1 & 0 \end{bmatrix}$ with respect to the ordered bases
$$S = \{[1 \quad 0 \quad -1], [0 \quad 2 \quad 0], [1 \quad 2 \quad 3]\}$$
and
$$T = \{[1 \quad -1], [2 \quad 0]\}.$$
Find the representation of L with respect to the natural bases for R_3 and R_2.

10. Let $L: R^3 \to R^3$ be the linear transformation whose representation with respect to the natural basis for R^3 is $A = \begin{bmatrix} a_{ij} \end{bmatrix}$. Let $P = \begin{bmatrix} 0 & 1 & 1 \\ 1 & 0 & 1 \\ 1 & 1 & 0 \end{bmatrix}$. Find a basis T for R^3 with respect to which $B = P^{-1}AP$ represents L. (*Hint*: See the solution of Example 2.)

11. Let A and B be similar. Show:
   (a) If A is nonsingular, then B is nonsingular.
   (b) If A is nonsingular, then A^{-1} and B^{-1} are similar.

12. Do Exercise 13(b) of Section 4.3 using transition matrices.

13. Do Exercise 17(b) of Section 4.3 using transition matrices.

14. Do Exercise 10(b) of Section 4.3 using transition matrices.

15. Do Exercise 20(b) of Section 4.3 using transition matrices.

16. Prove that A and O_n are similar if and only if $A = O_n$.

17. Let A and B be similar matrices. Show that:

(a) A^T and B^T are similar.

(b) rank A = rank B.

(c) A is nonsingular if and only if B is nonsingular.

(d) If A and B are nonsingular, then A^{-1} and B^{-1} are similar.

(e) $\text{Tr}(A) = \text{Tr}(B)$.

4.6 COMPUTER GRAPHICS (OPTIONAL)

We are all familiar with the astounding results being developed with computer graphics in the areas of video games and special effects in the film industry. Computer graphics also play a major role in the manufacturing world. *Computer-aided design* (CAD) is used to design a computer model of a product and then, by subjecting the computer model to a variety of tests (carried out on the computer), changes to the current design, can be implemented to obtain an improved design. One of the notable successes of this approach has been in the automobile industry, where the computer model can be viewed from different angles to obtain a most pleasing and popular style; can be tested for strength of components, for roadability, for seating comfort, and for safety in a crash.

In this section we give illustrations of linear operators $L: R^2 \to R^2$ that are useful in two-dimensional graphics. We will assume that the natural basis is used in R^2 to obtain a matrix representation of L. In each of the following four examples, we describe a geometric transformation on a set of points in R^2 and show the construction of the matrix representing the corresponding linear operator $L: R^2 \to R^2$.

EXAMPLE 1

A reflection with respect to the x-axis of a vector $\mathbf{v}$ in R^2 is defined by the linear operator

$$L(\mathbf{v}) = L\left(\begin{bmatrix} x \\ y \end{bmatrix}\right) = \begin{bmatrix} x \\ -y \end{bmatrix}$$

(see Example 8 in Section 4.1). Then

$$L\left(\begin{bmatrix} 1 \\ 0 \end{bmatrix}\right) = \begin{bmatrix} 1 \\ 0 \end{bmatrix} \quad \text{and} \quad L\left(\begin{bmatrix} 0 \\ 1 \end{bmatrix}\right) = \begin{bmatrix} 0 \\ -1 \end{bmatrix}.$$

Hence the standard matrix representing L with respect to the natural basis is

$$A = \begin{bmatrix} 1 & 0 \\ 0 & -1 \end{bmatrix}.$$

Thus we have

$$L(\mathbf{v}) = A\mathbf{v} = \begin{bmatrix} 1 & 0 \\ 0 & -1 \end{bmatrix}\begin{bmatrix} x \\ y \end{bmatrix} = \begin{bmatrix} x \\ -y \end{bmatrix}.$$

To illustrate a reflection with respect to the x-axis in computer graphics, let the triangle T in Figure 4.8(a) have vertices

$$(-1, 4), \quad (3, 1), \quad \text{and} \quad (2, 6).$$

To reflect T with respect to the x-axis, we let

$$\mathbf{v}_1 = \begin{bmatrix} -1 \\ 4 \end{bmatrix}, \qquad \mathbf{v}_2 = \begin{bmatrix} 3 \\ 1 \end{bmatrix}, \qquad \mathbf{v}_3 = \begin{bmatrix} 2 \\ 6 \end{bmatrix}$$

and compute the images $L(\mathbf{v}_1)$, $L(\mathbf{v}_2)$, and $L(\mathbf{v}_3)$ by forming the products

$$A\mathbf{v}_1 = \begin{bmatrix} 1 & 0 \\ 0 & -1 \end{bmatrix}\begin{bmatrix} -1 \\ 4 \end{bmatrix} = \begin{bmatrix} -1 \\ -4 \end{bmatrix},$$

$$A\mathbf{v}_2 = \begin{bmatrix} 1 & 0 \\ 0 & -1 \end{bmatrix}\begin{bmatrix} 3 \\ 1 \end{bmatrix} = \begin{bmatrix} 3 \\ -1 \end{bmatrix},$$

$$A\mathbf{v}_3 = \begin{bmatrix} 1 & 0 \\ 0 & -1 \end{bmatrix}\begin{bmatrix} 2 \\ 6 \end{bmatrix} = \begin{bmatrix} 2 \\ -6 \end{bmatrix}.$$

These three products can be written in terms of partitioned matrices as

$$A\begin{bmatrix} \mathbf{v}_1 & \mathbf{v}_2 & \mathbf{v}_3 \end{bmatrix} = \begin{bmatrix} -1 & 3 & 2 \\ -4 & -1 & -6 \end{bmatrix}.$$

Thus the image of T has vertices

$$(-1, -4), \quad (3, -1), \quad \text{and} \quad (2, -6)$$

and is displayed in Figure 4.8(b). ■

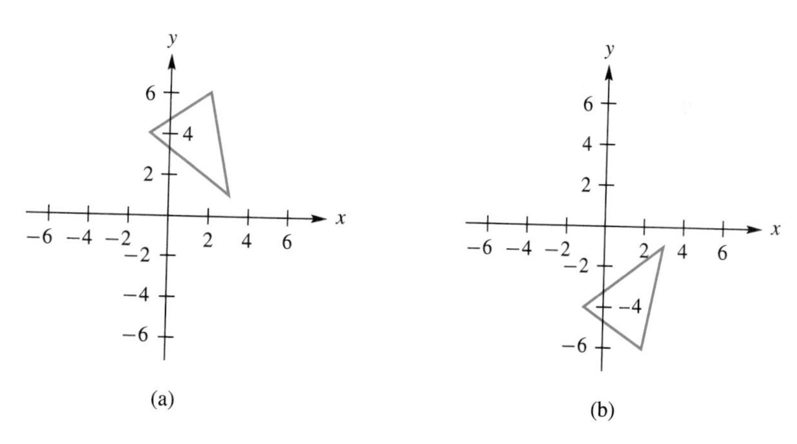

FIGURE 4.8

EXAMPLE 2

The reflection with respect to the line $y = -x$ of a vector $\mathbf{v}$ in R^2 is defined by the linear operator

$$L(\mathbf{v}) = L\left(\begin{bmatrix} x \\ y \end{bmatrix}\right) = \begin{bmatrix} -y \\ -x \end{bmatrix}.$$

Then

$$L\left(\begin{bmatrix}1\\0\end{bmatrix}\right)=\begin{bmatrix}0\\-1\end{bmatrix} \quad \text{and} \quad L\left(\begin{bmatrix}0\\1\end{bmatrix}\right)=\begin{bmatrix}-1\\0\end{bmatrix}.$$

Hence the matrix representation of L with respect to the natural basis is

$$B=\begin{bmatrix}0 & -1\\-1 & 0\end{bmatrix}.$$

To illustrate reflection with respect to the line $y=-x$, we use triangle T as defined in Example 1 and compute the products

$$A\begin{bmatrix}\mathbf{v}_1 & \mathbf{v}_2 & \mathbf{v}_3\end{bmatrix}=\begin{bmatrix}0 & -1\\-1 & 0\end{bmatrix}\begin{bmatrix}-1 & 3 & 2\\4 & 1 & 6\end{bmatrix}=\begin{bmatrix}-4 & -1 & -6\\1 & -3 & -2\end{bmatrix}.$$

Thus the image of T has vertices

$$(-4,1), \quad (-1,-3), \quad \text{and} \quad (-6,-2)$$

and is displayed in Figure 4.9. ■

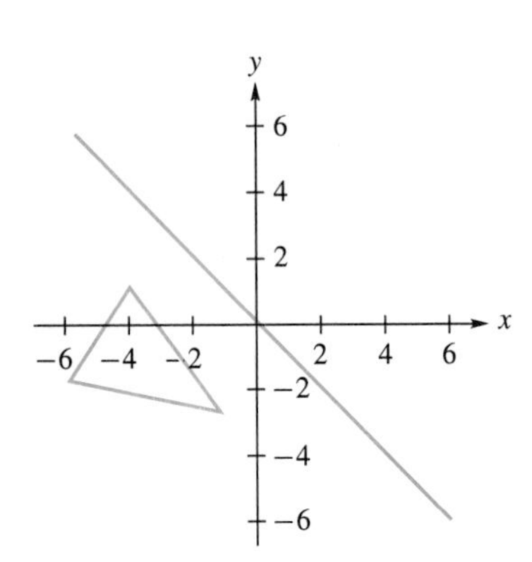

FIGURE 4.9

To perform a reflection with respect to the x-axis on triangle T of Example 1 followed by a reflection with respect to the line $y=-x$, we compute

$$B(A\mathbf{v}_1), \quad B(A\mathbf{v}_2), \quad \text{and} \quad B(A\mathbf{v}_3).$$

It is not difficult to show that reversing the order of these linear operators produces a different image (verify). Thus the order in which graphics transformations are performed is important.

EXAMPLE 3

Rotations in a plane have been defined in Example 9 of Section 4.1. A plane figure is rotated counterclockwise through an angle ϕ by using the linear operator $L: R^2 \to R^2$ whose standard matrix representation with respect to the natural basis in R^2 is given by

$$A=\begin{bmatrix}\cos\phi & -\sin\phi\\ \sin\phi & \cos\phi\end{bmatrix}.$$

Now suppose that we wish to rotate the parabola $y=x^2$ counterclockwise through 50°. We start by choosing a sample of points from the parabola, say,

$$(-2,4), \quad (-1,1), \quad (0,0), \quad \left(\tfrac{1}{2},\tfrac{1}{4}\right), \quad \text{and} \quad (3,9)$$

[see Figure 4.10(a)]. We then compute the images of these points. Thus letting

$$\mathbf{v}_1=\begin{bmatrix}-2\\4\end{bmatrix}, \quad \mathbf{v}_2=\begin{bmatrix}-1\\1\end{bmatrix}, \quad \mathbf{v}_3=\begin{bmatrix}0\\0\end{bmatrix}, \quad \mathbf{v}_4=\begin{bmatrix}\frac{1}{2}\\[2pt]\frac{1}{4}\end{bmatrix}, \quad \mathbf{v}_5=\begin{bmatrix}3\\9\end{bmatrix},$$

we compute the products (to four decimal places) (verify)

$$A\begin{bmatrix}\mathbf{v}_1 & \mathbf{v}_2 & \mathbf{v}_3 & \mathbf{v}_4 & \mathbf{v}_5\end{bmatrix} = \begin{bmatrix} -4.3498 & -1.4088 & 0 & 0.1299 & -4.9660 \\ 1.0391 & -0.1233 & 0 & 0.5437 & 8.0832 \end{bmatrix}.$$

The image points

$$(-4.3498, 1.0391), \quad (-1.4088, -0.1233), \quad (0, 0),$$
$$(0.1299, 0.5437), \quad \text{and} \quad (-4.9660, 8.0832)$$

are plotted, as shown in Figure 4.10(b), and successive points are connected showing the image of the parabola. ■

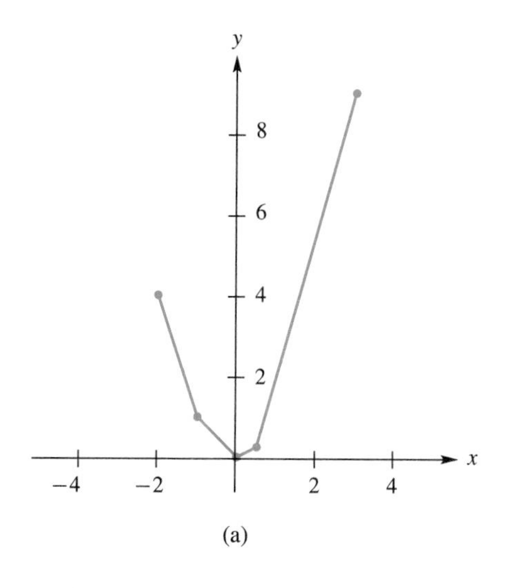

(a)

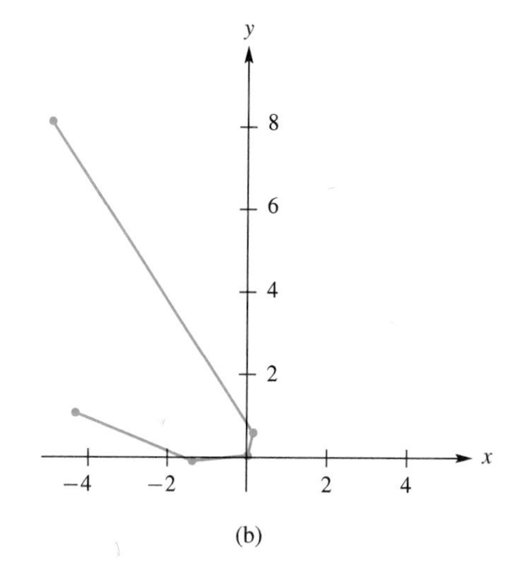

(b)

FIGURE 4.10

EXAMPLE 4

A **shear in the *x*-direction** is defined by the linear operator

$$L(\mathbf{v}) = L\left(\begin{bmatrix} x \\ y \end{bmatrix}\right) = \begin{bmatrix} x + ky \\ y \end{bmatrix},$$

where k is a scalar. Hence the standard matrix representing L with respect to the natural basis is

$$A = \begin{bmatrix} 1 & k \\ 0 & 1 \end{bmatrix}.$$

A shear in the x-direction takes the point (x, y) to the point $(x + ky, y)$. That is, the point (x, y) is moved parallel to the x-axis by the amount ky.

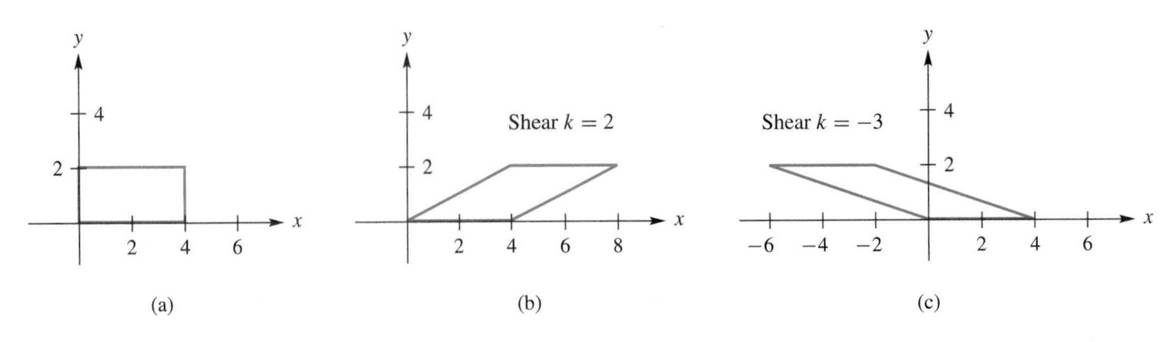

FIGURE 4.11

Consider now the rectangle R, shown in Figure 4.11(a), with vertices

$$(0, 0), \quad (0, 2), \quad (4, 0), \quad \text{and} \quad (4, 2).$$

If we apply the shear in the x-direction with $k = 2$, then the image of R is the parallelogram with vertices

$$(0, 0), \quad (4, 2), \quad (4, 0), \quad \text{and} \quad (8, 2),$$

shown in Figure 4.11(b). If we apply the shear in the x-direction with $k = -3$, then the image of R is the parallelogram with vertices

$$(0, 0), \quad (-6, 2), \quad (4, 0), \quad \text{and} \quad (-2, 2),$$

shown in Figure 4.11(c).

In the exercises we consider shears in the y-direction. ■

Other linear operators used in two-dimensional computer graphics are considered in the exercises at the end of this section. For a detailed discussion of computer graphics, the reader is referred to the books listed in the Further Readings section that follows.

FURTHER READINGS

FOLEY, J. D., A. VAN DAM, S. K. FEINER, J. F. HUGHES, and R. L. PHILLIPS. *Introduction to Computer Graphics.* Reading, Mass.: Addison-Wesley, 1994.

ROGERS, D. F., and J. A. ADAMS. *Mathematical Elements for Computer Graphics*, 2nd ed. New York: McGraw-Hill, 1989.

4.6 Exercises

1. Determine the matrix representation with respect to the natural basis of each of the following linear operators $L: R^2 \to R^2$.

(a) Reflection with respect to the y-axis.

(b) Reflection with respect to the line $y = x$.

(c) Rotation of 90° counterclockwise.

(d) Rotation of 30° clockwise.

2. Let R be the rectangle with vertices $(1, 1)$, $(1, 4)$, $(3, 1)$, and $(3, 4)$. Let L be the shear in the x-direction with $k = 3$. Find and sketch the image of R.

3. A **shear in the y-direction** is defined by the linear operator

$$L(\mathbf{v}) = L\left(\begin{bmatrix} x \\ y \end{bmatrix}\right) = \begin{bmatrix} x \\ y + kx \end{bmatrix},$$

where k is a scalar.

(a) Determine the standard matrix representing L with respect to the natural basis.

(b) Let R be the rectangle defined in Exercise 2 and let L be the shear in the y-direction with $k = -2$. Find and sketch the image of R.

4. A linear operator $L: R^2 \to R^2$ defined by

$$L(\mathbf{v}) = L\left(\begin{bmatrix} x \\ y \end{bmatrix}\right) = k\begin{bmatrix} x \\ y \end{bmatrix},$$

where k is a scalar, is called a **dilation** if $k > 1$ and it is called a **contraction** if $0 < k < 1$. Thus dilation stretches a vector, whereas contraction shrinks it.

(a) Determine the standard matrix representing L with respect to the natural basis.

(b) Let R be the rectangle defined in Exercise 2 and let L be the dilation with $k = 3$. Find and sketch the image of R.

5. A linear operator $L: R^2 \to R^2$ defined by

$$L(\mathbf{v}) = L\left(\begin{bmatrix} x \\ y \end{bmatrix}\right) = \begin{bmatrix} kx \\ y \end{bmatrix},$$

where k is a scalar, is called a **dilation in the x-direction** if $k > 1$; it is called a **contraction in the x-direction** if $0 < k < 1$.

(a) Determine the matrix representation of L with respect to the natural basis.

(b) Let R be the unit square and let L be the dilation in the x-direction with $k = 2$. Find and sketch the image of R.

6. A linear operator $L: R^2 \to R^2$ defined by

$$L(\mathbf{v}) = L\left(\begin{bmatrix} x \\ y \end{bmatrix}\right) = \begin{bmatrix} x \\ ky \end{bmatrix},$$

where k is a scalar, is called a **dilation in the y-direction** if $k > 1$; it is called a **contraction in the y-direction** if $0 < k < 1$.

(a) Determine the matrix representation of L with respect to the natural basis.

(b) Let R be the unit square and let L be the dilation in the y-direction with $k = \frac{1}{2}$. Find and sketch the image of R.

7. Let T be the triangle with vertices $(5, 0)$, $(0, 3)$, and $(2, -1)$. Find the coordinates of the vertices of the image of T under the linear operator with matrix representation

$$\begin{bmatrix} -2 & 1 \\ 3 & 4 \end{bmatrix}.$$

8. Let T be the triangle with vertices $(1, 1)$, $(-3, -3)$, and $(2, -1)$. Find the coordinates of the vertices of the image of T under the linear operator with matrix representation

$$\begin{bmatrix} 4 & -3 \\ -4 & 2 \end{bmatrix}.$$

9. Let L be the counterclockwise rotation through 60°. If T is the triangle defined in Exercise 8, find and sketch the image of T under L.

10. Let L_1 be reflection with respect to the y-axis and let L_2 be counterclockwise rotation.

(a) Show that the composite operator (see Section 4.4) $L_1 \circ L_2$ is different from $L_2 \circ L_1$.

(b) Find the standard matrices representing $L_1 \circ L_2$ and $L_2 \circ L_1$.

11. Let A be the singular matrix $\begin{bmatrix} 1 & 2 \\ 2 & 4 \end{bmatrix}$ and let T be the triangle defined in Exercise 8. Describe the image of T under L. (See also Exercise 16.)

12. Let $A = \begin{bmatrix} 5 & 0 \\ 0 & 2 \end{bmatrix}$ and let T be the triangle with vertices $(0, 0)$, $(0, 1)$, and $(2, 0)$. Compare the area of the image of T under L with the area of T.

Exercises 13 through 16 require the use of software that supports computer graphics.

13. Define a triangle T by identifying its vertices and sketch it on paper.

(a) Reflect T about the y-axis and record the resulting figure on paper, as Figure 1.

(b) Rotate Figure 1 counterclockwise through 30° and record the resulting figure on paper, as Figure 2.

(c) Reflect T about the line $y = x$, then dilate the resulting figure in the x-direction by a factor of 2 and record the new figure on paper, as Figure 3.

(d) Repeat the experiment in part (c) but interchange the order of the linear transformations. Record the resulting figure on paper, as Figure 4.

(e) Compare Figures 3 and 4.

(f) What does your answer in part (e) imply about the order of the linear transformations as applied to the triangle?

14. Consider the triangle T defined in Exercise 13. Record T on paper.

(a) Reflect T about the x-axis. Predict the result before pressing ENTER. Call this linear transformation L_1.

(b) Reflect the figure obtained in part (a) about the y-axis. Predict the result before pressing ENTER. Call this linear transformation L_2.

(c) Record on paper the figure that resulted from parts (a) and (b).

(d) Examine the relationship between the figure obtained in part (b) and T. What single linear transformation L_3 will accomplish the same result?

(e) Write a formula involving L_1, L_2, and L_3 that expresses the relationship you saw in part (d).

(f) Experiment with the formula in part (e) on several other figures until you can determine whether this formula is correct in general. Write a brief summary of your experiments, observations, and conclusions.

15. Consider the unit square S and record S on paper.

(a) Reflect S about the x-axis to obtain Figure 1. Now reflect Figure 1 about the y-axis to obtain Figure 2. Finally, reflect Figure 2 about the line $y = -x$ to obtain Figure 3. Record Figure 3 on paper.

(b) Compare S and Figure 3. Denote the reflection about the x-axis as L_1, the reflection about the y-axis as L_2, and the reflection about the line $y = -x$ as L_3. What formula does your comparison suggest when L_1 is followed by L_2, and then by L_3 on S?

(c) If M_i, $i = 1, 2, 3$, denotes the matrix representing L_i with respect to the natural basis for R^2, to what matrix is $M_3M_2M_1$ equal? Does this result agree with your conclusion in part (b)?

(d) Experiment with the successive application of these three linear transformations on other figures.

16. If your computer graphics software allows you to select any 2×2 matrix to use as a linear transformation, perform the following experiment. Choose a singular matrix and apply it to a triangle, unit square, rectangle, and pentagon. Write a brief summary of your experiments, observations, and conclusions indicating the behavior of "singular" linear transformations.

Supplementary Exercises

1. For an $n \times n$ matrix A, the trace of A, $\mathrm{Tr}(A)$, is defined as the sum of the diagonal entries of A (see Exercise 29 in Section 1.2). Prove that the trace defines a linear transformation from M_{nn} to the vector space of all real numbers.

2. Let $L\colon M_{nm} \to M_{mn}$ be the function defined by $L(A) = A^T$ (the transpose of A), for A in V. Is L a linear transformation? Justify your answer.

3. Let V be the vector space of all $n \times n$ matrices and let $L\colon V \to V$ be the function defined by

$$L(A) = \begin{cases} A^{-1} & \text{if } A \text{ is nonsingular} \\ O & \text{if } A \text{ is singular} \end{cases}$$

for A in V. Is L a linear transformation? Justify your answer.

4. Let $L\colon R_3 \to R_3$ be a linear transformation for which we know that

$$L\left(\begin{bmatrix}1 & 0 & 1\end{bmatrix}\right) = \begin{bmatrix}1 & 2 & 3\end{bmatrix},$$
$$L\left(\begin{bmatrix}0 & 1 & 2\end{bmatrix}\right) = \begin{bmatrix}1 & 0 & 0\end{bmatrix}, \quad \text{and}$$
$$L\left(\begin{bmatrix}1 & 1 & 0\end{bmatrix}\right) = \begin{bmatrix}1 & 0 & 1\end{bmatrix}.$$

(a) What is $L\left(\begin{bmatrix}4 & 1 & 0\end{bmatrix}\right)$?

(b) What is $L\left(\begin{bmatrix}0 & 0 & 0\end{bmatrix}\right)$?

(c) What is $L\left(\begin{bmatrix}a_1 & a_2 & a_3\end{bmatrix}\right)$?

5. Let $L\colon P_1 \to P_1$ be a linear transformation defined by

$$L(t-1) = t+2 \quad \text{and} \quad L(t+1) = 2t+1.$$

(a) What is $L(5t+1)$?

(b) What is $L(at+b)$?

6. Let $L\colon P_2 \to P_2$ be a linear transformation defined by

$$L(at^2 + bt + c) = (a+c)t^2 + (b+c)t.$$

(a) Is $t^2 - t - 1$ in $\ker L$?

(b) Is $t^2 + t - 1$ in $\ker L$?

(c) Is $2t^2 - t$ in range L?

(d) Is $t^2 - t + 2$ in range L?

(e) Find a basis for $\ker L$.

(f) Find a basis for range L.

7. Let $L: P_3 \to P_3$ be the linear transformation defined by

$$L(at^3 + bt^2 + ct + d) = (a - b)t^3 + (c - d)t.$$

(a) Is $t^3 + t^2 + t - 1$ in $\ker L$?

(b) Is $t^3 - t^2 + t - 1$ in $\ker L$?

(c) Is $3t^3 + t$ in range L?

(d) Is $3t^3 - t^2$ in range L?

(e) Find a basis for $\ker L$.

(f) Find a basis for range L.

8. Let $L: M_{22} \to M_{22}$ be the linear transformation defined by $L(A) = A^T$.

(a) Find a basis for $\ker L$.

(b) Find a basis for range L.

9. Let $L: M_{22} \to M_{22}$ be defined by

$$L(A) = A^T.$$

Let

$$S = \left\{ \begin{bmatrix} 1 & 0 \\ 0 & 0 \end{bmatrix}, \begin{bmatrix} 0 & 1 \\ 0 & 0 \end{bmatrix}, \begin{bmatrix} 0 & 0 \\ 1 & 0 \end{bmatrix}, \begin{bmatrix} 0 & 0 \\ 0 & 1 \end{bmatrix} \right\}$$

and

$$T = \left\{ \begin{bmatrix} 1 & 1 \\ 0 & 0 \end{bmatrix}, \begin{bmatrix} 0 & 1 \\ 0 & 0 \end{bmatrix}, \begin{bmatrix} 0 & 0 \\ 1 & 1 \end{bmatrix}, \begin{bmatrix} 1 & 0 \\ 0 & 1 \end{bmatrix} \right\}$$

be bases for M_{22}. Find the matrix of L with respect to:

(a) S. (b) S and T. (c) T and S. (d) T.

10. For the vector space V and linear transformation L of Exercise 22 in Section 4.1, find a basis for $\ker L$.

11. Let V be the vector space of real-valued continuous functions on $[0, 1]$ and let $L: V \to R$ be given by $L(f) = f(0)$, for f in V.

(a) Show that L is a linear transformation.

(b) Describe the kernel of L and give examples of polynomials, quotients of polynomials, and trigonometric functions that belong to $\ker L$.

(c) If we redefine L by $L(f) = f(\frac{1}{2})$, is it still a linear transformation? Explain.

12. Let $L: P_1 \to R$ be the linear transformation defined by

$$L(p(t)) = \int_0^1 p(t)\,dt.$$

(a) Find a basis for $\ker L$.

(b) Find a basis for range L.

(c) Verify Theorem 4.6 for L.

13. Let $L: P_2 \to P_2$ be the linear transformation defined by

$$L(at^2 + bt + c) = (a + 2c)t^2 + (b - c)t + (a - c).$$

Let $S = \{1, t, t^2\}$ and $T = \{t^2 - 1, t, t - 1\}$ be ordered bases for P_2.

(a) Find the matrix of L with respect to S and T.

(b) If $p(t) = 2t^2 - 3t + 1$, compute $L(p(t))$ using the matrix obtained in part (a).

14. Let $L: P_1 \to P_1$ be a linear transformation which is represented by the matrix

$$A = \begin{bmatrix} 2 & -3 \\ 1 & 2 \end{bmatrix}$$

with respect to the basis $S = \{p_1(t), p_2(t)\}$, where

$$p_1(t) = t - 2 \quad \text{and} \quad p_2(t) = t + 1.$$

(a) Compute $L(p_1(t))$ and $L(p_2(t))$.

(b) Compute $\left[L(p_1(t))\right]_S$ and $\left[L(p_2(t))\right]_S$.

(c) Compute $L(t + 2)$.

15. Let $L: P_3 \to P_3$ be defined by

$$L(at^3 + bt^2 + ct + d) = 3at^2 + 2bt + c.$$

Find the matrix of L with respect to the basis $S = \{t^3, t^2, t, 1\}$ for P_3.

16. Consider R^n as an inner product space with the standard inner product and let $L: R^n \to R^n$ be a linear transformation. Prove that for any vector $\mathbf{u}$ in R^n, $\|L(\mathbf{u}\| = \|\mathbf{u}\|$ if and only if

$$(L(\mathbf{u}), L(\mathbf{v})) = (\mathbf{u}, \mathbf{v})$$

for any vectors $\mathbf{u}$ and $\mathbf{v}$ in R^n. Such a linear transformation is said to **preserve inner products**.

17. Let $L_1: V \to V$ and $L_2: V \to V$ be linear transformations on a vector space V. Prove that

$$(L_1 + L_2)^2 = L_1^2 + 2L_1 \circ L_2 + L_2^2$$

if and only if $L_1 \circ L_2 = L_2 \circ L_1$ (see Exercise 23 in Section 4.4).

18. Let $\mathbf{u}$ and $\mathbf{v}$ be nonzero vectors in R^n. In Section 3.3 we defined the angle between $\mathbf{u}$ and $\mathbf{v}$ to be the angle θ such that

$$\cos\theta = \frac{(\mathbf{u}, \mathbf{v})}{\|\mathbf{u}\| \, \|\mathbf{v}\|} \qquad 0 \leq \theta \leq \pi.$$

A linear operator $L: R^n \to R^n$ is called **angle preserving** if the angle between $\mathbf{u}$ and $\mathbf{v}$ is the same as that between $L(\mathbf{u})$ and $L(\mathbf{v})$. Prove that if L is inner product preserving (see Exercise 16), then it is angle preserving.

19. Let $L: R^n \to R^n$ be a linear operator that preserves inner products (see Exercise 16), and let the $n \times n$ matrix A represent L with respect to some ordered basis S for R^n.

(a) Prove that $\ker L = \{\mathbf{0}\}$.

(b) Prove that $AA^T = I_n$ (*Hint*: Use Supplementary Exercise 17 in Chapter 3.)

20. Let $L: V \to W$ be a linear transformation. If $\{\mathbf{v}_1, \mathbf{v}_2, \ldots, \mathbf{v}_k\}$ spans V, show that $\{L(\mathbf{v}_1), L(\mathbf{v}_2), \ldots, L(\mathbf{v}_k)\}$ spans range L.

21. Let V be an n-dimensional vector space and $S = \{\mathbf{v}_1, \mathbf{v}_2, \ldots, \mathbf{v}_n\}$ a basis for V. Define $L: R^n \to V$ as follows: If

$$\mathbf{w} = \begin{bmatrix} a_1 \\ a_2 \\ \vdots \\ a_n \end{bmatrix}$$

is a vector in R^n, let

$$L(\mathbf{w}) = a_1\mathbf{v}_1 + a_2\mathbf{v}_2 + \cdots + a_n\mathbf{v}_n.$$

Show that:

(a) L is a linear transformation.

(b) L is one-to-one.

(c) L is onto.

22. Let V be an n-dimensional vector space. The vector space of all linear transformations from V into R^1 is called the **dual space** and is denoted by V^*. Prove that $\dim V = \dim V^*$. What does this imply?

Chapter 5

Determinants

5.1 DEFINITION

In Exercise 29 of Section 1.2, we defined the trace of a square ($n \times n$) matrix $A = \left[a_{ij}\right]$ by $\mathrm{Tr}(A) = \sum_{i=1}^{n} a_{ii}$. Another very important number associated with a square matrix A is the determinant of A, which we now define. Determinants first arose in the solution of linear systems. Although the methods given in Chapter 1 for solving such systems are more efficient than those involving determinants, determinants will be vital for our further study of a linear transformation $L: V \to V$. First, we deal briefly with permutations, which are used in our definition of determinant. Throughout this chapter, when we use the term *matrix* we mean *square matrix*.

Definition 5.1

Let $S = \{1, 2, \ldots, n\}$ be the set of integers from 1 to n, arranged in ascending order. A rearrangement $j_1 j_2 \cdots j_n$ of the elements of S is called a **permutation** of S. We can consider a permutation of S to be a one-to-one mapping of S onto itself. ▲

To illustrate the definition above, let $S = \{1, 2, 3, 4\}$. Then 4231 is a permutation of S. It corresponds to the function $f: S \to S$ defined by

$$f(1) = 4$$
$$f(2) = 2$$
$$f(3) = 3$$
$$f(4) = 1.$$

We can put any one of the n elements of S in first position, any one of the remaining $n-1$ elements in second position, any one of the remaining $n-2$ elements in third position, and so on until the nth position can only be filled by the last remaining element. Thus there are $n(n-1)(n-2)\cdots 2 \cdot 1 = n!$ (n factorial) permutations of S; we denote the set of all permutations of S by S_n.

EXAMPLE 1

Let $S = \{1, 2, 3\}$. The set S_3 of all permutations of S consists of the $3! = 6$ permutations 123, 132, 213, 231, 312, and 321. ■

A permutation $j_1 j_2 \ldots j_n$ is said to have an **inversion** if a larger integer, j_r, precedes a smaller one, j_s. A permutation is called **even** if the total number of inversions in it is even, or **odd** if the total number of inversions in it is odd. If $n \geq 2$, there are $n!/2$ even and $n!/2$ odd permutations in S_n.

EXAMPLE 2

S_1 has only $1! = 1$ permutation: 1, which is even because there are no inversions. ■

EXAMPLE 3

S_2 has $2! = 2$ permutations: 12, which is even (no inversions), and 21, which is odd (one inversion). ■

EXAMPLE 4

In the permutation 4312 in S_4, 4 precedes 3, 4 precedes 1, 4 precedes 2, 3 precedes 1, and 3 precedes 2. Thus the total number of inversions in this permutation is 5, and 4312 is odd. ■

EXAMPLE 5

S_3 has $3! = 3 \cdot 2 \cdot 1 = 6$ permutations: 123, 231, and 312, which are even, and 132, 213, and 321, which are odd. ■

Definition 5.2

Let $A = \begin{bmatrix} a_{ij} \end{bmatrix}$ be an $n \times n$ matrix. The determinant function, denoted by **det**, is defined by

$$\det(A) = \sum (\pm) a_{1j_1} a_{2j_2} \cdots a_{nj_n},$$

where the summation is over all permutations $j_1 j_2 \cdots j_n$ of the set $S = \{1, 2, \ldots, n\}$. The sign is taken as $+$ or $-$ according to whether the permutation $j_1 j_2 \cdots j_n$ is even or odd. ▲

In each term $(\pm) a_{1j_1} a_{2j_2} \cdots a_{nj_n}$ of $\det(A)$, the row subscripts are in natural order and the column subscripts are in the order $j_1 j_2 \cdots j_n$. Thus each term in $\det(A)$, with its appropriate sign, is a product of n entries of A, with exactly one entry from each row and exactly one entry from each column. Since we sum over all permutations of S, $\det(A)$ has $n!$ terms in the sum.

Another notation for $\det(A)$ is $|A|$. We shall use both $\det(A)$ and $|A|$.

EXAMPLE 6

If $A = \begin{bmatrix} a_{11} \end{bmatrix}$ is a 1×1 matrix, then $\det(A) = a_{11}$. ■

EXAMPLE 7

If

$$A = \begin{bmatrix} a_{11} & a_{12} \\ a_{21} & a_{22} \end{bmatrix},$$

then to obtain $\det(A)$ we write down the terms $a_{1-}a_{2-}$ and replace the dashes with all possible elements of S_2: The subscripts become 12 and 21. Now 12 is an even permutation and 21 is an odd permutation. Thus

$$\det(A) = a_{11}a_{22} - a_{12}a_{21}.$$

Hence we see that $\det(A)$ can be obtained by forming the product of the entries on the line from left to right and subtracting from this number the product of the entries on the line from right to left.

$$\begin{matrix} a_{11} & & a_{12} \\ a_{21} & & a_{22} \end{matrix}.$$

Thus, if $A = \begin{bmatrix} 2 & -3 \\ 4 & 5 \end{bmatrix}$, then $|A| = (2)(5) - (-3)(4) = 22$. ■

EXAMPLE 8

If

$$A = \begin{bmatrix} a_{11} & a_{12} & a_{13} \\ a_{21} & a_{22} & a_{23} \\ a_{31} & a_{32} & a_{33} \end{bmatrix},$$

then to compute $\det(A)$ we write down the six terms $a_{1-}a_{2-}a_{3-}$, $a_{1-}a_{2-}a_{3-}$, $a_{1-}a_{2-}a_{3-}$, $a_{1-}a_{2-}a_{3-}$, $a_{1-}a_{2-}a_{3-}$, $a_{1-}a_{2-}a_{3-}$. All the elements of S_3 are used to replace the dashes, and if we prefix each term by $+$ or $-$ according to whether the permutation is even or odd, we find that (verify)

$$\begin{aligned} \det(A) = a_{11}a_{22}a_{33} + a_{12}a_{23}a_{31} + a_{13}a_{21}a_{32} - a_{11}a_{23}a_{32} \\ - a_{12}a_{21}a_{33} - a_{13}a_{22}a_{31}. \end{aligned} \tag{1}$$

We can also obtain $|A|$ as follows. Repeat the first and second columns of A, as shown below. Form the sum of the products of the entries on the lines from left to right, and subtract from this number the products of the entries on the lines from right to left (verify):

$$\begin{matrix} a_{11} & a_{12} & a_{13} & a_{11} & a_{12} \\ a_{21} & a_{22} & a_{23} & a_{21} & a_{22} \\ a_{31} & a_{32} & a_{33} & a_{31} & a_{32} \end{matrix}.$$

■

EXAMPLE 9

Let

$$A = \begin{bmatrix} 1 & 2 & 3 \\ 2 & 1 & 3 \\ 3 & 1 & 2 \end{bmatrix}.$$

Evaluate $|A|$.

Solution Substituting in (1), we find that

$$\begin{vmatrix} 1 & 2 & 3 \\ 2 & 1 & 3 \\ 3 & 1 & 2 \end{vmatrix} = (1)(1)(2) + (2)(3)(3) + (3)(2)(1) - (1)(3)(1) \\ - (2)(2)(2) - (3)(1)(3) = 6.$$

We could obtain the same result by using the easy method illustrated above, as follows:

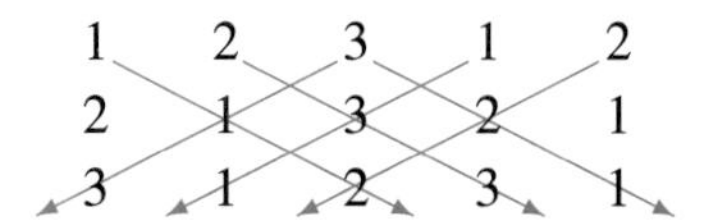

$$|A| = (1)(1)(2) + (2)(3)(3) + (3)(2)(1) - (3)(1)(3) - (1)(3)(1) \\ - (2)(2)(2) = 6.$$

■

It may already have struck the reader that Definition 5.2 is an extremely tedious way of computing determinants for a sizable value of n. In fact, $10! = 3.6288 \times 10^6$ and $20! = 2.4329 \times 10^{18}$, each an enormous number. In Section 5.2 we develop properties of determinants that will greatly reduce the computational effort.

Permutations are studied at some depth in abstract algebra courses and in courses dealing with group theory. As we just noted, we shall develop methods for evaluating determinants other than those involving permutations. However, we do require the following important property of permutations. If we interchange two numbers in the permutation $j_1 j_2 \cdots j_n$, then the number of inversions is either increased or decreased by an odd number.

A proof of this fact can be given by first noting that if two adjacent numbers in the permutation $j_1 j_2 \cdots j_n$ are interchanged, then the number of inversions is either increased or decreased by 1. Thus consider the permutations $j_1 j_2 \cdots j_e j_f \cdots j_n$ and $j_1 j_2 \cdots j_f j_e \cdots j_n$. If $j_e j_f$ is an inversion, then $j_f j_e$ is not an inversion, and the second permutation has one fewer inversion than the first one; if $j_e j_f$ is not an inversion, then $j_f j_e$ is, and so the second permutation has one more inversion than the first. Now an interchange of any two numbers in a permutation $j_1 j_2 \cdots j_n$ can always be achieved by an odd number of successive interchanges of adjacent numbers. Thus, if we wish to interchange j_c and j_k $(c < k)$ and there are s numbers between j_c and j_k, we move j_c to the right, by interchanging adjacent numbers, until j_c follows j_k. This requires $s + 1$ steps. Next, we move j_k to the left, by interchanging adjacent numbers until it is where j_c was. This requires s steps. Thus the total number of adjacent interchanges required is $(s + 1) + s = 2s + 1$, which is always odd. Since each adjacent interchange changes the number of inversions by 1 or -1, and since a sum of an odd number of numbers each of which is 1 or -1 is always odd, we conclude that the number of inversions is changed by an odd number. Thus the number of inversions in 54132 is 8 and the number of inversions in 52134 (obtained by interchanging 2 and 4) is 5.

5.1 Exercises

1. Find the number of inversions in each of the following permutations of $S = \{1, 2, 3, 4, 5\}$.

(a) 52134. (b) 45213. (c) 42135.

2. Find the number of inversions in each of the following permutations of $S = \{1, 2, 3, 4, 5\}$.

(a) 13542. (b) 35241. (c) 12345.

3. Determine whether each of the following permutations of $S = \{1, 2, 3, 4\}$ is even or odd.

(a) 4213. (b) 1243. (c) 1234.

4. Determine whether each of the following permutations of $S = \{1, 2, 3, 4\}$ is even or odd.

(a) 3214. (b) 1423. (c) 2143.

5. Determine the sign associated with each of the following permutations of the column indices of a 5×5 matrix.

(a) 25431. (b) 31245. (c) 21345.

6. Determine the sign associated with each of the following permutations of the column indices of a 5×5 matrix.

(a) 52341. (b) 34125. (c) 14523.

7. (a) Find the number of inversions in the permutation 436215.

(b) Verify that the number of inversions in the permutation 416235, obtained from that in part (a) by interchanging two numbers, differs from the answer in part (a) by an odd number.

8. Evaluate:

(a) $\begin{vmatrix} 2 & -1 \\ 3 & 2 \end{vmatrix}$. (b) $\begin{vmatrix} 2 & 1 \\ 4 & 3 \end{vmatrix}$.

9. Evaluate:

(a) $\begin{vmatrix} 1 & 2 \\ 2 & 4 \end{vmatrix}$. (b) $\begin{vmatrix} 3 & 1 \\ -3 & -1 \end{vmatrix}$.

10. Let $A = \begin{bmatrix} a_{ij} \end{bmatrix}$ be a 4×4 matrix. Develop the general expression for $\det(A)$.

11. Evaluate:

(a) $\det\left(\begin{bmatrix} 2 & 1 & 3 \\ 3 & 2 & 1 \\ 0 & 1 & 2 \end{bmatrix}\right)$. (b) $\begin{vmatrix} 2 & 1 & 3 \\ -3 & 2 & 1 \\ -1 & 3 & 4 \end{vmatrix}$.

(c) $\det\left(\begin{bmatrix} 0 & 0 & 0 & 3 \\ 0 & 0 & 4 & 0 \\ 0 & 2 & 0 & 0 \\ 6 & 0 & 0 & 0 \end{bmatrix}\right)$.

12. Evaluate:

(a) $\det\left(\begin{bmatrix} 2 & 0 & 0 \\ 0 & -3 & 0 \\ 0 & 0 & 4 \end{bmatrix}\right)$.

(b) $\det\left(\begin{bmatrix} 2 & 4 & 5 \\ 0 & -6 & 2 \\ 0 & 0 & 3 \end{bmatrix}\right)$.

(c) $\begin{vmatrix} 0 & 0 & 2 & 0 \\ 0 & 3 & 0 & 0 \\ 6 & 0 & 0 & 0 \\ 0 & 0 & 0 & 5 \end{vmatrix}$.

13. Evaluate:

(a) $\det\left(\begin{bmatrix} t-1 & 2 \\ 3 & t-2 \end{bmatrix}\right)$.

(b) $\begin{vmatrix} t-1 & -1 & -2 \\ 0 & t & 2 \\ 0 & 0 & t-3 \end{vmatrix}$.

14. Evaluate:

(a) $\begin{vmatrix} t & 4 \\ 5 & t-8 \end{vmatrix}$.

(b) $\det\left(\begin{bmatrix} t-1 & 0 & 1 \\ -2 & t & -1 \\ 0 & 0 & t+1 \end{bmatrix}\right)$.

15. For each of the matrices in Exercise 13, find values of t for which the determinant is 0.

16. For each of the matrices in Exercise 14, find values of t for which the determinant is 0.

5.2 PROPERTIES OF DETERMINANTS

In this section we examine properties of determinants that simplify their computation.

Theorem 5.1

If A is a matrix, then $\det(A) = \det(A^T)$.

Proof

Let $A = [a_{ij}]$ and $A^T = [b_{ij}]$, where $b_{ij} = a_{ji}$. We have

$$\det(A^T) = \sum(\pm) b_{1j_1} b_{2j_2} \cdots b_{nj_n} = \sum(\pm) a_{j_1 1} a_{j_2 2} \cdots a_{j_n n}.$$

We can then write $b_{1j_1} b_{2j_2} \cdots b_{nj_n} = a_{j_1 1} a_{j_2 2} \cdots a_{j_n n} = a_{1k_1} a_{2k_2} \cdots a_{nk_n}$, which is a term of $\det(A)$. Thus the terms in $\det(A^T)$ and $\det(A)$ are identical. We must now check that the signs of corresponding terms are also identical. It can be shown, by the properties of permutations discussed in an abstract algebra course,† that the number of inversions in the permutation $k_1 k_2 \ldots k_n$, which determines the sign associated with the term $a_{1k_1} a_{2k_2} \cdots a_{nk_n}$, is the same as the number of inversions in the permutation $j_1 j_2 \ldots j_n$, which determines the sign associated with the term $b_{1j_1} b_{2j_2} \cdots b_{nj_n}$. As an example,

$$b_{13} b_{24} b_{35} b_{41} b_{52} = a_{31} a_{42} a_{53} a_{14} a_{25} = a_{14} a_{25} a_{31} a_{42} a_{53};$$

the number of inversions in the permutation 45123 is 6 and the number of inversions in the permutation 34512 is also 6. Since the signs of corresponding terms are identical, we conclude that $\det(A^T) = \det(A)$. ●

EXAMPLE 1

Let A be the matrix in Example 9 of Section 5.1. Then

$$A^T = \begin{bmatrix} 1 & 2 & 3 \\ 2 & 1 & 1 \\ 3 & 3 & 2 \end{bmatrix}.$$

Substituting in (1) of Section 5.1 (or using the method of lines given in Example 8 of Section 5.1), we find that

$$\begin{aligned} |A^T| &= (1)(1)(2) + (2)(1)(3) + (3)(2)(3) \\ &\quad - (1)(1)(3) - (2)(2)(2) - (3)(1)(3) = 6 = |A|. \end{aligned}$$

■

Theorem 5.1 will enable us to replace "row" by "column" in many of the additional properties of determinants; we see how to do this in the following theorem.

†See J. Fraleigh, *A First Course in Abstract Algebra*, 6th ed., Reading, Mass.: Addison-Wesley Publishing Company, Inc., 1999; and J. Gallian, *Contemporary Abstract Algebra*, 4th ed., Lexington, Mass., D.C. Heath and Company, 1998.

Theorem 5.2

If matrix B results from matrix A by interchanging two rows (columns) of A, then $\det(B) = -\det(A)$.

Proof

Suppose that B arises from A by interchanging rows r and s of A, say $r < s$. Then we have $b_{rj} = a_{sj}$, $b_{sj} = a_{rj}$, and $b_{ij} = a_{ij}$ for $i \neq r, i \neq s$. Now

$$\begin{aligned}\det(B) &= \sum(\pm)b_{1j_1}b_{2j_2}\cdots b_{rj_r}\cdots b_{sj_s}\cdots b_{nj_n}\\ &= \sum(\pm)a_{1j_1}a_{2j_2}\cdots a_{sj_r}\cdots a_{rj_s}\cdots a_{nj_n}\\ &= \sum(\pm)a_{1j_1}a_{2j_2}\cdots a_{rj_s}\cdots a_{sj_r}\cdots a_{nj_n}.\end{aligned}$$

The permutation $j_1 j_2 \dots j_s \dots j_r \dots j_n$ results from the permutation $j_1 j_2 \dots j_r \dots j_s \dots j_n$ by an interchange of two numbers, and the number of inversions in the former differs by an odd number from the number of inversions in the latter. This means that the sign of each term in $\det(B)$ is the negative of the sign of the corresponding term in $\det(A)$. Hence $\det(B) = -\det(A)$.

Now let B arise from A by interchanging two columns of A. Then B^T arises from A^T by interchanging two rows of A^T. So $\det(B^T) = -\det(A^T)$, but $\det(B^T) = \det(B)$ and $\det(A^T) = \det(A)$. Hence $\det(B) = -\det(A)$. ●

In the results to follow, proofs will be given only for the rows of A; the proofs for the corresponding column cases proceed as at the end of the proof of Theorem 5.2.

EXAMPLE 2

We have $\begin{vmatrix} 2 & -1 \\ 3 & 2 \end{vmatrix} = -\begin{vmatrix} 3 & 2 \\ 2 & -1 \end{vmatrix} = \begin{vmatrix} 2 & 3 \\ -1 & 2 \end{vmatrix} = 7.$ ■

Theorem 5.3

If two rows (columns) of A are equal, then $\det(A) = 0$.

Proof

Suppose that rows r and s of A are equal. Interchange rows r and s of A to obtain a matrix B. Then $\det(B) = -\det(A)$. On the other hand, $B = A$, so $\det(B) = \det(A)$. Thus $\det(A) = -\det(A)$, and so $\det(A) = 0$. ●

EXAMPLE 3

We have $\begin{vmatrix} 1 & 2 & 3 \\ -1 & 0 & 7 \\ 1 & 2 & 3 \end{vmatrix} = 0$ (verify by the use of Definition 5.2). ■

Theorem 5.4

If a row (column) of A consists entirely of zeros, then $\det(A) = 0$.

Proof

Let the ith row of A consist entirely of zeros. Since each term in Definition 5.2 for the determinant of A contains a factor from the ith row, each term in $\det(A)$ is zero. Hence $\det(A) = 0$. ●

EXAMPLE 4

We have $\begin{vmatrix} 1 & 2 & 3 \\ 4 & 5 & 6 \\ 0 & 0 & 0 \end{vmatrix} = 0$ (verify by the use of Definition 5.2). ■

Theorem 5.5

If B is obtained from A by multiplying a row (column) of A by a real number c, then $\det(B) = c\det(A)$.

Proof

Suppose that the rth row of $A = \left[a_{ij}\right]$ is multiplied by c to obtain $B = \left[b_{ij}\right]$. Then $b_{ij} = a_{ij}$ if $i \neq r$ and $b_{rj} = ca_{rj}$. Using Definition 5.2, we obtain $\det(B)$ as

$$\begin{aligned}\det(B) &= \sum(\pm)b_{1j_1}b_{2j_2}\cdots b_{rj_r}\cdots b_{nj_n} \\ &= \sum(\pm)a_{1j_1}a_{2j_2}\cdots(ca_{rj_r})\cdots a_{nj_n} \\ &= c\left(\sum(\pm)a_{1j_1}a_{2j_2}\cdots a_{rj_r}\cdots a_{nj_n}\right) = c\det(A).\end{aligned}$$ ●

EXAMPLE 5

We have $\begin{vmatrix} 2 & 6 \\ 1 & 12 \end{vmatrix} = 2\begin{vmatrix} 1 & 3 \\ 1 & 12 \end{vmatrix} = (2)(3)\begin{vmatrix} 1 & 1 \\ 1 & 4 \end{vmatrix} = 6(4 - 1) = 18.$ ■

We can use Theorem 5.5 to simplify the computation of $\det(A)$ by factoring out common factors from rows and columns of A.

EXAMPLE 6

We have

$$\begin{vmatrix} 1 & 2 & 3 \\ 1 & 5 & 3 \\ 2 & 8 & 6 \end{vmatrix} = 2\begin{vmatrix} 1 & 2 & 3 \\ 1 & 5 & 3 \\ 1 & 4 & 3 \end{vmatrix} = (2)(3)\begin{vmatrix} 1 & 2 & 1 \\ 1 & 5 & 1 \\ 1 & 4 & 1 \end{vmatrix} = (2)(3)(0) = 0.$$

Here we first factored out 2 from the third row and 3 from the third column, and then used Theorem 5.3, since the first and third columns are equal. ■

Theorem 5.6

If $B = \left[b_{ij}\right]$ is obtained from $A = \left[a_{ij}\right]$ by adding to each element of the rth *row (column) of A, c times the corresponding element of the* sth *row (column), $r \neq s$, of A, then* $\det(B) = \det(A)$.

Proof

We prove the theorem for rows. We have $b_{ij} = a_{ij}$ for $i \neq r$, and $b_{rj} = a_{rj} + ca_{sj}$, $r \neq s$, say $r < s$. Then

$$\begin{aligned}\det(B) &= \sum(\pm)b_{1j_1}b_{2j_2}\cdots b_{rj_r}\cdots b_{nj_n}\\ &= \sum(\pm)a_{1j_1}a_{2j_2}\cdots(a_{rj_r} + ca_{sj_r})\cdots a_{sj_s}\cdots a_{nj_n}\\ &= \sum(\pm)a_{1j_1}a_{2j_2}\cdots a_{rj_r}\cdots a_{sj_s}\cdots a_{nj_n}\\ &\quad+\sum(\pm)a_{1j_1}a_{2j_2}\cdots(ca_{sj_r})\cdots a_{sj_s}\cdots a_{nj_n}.\end{aligned}$$

Now the first term in this last expression is $\det(A)$ while second is

$$c\left[\sum(\pm)a_{1j_1}a_{2j_2}\cdots a_{sj_r}\cdots a_{sj_s}\cdots a_{nj_n}\right].$$

Note that

$$\begin{aligned}&\sum(\pm)a_{1j_1}a_{2j_2}\cdots a_{sj_r}\cdots a_{sj_s}\cdots a_{nj_n}\\ &= \begin{vmatrix} a_{11} & a_{12} & \cdots & a_{1n} \\ a_{21} & a_{22} & \cdots & a_{2n} \\ \vdots & \vdots & & \vdots \\ a_{s1} & a_{s2} & \cdots & a_{sn} \\ \vdots & \vdots & & \vdots \\ a_{s1} & a_{s2} & \cdots & a_{sn} \\ \vdots & \vdots & & \vdots \\ a_{n1} & a_{n2} & \cdots & a_{nn} \end{vmatrix} \begin{matrix} \\ \\ \\ \leftarrow r\text{th row} \\ \\ \leftarrow s\text{th row} \\ \\ \\ \end{matrix}\\ &= 0,\end{aligned}$$

because this matrix has two equal rows. Hence $\det(B) = \det(A) + 0 = \det(A)$. ●

EXAMPLE 7

We have

$$\begin{vmatrix} 1 & 2 & 3 \\ 2 & -1 & 3 \\ 1 & 0 & 1 \end{vmatrix} = \begin{vmatrix} 5 & 0 & 9 \\ 2 & -1 & 3 \\ 1 & 0 & 1 \end{vmatrix},$$

obtained by adding twice the second row to the first row. By applying the definition of determinant to the second determinant, both are seen to have the value 4. ■

Theorem 5.7

If a matrix $A = \left[a_{ij}\right]$ is upper (lower) triangular, then $\det(A) = a_{11}a_{22}\cdots a_{nn}$; that is, the determinant of a triangular matrix is the product of the elements on the main diagonal.

Proof

Let $A = \left[a_{ij}\right]$ be upper triangular (that is, $a_{ij} = 0$ for $i > j$). Then a term $a_{1j_1}a_{2j_2}\cdots a_{nj_n}$ in the expression for $\det(A)$ can be nonzero only for $1 \leq j_1$, $2 \leq j_2$, ..., $n \leq j_n$. Now $j_1 j_2 \ldots j_n$ must be a permutation, or rearrangement, of $\{1, 2, \ldots, n\}$. Hence we must have $j_1 = 1$, $j_2 = 2$, ..., $j_n = n$. Thus the only term of $\det(A)$ that can be nonzero is the product of the elements on the main diagonal of A. Hence $\det(A) = a_{11}a_{22}\cdots a_{nn}$.

We leave the proof of the lower triangular case to the reader. ●

We now introduce the following notation for elementary row and elementary column operations on matrices and determinants.

- Interchange rows (columns) i and j:

$$\mathbf{r}_i \leftrightarrow \mathbf{r}_j \quad (\mathbf{c}_i \leftrightarrow \mathbf{c}_j).$$

- Replace row (column) i by k times row (column) i:

$$k\mathbf{r}_i \to \mathbf{r}_i \quad (k\mathbf{c}_i \to \mathbf{c}_i).$$

- Replace row (column) j by k times row (column) i + row (column) j:

$$k\mathbf{r}_i + \mathbf{r}_j \to \mathbf{r}_j \quad (k\mathbf{c}_i + \mathbf{c}_j \to \mathbf{c}_j).$$

We can now interpret Theorems 5.2, 5.5, and 5.6 in terms of this notation as follows.

$$|A|_{\mathbf{r}_i \leftrightarrow \mathbf{r}_j} = -|A|,$$

$$|A|_{k\mathbf{r}_i \to \mathbf{r}_i} = \frac{1}{k}|A|,$$

$$|A|_{k\mathbf{r}_i + \mathbf{r}_j \to \mathbf{r}_j} = |A|.$$

Similarly for column operations.

Theorems 5.2, 5.5, and 5.6 are very useful in evaluating determinants. What we do is transform A by means of our elementary row or column operations to a triangular matrix. Of course, we must keep track of how the determinant of the resulting matrices changes as we perform the elementary row or column operations.

EXAMPLE 8

We have

$$\begin{vmatrix} 4 & 3 & 2 \\ 3 & -2 & 5 \\ 2 & 4 & 6 \end{vmatrix}_{\frac{1}{2}\mathbf{r}_3\to\mathbf{r}_3} = 2\begin{vmatrix} 4 & 3 & 2 \\ 3 & -2 & 5 \\ 1 & 2 & 3 \end{vmatrix}_{\mathbf{r}_1\leftrightarrow\mathbf{r}_3} = -2\begin{vmatrix} 1 & 2 & 3 \\ 3 & -2 & 5 \\ 4 & 3 & 2 \end{vmatrix}_{\mathbf{r}_2-3\mathbf{r}_1\to\mathbf{r}_2}$$

$$= -2\begin{vmatrix} 1 & 2 & 3 \\ 0 & -8 & -4 \\ 4 & 3 & 2 \end{vmatrix}_{\mathbf{r}_3-4\mathbf{r}_1\to\mathbf{r}_3} = -2\begin{vmatrix} 1 & 2 & 3 \\ 0 & -8 & -4 \\ 0 & -5 & -10 \end{vmatrix}_{\frac{1}{4}\mathbf{r}_2\to\mathbf{r}_2}$$

$$= (-2)(4)\begin{vmatrix} 1 & 2 & 3 \\ 0 & -2 & -1 \\ 0 & -5 & -10 \end{vmatrix}_{\frac{1}{5}\mathbf{r}_3\to\mathbf{r}_3} = (-2)(4)(5)\begin{vmatrix} 1 & 2 & 3 \\ 0 & -2 & -1 \\ 0 & -1 & -2 \end{vmatrix}_{\mathbf{r}_3-\frac{1}{2}\mathbf{r}_2\to\mathbf{r}_3}$$

$$= (-2)(4)(5)\begin{vmatrix} 1 & 2 & 3 \\ 0 & -2 & -1 \\ 0 & 0 & -\frac{3}{2} \end{vmatrix} = (-2)(4)(5)(1)(-2)\left(-\tfrac{3}{2}\right) = -120.$$

■

Remark The method used to compute a determinant in Example 8 will be referred to as the computation **via reduction to triangular form**.

We can now compute the determinant of the identity matrix I_n: $\det(I_n) = 1$. We can also compute the determinants of the elementary matrices discussed in Section 1.6, as follows.

Let E_1 be an elementary matrix of type I; that is, E_1 is obtained from I_n by interchanging, say, the ith and jth rows of I_n. By Theorem 5.2 we have that $\det(E_1) = -\det(I_n) = -1$. Now let E_2 be an elementary matrix of type II; that is, E_2 is obtained from I_n by multiplying, say, the ith row of I_n by $c \neq 0$. By Theorem 5.5 we have that $\det(E_2) = c\det(I_n) = c$. Finally, let E_3 be an elementary matrix of type III; that is, E_3 is obtained from I_n by adding c times the sth row of I_n to the rth row of I_n $(r \neq s)$. By Theorem 5.6 we have that $\det(E_3) = \det(I_n) = 1$. Thus the determinant of an elementary matrix is never zero.

Next, we prove that the determinant of a product of two matrices is the product of their determinants and that A is nonsingular if and only if $\det(A) \neq 0$.

Lemma 5.1

If E is an elementary matrix, then $\det(EA) = \det(E)\det(A)$, and $\det(AE) = \det(A)\det(E)$.

Proof

If E is an elementary matrix of type I, then EA is obtained from A by interchanging two rows of A, so $\det(EA) = -\det(A)$. Also $\det(E) = -1$. Thus $\det(EA) = \det(E)\det(A)$.

If E is an elementary matrix of type II, then EA is obtained from A by multiplying a given row of A by $c \neq 0$. Then $\det(EA) = c\det(A)$ and $\det(E) = c$, so $\det(EA) = \det(E)\det(A)$.

Finally, if E is an elementary matrix of type III, then EA is obtained from A by adding a multiple of a row of A to a different row of A. Then $\det(EA) = \det(A)$ and $\det(E) = 1$, so $\det(EA) = \det(E)\det(A)$.

Thus, in all cases, $\det(EA) = \det(E)\det(A)$. By a similar proof, we can show that $\det(AE) = \det(A)\det(E)$. ●

It also follows from Lemma 5.1 that if $B = E_r E_{r-1} \cdots E_2 E_1 A$, then

$$\begin{aligned}\det(B) &= \det(E_r(E_{r-1} \cdots E_2 E_1 A)) \\ &= \det(E_r)\det(E_{r-1} E_{r-2} \cdots E_2 E_1 A) \\ &\;\;\vdots \\ &= \det(E_r)\det(E_{r-1}) \cdots \det(E_2)\det(E_1)\det(A).\end{aligned}$$

Theorem 5.8

If A is an $n \times n$ matrix, then A is nonsingular if and only if $\det(A) \neq 0$.

Proof

If A is nonsingular, then A is a product of elementary matrices (Theorem 1.16). Thus let $A = E_1 E_2 \cdots E_k$. Then

$$\det(A) = \det(E_1 E_2 \cdots E_k) = \det(E_1)\det(E_2) \cdots \det(E_k) \neq 0.$$

If A is singular, then A is row equivalent to a matrix B that has a row of zeros (Theorem 1.18). Then $A = E_1 E_2 \cdots E_r B$, where $E_1, E_2, \ldots, E_r$ are elementary matrices. It then follows by the observation following Lemma 5.1 that

$$\det(A) = \det(E_1 E_2 \cdots E_r B) = \det(E_1)\det(E_2) \cdots \det(E_r)\det(B) = 0,$$

since $\det(B) = 0$. ●

Corollary 5.1

If A is an $n \times n$ matrix, then rank $A = n$ *if and only if* $\det(A) \neq 0$.

Proof

By Corollary 2.7 we know that A is nonsingular if and only if rank $A = n$. ●

Corollary 5.2

If A is an $n \times n$ matrix, then $\mathbf{Ax} = \mathbf{0}$ has a nontrivial solution if and only if $\det(A) = 0$.

Proof

Exercise. ●

Theorem 5.9

If A and B are $n \times n$ matrices, then $\det(AB) = \det(A)\det(B)$.

Proof

If A is nonsingular, then A is row equivalent to I_n. Thus $A = E_k E_{k-1} \cdots E_2 E_1 I_n = E_k E_{k-1} \cdots E_2 E_1$, where $E_1, E_2, \ldots, E_k$ are elementary matrices. Then

$$\det(A) = \det(E_k E_{k-1} \cdots E_2 E_1) = \det(E_k)\det(E_{k-1}) \cdots \det(E_2)\det(E_1).$$

Now

$$\begin{aligned}\det(AB) &= \det(E_k E_{k-1} \cdots E_2 E_1 B)\\ &= \det(E_k)\det(E_{k-1}) \cdots \det(E_2)\det(E_1)\det(B)\\ &= \det(A)\det(B).\end{aligned}$$

If A is singular, then $\det(A) = 0$ by Theorem 5.8. Moreover, if A is singular, then A is row equivalent to a matrix C that has a row consisting entirely of zeros (Theorem 1.18). Thus $C = E_k E_{k-1} \cdots E_2 E_1 A$, so

$$CB = E_k E_{k-1} \cdots E_2 E_1 AB.$$

This means that AB is row equivalent to CB, and since CB has a row consisting entirely of zeros, it follows that AB is singular. Hence $\det(AB) = 0$, and in this case we also have $\det(AB) = \det(A)\det(B)$. ●

EXAMPLE 9

Let

$$A = \begin{bmatrix} 1 & 2 \\ 3 & 4 \end{bmatrix} \quad \text{and} \quad B = \begin{bmatrix} 2 & -1 \\ 1 & 2 \end{bmatrix}.$$

Then

$$|A| = -2 \quad \text{and} \quad |B| = 5.$$

On the other hand, $AB = \begin{bmatrix} 4 & 3 \\ 10 & 5 \end{bmatrix}$, and $|AB| = -10 = |A||B|$. ■

Corollary 5.3

If A is nonsingular, then $\det(A^{-1}) = \dfrac{1}{\det(A)}$.

Proof

Exercise. ●

Corollary 5.4

If A and B are similar matrices, then $\det(A) = \det(B)$.

Proof

Exercise. ●

The determinant of a sum of two $n \times n$ matrices A and B is, in general, not the sum of the determinants of A and B. The best result we can give along these lines is that if A, B, and C are $n \times n$ matrices all of whose entries are equal except for the kth row (column), and the kth row (column) of C is the sum of the kth rows (columns) of A and B, then $\det(C) = \det(A) + \det(B)$. We shall not prove this result, but will consider an example.

EXAMPLE 10

Let

$$A = \begin{bmatrix} 2 & 2 & 3 \\ 0 & 3 & 4 \\ 0 & 2 & 4 \end{bmatrix}, \quad B = \begin{bmatrix} 2 & 2 & 3 \\ 0 & 3 & 4 \\ 1 & -2 & -4 \end{bmatrix},$$

and

$$C = \begin{bmatrix} 2 & 2 & 3 \\ 0 & 3 & 4 \\ 1 & 0 & 0 \end{bmatrix}.$$

Then $|A| = 8$, $|B| = -9$, and $|C| = -1$, so $|C| = |A| + |B|$. ■

5.2 Exercises

1. Compute the following determinants via reduction to triangular form or by citing a particular theorem or corollary.

(a) $\begin{vmatrix} 3 & 0 \\ 2 & 1 \end{vmatrix}$.

(b) $\begin{vmatrix} 2 & 1 \\ 4 & 3 \end{vmatrix}$.

(c) $\begin{vmatrix} 4 & 0 & 0 \\ 0 & 2 & 0 \\ 0 & 0 & 3 \end{vmatrix}$.

(d) $\begin{vmatrix} 4 & 1 & 3 \\ 2 & 3 & 0 \\ 1 & 3 & 2 \end{vmatrix}$.

(e) $\begin{vmatrix} 4 & 2 & 2 & 0 \\ 2 & 0 & 0 & 0 \\ 3 & 0 & 0 & 1 \\ 0 & 0 & 1 & 0 \end{vmatrix}$.

(f) $\begin{vmatrix} 4 & 2 & 3 & -4 \\ 3 & -2 & 1 & 5 \\ -2 & 0 & 1 & -3 \\ 8 & -2 & 6 & 4 \end{vmatrix}$.

2. Compute the following determinants via reduction to triangular form or by citing a particular theorem or corollary.

(a) $\begin{vmatrix} 2 & -2 \\ 3 & -1 \end{vmatrix}$. (b) $\begin{vmatrix} 4 & 2 & 0 \\ 0 & -2 & 5 \\ 0 & 0 & 3 \end{vmatrix}$.

(c) $\begin{vmatrix} 3 & 4 & 2 \\ 2 & 5 & 0 \\ 3 & 0 & 0 \end{vmatrix}$. (d) $\begin{vmatrix} 4 & -3 & 5 \\ 5 & 2 & 0 \\ 2 & 0 & 4 \end{vmatrix}$.

(e) $\begin{vmatrix} 4 & 0 & 0 & 0 \\ -1 & 2 & 0 & 0 \\ 1 & 2 & -3 & 0 \\ 1 & 5 & 3 & 5 \end{vmatrix}$.

(f) $\begin{vmatrix} 2 & 0 & 1 & 4 \\ 3 & 2 & -4 & -2 \\ 2 & 3 & -1 & 0 \\ 11 & 8 & -4 & 6 \end{vmatrix}$.

3. If $\begin{vmatrix} a_1 & a_2 & a_3 \\ b_1 & b_2 & b_3 \\ c_1 & c_2 & c_3 \end{vmatrix} = 3$, find

$$\begin{vmatrix} a_1 + 2b_1 - 3c_1 & a_2 + 2b_2 - 3c_2 & a_3 + 2b_3 - 3c_3 \\ b_1 & b_2 & b_3 \\ c_1 & c_2 & c_3 \end{vmatrix}.$$

4. If $\begin{vmatrix} a_1 & a_2 & a_3 \\ b_1 & b_2 & b_3 \\ c_1 & c_2 & c_3 \end{vmatrix} = -2$, find

$$\begin{vmatrix} a_1 - \frac{1}{2}a_3 & a_2 & a_3 \\ b_1 - \frac{1}{2}b_3 & b_2 & b_3 \\ c_1 - \frac{1}{2}c_3 & c_2 & c_3 \end{vmatrix}.$$

5. If $\begin{vmatrix} a_1 & a_2 & a_3 \\ b_1 & b_2 & b_3 \\ c_1 & c_2 & c_3 \end{vmatrix} = 4$, find

$$\begin{vmatrix} a_1 & a_2 & 4a_3 - 2a_2 \\ b_1 & b_2 & 4b_3 - 2b_2 \\ \frac{1}{2}c_1 & \frac{1}{2}c_2 & 2c_3 - c_2 \end{vmatrix}.$$

6. Verify that $\det(AB) = \det(A)\det(B)$ for the following.

(a) $A = \begin{bmatrix} 1 & -2 & 3 \\ -2 & 3 & 1 \\ 0 & 1 & 0 \end{bmatrix}$, $B = \begin{bmatrix} 1 & 0 & 2 \\ 3 & -2 & 5 \\ 2 & 1 & 3 \end{bmatrix}$.

(b) $A = \begin{bmatrix} 2 & 3 & 6 \\ 0 & 3 & 2 \\ 0 & 0 & -4 \end{bmatrix}$, $B = \begin{bmatrix} 3 & 0 & 0 \\ 4 & 5 & 0 \\ 2 & 1 & -2 \end{bmatrix}$.

7. Evaluate:

(a) $\begin{vmatrix} -4 & 2 & 0 & 0 \\ 2 & 3 & 1 & 0 \\ 3 & 1 & 0 & 2 \\ 1 & 3 & 0 & 3 \end{vmatrix}$.

(b) $\begin{vmatrix} 2 & 0 & 0 & 0 \\ -5 & 3 & 0 & 0 \\ 3 & 2 & 4 & 0 \\ 4 & 2 & 1 & -5 \end{vmatrix}$.

(c) $\begin{vmatrix} t-1 & -1 & -2 \\ 0 & t-2 & 2 \\ 0 & 0 & t-3 \end{vmatrix}$.

(d) $\begin{vmatrix} t+1 & 4 \\ 2 & t-3 \end{vmatrix}$.

8. Is $\det(AB) = \det(BA)$? Justify your answer.

9. If $\det(AB) = 0$, is $\det(A) = 0$ or $\det(B) = 0$? Give reasons for your answer.

10. Show that if c is a scalar and A is $n \times n$, then $\det(cA) = c^n \det(A)$.

11. Show that if A is $n \times n$ with n odd and skew symmetric, then $\det(A) = 0$.

12. Show that if A is a matrix such that in each row and in each column one and only one element is $\neq 0$, then $\det(A) \neq 0$.

13. Prove Corollary 5.2.

14. Show that if $AB = I_n$, then $\det(A) \neq 0$ and $\det(B) \neq 0$.

15. (a) Show that if $A = A^{-1}$, then $\det(A) = \pm 1$.

(b) If $A^T = A^{-1}$, what is $\det(A)$?

16. Show that if A and B are square matrices, then $\begin{bmatrix} A & O \\ O & B \end{bmatrix} = |A|\,|B|$.

17. If A is a nonsingular matrix such that $A^2 = A$, what is $\det(A)$?

18. Prove Corollary 5.3.

19. Show that if A, B, and C are square matrices, then $\begin{bmatrix} A & O \\ C & B \end{bmatrix} = |A|\,|B|$.

20. Show that if A and B are both $n \times n$, then:

(a) $\det(A^T B^T) = \det(A)\det(B^T)$.

(b) $\det(A^T B^T) = \det(A^T)\det(B)$.

21. Verify the result in Exercise 16 for $A = \begin{bmatrix} 1 & 2 \\ 3 & 4 \end{bmatrix}$ and $B = \begin{bmatrix} 2 & 1 \\ -3 & 2 \end{bmatrix}$.

22. Use the properties of Section 5.2 to prove that

$$\begin{vmatrix} 1 & a & a^2 \\ 1 & b & b^2 \\ 1 & c & c^2 \end{vmatrix} = (b-a)(c-a)(c-b).$$

(*Hint*: Use factorization.) This determinant is called a **Vandermonde determinant**.

23. If $\det(A) = 2$, find $\det(A^5)$.

24. Use Theorem 5.8 to determine which of the following matrices are nonsingular.

(a) $\begin{bmatrix} 1 & 2 & 3 \\ 0 & 1 & 2 \\ 2 & -3 & 1 \end{bmatrix}$. (b) $\begin{bmatrix} 1 & 2 \\ 3 & 4 \end{bmatrix}$.

25. Use Theorem 5.8 to determine which of the following matrices are nonsingular.

(a) $\begin{bmatrix} 1 & 3 & 2 \\ 2 & 1 & 4 \\ 1 & -7 & 2 \end{bmatrix}$.

(b) $\begin{bmatrix} 1 & 2 & 0 & 5 \\ 3 & 4 & 1 & 7 \\ -2 & 5 & 2 & 0 \\ 0 & 1 & 2 & -7 \end{bmatrix}$.

26. Use Corollary 5.1 to find out whether rank $A = 3$ for the following matrices.

(a) $A = \begin{bmatrix} 1 & 2 & 3 \\ 2 & 1 & 0 \\ -3 & 1 & 2 \end{bmatrix}$.

(b) $A = \begin{bmatrix} 1 & 3 & -4 \\ -2 & 1 & 2 \\ -9 & 15 & 0 \end{bmatrix}$.

(c) $A = \begin{bmatrix} 1 & 0 & 1 \\ 1 & 1 & 0 \\ 2 & 1 & 0 \end{bmatrix}$.

27. Use Corollary 5.2 to find out whether the following homogeneous system has a nontrivial solution (do *not* solve).

$$\begin{aligned} x_1 - 2x_2 + x_3 &= 0 \\ 2x_1 + 3x_2 + x_3 &= 0 \\ 3x_1 + x_2 + 2x_3 &= 0. \end{aligned}$$

28. Repeat Exercise 27 for the following homogeneous system:

$$\begin{bmatrix} 1 & 2 & 0 & 1 \\ 0 & 1 & 2 & 3 \\ 0 & 0 & 1 & 2 \\ 0 & 1 & 2 & -1 \end{bmatrix} \begin{bmatrix} x_1 \\ x_2 \\ x_3 \\ x_4 \end{bmatrix} = \begin{bmatrix} 0 \\ 0 \\ 0 \\ 0 \end{bmatrix}.$$

29. Let $A = \begin{bmatrix} a_{ij} \end{bmatrix}$ be an upper triangular matrix. Prove that A is nonsingular if and only if $a_{ii} \neq 0$ for $i = 1, 2, \ldots, n$.

30. Let $A^2 = A$. Prove that either A is singular or $\det(A) = 1$.

31. Prove Corollary 5.4.

32. Let $AB = AC$. Prove that if $\det(A) \neq 0$, then $B = C$.

33. Determine if the software you are using has a command for computing the determinant of a matrix. If it does, verify the computations in Examples 8, 9, and 10. Experiment further by using the matrices in Exercises 1 and 2.

34. Assuming that your software has a command to compute the determinant of a matrix, read the accompanying software documentation to determine the method used. Is the description closest to that in Section 5.1, Example 8 in Section 5.2, or the material in Section 1.8?

35. **Warning:** Theorem 5.8 assumes that all calculations for $\det(A)$ are done using exact arithmetic. As noted previously, this is usually not the case in software. Hence, computationally the determinant may not be a valid text for nonsingularity. Perform the following experiment. Let $A = \begin{bmatrix} 1 & 2 & 3 \\ 4 & 5 & 6 \\ 7 & 8 & 9 \end{bmatrix}$. Show that $\det(A)$ is 0 either by hand or using your software. Next show by hand computation that $\det(B) = -3\epsilon$, where $B = \begin{bmatrix} 1 & 2 & 3 \\ 4 & 5 & 6 \\ 7 & 8 & 9+\epsilon \end{bmatrix}$. Hence, theoretically, for any $\epsilon \neq 0$ matrix B is nonsingular. Let your software compute $\det(B)$ for $\epsilon = \pm 10^{-k}$, $k = 5, 6, \ldots, 20$. Do the computational results match the theoretical result? If not, formulate a conjecture to explain why not.

5.3 COFACTOR EXPANSION

Thus far we have evaluated determinants by using Definition 5.2 and the properties established in Section 5.2. We now develop a method for evaluating the determinant of an $n \times n$ matrix which reduces the problem to the evaluation of determinants of matrices of order $n - 1$. We can then repeat the process for these $(n-1) \times (n-1)$ matrices until we get to 2×2 matrices.

Definition 5.3

Let $A = \left[a_{ij}\right]$ be an $n \times n$ matrix. Let M_{ij} be the $(n-1) \times (n-1)$ submatrix of A obtained by deleting the ith row and jth column of A. The determinant $\det(M_{ij})$ is called the **minor** of a_{ij}. ▲

Definition 5.4

Let $A = \left[a_{ij}\right]$ be an $n \times n$ matrix. The **cofactor** A_{ij} of a_{ij} is defined as $A_{ij} = (-1)^{i+j} \det(M_{ij})$. ▲

EXAMPLE 1

Let

$$A = \begin{bmatrix} 3 & -1 & 2 \\ 4 & 5 & 6 \\ 7 & 1 & 2 \end{bmatrix}.$$

Then

$$\det(M_{12}) = \begin{vmatrix} 4 & 6 \\ 7 & 2 \end{vmatrix} = 8 - 42 = -34, \quad \det(M_{23}) = \begin{vmatrix} 3 & -1 \\ 7 & 1 \end{vmatrix} = 3 + 7 = 10,$$

and

$$\det(M_{31}) = \begin{vmatrix} -1 & 2 \\ 5 & 6 \end{vmatrix} = -6 - 10 = -16.$$

Also,

$$A_{12} = (-1)^{1+2} \det(M_{12}) = (-1)(-34) = 34,$$
$$A_{23} = (-1)^{2+3} \det(M_{23}) = (-1)(10) = -10,$$

and

$$A_{31} = (-1)^{3+1} \det(M_{31}) = (1)(-16) = -16.$$ ■

If we think of the sign $(-1)^{i+j}$ as being located in position (i, j) of an $n \times n$ matrix, then the signs form a checkerboard pattern that has a $+$ in the $(1, 1)$ position.

The patterns for $n = 3$ and $n = 4$ are as follows:

$$\begin{bmatrix} + & - & + \\ - & + & - \\ + & - & + \end{bmatrix} \qquad \begin{bmatrix} + & - & + & - \\ - & + & - & + \\ + & - & + & - \\ - & + & - & + \end{bmatrix}$$

$n = 3$ $\qquad$ $n = 4$

Theorem 5.10

Let $A = [a_{ij}]$ be an $n \times n$ matrix. Then

$$\det(A) = a_{i1}A_{i1} + a_{i2}A_{i2} + \cdots + a_{in}A_{in}$$
$$[\textit{expansion of } \det(A) \textit{ along the } i\text{th } \textit{row}]$$

and

$$\det(A) = a_{1j}A_{1j} + a_{2j}A_{2j} + \cdots + a_{nj}A_{nj}$$
$$[\textit{expansion of } \det(A) \textit{ along the } j\text{th } \textit{column}].$$

Proof

The first formula follows from the second by Theorem 5.1, that is, from the fact that $\det(A^T) = \det(A)$. We omit the general proof and consider the 3×3 matrix $A = [a_{ij}]$. From (1) in Section 5.1,

$$\begin{aligned} \det(A) = &\ a_{11}a_{22}a_{33} + a_{12}a_{23}a_{31} + a_{13}a_{21}a_{32} \\ &- a_{11}a_{23}a_{32} - a_{12}a_{21}a_{33} - a_{13}a_{22}a_{31}. \end{aligned} \tag{1}$$

We can write this expression as

$$\begin{aligned} \det(A) = &\ a_{11}(a_{22}a_{33} - a_{23}a_{32}) + a_{12}(a_{23}a_{31} - a_{21}a_{33}) \\ &+ a_{13}(a_{21}a_{32} - a_{22}a_{31}). \end{aligned}$$

Now,

$$A_{11} = (-1)^{1+1}\begin{vmatrix} a_{22} & a_{23} \\ a_{32} & a_{33} \end{vmatrix} = (a_{22}a_{33} - a_{23}a_{32}),$$

$$A_{12} = (-1)^{1+2}\begin{vmatrix} a_{21} & a_{23} \\ a_{31} & a_{33} \end{vmatrix} = (a_{23}a_{31} - a_{21}a_{33}),$$

$$A_{13} = (-1)^{1+3}\begin{vmatrix} a_{21} & a_{22} \\ a_{31} & a_{32} \end{vmatrix} = (a_{21}a_{32} - a_{22}a_{31}).$$

Hence

$$\det(A) = a_{11}A_{11} + a_{12}A_{12} + a_{13}A_{13},$$

which is the expansion of $\det(A)$ along the first row.

If we now write (1) as

$$\det(A) = a_{13}(a_{21}a_{32} - a_{22}a_{31}) + a_{23}(a_{12}a_{31} - a_{11}a_{32}) + a_{33}(a_{11}a_{22} - a_{12}a_{21}),$$

we can verify that

$$\det(A) = a_{13}A_{13} + a_{23}A_{23} + a_{33}A_{33},$$

which is the expansion of $\det(A)$ along the third column. ●

EXAMPLE 2

To evaluate the determinant

$$\begin{vmatrix} 1 & 2 & -3 & 4 \\ -4 & 2 & 1 & 3 \\ 3 & 0 & 0 & -3 \\ 2 & 0 & -2 & 3 \end{vmatrix},$$

it is best to expand along either the second column or the third row because they each have two zeros. Obviously, the optimal course of action is to expand along the row or column that has the largest number of zeros, because in that case the cofactors A_{ij} of those a_{ij} which are zero do not have to be evaluated since $a_{ij}A_{ij} = (0)(A_{ij}) = 0$. Thus, expanding along the third row, we have

$$\begin{aligned}
\begin{vmatrix} 1 & 2 & -3 & 4 \\ -4 & 2 & 1 & 3 \\ 3 & 0 & 0 & -3 \\ 2 & 0 & -2 & 3 \end{vmatrix}
&= (-1)^{3+1}(3)\begin{vmatrix} 2 & -3 & 4 \\ 2 & 1 & 3 \\ 0 & -2 & 3 \end{vmatrix} + (-1)^{3+2}(0)\begin{vmatrix} 1 & -3 & 4 \\ -4 & 1 & 3 \\ 2 & -2 & 3 \end{vmatrix} \\
&\quad + (-1)^{3+3}(0)\begin{vmatrix} 1 & 2 & 4 \\ -4 & 2 & 3 \\ 2 & 0 & 3 \end{vmatrix} + (-1)^{3+4}(-3)\begin{vmatrix} 1 & 2 & -3 \\ -4 & 2 & 1 \\ 2 & 0 & -2 \end{vmatrix} \\
&= (+1)(3)(20) + 0 + 0 + (-1)(-3)(-4) = 48.
\end{aligned}$$

■

We can use the properties of Section 5.2 to introduce many zeros in a given row or column and then expand along that row or column. Consider the following example.

EXAMPLE 3

We have

$$\begin{vmatrix} 1 & 2 & -3 & 4 \\ -4 & 2 & 1 & 3 \\ 1 & 0 & 0 & -3 \\ 2 & 0 & -2 & 3 \end{vmatrix}_{\mathbf{c}_4+3\mathbf{c}_1\to\mathbf{c}_4} = \begin{vmatrix} 1 & 2 & -3 & 7 \\ -4 & 2 & 1 & -9 \\ 1 & 0 & 0 & 0 \\ 2 & 0 & -2 & 9 \end{vmatrix}$$

$$= (-1)^{3+1}(1) \begin{vmatrix} 2 & -3 & 7 \\ 2 & 1 & -9 \\ 0 & -2 & 9 \end{vmatrix}_{\mathbf{r}_1-\mathbf{r}_2\to\mathbf{r}_1}$$

$$= (-1)^4(1) \begin{vmatrix} 0 & -4 & 16 \\ 2 & 1 & -9 \\ 0 & -2 & 9 \end{vmatrix}$$

$$= (-1)^4(1)(8) = 8.$$

■

Application to Computing Areas

Consider the triangle with vertices (x_1, y_1), (x_2, y_2), and (x_3, y_3), which we show in Figure 5.1.

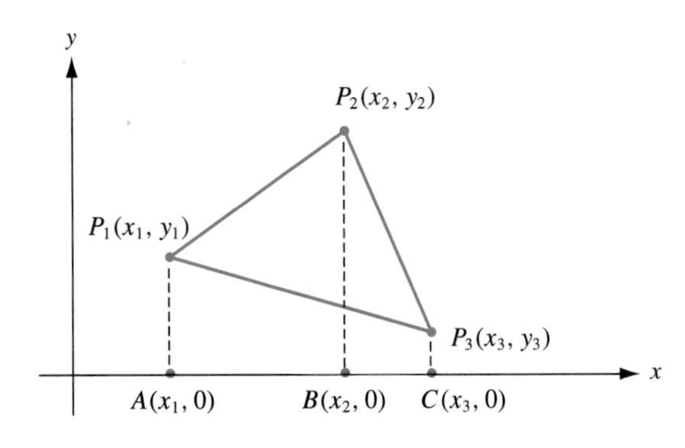

FIGURE 5.1

We may compute the area of this triangle as

$$\text{area of trapezoid } AP_1P_2B + \text{area of trapezoid } BP_2P_3C$$
$$- \text{ area of trapezoid } AP_1P_3C.$$

Now recall that the area of a trapezoid is $\frac{1}{2}$ the distance between the parallel sides of the trapezoid times the sum of the lengths of the parallel sides. Thus

area of triangle $P_1P_2P_3$

$$= \frac{1}{2}(x_2 - x_1)(y_1 + y_2) + \frac{1}{2}(x_3 - x_2)(y_2 + y_3) - \frac{1}{2}(x_3 - x_1)(y_1 + y_3)$$

$$= \frac{1}{2}x_2y_1 - \frac{1}{2}x_1y_2 + \frac{1}{2}x_3y_2 - \frac{1}{2}x_2y_3 - \frac{1}{2}x_3y_1 + \frac{1}{2}x_1y_3$$

$$= \frac{1}{2}\det\left(\begin{bmatrix} x_1 & y_1 & 1 \\ x_2 & y_2 & 1 \\ x_3 & y_3 & 1 \end{bmatrix}\right).$$

When the points are in the other quadrants or the points are labeled in a different order, the formula just obtained may yield the negative of the area of the triangle. Thus, for a triangle with vertices (x_1, y_1), (x_2, y_2), and (x_3, y_3), we have

$$\text{area of triangle} = \frac{1}{2}\left|\det\left(\begin{bmatrix} x_1 & y_1 & 1 \\ x_2 & y_2 & 1 \\ x_3 & y_3 & 1 \end{bmatrix}\right)\right| \tag{2}$$

(the area is $\frac{1}{2}$ the absolute value of the determinant).

EXAMPLE 4

Compute the area of the triangle T, shown in Figure 5.2, with vertices $(-1, 4)$, $(3, 1)$, and $(2, 6)$.

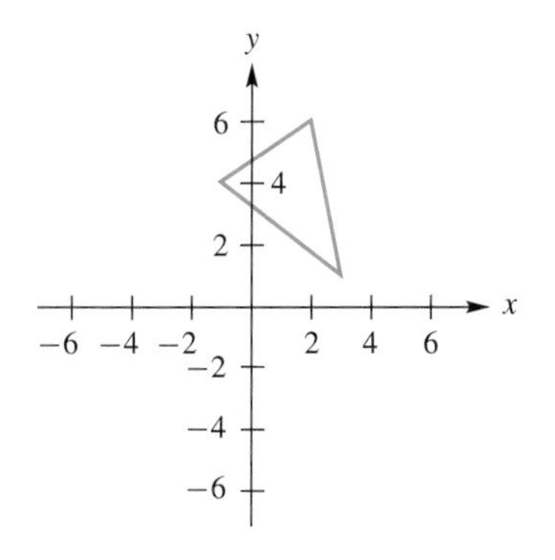

FIGURE 5.2

Solution By Equation (2), the area of T is

$$\frac{1}{2}\left|\det\left(\begin{bmatrix} -1 & 4 & 1 \\ 3 & 1 & 1 \\ 2 & 6 & 1 \end{bmatrix}\right)\right| = \frac{1}{2}|17| = 8.5.$$

■

Suppose we now have the parallelogram shown in Figure 5.3. Since a diagonal divides the parallelogram into two equal triangles, it follows from Equation (2) that

$$\text{area of parallelogram} = \left|\det\left(\begin{bmatrix} x_1 & y_1 & 1 \\ x_2 & y_2 & 1 \\ x_3 & y_3 & 1 \end{bmatrix}\right)\right|.$$

An arbitrary 2×2 matrix defines a linear operator $L\colon R^2 \to R^2$ by

$$L(\mathbf{v}) = A\mathbf{v}$$

for a vector $\mathbf{v}$ in R^2. How does the area of the image of a closed figure, such as a triangle T that is obtained by applying L to T, compare with the area of T? Thus

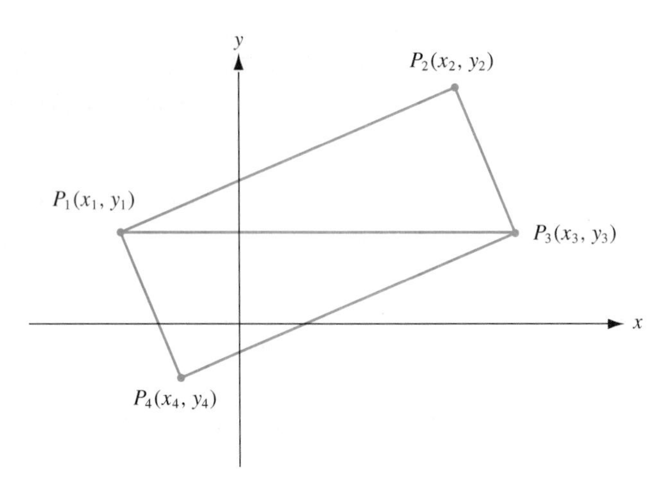

FIGURE 5.3

suppose that T is the triangle with vertices $(x_1, y_1, (x_2, y_2)$, and (x_3, y_3) and let

$$A = \begin{bmatrix} a & b \\ c & d \end{bmatrix}.$$

In Exercise 19, we ask you to first compute the coordinates of the image $L(T)$ and then show that

$$\text{area of } L(T) = |\det(A)| \cdot \text{area of } T.$$

EXAMPLE 5

Consider the triangle T defined in Example 4 and let

$$A = \begin{bmatrix} 6 & -3 \\ -1 & 1 \end{bmatrix}.$$

The image of T, using the linear operator defined by the matrix A, is the triangle with vertices $(-18, 5)$, $(15, -2)$, and $(-6, 4)$ (verify). See Figure 5.4. The area of this triangle is, by (2),

$$\frac{1}{2}\left|\det\left(\begin{bmatrix} -18 & 5 & 1 \\ 15 & -2 & 1 \\ -6 & 4 & 1 \end{bmatrix}\right)\right| = \frac{1}{2}|51| = 25.5.$$

Since $\det(A) = 3$, we see that $3 \times$ area of triangle $T =$ area of the image. ■

Remark The result discussed in Example 5 is true in general; that is, if S is a closed figure in 2-space (3-space), A is a matrix of appropriate dimension, and L is the linear transformation defined by A, then the area of the image $= |\det(A)| \cdot$ area of S (volume of the image $= |\det(A)| \cdot$ volume of S).

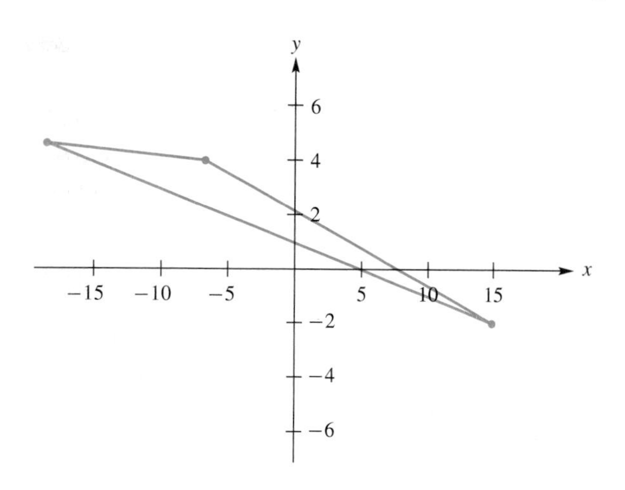

FIGURE 5.4

5.3 Exercises

1. Let $A = \begin{bmatrix} 1 & 0 & -2 \\ 3 & 1 & 4 \\ 5 & 2 & -3 \end{bmatrix}$. Find the following minors.

(a) $\det(M_{13})$. (b) $\det(M_{22})$.
(c) $\det(M_{31})$. (d) $\det(M_{32})$.

2. Let $A = \begin{bmatrix} 2 & -1 & 0 & 3 \\ 1 & 2 & -2 & 4 \\ -1 & 1 & -3 & -2 \\ 0 & 2 & -1 & 5 \end{bmatrix}$. Find the following minors.

(a) $\det(M_{12})$. (b) $\det(M_{23})$.
(c) $\det(M_{33})$. (d) $\det(M_{41})$.

3. Let $A = \begin{bmatrix} -1 & 2 & 3 \\ -2 & 5 & 4 \\ 0 & 1 & -3 \end{bmatrix}$. Find the following cofactors.

(a) A_{13}. (b) A_{21}.
(c) A_{32}. (d) A_{33}.

4. Let $A = \begin{bmatrix} 1 & 0 & 3 & 0 \\ 2 & 1 & -4 & -1 \\ 3 & 2 & 4 & 0 \\ 0 & 3 & -1 & 0 \end{bmatrix}$. Find the following cofactors.

(a) A_{12}. (b) A_{23}.
(c) A_{33}. (d) A_{41}.

5. Use Theorem 5.10 to evaluate the determinants in Exercise 1(a), (d), and (e) of Section 5.2.

6. Use Theorem 5.10 to evaluate the determinants in Exercise 1(b), (c), and (f) of Section 5.2.

7. Use Theorem 5.10 to evaluate the determinants in Exercise 2(a), (c), and (f) of Section 5.2.

8. Use Theorem 5.10 to evaluate the determinants in Exercise 2(b), (d), and (e) of Section 5.2.

9. Show by a column (row) expansion that if $A = \left[a_{ij}\right]$ is upper (lower) triangular, then $\det(A) = a_{11}a_{22}\cdots a_{nn}$.

10. If $A = \left[a_{ij}\right]$ is a 3×3 matrix, develop the general expression for $\det(A)$ by expanding:

(a) Along the second column.

(b) Along the third row.

Compare these answers with those obtained for Example 8 in Section 5.1.

11. Find all values of t for which:

(a) $\begin{vmatrix} t-2 & 2 \\ 3 & t-3 \end{vmatrix} = 0$.

(b) $\begin{vmatrix} t-1 & -4 \\ 0 & t-4 \end{vmatrix} = 0$.

12. Find all values of t for which

$$\begin{vmatrix} t-1 & 0 & 1 \\ -2 & t+2 & -1 \\ 0 & 0 & t+1 \end{vmatrix} = 0.$$

13. Let A be an $n \times n$ matrix.

(a) Show that $f(t) = \det(tI_n - A)$ is a polynomial in t of degree n.

(b) What is the coefficient of t^n in $f(t)$?

(c) What is the constant term in $f(t)$?

14. Verify your answers to Exercise 13 with the following matrices.

(a) $\begin{bmatrix} 1 & 2 \\ 3 & 4 \end{bmatrix}$. (b) $\begin{bmatrix} 1 & 3 & 2 \\ 2 & -1 & 3 \\ 3 & 0 & 1 \end{bmatrix}$.

(c) $\begin{bmatrix} 1 & 1 \\ 1 & 1 \end{bmatrix}$.

15. Let T be the triangle with vertices $(3, 3)$, $(-1, -1)$, $(4, 1)$.

(a) Find the area of the triangle T.

(b) Find the coordinates of the vertices of the image of T under the linear operator with matrix representation

$$A = \begin{bmatrix} 4 & -3 \\ -4 & 2 \end{bmatrix}.$$

(c) Find the area of the triangle whose vertices are obtained in part (b).

16. Find the area of the parallelogram with vertices $(2, 3)$, $(5, 3)$, $(4, 5)$, $(7, 5)$.

17. Let Q be the quadrilateral with vertices $(-2, 3)$, $(1, 4)$, $(3, 0)$, and $(-1, -3)$. Find the area of Q.

18. Prove that a rotation leaves the area of a triangle unchanged.

19. Let T be the triangle with vertices (x_1, y_1), (x_2, y_2), and (x_3, y_3), and let

$$A = \begin{bmatrix} a & b \\ c & d \end{bmatrix}.$$

Let L be the linear operator defined by $L(\mathbf{v}) = A\mathbf{v}$ for a vector $\mathbf{v}$ in R^2. First, compute the vertices of $L(T)$, the image of T under L, and then show that the area of $L(T)$ is $|\det(A)| \cdot$ area of T.

5.4 INVERSE OF A MATRIX

We saw in Section 5.3 that Theorem 5.10 provides formulas for expanding $\det(A)$ along either a row or a column of A. Thus $\det(A) = a_{i1}A_{i1} + a_{i2}A_{i2} + \cdots + a_{in}A_{in}$ is the expansion of $\det(A)$ along the ith row. It is interesting to ask what $a_{i1}A_{k1} + a_{i2}A_{k2} + \cdots + a_{in}A_{kn}$ is for $i \neq k$, because as soon as we answer this question we shall obtain another method for finding the inverse of a nonsingular matrix.

Theorem 5.11

If $A = \begin{bmatrix} a_{ij} \end{bmatrix}$ is an $n \times n$ matrix, then

$$a_{i1}A_{k1} + a_{i2}A_{k2} + \cdots + a_{in}A_{kn} = 0 \quad \text{for } i \neq k;$$
$$a_{1j}A_{1k} + a_{2j}A_{2k} + \cdots + a_{nj}A_{nk} = 0 \quad \text{for } j \neq k.$$

Proof

We prove only the first formula. The second follows from the first one by Theorem 5.1.

Consider the matrix B obtained from A by replacing the kth row of A by its ith row. Thus B is a matrix having two identical rows—the ith and kth, so $\det(B) = 0$. Now expand $\det(B)$ along the kth row. The elements of the kth row of B are $a_{i1}, a_{i2}, \ldots, a_{in}$. The cofactors of the kth row are $A_{k1}, A_{k2}, \ldots, A_{kn}$. Thus

$$0 = \det(B) = a_{i1}A_{k1} + a_{i2}A_{k2} + \cdots + a_{in}A_{kn},$$

which is what we wanted to show. ●

This theorem says that if we sum the products of the elements of any row (column) times the corresponding cofactors of any other row (column), then we obtain zero.

EXAMPLE 1

Let $A = \begin{bmatrix} 1 & 2 & 3 \\ -2 & 3 & 1 \\ 4 & 5 & -2 \end{bmatrix}$. Then

$$A_{21} = (-1)^{2+1} \begin{vmatrix} 2 & 3 \\ 5 & -2 \end{vmatrix} = 19,$$

$$A_{22} = (-1)^{2+2} \begin{vmatrix} 1 & 3 \\ 4 & -2 \end{vmatrix} = -14, \quad \text{and} \quad A_{23} = (-1)^{2+3} \begin{vmatrix} 1 & 2 \\ 4 & 5 \end{vmatrix} = 3.$$

Now

$$a_{31}A_{21} + a_{32}A_{22} + a_{33}A_{23} = (4)(19) + (5)(-14) + (-2)(3) = 0,$$

and

$$a_{11}A_{21} + a_{12}A_{22} + a_{13}A_{23} = (1)(19) + (2)(-14) + (3)(3) = 0.$$ ■

We may summarize our expansion results by writing

$$\begin{aligned} a_{i1}A_{k1} + a_{i2}A_{k2} + \cdots + a_{in}A_{kn} &= \det(A) && \text{if } i = k \\ &= 0 && \text{if } i \neq k \end{aligned}$$

and

$$\begin{aligned} a_{1j}A_{1k} + a_{2j}A_{2k} + \cdots + a_{nj}A_{nk} &= \det(A) && \text{if } j = k \\ &= 0 && \text{if } j \neq k. \end{aligned}$$

Definition 5.5

Let $A = \left[a_{ij}\right]$ be an $n \times n$ matrix. The $n \times n$ matrix adj A, called the **adjoint** of A, is the matrix whose (i, j)th entry is the cofactor A_{ji} of a_{ji}. Thus

$$\text{adj } A = \begin{bmatrix} A_{11} & A_{21} & \cdots & A_{n1} \\ A_{12} & A_{22} & \cdots & A_{n2} \\ \vdots & \vdots & & \vdots \\ A_{1n} & A_{2n} & \cdots & A_{nn} \end{bmatrix}.$$ ▲

Remark It should be noted that the term *adjoint* has other meanings in linear algebra in addition to its use in the definition above.

EXAMPLE 2

Let $A = \begin{bmatrix} 3 & -2 & 1 \\ 5 & 6 & 2 \\ 1 & 0 & -3 \end{bmatrix}$. Compute adj A.

Solution We first compute the cofactors of A. We have

$$A_{11} = (-1)^{1+1} \begin{vmatrix} 6 & 2 \\ 0 & -3 \end{vmatrix} = -18,$$

$$A_{12} = (-1)^{1+2} \begin{vmatrix} 5 & 2 \\ 1 & -3 \end{vmatrix} = 17, \quad A_{13} = (-1)^{1+3} \begin{vmatrix} 5 & 6 \\ 1 & 0 \end{vmatrix} = -6,$$

$$A_{21} = (-1)^{2+1} \begin{vmatrix} -2 & 1 \\ 0 & -3 \end{vmatrix} = -6,$$

$$A_{22} = (-1)^{2+2} \begin{vmatrix} 3 & 1 \\ 1 & -3 \end{vmatrix} = -10, \quad A_{23} = (-1)^{2+3} \begin{vmatrix} 3 & -2 \\ 1 & 0 \end{vmatrix} = -2,$$

$$A_{31} = (-1)^{3+1} \begin{vmatrix} -2 & 1 \\ 6 & 2 \end{vmatrix} = -10,$$

$$A_{32} = (-1)^{3+2} \begin{vmatrix} 3 & 1 \\ 5 & 2 \end{vmatrix} = -1, \quad A_{33} = (-1)^{3+3} \begin{vmatrix} 3 & -2 \\ 5 & 6 \end{vmatrix} = 28.$$

Then

$$\text{adj}\, A = \begin{bmatrix} -18 & -6 & -10 \\ 17 & -10 & -1 \\ -6 & -2 & 28 \end{bmatrix}.$$

■

Theorem 5.12

If $A = \left[a_{ij} \right]$ is an $n \times n$ matrix, then $A(\text{adj}\, A) = (\text{adj}\, A)A = \det(A)I_n$.

Proof

We have

$$A(\text{adj}\, A) = \begin{bmatrix} a_{11} & a_{12} & \cdots & a_{1n} \\ a_{21} & a_{22} & \cdots & a_{2n} \\ \vdots & \vdots & & \vdots \\ a_{i1} & a_{i2} & \cdots & a_{in} \\ \vdots & \vdots & & \vdots \\ a_{n1} & a_{n2} & \cdots & a_{nn} \end{bmatrix} \begin{bmatrix} A_{11} & A_{21} & \cdots & A_{j1} & \cdots & A_{n1} \\ A_{12} & A_{22} & \cdots & A_{j2} & \cdots & A_{n2} \\ \vdots & \vdots & & \vdots & & \vdots \\ A_{1n} & A_{2n} & \cdots & A_{jn} & \cdots & A_{nn} \end{bmatrix}.$$

The (i, j)th element in the product matrix $A(\text{adj } A)$ is, by Theorem 5.10,

$$\begin{aligned} a_{i1}A_{j1} + a_{i2}A_{j2} + \cdots + a_{in}A_{jn} &= \det(A) \quad \text{if } i = j \\ &= 0 \quad \text{if } i \neq j. \end{aligned}$$

This means that

$$A(\text{adj } A) = \begin{bmatrix} \det(A) & 0 & \cdots & 0 \\ 0 & \det(A) & & 0 \\ \vdots & \vdots & \vdots & \vdots \\ 0 & \cdots & 0 & \det(A) \end{bmatrix} = \det(A)I_n.$$

The (i, j)th element in the product matrix $(\text{adj } A)A$ is, by Theorem 5.10,

$$\begin{aligned} A_{1i}a_{1j} + A_{2i}a_{2j} + \cdots + A_{ni}a_{nj} &= \det(A) \quad \text{if } i = j \\ &= 0 \quad \text{if } i \neq j. \end{aligned}$$

Thus $(\text{adj } A)A = \det(A)I_n$.

EXAMPLE 3

Consider the matrix of Example 2. Then

$$\begin{bmatrix} 3 & -2 & 1 \\ 5 & 6 & 2 \\ 1 & 0 & -3 \end{bmatrix} \begin{bmatrix} -18 & -6 & -10 \\ 17 & -10 & -1 \\ -6 & -2 & 28 \end{bmatrix} = \begin{bmatrix} -94 & 0 & 0 \\ 0 & -94 & 0 \\ 0 & 0 & -94 \end{bmatrix}$$

$$= -94 \begin{bmatrix} 1 & 0 & 0 \\ 0 & 1 & 0 \\ 0 & 0 & 1 \end{bmatrix}$$

and

$$\begin{bmatrix} -18 & -6 & -10 \\ 17 & -10 & -1 \\ -6 & -2 & 28 \end{bmatrix} \begin{bmatrix} 3 & -2 & 1 \\ 5 & 6 & 2 \\ 1 & 0 & -3 \end{bmatrix} = -94 \begin{bmatrix} 1 & 0 & 0 \\ 0 & 1 & 0 \\ 0 & 0 & 1 \end{bmatrix}.$$

We now have a new method for finding the inverse of a nonsingular matrix, and we state this result as the following corollary.

Corollary 5.5

If A is an $n \times n$ matrix and $\det(A) \neq 0$, *then*

$$A^{-1} = \frac{1}{\det(A)}(\operatorname{adj} A) = \begin{bmatrix} \frac{A_{11}}{\det(A)} & \frac{A_{21}}{\det(A)} & \cdots & \frac{A_{n1}}{\det(A)} \\ \frac{A_{12}}{\det(A)} & \frac{A_{22}}{\det(A)} & \cdots & \frac{A_{n2}}{\det(A)} \\ \vdots & \vdots & & \vdots \\ \frac{A_{1n}}{\det(A)} & \frac{A_{2n}}{\det(A)} & \cdots & \frac{A_{nn}}{\det(A)} \end{bmatrix}.$$

Proof

By Theorem 5.12, $A(\operatorname{adj} A) = \det(A)I_n$, so if $\det(A) \neq 0$, then

$$A\left(\frac{1}{\det(A)}(\operatorname{adj} A)\right) = \frac{1}{\det(A)}(A(\operatorname{adj} A)) = \frac{1}{\det(A)}(\det(A)I_n) = I_n.$$

Hence

$$A^{-1} = \frac{1}{\det(A)}(\operatorname{adj} A).$$

●

EXAMPLE 4

Again consider the matrix of Example 2. Then $\det(A) = -94$, and

$$A^{-1} = \frac{1}{\det(A)}(\operatorname{adj} A) = \begin{bmatrix} \frac{18}{94} & \frac{6}{94} & \frac{10}{94} \\ -\frac{17}{94} & \frac{10}{94} & \frac{1}{94} \\ \frac{6}{94} & \frac{2}{94} & -\frac{28}{94} \end{bmatrix}.$$

■

We might note that the method of inverting a nonsingular matrix given in Corollary 5.5 is much less efficient than the method given in Chapter 1. In fact, the computation of A^{-1} using determinants, as given in Corollary 5.5, becomes too expensive for $n > 4$. We discuss these matters in Section 5.6, where we deal with determinants from a computational point of view. However, Corollary 5.5 is still a useful result on other grounds.

5.4 Exercises

1. Verify Theorem 5.11 for the matrix

$$A = \begin{bmatrix} -2 & 3 & 0 \\ 4 & 1 & -3 \\ 2 & 0 & 1 \end{bmatrix}$$

by computing $a_{11}A_{12} + a_{21}A_{22} + a_{31}A_{32}$.

2. Let $A = \begin{bmatrix} 2 & 1 & 3 \\ -1 & 2 & 0 \\ 3 & -2 & 1 \end{bmatrix}$.

(a) Find adj A.

(b) Compute $\det(A)$.

(c) Verify Theorem 5.12; that is, show that

$$A(\text{adj } A) = (\text{adj } A)A = \det(A)I_3.$$

3. Let $A = \begin{bmatrix} 6 & 2 & 8 \\ -3 & 4 & 1 \\ 4 & -4 & 5 \end{bmatrix}$. Follow the directions of Exercise 2.

4. Find the inverse of the matrix in Exercise 2 by the method given in Corollary 5.5.

5. Repeat Exercise 11 of Section 1.6 by the method given in Corollary 5.5. Compare your results with those obtained earlier.

6. Prove that if A is a symmetric matrix, then adj A is symmetric.

7. Use the method given in Corollary 5.5 to find the inverse, if it exists, of:

(a) $\begin{bmatrix} 0 & 2 & 1 & 3 \\ 2 & -1 & 3 & 4 \\ -2 & 1 & 5 & 2 \\ 0 & 1 & 0 & 2 \end{bmatrix}$.

(b) $\begin{bmatrix} 4 & 2 & 2 \\ 0 & 1 & 2 \\ 1 & 0 & 3 \end{bmatrix}$.

(c) $\begin{bmatrix} 3 & 2 \\ -3 & 4 \end{bmatrix}$.

8. Prove that if A is a nonsingular upper triangular matrix, then A^{-1} is upper triangular.

9. Use the method given in Corollary 5.5 to find the inverse of

$$A = \begin{bmatrix} a & b \\ c & d \end{bmatrix} \quad \text{if } ad - bc \neq 0.$$

10. Use the method given in Corollary 5.5 to find the inverse of

$$A = \begin{bmatrix} 1 & a & a^2 \\ 1 & b & b^2 \\ 1 & c & c^2 \end{bmatrix}.$$

[*Hint*: See Exercise 22 in Section 5.2, where $\det(A)$ is computed.]

11. Use the method given in Corollary 5.5 to find the inverse of

$$A = \begin{bmatrix} 4 & 0 & 0 \\ 0 & -3 & 0 \\ 0 & 0 & 2 \end{bmatrix}.$$

12. Use the method given in Corollary 5.5 to find the inverse of

$$A = \begin{bmatrix} 4 & 1 & 2 \\ 0 & -3 & 3 \\ 0 & 0 & 2 \end{bmatrix}.$$

13. Prove that if A is singular, then adj A is singular. [*Hint*: First show that if A is singular, then $A(\text{adj } A) = O$.]

14. Prove that if A is an $n \times n$ matrix, then $\det(\text{adj } A) = [\det(A)]^{n-1}$.

15. Assuming that your software has a command for computing the inverse of a matrix (see Exercise 47 in Section 1.4), read the accompanying software documentation to determine the method used. Is the description closer to that in Section 1.6 or Corollary 5.5? See also the comments in Section 5.6.

5.5 OTHER APPLICATIONS OF DETERMINANTS

We can use the results developed in Theorem 5.12 to obtain another method for solving a linear system of n equations in n unknowns. This method is known as **Cramer's rule**.

Theorem 5.13 (*Cramer's*[†] *Rule*).

Let

$$\begin{aligned} a_{11}x_1 + a_{12}x_2 + \cdots + a_{1n}x_n &= b_1 \\ a_{21}x_1 + a_{22}x_2 + \cdots + a_{2n}x_n &= b_2 \\ &\vdots \\ a_{n1}x_1 + a_{n2}x_2 + \cdots + a_{nn}x_n &= b_n \end{aligned}$$

be a linear system of n equations in n unknowns, and let $A = \left[a_{ij} \right]$ *be the coefficient matrix so that we can write the given system as* $A\mathbf{x} = \mathbf{b}$, *where*

$$\mathbf{b} = \begin{bmatrix} b_1 \\ b_2 \\ \vdots \\ b_n \end{bmatrix}.$$

If $\det(A) \neq 0$, *then the system has the unique solution*

$$x_1 = \frac{\det(A_1)}{\det(A)}, \quad x_2 = \frac{\det(A_2)}{\det(A)}, \quad \ldots, \quad x_n = \frac{\det(A_n)}{\det(A)},$$

where A_i *is the matrix obtained from A by replacing the ith column of A by* $\mathbf{b}$.

Proof

If $\det(A) \neq 0$, then, by Theorem 5.8, A is nonsingular. Hence

$$\mathbf{x} = \begin{bmatrix} x_1 \\ x_2 \\ \vdots \\ x_n \end{bmatrix} = A^{-1}\mathbf{b} = \begin{bmatrix} \frac{A_{11}}{\det(A)} & \frac{A_{21}}{\det(A)} & \cdots & \frac{A_{n1}}{\det(A)} \\ \frac{A_{12}}{\det(A)} & \frac{A_{22}}{\det(A)} & \cdots & \frac{A_{n2}}{\det(A)} \\ \vdots & \vdots & & \vdots \\ \frac{A_{1i}}{\det(A)} & \frac{A_{2i}}{\det(A)} & \cdots & \frac{A_{ni}}{\det(A)} \\ \vdots & \vdots & & \vdots \\ \frac{A_{1n}}{\det(A)} & \frac{A_{2n}}{\det(A)} & \cdots & \frac{A_{nn}}{\det(A)} \end{bmatrix} \begin{bmatrix} b_1 \\ b_2 \\ \vdots \\ b_n \end{bmatrix}.$$

This means that

$$x_i = \frac{A_{1i}}{\det(A)} b_1 + \frac{A_{2i}}{\det(A)} b_2 + \cdots + \frac{A_{ni}}{\det(A)} b_n \qquad \text{for } i = 1, 2, \ldots, n.$$

[†]Gabriel Cramer (1704–1752) was born in Geneva, Switzerland, and lived there all his life. Remaining single, he traveled extensively, taught at the Académie de Calvin, and participated actively in civic affairs.

The rule for solving systems of linear equations appeared in an appendix to his 1750 book, *Introduction à l'analyse des lignes courbes algébriques*. It was known previously by other mathematicians but was not widely known or clearly explained until its appearance in Cramer's influential work.

Now let

$$A_i = \begin{bmatrix} a_{11} & a_{12} & \cdots & a_{1\,i-1} & b_1 & a_{1\,i+1} & \cdots & a_{1n} \\ a_{21} & a_{22} & \cdots & a_{2\,i-1} & b_2 & a_{2\,i+1} & \cdots & a_{2n} \\ \vdots & \vdots & & \vdots & \vdots & \vdots & & \vdots \\ a_{n1} & a_{n2} & \cdots & a_{n\,i-1} & b_n & a_{n\,i+1} & \cdots & a_{nn} \end{bmatrix}.$$

If we evaluate $\det(A_i)$ by expanding along the cofactors of the ith column, we find that

$$\det(A_i) = A_{1i}b_1 + A_{2i}b_2 + \cdots + A_{ni}b_n.$$

Hence

$$x_i = \frac{\det(A_i)}{\det(A)} \qquad \text{for } i = 1, 2, \ldots, n.$$

In the expression for x_i given in Equation (1), the determinant, $\det(A_i)$, of A_i can be calculated by any method desired. It was only in the *derivation* of the expression for x_i that we had to evaluate $\det(A_i)$ by expanding along the ith column.

EXAMPLE 1

Consider the following linear system:

$$\begin{aligned} -2x_1 + 3x_2 - x_3 &= 1 \\ x_1 + 2x_2 - x_3 &= 4 \\ -2x_1 - x_2 + x_3 &= -3. \end{aligned}$$

We have $|A| = \begin{vmatrix} -2 & 3 & -1 \\ 1 & 2 & -1 \\ -2 & -1 & 1 \end{vmatrix} = -2$. Then

$$x_1 = \frac{\begin{vmatrix} 1 & 3 & -1 \\ 4 & 2 & -1 \\ -3 & -1 & 1 \end{vmatrix}}{|A|} = \frac{-4}{-2} = 2,$$

$$x_2 = \frac{\begin{vmatrix} -2 & 1 & -1 \\ 1 & 4 & -1 \\ -2 & -3 & 1 \end{vmatrix}}{|A|} = \frac{-6}{-2} = 3,$$

and

$$x_3 = \frac{\begin{vmatrix} -2 & 3 & 1 \\ 1 & 2 & 4 \\ -2 & -1 & -3 \end{vmatrix}}{|A|} = \frac{-8}{-2} = 4.$$

We note that Cramer's rule is applicable only to the case in which we have n equations in n unknowns and the coefficient matrix A is nonsingular. If we have to solve a linear system of n equations in n unknowns whose coefficient matrix is singular, then we must use the Gaussian elimination or Gauss–Jordan reduction methods as discussed in Section 1.5. Cramer's rule becomes computationally inefficient for $n \geq 4$, and it is then better to also use the Gaussian elimination or Gauss–Jordan reduction methods.

Our next application of determinants will enable us to tell whether a set of n vectors in R^n or R_n is linearly independent.

Theorem 5.14

Let $S = \{\mathbf{v}_1, \mathbf{v}_2, \ldots, \mathbf{v}_n\}$ be a set of n vectors in R^n (R_n). Let A be the matrix whose columns (rows) are the elements of S. Then S is linearly independent if and only if $\det(A) \neq 0$.

Proof

We shall prove the result for columns only; the proof for rows is analogous.

If S is linearly independent, then the dimension of the column space of A = rank $A = n$, and from Corollary 5.1 it follows that $\det(A) \neq 0$. Conversely, if $\det(A) \neq 0$, by Corollary 5.1 we know that rank $A = n$, so the columns of A are linearly independent. ●

EXAMPLE 2

Is $S = \{\begin{bmatrix}1 & 2 & 3\end{bmatrix}, \begin{bmatrix}0 & 1 & 2\end{bmatrix}, \begin{bmatrix}3 & 0 & -1\end{bmatrix}\}$ a linearly independent set of vectors in R^3?

Solution We form the matrix A whose rows are the vectors in S:

$$A = \begin{bmatrix} 1 & 2 & 3 \\ 0 & 1 & 2 \\ 3 & 0 & -1 \end{bmatrix}.$$

Since $\det(A) = 2$ (verify), we conclude that S is linearly independent. ■

EXAMPLE 3

To find out if $S = \{t^2 + t, t + 1, t - 1\}$ is a basis for P_2, we note that P_2 is a three-dimensional vector space isomorphic to R^3 under the mapping $L\colon P_2 \to R^3$ defined by $L(at^2 + bt + c) = \begin{bmatrix} a \\ b \\ c \end{bmatrix}$. Therefore, S is a basis for P_2 if and only if $T = \{L(t^2 + t), L(t + 1), L(t - 1)\}$ is a basis for R^3. To decide whether this is so, we apply Theorem 5.14. Thus let A be the matrix whose columns are $L(t^2 + t)$,

$L(t+1)$, $L(t-1)$, respectively. Now

$$L(t^2+t) = \begin{bmatrix} 1 \\ 1 \\ 0 \end{bmatrix}, \quad L(t+1) = \begin{bmatrix} 0 \\ 1 \\ 1 \end{bmatrix}, \quad \text{and} \quad L(t-1) = \begin{bmatrix} 0 \\ 1 \\ -1 \end{bmatrix},$$

so

$$A = \begin{bmatrix} 1 & 0 & 0 \\ 1 & 1 & 1 \\ 0 & 1 & -1 \end{bmatrix}.$$

Since $\det(A) = -2$ (verify), we conclude that T is linearly independent. Hence S is linearly independent, and since $\dim P_2 = 3$, S is a basis for P_2. ■

We now recall that if a set S of n vectors in R^n (R_n) is linearly independent, then S spans R^n (R_n), and conversely, if S spans R^n (R_n), then S is linearly independent (Theorem 2.11 in Section 2.5). Thus the condition in Theorem 5.14 (that $\det(A) \neq 0$) is also necessary and sufficient for S to span R^n (R_n).

We may summarize our results on the application of determinants by noting that the following statements are equivalent for an $n \times n$ matrix A:

1. $\det(A) \neq 0$.
2. A is nonsingular.
3. The rows (columns) of A are linearly independent.
4. $A\mathbf{x} = \mathbf{0}$ has only the trivial solution.
5. $\text{rank}\, A = n$.

Determinants and Cross Product (Optional)

Our final application of determinants is to cross products. Recall the definition given in Section 3.2 for the cross product $\mathbf{u} \times \mathbf{v}$ of the vectors $\mathbf{u} = u_1\mathbf{i} + u_2\mathbf{j} + u_3\mathbf{k}$ and $\mathbf{v} = v_1\mathbf{i} + v_2\mathbf{j} + v_3\mathbf{k}$ in R^3:

$$\mathbf{u} \times \mathbf{v} = (u_2v_3 - u_3v_2)\mathbf{i} + (u_3v_1 - u_1v_3)\mathbf{j} + (u_1v_2 - u_2v_1)\mathbf{k}.$$

If we formally write the matrix

$$C = \begin{bmatrix} \mathbf{i} & \mathbf{j} & \mathbf{k} \\ u_1 & u_2 & u_3 \\ v_1 & v_2 & v_3 \end{bmatrix},$$

then the determinant of C, obtained by expanding along the cofactors of the first row, is $\mathbf{u} \times \mathbf{v}$; that is,

$$\mathbf{u} \times \mathbf{v} = \det(C) = \begin{vmatrix} u_2 & u_3 \\ v_2 & v_3 \end{vmatrix} \mathbf{i} - \begin{vmatrix} u_1 & u_3 \\ v_1 & v_3 \end{vmatrix} \mathbf{j} + \begin{vmatrix} u_1 & u_2 \\ v_1 & v_2 \end{vmatrix} \mathbf{k}.$$

Of course, C is not really a matrix and $\det(C)$ is not really a determinant, but it is convenient to think of the computation in this way.

EXAMPLE 4

If $\mathbf{u} = 2\mathbf{i} + \mathbf{j} + 2\mathbf{k}$ and $\mathbf{v} = 3\mathbf{i} - \mathbf{j} - 3\mathbf{k}$, as in Example 1 of Section 3.2, then

$$C = \begin{bmatrix} \mathbf{i} & \mathbf{j} & \mathbf{k} \\ 2 & 1 & 2 \\ 3 & -1 & -3 \end{bmatrix},$$

and $\det(C) = \mathbf{u} \times \mathbf{v} = -\mathbf{i} + 12\mathbf{j} - 5\mathbf{k}$, when expanded along its first row. ■

5.5 Exercises

1. If possible, solve the following linear systems by Cramer's rule:

$$\begin{aligned} 2x_1 + 4x_2 + 6x_3 &= 2 \\ x_1 \qquad\quad + 2x_3 &= 0 \\ 2x_1 + 3x_2 - x_3 &= -5. \end{aligned}$$

2. Repeat Exercise 1 for the linear system

$$\begin{bmatrix} 1 & 1 & 1 & -2 \\ 0 & 2 & 1 & 3 \\ 2 & 1 & -1 & 2 \\ 1 & -1 & 0 & 1 \end{bmatrix} \begin{bmatrix} x_1 \\ x_2 \\ x_3 \\ x_4 \end{bmatrix} = \begin{bmatrix} -4 \\ 4 \\ 5 \\ 4 \end{bmatrix}.$$

3. Is $S = \left\{ \begin{bmatrix} 2 \\ 2 \\ 3 \end{bmatrix}, \begin{bmatrix} 1 \\ 0 \\ 2 \end{bmatrix}, \begin{bmatrix} 0 \\ 1 \\ 3 \end{bmatrix} \right\}$ a linearly independent set of vectors in R^3? Use Theorem 5.14.

4. Solve the following linear system for x_3, by Cramer's rule:

$$\begin{aligned} 2x_1 + x_2 + x_3 &= 6 \\ 3x_1 + 2x_2 - 2x_3 &= -2 \\ x_1 + x_2 + 2x_3 &= -4. \end{aligned}$$

5. Repeat Exercise 5 of Section 1.5; use Cramer's rule.

6. Is $S = \{ \begin{bmatrix} 0 & -1 & 4 \end{bmatrix}, \begin{bmatrix} 4 & 1 & 2 \end{bmatrix}, \begin{bmatrix} 2 & 0 & 3 \end{bmatrix} \}$ a basis for R_3? Use Theorem 5.14.

7. Repeat Exercise 1 for the following linear system:

$$\begin{aligned} 2x_1 - x_2 + 3x_3 &= 0 \\ x_1 + 2x_2 - 3x_3 &= 0 \\ 4x_1 + 2x_2 + x_3 &= 0. \end{aligned}$$

8. Does the set $S = \left\{ \begin{bmatrix} 3 \\ 1 \\ 0 \end{bmatrix}, \begin{bmatrix} 0 \\ -1 \\ 1 \end{bmatrix}, \begin{bmatrix} 1 \\ 2 \\ -1 \end{bmatrix} \right\}$ span R^3? Use Theorem 5.14.

9. Repeat Exercise 6(b) of Section 1.5; use Cramer's rule.

10. Repeat Exercise 1 for the following linear systems:

$$\begin{aligned} 2x_1 + 3x_2 + 7x_3 &= 0 \\ -2x_1 \qquad\quad - 4x_3 &= 0 \\ x_1 + 2x_2 + 4x_3 &= 0. \end{aligned}$$

11. Is $S = \{t^3 + t + 1, 2t^2 + 3, t - 1, 2t^3 - 2t^2\}$ a basis for P_3? Use Theorem 5.14.

12. Is

$$S = \left\{ \begin{bmatrix} 1 & 1 \\ 0 & 1 \end{bmatrix}, \begin{bmatrix} 0 & 2 \\ 1 & 3 \end{bmatrix}, \begin{bmatrix} 0 & 0 \\ -1 & 2 \end{bmatrix}, \begin{bmatrix} 0 & -2 \\ 0 & 3 \end{bmatrix} \right\}$$

a basis for M_{22}? Use Theorem 5.14.

13. For what values of c is the set $\{t + 3, 2t + c^2 + 2\}$ linearly independent?

14. Let $\mathbf{u} = u_1\mathbf{i} + u_2\mathbf{j} + u_3\mathbf{k}$, $\mathbf{v} = v_1\mathbf{i} + v_2\mathbf{j} + v_3\mathbf{k}$, and $\mathbf{w} = w_1\mathbf{i} + w_2\mathbf{j} + w_3\mathbf{k}$ be vectors in R^3. Show that

$$(\mathbf{u} \times \mathbf{v}) \cdot \mathbf{w} = \begin{vmatrix} u_1 & u_2 & u_3 \\ v_1 & v_2 & v_3 \\ w_1 & w_2 & w_3 \end{vmatrix}.$$

15. Compute $\mathbf{u} \times \mathbf{v}$ by the method of Example 4.

(a) $\mathbf{u} = 2\mathbf{i} + 3\mathbf{j} + 4\mathbf{k}$, $\mathbf{v} = -\mathbf{i} + 3\mathbf{j} - \mathbf{k}$.

(b) $\mathbf{u} = \mathbf{i} + \mathbf{k}$, $\mathbf{v} = 2\mathbf{i} + 3\mathbf{j} - \mathbf{k}$.

(c) $\mathbf{u} = \mathbf{i} - \mathbf{j} + 2\mathbf{k}$, $\mathbf{v} = 3\mathbf{i} - 4\mathbf{j} + \mathbf{k}$.

(d) $\mathbf{u} = 2\mathbf{i} + \mathbf{j} - 2\mathbf{k}$, $\mathbf{v} = \mathbf{i} + 3\mathbf{k}$.

5.6 DETERMINANTS FROM A COMPUTATIONAL POINT OF VIEW

In Chapter 1 we discussed three methods for solving a linear system: Gaussian elimination, Gauss–Jordan reduction, and LU-factorization. In this chapter, we presented one more way: Cramer's rule. We also have two methods for inverting a nonsingular matrix: the method developed in Section 1.6, which uses elementary matrices, and the method involving determinants, presented in Section 5.4. In this section we discuss criteria to be considered when selecting one or another of these methods.

In general, if we are seeking numerical answers, then any method involving determinants can be used for $n \leq 4$. Gaussian elimination, Gauss–Jordan reduction, and LU-factorization all require approximately $n^3/3$ operations to solve the linear system $A\mathbf{x} = \mathbf{b}$, where A is an $n \times n$ matrix. We now compare these methods with Cramer's rule, when A is 25×25, which in the world of real applications is a small problem (in some applications A can be as large as $100{,}000 \times 100{,}000$).

If we find $\mathbf{x}$ by Cramer's rule, then we must first obtain $\det(A)$. Suppose that we compute $\det(A)$ by cofactor expansion, say

$$\det(A) = a_{11}A_{11} + a_{21}A_{21} + \cdots + a_{n1}A_{n1},$$

where we have expanded along the first column of A. If each cofactor A_{ij} is available, we need 25 multiplications to compute $\det(A)$. Now each cofactor A_{ij} is the determinant of a 24×24 matrix, and it can be expanded along a particular row or column, requiring 24 multiplications. Thus the computation of $\det(A)$ requires more than $25 \times 24 \times \cdots \times 2 \times 1 = 25!$ (approximately 1.55×10^{25}) multiplications. Even if we were to use a futuristic (not very far into the future) supercomputer capable of performing *ten trillion* (1×10^{12}) multiplications *per second* (3.15×10^{19} *per year*), it would take *49,000 years* to evaluate $\det(A)$. However, Gaussian elimination takes approximately $25^3/3$ multiplications, and we obtain the solution in less than *one second.* Of course. $\det(A)$ can be computed in a much more efficient way, by using elementary row operations to reduce A to triangular form and then using Theorem 5.7 (see Example 8 in Section 5.2). When implemented this way, Cramer's rule will require approximately n^4 multiplications for an $n \times n$ matrix, compared to $n^3/3$ multiplications for Gaussian elimination. The most widely used method in practice is LU-factorization because it is cheapest, especially when we need to solve many linear systems with different right sides.

The importance of determinants obviously does not lie in their computational use; determinants enable one to express the inverse of a matrix and the solutions to a system of n linear equations in n unknowns by means of *expressions* or *formulas*. The other methods mentioned above for solving a linear system, and the method for finding A^{-1} using elementary matrices, have the property that we cannot write a *formula* for the answer; we must proceed algorithmically to obtain the answer. Sometimes we do not need a numerical answer but merely an expression for the answer, because we may wish to further manipulate the answer, for example, integrate it. However, the most important reason for studying determinants is that they play a key role in the further study of the properties of a linear transformation mapping a vector space V into itself. This study will be undertaken in Chapter 6.

Supplementary Exercises

1. Compute $|A|$ for each of the following.

(a) $A = \begin{bmatrix} 2 & 3 & 4 \\ 1 & 2 & 4 \\ 4 & 3 & 1 \end{bmatrix}$.

(b) $A = \begin{bmatrix} 2 & 1 & 0 \\ 1 & 2 & 1 \\ 0 & 1 & 2 \end{bmatrix}$.

(c) $A = \begin{bmatrix} 2 & 1 & -1 & 2 \\ 2 & -3 & -1 & 4 \\ 1 & 3 & 2 & -3 \\ 1 & -2 & -1 & 1 \end{bmatrix}$.

(d) $A = \begin{bmatrix} 2 & 1 & 0 & 0 \\ 1 & 2 & 1 & 0 \\ 0 & 1 & 2 & 1 \\ 0 & 0 & 1 & 2 \end{bmatrix}$.

2. Find all values of t for which $\det(tI_3 - A) = 0$ for each of the following.

(a) $A = \begin{bmatrix} 1 & 4 & 0 \\ 0 & 4 & 0 \\ 0 & 0 & 1 \end{bmatrix}$.

(b) $A = \begin{bmatrix} 2 & -2 & 0 \\ -3 & 1 & 0 \\ 0 & 0 & 3 \end{bmatrix}$.

(c) $A = \begin{bmatrix} 0 & 1 & 0 \\ 0 & 0 & 1 \\ 6 & -11 & 6 \end{bmatrix}$.

(d) $A = \begin{bmatrix} 0 & 1 & 0 \\ 0 & 0 & 1 \\ 3 & 1 & -3 \end{bmatrix}$.

3. Show that if $A^n = O$ for some positive integer n (i.e., if A is a nilpotent matrix), then $\det(A) = 0$.

4. Using only elementary row or elementary column operations and Theorems 5.2, 5.5, and 5.6 (do not expand the determinants), verify the following.

(a) $\begin{vmatrix} a-b & 1 & a \\ b-c & 1 & b \\ c-a & 1 & c \end{vmatrix} = \begin{vmatrix} a & 1 & b \\ b & 1 & c \\ c & 1 & a \end{vmatrix}$.

(b) $\begin{vmatrix} 1 & a & bc \\ 1 & b & ca \\ 1 & c & ab \end{vmatrix} = \begin{vmatrix} 1 & a & a^2 \\ 1 & b & b^2 \\ 1 & c & c^2 \end{vmatrix}$.

5. If (x_1, y_1) and (x_2, y_2) are distinct points in the plane, show that

$$\begin{vmatrix} x & y & 1 \\ x_1 & y_1 & 1 \\ x_2 & y_2 & 1 \end{vmatrix} = 0$$

is the equation of the line through (x_1, y_1) and (x_2, y_2). Use this result to develop a test for collinearity of three points.

6. Let $P_i(x_i, y_i, z_i)$, $i = 1, 2, 3$ be three points in 3-space. Show that

$$\begin{vmatrix} x & y & z & 1 \\ x_1 & y_1 & z_1 & 1 \\ x_2 & y_2 & z_2 & 1 \\ x_3 & y_3 & z_3 & 1 \end{vmatrix} = 0$$

is the equation of a plane (see Section 3.2) through points P_i, $i = 1, 2, 3$.

7. Show that if A is an $n \times n$ matrix, then $\det(AA^T) \geq 0$.

8. Prove or disprove that the determinant function is a linear transformation of M_{nn} into R^1.

9. Show that if A is a nonsingular matrix, then adj A is nonsingular and

$$(\text{adj } A)^{-1} = \frac{1}{\det(A)} A = \text{adj}(A^{-1}).$$

10. Prove that if two rows (columns) of the $n \times n$ matrix A are proportional, then $\det(A) = 0$.

11. Let Q be the $n \times n$ real matrix in which each entry is 1. Show that $\det(Q - nI_n) = 0$.

12. Let A be an $n \times n$ matrix with integer entries. Prove that A is nonsingular and A^{-1} has integer entries if and only if $\det(A) = \pm 1$.

13. Let A be an $n \times n$ matrix with integer entries and $\det(A) = \pm 1$. Show that if $\mathbf{b}$ has all integer entries, then every solution to $A\mathbf{x} = \mathbf{b}$ consists of integers.

Chapter 6

Eigenvalues and Eigenvectors

6.1 EIGENVALUES AND EIGENVECTORS

Definitions and Examples

Let $L: V \to V$ be a linear transformation of an n-dimensional vector space V into itself (a linear operator on V). Then L maps a vector $\mathbf{v}$ in V to another vector $L(\mathbf{v})$ in V. A question that arises in a wide variety of applications is that of determining whether $L(\mathbf{v})$ can be a multiple of $\mathbf{v}$. If V is R^n, then this question becomes that of determining whether $L(\mathbf{v})$ can be parallel to $\mathbf{v}$. Note that if $\mathbf{v} = \mathbf{0}$, then $L(\mathbf{0}) = \mathbf{0}$, so $L(\mathbf{0})$ is a multiple of $\mathbf{0}$ and is then parallel to $\mathbf{0}$. Thus we need only consider nonzero vectors in V.

EXAMPLE 1

Let $L: R^2 \to R^2$ be a reflection with respect to the x-axis, defined by

$$L(\mathbf{v}) = L\left(\begin{bmatrix} x_1 \\ x_2 \end{bmatrix}\right) = \begin{bmatrix} x_1 \\ -x_2 \end{bmatrix}$$

which we considered in Example 8 of Section 4.1. See Figure 6.1. If we want $L(\mathbf{v})$ to be parallel to a nonzero vector $\mathbf{v}$, we must have

$$L\left(\begin{bmatrix} x_1 \\ x_2 \end{bmatrix}\right) = \lambda \begin{bmatrix} x_1 \\ x_2 \end{bmatrix},$$

where λ is a scalar. Thus,

$$\lambda \begin{bmatrix} x_1 \\ x_2 \end{bmatrix} = \begin{bmatrix} x_1 \\ -x_2 \end{bmatrix}$$

or

$$\begin{aligned} \lambda x_1 \quad &= \quad x_1 \\ \lambda x_2 &= -x_2. \end{aligned}$$

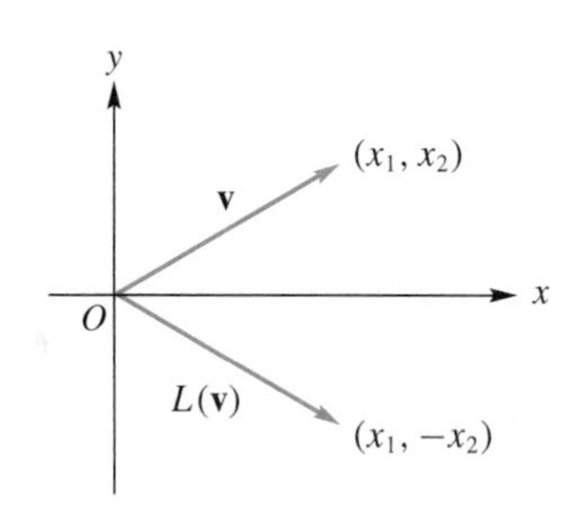

FIGURE 6.1

Since $\mathbf{v}$ is not the zero vector, both x_1 and x_2 cannot be zero. If $x_1 \neq 0$, then from the first equation it follows that $\lambda = 1$, and from the second equation we conclude that $x_2 = 0$. Thus, $\mathbf{v} = \begin{bmatrix} r \\ 0 \end{bmatrix}$, $r \neq 0$, which represents any vector along the x-axis. If $x_2 \neq 0$, then from the second equation it follows that $\lambda = -1$, and from the first equation we have $x_1 = 0$. Thus, $\mathbf{v} = \begin{bmatrix} 0 \\ s \end{bmatrix}$, $s \neq 0$, which represents any vector along the y-axis. Hence, for any vector $\mathbf{v}$ along the x-axis or along the y-axis, $L(\mathbf{v})$ will be parallel to $\mathbf{v}$. ■

EXAMPLE 2

Let $L: R^2 \to R^2$ be the linear operator defined by

$$L\left(\begin{bmatrix} x \\ y \end{bmatrix}\right) = \begin{bmatrix} \cos\phi & -\sin\phi \\ \sin\phi & \cos\phi \end{bmatrix} \begin{bmatrix} x \\ y \end{bmatrix},$$

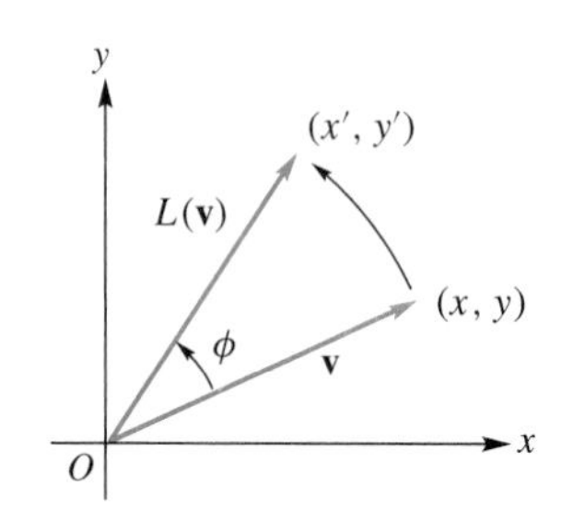

FIGURE 6.2

a counterclockwise rotation through the angle ϕ, $0 \leq \phi < 2\pi$, as defined in Example 9 of Section 4.1. See Figure 6.2. It follows that if $\phi \neq 0$ and $\phi \neq \pi$, then for every vector $\mathbf{v} = \begin{bmatrix} x \\ y \end{bmatrix}$, $L(\mathbf{v})$ is oriented in a direction different from that of $\mathbf{v}$, so $L(\mathbf{v})$ and $\mathbf{v}$ are never parallel. If $\phi = 0$, then $L(\mathbf{v}) = \mathbf{v}$ (verify), which means that $L(\mathbf{v})$ and $\mathbf{v}$ are in the same direction and are thus parallel. If $\phi = \pi$, then $L(\mathbf{v}) = -\mathbf{v}$ (verify), so $L(\mathbf{v})$ and $\mathbf{v}$ are in opposite directions and are thus parallel. ■

As we can see from Examples 1 and 2, the problem of determining all vectors $\mathbf{v}$ in an n-dimensional vector space V that are mapped by a given linear operator $L: V \to V$ to a multiple of $\mathbf{v}$ is one that is not seemingly simple. To further study this important problem we now formulate some terminology for it.

Definition 6.1

Let $L: V \to V$ be a linear transformation of an n-dimensional vector space V into itself (a linear operator on V). The real number λ is called an **eigenvalue** of L if there exists a *nonzero* vector $\mathbf{x}$ in V such that

$$L(\mathbf{x}) = \lambda\mathbf{x}. \tag{1}$$

Every nonzero vector $\mathbf{x}$ satisfying this equation is then called an **eigenvector** of L **associated with the eigenvalue λ**. We might mention that the word *eigenvalue* is a hybrid (*eigen* in German means *proper*). Eigenvalues are also called **proper**, **characteristic**, or **latent vectors**. ▲

Note that if we do not require that $\mathbf{x}$ be nonzero in Definition 6.1, then *every* real number λ would be an eigenvalue, since $L(\mathbf{0}) = \mathbf{0} = \lambda\mathbf{0}$. Such a definition would be of no interest. This is why we insist that $\mathbf{x}$ be nonzero.

In some applications we encounter vector spaces over the complex numbers (see Sections B.1 and B.2, respectively). In such settings the preceding definition of eigenvalue is modified so that an eigenvalue can be a real *or* a complex number. An introduction to this approach, a treatment usually presented in more advanced books, is given in Section B.2.

Throughout the rest of this book, unless stated otherwise, we require that an eigenvalue be a real number.

EXAMPLE 3

Let $L\colon V \to V$ be the linear operator defined by $L(\mathbf{x}) = 2\mathbf{x}$. We can see that the only eigenvalue of L is $\lambda = 2$ and that every nonzero vector in V is an eigenvector of L associated with the eigenvalue $\lambda = 2$. ■

Example 3 shows that an eigenvalue λ can have associated with it many different eigenvectors. In fact, if $\mathbf{x}$ is an eigenvector of L associated with the eigenvalue λ [i.e., $L(\mathbf{x}) = \lambda\mathbf{x}$], then

$$L(r\mathbf{x}) = rL(\mathbf{x}) = r(\lambda\mathbf{x}) = \lambda(r\mathbf{x}),$$

for any real number r. Thus, if $r \neq 0$, then $r\mathbf{x}$ is also an eigenvector of L associated with λ so that eigenvectors are never unique.

EXAMPLE 4

Let $L\colon R^2 \to R^2$ be the linear transformation that is defined by $L\left(\begin{bmatrix} a_1 \\ a_2 \end{bmatrix}\right) = \begin{bmatrix} a_2 \\ a_1 \end{bmatrix}$.

It follows that

$$L\left(\begin{bmatrix} r \\ r \end{bmatrix}\right) = 1\begin{bmatrix} r \\ r \end{bmatrix} \quad \text{and} \quad L\left(\begin{bmatrix} r \\ -r \end{bmatrix}\right) = -1\begin{bmatrix} r \\ -r \end{bmatrix}.$$

Thus any vector of the form $\begin{bmatrix} r \\ r \end{bmatrix}$, where r is any nonzero real number, for example, $\mathbf{x}_1 = \begin{bmatrix} 1 \\ 1 \end{bmatrix}$, is an eigenvector of L associated with the eigenvalue $\lambda = 1$; any vector of the form $\begin{bmatrix} r \\ -r \end{bmatrix}$, where r is any nonzero real number, such as $\mathbf{x}_2 = \begin{bmatrix} 1 \\ -1 \end{bmatrix}$, is an eigenvector of L associated with the eigenvalue $\lambda = -1$ (see Figure 6.3). ■

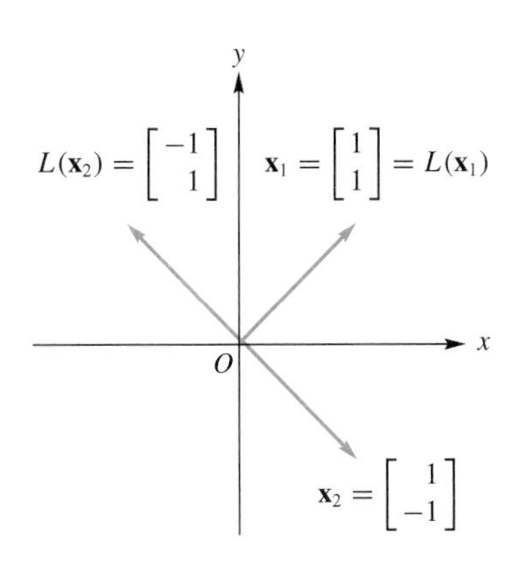

FIGURE 6.3

EXAMPLE 5

Let $L\colon R^2 \to R^2$ be counterclockwise rotation through the angle ϕ, $0 \leq \phi < 2\pi$, as defined in Example 2. It follows from our discussion in Example 2 that $\phi = 0$ and $\phi = \pi$ are the only angles for which L has eigenvalues (recall that we require eigenvalues to be real numbers). Thus, if $\phi = 0$ then $\lambda = 1$ is the only eigenvalue of L and every nonzero vector in R^2 is an eigenvector of L associated with the eigenvalue $\lambda = 1$. If we allow eigenvalues to be complex numbers, then for every angle ϕ, L has eigenvalues and associated eigenvectors. For example, if $\phi = \pi/4$, then it can be shown that $\lambda = \sqrt{2} + i\sqrt{2}$ is an eigenvalue of L with associated eigenvector $\mathbf{x} = \begin{bmatrix} i \\ 1 \end{bmatrix}$. ■

EXAMPLE 6

Let $L: R_2 \to R_2$ be defined by $L\left(\begin{bmatrix} x_1 & x_2 \end{bmatrix}\right) = \begin{bmatrix} 0 & x_2 \end{bmatrix}$. We can then see that

$$L\left(\begin{bmatrix} r & 0 \end{bmatrix}\right) = \begin{bmatrix} 0 & 0 \end{bmatrix} = 0\begin{bmatrix} r & 0 \end{bmatrix}$$

so that a vector of the form $\begin{bmatrix} r & 0 \end{bmatrix}$, where r is any nonzero real number (such as $\begin{bmatrix} 2 & 0 \end{bmatrix}$), is an eigenvector of L associated with the eigenvalue $\lambda = 0$. Also,

$$L\left(\begin{bmatrix} 0 & r \end{bmatrix}\right) = \begin{bmatrix} 0 & r \end{bmatrix} = 1\begin{bmatrix} 0 & r \end{bmatrix}$$

so that a vector of the form $\begin{bmatrix} 0 & r \end{bmatrix}$ (such as $\begin{bmatrix} 0 & 1 \end{bmatrix}$), where r is any nonzero real number, is an eigenvector of L associated with the eigenvalue $\lambda = 1$. ■

By definition the zero vector cannot be an eigenvector. However, Example 6 shows that the scalar zero can be an eigenvalue.

EXAMPLE 7

Although we introduced this chapter with the requirement that V be an n-dimensional vector space, the notions of eigenvalues and eigenvectors can be considered for infinite-dimensional vector spaces. In this example we look at such a situation.

Let V be the vector space of all real-valued functions of a single variable that have derivatives of all orders. Let $L: V \to V$ be the linear operator defined by

$$L(f) = f'.$$

Then the problem presented in Definition 6.1 can be stated as follows: Can we find a real number λ and a function $f \neq 0$ in V so that

$$L(f) = \lambda f? \tag{2}$$

If $y = f(x)$, then (2) can be written as

$$\frac{dy}{dx} = \lambda y. \tag{3}$$

Equation (3) states that the quantity y is one whose rate of change, with respect to x, is proportional to y itself. Examples of physical phenomena in which a quantity satisfies (3) include growth of human population, growth of bacteria and other organisms, investment problems, radioactive decay, carbon dating, and concentration of a drug in the body.

For each real number λ (an eigenvalue of L) we obtain, by using calculus, an associated eigenvector given by

$$f(x) = Ke^{\lambda x},$$

where K is an arbitrary nonzero constant. ■

Equation (3) is a simple example of a differential equation. The subject of differential equations is a major area in mathematics. In Section 7.1 we provide a brief introduction to homogeneous linear systems of differential equations.

Let L be a linear transformation of an n-dimensional vector space V into itself. If $S = \{\mathbf{x}_1, \mathbf{x}_2, \ldots, \mathbf{x}_n\}$ is a basis for V, then there is an $n \times n$ matrix A that represents L with respect to S (see Section 4.3). To determine an eigenvalue λ of L and an eigenvector $\mathbf{x}$ of L associated with the eigenvalue λ, we solve the equation

$$L(\mathbf{x}) = \lambda \mathbf{x}.$$

Using Theorem 4.9, we see that an equivalent matrix equation is

$$A\left[\mathbf{x}\right]_S = \lambda\left[\mathbf{x}\right]_S.$$

This formulation allows us to use techniques for solving linear systems in R^n to determine eigenvalue-eigenvector pairs of L.

EXAMPLE 8

Let $L\colon P_2 \to P_2$ be a linear operator defined by

$$L(at^2 + bt + c) = -bt - 2c.$$

The eigen-problem for L can be formulated in terms of a matrix representing L with respect to a specific basis for P_2. Find the corresponding matrix eigen-problem for each of the bases $S = \{1 - t, 1 + t, t^2\}$ and $T = \{t - 1, 1, t^2\}$ for P_2.

Solution To find the matrix A that represents L with respect to the basis S, we compute (verify)

$$L(1 - t) = t - 2 = -\tfrac{3}{2}(1 - t) - \tfrac{1}{2}(1 + t) + 0t^2, \qquad \text{so } \left[L(1 - t)\right]_S = \begin{bmatrix} -\frac{3}{2} \\ -\frac{1}{2} \\ 0 \end{bmatrix},$$

$$L(1 + t) = -t - 2 = -\tfrac{1}{2}(1 - t) - \tfrac{3}{2}(1 + t) + 0t^2, \quad \text{so } \left[L(1 + t)\right]_S = \begin{bmatrix} -\frac{1}{2} \\ -\frac{3}{2} \\ 0 \end{bmatrix},$$

$$L(t^2) = 0 = 0(1 - t) + 0(1 + t) + 0t^2, \qquad \text{so } \left[L(t^2)\right]_S = \begin{bmatrix} 0 \\ 0 \\ 0 \end{bmatrix},$$

Then

$$A = \begin{bmatrix} -\frac{3}{2} & -\frac{1}{2} & 0 \\ -\frac{1}{2} & -\frac{3}{2} & 0 \\ 0 & 0 & 0 \end{bmatrix}$$

and the matrix eigen-problem for L with respect to S is that of finding a real number λ and a nonzero vector $\mathbf{x}$ in R^3 so that

$$A\mathbf{x} = \lambda\mathbf{x}.$$

In a similar fashion we can show that the matrix B which represents L with respect to the basis T is

$$B = \begin{bmatrix} -1 & 0 & 0 \\ 1 & -2 & 0 \\ 0 & 0 & 0 \end{bmatrix}$$

(verify) and the corresponding matrix eigen-problem for L with respect to T is

$$B\mathbf{x} = \lambda\mathbf{x}.$$

Thus the matrix eigen-problem for L depends upon the basis selected for V. We show in Section 6.2 that the eigenvalues of L will not depend upon the matrix representing L. ■

As we have seen in Example 8, the eigen-problem for a linear transformation can be expressed in terms of a matrix representing L. We now formulate the notions of eigenvalue and eigenvector for *any* square matrix. If A is an $n \times n$ matrix, we can consider, as in Section 4.1, the linear operator $L: R^n \to R^n$ defined by $L(\mathbf{x}) = A\mathbf{x}$ for $\mathbf{x}$ in R^n. If λ is a scalar and $\mathbf{x} \neq \mathbf{0}$ a vector in R^n such that

$$A\mathbf{x} = \lambda\mathbf{x}, \tag{4}$$

then we shall say that λ is an **eigenvalue** of A and $\mathbf{x}$ is an **eigenvector** of A **associated with** $\boldsymbol{\lambda}$. That is, λ is an eigenvalue of L and $\mathbf{x}$ is an eigenvector of L associated with λ.

Computing Eigenvalues and Eigenvectors

Thus far we have found the eigenvalues and associated eigenvectors of a given linear transformation by inspection, geometric arguments, or very simple algebraic approaches. In the following example, we compute the eigenvalues and associated eigenvectors of a matrix by a somewhat more systematic method.

EXAMPLE 9

Let $A = \begin{bmatrix} 1 & 1 \\ -2 & 4 \end{bmatrix}$. We wish to find the eigenvalues of A and their associated eigenvectors. Thus we wish to find all real numbers λ and all nonzero vectors

$\mathbf{x} = \begin{bmatrix} x_1 \\ x_2 \end{bmatrix}$ that satisfy Equation (4):

$$\begin{bmatrix} 1 & 1 \\ -2 & 4 \end{bmatrix} \begin{bmatrix} x_1 \\ x_2 \end{bmatrix} = \lambda \begin{bmatrix} x_1 \\ x_2 \end{bmatrix}, \tag{5a}$$

which yields

$$\begin{aligned} x_1 + x_2 &= \lambda x_1 \\ -2x_1 + 4x_2 &= \lambda x_2, \end{aligned} \quad \text{or} \quad \begin{aligned} (\lambda - 1)x_1 - x_2 &= 0 \\ 2x_1 + (\lambda - 4)x_2 &= 0. \end{aligned} \tag{5b}$$

This homogeneous system of two equations in two unknowns has a nontrivial solution if and only if the determinant of the coefficient matrix is zero. Thus

$$\begin{vmatrix} \lambda - 1 & -1 \\ 2 & \lambda - 4 \end{vmatrix} = 0.$$

This means that

$$\lambda^2 - 5\lambda + 6 = 0 = (\lambda - 3)(\lambda - 2),$$

and so $\lambda_1 = 2$ and $\lambda_2 = 3$ are the eigenvalues of A. That is, Equation (5b) will have a nontrivial solution only when $\lambda_1 = 2$ or $\lambda_2 = 3$. To find all eigenvectors of A associated with $\lambda_1 = 2$, we substitute $\lambda_1 = 2$ in Equation (5a):

$$\begin{bmatrix} 1 & 1 \\ -2 & 4 \end{bmatrix} \begin{bmatrix} x_1 \\ x_2 \end{bmatrix} = 2 \begin{bmatrix} x_1 \\ x_2 \end{bmatrix},$$

which yields

$$\begin{aligned} x_1 + x_2 &= 2x_1 \\ -2x_1 + 4x_2 &= 2x_2 \end{aligned} \quad \text{or} \quad \begin{aligned} (2 - 1)x_1 - x_2 &= 0 \\ 2x_1 + (2 - 4)x_2 &= 0 \end{aligned}$$

$$\text{or} \quad \begin{aligned} x_1 - x_2 &= 0 \\ 2x_1 - 2x_2 &= 0. \end{aligned}$$

Note that we could have obtained this last homogeneous system by merely substituting $\lambda_1 = 2$ in (5b). All solutions to this last system are given by

$$\begin{aligned} x_1 &= x_2 \\ x_2 &= \text{any real number } r. \end{aligned}$$

Hence all eigenvectors associated with the eigenvalue $\lambda_1 = 2$ are given by $\begin{bmatrix} r \\ r \end{bmatrix}$, r any nonzero real number. In particular, for $r = 1$, $\mathbf{x}_1 = \begin{bmatrix} 1 \\ 1 \end{bmatrix}$ is an eigenvector associated with $\lambda_1 = 2$. Similarly, substituting $\lambda_2 = 3$ in Equation (5b), we obtain

$$\begin{aligned} (3 - 1)x_1 - x_2 &= 0 \\ 2x_1 + (3 - 4)x_2 &= 0 \end{aligned} \quad \text{or} \quad \begin{aligned} 2x_1 - x_2 &= 0 \\ 2x_1 - x_2 &= 0. \end{aligned}$$

All solutions to this last homogeneous system are given by

$$x_1 = \tfrac{1}{2}x_2$$
$$x_2 = \text{any real number } r.$$

Hence all eigenvectors associated with the eigenvalue $\lambda_2 = 3$ are given by $\begin{bmatrix} \frac{r}{2} \\ r \end{bmatrix}$, r any nonzero real number. In particular, for $r = 2$, $\mathbf{x}_2 = \begin{bmatrix} 1 \\ 2 \end{bmatrix}$ is an eigenvector associated with the eigenvalue $\lambda_2 = 3$. ■

We now use the method followed in Example 9 as our standard method for finding the eigenvalues and associated eigenvectors of a given matrix. We first state some terminology.

Definition 6.2

Let $A = \begin{bmatrix} a_{11} & a_{12} & \cdots & a_{1n} \\ a_{21} & a_{22} & \cdots & a_{2n} \\ \vdots & \vdots & & \vdots \\ a_{n1} & a_{n2} & \cdots & a_{nn} \end{bmatrix}$ be an $n \times n$ matrix. Then the determinant of the matrix

$$\lambda I_n - A = \begin{bmatrix} \lambda - a_{11} & -a_{12} & \cdots & -a_{1n} \\ -a_{21} & \lambda - a_{22} & \cdots & -a_{2n} \\ \vdots & \vdots & & \vdots \\ -a_{n1} & -a_{n2} & \cdots & \lambda - a_{nn} \end{bmatrix}$$

is called the **characteristic polynomial** of A. The equation

$$p(\lambda) = \det(\lambda I_n - A) = 0$$

is called the **characteristic equation** of A. ▲

Recall from Chapter 5 that each term in the expansion of the determinant of an $n \times n$ matrix is a product of n entries of the matrix, containing exactly one entry from each row and exactly one entry from each column. Thus, if we expand $\det(\lambda I_n - A)$, we obtain a polynomial of degree n. The expression involving λ^n in the characteristic polynomial of A comes from the product

$$(\lambda - a_{11})(\lambda - a_{22}) \cdots (\lambda - a_{nn}),$$

and so the coefficient of λ^n is 1. We can then write

$$\det(\lambda I_n - A) = p(\lambda) = \lambda^n + a_1\lambda^{n-1} + a_2\lambda^{n-2} + \cdots + a_{n-1}\lambda + a_n.$$

Note that if we let $\lambda = 0$ in $\det(\lambda I_n - A)$ as well as in the expression on the right, then we get $\det(-A) = a_n$, and thus the constant term of the characteristic polynomial of A is $a_n = (-1)^n \det(A)$.

EXAMPLE 10

Let $A = \begin{bmatrix} 1 & 2 & -1 \\ 1 & 0 & 1 \\ 4 & -4 & 5 \end{bmatrix}$. The characteristic polynomial of A is

$$p(\lambda) = \det(\lambda I_3 - A) = \begin{vmatrix} \lambda - 1 & -2 & 1 \\ -1 & \lambda & -1 \\ -4 & 4 & \lambda - 5 \end{vmatrix} = \lambda^3 - 6\lambda^2 + 11\lambda - 6.$$

(verify). ■

We now connect the characteristic polynomial of a matrix with its eigenvalues in the following theorem.

Theorem 6.1

Let A be an $n \times n$ matrix. The eigenvalues of A are the real roots of the characteristic polynomial of A.

Proof

Let $\mathbf{x}$ in R^n be an eigenvector of A associated with the eigenvalue λ. Then

$$A\mathbf{x} = \lambda\mathbf{x} \quad \text{or} \quad A\mathbf{x} = (\lambda I_n)\mathbf{x} \quad \text{or} \quad (\lambda I_n - A)\mathbf{x} = \mathbf{0}.$$

This is a homogeneous system of n equations in n unknowns; a nontrivial solution exists if and only if $\det(\lambda I_n - A) = 0$. Hence λ is a real root of the characteristic polynomial of A.

Conversely, if λ is a real root of the characteristic polynomial of A, then $\det(\lambda I_n - A) = 0$, so the homogeneous system $(\lambda I_n - A)\mathbf{x} = \mathbf{0}$ has a nontrivial solution. Hence λ is an eigenvalue of A. ●

Thus, to find the eigenvalues of a given matrix A, we must find the real roots of its characteristic polynomial $p(\lambda)$. There are many methods for finding approximations to the roots of a polynomial, some of them more effective than others. Two results that are sometimes useful in this connection are: (1) the product of all the roots of the polynomial

$$p(\lambda) = \lambda^n + a_1\lambda^{n-1} + \cdots + a_{n-1}\lambda + a_n$$

is $(-1)^n a_n$, and (2) if $a_1, a_2, \ldots, a_n$ are integers, then $p(\lambda)$ cannot have a rational root that is not already an integer. Thus as possible rational roots of $p(\lambda)$ one need only try the integer factors of a_n. Of course, $p(\lambda)$ might well have irrational roots.

To minimize the computational effort and as a convenience to the reader, *almost all of the characteristic polynomials to be solved in the rest of this chapter have only integer roots*, and each of these roots is a factor of the constant term of the characteristic polynomial of A. The corresponding eigenvectors are obtained by substituting for λ in the matrix equation

$$(\lambda I_n - A)\mathbf{x} = \mathbf{0} \tag{6}$$

and solving the resulting homogeneous system. The solution to these types of problems has been studied in Section 2.6.

EXAMPLE 11

Compute the eigenvalues and associated eigenvectors of the matrix A defined in Example 10.

Solution In Example 10 we found the characteristic polynomial of A to be

$$p(\lambda) = \lambda^3 - 6\lambda^2 + 11\lambda - 6.$$

The possible integer roots of $p(\lambda)$ are ± 1, ± 2, ± 3, and ± 6. By substituting these values in $p(\lambda)$, we find that $p(1) = 0$, so that $\lambda = 1$ is a root of $p(\lambda)$. Hence $(\lambda - 1)$ is a factor of $p(\lambda)$. Dividing $p(\lambda)$ by $(\lambda - 1)$, we obtain

$$p(\lambda) = (\lambda - 1)(\lambda^2 - 5\lambda + 6) \quad \text{(verify)}.$$

Factoring $\lambda^2 - 5\lambda + 6$, we have

$$p(\lambda) = (\lambda - 1)(\lambda - 2)(\lambda - 3).$$

The eigenvalues of A are then $\lambda_1 = 1$, $\lambda_2 = 2$, and $\lambda_3 = 3$. To find an eigenvector $\mathbf{x}_1$ associated with $\lambda_1 = 1$, we substitute $\lambda = 1$ in (6), obtaining

$$\begin{bmatrix} 1-1 & -2 & 1 \\ -1 & 1 & -1 \\ -4 & 4 & 1-5 \end{bmatrix} \begin{bmatrix} x_1 \\ x_2 \\ x_3 \end{bmatrix} = \begin{bmatrix} 0 \\ 0 \\ 0 \end{bmatrix}$$

or

$$\begin{bmatrix} 0 & -2 & 1 \\ -1 & 1 & -1 \\ -4 & 4 & -4 \end{bmatrix} \begin{bmatrix} x_1 \\ x_2 \\ x_3 \end{bmatrix} = \begin{bmatrix} 0 \\ 0 \\ 0 \end{bmatrix}.$$

The vector $\begin{bmatrix} -\frac{r}{2} \\ \frac{r}{2} \\ r \end{bmatrix}$ is a solution for any real number r. Thus $\mathbf{x}_1 = \begin{bmatrix} -1 \\ 1 \\ 2 \end{bmatrix}$ is an eigenvector of A associated with $\lambda_1 = 1$ (r was taken as 2).

To find an eigenvector $\mathbf{x}_2$ associated with $\lambda_2 = 2$, we substitute $\lambda = 2$ in (6), obtaining

$$\begin{bmatrix} 2-1 & -2 & 1 \\ -1 & 2 & -1 \\ -4 & 4 & 2-5 \end{bmatrix} \begin{bmatrix} x_1 \\ x_2 \\ x_3 \end{bmatrix} = \begin{bmatrix} 0 \\ 0 \\ 0 \end{bmatrix}$$

or

$$\begin{bmatrix} 1 & -2 & 1 \\ -1 & 2 & -1 \\ -4 & 4 & -3 \end{bmatrix} \begin{bmatrix} x_1 \\ x_2 \\ x_3 \end{bmatrix} = \begin{bmatrix} 0 \\ 0 \\ 0 \end{bmatrix}.$$

The vector $\begin{bmatrix} -\frac{r}{2} \\ \frac{r}{4} \\ r \end{bmatrix}$ is a solution for any real number r. Thus $\mathbf{x}_2 = \begin{bmatrix} -2 \\ 1 \\ 4 \end{bmatrix}$ is an eigenvector of A associated with $\lambda_2 = 2$ (r was taken as 4).

To find an eigenvector $\mathbf{x}_3$ associated with $\lambda_3 = 3$, we substitute $\lambda = 3$ in (6), obtaining

$$\begin{bmatrix} 3-1 & -2 & 1 \\ -1 & 3 & -1 \\ -4 & 4 & 3-5 \end{bmatrix} \begin{bmatrix} x_1 \\ x_2 \\ x_3 \end{bmatrix} = \begin{bmatrix} 0 \\ 0 \\ 0 \end{bmatrix}$$

or

$$\begin{bmatrix} 2 & -2 & 1 \\ -1 & 3 & -1 \\ -4 & 4 & -2 \end{bmatrix} \begin{bmatrix} x_1 \\ x_2 \\ x_3 \end{bmatrix} = \begin{bmatrix} 0 \\ 0 \\ 0 \end{bmatrix}.$$

The vector $\begin{bmatrix} -\frac{r}{4} \\ \frac{r}{4} \\ r \end{bmatrix}$ is a solution for any real number r. Thus $\mathbf{x}_3 = \begin{bmatrix} -1 \\ 1 \\ 4 \end{bmatrix}$ is an eigenvector of A associated with $\lambda_3 = 3$ (r was taken as 4). ■

EXAMPLE 12

Let L be the linear operator on P_2 defined in Example 8. Using the matrix B obtained there representing L with respect to the basis $\{t-1, 1, t^2\}$ for P_2, find the eigenvalues and associated eigenvectors of L.

Solution The characteristic polynomial of

$$B = \begin{bmatrix} -1 & 0 & 0 \\ 1 & -2 & 0 \\ 0 & 0 & 0 \end{bmatrix}$$

is $p(\lambda) = \lambda(\lambda+2)(\lambda+1)$ (verify), so the eigenvalues of L are $\lambda_1 = 0$, $\lambda_2 = -2$, and $\lambda_3 = -1$. Associated eigenvectors are (verify)

$$\mathbf{x}_1 = \begin{bmatrix} 0 \\ 0 \\ 1 \end{bmatrix}, \quad \mathbf{x}_2 = \begin{bmatrix} 1 \\ 1 \\ 0 \end{bmatrix}, \quad \mathbf{x}_3 = \begin{bmatrix} 0 \\ 1 \\ 0 \end{bmatrix}.$$

These are the coordinate vectors of the eigenvectors of L, so the corresponding eigenvectors of L are

$$0(t-1) + 0(1) + 1(t^2) = t^2$$

$$1(t-1) + 1(1) + 0(t^2) = t$$

and

$$0(t-1) + 1(1) + 0(t^2) = 1,$$

respectively. ■

EXAMPLE 13

The characteristic polynomial of $A = \begin{bmatrix} 0 & 1 \\ -1 & 0 \end{bmatrix}$ is $p(\lambda) = \lambda^2 + 1$ (verify), which has no real roots. Thus, according to our definition, A has no eigenvalues. Note that this is just the matrix representing a counterclockwise rotation through the angle $\pi/2$. Compare with Example 2. ■

The procedure for finding the eigenvalues and associated eigenvectors of a matrix is as follows.

STEP 1 Determine the real roots of the characteristic polynomial $p(\lambda) = \det(\lambda I_n - A)$. These are the eigenvalues of A.

STEP 2 For each eigenvalue λ, find all the nontrivial solutions to the homogeneous system $(\lambda I_n - A)\mathbf{x} = \mathbf{0}$. These are the eigenvectors of A associated with the eigenvalue λ.

Of course, the characteristic polynomial of a matrix may have some complex roots and it may even have no real roots (see Example 13). However, in the important case of symmetric matrices, all the roots of the characteristic polynomial are real. We shall prove this in Section 6.4 (Theorem 6.6).

Eigenvalues and eigenvectors satisfy many important and interesting properties. For example, if A is an upper (lower) triangular matrix, then the eigenvalues of A are the elements on the main diagonal of A (Exercise 9). Other properties are developed in the exercises for this section.

It must be pointed out that the method for finding the eigenvalues of a linear transformation or matrix by obtaining the real roots of the characteristic polynomial is not practical for $n > 4$, since it involves evaluating a determinant. Efficient numerical methods for finding eigenvalues are studied in numerical analysis courses.

6.1 Exercises

1. Let $L: R^2 \to R^2$ be counterclockwise rotation through an angle π. Find the eigenvalues and associated eigenvectors of L.

2. Let $L: P_1 \to P_1$ be the linear operator defined by $L(at + b) = bt - a$. Using the matrix representing L with respect to the basis $\{1, t\}$ for P_1, find the eigenvalues and associated eigenvectors of L.

3. Let $L: P_2 \to P_2$ be the linear operator defined by $L(at^2 + bt + c) = c - at^2$. Using the matrix representing L with respect to the basis $\{t^2 + 1, t, 1\}$ for P_2, find the eigenvalues and associated eigenvectors of L.

4. Let $L: R_3 \to R_3$ be defined by

$$L\left(\begin{bmatrix} a_1 & a_2 & a_3 \end{bmatrix}\right) = \begin{bmatrix} 2a_1 + 3a_2 & -a_2 + 4a_3 & 3a_3 \end{bmatrix}.$$

Using the natural basis for R_3, find the eigenvalues and associated eigenvectors of L.

5. Find the characteristic polynomial of each of the following matrices.

(a) $\begin{bmatrix} 2 & 1 \\ -1 & 3 \end{bmatrix}$.

(b) $\begin{bmatrix} 1 & 2 & 1 \\ 0 & 1 & 2 \\ -1 & 3 & 2 \end{bmatrix}$.

(c) $\begin{bmatrix} 4 & -1 & 3 \\ 0 & 2 & 1 \\ 0 & 0 & 3 \end{bmatrix}$. (d) $\begin{bmatrix} 4 & 2 \\ 3 & 3 \end{bmatrix}$.

6. Find the characteristic polynomial, the eigenvalues, and associated eigenvectors of each of the following matrices.

(a) $\begin{bmatrix} 1 & 1 \\ 1 & 1 \end{bmatrix}$. (b) $\begin{bmatrix} 1 & 0 & 0 \\ -1 & 3 & 0 \\ 3 & 2 & -2 \end{bmatrix}$.

(c) $\begin{bmatrix} 0 & 1 & 2 \\ 0 & 0 & 3 \\ 0 & 0 & 0 \end{bmatrix}$. (d) $\begin{bmatrix} 2 & 1 & 2 \\ 2 & 2 & -2 \\ 3 & 1 & 1 \end{bmatrix}$.

7. Find the characteristic polynomial, the eigenvalues, and associated eigenvectors of each of the following matrices.

(a) $\begin{bmatrix} 1 & -1 \\ 2 & 4 \end{bmatrix}$. (b) $\begin{bmatrix} 2 & -2 & 3 \\ 0 & 3 & -2 \\ 0 & -1 & 2 \end{bmatrix}$.

(c) $\begin{bmatrix} 2 & 2 & 3 \\ 1 & 2 & 1 \\ 2 & -2 & 1 \end{bmatrix}$. (d) $\begin{bmatrix} 2 & -2 & 3 \\ 0 & 3 & -2 \\ 0 & -1 & 2 \end{bmatrix}$.

8. Find all the eigenvalues and associated eigenvectors of each of the following matrices.

(a) $\begin{bmatrix} 1 & 4 \\ 1 & -2 \end{bmatrix}$. (b) $\begin{bmatrix} 0 & -9 \\ 1 & 0 \end{bmatrix}$.

(c) $\begin{bmatrix} 4 & 2 & -4 \\ 1 & 5 & -4 \\ 0 & 0 & 6 \end{bmatrix}$. (d) $\begin{bmatrix} 0 & -1 & 0 \\ 1 & 0 & 0 \\ 0 & 1 & 0 \end{bmatrix}$.

9. Prove that if A is an upper (lower) triangular matrix, then the eigenvalues of A are the elements on the main diagonal of A.

10. Prove that A and A^T have the same eigenvalues. What, if anything, can we say about the associated eigenvectors of A and A^T?

11. Let

$$A = \begin{bmatrix} 1 & 2 & 3 & 4 \\ 0 & -1 & 3 & 2 \\ 0 & 0 & 3 & 3 \\ 0 & 0 & 0 & 2 \end{bmatrix}$$

represent the linear transformation $L: M_{22} \to M_{22}$ with respect to the basis

$$S = \left\{ \begin{bmatrix} 1 & 0 \\ 0 & 0 \end{bmatrix}, \begin{bmatrix} 0 & 1 \\ 0 & 0 \end{bmatrix}, \begin{bmatrix} 0 & 0 \\ 1 & 0 \end{bmatrix}, \begin{bmatrix} 0 & 0 \\ 0 & 1 \end{bmatrix} \right\}.$$

Find the eigenvalues and associated eigenvectors of L.

12. Let $L: V \to V$ be a linear operator, where V is an n-dimensional vector space. Let λ be an eigenvalue of L. Prove that the subset of V consisting of $\mathbf{0}_V$ and all eigenvectors of L associated with λ is a subspace of V. This subspace is called the **eigenspace** associated with λ.

13. Let λ be an eigenvalue of the $n \times n$ matrix A. Prove that the subset of R^n consisting of the zero vector and all eigenvectors of A associated with λ is a subspace of R^n. This subspace is called the **eigenspace** associated with λ. (This result is a corollary to the result in Exercise 12.)

14. In Exercises 12 and 13, why do we have to include $\mathbf{0}_V$ in the set of all eigenvectors associated with λ?

15. Find a basis for the eigenspace (see Exercise 13) associated with λ in each of the following.

(a) $\begin{bmatrix} 0 & 0 & 1 \\ 0 & 1 & 0 \\ 1 & 0 & 0 \end{bmatrix}$, $\lambda = 1$.

(b) $\begin{bmatrix} 2 & 1 & 0 \\ 1 & 2 & 1 \\ 0 & 1 & 2 \end{bmatrix}$, $\lambda = 2$.

(c) $\begin{bmatrix} 3 & 0 & 0 \\ -2 & 3 & -2 \\ 2 & 0 & 5 \end{bmatrix}$, $\lambda = 3$.

(d) $\begin{bmatrix} 4 & 2 & 0 & 0 \\ 3 & 3 & 0 & 0 \\ 0 & 0 & 2 & 5 \\ 0 & 0 & 0 & 2 \end{bmatrix}$, $\lambda = 2$.

16. Let $A = \begin{bmatrix} 2 & 2 & 3 & 4 \\ 0 & 2 & 3 & 2 \\ 0 & 0 & 1 & 1 \\ 0 & 0 & 0 & 1 \end{bmatrix}$.

(a) Find a basis for the eigenspace associated with the eigenvalue $\lambda_1 = 1$.

(b) Find a basis for the eigenspace associated with the eigenvalue $\lambda_2 = 2$.

17. Prove that if λ is an eigenvalue of a matrix A with associated eigenvector $\mathbf{x}$, and k is a positive integer, then λ^k is an eigenvalue of the matrix $A^k = A \cdot A \cdot \cdots \cdot A$ (k factors) with associated eigenvector $\mathbf{x}$.

18. Let $A = \begin{bmatrix} 1 & 4 \\ 1 & -2 \end{bmatrix}$ be the matrix of Exercise 8(a). Find the eigenvalues and eigenvectors of A^2 and verify Exercise 17.

19. Prove that if $A^k = O$ for some positive integer k (i.e., if A is a nilpotent matrix (see Supplementary Exercise 22 in Chapter 1)), then 0 is the only eigenvalue of A. (*Hint*: Use Exercise 17.)

20. Let A be an $n \times n$ matrix.

(a) Show that $\det(A)$ is the product of all the roots of the characteristic polynomial of A.

(b) Show that A is singular if and only if 0 is an eigenvalue of A.

(c) Also prove the analogous statement for a linear transformation: If $L: V \to V$ is a linear transformation, show that L is not one-to-one if and only if 0 is an eigenvalue of L.

(d) Show that if A is nilpotent (see Supplementary Exercise 22 in Chapter 1), then A is singular.

21. Let $L: V \to V$ be an invertible linear operator and let λ be an eigenvalue of L with associated eigenvector $\mathbf{x}$.

(a) Show that $1/\lambda$ is an eigenvalue of L^{-1} with associated eigenvector $\mathbf{x}$.

(b) State and prove the analogous statement for matrices.

22. Let A be an $n \times n$ matrix with eigenvalues λ_1 and λ_2, where $\lambda_1 \neq \lambda_2$. Let S_1 and S_2 be the eigenspaces associated with λ_1 and λ_2, respectively. Explain why the zero vector is the only vector that is in both S_1 and S_2.

23. Let λ be an eigenvalue of A with associated eigenvector $\mathbf{x}$. Show that $\lambda + r$ is an eigenvalue of $A + rI_n$ with associated eigenvector $\mathbf{x}$. Thus, adding a scalar multiple of the identity matrix to A merely shifts the eigenvalues by the scalar multiple.

24. Let A be an $n \times n$ matrix and consider the linear operator on R^n defined by $L(\mathbf{u}) = A\mathbf{u}$, for $\mathbf{u}$ in R^n. A subspace W of R^n is called **invariant** under L if for any $\mathbf{w}$ in W, $L(\mathbf{w})$ is also in W. Show that an eigenspace of A is invariant under L.

25. Let A and B be $n \times n$ matrices such that $A\mathbf{x} = \lambda\mathbf{x}$ and $B\mathbf{x} = \mu\mathbf{x}$. Show that:

(a) $(A + B)\mathbf{x} = (\lambda + \mu)\mathbf{x}$.

(b) $(AB)\mathbf{x} = (\lambda\mu)\mathbf{x}$.

26. The **Cayley–Hamilton theorem** states that a matrix satisfies its characteristic equation; that is, if A is an $n \times n$ matrix with characteristic polynomial

$$p(\lambda) = \lambda^n + a_1\lambda^{n-1} + \cdots + a_{n-1}\lambda + a_n,$$

then

$$A^n + a_1A^{n-1} + \cdots + a_{n-1}A + a_nI_n = O.$$

The proof and applications of this result, unfortunately, lie beyond the scope of this book. Verify the Cayley–Hamilton theorem for the following matrices.

(a) $\begin{bmatrix} 1 & 2 & 3 \\ 2 & -1 & 5 \\ 3 & 2 & 1 \end{bmatrix}$. (b) $\begin{bmatrix} 1 & 2 & 3 \\ 0 & 2 & 2 \\ 0 & 0 & -3 \end{bmatrix}$.

(c) $\begin{bmatrix} 3 & 3 \\ 2 & 4 \end{bmatrix}$.

27. Let A be an $n \times n$ matrix whose characteristic polynomial is

$$p(\lambda) = \lambda^n + a_1\lambda^{n-1} + \cdots + a_{n-1}\lambda + a_n.$$

If A is nonsingular, show that

$$A^{-1} = -\frac{1}{a_n}(A^{n-1} + a_1A^{n-2} + \cdots + a_{n-2}A + a_{n-1}I_n).$$

[*Hint*: Use the Cayley–Hamilton theorem (Exercise 26).]

28. Let

$$A = \begin{bmatrix} a & b \\ c & d \end{bmatrix}.$$

Prove that the characteristic polynomial $p(\lambda)$ of A is given by

$$p(\lambda) = \lambda^2 - \mathrm{Tr}(A)\lambda + \det(A),$$

where $\mathrm{Tr}(A)$ denotes the trace of A (see Exercise 29 in Section 1.2).

29. Show that if A is a matrix all of whose columns add up to 1, then $\lambda = 1$ is an eigenvalue of A. (*Hint*: Consider the product $A^T\mathbf{x}$, where $\mathbf{x}$ is a vector all of whose entries are 1 and use Exercise 10.)

30. Determine if your software has a command for finding the characteristic polynomial of a matrix A. If it does, compare the output from your software with the results in Examples 9 and 11. Software output for a characteristic polynomial often is just the set of coefficients of the polynomial with the powers of λ omitted. Carefully determine the order in which the coefficients are listed. Experiment further with the matrices in Exercises 5 and 6.

31. If your software has a command for finding the characteristic polynomial of a matrix A (see Exercise 30), it probably has another command for finding the roots of polynomials. Investigate the use of these commands in your software. The roots of the characteristic polynomial of A are the eigenvalues of A. (In this book, eigenvalues are defined to be real numbers.)

32. Assuming that your software has the commands discussed in Exercises 30 and 31, apply them to find the eigenvalues of $A = \begin{bmatrix} 0 & 1 \\ -1 & 0 \end{bmatrix}$. If your software is successful, the results should be $\lambda = i, -i$, where $i = \sqrt{-1}$. (See Appendix B.) (**Caution:** Some software does not handle complex roots and may not permit complex elements in a matrix. Determine the situation for the software you use.)

33. Most linear algebra software has a command for automatically determining the eigenvalues of a matrix. Determine the command available in your software. Test its behavior on Examples 11 and 13. Often such a command uses techniques which are different than finding the roots of the characteristic polynomial. Use the documentation accompanying your software to find the method used. (**Warning:** It may involve ideas from Section 6.4 or more sophisticated procedures.)

34. Following the ideas in Exercise 33, determine the command in your software for obtaining the eigenvectors of a matrix. Often it is a variation of the eigenvalue command. Test it on the matrices in Examples 9 and 11. These examples cover the types of cases for eigenvectors that you will encounter in this course.

6.2 DIAGONALIZATION AND SIMILAR MATRICES

If $L: V \to V$ is a linear operator on an n-dimensional vector space V, we have already seen in Section 6.1 that we can find the eigenvalues of L and associated eigenvectors by using a matrix representing L with respect to a basis for V. The computational steps involved depend upon the matrix selected to represent L. An ideal situation would be the following one: Suppose that L is represented by a matrix A with respect to a certain basis for V. Find a basis for V with respect to which L is represented by a diagonal matrix D whose eigenvalues are the same as the eigenvalues of A. Of course, this is a very desirable situation, since the eigenvalues of D are merely its entries on the main diagonal. Now recall from Theorem 4.14 in Section 4.5 that A and D represent the same linear operator $L: V \to V$ with respect to two bases for V if and only if they are similar. That is, if and only if there exists a nonsingular matrix P such that $D = P^{-1}AP$. In this section we examine the type of linear transformations and matrices for which this situation is possible.

Definition 6.3

Let $L: V \to V$ be a linear operator on an n-dimensional vector space V. We say that L is **diagonalizable** or can be **diagonalized** if there exists a basis S for V such that L is represented with respect to S by a diagonal matrix D. ▲

EXAMPLE 1

In Example 2 of Section 4.5 we considered the linear transformation $L: R_3 \to R_3$ defined by

$$L\left(\begin{bmatrix} a_1 & a_2 & a_3 \end{bmatrix}\right) = \begin{bmatrix} 2a_1 - a_3 & a_1 + a_2 - a_3 & a_3 \end{bmatrix}.$$

In that example we used the basis

$$S' = \{\begin{bmatrix} 1 & 0 & 1 \end{bmatrix}, \begin{bmatrix} 0 & 1 & 0 \end{bmatrix}, \begin{bmatrix} 1 & 1 & 0 \end{bmatrix}\}$$

for R_3 and showed that the representation of L with respect to S' is

$$B = \begin{bmatrix} 1 & 0 & 0 \\ 0 & 1 & 0 \\ 0 & 0 & 2 \end{bmatrix}.$$

Hence L is a diagonalizable linear transformation. ■

We next show that similar matrices have the same eigenvalues.

Theorem 6.2

Similar matrices have the same eigenvalues.

Proof

Let A and B be similar. Then $B = P^{-1}AP$, for some nonsingular matrix P. We prove that A and B have the same characteristic polynomials, $p_A(\lambda)$ and $p_B(\lambda)$, respectively. We have

$$\begin{aligned} p_B(\lambda) &= \det(\lambda I_n - B) = \det(\lambda I_n - P^{-1}AP) \\ &= \det(P^{-1}\lambda I_n P - P^{-1}AP) = \det(P^{-1}(\lambda I_n - A)P) \\ &= \det(P^{-1}) \det(\lambda I_n - A) \det(P) \\ &= \det(P^{-1}) \det(P) \det(\lambda I_n - A) \\ &= \det(\lambda I_n - A) = p_A(\lambda). \end{aligned} \tag{1}$$

Since $p_A(\lambda) = p_B(\lambda)$, it follows that A and B have the same eigenvalues. ●

Note that in the proof of Theorem 6.2 we have used the fact that the product of $\det(P^{-1})$ and $\det(P)$ is 1, and that determinants are numbers, so their order as factors in multiplication does not matter.

Let $L: V \to V$ be a diagonalizable linear operator on an n-dimensional vector space V and let $S = \{\mathbf{x}_1, \mathbf{x}_2, \ldots, \mathbf{x}_n\}$ be a basis for V such that L is represented with respect to S by a diagonal matrix

$$D = \begin{bmatrix} \lambda_1 & 0 & \cdots & & 0 \\ 0 & \lambda_2 & \cdots & & 0 \\ \vdots & \vdots & & & \vdots \\ & & & & 0 \\ 0 & 0 & \cdots & 0 & \lambda_n \end{bmatrix},$$

where $\lambda_1, \lambda_2, \ldots, \lambda_n$ are real scalars. Now recall that if D represents L with respect to S, then the jth column of D is the coordinate vector $\left[L(\mathbf{x}_j)\right]_S$ of $L(\mathbf{x}_j)$ with respect to S. Thus we have

$$\left[L(\mathbf{x}_j)\right]_S = \begin{bmatrix} 0 \\ 0 \\ \vdots \\ 0 \\ \lambda_j \\ 0 \\ \vdots \\ 0 \end{bmatrix} \quad \leftarrow j\text{th row},$$

which means that

$$L(\mathbf{x}_j) = 0\mathbf{x}_1 + 0\mathbf{x}_2 + \cdots + 0\mathbf{x}_{j-1} + \lambda_j\mathbf{x}_j + 0\mathbf{x}_{j+1} + \cdots + 0\mathbf{x}_n = \lambda_j\mathbf{x}_j.$$

Conversely, let $S = \{\mathbf{x}_1, \mathbf{x}_2, \dots, \mathbf{x}_n\}$ be a basis for V such that

$$L(\mathbf{x}_j) = \lambda_j\mathbf{x}_j = 0\mathbf{x}_1 + 0\mathbf{x}_2 + \cdots + 0\mathbf{x}_{j-1} + \lambda_j\mathbf{x}_j + 0\mathbf{x}_{j+1} + \cdots + 0\mathbf{x}_n \quad \text{for } j = 1, 2, \dots, n.$$

We now find the matrix representing L with respect to S. The jth column of this matrix is

$$\left[L(\mathbf{x}_j)\right]_S = \begin{bmatrix} 0 \\ 0 \\ \vdots \\ 0 \\ \lambda_j \\ 0 \\ \vdots \\ 0 \end{bmatrix}.$$

Hence

$$D = \begin{bmatrix} \lambda_1 & 0 & \cdots & & 0 \\ 0 & \lambda_2 & \cdots & & 0 \\ \vdots & \vdots & & & \vdots \\ & & & & 0 \\ 0 & 0 & \cdots & 0 & \lambda_n \end{bmatrix},$$

a diagonal matrix, represents L with respect to S, so L is diagonalizable.

We can now state the following theorem, whose proof has been given above.

Theorem 6.3

Let $L: V \to V$ be a linear operator on an n-dimensional vector space V. Then L is diagonalizable if and only if V has a basis S of eigenvectors of L. Moreover, if D is the diagonal matrix representing L with respect to S, then the entries on the main diagonal of D are the eigenvalues of L. ●

In terms of matrices, Theorem 6.3 can be stated as follows.

Theorem 6.4

An $n \times n$ matrix A is similar to a diagonal matrix D if and only if R^n has a basis of eigenvectors of A. Moreover, the elements on the main diagonal of D are the eigenvalues of A.

Remark If a matrix A is similar to a diagonal matrix, we say that A is **diagonalizable** or can be diagonalized.

To use Theorem 6.2, we need only show that there is a set of n eigenvectors of A that are linearly independent, since n linearly independent vectors in R^n form a basis for R^n (Theorem 2.11 in Section 2.5).

EXAMPLE 2

Let $A = \begin{bmatrix} 1 & 1 \\ -2 & 4 \end{bmatrix}$. In Example 9 of Section 6.1, we found that the eigenvalues of A are $\lambda_1 = 2$ and $\lambda_2 = 3$, with associated eigenvectors $\mathbf{x}_1 = \begin{bmatrix} 1 \\ 1 \end{bmatrix}$ and $\mathbf{x}_2 = \begin{bmatrix} 1 \\ 2 \end{bmatrix}$, respectively. Since $S = \left\{ \begin{bmatrix} 1 \\ 1 \end{bmatrix}, \begin{bmatrix} 1 \\ 2 \end{bmatrix} \right\}$ is linearly independent (verify), R^2 has a basis of two eigenvectors of A and hence A can be diagonalized. From Theorem 6.4, we conclude that A is similar to $D = \begin{bmatrix} 2 & 0 \\ 0 & 3 \end{bmatrix}$. ■

EXAMPLE 3

Let $A = \begin{bmatrix} 1 & 1 \\ 0 & 1 \end{bmatrix}$. Can A be diagonalized?

Solution Since A is upper triangular, its eigenvalues are the entries on its main diagonal (Exercise 9 in Section 6.1). Thus, the eigenvalues of A are $\lambda_1 = 1$ and $\lambda_2 = 1$.

We now find eigenvectors of A associated with $\lambda_1 = 1$. Equation (6) of Section 6.1, $(\lambda I_n - A)\mathbf{x} = \mathbf{0}$, becomes, with $\lambda = 1$, the homogeneous system

$$\begin{aligned} (1-1)x_1 - \quad x_2 &= 0 \\ (1-1)x_2 &= 0. \end{aligned}$$

The vector $\begin{bmatrix} r \\ 0 \end{bmatrix}$, for any real number r, is a solution. Thus, all eigenvectors of A are multiples of the vector $\begin{bmatrix} 1 \\ 0 \end{bmatrix}$. Since A does not have two linearly independent eigenvectors, it cannot be diagonalized. ■

If an $n \times n$ matrix A is similar to a diagonal matrix D, then $P^{-1}AP = D$ for some nonsingular matrix P. We now discuss how to construct such a matrix P. We have $AP = PD$. Let

$$D = \begin{bmatrix} \lambda_1 & 0 & \cdots & & 0 \\ 0 & \lambda_2 & \cdots & & 0 \\ \vdots & 0 & \cdots & & \vdots \\ & \vdots & & & 0 \\ 0 & 0 & \cdots & 0 & \lambda_n \end{bmatrix},$$

and let $\mathbf{x}_j$, $j = 1, 2, \ldots, n$, be the jth column of P. Note that the jth column of AP is $A\mathbf{x}_j$, and the jth column of PD is $\lambda_j \mathbf{x}_j$ (see Exercise 32 in Section 1.2). Thus we have

$$A\mathbf{x}_j = \lambda_j \mathbf{x}_j,$$

which means that λ_j is an eigenvalue of A and $\mathbf{x}_j$ is an associated eigenvector.

Conversely, if $\lambda_1, \lambda_2, \ldots, \lambda_n$ are n eigenvalues of an $n \times n$ matrix A and $\mathbf{x}_1, \mathbf{x}_2, \ldots, \mathbf{x}_n$ are associated eigenvectors forming a linearly independent set in R^n, we let P be the matrix whose jth column is $\mathbf{x}_j$. Then rank $P = n$, so by Corollary 2.7 P is nonsingular. Since $A\mathbf{x}_j = \lambda_j \mathbf{x}_j$, $j = 1, 2, \ldots, n$, we have $AP = PD$, or $P^{-1}AP = D$, which means that A is diagonalizable. Thus, if n eigenvectors $\mathbf{x}_1, \mathbf{x}_2, \ldots, \mathbf{x}_n$ of the $n \times n$ matrix A form a linearly independent set in R^n, we can diagonalize A by letting P be the matrix whose jth column is $\mathbf{x}_j$, and we find that $P^{-1}AP = D$, a diagonal matrix whose entries on the main diagonal are the associated eigenvalues of A. Of course, the order of columns of P determines the order of the diagonal entries of D.

EXAMPLE 4

Let A be as in Example 2. The eigenvalues of A are $\lambda_1 = 2$ and $\lambda_2 = 3$, and associated eigenvectors are $\mathbf{x}_1 = \begin{bmatrix} 1 \\ 1 \end{bmatrix}$ and $\mathbf{x}_2 = \begin{bmatrix} 1 \\ 2 \end{bmatrix}$, respectively. Thus $P = \begin{bmatrix} 1 & 1 \\ 1 & 2 \end{bmatrix}$ and $P^{-1} = \begin{bmatrix} 2 & -1 \\ -1 & 1 \end{bmatrix}$ (verify). Hence

$$P^{-1}AP = \begin{bmatrix} 2 & -1 \\ -1 & 1 \end{bmatrix} \begin{bmatrix} 1 & 1 \\ -2 & 4 \end{bmatrix} \begin{bmatrix} 1 & 1 \\ 1 & 2 \end{bmatrix} = \begin{bmatrix} 2 & 0 \\ 0 & 3 \end{bmatrix}.$$

On the other hand, if we let $\lambda_1 = 3$ and $\lambda_2 = 2$, then $\mathbf{x}_1 = \begin{bmatrix} 1 \\ 2 \end{bmatrix}$ and $\mathbf{x}_2 = \begin{bmatrix} 1 \\ 1 \end{bmatrix}$; $P = \begin{bmatrix} 1 & 1 \\ 2 & 1 \end{bmatrix}$ and $P^{-1} = \begin{bmatrix} -1 & 1 \\ 2 & -1 \end{bmatrix}$, and

$$P^{-1}AP = \begin{bmatrix} -1 & 1 \\ 2 & -1 \end{bmatrix} \begin{bmatrix} 1 & 1 \\ -2 & 4 \end{bmatrix} \begin{bmatrix} 1 & 1 \\ 2 & 1 \end{bmatrix} = \begin{bmatrix} 3 & 0 \\ 0 & 2 \end{bmatrix}.$$

■

EXAMPLE 5

The characteristic polynomial of $A = \begin{bmatrix} 0 & 1 \\ -1 & 0 \end{bmatrix}$ is $p(\lambda) = \lambda^2 + 1$, which has no real roots. Thus, according to our definition, A cannot be diagonalized. If we permit eigenvalues to be complex numbers then A can be diagonalized (see Appendix B.2). Note that this is just the matrix representing a counterclockwise rotation through the angle $\pi/2$. Compare with Example 2 in Section 6.1. ■

The following useful theorem identifies a large class of matrices that can be diagonalized.

Theorem 6.5

If the roots of the characteristic polynomial of an $n \times n$ matrix A are real and all are different from each other (i.e., distinct), then A is diagonalizable.

Proof

Let $\{\lambda_1, \lambda_2, \ldots, \lambda_n\}$ be the set of distinct eigenvalues of A, and let $S = \{\mathbf{x}_1, \mathbf{x}_2, \ldots, \mathbf{x}_n\}$ be a set of associated eigenvectors. We wish to prove that S is a basis for R^n, and it suffices to show that S is linearly independent.

Suppose that S is linearly dependent. Then Theorem 2.6 implies that some vector $\mathbf{x}_j$ is a linear combination of the preceding vectors in S. We can assume that $S_1 = \{\mathbf{x}_1, \mathbf{x}_2, \ldots, \mathbf{x}_{j-1}\}$ is linearly independent, for otherwise one of the vectors in S_1 is a linear combination of the preceding ones, and we can choose a new set S_2, and so on. We thus have that S_1 is linearly independent and that

$$\mathbf{x}_j = a_1\mathbf{x}_1 + a_2\mathbf{x}_2 + \cdots + a_{j-1}\mathbf{x}_{j-1}, \tag{2}$$

where $a_1, a_2, \ldots, a_{j-1}$ are real numbers. This means that

$$\begin{aligned} A\mathbf{x}_j &= A(a_1\mathbf{x}_1 + a_2\mathbf{x}_2 + \cdots + a_{j-1}\mathbf{x}_{j-1}) \\ &= a_1A\mathbf{x}_1 + a_2A\mathbf{x}_2 + \cdots + a_{j-1}A\mathbf{x}_{j-1}. \end{aligned} \tag{3}$$

Since $\lambda_1, \lambda_2, \ldots, \lambda_j$ are eigenvalues and $\mathbf{x}_1, \mathbf{x}_2, \ldots, \mathbf{x}_j$ are associated eigenvectors, we know that $A\mathbf{x}_i = \lambda_i\mathbf{x}_i$ for $i = 1, 2, \ldots, n$. Substituting in (3), we have

$$\lambda_j\mathbf{x}_j = a_1\lambda_1\mathbf{x}_1 + a_2\lambda_2\mathbf{x}_2 + \cdots + a_{j-1}\lambda_{j-1}\mathbf{x}_{j-1}. \tag{4}$$

Multiplying (1) by λ_j, we get

$$\lambda_j\mathbf{x}_j = \lambda_j a_1\mathbf{x}_1 + \lambda_j a_2\mathbf{x}_2 + \cdots + \lambda_j a_{j-1}\mathbf{x}_{j-1}. \tag{5}$$

Subtracting (4) from (3), we have

$$\begin{aligned} \mathbf{0} &= \lambda_j\mathbf{x}_j - \lambda_j\mathbf{x}_j \\ &= a_1(\lambda_1 - \lambda_j)\mathbf{x}_1 + a_2(\lambda_2 - \lambda_j)\mathbf{x}_2 + \cdots + a_{j-1}(\lambda_{j-1} - \lambda_j)\mathbf{x}_{j-1}. \end{aligned}$$

Since S_1 is linearly independent, we must have

$$a_1(\lambda_1 - \lambda_j) = 0, \quad a_2(\lambda_2 - \lambda_j) = 0, \quad \ldots, \quad a_{j-1}(\lambda_{j-1} - \lambda_j) = 0.$$

Now $(\lambda_1 - \lambda_j) \neq 0$, $(\lambda_2 - \lambda_j) \neq 0, \ldots, (\lambda_{j-1} - \lambda_j) \neq 0$, since the λ's are distinct, which implies that

$$a_1 = a_2 = \cdots = a_{j-1} = 0.$$

This means that $\mathbf{x}_j = \mathbf{0}$, which is impossible if $\mathbf{x}_j$ is an eigenvector. Hence S is linearly independent, so A is diagonalizable. ●

Remark In the proof of Theorem 6.5 we have actually established the following somewhat stronger result: Let A be an $n \times n$ matrix and let $\lambda_1, \lambda_2, \ldots, \lambda_k$ be k distinct eigenvalues of A with associated eigenvectors $\mathbf{x}_1, \mathbf{x}_2, \ldots, \mathbf{x}_k$. Then $\mathbf{x}_1, \mathbf{x}_2, \ldots, \mathbf{x}_k$ are linearly independent (Exercise 25).

If all the roots of the characteristic polynomial of A are real and not all distinct, then A may or may not be diagonalizable. The characteristic polynomial of A can be written as the product of n factors, each of the form $\lambda - \lambda_0$, where λ_0 is a root of the characteristic polynomial. Now the eigenvalues of A are the real roots of the characteristic polynomial of A. Thus the characteristic polynomial can be written as

$$(\lambda - \lambda_1)^{k_1}(\lambda - \lambda_2)^{k_2} \cdots (\lambda - \lambda_r)^{k_r},$$

where $\lambda_1, \lambda_2, \ldots, \lambda_r$ are the distinct eigenvalues of A, and $k_1, k_2, \ldots, k_r$ are integers whose sum is n. The integer k_i is called the **multiplicity** of λ_i. Thus, in Example 3, $\lambda = 1$ is an eigenvalue of $A = \begin{bmatrix} 1 & 1 \\ 0 & 1 \end{bmatrix}$ of multiplicity 2. It can be shown that if the roots of the characteristic polynomial of A are all real, then A can be diagonalized if and only if for each eigenvalue λ of multiplicity k, we can find k linearly independent eigenvectors. This means that the solution space of the homogeneous system $(\lambda I_n - A)\mathbf{x} = \mathbf{0}$ has dimension k. It can also be shown that if λ is an eigenvalue of A of multiplicity k, then we can never find more than k linearly independent eigenvectors associated with λ.

EXAMPLE 6

Let $A = \begin{bmatrix} 0 & 0 & 1 \\ 0 & 1 & 2 \\ 0 & 0 & 1 \end{bmatrix}$. Then the characteristic polynomial of A is $p(\lambda) = \lambda(\lambda - 1)^2$ (verify), so the eigenvalues of A are $\lambda_1 = 0$, $\lambda_2 = 1$, and $\lambda_3 = 1$; thus $\lambda_2 = 1$ is an eigenvalue of multiplicity 2. We now consider the eigenvectors associated with the eigenvalues $\lambda_2 = \lambda_3 = 1$. They are obtained by solving the homogeneous system $(1I_3 - A)\mathbf{x} = \mathbf{0}$ (Equation (6) in Section 6.1):

$$\begin{bmatrix} 1 & 0 & -1 \\ 0 & 0 & -2 \\ 0 & 0 & 0 \end{bmatrix} \begin{bmatrix} x_1 \\ x_2 \\ x_3 \end{bmatrix} = \begin{bmatrix} 0 \\ 0 \\ 0 \end{bmatrix}.$$

The solutions are the vectors of the form $\begin{bmatrix} 0 \\ r \\ 0 \end{bmatrix}$, where r is any real number, so the dimension of the solution space of $(1I_3 - A)\mathbf{x} = \mathbf{0}$ is 1 (why?), and we cannot find two linearly independent eigenvectors. Thus A cannot be diagonalized. ■

EXAMPLE 7

Let $A = \begin{bmatrix} 0 & 0 & 0 \\ 0 & 1 & 0 \\ 1 & 0 & 1 \end{bmatrix}$. The characteristic polynomial of A is $p(\lambda) = \lambda(\lambda - 1)^2$ (verify), so the eigenvalues of A are $\lambda_1 = 0$, $\lambda_2 = 1$, and $\lambda_3 = 1$; thus $\lambda_2 = 1$ is again an eigenvalue of multiplicity 2. Now we consider the eigenvectors associated with the eigenvalues $\lambda_2 = \lambda_3 = 1$. They are obtained by solving the homogeneous system $(1I_3 - A)\mathbf{x} = \mathbf{0}$ (Equation (6) in Section 6.1):

$$\begin{bmatrix} 1 & 0 & 0 \\ 0 & 0 & 0 \\ -1 & 0 & 0 \end{bmatrix} \begin{bmatrix} x_1 \\ x_2 \\ x_3 \end{bmatrix} = \begin{bmatrix} 0 \\ 0 \\ 0 \end{bmatrix}.$$

The solutions are the vectors of the form $\begin{bmatrix} 0 \\ r \\ s \end{bmatrix}$ for any real numbers r and s. Thus $\mathbf{x}_2 = \begin{bmatrix} 0 \\ 1 \\ 0 \end{bmatrix}$ and $\mathbf{x}_3 = \begin{bmatrix} 0 \\ 0 \\ 1 \end{bmatrix}$ are eigenvectors.

Next we look for an eigenvector associated with $\lambda_1 = 0$. We have to solve the homogeneous system

$$\begin{bmatrix} 0 & 0 & 0 \\ 0 & -1 & 0 \\ -1 & 0 & -1 \end{bmatrix} \begin{bmatrix} x_1 \\ x_2 \\ x_3 \end{bmatrix} = \begin{bmatrix} 0 \\ 0 \\ 0 \end{bmatrix}.$$

The solutions are the vectors of the form $\begin{bmatrix} r \\ 0 \\ -r \end{bmatrix}$ for any real number r. Thus $\mathbf{x}_1 = \begin{bmatrix} 1 \\ 0 \\ -1 \end{bmatrix}$ is an eigenvector associated with $\lambda_1 = 0$. Now $S = \{\mathbf{x}_1, \mathbf{x}_2, \mathbf{x}_3\}$ is linearly independent and so A can be diagonalized. ■

Thus, a matrix may fail to be diagonalizable either because not all the roots of its characteristic polynomial are real numbers, or because its eigenvectors do not form a basis for R^n.

Now define the **characteristic polynomial** of a linear operator $L: V \to V$ as the characteristic polynomial of any matrix representing L; by Theorem 6.2 all representations of L will give the same characteristic polynomial. It follows that a scalar λ is an eigenvalue of L if and only if λ is a real root of the characteristic polynomial of L.

EXAMPLE 8

In Example 12 of Section 6.1, we derived the matrix eigen-problem for the linear operator $L: P_2 \to P_2$, defined in Example 8 of that section, by $L(at^2 + bt + c) = -bt - 2c$ by using the matrix

$$B = \begin{bmatrix} -1 & 0 & 0 \\ 1 & -2 & 0 \\ 0 & 0 & 0 \end{bmatrix},$$

which represents L with respect to the basis $\{t - 1, 1, t^2\}$ for P_2. We computed the characteristic polynomial of B to be $p(\lambda) = \lambda(\lambda + 2)(\lambda + 1)$, so this is also the characteristic polynomial of L. Since the eigenvalues are distinct, it follows that B and L are diagonalizable. Of course, any other matrix representing L could be used in place of B. ■

6.2 Exercises

1. Let $L: P_2 \to P_2$ be the linear operator defined by $L(p(t)) = p'(t)$ for $p(t)$ in P_2. Is L diagonalizable? If it is, find a basis S for P_2 with respect to which L is represented by a diagonal matrix.

2. Let $L: P_1 \to P_1$ be the linear operator defined by

$$L(at + b) = -bt - a.$$

Find, if possible, a basis for P_1 with respect to which L is represented by a diagonal matrix.

3. Let $L: P_2 \to P_2$ be the linear operator defined by

$$L(at^2 + bt + c) = at^2 - c.$$

Find, if possible, a basis for P_2 with respect to which L is represented by a diagonal matrix.

4. Let V be the vector space of continuous functions with basis $\{\sin t, \cos t\}$, and let $L: V \to V$ be defined as $L(g(t)) = g'(t)$. Is L diagonalizable?

5. Let $L: P_2 \to P_2$ be the linear operator defined by

$$L(at^2 + bt + c) = (2a + b + c)t^2 + (2c - 3b)t + 4c.$$

Find the eigenvalues and eigenvectors of L. Is L diagonalizable?

6. Which of the following matrices are diagonalizable?

(a) $\begin{bmatrix} 1 & 4 \\ 1 & -2 \end{bmatrix}$. (b) $\begin{bmatrix} 1 & 0 \\ -2 & 1 \end{bmatrix}$.

(c) $\begin{bmatrix} 1 & 1 & -2 \\ 4 & 0 & 4 \\ 1 & -1 & 4 \end{bmatrix}$. (d) $\begin{bmatrix} 1 & 2 & 3 \\ 0 & -1 & 2 \\ 0 & 0 & 2 \end{bmatrix}$.

7. Which of the following matrices are diagonalizable?

(a) $\begin{bmatrix} 3 & 1 & 0 \\ 0 & 3 & 1 \\ 0 & 0 & 3 \end{bmatrix}$. (b) $\begin{bmatrix} -2 & 2 \\ 5 & 1 \end{bmatrix}$.

(c) $\begin{bmatrix} 2 & 0 & 3 \\ 0 & 1 & 0 \\ 0 & 1 & 2 \end{bmatrix}$. (d) $\begin{bmatrix} 2 & 3 & 3 & 5 \\ 3 & 2 & 2 & 3 \\ 0 & 0 & 2 & 2 \\ 0 & 0 & 0 & 2 \end{bmatrix}$.

8. Find a 2×2 nondiagonal matrix whose eigenvalues are 2 and -3, and associated eigenvectors are $\begin{bmatrix} -1 \\ 2 \end{bmatrix}$ and $\begin{bmatrix} 1 \\ 1 \end{bmatrix}$, respectively.

9. Find a 3×3 nondiagonal matrix whose eigenvalues are -2, -2, and 3, and associated eigenvectors are $\begin{bmatrix} 1 \\ 0 \\ 1 \end{bmatrix}$, $\begin{bmatrix} 0 \\ 1 \\ 1 \end{bmatrix}$, and $\begin{bmatrix} 1 \\ 1 \\ 1 \end{bmatrix}$, respectively.

10. For each of the following matrices find, if possible, a nonsingular matrix P such that $P^{-1}AP$ is diagonal.

(a) $\begin{bmatrix} 4 & 2 & 3 \\ 2 & 1 & 2 \\ -1 & -2 & 0 \end{bmatrix}$. (b) $\begin{bmatrix} 1 & 1 & 2 \\ 0 & 1 & 0 \\ 0 & 1 & 3 \end{bmatrix}$.

(c) $\begin{bmatrix} 1 & 2 & 3 \\ 0 & 1 & 0 \\ 2 & 1 & 2 \end{bmatrix}$. (d) $\begin{bmatrix} 0 & -1 \\ 2 & 3 \end{bmatrix}$.

11. For each of the following matrices find, if possible, a nonsingular matrix P such that $P^{-1}AP$ is diagonal.

(a) $\begin{bmatrix} 3 & -2 & 1 \\ 0 & 2 & 0 \\ 0 & 0 & 0 \end{bmatrix}$. (b) $\begin{bmatrix} 2 & 2 & 2 \\ 2 & 2 & 2 \\ 2 & 2 & 2 \end{bmatrix}$.

(c) $\begin{bmatrix} 3 & 0 & 0 \\ 2 & 3 & 0 \\ 0 & 0 & 3 \end{bmatrix}$. (d) $\begin{bmatrix} 1 & 0 & 1 \\ 0 & 1 & 0 \\ 0 & 1 & 2 \end{bmatrix}$.

12. Let A be a 2×2 matrix whose eigenvalues are 3 and 4, and associated eigenvectors are $\begin{bmatrix} -1 \\ 1 \end{bmatrix}$ and $\begin{bmatrix} 2 \\ 1 \end{bmatrix}$, respectively. Without computation, find a diagonal matrix D that is similar to A and nonsingular matrix P such that $P^{-1}AP = D$.

13. Let A be a 3×3 matrix whose eigenvalues are -3, 4, and 4, and associated eigenvectors are

$$\begin{bmatrix} -1 \\ 0 \\ 1 \end{bmatrix}, \quad \begin{bmatrix} 0 \\ 0 \\ 1 \end{bmatrix}, \quad \text{and} \quad \begin{bmatrix} 0 \\ 1 \\ 1 \end{bmatrix},$$

respectively. Without computation, find a diagonal matrix D that is similar to A and nonsingular matrix P such that $P^{-1}AP = D$.

14. Which of the following matrices are similar to a diagonal matrix?

(a) $\begin{bmatrix} 2 & 3 & 0 \\ 0 & 1 & 0 \\ 0 & 0 & 2 \end{bmatrix}$. (b) $\begin{bmatrix} 2 & 3 & 1 \\ 0 & 1 & 0 \\ 0 & 0 & 2 \end{bmatrix}$.

(c) $\begin{bmatrix} -3 & 0 \\ 1 & 2 \end{bmatrix}$. (d) $\begin{bmatrix} 1 & 1 & 0 \\ 2 & 2 & 0 \\ 3 & 3 & 3 \end{bmatrix}$.

15. Show that each of the following matrices is diagonalizable and find a diagonal matrix similar to the given matrix.

(a) $\begin{bmatrix} 4 & 2 \\ 3 & 3 \end{bmatrix}$. (b) $\begin{bmatrix} 3 & 2 \\ 6 & 4 \end{bmatrix}$.

(c) $\begin{bmatrix} 2 & -2 & 3 \\ 0 & 3 & -2 \\ 0 & -1 & 2 \end{bmatrix}$. (d) $\begin{bmatrix} 0 & -2 & 1 \\ 1 & 3 & -1 \\ 0 & 0 & 1 \end{bmatrix}$.

16. Show that each of the following matrices is diagonalizable.

(a) $\begin{bmatrix} 1 & 1 \\ 0 & 1 \end{bmatrix}$. (b) $\begin{bmatrix} 2 & 0 & 0 \\ 3 & 2 & 0 \\ 0 & 0 & 5 \end{bmatrix}$.

(c) $\begin{bmatrix} 10 & 11 & 3 \\ -3 & -4 & -3 \\ -8 & -8 & -1 \end{bmatrix}$. (d) $\begin{bmatrix} 2 & 3 & 3 & 5 \\ 3 & 2 & 2 & 3 \\ 0 & 0 & 1 & 1 \\ 0 & 0 & 0 & 1 \end{bmatrix}$.

17. A matrix A is called **defective** if A has an eigenvalue λ of multiplicity $m > 1$ for which the associated eigenspace has a basis of fewer than m vectors; that is, the dimension of the eigenspace associated with λ is less than m. Use the eigenvalues of the following matrices to determine which matrices are defective.

(a) $\begin{bmatrix} 8 & 7 \\ 0 & 8 \end{bmatrix}$, $\lambda = 8, 8$.

(b) $\begin{bmatrix} 3 & 0 & 0 \\ -2 & 3 & -2 \\ 2 & 0 & 5 \end{bmatrix}$, $\lambda = 3, 3, 5$.

(c) $\begin{bmatrix} 3 & 3 & 3 \\ 3 & 3 & 3 \\ -3 & -3 & -3 \end{bmatrix}$, $\lambda = 0, 0, 3$.

(d) $\begin{bmatrix} 0 & 0 & 1 & 0 \\ 0 & 0 & 0 & -1 \\ 1 & 0 & 0 & 0 \\ 0 & -1 & 0 & 0 \end{bmatrix}$, $\lambda = 1, 1, -1, -1$.

18. Let $D = \begin{bmatrix} 2 & 0 \\ 0 & -2 \end{bmatrix}$. Compute D^9.

19. Let $A = \begin{bmatrix} 3 & -5 \\ 1 & -3 \end{bmatrix}$. Compute A^9. (*Hint*: Find a matrix P such that $P^{-1}AP$ is a diagonal matrix D and show that $A^9 = PD^9P^{-1}$.)

20. Let $A = \begin{bmatrix} a & b \\ c & d \end{bmatrix}$. Find necessary and sufficient conditions for A to be diagonalizable.

21. Let A and B be nonsingular $n \times n$ matrices. Prove that AB and BA have the same eigenvalues.

22. Let V be the vector space of continuous functions with basis $\{e^t, e^{-t}\}$. Let $L\colon V \to V$ be defined by $L(g(t)) = g'(t)$ for $g(t)$ in V. Show that L is diagonalizable.

23. Prove that if A is diagonalizable, then (a) A^T is diagonalizable, and (b) A^k is diagonalizable, where k is a positive integer.

24. Show that if A is nonsingular and diagonalizable, then A^{-1} is diagonalizable.

25. Let $\lambda_1, \lambda_2, \dots, \lambda_k$ be distinct eigenvalues of a matrix A with associated eigenvectors $\mathbf{x}_1, \mathbf{x}_2, \dots, \mathbf{x}_k$. Prove that $\mathbf{x}_1, \mathbf{x}_2, \dots, \mathbf{x}_k$ are linearly independent. (*Hint*: See the proof of Theorem 6.4.)

26. Let A and B be nonsingular $n \times n$ matrices. Prove that AB and BA have the same eigenvalues.

27. Show that if a matrix A is similar to a diagonal matrix D, then $\text{Tr}(A) = \text{Tr}(D)$, where $\text{Tr}(A)$ is the trace of A.[*Hint*: See Exercise 29, Section 1.2, where part (c) establishes $\text{Tr}(AB) = \text{Tr}(BA)$.]

6.3 STABLE AGE DISTRIBUTION IN A POPULATION; MARKOV PROCESSES (OPTIONAL)

In this section we deal with two applications of eigenvalues and eigenvectors. These applications find use in a wide variety of everyday situations, including harvesting of animal resources and planning of mass transportation systems.

Stable Age Distribution in a Population

Consider a population of animals that can live to a maximum age of n years (or any other time unit). Suppose that the number of males in the population is always a fixed percentage of the female population. Thus, in studying the growth of the entire population, we can ignore the male population and concentrate our attention on the female population. We divide the female population into $n+1$ age groups as follows:

$$
\begin{aligned}
x_i^{(k)} &= \text{number of females of age } i \text{ who are alive at time } k,\ 0 \le i \le n;\\
f_i &= \text{fraction of females of age } i \text{ who will be alive a year later;}\\
b_i &= \text{average number of females born to a female of age } i.
\end{aligned}
$$

Let

$$
\mathbf{x}^{(k)} = \begin{bmatrix} x_0^{(k)} \\ x_1^{(k)} \\ \vdots \\ x_n^{(k)} \end{bmatrix} \qquad (k \ge 0)
$$

denote the age distribution vector at time k.

The number of females in the first age group (age zero) at time $k+1$ is merely the total number of females born from time k to time $k+1$. There are $x_0^{(k)}$ females in the first age group at time k and each of these females, on the average, produces b_0 female offspring, so the first age group produces a total of $b_0 x_0^{(k)}$ females. Similarly, the $x_1^{(k)}$ females in the second age group (age 1) produces a total of $b_1 x_1^{(k)}$ females. Thus

$$
x_0^{(k+1)} = b_0 x_0^{(k)} + b_1 x_1^{(k)} + \cdots + b_n x_n^{(k)}. \tag{1}
$$

The number $x_1^{(k+1)}$ of females in the second age group at time $k+1$ is the number of females from the first age group at time k who are alive a year later. Thus

$$x_1^{(k+1)} = \begin{pmatrix} \text{fraction of females in} \\ \text{first age group who are} \\ \text{alive a year later} \end{pmatrix} \times \begin{pmatrix} \text{number of females in} \\ \text{first age group} \end{pmatrix},$$

or

$$x_1^{(k+1)} = f_0 x_0^{(k)},$$

and, in general,

$$x_j^{(k+1)} = f_{j-1} x_{j-1}^{(k)} \qquad (1 \leq j \leq n). \tag{2}$$

We can write (1) and (2), using matrix notation, as

$$\mathbf{x}^{(k+1)} = A\mathbf{x}^{(k)} \qquad (k \geq 1) \tag{3}$$

where

$$A = \begin{bmatrix} b_0 & b_1 & b_2 & \cdots & b_{n-1} & b_n \\ f_0 & 0 & 0 & \cdots & 0 & 0 \\ 0 & f_1 & 0 & \cdots & 0 & 0 \\ \vdots & \vdots & \vdots & & \vdots & \vdots \\ 0 & 0 & 0 & \cdots & f_{n-1} & 0 \end{bmatrix}.$$

We can use Equation (3) to try to determine a distribution of the population by age groups at time $k+1$ so that the number of females in each age group at time $k+1$ will be a fixed multiple of the number in the corresponding age group at time k. That is, if λ is the multiplier, we want

$$\mathbf{x}^{(k+1)} = \lambda\mathbf{x}^{(k)}$$

or

$$A\mathbf{x}^{(k)} = \lambda\mathbf{x}^{(k)}.$$

Thus λ is an eigenvalue of A and $\mathbf{x}^{(k)}$ is a corresponding eigenvector. If $\lambda = 1$, the number of females in each age group will be the same year after year. If we can find an eigenvector $\mathbf{x}^{(k)}$ corresponding to the eigenvalue $\lambda = 1$, we say that we have a **stable age distribution**.

EXAMPLE 1

Consider a beetle that can live to a maximum age of two years and whose population dynamics are represented by the matrix

$$A = \begin{bmatrix} 0 & 0 & 6 \\ \frac{1}{2} & 0 & 0 \\ 0 & \frac{1}{3} & 0 \end{bmatrix}.$$

We find that $\lambda = 1$ is an eigenvalue of A with corresponding eigenvector

$$\begin{bmatrix} 6 \\ 3 \\ 1 \end{bmatrix}.$$

Thus, if the numbers of females in the three groups are proportional to 6:3:1, we have a stable age distribution. That is, if we have 600 females in the first age group, 300 in the second, and 100 in the third, then year after year the number of females in each age group will remain the same. ■

Remark Population growth problems of the type considered in Example 1 have applications to animal harvesting.

Markov Processes

A **Markov chain**, or **Markov process**, is a process in which the probability of the system being in a particular state at a given observation period depends only on its state at the immediately preceding observation period.

Suppose that the system has n possible states. For each $i = 1, 2, \ldots, n$, and $j = 1, 2, \ldots, n$, let t_{ij} be the probability that if the system is in state j at a certain observation period, it will be in state i at the next observation period; t_{ij} is called a **transition probability**. Moreover, t_{ij} applies to every time period; that is, it does not change with time.

Since t_{ij} is a probability, we must have

$$0 \leq t_{ij} \leq 1 \qquad (1 \leq i, j \leq n).$$

Also, if the system is in state j at a certain observation period, then it must be in one of the n states (it may remain in state j) at the next observation period. Thus we have

$$t_{1j} + t_{2j} + \cdots + t_{nj} = 1. \tag{4}$$

It is convenient to arrange the transition probabilities as the $n \times n$ matrix $T = \left[t_{ij} \right]$, which is called the **transition matrix** of the Markov chain. Other names for a transition matrix are **Markov matrix**, **stochastic matrix**, and **probability matrix**. We see that the entries in each column of T are nonnegative and, from Equation (4), add up to 1.

We shall now use the transition matrix of the Markov process to determine the probability of the system being in any of the n states at future times.

Let

$$\mathbf{x}^{(k)} = \begin{bmatrix} p_1^{(k)} \\ p_2^{(k)} \\ \vdots \\ p_n^{(k)} \end{bmatrix} \qquad (k \geq 0)$$

denote the **state vector** of the Markov process at the observation period k, where $p_j^{(k)}$ is the probability that the system is in state j at the observation period k. The state vector $\mathbf{x}^{(0)}$, at the observation period 0, is called the **initial state vector**.

It follows from the basic properties of probability theory that if T is the transition matrix of a Markov process, then the state vector $\mathbf{x}^{(k+1)}$, at the $(k+1)$th observation period, can be determined from the state vector $\mathbf{x}^{(k)}$, at the kth observation period, as

$$\mathbf{x}^{(k+1)} = T\mathbf{x}^{(k)}. \tag{5}$$

From (5), we have

$$\begin{aligned}\mathbf{x}^{(1)} &= T\mathbf{x}^{(0)}\\ \mathbf{x}^{(2)} &= T\mathbf{x}^{(1)} = T(T\mathbf{x}^{(0)}) = T^2\mathbf{x}^{(0)}\\ \mathbf{x}^{(3)} &= T\mathbf{x}^{(2)} = T(T^2\mathbf{x}^{(0)}) = T^3\mathbf{x}^{(0)},\end{aligned}$$

and, in general,

$$\mathbf{x}^{(n)} = T^n\mathbf{x}^{(0)}.$$

Thus the transition matrix and the initial state vector completely determine every other state vector.

For certain types of Markov processes, as the number of observation periods increases, the state vectors converge to a fixed vector. In this case, we say that the Markov process has reached **equilibrium**. The fixed vector is called the **steady-state vector**. Markov processes are generally used to determine the behavior of a system in the long run; for example, the share of the market that a certain manufacturer can expect to retain on a somewhat permanent basis. Thus, the question of whether or not a Markov process reaches equilibrium is quite important. To identify a class of Markov processes that reach equilibrium, we need several additional notions.

The vector

$$\mathbf{u} = \begin{bmatrix} u_1 \\ u_2 \\ \vdots \\ u_n \end{bmatrix}$$

is called a **probability vector** if $u_i \geq 0$ $(1 \leq i \leq n)$ and

$$u_1 + u_2 + \cdots + u_n = 1.$$

A Markov process is called **regular** if its transition matrix T has the property that all the entries in some power of T are positive. It can be shown that a regular Markov process always reaches equilibrium and the steady-state vector is a probability vector. The steady-state vector can be found by obtaining the limit of the successive powers $T^n\mathbf{x}$ for an arbitrary probability vector $\mathbf{x}$. Observe that if $\mathbf{u}$ is the steady-state vector of a Markov process with transition matrix T, then $T\mathbf{u} = \mathbf{u}$, so that $\lambda = 1$ is an eigenvalue of T with associated eigenvector $\mathbf{u}$ (a probability vector).

EXAMPLE 2

Suppose that the weather in a certain city is either rainy or dry. As a result of extensive record keeping it has been determined that the probability of a rainy day following a dry day is $\frac{1}{3}$, and the probability of a rainy day following a rainy day is $\frac{1}{2}$. Let state D be a dry day and state R be a rainy day. Then the transition matrix of this Markov process is

$$T = \begin{array}{c} \\ \\ \end{array}\begin{array}{cc} \text{D} & \text{R} \\ \left[\begin{array}{cc} \frac{2}{3} & \frac{1}{2} \\ \frac{1}{3} & \frac{1}{2} \end{array}\right] \end{array}\begin{array}{c} \\ \text{D} \\ \text{R} \end{array}.$$

Since all the entries in T are positive, we are dealing with a regular Markov process, so the process reaches equilibrium. Suppose that when we begin our observations (day 0), it is dry, so the initial state vector is

$$\mathbf{x}^{(0)} = \begin{bmatrix} 1 \\ 0 \end{bmatrix},$$

a probability vector. Then the state vector on day 1 (the day after we begin our observations) is

$$\mathbf{x}^{(1)} = T\mathbf{x}^{(0)} = \begin{bmatrix} 0.67 & 0.5 \\ 0.33 & 0.5 \end{bmatrix}\begin{bmatrix} 1 \\ 0 \end{bmatrix} = \begin{bmatrix} 0.67 \\ 0.33 \end{bmatrix},$$

where for convenience we have written $\frac{2}{3}$ and $\frac{1}{3}$ as 0.67 and 0.33, respectively. Moreover, to simplify matters the output of calculations is recorded to three digits of accuracy. Thus the probability of no rain on day 1 is 0.67 and the probability of rain on that day is 0.33. Similarly,

$$\mathbf{x}^{(2)} = T\mathbf{x}^{(1)} = \begin{bmatrix} 0.67 & 0.5 \\ 0.33 & 0.5 \end{bmatrix}\begin{bmatrix} 0.67 \\ 0.33 \end{bmatrix} = \begin{bmatrix} 0.614 \\ 0.386 \end{bmatrix}$$

$$\mathbf{x}^{(3)} = T\mathbf{x}^{(2)} = \begin{bmatrix} 0.67 & 0.5 \\ 0.33 & 0.5 \end{bmatrix}\begin{bmatrix} 0.614 \\ 0.386 \end{bmatrix} = \begin{bmatrix} 0.604 \\ 0.396 \end{bmatrix}$$

$$\mathbf{x}^{(4)} = T\mathbf{x}^{(3)} = \begin{bmatrix} 0.67 & 0.5 \\ 0.33 & 0.5 \end{bmatrix}\begin{bmatrix} 0.604 \\ 0.396 \end{bmatrix} = \begin{bmatrix} 0.603 \\ 0.397 \end{bmatrix}$$

$$\mathbf{x}^{(5)} = T\mathbf{x}^{(4)} = \begin{bmatrix} 0.67 & 0.5 \\ 0.33 & 0.5 \end{bmatrix}\begin{bmatrix} 0.603 \\ 0.397 \end{bmatrix} = \begin{bmatrix} 0.603 \\ 0.397 \end{bmatrix}.$$

From the fourth day on, the state vector is always the same,

$$\begin{bmatrix} 0.603 \\ 0.397 \end{bmatrix},$$

so this is the steady-state vector.

This means that from the fourth day on, it is dry about 60 percent of the time, and it rains about 40 percent of the time.

The steady-state vector can also be found as follows. Since $\lambda = 1$ is an eigenvalue of T, we find an associated eigenvector $\mathbf{u} = \begin{bmatrix} u_1 \\ u_2 \end{bmatrix}$ by solving the equation

$$T\mathbf{u} = \mathbf{u}$$

or

$$(I_2 - T)\mathbf{u} = \mathbf{0}.$$

From the infinitely many solutions obtained by solving the resulting homogeneous system, we determine a unique solution $\mathbf{u}$ by requiring that its components add up to 1 (since $\mathbf{u}$ is a probability vector). In this case we have to solve the homogeneous system

$$\begin{aligned} \tfrac{1}{3}u_1 - \tfrac{1}{2}u_2 &= 0 \\ -\tfrac{1}{3}u_1 + \tfrac{1}{2}u_2 &= 0. \end{aligned}$$

Then $u_2 = \frac{2}{3}u_1$. Substituting in the equation

$$u_1 + u_2 = 1,$$

we obtain $u_1 = 0.6$ and $u_2 = 0.4$. ■

6.3 Exercises

1. Consider a living organism that can live to a maximum age of 2 years and whose matrix is

$$A = \begin{bmatrix} 0 & 0 & 8 \\ \frac{1}{4} & 0 & 0 \\ 0 & \frac{1}{2} & 0 \end{bmatrix}.$$

Find a stable age distribution.

2. Consider a living organism that can live to a maximum age of 2 years and whose matrix is

$$A = \begin{bmatrix} 0 & 4 & 0 \\ \frac{1}{4} & 0 & 0 \\ 0 & \frac{1}{2} & 0 \end{bmatrix}.$$

Find a stable age distribution.

3. Which of the following can be transition matrices of a Markov process?

(a) $\begin{bmatrix} 0.3 & 0.7 \\ 0.4 & 0.6 \end{bmatrix}$. (b) $\begin{bmatrix} 0.2 & 0.3 & 0.1 \\ 0.8 & 0.5 & 0.7 \\ 0.0 & 0.2 & 0.2 \end{bmatrix}$.

(c) $\begin{bmatrix} 0.55 & 0.33 \\ 0.45 & 0.67 \end{bmatrix}$. (d) $\begin{bmatrix} 0.3 & 0.4 & 0.2 \\ 0.2 & 0.0 & 0.8 \\ 0.1 & 0.3 & 0.6 \end{bmatrix}$.

4. Which of the following are probability vectors?

(a) $\begin{bmatrix} \frac{1}{2} \\ \frac{1}{3} \\ \frac{2}{3} \end{bmatrix}$. (b) $\begin{bmatrix} 0 \\ 1 \\ 0 \end{bmatrix}$.

(c) $\begin{bmatrix} \frac{1}{4} \\ \frac{1}{6} \\ \frac{1}{3} \\ \frac{1}{4} \end{bmatrix}$. (d) $\begin{bmatrix} \frac{1}{5} \\ \frac{2}{5} \\ \frac{1}{10} \\ \frac{2}{10} \end{bmatrix}$.

5. Consider the transition matrix

$$T = \begin{bmatrix} 0.7 & 0.4 \\ 0.3 & 0.6 \end{bmatrix}.$$

(a) If $\mathbf{x}^{(0)} = \begin{bmatrix} 1 \\ 0 \end{bmatrix}$, compute $\mathbf{x}^{(1)}$, $\mathbf{x}^{(2)}$, and $\mathbf{x}^{(3)}$ to three decimal places.

(b) Show that T is regular and find its steady-state vector.

6. Consider the transition matrix

$$T = \begin{bmatrix} 0 & 0.2 & 0.0 \\ 0 & 0.3 & 0.3 \\ 1 & 0.5 & 0.7 \end{bmatrix}.$$

(a) If

$$\mathbf{x}^{(0)} = \begin{bmatrix} 0 \\ 1 \\ 0 \end{bmatrix},$$

compute $\mathbf{x}^{(1)}$, $\mathbf{x}^{(2)}$, $\mathbf{x}^{(3)}$, and $\mathbf{x}^{(4)}$ to three decimal places.

(b) Show that T is regular and find its steady-state vector.

7. Which of the following transition matrices are regular?

(a) $\begin{bmatrix} 0 & \frac{1}{2} \\ 1 & \frac{1}{2} \end{bmatrix}$. (b) $\begin{bmatrix} \frac{1}{2} & 0 & 0 \\ 0 & 1 & \frac{1}{2} \\ \frac{1}{2} & 0 & \frac{1}{2} \end{bmatrix}$.

(c) $\begin{bmatrix} 1 & \frac{1}{3} & 0 \\ 0 & \frac{1}{3} & 1 \\ 0 & \frac{1}{3} & 0 \end{bmatrix}$. (d) $\begin{bmatrix} \frac{1}{4} & \frac{3}{5} & \frac{1}{2} \\ \frac{1}{2} & 0 & 0 \\ \frac{1}{4} & \frac{2}{5} & \frac{1}{2} \end{bmatrix}$.

8. Show that each of the following transition matrices reaches a state of equilibrium.

(a) $\begin{bmatrix} \frac{1}{2} & 1 \\ \frac{1}{2} & 0 \end{bmatrix}$. (b) $\begin{bmatrix} 0.4 & 0.2 \\ 0.6 & 0.8 \end{bmatrix}$.

(c) $\begin{bmatrix} \frac{1}{3} & 1 & \frac{1}{2} \\ \frac{1}{3} & 0 & \frac{1}{4} \\ \frac{1}{3} & 0 & \frac{1}{4} \end{bmatrix}$. (d) $\begin{bmatrix} 0.3 & 0.1 & 0.4 \\ 0.2 & 0.4 & 0.0 \\ 0.5 & 0.5 & 0.6 \end{bmatrix}$.

9. Find the steady-state vector of each of the following regular matrices.

(a) $\begin{bmatrix} \frac{1}{3} & \frac{1}{2} \\ \frac{2}{3} & \frac{1}{2} \end{bmatrix}$. (b) $\begin{bmatrix} 0.3 & 0.1 \\ 0.7 & 0.9 \end{bmatrix}$.

(c) $\begin{bmatrix} \frac{1}{4} & \frac{1}{2} & \frac{1}{3} \\ 0 & \frac{1}{2} & \frac{2}{3} \\ \frac{3}{4} & 0 & 0 \end{bmatrix}$. (d) $\begin{bmatrix} 0.4 & 0.0 & 0.1 \\ 0.2 & 0.5 & 0.3 \\ 0.4 & 0.5 & 0.6 \end{bmatrix}$.

10. (**Psychology**) A behavioral psychologist places a rat each day in a cage with two doors, A and B. The rat can go through door A, where it receives an electric shock, or through door B, where it receives some food. A record is made of the door through which the rat passes. At the start of the experiment, on a Monday, the rat is equally likely to go through door A as through door B. After going through door A, and receiving a shock, the probability of going through the same door on the next day is 0.3. After going through door B, and receiving food, the probability of going through the same door on the next day is 0.6.

(a) Write the transition matrix for the Markov process.

(b) What is the probability of the rat going through door A on Thursday (the third day after starting the experiment)?

(c) What is the steady-state vector?

11. (**Sociology**) A study has determined that the occupation of a boy, as an adult, depends upon the occupation of his father and is given by the following transition matrix where P = professional, F = farmer, and L = laborer.

		Father's occupation		
		P	F	L
Son's occupation	P	0.8	0.3	0.2
	F	0.1	0.5	0.2
	L	0.1	0.2	0.6

Thus the probability that the son of a professional will also be a professional is 0.8, and so on.

(a) What is the probability that the grandchild of a professional will also be a professional?

(b) In the long run, what proportion of the population will be farmers?

12. (**Genetics**) Consider a plant that can have red flowers (R), pink flowers (P), or white flowers (W), depending upon the genotypes RR, RW, and WW. When we cross each of these genotypes with a genotype RW, we obtain the transition matrix

		Flowers of parent plant		
		R	P	W
Flowers of offspring plant	R	0.5	0.25	0.0
	P	0.5	0.50	0.5
	W	0.0	0.25	0.5

Suppose that each successive generation is produced by crossing only with plants of RW genotype. When the process reaches equilibrium, what percentage of the plants will have red, pink, or white flowers?

13. (**Mass Transit**) A new mass transit system has just gone into operation. The transit authority has made studies that predict the percentage of commuters who will change to mass transit (M) or continue driving their automobile (A). The following transition matrix has been obtained:

$$\text{Next year}\quad \begin{matrix} & \text{This year} \\ & \begin{matrix} \text{M} & \text{A} \end{matrix} \\ \begin{matrix} \text{M} \\ \text{A} \end{matrix} & \begin{bmatrix} 0.7 & 0.2 \\ 0.3 & 0.8 \end{bmatrix} \end{matrix}$$

Suppose that the population of the area remains constant, and that initially 30 percent of the commuters use mass transit and 70 percent use their automobiles.

(a) What percentage of the commuters will be using the mass transit system after 1 year? After 2 years?

(b) What percentage of the commuters will be using the mass transit system in the long run?

6.4 DIAGONALIZATION OF SYMMETRIC MATRICES

In this section we consider the diagonalization of a symmetric matrix (i.e., a matrix A for which $A = A^T$). We restrict our attention to symmetric matrices, because they are easier to handle than general matrices and because they arise in many applied problems. One of these applications will be discussed in Section 6.6.

Theorem 6.5 assures us that an $n \times n$ matrix A is diagonalizable if it has n distinct eigenvalues: if this is not so, then A may fail to be diagonalizable. However, every symmetric matrix can be diagonalized; that is, if A is symmetric, there exists a nonsingular matrix P such that $P^{-1}AP = D$, where D is a diagonal matrix. Moreover, P has some noteworthy properties that we remark upon. We thus turn to the study of symmetric matrices in this section.

We first prove that all the roots of the characteristic polynomial of a symmetric matrix are real. Section B.2 contains examples of matrices with complex eigenvalues and provides more background and motivation for considering the case of symmetric matrices. A review of complex arithmetic appears in Section B.1.

Theorem 6.6

All the roots of the characteristic polynomial of a real symmetric matrix are real numbers.

Proof

We give two proofs of this result. They both require some facts about complex numbers, which are covered in Appendix B.1. The first proof requires fewer of these facts, but is more computational and longer. Let $\lambda = a + bi$ be any root of the characteristic polynomial of A. We shall prove that $b = 0$, so that λ is a real number. Now

$$\det(\lambda I_n - A) = 0 = \det((a + bi)I_n - A).$$

This means that the homogeneous system

$$((a + bi)I_n - A)(\mathbf{x} + \mathbf{y}i) = \mathbf{0} = \mathbf{0} + \mathbf{0}i \tag{1}$$

has a nontrivial solution $\mathbf{x} + \mathbf{y}i$, where $\mathbf{x}$ and $\mathbf{y}$ are vectors in R^n that are not both the zero vector. Carrying out the multiplication in (1), we obtain

$$(aI_n\mathbf{x} - A\mathbf{x} - bI_n\mathbf{y}) + i(aI_n\mathbf{y} + bI_n\mathbf{x} - A\mathbf{y}) = \mathbf{0} + \mathbf{0}i. \tag{2}$$

Setting the real and imaginary parts equal to $\mathbf{0}$, we have

$$\begin{aligned} aI_n\mathbf{x} - A\mathbf{x} - bI_n\mathbf{y} &= \mathbf{0} \\ aI_n\mathbf{y} - A\mathbf{y} + bI_n\mathbf{x} &= \mathbf{0}. \end{aligned} \tag{3}$$

Forming the inner products of both sides of the first equation in (3) with $\mathbf{y}$ and of both sides of the second equation of (3) with $\mathbf{x}$, we have

$$\begin{aligned} (\mathbf{y}, aI_n\mathbf{x} - A\mathbf{x} - bI_n\mathbf{y}) &= (\mathbf{y}, \mathbf{0}) = 0 \\ (aI_n\mathbf{y} - A\mathbf{y} + bI_n\mathbf{x}, \mathbf{x}) &= (\mathbf{0}, \mathbf{x}) = 0 \end{aligned}$$

or

$$\begin{aligned} a(\mathbf{y}, I_n\mathbf{x}) - (\mathbf{y}, A\mathbf{x}) - b(\mathbf{y}, I_n\mathbf{y}) &= 0 \\ a(I_n\mathbf{y}, \mathbf{x}) - (A\mathbf{y}, \mathbf{x}) + b(I_n\mathbf{x}, \mathbf{x}) &= 0. \end{aligned} \tag{4}$$

Now, by Equation (3) in Section 3.3, we see that $(I_n\mathbf{y}, \mathbf{x}) = (\mathbf{y}, I_n^T\mathbf{x}) = (\mathbf{y}, I_n\mathbf{x})$ and that $(A\mathbf{y}, \mathbf{x}) = (\mathbf{y}, A^T\mathbf{x}) = (\mathbf{y}, A\mathbf{x})$. Note that we have used the fact that $I_n^T = I_n$ and that, since A is symmetric, we have $A^T = A$. Subtracting the two equations in (4), we now get

$$-b(\mathbf{y}, I_n\mathbf{y}) - b(I_n\mathbf{x}, \mathbf{x}) = 0 \tag{5}$$

or

$$-b[(\mathbf{y}, \mathbf{y}) + (\mathbf{x}, \mathbf{x})] = 0. \tag{6}$$

Since $\mathbf{x}$ and $\mathbf{y}$ are not both the zero vector, $(\mathbf{x}, \mathbf{x}) > 0$ or $(\mathbf{y}, \mathbf{y}) > 0$. From (6) we conclude that $b = 0$. Hence every root of the characteristic polynomial of A is a real number.

Alternative Proof

Let λ be any root of the characteristic polynomial of A. We will prove that λ is real by showing that $\lambda = \overline{\lambda}$, its complex conjugate. We have

$$A\mathbf{x} = \lambda\mathbf{x}.$$

Multiplying both sides of this equation by $\overline{\mathbf{x}}^T$ on the left, we obtain

$$\overline{\mathbf{x}}^T A\mathbf{x} = \overline{\mathbf{x}}^T \lambda\mathbf{x}.$$

Taking the conjugate transpose of both sides yields

$$\overline{\mathbf{x}}^T \overline{A}^T \mathbf{x} = \overline{\lambda}\,\overline{\mathbf{x}}^T\mathbf{x}$$

or

$$\begin{aligned} \overline{\mathbf{x}}^T A\mathbf{x} &= \overline{\lambda}\,\overline{\mathbf{x}}^T\mathbf{x} \qquad (\text{since } A = A^T) \\ \lambda\,\overline{\mathbf{x}}^T\mathbf{x} &= \overline{\lambda}\,\overline{\mathbf{x}}^T\mathbf{x}, \end{aligned}$$

so

$$(\lambda - \overline{\lambda})(\overline{\mathbf{x}}^T\mathbf{x}) = 0.$$

Since $\mathbf{x} \neq \mathbf{0}$, $\overline{\mathbf{x}}^T\mathbf{x} \neq 0$. Hence $\lambda - \overline{\lambda} = 0$ or $\lambda = \overline{\lambda}$. ●

Once we have established this result, we know that complex numbers do not enter into the study of the diagonalization problem for real symmetric matrices. Thus, throughout the remainder of this book, except for Appendix B, we again deal only with real numbers.

Corollary 6.1

If A is a symmetric matrix and all the eigenvalues of A are distinct, then A is diagonalizable.

Proof

Since A is symmetric, all the roots of its characteristic polynomial are real. From Theorem 6.5 it now follows that A can be diagonalized. Moreover, if D is the diagonal matrix that is similar to A, then the elements on the main diagonal of D are the eigenvalues of A. ●

Theorem 6.7

If A is a symmetric matrix, then eigenvectors that belong to distinct eigenvalues of A are orthogonal.

Proof

Let $\mathbf{x}_1$ and $\mathbf{x}_2$ be eigenvectors of A that are associated with the distinct eigenvalues λ_1 and λ_2 of A. We then have

$$A\mathbf{x}_1 = \lambda_1\mathbf{x}_1 \quad \text{and} \quad A\mathbf{x}_2 = \lambda_2\mathbf{x}_2.$$

Now using Equation (3) of Section 3.3 and the fact that $A^T = A$, since A is symmetric, we have

$$\begin{aligned}\lambda_1(\mathbf{x}_1, \mathbf{x}_2) = (\lambda_1\mathbf{x}_1, \mathbf{x}_2) &= (A\mathbf{x}_1, \mathbf{x}_2)\\ = (\mathbf{x}_1, A^T\mathbf{x}_2) &= (\mathbf{x}_1, A\mathbf{x}_2)\\ = (\mathbf{x}_1, \lambda_2\mathbf{x}_2) &= \lambda_2(\mathbf{x}_1, \mathbf{x}_2).\end{aligned}$$

Thus

$$\lambda_1(\mathbf{x}_1, \mathbf{x}_2) = \lambda_2(\mathbf{x}_1, \mathbf{x}_2)$$

and subtracting, we obtain

$$\begin{aligned}0 &= \lambda_1(\mathbf{x}_1, \mathbf{x}_2) - \lambda_2(\mathbf{x}_1, \mathbf{x}_2)\\ &= (\lambda_1 - \lambda_2)(\mathbf{x}_1, \mathbf{x}_2).\end{aligned}$$

Since $\lambda_1 \neq \lambda_2$, we conclude that $(\mathbf{x}_1, \mathbf{x}_2) = 0$. ●

EXAMPLE 1

Let $A = \begin{bmatrix} 0 & 0 & -2 \\ 0 & -2 & 0 \\ -2 & 0 & 3 \end{bmatrix}$. The characteristic polynomial of A is

$$p(\lambda) = (\lambda + 2)(\lambda - 4)(\lambda + 1)$$

(verify), so the eigenvalues of A are $\lambda_1 = -2$, $\lambda_2 = 4$, $\lambda_3 = -1$. Associated eigenvectors are the nontrivial solutions of the homogeneous system [Equation (6) in Section 6.1]

$$\begin{bmatrix} \lambda & 0 & 2 \\ 0 & \lambda+2 & 0 \\ 2 & 0 & \lambda-3 \end{bmatrix} \begin{bmatrix} x_1 \\ x_2 \\ x_3 \end{bmatrix} = \begin{bmatrix} 0 \\ 0 \\ 0 \end{bmatrix}.$$

For $\lambda_1 = -2$, we find that $\mathbf{x}_1$ is any vector of the form $\begin{bmatrix} 0 \\ r \\ 0 \end{bmatrix}$, where r is any nonzero real number (verify). Thus we may take $\mathbf{x}_1 = \begin{bmatrix} 0 \\ 1 \\ 0 \end{bmatrix}$. For $\lambda_2 = 4$, we find that $\mathbf{x}_2$ is any vector of the form $\begin{bmatrix} -\frac{r}{2} \\ 0 \\ r \end{bmatrix}$, where r is any nonzero real number (verify). Thus we may take $\mathbf{x}_2 = \begin{bmatrix} -1 \\ 0 \\ 2 \end{bmatrix}$. For $\lambda_3 = -1$, we find that $\mathbf{x}_3$ is any vector of the form $\begin{bmatrix} 2r \\ 0 \\ r \end{bmatrix}$, where r is any nonzero real number (verify). Thus we may take $\mathbf{x}_3 = \begin{bmatrix} 2 \\ 0 \\ 1 \end{bmatrix}$. It is clear that $\{\mathbf{x}_1, \mathbf{x}_2, \mathbf{x}_3\}$ is orthogonal and linearly independent. Thus A is similar to $D = \begin{bmatrix} -2 & 0 & 0 \\ 0 & 4 & 0 \\ 0 & 0 & -1 \end{bmatrix}$. ■

If A can be diagonalized, then there exists a nonsingular matrix P such that $P^{-1}AP$ is diagonal. Moreover, the columns of P are eigenvectors of A. Now, if the eigenvectors of A form an orthogonal set S, as happens when A is symmetric and the eigenvalues of A are distinct, then since any nonzero scalar multiple of an eigenvector of A is also an eigenvector of A, we can normalize S to obtain an orthonormal set $T = \{\mathbf{x}_1, \mathbf{x}_2, \ldots, \mathbf{x}_n\}$ of eigenvectors of A. Let the jth column of P be the eigenvector $\mathbf{x}_j$, and we now see what type of matrix P must be. We can write P as a partitioned matrix in the form $P = \begin{bmatrix} \mathbf{x}_1 & \mathbf{x}_2 & \cdots & \mathbf{x}_n \end{bmatrix}$. Then $P^T = \begin{bmatrix} \mathbf{x}_1^T \\ \mathbf{x}_2^T \\ \vdots \\ \mathbf{x}_n^T \end{bmatrix}$, where $\mathbf{x}_i^T$ is the transpose of the $n \times 1$ matrix (or vector) $\mathbf{x}_i$. We find that the (i, j) entry in $P^T P$ is $(\mathbf{x}_i, \mathbf{x}_j)$. Since $(\mathbf{x}_i, \mathbf{x}_j) = 1$ if $i = j$ and $(\mathbf{x}_i, \mathbf{x}_j) = 0$ if $i \neq j$, we have $P^T P = I_n$, which means that $P^T = P^{-1}$. Such matrices are important enough to have a special name.

Definition 6.4

A real square matrix A is called **orthogonal** if $A^{-1} = A^T$. Of course, we can also say that A is orthogonal if $A^T A = I_n$. ▲

EXAMPLE 2

Let

$$A = \begin{bmatrix} \frac{2}{3} & -\frac{2}{3} & \frac{1}{3} \\ \frac{2}{3} & \frac{1}{3} & -\frac{2}{3} \\ \frac{1}{3} & \frac{2}{3} & \frac{2}{3} \end{bmatrix}.$$

It is easy to check that $A^T A = I_n$, so A is an orthogonal matrix. ■

EXAMPLE 3

Let A be the matrix defined in Example 1. We already know that the set of eigenvectors

$$\left\{ \begin{bmatrix} 0 \\ 1 \\ 0 \end{bmatrix}, \begin{bmatrix} -1 \\ 0 \\ 2 \end{bmatrix}, \begin{bmatrix} 2 \\ 0 \\ 1 \end{bmatrix} \right\}$$

is orthogonal. If we normalize these vectors, we find that

$$T = \left\{ \begin{bmatrix} 0 \\ 1 \\ 0 \end{bmatrix}, \begin{bmatrix} -\frac{1}{\sqrt{5}} \\ 0 \\ \frac{2}{\sqrt{5}} \end{bmatrix}, \begin{bmatrix} \frac{2}{\sqrt{5}} \\ 0 \\ \frac{1}{\sqrt{5}} \end{bmatrix} \right\}$$

is an orthonormal basis for R^3. A matrix P such that $P^{-1}AP$ is diagonal is the matrix whose columns are the vectors in T. Thus

$$P = \begin{bmatrix} 0 & -\frac{1}{\sqrt{5}} & \frac{2}{\sqrt{5}} \\ 1 & 0 & 0 \\ 0 & \frac{2}{\sqrt{5}} & \frac{1}{\sqrt{5}} \end{bmatrix}.$$

We leave it to the reader to verify that P is an orthogonal matrix and that

$$P^{-1}AP = P^T AP = \begin{bmatrix} -2 & 0 & 0 \\ 0 & 4 & 0 \\ 0 & 0 & -1 \end{bmatrix}.$$ ■

The following theorem is not difficult to prove.

Theorem 6.8

The $n \times n$ matrix A is orthogonal if and only if the columns (rows) of A form an orthonormal set.

Proof

Exercise. ●

If A is an orthogonal matrix, then we can show that $\det(A) = \pm 1$ (Exercise 8). We now look at some of the geometric properties of orthogonal matrices. If A is an orthogonal $n \times n$ matrix, let $L\colon R^n \to R^n$ be the linear operator defined by $L(\mathbf{x}) = A\mathbf{x}$, for $\mathbf{x}$ in R^n (recall Chapter 4). If $\det(A) = 1$ and $n = 2$, it then follows that L is a counterclockwise rotation. It can also be shown that if $\det(A) = -1$, then L is a reflection about the x-axis followed by a counterclockwise rotation (see Exercise 32).

Again, let A be an orthogonal $n \times n$ matrix and let $L\colon R^n \to R^n$ be defined by $L(\mathbf{x}) = A\mathbf{x}$ for $\mathbf{x}$ in R^n. We now compute $(L(\mathbf{x}), L(\mathbf{y}))$ for any vectors $\mathbf{x}$, $\mathbf{y}$ in R^n, using the standard inner product on R^n. We have

$$(L(\mathbf{x}), L(\mathbf{y})) = (A\mathbf{x}, A\mathbf{y}) = (\mathbf{x}, A^T A\mathbf{y}) = (\mathbf{x}, A^{-1} A\mathbf{y}) = (\mathbf{x}, I_n\mathbf{y}) = (\mathbf{x}, \mathbf{y}), \quad (7)$$

where we have used Equation (3) in Section 3.3. This means that L preserves the inner product of two vectors and, consequently, L preserves length (why?). It also follows that if θ is the angle between vectors $\mathbf{x}$ and $\mathbf{y}$ in R^n, then the angle between $L(\mathbf{x})$ and $L(\mathbf{y})$ is also θ. A linear transformation satisfying Equation (7), $(L(\mathbf{x}), L(\mathbf{y})) = (\mathbf{x}, \mathbf{y})$, is called an **isometry** (from the Greek meaning *equal length*). Conversely, let $L\colon R^n \to R^n$ be an isometry, so that $(L(\mathbf{x}), L(\mathbf{y})) = (\mathbf{x}, \mathbf{y})$ for any $\mathbf{x}$ and $\mathbf{y}$ in R^n. Let A be the standard matrix representing L. Then $L(\mathbf{x}) = A\mathbf{x}$. We now have

$$(\mathbf{x}, \mathbf{y}) = (L(\mathbf{x}), L(\mathbf{y})) = (A\mathbf{x}, A\mathbf{y}) = (\mathbf{x}, A^T A\mathbf{y}).$$

Since this holds for all $\mathbf{x}$ in R^n, then, by Exercise 7(e) in Section 3.3, we conclude that $A^T A\mathbf{y} = \mathbf{y}$ for any $\mathbf{y}$ in R^n. It follows that $A^T A = I_n$ (Exercise 36), so A is an orthogonal matrix. Other properties of orthogonal matrices and isometries are examined in the exercises. (See also Supplementary Exercises 16 and 18 in Chapter 4.)

We now turn to the general situation for a symmetric matrix; even if A has eigenvalues whose multiplicities are greater than one, it turns out that we can still diagonalize A. We omit the proof of the following theorem. For a proof, see Ortega, J.M., *Matrix Theory: A Second Course*, New York: Plenum Press, 1987.

Theorem 6.9

If A is a symmetric $n \times n$ matrix, then there exists an orthogonal matrix P such that $P^{-1}AP = P^TAP = D$, a diagonal matrix. The eigenvalues of A lie on the main diagonal of D. ●

It can be shown (see the book by Ortega cited above) that if a symmetric matrix A has an eigenvalue λ of multiplicity k, then the solution space of the homogeneous system $(\lambda I_n - A)\mathbf{x} = \mathbf{0}$ [Equation (6) in Section 6.1] has dimension k. This means that there exist k linearly independent eigenvectors of A associated with the eigenvalue λ. By the Gram–Schmidt process, we can choose an orthonormal basis for this solution space. Thus we obtain a set of k orthonormal eigenvectors associated with the eigenvalue λ. Since eigenvectors associated with distinct eigenvalues are orthogonal, if we form the set of all eigenvectors we get an orthonormal set. Hence the matrix P whose columns are the eigenvectors is orthogonal.

EXAMPLE 4

Let

$$A = \begin{bmatrix} 0 & 2 & 2 \\ 2 & 0 & 2 \\ 2 & 2 & 0 \end{bmatrix}.$$

The characteristic polynomial of A is

$$p(\lambda) = (\lambda + 2)^2(\lambda - 4),$$

(verify), so its eigenvalues are

$$\lambda_1 = -2, \quad \lambda_2 = -2, \quad \text{and} \quad \lambda_3 = 4.$$

That is, -2 is an eigenvalue of multiplicity 2. To find eigenvectors associated with -2, we solve the homogeneous system $(-2I_3 - A)\mathbf{x} = \mathbf{0}$:

$$\begin{bmatrix} -2 & -2 & -2 \\ -2 & -2 & -2 \\ -2 & -2 & -2 \end{bmatrix} \begin{bmatrix} x_1 \\ x_2 \\ x_3 \end{bmatrix} = \begin{bmatrix} 0 \\ 0 \\ 0 \end{bmatrix}. \tag{8}$$

A basis for the solution space of (8) consists of the eigenvectors

$$\mathbf{x}_1 = \begin{bmatrix} -1 \\ 1 \\ 0 \end{bmatrix} \quad \text{and} \quad \mathbf{x}_2 = \begin{bmatrix} -1 \\ 0 \\ 1 \end{bmatrix}$$

(verify). Now $\mathbf{x}_1$ and $\mathbf{x}_2$ are not orthogonal, since $(\mathbf{x}_1, \mathbf{x}_2) \neq 0$. We can use the Gram–Schmidt process to obtain an orthonormal basis for the solution space of (8) (the eigenspace associated with -2) as follows. Let $\mathbf{y}_1 = \mathbf{x}_1$ and

$$\mathbf{y}_2 = \mathbf{x}_2 - \left(\frac{\mathbf{x}_2 \cdot \mathbf{y}_1}{\mathbf{y}_1 \cdot \mathbf{y}_1}\right)\mathbf{x}_1 = \begin{bmatrix} -\frac{1}{2} \\ -\frac{1}{2} \\ 1 \end{bmatrix}.$$

To eliminate fractions, we let

$$\mathbf{y}_2^* = 2\mathbf{y}_2 = \begin{bmatrix} -1 \\ -1 \\ 2 \end{bmatrix}.$$

The set $\{\mathbf{y}_1, \mathbf{y}_2^*\}$ is an orthogonal set of vectors. Normalizing, we obtain

$$\mathbf{z}_1 = \frac{1}{\|\mathbf{y}_1\|}\mathbf{y}_1 = \frac{1}{\sqrt{2}}\begin{bmatrix} -1 \\ 1 \\ 0 \end{bmatrix} \quad \text{and} \quad \mathbf{z}_2 = \frac{1}{\|\mathbf{y}_2^*\|}\mathbf{y}_2^* = \frac{1}{\sqrt{6}}\begin{bmatrix} -1 \\ -1 \\ 2 \end{bmatrix}.$$

The set $\{\mathbf{z}_1, \mathbf{z}_2\}$ is an orthonormal basis for the eigenspace associated with $\lambda = -2$. Now we find a basis for the eigenspace associated with $\lambda = 4$ by solving the homogeneous system $(4I_3 - A)\mathbf{x} = \mathbf{0}$:

$$\begin{bmatrix} 4 & -2 & -2 \\ -2 & 4 & -2 \\ -2 & -2 & 4 \end{bmatrix}\begin{bmatrix} x_1 \\ x_2 \\ x_3 \end{bmatrix} = \begin{bmatrix} 0 \\ 0 \\ 0 \end{bmatrix}, \tag{9}$$

A basis for this eigenspace consists of the vector

$$\mathbf{x}_3 = \begin{bmatrix} 1 \\ 1 \\ 1 \end{bmatrix}.$$

(verify). Normalizing this vector, we have the eigenvector

$$\mathbf{z}_3 = \frac{1}{\sqrt{3}}\begin{bmatrix} 1 \\ 1 \\ 1 \end{bmatrix}$$

as a basis for the eigenspace associated with $\lambda = 4$. Since eigenvectors associated with distinct eigenvalues are orthogonal, we observe that $\mathbf{z}_3$ is orthogonal to both $\mathbf{z}_1$ and $\mathbf{z}_2$. Thus the set $\{\mathbf{z}_1, \mathbf{z}_2, \mathbf{z}_3\}$ is an orthonormal basis for R^3 consisting of eigenvectors of A. The matrix P is the matrix whose jth column is $\mathbf{z}_j$:

$$P = \begin{bmatrix} -\frac{1}{\sqrt{2}} & -\frac{1}{\sqrt{6}} & \frac{1}{\sqrt{3}} \\ \frac{1}{\sqrt{2}} & -\frac{1}{\sqrt{6}} & \frac{1}{\sqrt{3}} \\ 0 & \frac{2}{\sqrt{6}} & \frac{1}{\sqrt{3}} \end{bmatrix}.$$

We leave it to the reader to verify that

$$P^{-1}AP = P^TAP = \begin{bmatrix} -2 & 0 & 0 \\ 0 & -2 & 0 \\ 0 & 0 & 4 \end{bmatrix}.$$

■

EXAMPLE 5

Let

$$A = \begin{bmatrix} 1 & 2 & 0 & 0 \\ 2 & 1 & 0 & 0 \\ 0 & 0 & 1 & 2 \\ 0 & 0 & 2 & 1 \end{bmatrix}.$$

Either by straightforward computation, or by Exercise 16 in Section 5.2, we find that the characteristic polynomial of A is

$$p(\lambda) = (\lambda + 1)^2(\lambda - 3)^2,$$

so its eigenvalues are

$$\lambda_1 = -1, \quad \lambda_2 = -1, \quad \lambda_3 = 3, \quad \text{and} \quad \lambda_4 = 3.$$

We now compute associated eigenvectors and the orthogonal matrix P. The eigenspace associated with the eigenvalue -1, of multiplicity 2, is the solution space of the homogeneous system $(-1I_4 - A)\mathbf{x} = \mathbf{0}$:

$$\begin{bmatrix} -2 & -2 & 0 & 0 \\ 2 & 2 & 0 & 0 \\ 0 & 0 & -2 & -2 \\ 0 & 0 & -2 & -2 \end{bmatrix} \begin{bmatrix} x_1 \\ x_2 \\ x_3 \\ x_4 \end{bmatrix} = \begin{bmatrix} 0 \\ 0 \\ 0 \\ 0 \end{bmatrix},$$

which is the set of all vectors of the form

$$\begin{bmatrix} r \\ -r \\ s \\ -s \end{bmatrix} = r \begin{bmatrix} 1 \\ -1 \\ 0 \\ 0 \end{bmatrix} + s \begin{bmatrix} 0 \\ 0 \\ 1 \\ -1 \end{bmatrix},$$

where r and s are any real numbers. Thus the eigenvectors

$$\begin{bmatrix} 1 \\ -1 \\ 0 \\ 0 \end{bmatrix} \quad \text{and} \quad \begin{bmatrix} 0 \\ 0 \\ 1 \\ -1 \end{bmatrix}$$

form a basis for the eigenspace associated with -1, and the dimension of this eigenspace is 2. Note that the eigenvectors

$$\begin{bmatrix} 1 \\ -1 \\ 0 \\ 0 \end{bmatrix} \quad \text{and} \quad \begin{bmatrix} 0 \\ 0 \\ 1 \\ -1 \end{bmatrix}$$

happen to be orthogonal. Since we are looking for an orthonormal basis for this eigenspace, we take

$$\mathbf{x}_1 = \begin{bmatrix} \frac{1}{\sqrt{2}} \\ -\frac{1}{\sqrt{2}} \\ 0 \\ 0 \end{bmatrix} \quad \text{and} \quad \mathbf{x}_2 = \begin{bmatrix} 0 \\ 0 \\ \frac{1}{\sqrt{2}} \\ -\frac{1}{\sqrt{2}} \end{bmatrix}$$

as eigenvectors associated with λ_1 and λ_2, respectively. Then $\{\mathbf{x}_1, \mathbf{x}_2\}$ is an orthonormal basis for the eigenspace associated with -1. The eigenspace associated with the eigenvalue 3, of multiplicity 2, is the solution space of the homogeneous system $(3I_4 - A)\mathbf{x} = \mathbf{0}$:

$$\begin{bmatrix} 2 & -2 & 0 & 0 \\ -2 & 2 & 0 & 0 \\ 0 & 0 & 2 & -2 \\ 0 & 0 & -2 & 2 \end{bmatrix} \begin{bmatrix} x_1 \\ x_2 \\ x_3 \\ x_4 \end{bmatrix} = \begin{bmatrix} 0 \\ 0 \\ 0 \\ 0 \end{bmatrix},$$

which is the set of all vectors of the form

$$\begin{bmatrix} r \\ r \\ s \\ s \end{bmatrix} = r \begin{bmatrix} 1 \\ 1 \\ 0 \\ 0 \end{bmatrix} + s \begin{bmatrix} 0 \\ 0 \\ 1 \\ 1 \end{bmatrix},$$

where r and s are any real numbers. Thus the eigenvectors

$$\begin{bmatrix} 1 \\ 1 \\ 0 \\ 0 \end{bmatrix} \quad \text{and} \quad \begin{bmatrix} 0 \\ 0 \\ 1 \\ 1 \end{bmatrix}$$

form a basis for the eigenspace associated with 3, and the dimension of this eigenspace is 2. Since these eigenvectors are orthogonal, we normalize them and let

$$\mathbf{x}_3 = \begin{bmatrix} \frac{1}{\sqrt{2}} \\ \frac{1}{\sqrt{2}} \\ 0 \\ 0 \end{bmatrix} \quad \text{and} \quad \mathbf{x}_4 = \begin{bmatrix} 0 \\ 0 \\ \frac{1}{\sqrt{2}} \\ \frac{1}{\sqrt{2}} \end{bmatrix}$$

be eigenvectors associated with λ_3 and λ_4, respectively. Then $\{\mathbf{x}_3, \mathbf{x}_4\}$ is an orthonormal basis for the eigenspace associated with 3. Now eigenvectors associated with distinct eigenvalues are orthogonal, so $\{\mathbf{x}_1, \mathbf{x}_2, \mathbf{x}_3, \mathbf{x}_4\}$ is an orthonormal basis for R^4. The matrix P is the matrix whose jth column is $\mathbf{x}_j$, $j = 1, 2, 3, 4$. Thus

$$P = \begin{bmatrix} \frac{1}{\sqrt{2}} & 0 & \frac{1}{\sqrt{2}} & 0 \\ -\frac{1}{\sqrt{2}} & 0 & \frac{1}{\sqrt{2}} & 0 \\ 0 & \frac{1}{\sqrt{2}} & 0 & \frac{1}{\sqrt{2}} \\ 0 & -\frac{1}{\sqrt{2}} & 0 & \frac{1}{\sqrt{2}} \end{bmatrix}.$$

We leave it to the reader to verify that P is an orthogonal matrix and that

$$P^{-1}AP = P^TAP = \begin{bmatrix} -1 & 0 & 0 & 0 \\ 0 & -1 & 0 & 0 \\ 0 & 0 & 3 & 0 \\ 0 & 0 & 0 & 3 \end{bmatrix}.$$

■

The procedure for diagonalizing a matrix A is as follows.

STEP 1 Form the characteristic polynomial $p(\lambda) = \det(\lambda I_n - A)$ of A.

STEP 2 Find the roots of the characteristic polynomial of A. If the roots are not real, then A cannot be diagonalized.

STEP 3 For each eigenvalue λ_j of A of multiplicity k_j, find a basis for the solution space of $(\lambda_j I_n - A)\mathbf{x} = \mathbf{0}$ (the eigenspace associated with λ_j). If the dimension of the eigenspace is less than k_j, then A is not diagonalizable. We thus determine n linearly independent eigenvectors of A. In Section 2.6 we solved the problem of finding a basis for the solution space of a homogeneous system.

STEP 4 Let P be the matrix whose columns are the n linearly independent eigenvectors determined in Step 3. Then $P^{-1}AP = D$, a diagonal matrix whose diagonal elements are the eigenvalues of A that correspond to the columns of P.

If A is an $n \times n$ symmetric matrix, we know that we can find an orthogonal matrix P such that $P^{-1}AP$ is diagonal. Conversely, suppose that A is a matrix for which we can find an orthogonal matrix P such that $P^{-1}AP = D$ is a diagonal matrix. What type of matrix is A? Since $P^{-1}AP = D$, $A = PDP^{-1}$. Also, $P^{-1} = P^T$ since P is orthogonal. Then

$$A^T = (PDP^T)^T = (P^T)^T D^T P^T = PDP^T = A,$$

which means that A is symmetric.

Some remarks about nonsymmetric matrices are in order at this point. Theorem 6.5 assures us that A is diagonalizable if all the roots of its characteristic polynomial are real and distinct. We also studied examples, in Section 6.2, of nonsymmetric matrices that had repeated eigenvalues which were diagonalizable and others that were not diagonalizable. There are some striking differences between the symmetric and nonsymmetric cases, which we now summarize. If A is nonsymmetric, then the roots of its characteristic polynomial need not all be real numbers; if an eigenvalue λ has multiplicity k, then the solution space of $(\lambda I_n - A)\mathbf{x} = \mathbf{0}$ may have dimension less than k; if the roots of the characteristic polynomial of A are all real, it is still possible for the eigenvectors to not form a basis for R^n; eigenvectors associated with distinct eigenvalues need not be orthogonal. Thus, in Example 7 of Section 6.2, the eigenvectors $\mathbf{x}_1$ and $\mathbf{x}_3$ associated with the eigenvalues $\lambda_1 = 0$ and $\lambda_3 = 1$ are not orthogonal. If a matrix A cannot be diagonalized, then we can often find a matrix B similar to A that is "nearly diagonal." The matrix B is said to be in **Jordan canonical form**. The study of such matrices lies beyond the scope of this book, but they are studied in advanced books on linear algebra [e.g., K. Hoffman and R. Kunze, *Linear Algebra*, 2nd ed. (Upper Saddle River, N.J.: Prentice Hall, 1971)]; they play a key role in many applications of linear algebra.

It should be noted that, in many applications, we need only find a diagonal matrix D that is similar to the given matrix A; that is, we do not explicitly have to know the matrix P such that $P^{-1}AP = D$.

Eigenvalue problems arise in all applications involving vibrations; they occur in aerodynamics, elasticity, nuclear physics, mechanics, chemical engineering, biology, differential equations, and so on. Many of the matrices to be diagonalized in applied problems are either symmetric or all the roots of their characteristic polynomial are real. Of course, the methods for finding eigenvalues that have been presented in this chapter are not recommended for matrices of large order because of the need to evaluate determinants.

6.4 Exercises

1. Verify that

$$P = \begin{bmatrix} \frac{2}{3} & -\frac{2}{3} & \frac{1}{3} \\ \frac{2}{3} & \frac{1}{3} & -\frac{2}{3} \\ \frac{1}{3} & \frac{2}{3} & \frac{2}{3} \end{bmatrix}$$

is an orthogonal matrix.

2. Find the inverse of each of the following orthogonal matrices.

(a) $A = \begin{bmatrix} 1 & 0 & 0 \\ 0 & \cos\phi & \sin\phi \\ 0 & -\sin\phi & \cos\phi \end{bmatrix}$.

(b) $B = \begin{bmatrix} 1 & 0 & 0 \\ 0 & \frac{1}{\sqrt{2}} & -\frac{1}{\sqrt{2}} \\ 0 & -\frac{1}{\sqrt{2}} & -\frac{1}{\sqrt{2}} \end{bmatrix}$.

3. Show that if A and B are orthogonal matrices, then AB is an orthogonal matrix.

4. Show that if A is an orthogonal matrix, then A^{-1} is orthogonal.

5. Prove Theorem 6.8.

6. Verify Theorem 6.8 for the matrices in Exercise 2.

7. Verify that the matrix P in Example 3 is an orthogonal matrix and that

$$P^{-1}AP = P^TAP = \begin{bmatrix} -2 & 0 & 0 \\ 0 & 4 & 0 \\ 0 & 0 & -1 \end{bmatrix}.$$

8. Show that if A is an orthogonal matrix, then $\det(A) = \pm 1$.

9. (a) Verify that the matrix

$$\begin{bmatrix} \cos\phi & -\sin\phi \\ \sin\phi & \cos\phi \end{bmatrix}$$

is orthogonal.

(b) Prove that if A is an orthogonal 2×2 matrix, then there exists a real number ϕ such that either

$$A = \begin{bmatrix} \cos\phi & -\sin\phi \\ \sin\phi & \cos\phi \end{bmatrix}$$

or

$$A = \begin{bmatrix} \cos\phi & \sin\phi \\ \sin\phi & -\cos\phi \end{bmatrix}.$$

10. For the orthogonal matrix

$$A = \begin{bmatrix} \frac{1}{\sqrt{2}} & -\frac{1}{\sqrt{2}} \\ -\frac{1}{\sqrt{2}} & -\frac{1}{\sqrt{2}} \end{bmatrix},$$

verify that $(A\mathbf{x}, A\mathbf{y}) = (\mathbf{x}, \mathbf{y})$ for any vectors $\mathbf{x}$ and $\mathbf{y}$ in R^2.

11. Let A be an $n \times n$ orthogonal matrix, and let $L: R^n \to R^n$ be the linear operator associated with A; that is, $L(\mathbf{x}) = A\mathbf{x}$ for $\mathbf{x}$ in R^n. Let θ be the angle between vectors $\mathbf{x}$ and $\mathbf{y}$ in R^n. Prove that the angle between $L(\mathbf{x})$ and $L(\mathbf{y})$ is also θ.

12. A linear operator $L: V \to V$, where V is an n-dimensional Euclidean space, is called **orthogonal** if $(L(\mathbf{x}), L(\mathbf{y})) = (\mathbf{x}, \mathbf{y})$. Let S be an orthonormal basis for V, and let the matrix A represent the orthogonal linear operator L with respect to S. Prove that A is an orthogonal matrix.

13. Let $L: R^2 \to R^2$ be the linear operator performing a counterclockwise rotation through $\pi/4$, and let A be the matrix representing L with respect to the natural basis for R^2. Prove that A is orthogonal.

14. Let A be an $n \times n$ matrix and let $B = P^{-1}AP$ be similar to A. Prove that if $\mathbf{x}$ is an eigenvector of A associated with the eigenvalue λ of A, then $P^{-1}\mathbf{x}$ is an eigenvector of B associated with the eigenvalue λ of B.

In Exercises 15 *through* 20, *diagonalize each given matrix and find an orthogonal matrix P such that $P^{-1}AP$ is diagonal.*

15. $A = \begin{bmatrix} 2 & 2 \\ 2 & 2 \end{bmatrix}$.

16. $A = \begin{bmatrix} 0 & 0 & 1 \\ 0 & 0 & 0 \\ 1 & 0 & 0 \end{bmatrix}$.

17. $A = \begin{bmatrix} 0 & 0 & 0 \\ 0 & 2 & 2 \\ 0 & 2 & 2 \end{bmatrix}$.

18. $A = \begin{bmatrix} 0 & 0 & 0 & 0 \\ 0 & 0 & 0 & 0 \\ 0 & 0 & 0 & 1 \\ 0 & 0 & 1 & 0 \end{bmatrix}$.

19. $A = \begin{bmatrix} 0 & -1 & -1 \\ -1 & 0 & -1 \\ -1 & -1 & 0 \end{bmatrix}$.

20. $A = \begin{bmatrix} -1 & 2 & 2 \\ 2 & -1 & 2 \\ 2 & 2 & -1 \end{bmatrix}$.

In Exercises 21 *through* 28, *diagonalize each given matrix.*

21. $A = \begin{bmatrix} 2 & 1 \\ 1 & 2 \end{bmatrix}$.

22. $A = \begin{bmatrix} 2 & 2 & 0 & 0 \\ 2 & 2 & 0 & 0 \\ 0 & 0 & 2 & 2 \\ 0 & 0 & 2 & 2 \end{bmatrix}$.

23. $A = \begin{bmatrix} 1 & 1 & 0 \\ 1 & 1 & 0 \\ 0 & 0 & 1 \end{bmatrix}$.

24. $A = \begin{bmatrix} 1 & 0 & 0 \\ 0 & 3 & -2 \\ 0 & -2 & 3 \end{bmatrix}$.

25. $A = \begin{bmatrix} 1 & 0 & 0 \\ 0 & 1 & 1 \\ 0 & 1 & 1 \end{bmatrix}$.

26. $A = \begin{bmatrix} 0 & 0 & 0 & 1 \\ 0 & 0 & 0 & 0 \\ 0 & 0 & 0 & 0 \\ 1 & 0 & 0 & 0 \end{bmatrix}$.

27. $A = \begin{bmatrix} 1 & -1 & 2 \\ -1 & 1 & 2 \\ 2 & 2 & 2 \end{bmatrix}$.

28. $A = \begin{bmatrix} -3 & 0 & -1 \\ 0 & -2 & 0 \\ -1 & 0 & -3 \end{bmatrix}$.

29. Prove Theorem 6.9 for the 2×2 case by studying the two possible cases for the roots of the characteristic polynomial of A.

30. Let $L: V \to V$ be an orthogonal linear operator (see Exercise 12), where V is an n-dimensional Euclidean space. Show that if λ is an eigenvalue of L, then $|\lambda| = 1$.

31. Let $L\colon R^2 \to R^2$ be defined by

$$L\left(\begin{bmatrix} x \\ y \end{bmatrix}\right) = \begin{bmatrix} \frac{1}{\sqrt{2}} & \frac{1}{\sqrt{2}} \\ \frac{1}{\sqrt{2}} & -\frac{1}{\sqrt{2}} \end{bmatrix} \begin{bmatrix} x \\ y \end{bmatrix}.$$

Show that L is an isometry of R^2.

32. Let $L\colon R^2 \to R^2$ be defined by $L(\mathbf{x}) = A\mathbf{x}$, for $\mathbf{x}$ in R^2, where A is an orthogonal matrix.

(a) Prove that if $\det(A) = 1$, then L is a counterclockwise rotation.

(b) Prove that if $\det(A) = -1$, then L is a reflection about the x-axis followed by a counterclockwise rotation.

33. Let $L\colon R^n \to R^n$ be a linear operator.

(a) Prove that if L is an isometry, then $\|L(\mathbf{x})\| = \|\mathbf{x}\|$, for $\mathbf{x}$ in R^n.

(b) Prove that if L is an isometry and θ is the angle between vectors $\mathbf{x}$ and $\mathbf{y}$ in R^n, then the angle between $L(\mathbf{x})$ and $L(\mathbf{y})$ is also θ.

34. Let $L\colon R^n \to R^n$ be a linear operator defined by $L(\mathbf{x}) = A\mathbf{x}$ for $\mathbf{x}$ in R^n. Prove that if L is an isometry, then L^{-1} is an isometry.

35. Let $L\colon R^n \to R^n$ be a linear operator and $S = \{\mathbf{v}_1, \mathbf{v}_2, \ldots, \mathbf{v}_n\}$ an orthonormal basis for R^n. Prove that L is an isometry if and only if $T = \{L(\mathbf{v}_1), L(\mathbf{v}_2), \ldots, L(\mathbf{v}_n)\}$ is an orthonormal basis for R^n.

36. Show that if $A^T A\mathbf{y} = \mathbf{y}$ for all $\mathbf{y}$ in R^n, then $A^T A = I_n$.

37. Show that if A is an orthogonal matrix, then A^T is also orthogonal.

38. Let A be an orthogonal matrix. Show that cA is orthogonal if and only if $c = \pm 1$.

39. Assuming that the software you use has a command for eigenvalues and eigenvectors (see Exercises 33 and 34 in Section 6.1), determine if a set of orthonormal eigenvectors is returned when the input matrix A is symmetric (see Theorem 6.9). Experiment with the matrices in Examples 4 and 5.

40. If the answer to Exercise 39 is no, you can use the Gram–Schmidt procedure to obtain an orthonormal set of eigenvectors (see Exercise 38 in Section 3.4). Experiment with the matrices in Examples 4 and 5 if necessary.

6.5 SPECTRAL DECOMPOSITION AND SINGULAR VALUE DECOMPOSITION (OPTIONAL)

Theorem 6.9 tells us that an $n \times n$ symmetric matrix A can be expressed as the matrix product

$$A = PDP^T, \tag{1}$$

where D is a diagonal matrix and P is an orthogonal matrix. The diagonal entries of D are the eigenvalues of A, $\lambda_1, \lambda_2, \ldots, \lambda_n$, and the columns of P are associated orthonormal eigenvectors $\mathbf{x}_1, \mathbf{x}_2, \ldots, \mathbf{x}_n$. The expression in (1) is called the **spectral decomposition** of A. It is helpful to write (1) in the following form:

$$A = \begin{bmatrix} \mathbf{x}_1 & \mathbf{x}_2 & \cdots & \mathbf{x}_n \end{bmatrix} \begin{bmatrix} \lambda_1 & 0 & \cdots & \cdots & 0 \\ 0 & \lambda_2 & 0 & \cdots & 0 \\ \vdots & 0 & \ddots & \ddots & 0 \\ \vdots & \vdots & \ddots & \ddots & 0 \\ 0 & 0 & \cdots & 0 & \lambda_n \end{bmatrix} \begin{bmatrix} \mathbf{x}_1^T \\ \mathbf{x}_2^T \\ \vdots \\ \mathbf{x}_n^T \end{bmatrix}. \tag{2}$$

The expression in (2) can be used to gain information about quadratic forms as shown in Section 6.6, and the nature of the eigenvalues of A can be utilized in other applications. However, we can use (1) and (2) to reveal aspects of the information contained in the matrix A. In particular, we can express A as a linear combination of simple symmetric matrices which are fundamental building blocks of the total information within A. The product DP^T can be computed and gives

$$DP^T = \begin{bmatrix} \lambda_1 & 0 & \cdots & \cdots & 0 \\ 0 & \lambda_2 & 0 & \cdots & 0 \\ \vdots & 0 & \ddots & \ddots & 0 \\ \vdots & \vdots & \ddots & \ddots & 0 \\ 0 & 0 & \cdots & 0 & \lambda_n \end{bmatrix} \begin{bmatrix} \mathbf{x}_1^T \\ \mathbf{x}_2^T \\ \vdots \\ \mathbf{x}_n^T \end{bmatrix} = \begin{bmatrix} \lambda_1 \mathbf{x}_1^T \\ \lambda_2 \mathbf{x}_2^T \\ \vdots \\ \lambda_n \mathbf{x}_n^T \end{bmatrix},$$

and hence (2) becomes

$$A = \begin{bmatrix} \mathbf{x}_1 & \mathbf{x}_2 & \cdots & \mathbf{x}_n \end{bmatrix} \begin{bmatrix} \lambda_1 \mathbf{x}_1^T \\ \lambda_2 \mathbf{x}_2^T \\ \vdots \\ \lambda_n \mathbf{x}_n^T \end{bmatrix}. \tag{3}$$

A careful analysis of the product on the right side of (3) reveals that we can express A as a linear combination of the matrices $\mathbf{x}_j \mathbf{x}_j^T$ and the coefficients are the eigenvalues of A. That is,

$$A = \sum_{j=1}^{n} \lambda_j \mathbf{x}_j \mathbf{x}_j^T = \lambda_1 \mathbf{x}_1 \mathbf{x}_1^T + \lambda_2 \mathbf{x}_2 \mathbf{x}_2^T + \cdots + \lambda_n \mathbf{x}_n \mathbf{x}_n^T. \tag{4}$$

A formal proof of (4) is quite tedious. In Example 1 we show the process for obtaining (4) in the case $n = 2$. The steps involved reveal the pattern that can be followed for the general case.

EXAMPLE 1

Let A be a 2×2 symmetric matrix with eigenvalues λ_1 and λ_2 and associated orthonormal eigenvectors $\mathbf{x}_1$ and $\mathbf{x}_2$. Let $P = \begin{bmatrix} \mathbf{x}_1 & \mathbf{x}_2 \end{bmatrix}$ and, to make the manipulations easier to see, let $\mathbf{x}_1 = \begin{bmatrix} a \\ b \end{bmatrix}$ and $\mathbf{x}_2 = \begin{bmatrix} c \\ d \end{bmatrix}$. Since A is symmetric it is diagonalizable using an orthogonal matrix. We have

$$P^T A P = D = \begin{bmatrix} \lambda_1 & 0 \\ 0 & \lambda_2 \end{bmatrix} \quad \text{and so} \quad A = P \begin{bmatrix} \lambda_1 & 0 \\ 0 & \lambda_2 \end{bmatrix} P^T.$$

Next, perform the following matrix and algebraic operations.

$$
\begin{aligned}
A &= \begin{bmatrix} a & c \\ b & d \end{bmatrix} \begin{bmatrix} \lambda_1 a & \lambda_1 b \\ \lambda_2 c & \lambda_2 d \end{bmatrix} = \begin{bmatrix} \lambda_1 a^2 + \lambda_2 c^2 & \lambda_1 ab + \lambda_2 cd \\ \lambda_1 ab + \lambda_2 cd & \lambda_1 b^2 + \lambda_2 d^2 \end{bmatrix} \\
&= \begin{bmatrix} \lambda_1 a^2 & \lambda_1 ab \\ \lambda_1 ab & \lambda_1 b^2 \end{bmatrix} + \begin{bmatrix} \lambda_2 c^2 & \lambda_2 cd \\ \lambda_2 cd & \lambda_2 d^2 \end{bmatrix} \\
&= \lambda_1 \begin{bmatrix} a^2 & ab \\ ab & b^2 \end{bmatrix} + \lambda_2 \begin{bmatrix} c^2 & cd \\ cd & d^2 \end{bmatrix} \\
&= \lambda_1 \begin{bmatrix} a\begin{bmatrix} a & b \end{bmatrix} \\ b\begin{bmatrix} a & b \end{bmatrix} \end{bmatrix} + \lambda_2 \begin{bmatrix} c\begin{bmatrix} c & d \end{bmatrix} \\ d\begin{bmatrix} c & d \end{bmatrix} \end{bmatrix} \\
&= \lambda_1 \begin{bmatrix} a \\ b \end{bmatrix} \begin{bmatrix} a & b \end{bmatrix} + \lambda_2 \begin{bmatrix} c \\ d \end{bmatrix} \begin{bmatrix} c & d \end{bmatrix} \\
&= \lambda_1 \mathbf{x}_1 \mathbf{x}_1^T + \lambda_2 \mathbf{x}_2 \mathbf{x}_2^T. \qquad \blacksquare
\end{aligned}
$$

The expression in (4) is equivalent to the spectral decomposition in (1), but displays the eigenvalue and eigenvector information in a different form. Example 2 illustrates the spectral decomposition in both forms.

EXAMPLE 2

Let

$$A = \begin{bmatrix} 2 & 1 & 0 \\ 1 & 2 & 0 \\ 0 & 0 & -1 \end{bmatrix}.$$

To determine the spectral representation of A we first obtain its eigenvalues and eigenvectors. We find that A has three distinct eigenvalues $\lambda_1 = 1$, $\lambda_2 = 3$, and $\lambda_3 = -1$ and that associated eigenvectors are respectively (verify)

$$\begin{bmatrix} 1 \\ -1 \\ 0 \end{bmatrix}, \quad \begin{bmatrix} 1 \\ 1 \\ 0 \end{bmatrix}, \quad \text{and} \quad \begin{bmatrix} 0 \\ 0 \\ 1 \end{bmatrix}.$$

Since the eigenvalues are distinct we are assured that the corresponding eigenvectors form an orthogonal set. (See Theorem 6.7.) Normalizing these vectors we obtain eigenvectors of unit length that are an orthonormal set:

$$\frac{1}{\sqrt{2}}\begin{bmatrix} 1 \\ -1 \\ 0 \end{bmatrix}, \quad \frac{1}{\sqrt{2}}\begin{bmatrix} 1 \\ 1 \\ 0 \end{bmatrix}, \quad \text{and} \quad \begin{bmatrix} 0 \\ 0 \\ 1 \end{bmatrix}.$$

Then the spectral representation of A is

$$A = \lambda_1 \mathbf{x}_1 \mathbf{x}_1^T + \lambda_2 \mathbf{x}_2 \mathbf{x}_2^T + \lambda_3 \mathbf{x}_3 \mathbf{x}_3^T$$

$$= 1\left(\frac{1}{\sqrt{2}}\right)^2 \begin{bmatrix} 1 \\ -1 \\ 0 \end{bmatrix} \begin{bmatrix} 1 & -1 & 0 \end{bmatrix} + 3\left(\frac{1}{\sqrt{2}}\right)^2 \begin{bmatrix} 1 \\ 1 \\ 0 \end{bmatrix} \begin{bmatrix} 1 & 1 & 0 \end{bmatrix}$$

$$+ (-1) \begin{bmatrix} 0 \\ 0 \\ 1 \end{bmatrix} \begin{bmatrix} 0 & 0 & 1 \end{bmatrix}$$

$$= 1\left(\frac{1}{2}\right) \begin{bmatrix} 1 & -1 & 0 \\ -1 & 1 & 0 \\ 0 & 0 & 0 \end{bmatrix} + 3\left(\frac{1}{2}\right) \begin{bmatrix} 1 & 1 & 0 \\ 1 & 1 & 0 \\ 0 & 0 & 0 \end{bmatrix} + (-1) \begin{bmatrix} 0 & 0 & 0 \\ 0 & 0 & 0 \\ 0 & 0 & 1 \end{bmatrix}.$$

■

In Example 2 the eigenvalues of A were distinct, so the associated eigenvectors formed an orthogonal set. If a symmetric matrix has an eigenvalue that is repeated, then the linearly independent eigenvectors associated with the repeated eigenvalue need not form an orthogonal set. However, we can apply the Gram–Schmidt process to the linearly independent eigenvectors associated with a repeated eigenvalue to obtain a set of orthogonal eigenvectors.

We note that (4) expresses the symmetric matrix A as a linear combination of matrices, $\mathbf{x}_j \mathbf{x}_j^T$, which are $n \times n$ since $\mathbf{x}_j$ is $n \times 1$ and $\mathbf{x}_j^T$ is $1 \times n$. The matrix $\mathbf{x}_j \mathbf{x}_j^T$ has a simple construction as shown in Figure 6.4.

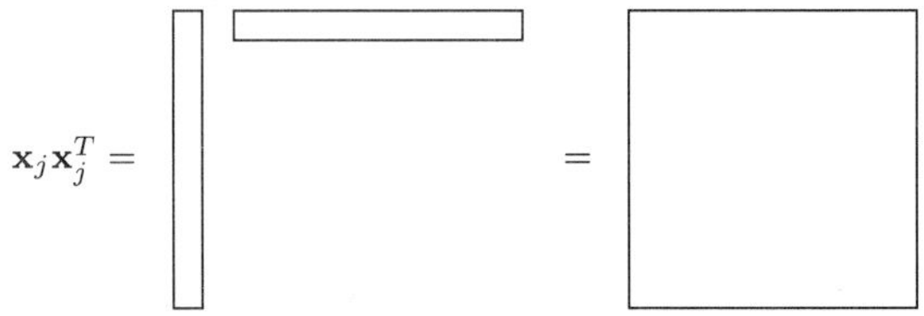

$$\mathbf{x}_j \mathbf{x}_j^T = \boxed{}\ \boxed{\qquad\qquad} = \boxed{\qquad}$$

FIGURE 6.4

We call $\mathbf{x}_j \mathbf{x}_j^T$ an **outer product** (see Supplementary Exercise 27 in Chapter 1). It can be shown that each row is a multiple of $\mathbf{x}_j^T$. Hence the reduced row echelon form of $\mathbf{x}_j \mathbf{x}_j^T$ [denoted $\mathbf{rref}(\mathbf{x}_j \mathbf{x}_j^T)$] has one nonzero row and thus has rank one. We interpret this in (4) to mean that each outer product $\mathbf{x}_j \mathbf{x}_j^T$ contributes just one piece of information to the construction of matrix A. Thus we can say that the spectral decomposition certainly reveals basic information about matrix A.

The spectral decomposition in (4) expresses A as a linear combination of outer products of the eigenvectors of A with coefficients that are the corresponding eigenvalues. Since each outer product has rank one we could say that they have "equal value" in building matrix A. However, the magnitude of the eigenvalue λ_j determines the "weight" given to the information contained in the outer product $\mathbf{x}_j \mathbf{x}_j^T$. Intuitively, if we labeled the eigenvalues so that $|\lambda_1| \geq |\lambda_2| \geq \cdots \geq |\lambda_n|$, then the beginning terms in the sum

$$A = \sum_{j=1}^{n} \lambda_j \mathbf{x}_j \mathbf{x}_j^T = \lambda_1 \mathbf{x}_1 \mathbf{x}_1^T + \lambda_2 \mathbf{x}_2 \mathbf{x}_2^T + \cdots + \lambda_n \mathbf{x}_n \mathbf{x}_n^T$$

contribute more information than the later terms, which correspond to the smaller eigenvalues. Note that if any eigenvalue is zero, then its eigenvector contributes no information to the construction of A. This is the case for a singular symmetric matrix.

EXAMPLE 3

From Example 2 we have that the eigenvalues are ordered as $|3| \geq |-1| \geq |1|$. Thus the contribution of the eigenvectors corresponding to eigenvalues -1 and 1 can be considered equal, whereas the one corresponding to 3 is dominant. Rewriting the spectral decomposition, using the terms in the order of the magnitude of the eigenvalues, we obtain the following.

$$A = 3\left(\frac{1}{\sqrt{2}}\right)^2 \begin{bmatrix} 1 \\ 1 \\ 0 \end{bmatrix} \begin{bmatrix} 1 & 1 & 0 \end{bmatrix} + (-1) \begin{bmatrix} 0 \\ 0 \\ 1 \end{bmatrix} \begin{bmatrix} 0 & 0 & 1 \end{bmatrix}$$
$$+ 1\left(\frac{1}{\sqrt{2}}\right)^2 \begin{bmatrix} 1 \\ -1 \\ 0 \end{bmatrix} \begin{bmatrix} 1 & -1 & 0 \end{bmatrix}$$
$$= 3\left(\frac{1}{2}\right) \begin{bmatrix} 1 & 1 & 0 \\ 1 & 1 & 0 \\ 0 & 0 & 0 \end{bmatrix} + (-1) \begin{bmatrix} 0 & 0 & 0 \\ 0 & 0 & 0 \\ 0 & 0 & 1 \end{bmatrix} + 1\left(\frac{1}{2}\right) \begin{bmatrix} 1 & -1 & 0 \\ -1 & 1 & 0 \\ 0 & 0 & 0 \end{bmatrix}. \quad \blacksquare$$

Looking at the terms of the partial sums in Example 3 individually we have the following matrices:

$$\begin{bmatrix} \frac{3}{2} & \frac{3}{2} & 0 \\ \frac{3}{2} & \frac{3}{2} & 0 \\ 0 & 0 & 0 \end{bmatrix}$$

$$\begin{bmatrix} \frac{3}{2} & \frac{3}{2} & 0 \\ \frac{3}{2} & \frac{3}{2} & 0 \\ 0 & 0 & 0 \end{bmatrix} + \begin{bmatrix} 0 & 0 & 0 \\ 0 & 0 & 0 \\ 0 & 0 & -1 \end{bmatrix} = \begin{bmatrix} \frac{3}{2} & \frac{3}{2} & 0 \\ \frac{3}{2} & \frac{3}{2} & 0 \\ 0 & 0 & -1 \end{bmatrix}$$

$$\begin{bmatrix} \frac{3}{2} & \frac{3}{2} & 0 \\ \frac{3}{2} & \frac{3}{2} & 0 \\ 0 & 0 & -1 \end{bmatrix} + \begin{bmatrix} \frac{1}{2} & -\frac{1}{2} & 0 \\ -\frac{1}{2} & \frac{1}{2} & 0 \\ 0 & 0 & 0 \end{bmatrix} = \begin{bmatrix} 2 & 1 & 0 \\ 1 & 2 & 0 \\ 0 & 0 & -1 \end{bmatrix} = A$$

This suggests that we can approximate the information in matrix A, using the partial sums of the spectral decomposition in (4). In fact, this type of development is the foundation of a number of approximation procedures in mathematics.

While the utility of a symmetric matrix is somewhat limited in terms of the representation of information, a generalization of a spectral style decomposition to arbitrary size matrices dramatically extends the usefulness of the approximation procedure by partial sums. We indicate this more general decomposition next.

We state the following result, whose proof can be found in the references listed at the end of the section.

The Singular Value Decomposition of a Matrix

Let A be an $m \times n$ real matrix. Then there exist orthogonal matrices U of size $m \times m$ and V of size $n \times n$ such that

$$A = USV^T, \tag{5}$$

where S is an $m \times n$ matrix with nondiagonal entries all zero and $s_{11} \geq s_{22} \geq \cdots \geq s_{pp} \geq 0$, where $p = \min\{m, n\}$.

The diagonal entries of S are called the **singular values** of A, the columns of U are called the **left singular vectors** of A, and the columns of V are called the **right singular vectors** of A. The singular value decomposition of A in (5) can be expressed as the following linear combination:

$$\begin{aligned} A = {} & \text{col}_1(U)s_{11}\text{col}_1(V)^T + \text{col}_2(U)s_{22}\text{col}_2(V)^T \\ & + \cdots + \text{col}_p(U)s_{pp}\text{col}_p(V)^T, \end{aligned} \tag{6}$$

which has the same form as the spectral representation of a symmetric matrix as given in Equation (4).

To determine the matrices U, S, and V in the singular value decomposition given in (5) we start as follows. An $n \times n$ symmetric matrix related to A is A^TA. By Theorem 6.9 there exists an orthogonal $n \times n$ matrix V such that

$$V(A^TA)V^T = D,$$

where D is a diagonal matrix whose diagonal entries $\lambda_1, \lambda_2, \ldots, \lambda_n$ are the eigenvalues of A^TA. If $\mathbf{v}_j$ denotes column j of V, then $(A^TA)\mathbf{v}_j = \lambda_j\mathbf{v}_j$. Multiply both sides of this expression on the left by $\mathbf{v}_j^T$; then we can rearrange the expression as

$$\mathbf{v}_j^T(A^TA)\mathbf{v}_j = \lambda_j\mathbf{v}_j^T\mathbf{v}_j \quad \text{or} \quad (A\mathbf{v}_j)^T(A\mathbf{v}_j) = \lambda_j\mathbf{v}_j^T\mathbf{v}_j \quad \text{or} \quad \|A\mathbf{v}_j\|^2 = \lambda_j\|\mathbf{v}_j\|^2.$$

Since the length of a vector is nonnegative, the last expression implies that $\lambda_j \geq 0$. Hence, each eigenvalue of A^TA is nonnegative. If necessary, renumber the eigenvalues of A^TA so that $\lambda_1 \geq \lambda_2 \geq \cdots \geq \lambda_n$; then define $s_{jj} = \sqrt{\lambda_j}$. We note that since V is an orthogonal matrix, each of its columns is a unit vector, that is, $\|\mathbf{v}_j\| = 1$. Hence $s_{jj} = \|A\mathbf{v}_j\|$. (Verify.) Thus the singular values of A are the square roots of the eigenvalues of A^TA.

Finally we determine the $m \times m$ orthogonal matrix U. Assuming the matrix equation in (5), let us see what the columns of U should look like.

- Since U is to be orthogonal, its columns must be an orthonormal set, hence they are linearly independent $m \times 1$ vectors.
- The matrix S has the form (block diagonal)

$$S = \left[\begin{array}{cccc|c} s_{11} & 0 & \cdots & 0 & \\ 0 & s_{22} & \ddots & 0 & O_{p,n-p} \\ \vdots & \ddots & \ddots & 0 & \\ 0 & \cdots & 0 & s_{pp} & \\ \hline & O_{m-p,p} & & & O_{m-p,n-p} \end{array}\right].$$

- From (5), $AV = US$ so

$$A\begin{bmatrix}\mathbf{v}_1 & \mathbf{v}_2 & \cdots & \mathbf{v}_n\end{bmatrix}$$

$$= \begin{bmatrix}\mathbf{u}_1 & \mathbf{u}_2 & \cdots & \mathbf{u}_m\end{bmatrix} \left[\begin{array}{cccc|c} s_{11} & 0 & \cdots & 0 & \\ 0 & s_{22} & \ddots & 0 & O_{p,n-p} \\ \vdots & \ddots & \ddots & 0 & \\ 0 & \cdots & 0 & s_{pp} & \\ \hline & O_{m-p,p} & & & O_{m-p,n-p} \end{array}\right].$$

  This implies that we need to require that $A\mathbf{v}_j = s_{jj}\mathbf{u}_j$ for $j = 1, 2, \ldots, p$. (Verify.)
- However, U must have m orthonormal columns and $m \geq p$. In order to construct U we need an orthonormal basis for R^m whose first p vectors are $\mathbf{u}_j = \dfrac{1}{s_{jj}} A\mathbf{v}_j$. In Theorem 2.10 we showed how to extend any linearly independent subset of a vector space to a basis. We use that technique here to obtain the remaining $m - p$ columns of U. (This is necessary only if $m > p$.) Since these $m - p$ columns are not unique, matrix U is not unique. (Neither is V if any of the eigenvalues of $A^T A$ are repeated.)

It can be shown that the preceding construction gives matrices U, S, and V, so that $A = USV^T$. We illustrate the process in Example 4.

EXAMPLE 4

To find the singular value decomposition of $A = \begin{bmatrix} 2 & -4 \\ 2 & 2 \\ -4 & 0 \\ 1 & 4 \end{bmatrix}$ we follow the steps outlined above. First, we compute $A^T A$, and then compute its eigenvalues and eigenvectors. We obtain (verify)

$$A^T A = \begin{bmatrix} 25 & 0 \\ 0 & 36 \end{bmatrix},$$

and since it is diagonal we know that its eigenvalues are its diagonal entries. It follows that $\begin{bmatrix} 1 \\ 0 \end{bmatrix}$ is an eigenvector associated with eigenvalue 25, and that $\begin{bmatrix} 0 \\ 1 \end{bmatrix}$ is an eigenvector associated with eigenvalue 36. (Verify.) We label the eigenvalues in decreasing magnitude as $\lambda_1 = 36$ and $\lambda_2 = 25$ with corresponding eigenvectors $\mathbf{v}_1 = \begin{bmatrix} 0 \\ 1 \end{bmatrix}$ and $\mathbf{v}_2 = \begin{bmatrix} 1 \\ 0 \end{bmatrix}$, respectively. Hence

$$V = \begin{bmatrix} \mathbf{v}_1 & \mathbf{v}_2 \end{bmatrix} = \begin{bmatrix} 0 & 1 \\ 1 & 0 \end{bmatrix} \quad \text{and} \quad s_{11} = 6 \quad \text{and} \quad s_{22} = 5.$$

It follows that

$$S = \begin{bmatrix} 6 & 0 \\ 0 & 5 \\ \hline \multicolumn{2}{c}{O_2} \end{bmatrix},$$

Next, we determine the matrix U, starting with the first two columns:

$$\mathbf{u}_1 = \frac{1}{s_{11}} A\mathbf{v}_1 = \frac{1}{6} \begin{bmatrix} -4 \\ 2 \\ 0 \\ 4 \end{bmatrix} \quad \text{and} \quad \mathbf{u}_2 = \frac{1}{s_{22}} A\mathbf{v}_2 = \frac{1}{5} \begin{bmatrix} 2 \\ 2 \\ -4 \\ 1 \end{bmatrix}.$$

The remaining two columns are found by extending the linearly independent vectors in $\{\mathbf{u}_1, \mathbf{u}_2\}$ to a basis for R^4 and then applying the Gram–Schmidt process. We proceed as follows. We first compute the reduced row echelon form (verify):

$$\mathbf{rref}\left(\begin{bmatrix} \mathbf{u}_1 & \mathbf{u}_2 & I_4 \end{bmatrix}\right) = \begin{bmatrix} 1 & 0 & 0 & 0 & \frac{3}{8} & \frac{3}{2} \\ 0 & 1 & 0 & 0 & -\frac{3}{2} & 0 \\ 0 & 0 & 1 & 0 & \frac{3}{4} & 1 \\ 0 & 0 & 0 & 1 & \frac{3}{8} & -\frac{1}{2} \end{bmatrix}.$$

This tells us that the set $\{\mathbf{u}_1, \mathbf{u}_2, \mathbf{e}_1, \mathbf{e}_2\}$ is a basis for R^4. (Explain why.) Next, we apply the Gram–Schmidt process to this set to find the orthogonal matrix U. Recording the results to six decimals we obtain

$$U = \begin{bmatrix} & & 0.628932 & 0 \\ \mathbf{u}_1 & \mathbf{u}_2 & 0.098933 & 0.847998 \\ & & 0.508798 & 0.317999 \\ & & 0.579465 & -0.423999 \end{bmatrix}.$$

Thus we have the singular value decomposition of A and

$$A = USV^T = \left[\begin{array}{cc|cc} \mathbf{u}_1 & \mathbf{u}_2 & \begin{array}{c} 0.628932 \\ 0.098933 \\ 0.508798 \\ 0.579465 \end{array} & \begin{array}{c} 0 \\ 0.847998 \\ 0.317999 \\ -0.423999 \end{array} \end{array}\right] \left[\begin{array}{cc} 6 & 0 \\ 0 & 5 \\ \hline \multicolumn{2}{c}{O_2} \end{array}\right] \begin{bmatrix} \mathbf{v}_1 & \mathbf{v}_2 \end{bmatrix}^T$$

$$= \begin{bmatrix} -\frac{2}{3} & \frac{2}{5} & 0.628932 & 0 \\ \frac{1}{3} & \frac{2}{5} & 0.098933 & 0.847998 \\ 0 & -\frac{4}{5} & 0.508798 & 0.317999 \\ \frac{2}{3} & \frac{1}{5} & 0.579465 & -0.423999 \end{bmatrix} \begin{bmatrix} 6 & 0 \\ 0 & 5 \\ 0 & 0 \\ 0 & 0 \end{bmatrix} \begin{bmatrix} 0 & 1 \\ 1 & 0 \end{bmatrix}^T .$$

■

The singular decomposition of a matrix A has been called one of the most useful tools in terms of the information that it reveals. For example, previously in Section 2.8, we indicated that we could compute the rank A by determining the number of nonzero rows in the reduced row echelon form of A. An implicit assumption in this statement is that all the computational steps in the row operations would use exact arithmetic. Unfortunately, in most computing environments, when we perform row operations exact arithmetic is not used. Rather, floating point arithmetic, which is a model of exact arithmetic, is used. In doing so, we may lose accuracy because of the accumulation of small errors in the arithmetic steps. In some cases this loss of accuracy is enough to introduce doubt into the computation of rank. The following two results, which we state without proof, indicate how the singular value decomposition of a matrix can be used to compute its rank.

Theorem 6.10

Let A be an $m \times n$ matrix and let B and C be nonsingular matrices of sizes $m \times m$ and $n \times n$, respectively. Then

$$\text{rank } BA = \text{rank } A = \text{rank } AC.$$

●

Corollary 6.2

The rank of A is the number of nonzero singular values of A. (Multiple singular values are counted according to their multiplicity.)

●

(See Exercise 7.)

Because matrices U and V of a singular value decomposition are orthogonal, it can be shown that most of the errors due to using floating point arithmetic occur in the computation of the singular values. The size of the matrix A and characteristics of the floating point arithmetic are often used to determine a threshold value below which singular values are considered zero. It has been argued that singular values and singular value decomposition gives us a computationally reliable way to compute the rank of A.

In addition to determining rank, the singular value decomposition of a matrix also provides orthonormal bases for the fundamental subspaces associated with a linear system of equations. We state without proof the following.

Theorem 6.11

Let A be an $m \times n$ real matrix with singular value decomposition USV^T. Then:

(a) *The first p columns of U are an orthonormal basis for the column space of A.*

(b) *The first p columns of V are an orthonormal basis for the row space of A.*

(c) *The last $n - p$ columns of V are an orthonormal basis for the null space of A.* ●

The singular value decomposition is also a valuable tool in the computation of least-squares approximations and can be used to approximate images which have been digitized. For examples of these applications see the references at the end of this section.

6.5 Exercises

1. Find the singular values of each of the following matrices.

(a) $A = \begin{bmatrix} 5 & 0 \\ 0 & 0 \\ 0 & -1 \end{bmatrix}$. (b) $A = \begin{bmatrix} 1 & 1 \\ 1 & 1 \end{bmatrix}$.

(c) $A = \begin{bmatrix} 1 & 2 \\ 0 & 1 \\ -2 & 1 \end{bmatrix}$.

(d) $A = \begin{bmatrix} 1 & 0 & 1 & -1 \\ 0 & 1 & -1 & 1 \end{bmatrix}$.

2. Determine the singular value decomposition of $A = \begin{bmatrix} 1 & -4 \\ -2 & 2 \\ 2 & 4 \end{bmatrix}$.

3. Determine the singular value decomposition of $A = \begin{bmatrix} 1 & 1 \\ -1 & 1 \end{bmatrix}$.

4. Determine the singular value decomposition of $A = \begin{bmatrix} 1 & 0 & 1 \\ 1 & 1 & -1 \\ -1 & 1 & 0 \\ 0 & 1 & 1 \end{bmatrix}$.

5. There is a MATLAB command, **svd**, that computes the singular value decomposition of a matrix A. Used in the form **svd(A)**, the output is a list of the singular values of A. In the form **[U, S, V] = svd(A)** we get matrices U, S, and V such that $A = USV^T$. Use this command to determine the singular values of each of the following matrices.

(a) $\begin{bmatrix} 5 & 2 \\ 8 & 5 \end{bmatrix}$. (b) $\begin{bmatrix} 3 & 12 & -10 \\ 1 & -1 & 5 \\ -2 & -8 & 5 \end{bmatrix}$.

(c) $\begin{bmatrix} 1 & 5 & 9 \\ 2 & 6 & 10 \\ 3 & 7 & 11 \\ 4 & 8 & 12 \end{bmatrix}$.

6. An $m \times n$ matrix A is said to have **full rank** if rank $A =$ minimum $\{m, n\}$. The singular value decomposition lets us measure how close A is to not having full rank. If any singular value is zero, then A does not have full rank. If $s_{\min}$ is the smallest singular value of A and $s_{\min} \neq 0$, then the distance from A to the set of matrices with rank $r = \min\{m, n\} - 1$ is $s_{\min}$. Determine the distance from each of the following matrices to the matrices of the same size with rank $\min\{m, n\} - 1$. (Use MATLAB to find the singular values.)

(a) $\begin{bmatrix} 1 & 3 & 2 \\ 4 & 1 & 0 \\ 2 & -5 & -4 \\ 1 & 2 & 1 \end{bmatrix}$. (b) $\begin{bmatrix} 1 & 2 & 3 \\ 1 & 0 & 1 \\ 3 & 6 & 9 \end{bmatrix}$.

(c) $\begin{bmatrix} 1 & 0 & 0 & -1 & 0 \\ -1 & 1 & 1 & 0 & 0 \\ 0 & 0 & 1 & -2 & -1 \\ 0 & 1 & 0 & 1 & 1 \end{bmatrix}$.

7. Prove Corollary 6.2 using Theorem 6.10 and the singular value decomposition.

REFERENCES

HERN, T. and C. LONG, Viewing Some Concepts and Applications in Linear Algebra, in *Visualization in Teaching and Learning Mathematics*, MAA Notes Number 19, pp. 173–190, The Mathematical Association of America, 1991.

HILL, D., *Experiments in Computational Matrix Algebra*, Random House, N.Y., 1988 (distributed by McGraw-Hill).

HILL, D., Nuggets of Information, in *Using MATLAB in the Classroom*, pp. 77–93, Prentice Hall, Upper Saddle River, N.J., 1993.

KAHANER, D., C. MOLER, and S. NASH, *Numerical Methods and Software*, Prentice Hall, Upper Saddle River, N.J., 1989.

STEWART, G., *Introduction to Matrix Computations*, Academic Press, N.Y., 1973.

6.6 REAL QUADRATIC FORMS

In your precalculus and calculus courses you have seen that the graph of the equation

$$ax^2 + 2bxy + cy^2 = d, \tag{1}$$

where a, b, c, and d are real numbers, is a **conic section** centered at the origin of a rectangular Cartesian coordinate system in two-dimensional space. Similarly, the graph of the equation

$$ax^2 + 2dxy + 2exz + by^2 + 2fyz + cz^2 = g, \tag{2}$$

where a, b, c, d, e, f, and g are real numbers, is a **quadric surface** centered at the origin of a rectangular Cartesian coordinate system in three-dimensional space. If a conic section or quadric surface is not centered at the origin, its equations are more complicated than those given in (1) and (2).

The identification of the conic section or quadric surface that is the graph of a given equation often requires the rotation and translation of the coordinate axes. These methods can best be understood as an application of eigenvalues and eigenvectors of matrices that are discussed in Sections 6.7 and 6.8.

The expressions on the left sides of Equations (1) and (2) are examples of quadratic forms. Quadratic forms arise in statistics, mechanics, and in other areas of physics; in quadratic programming; in the study of maxima and minima of functions of several variables; and in other applied problems. In this section we use our results on eigenvalues and eigenvectors of matrices to give a brief treatment of real quadratic forms in n variables. In Section 6.7 we apply these results to the classification of the conic sections, and in Section 6.8 to the classification of the quadric surfaces.

Definition 6.5

If A is a symmetric matrix, then the function $g: R^n \to R^1$ (a real-valued function on R^n) defined by

$$g(\mathbf{x}) = \mathbf{x}^T A\mathbf{x},$$

where

$$\mathbf{x} = \begin{bmatrix} x_1 \\ x_2 \\ \vdots \\ x_n \end{bmatrix},$$

is called a **real quadratic form in the n variables** $x_1, x_2, \ldots, x_n$. The matrix A is called the **matrix of the quadratic form** g. We shall also denote the quadratic form by $g(\mathbf{x})$. ▲

EXAMPLE 1

Write the left side of (1) as the quadratic form in the variables x and y.

Solution Let

$$\mathbf{x} = \begin{bmatrix} x \\ y \end{bmatrix} \quad \text{and} \quad A = \begin{bmatrix} a & b \\ b & c \end{bmatrix}.$$

Then the left side of (1) is the quadratic form

$$g(\mathbf{x}) = \mathbf{x}^T A\mathbf{x}.$$

■

EXAMPLE 2

Write the left side of (2) as the quadratic form in the variables x, y, and z.

Solution Let

$$\mathbf{x} = \begin{bmatrix} x \\ y \\ z \end{bmatrix} \quad \text{and} \quad A = \begin{bmatrix} a & d & e \\ d & b & f \\ e & f & c \end{bmatrix}.$$

Then the left side of (2) is the quadratic form

$$g(\mathbf{x}) = \mathbf{x}^T A\mathbf{x}.$$

■

EXAMPLE 3

The following expressions are quadratic forms:

(a) $3x^2 - 5xy - 7y^2 = \begin{bmatrix} x & y \end{bmatrix} \begin{bmatrix} 3 & -\frac{5}{2} \\ -\frac{5}{2} & -7 \end{bmatrix} \begin{bmatrix} x \\ y \end{bmatrix}.$

(b) $$3x^2-7xy+5xz+4y^2-4yz-3z^2 = \begin{bmatrix} x & y & z \end{bmatrix} \begin{bmatrix} 3 & -\frac{7}{2} & \frac{5}{2} \\ -\frac{7}{2} & 4 & -2 \\ \frac{5}{2} & -2 & -3 \end{bmatrix} \begin{bmatrix} x \\ y \\ z \end{bmatrix}.$$

■

Suppose now that $g(\mathbf{x}) = \mathbf{x}^T A\mathbf{x}$ is a quadratic form. To simplify the quadratic form, we change from the variables $x_1, x_2, \ldots, x_n$ to the variables $y_1, y_2, \ldots, y_n$, where we assume that the old variables are related to the new variables by $\mathbf{x} = P\mathbf{y}$ for some orthogonal matrix P. Then

$$g(\mathbf{x}) = \mathbf{x}^T A\mathbf{x} = (P\mathbf{y})^T A(P\mathbf{y}) = \mathbf{y}^T(P^T AP)\mathbf{y} = \mathbf{y}^T B\mathbf{y},$$

where $B = P^T AP$. We shall let you verify that if A is a symmetric matrix, then $P^T AP$ is also symmetric (Exercise 25). Thus

$$h(\mathbf{y}) = \mathbf{y}^T B\mathbf{y}$$

is another quadratic form and $g(\mathbf{x}) = h(\mathbf{y})$.

This situation is important enough to formulate the following definitions.

Definition 6.6

If A and B are $n \times n$ matrices, we say that B is **congruent** to A if $B = P^T AP$ for a nonsingular matrix P. ▲

In light of Exercise 26, "A is congruent to B," and "B is congruent to A" can both be replaced by "A and B are congruent."

Definition 6.7

Two quadratic forms g and h with matrices A and B, respectively, are said to be **equivalent** if A and B are congruent. ▲

The congruence of matrices and equivalence of forms are more general concepts than the notion of similarity since the matrix P is only required to be nonsingular (not necessarily orthogonal). We shall consider here the more restrictive situation with P orthogonal.

EXAMPLE 4

Consider the quadratic form in the variables x and y defined by

$$g(\mathbf{x}) = 2x^2 + 2xy + 2y^2 = \begin{bmatrix} x & y \end{bmatrix} \begin{bmatrix} 2 & 1 \\ 1 & 2 \end{bmatrix} \begin{bmatrix} x \\ y \end{bmatrix}. \tag{3}$$

We now change from the variables x and y to the variables x' and y'. Suppose that the old variables are related to the new variables by the equations

$$x = \frac{1}{\sqrt{2}}x' - \frac{1}{\sqrt{2}}y' \quad \text{and} \quad y = \frac{1}{\sqrt{2}}x' + \frac{1}{\sqrt{2}}y', \tag{4}$$

which can be written in matrix form as

$$\mathbf{x} = \begin{bmatrix} x \\ y \end{bmatrix} = \begin{bmatrix} \frac{1}{\sqrt{2}} & -\frac{1}{\sqrt{2}} \\ \frac{1}{\sqrt{2}} & \frac{1}{\sqrt{2}} \end{bmatrix} \begin{bmatrix} x' \\ y' \end{bmatrix} = P\mathbf{y},$$

where the orthogonal (hence nonsingular) matrix

$$P = \begin{bmatrix} \frac{1}{\sqrt{2}} & -\frac{1}{\sqrt{2}} \\ \frac{1}{\sqrt{2}} & \frac{1}{\sqrt{2}} \end{bmatrix} \quad \text{and} \quad \mathbf{y} = \begin{bmatrix} x' \\ y' \end{bmatrix}.$$

We shall soon see why and how this particular matrix P was selected. Substituting in (3), we obtain

$$\begin{aligned} g(\mathbf{x}) &= \mathbf{x}^T A\mathbf{x} = (P\mathbf{y})^T A(P\mathbf{y}) = \mathbf{y}^T P^T AP\mathbf{y} \\ &= \begin{bmatrix} x' & y' \end{bmatrix} \begin{bmatrix} \frac{1}{\sqrt{2}} & -\frac{1}{\sqrt{2}} \\ \frac{1}{\sqrt{2}} & \frac{1}{\sqrt{2}} \end{bmatrix}^T \begin{bmatrix} 2 & 1 \\ 1 & 2 \end{bmatrix} \begin{bmatrix} \frac{1}{\sqrt{2}} & -\frac{1}{\sqrt{2}} \\ \frac{1}{\sqrt{2}} & \frac{1}{\sqrt{2}} \end{bmatrix} \begin{bmatrix} x' \\ y' \end{bmatrix} \\ &= \begin{bmatrix} x' & y' \end{bmatrix} \begin{bmatrix} 3 & 0 \\ 0 & 1 \end{bmatrix} \begin{bmatrix} x' \\ y' \end{bmatrix} = h(\mathbf{y}) \\ &= 3x'^2 + y'^2 \end{aligned}$$

Thus the matrices

$$\begin{bmatrix} 2 & 1 \\ 1 & 2 \end{bmatrix} \quad \text{and} \quad \begin{bmatrix} 3 & 0 \\ 0 & 1 \end{bmatrix}$$

are congruent and the quadratic forms g and h are equivalent. ■

We now turn to the question of how to select the matrix P.

Theorem 6.12 (Principal Axes Theorem).
Any quadratic form in n variables $g(\mathbf{x}) = \mathbf{x}^T A\mathbf{x}$ is equivalent by means of an orthogonal matrix P to a quadratic form, $h(\mathbf{y}) = \lambda_1 y_1^2 + \lambda_2 y_2^2 + \cdots + \lambda_n y_n^2$, where

$$\mathbf{y} = \begin{bmatrix} y_1 \\ y_2 \\ \vdots \\ y_n \end{bmatrix}$$

and $\lambda_1, \lambda_2, \ldots, \lambda_n$ are the eigenvalues of the matrix A of g.

Proof
If A is the matrix of g, then, since A is symmetric, we know by Theorem 6.9 that A can be diagonalized by an orthogonal matrix. This means that there exists an orthogonal matrix P such that $D = P^{-1}AP$ is a diagonal matrix. Since P is orthogonal, $P^{-1} = P^T$, so $D = P^TAP$. Moreover, the elements on the main diagonal of D are the eigenvalues, $\lambda_1, \lambda_2, \ldots, \lambda_n$ of A, which are real numbers. The quadratic form h with matrix D is given by

$$h(\mathbf{y}) = \lambda_1 y_1^2 + \lambda_2 y_2^2 + \cdots + \lambda_n y_n^2;$$

g and h are equivalent. ●

EXAMPLE 5

Consider the quadratic form g in the variables x, y, and z defined by

$$g(\mathbf{x}) = 2x^2 + 4y^2 + 6yz - 4z^2.$$

Determine a quadratic form h of the form in Theorem 6.12 to which g is equivalent.

Solution The matrix of g is

$$A = \begin{bmatrix} 2 & 0 & 0 \\ 0 & 4 & 3 \\ 0 & 3 & -4 \end{bmatrix},$$

and the eigenvalues of A are

$$\lambda_1 = 2, \quad \lambda_2 = 5, \quad \text{and} \quad \lambda_3 = -5 \qquad \text{(verify)}.$$

Let h be the quadratic form in the variables x', y', and z' defined by

$$h(\mathbf{y}) = 2x'^2 + 5y'^2 - 5z'^2.$$

Then g and h are equivalent by means of some orthogonal matrix. Note that $\hat{h}(\mathbf{y}) = -5x'^2 + 2y'^2 + 5z'^2$ is also equivalent to g. ■

Observe that to apply Theorem 6.12 to diagonalize a given quadratic form, as shown in Example 5, we do not need to know the eigenvectors of A (nor the matrix P); we only require the eigenvalues of A.

To understand the significance of Theorem 6.12, we consider quadratic forms in two and three variables. As we have already observed at the beginning of this section, the graph of the equation

$$g(\mathbf{x}) = \mathbf{x}^T A\mathbf{x} = 1,$$

where $\mathbf{x}$ is a vector in R^2 and A is a symmetric 2×2 matrix, is a conic section centered at the origin of the xy-plane. From Theorem 6.12 it follows that there is a Cartesian coordinate system in the xy-plane with respect to which the equation of this conic section is

$$ax'^2 + by'^2 = 1,$$

where a and b are real numbers. Similarly, the graph of the equation

$$g(\mathbf{x}) = \mathbf{x}^T A\mathbf{x} = 1,$$

where $\mathbf{x}$ is a vector in R^3 and A is a symmetric 3×3 matrix, is a quadric surface centered at the origin of the xyz Cartesian coordinate system. From Theorem 6.12 it follows that there is a Cartesian coordinate system in 3-space with respect to which the equation of the quadric surface is

$$ax'^2 + by'^2 + cz'^2 = 1,$$

where a, b, and c are real numbers. The principal axes of the conic or surface lie along the new coordinate axes, and this is the reason for calling Theorem 6.12 the **principal axes theorem**.

EXAMPLE 6

Consider the conic section whose equation is

$$g(\mathbf{x}) = 2x^2 + 2xy + 2y^2 = 9.$$

From Example 4 it follows that this conic section can also be described by the equation

$$h(\mathbf{y}) = 3x'^2 + y'^2 = 9,$$

which can be rewritten as

$$\frac{x'^2}{3} + \frac{y'^2}{9} = 1.$$

The graph of this equation is an ellipse (Figure 6.5) whose major axis is along the y'-axis. The major axis is of length 6; the minor axis is of length $2\sqrt{3}$. We now note that there is a very close connection between the eigenvectors of the matrix of (3) and the location of the x'- and y'-axes.

Since $\mathbf{x} = P\mathbf{y}$, we have $\mathbf{y} = P^{-1}\mathbf{x} = P^T\mathbf{x} = P\mathbf{x}$ (P is orthogonal and, in this example, also symmetric). Thus

$$x' = \frac{1}{\sqrt{2}}x + \frac{1}{\sqrt{2}}y \quad \text{and} \quad y' = -\frac{1}{\sqrt{2}}x + \frac{1}{\sqrt{2}}y.$$

This means that, in terms of the x- and y-axes, the x'-axis lies along the vector

$$\mathbf{x}_1 = \begin{bmatrix} \frac{1}{\sqrt{2}} \\ \frac{1}{\sqrt{2}} \end{bmatrix},$$

and the y'-axis lies along the vector

$$\mathbf{x}_2 = \begin{bmatrix} -\frac{1}{\sqrt{2}} \\ \frac{1}{\sqrt{2}} \end{bmatrix}.$$

Now $\mathbf{x}_1$ and $\mathbf{x}_2$ are the columns of the matrix

$$P = \begin{bmatrix} \frac{1}{\sqrt{2}} & -\frac{1}{\sqrt{2}} \\ \frac{1}{\sqrt{2}} & \frac{1}{\sqrt{2}} \end{bmatrix},$$

which in turn are eigenvectors of the matrix of (3). Thus the x'- and y'-axes lie along the eigenvectors of the matrix of (3) (see Figure 6.5). ■

The situation described in Example 6 is true in general. That is, the principal axes of a conic section or quadric surface lie along the eigenvectors of the matrix of the quadratic form.

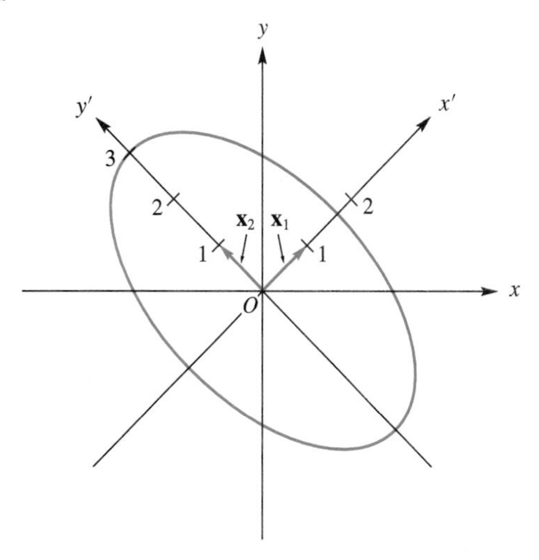

FIGURE 6.5

Let $g(\mathbf{x}) = \mathbf{x}^T A\mathbf{x}$ be a quadratic form in n variables. Then we know that g is equivalent to the quadratic form

$$h(\mathbf{y}) = \lambda_1 y_1^2 + \lambda_2 y_2^2 + \cdots + \lambda_n y_n^2,$$

where $\lambda_1, \lambda_2, \ldots, \lambda_n$ are eigenvalues of the symmetric matrix A of g, and hence are all real. We can label the eigenvalues so that all the positive eigenvalues of A, if any, are listed first, followed by all the negative eigenvalues, if any, followed by the zero eigenvalues, if any. Thus let $\lambda_1, \lambda_2, \ldots, \lambda_p$ be positive, $\lambda_{p+1}, \lambda_{p+2}, \ldots, \lambda_r$ be negative, and $\lambda_{r+1}, \lambda_{r+2}, \ldots, \lambda_n$ be zero. We now define the diagonal matrix H whose entries on the main diagonal are

$$\frac{1}{\sqrt{\lambda_1}}, \frac{1}{\sqrt{\lambda_2}}, \ldots, \frac{1}{\sqrt{\lambda_p}}, \frac{1}{\sqrt{-\lambda_{p+1}}}, \frac{1}{\sqrt{-\lambda_{p+2}}}, \ldots, \frac{1}{\sqrt{-\lambda_r}}, 1, 1, \ldots, 1,$$

with $n - r$ ones. Let D be the diagonal matrix whose entries on the main diagonal are $\lambda_1, \lambda_2, \ldots, \lambda_p, \lambda_{p+1}, \ldots, \lambda_r, \lambda_{r+1}, \ldots, \lambda_n$; A and D are congruent. Let $D_1 = H^T DH$ be the matrix whose diagonal elements are $1, 1, \ldots, 1, -1, \ldots, -1, 0, 0, \ldots, 0$ (p ones, $r - p - 1$ negative ones, and $n - r$ zeros); D and D_1 are then congruent. From Exercise 26 it follows that A and D_1 are congruent. In terms of quadratic forms, we have established Theorem 6.13.

Theorem 6.13

A quadratic form $g(\mathbf{x}) = \mathbf{x}^T A\mathbf{x}$ in n variables is equivalent to a quadratic form

$$h(\mathbf{y}) = y_1^2 + y_2^2 + \cdots + y_p^2 - y_{p+1}^2 - y_{p+2}^2 - \cdots - y_r^2.$$

It is clear that the rank of the matrix D_1 is r, the number of nonzero entries on its main diagonal. Now it can be shown (see J. L. Goldberg, *Matrix Theory with Applications*, New York: McGraw-Hill, Inc., 1991)that congruent matrices have equal ranks. Since the rank of D_1 is r, the rank of A is also r. We also refer to r as the **rank** of the quadratic form g whose matrix is A. It can be shown (see the book by Goldberg cited above) that the number p of positive terms in the quadratic form h of Theorem 6.13 is unique; that is, no matter how we simplify the given quadratic form g to obtain an equivalent quadratic form, the latter will always have p positive terms. Hence the quadratic form h in Theorem 6.13 is unique; it is often called the **canonical form** of a quadratic form in n variables. The difference between the number of positive eigenvalues and the number of negative eigenvalues is $s = p-(r-p) = 2p-r$ and is called the **signature** of the quadratic form. Thus, if g and h are equivalent quadratic forms, then they have equal ranks and signatures. However, it can also be shown (see the book by Goldberg cited above) that if g and h have equal ranks and signatures, then they are equivalent.

EXAMPLE 7

Consider the quadratic form in x_1, x_2, x_3, given by

$$\begin{aligned} g(\mathbf{x}) &= 3x_2^2 + 8x_2x_3 - 3x_3^2 \\ &= \mathbf{x}^T A\mathbf{x} = \begin{bmatrix} x_1 & x_2 & x_3 \end{bmatrix} \begin{bmatrix} 0 & 0 & 0 \\ 0 & 3 & 4 \\ 0 & 4 & -3 \end{bmatrix} \begin{bmatrix} x_1 \\ x_2 \\ x_3 \end{bmatrix}. \end{aligned}$$

The eigenvalues of A are (verify)

$$\lambda_1 = 5, \quad \lambda_2 = -5, \quad \text{and} \quad \lambda_3 = 0.$$

In this case A is congruent to

$$D = \begin{bmatrix} 5 & 0 & 0 \\ 0 & -5 & 0 \\ 0 & 0 & 0 \end{bmatrix}.$$

If we let

$$H = \begin{bmatrix} \frac{1}{\sqrt{5}} & 0 & 0 \\ 0 & \frac{1}{\sqrt{5}} & 0 \\ 0 & 0 & 1 \end{bmatrix},$$

then

$$D_1 = H^T DH = \begin{bmatrix} 1 & 0 & 0 \\ 0 & -1 & 0 \\ 0 & 0 & 0 \end{bmatrix},$$

and A are congruent, and the given quadratic form is equivalent to the canonical form

$$h(\mathbf{y}) = y_1^2 - y_2^2.$$

The rank of g is 2 and, since $p = 1$, the signature $s = 2p - r = 0$. ■

As a final application of quadratic forms we consider positive definite symmetric matrices. We recall that, in Section 3.3, a symmetric $n \times n$ matrix A was called positive definite if $\mathbf{x}^T A\mathbf{x} > 0$ for every nonzero vector $\mathbf{x}$ in R^n.

If A is a symmetric matrix, then $g(\mathbf{x}) = \mathbf{x}^T A\mathbf{x}$ is a quadratic form and, by Theorem 6.12, g is equivalent to h, where

$$h(\mathbf{y}) = \lambda_1 y_1^2 + \lambda_2 y_2^2 + \cdots + \lambda_p y_p^2 + \lambda_{p+1} y_{p+1}^2 + \lambda_{p+2} y_{p+2}^2 + \cdots + \lambda_r y_r^2.$$

Now A is positive definite if and only if $h(\mathbf{y}) > 0$ for each $\mathbf{y} \neq \mathbf{0}$. However, this can happen if and only if all summands in $h(\mathbf{y})$ are positive and $r = n$. These remarks have established the following theorem.

Theorem 6.14

A symmetric matrix A is positive definite if and only if all the eigenvalues of A are positive. ●

A quadratic form is then called **positive definite** if its matrix is positive definite.

6.6 Exercises

In Exercises 1 and 2, write each quadratic form as $\mathbf{x}^T A\mathbf{x}$, where A is a symmetric matrix.

1. (a) $-3x^2 + 5xy - 2y^2$.

(b) $2x_1^2 + 3x_1x_2 - 5x_1x_3 + 7x_2x_3$.

(c) $3x_1^2 + x_2^2 - 2x_3^2 + x_1x_2 - x_1x_3 - 4x_2x_3$.

2. (a) $x_1^2 - 3x_2^2 + 4x_3^2 - 4x_1x_2 + 6x_2x_3$.

(b) $4x^2 - 6xy - 2y^2$.

(c) $-2x_1x_2 + 4x_1x_3 + 6x_2x_3$.

In Exercises 3 and 4, for each given symmetric matrix A find a diagonal matrix D that is congruent to A.

3. (a) $A = \begin{bmatrix} -1 & 0 & 0 \\ 0 & 1 & 1 \\ 0 & 1 & 1 \end{bmatrix}$.

(b) $A = \begin{bmatrix} 1 & 1 & 1 \\ 1 & 1 & 1 \\ 1 & 1 & 1 \end{bmatrix}$.

(c) $A = \begin{bmatrix} 0 & 2 & 2 \\ 2 & 0 & 2 \\ 2 & 2 & 0 \end{bmatrix}$.

4. (a) $A = \begin{bmatrix} 3 & 4 & 0 \\ 4 & -3 & 0 \\ 0 & 0 & 5 \end{bmatrix}$.

(b) $A = \begin{bmatrix} 2 & 1 & 1 \\ 1 & 2 & 1 \\ 1 & 1 & 2 \end{bmatrix}$.

(c) $A = \begin{bmatrix} 0 & 0 & 1 \\ 0 & 1 & 0 \\ 1 & 0 & 0 \end{bmatrix}$.

In Exercises 5 through 10, find a quadratic form of the type in Theorem 6.12 that is equivalent to the given quadratic form.

5. $2x^2 - 4xy - y^2$.

6. $x_1^2 + x_2^2 + x_3^2 + 2x_2x_3$.

7. $2x_1x_3$.

8. $2x_2^2 + 2x_3^2 + 4x_2x_3$.

9. $-2x_1^2 - 4x_2^2 + 4x_3^2 - 6x_2x_3$.

10. $6x_1x_2 + 8x_2x_3$.

In Exercises 11 through 16, find a quadratic form of the type in Theorem 6.13 that is equivalent to the given quadratic form.

11. $2x^2 + 4xy + 2y^2$.

12. $x_1^2 + x_2^2 + x_3^2 + 2x_1x_2$.

13. $2x_1^2 + 4x_2^2 + 4x_3^2 + 10x_2x_3$.

14. $2x_1^2 + 3x_2^2 + 3x_3^2 + 4x_2x_3$.

15. $-3x_1^2 + 2x_2^2 + 2x_3^2 + 4x_2x_3$.

16. $-3x_1^2 + 5x_2^2 + 3x_3^2 - 8x_1x_3$.

17. Let $g(\mathbf{x}) = 4x_2^2 + 4x_3^2 - 10x_2x_3$ be a quadratic form in three variables. Find a quadratic form of the type in Theorem 6.13 that is equivalent to g. What is the rank of g? What is the signature of g?

18. Let $g(\mathbf{x}) = 3x_1^2 - 3x_2^2 - 3x_3^2 + 4x_2x_3$ be a quadratic form in three variables. Find a quadratic form of the type in Theorem 6.13 that is equivalent to g. What is the rank of g? What is the signature of g?

19. Find all quadratic forms $g(\mathbf{x}) = \mathbf{x}^T A\mathbf{x}$ in two variables of the type described in Theorem 6.13. What conics do the equations $\mathbf{x}^T A\mathbf{x} = 1$ represent?

20. Find all quadratic forms $g(\mathbf{x}) = \mathbf{x}^T A\mathbf{x}$ in two variables of rank 1 of the type described in Theorem 6.13. What conics do the equations $\mathbf{x}^T A\mathbf{x} = 1$ represent?

In Exercises 21 and 22, which of the given quadratic forms in three variables are equivalent?

21. $g_1(\mathbf{x}) = x_1^2 + x_2^2 + x_3^2 + 2x_1x_2$.
$g_2(\mathbf{x}) = 2x_2^2 + 2x_3^2 + 2x_2x_3$.
$g_3(\mathbf{x}) = 3x_2^2 - 3x_3^2 + 8x_2x_3$.
$g_4(\mathbf{x}) = 3x_2^2 + 3x_3^2 - 4x_2x_3$.

22. $g_1(\mathbf{x}) = x_2^2 + 2x_1x_3$.
$g_2(\mathbf{x}) = 2x_1^2 + 2x_2^2 + x_3^2 + 2x_1x_2 + 2x_1x_3 + 2x_2x_3$.
$g_3(\mathbf{x}) = 2x_1x_2 + 2x_1x_3 + 2x_2x_3$.
$g_4(\mathbf{x}) = 4x_1^2 + 3x_2^2 + 4x_3^2 + 10x_1x_3$.

In Exercises 23 *and* 24, *which of the given matrices are positive definite?*

23. (a) $\begin{bmatrix} 2 & -1 \\ -1 & 2 \end{bmatrix}$. (b) $\begin{bmatrix} 2 & 1 \\ 1 & 2 \end{bmatrix}$.

(c) $\begin{bmatrix} 3 & 1 & 0 \\ 1 & 3 & 0 \\ 0 & 0 & 3 \end{bmatrix}$. (d) $\begin{bmatrix} 1 & 0 & 0 \\ 0 & 2 & 0 \\ 0 & 0 & -3 \end{bmatrix}$.

(e) $\begin{bmatrix} 2 & 2 \\ 2 & 2 \end{bmatrix}$.

24. (a) $\begin{bmatrix} 0 & -1 \\ -1 & 0 \end{bmatrix}$. (b) $\begin{bmatrix} 1 & 1 \\ 1 & 1 \end{bmatrix}$.

(c) $\begin{bmatrix} 0 & 0 & 0 \\ 0 & 1 & 2 \\ 0 & 2 & 1 \end{bmatrix}$. (d) $\begin{bmatrix} 7 & 4 & 4 \\ 4 & 7 & 4 \\ 4 & 4 & 7 \end{bmatrix}$.

(e) $\begin{bmatrix} 2 & 0 & 0 & 0 \\ 0 & 1 & 0 & 0 \\ 0 & 0 & 3 & 4 \\ 0 & 0 & 4 & -3 \end{bmatrix}$.

25. Prove that if A is a symmetric matrix, then $P^T AP$ is also symmetric.

26. If A, B, and C are $n \times n$ symmetric matrices, prove the following.

(a) A is congruent to A.

(b) If B is congruent to A, then A is congruent to B.

(c) If C is congruent to B and B is congruent to A, then C is congruent to A.

27. Prove that if A is symmetric, then A is congruent to a diagonal matrix D.

28. Let $A = \begin{bmatrix} a & b \\ b & d \end{bmatrix}$ be a 2×2 symmetric matrix. Prove that A is positive definite if and only if $\det(A) > 0$ and $a > 0$.

29. Prove that a symmetric matrix A is positive definite if and only if $A = P^T P$ for a nonsingular matrix P.

6.7 CONIC SECTIONS

In this section we discuss the classification of the conic sections in the plane. A **quadratic equation** in the variables x and y has the form

$$ax^2 + 2bxy + cy^2 + dx + ey + f = 0, \tag{1}$$

where a, b, c, d, e, and f are real numbers. The graph of Equation (1) is a **conic section**, a curve so named because it is obtained by intersecting a plane with a right circular cone that has two nappes. In Figure 6.6 we show that a plane cuts the cone

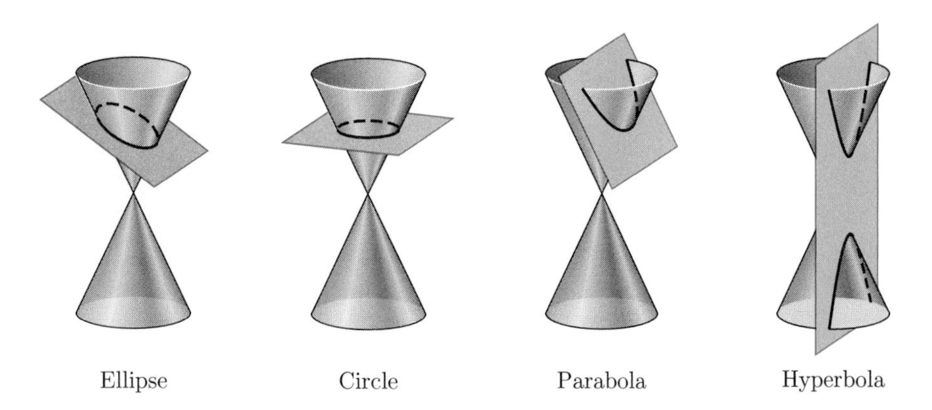

FIGURE 6.6

Nondegenerate conic sections.

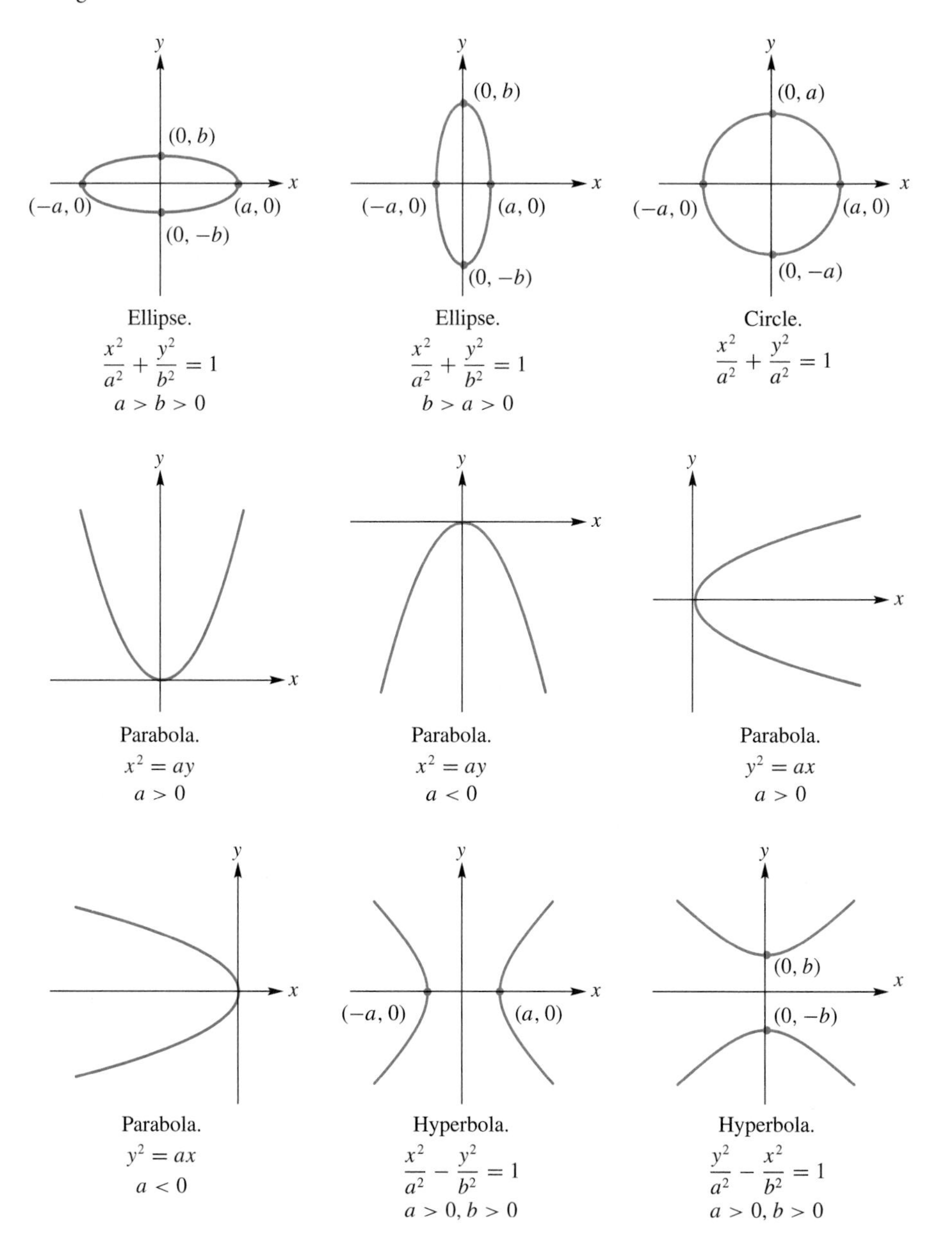

FIGURE 6.7

Conic sections in standard position.

in a circle, ellipse, parabola, or hyperbola. Degenerate cases of the conic sections are a point, a line, a pair of lines, or the empty set.

The nondegenerate conics are said to be in **standard position** if their graphs and equations are as given in Figure 6.7. The equation is said to be in **standard form**.

EXAMPLE 1

Identify the graph of the given equation.

(a) $4x^2 + 25y^2 - 100 = 0$.
(b) $9y^2 - 4x^2 = -36$.
(c) $x^2 + 4y = 0$.
(d) $y^2 = 0$.
(e) $x^2 + 9y^2 + 9 = 0$.
(f) $x^2 + y^2 = 0$.

Solution (a) We rewrite the given equation as

$$\frac{4}{100}x^2 + \frac{25}{100}y^2 = \frac{100}{100}$$

or

$$\frac{x^2}{25} + \frac{y^2}{4} = 1,$$

whose graph is an ellipse in standard position with $a = 5$ and $b = 2$. Thus the x-intercepts are $(5, 0)$ and $(-5, 0)$, and the y-intercepts are $(0, 2)$ and $(0, -2)$.
(b) Rewriting the given equation as

$$\frac{x^2}{9} - \frac{y^2}{4} = 1,$$

we see that its graph is a hyperbola in standard position with $a = 3$ and $b = 2$. The x-intercepts are $(3, 0)$ and $(-3, 0)$.
(c) Rewriting the given equation as

$$x^2 = -4y,$$

we see that its graph is a parabola in standard position with $a = -4$, so it opens downward.
(d) Every point satisfying the given equation must have a y-coordinate equal to zero. Thus the graph of this equation consists of all the points on the x-axis.
(e) Rewriting the given equation as

$$x^2 + 9y^2 = -9,$$

we conclude that there are no points in the plane whose coordinates satisfy the given equation.
(f) The only point satisfying the equation is the origin $(0, 0)$, so the graph of this equation is the single point consisting of the origin. ■

We next turn to the study of conic sections whose graphs are not in standard position. First, notice that the equations of the conic sections whose graphs are in standard position do not contain an xy-term (called a **cross-product term**). If a cross-product term appears in the equation, the graph is a conic section that has been rotated from its standard position [see Figure 6.8(a)]. Also notice that none of the equations in Figure 6.7 contain an x^2-term and an x-term or a y^2-term and a y-term. If either of these cases occurs and there is no xy-term in the equation, the graph is a conic section that has been translated from its standard position [see Figure 6.8(b)]. On the other hand, if an xy-term is present, the graph is a conic section that has been rotated and possibly also translated [see Figure 6.8(c)].

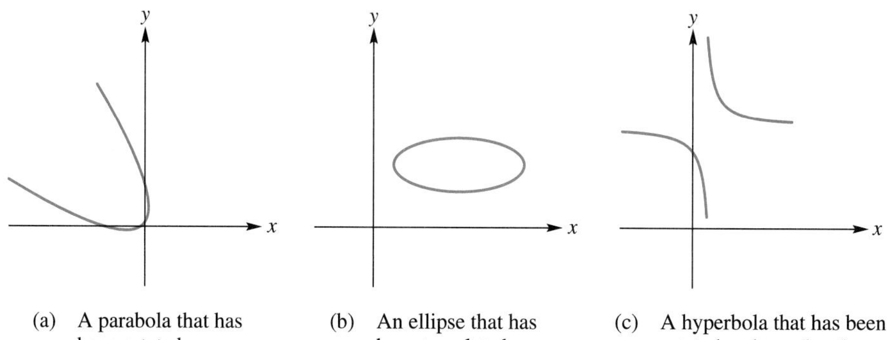

(a) A parabola that has been rotated.

(b) An ellipse that has been translated.

(c) A hyperbola that has been rotated and translated.

FIGURE 6.8

To identify a nondegenerate conic section whose graph is not in standard position, we proceed as follows:

1. If a cross-product term is present in the given equation, rotate the xy-coordinate axes by means of an orthogonal linear transformation so that in the resulting equation the xy-term no longer appears.
2. If an xy-term is not present in the given equation, but an x^2-term and an x-term, or a y^2-term and a y-term appear, translate the xy-coordinate axes by completing the square so that the graph of the resulting equation will be in standard position with respect to the origin of the new coordinate system.

Thus, if an xy-term appears in a given equation, we first rotate the xy-coordinate axes and then, if necessary, translate the rotated axes. In the next example, we deal with the case requiring only a translation of axes.

EXAMPLE 2

Identify and sketch the graph of the equation

$$x^2 - 4y^2 + 6x + 16y - 23 = 0. \tag{2}$$

Also, write its equation in standard form.

Solution Since there is no cross-product term, we only need to translate axes. Completing the squares in the x- and y-terms, we have

$$
\begin{aligned}
x^2 + 6x + 9 - 4(y^2 - 4y + 4) - 23 &= 9 - 16 \\
(x + 3)^2 - 4(y - 2)^2 &= 23 + 9 - 16 = 16.
\end{aligned} \tag{3}
$$

Letting

$$x' = x + 3 \quad \text{and} \quad y' = y - 2,$$

we can rewrite Equation (3) as

$$x'^2 - 4y'^2 = 16$$

or in standard form as

$$\frac{x'^2}{16} - \frac{y'^2}{4} = 1. \tag{4}$$

If we translate the xy-coordinate system to the $x'y'$-coordinate system, whose origin is at $(-3, 2)$, then the graph of Equation (4) is a hyperbola in standard position with respect to the $x'y'$-coordinate system (see Figure 6.9). ■

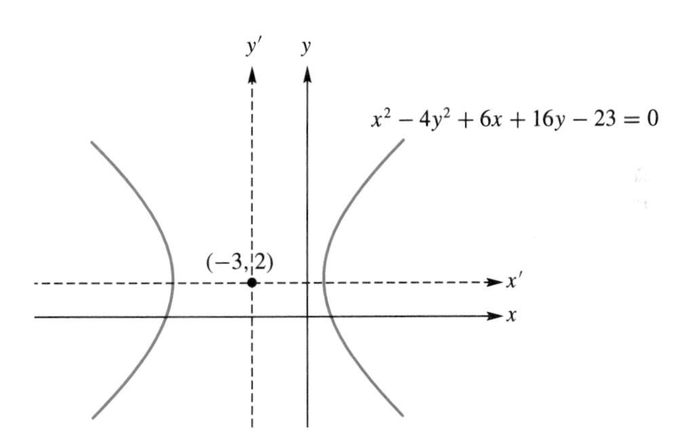

FIGURE 6.9

We now turn to the problem of identifying the graph of Equation (1), where we assume that $b \neq 0$, that is, a cross-product term is present. This equation can be written in matrix form as

$$\mathbf{x}^T A\mathbf{x} + B\mathbf{x} + f = 0, \tag{5}$$

where

$$\mathbf{x} = \begin{bmatrix} x \\ y \end{bmatrix}, \quad A = \begin{bmatrix} a & b \\ b & c \end{bmatrix}, \quad \text{and} \quad B = \begin{bmatrix} d & e \end{bmatrix}.$$

Since A is a symmetric matrix, we know from Section 6.4 that it can be diagonalized by an orthogonal matrix P. Thus

$$P^T A P = \begin{bmatrix} \lambda_1 & 0 \\ 0 & \lambda_2 \end{bmatrix},$$

where λ_1 and λ_2 are the eigenvalues of A and the columns of P are $\mathbf{x}_1$ and $\mathbf{x}_2$, orthonormal eigenvectors of A associated with λ_1 and λ_2, respectively.

Letting

$$\mathbf{x} = P\mathbf{y}, \quad \text{where} \quad \mathbf{y} = \begin{bmatrix} x' \\ y' \end{bmatrix},$$

we can rewrite Equation (5) as

$$(P\mathbf{y})^T A(P\mathbf{y}) + B(P\mathbf{y}) + f = 0$$
$$\mathbf{y}^T (P^T A P)\mathbf{y} + B(P\mathbf{y}) + f = 0$$

or

$$\begin{bmatrix} x' & y' \end{bmatrix} \begin{bmatrix} \lambda_1 & 0 \\ 0 & \lambda_2 \end{bmatrix} \begin{bmatrix} x' \\ y' \end{bmatrix} + B(P\mathbf{y}) + f = 0 \tag{6}$$

or

$$\lambda_1 x'^2 + \lambda_2 y'^2 + d'x' + e'y' + f = 0. \tag{7}$$

Equation (7) is the resulting equation for the given conic section and it has no cross-product term.

As discussed in Section 6.6, the x' and y' coordinate axes lie along the eigenvectors $\mathbf{x}_1$ and $\mathbf{x}_2$, respectively. Since P is an orthogonal matrix, $\det(P) = \pm 1$ and, if necessary, we can interchange the columns of P (the eigenvectors $\mathbf{x}_1$ and $\mathbf{x}_2$ of A) or multiply a column of P by -1, so that $\det(P) = 1$. As noted in Section 6.4, it then follows that P is the matrix of a counterclockwise rotation of R^2 through an angle θ that can be determined as follows. If

$$\mathbf{x}_1 = \begin{bmatrix} x_{11} \\ x_{21} \end{bmatrix},$$

then

$$\theta = \tan^{-1}\left(\frac{x_{21}}{x_{11}}\right),$$

a result that is frequently developed in a calculus course.

EXAMPLE 3

Identify and sketch the graph of the equation

$$5x^2 - 6xy + 5y^2 - 24x\sqrt{2} + 8y\sqrt{2} + 56 = 0. \tag{8}$$

Write the equation in standard form.

Solution Rewriting the given equation in matrix form, we obtain

$$\begin{bmatrix} x & y \end{bmatrix} \begin{bmatrix} 5 & -3 \\ -3 & 5 \end{bmatrix} \begin{bmatrix} x \\ y \end{bmatrix} + \begin{bmatrix} -24\sqrt{2} & 8\sqrt{2} \end{bmatrix} \begin{bmatrix} x \\ y \end{bmatrix} + 56 = 0.$$

We now find the eigenvalues of the matrix

$$A = \begin{bmatrix} 5 & -3 \\ -3 & 5 \end{bmatrix}.$$

Thus

$$\begin{aligned} |\lambda I_2 - A| &= \begin{bmatrix} \lambda - 5 & 3 \\ 3 & \lambda - 5 \end{bmatrix} \\ &= (\lambda - 5)(\lambda - 5) - 9 = \lambda^2 - 10\lambda + 16 \\ &= (\lambda - 2)(\lambda - 8), \end{aligned}$$

so the eigenvalues of A are

$$\lambda_1 = 2, \quad \lambda_2 = 8.$$

Associated eigenvectors are obtained by solving the homogeneous system

$$(\lambda I_2 - A)\mathbf{x} = \mathbf{0}.$$

Thus, for $\lambda_1 = 2$, we have

$$\begin{bmatrix} -3 & 3 \\ 3 & -3 \end{bmatrix} \mathbf{x} = \mathbf{0},$$

so an eigenvector of A associated with $\lambda_1 = 2$ is

$$\begin{bmatrix} 1 \\ 1 \end{bmatrix}.$$

For $\lambda_2 = 8$ we have

$$\begin{bmatrix} 3 & 3 \\ 3 & 3 \end{bmatrix} \mathbf{x} = \mathbf{0},$$

so an eigenvector of A associated with $\lambda_2 = 8$ is

$$\begin{bmatrix} -1 \\ 1 \end{bmatrix}.$$

Normalizing these eigenvectors, we obtain the orthogonal matrix

$$P = \begin{bmatrix} \frac{1}{\sqrt{2}} & -\frac{1}{\sqrt{2}} \\ \frac{1}{\sqrt{2}} & \frac{1}{\sqrt{2}} \end{bmatrix}.$$

Then

$$P^T A P = \begin{bmatrix} 2 & 0 \\ 0 & 8 \end{bmatrix}.$$

Letting $\mathbf{x} = P\mathbf{y}$, we write the transformed equation for the given conic section, Equation (6), as

$$2x'^2 + 8y'^2 - 16x' + 32y' + 56 = 0$$

or

$$x'^2 + 4y'^2 - 8x' + 16y' + 28 = 0.$$

To identify the graph of this equation, we need to translate axes, so we complete the squares, obtaining

$$(x' - 4)^2 + 4(y' + 2)^2 + 28 = 16 + 16$$

$$(x' - 4)^2 + 4(y' + 2)^2 = 4$$

$$\frac{(x' - 4)^2}{4} + \frac{(y' + 2)^2}{1} = 1. \tag{9}$$

Letting $x'' = x' - 4$ and $y'' = y' + 2$, we find that Equation (9) becomes

$$\frac{x''^2}{4} + \frac{y''^2}{1} = 1, \tag{10}$$

whose graph is an ellipse in standard position with respect to the $x''y''$-coordinate axes, as shown in Figure 6.10, where the origin of the $x''y''$-coordinate system in the $x'y'$-coordinate system is at $(4, -2)$, and at $(3\sqrt{2}, -\sqrt{2})$ in the xy-coordinate system. Equation (10) is the standard form of the equation of the ellipse. Since the eigenvector

$$\mathbf{x}_1 = \begin{bmatrix} \frac{1}{\sqrt{2}} \\ \frac{1}{\sqrt{2}} \end{bmatrix},$$

the xy-coordinate axes have been rotated through the angle θ, where

$$\theta = \tan^{-1}\left(\frac{\frac{1}{\sqrt{2}}}{\frac{1}{\sqrt{2}}}\right) = \tan^{-1} 1,$$

so $\theta = 45°$. The x' and y' axes lie along the respective eigenvectors

$$\mathbf{x}_1 = \begin{bmatrix} \frac{1}{\sqrt{2}} \\ \frac{1}{\sqrt{2}} \end{bmatrix} \quad \text{and} \quad \mathbf{x}_2 = \begin{bmatrix} -\frac{1}{\sqrt{2}} \\ \frac{1}{\sqrt{2}} \end{bmatrix},$$

as shown in Figure 6.11. ■

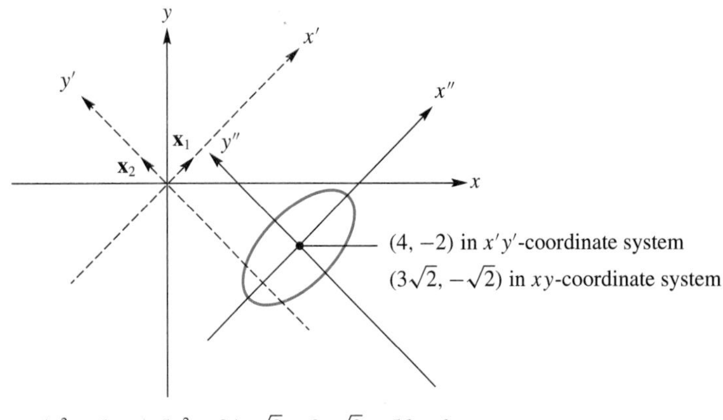

FIGURE 6.10

The graph of a given quadratic equation in x and y can be identified from the equation that is obtained after rotating axes, that is, from Equation (6) or (7). The identification of the conic section given by these equations is shown in Table 6.1.

TABLE 6.1 Identification of the Conic Sections

λ_1, λ_2 both nonzero		Exactly one of λ_1, λ_2 is zero
$\lambda_1\lambda_2 > 0$	$\lambda_1\lambda_2 < 0$	
Ellipse	Hyperbola	Parabola

6.7 Exercises

In Exercises 1 *through* 10, *identify the graph of the equation.*

1. $x^2 + 9y^2 - 9 = 0.$
2. $x^2 = 2y.$
3. $25y^2 - 4x^2 = 100.$
4. $y^2 - 16 = 0.$
5. $3x^2 - y^2 = 0.$
6. $y = 0.$
7. $4x^2 + 4y^2 - 9 = 0.$
8. $-25x^2 + 9y^2 + 225 = 0.$
9. $4x^2 + y^2 = 0.$
10. $9x^2 + 4y^2 + 36 = 0.$

In Exercises 11 *through* 18, *translate axes to identify the graph of the equation and write the equation in standard form.*

11. $x^2 + 2y^2 - 4x - 4y + 4 = 0.$
12. $x^2 - y^2 + 4x - 6y - 9 = 0.$
13. $x^2 + y^2 - 8x - 6y = 0.$
14. $x^2 - 4x + 4y + 4 = 0.$
15. $y^2 - 4y = 0.$
16. $4x^2 + 5y^2 - 30y + 25 = 0.$
17. $x^2 + y^2 - 2x - 6y + 10 = 0.$
18. $2x^2 + y^2 - 12x - 4y + 24 = 0.$

In Exercises 19 through 24, rotate axes to identify the graph of the equation and write the equation in standard form.

19. $x^2 + xy + y^2 = 6.$

20. $xy = 1.$

21. $9x^2 + y^2 + 6xy = 4.$

22. $x^2 + y^2 + 4xy = 9.$

23. $4x^2 + 4y^2 - 10xy = 0.$

24. $9x^2 + 6y^2 + 4xy - 5 = 0.$

In Exercises 25 through 30, identify the graph of the equation and write the equation in standard form.

25. $9x^2 + y^2 + 6xy - 10\sqrt{10}\,x + 10\sqrt{10}\,y + 90 = 0.$

26. $5x^2 + 5y^2 - 6xy - 30\sqrt{2}\,x + 18\sqrt{2}\,y + 82 = 0.$

27. $5x^2 + 12xy - 12\sqrt{13}\,x = 36.$

28. $6x^2 + 9y^2 - 4xy - 4\sqrt{5}\,x - 18\sqrt{5}\,y = 5.$

29. $x^2 - y^2 + 2\sqrt{3}\,xy + 6x = 0.$

30. $8x^2 + 8y^2 - 16xy + 33\sqrt{2}\,x - 31\sqrt{2}\,y + 70 = 0.$

6.8 QUADRIC SURFACES

In Section 6.7 conic sections were used to provide geometric models for quadratic forms in two variables. In this section we investigate quadratic forms in three variables and use particular surfaces called quadric surfaces as geometric models. Quadric surfaces are often studied and sketched in analytic geometry and calculus. Here we use Theorems 6.12 and 6.13 to develop a classification scheme for quadric surfaces.

A **second-degree polynomial equation** in three variables x, y, and z has the form

$$ax^2 + by^2 + cz^2 + 2dxy + 2exz + 2fyz + gx + hy + iz = j, \tag{1}$$

where coefficients a through j are real numbers with $a, b, \ldots, f$ not all zero. Equation (1) can be written in matrix form as

$$\mathbf{x}^T A\mathbf{x} + B\mathbf{x} = j, \tag{2}$$

where

$$A = \begin{bmatrix} a & d & e \\ d & b & f \\ e & f & c \end{bmatrix}, \quad B = \begin{bmatrix} g & h & i \end{bmatrix}, \quad \text{and} \quad \mathbf{x} = \begin{bmatrix} x \\ y \\ z \end{bmatrix}.$$

We call $\mathbf{x}^T A\mathbf{x}$ the **quadratic form (in three variables) associated with the second-degree polynomial** in (1). As in Section 6.6, the symmetric matrix A is called the matrix of the quadratic form.

The graph of (1) in R^3 is called a **quadric surface**. As in the case of the classification of conic sections in Section 6.7, the classification of (1) as to the type of surface represented depends on the matrix A. Using the ideas in Section 6.7, we have the following strategies to determine a simpler equation for a quadric surface.

1. If A is not diagonal, then a rotation of axes is used to eliminate any cross-product terms xy, xz, or yz.
2. If $B = \begin{bmatrix} g & h & i \end{bmatrix} \neq \mathbf{0}$, then a translation of axes is used to eliminate any first-degree terms.

The resulting equation will have the standard form

$$\lambda_1 x''^2 + \lambda_2 y''^2 + \lambda_3 z''^2 = k$$

or, in matrix form,

$$\mathbf{y}^T C \mathbf{y} = k, \tag{3}$$

where $\mathbf{y} = \begin{bmatrix} x'' \\ y'' \\ z'' \end{bmatrix}$, k is some real constant, and C is a diagonal matrix with diagonal entries $\lambda_1, \lambda_2, \lambda_3$, which are the eigenvalues of A.

We now turn to the classification of quadric surfaces.

Definition 6.8

Let A be an $n \times n$ symmetric matrix. The **inertia** of A, denoted $\text{In}(A)$, is an ordered triple of numbers

$$(\text{pos}, \text{neg}, \text{zer}),$$

where pos, neg, and zer are the number of positive, negative, and zero eigenvalues of A, respectively. ▲

EXAMPLE 1

Find the inertia of each of the following matrices:

$$A_1 = \begin{bmatrix} 2 & 2 \\ 2 & 2 \end{bmatrix}, \quad A_2 = \begin{bmatrix} 2 & 1 \\ 1 & 2 \end{bmatrix}, \quad A_3 = \begin{bmatrix} 0 & 2 & 2 \\ 2 & 0 & 2 \\ 2 & 2 & 0 \end{bmatrix}.$$

Solution We determine the eigenvalues of each of the matrices. It follows that (verify)

$\det(\lambda I_2 - A_1) = \lambda(\lambda - 4) = 0,$ so $\lambda_1 = 0, \lambda_2 = 4$, and $\text{In}(A_1) = (1, 0, 1)$.

$\det(\lambda I_2 - A_2) = (\lambda - 1)(\lambda - 3) = 0,$ so $\lambda_1 = 1, \lambda_2 = 3$, and $\text{In}(A_2) = (2, 0, 0)$.

$\det(\lambda I_3 - A_3) = (\lambda + 2)^2(\lambda - 4) = 0,$ so $\lambda_1 = \lambda_2 = -2, \lambda_3 = 4$, and $\text{In}(A_3) = (1, 2, 0)$. ■

From Section 6.6, the signature of a quadratic form $\mathbf{x}^T A\mathbf{x}$ is the difference between the number of positive eigenvalues and the number of negative eigenvalues of A. In terms of inertia, the signature of $\mathbf{x}^T A\mathbf{x}$ is $s = \text{pos} - \text{neg}$.

In order to use inertia for classification of quadric surfaces (or conic sections), we assume that the eigenvalues of an $n \times n$ symmetric matrix A of a quadratic form in n variables are denoted by

$$\begin{aligned} \lambda_1 \geq \cdots \geq \lambda_{\text{pos}} &> 0 \\ \lambda_{\text{pos}+1} \leq \cdots \leq \lambda_{\text{pos}+\text{neg}} &< 0 \\ \lambda_{\text{pos}+\text{neg}+1} = \cdots = \lambda_n &= 0. \end{aligned}$$

The largest positive eigenvalue is denoted by λ_1 and the smallest one by λ_{pos}. We also assume that $\lambda_1 > 0$ and $j \geq 0$ in (2), which eliminates redundant and impossible cases. For example, if

$$A = \begin{bmatrix} -1 & 0 & 0 \\ 0 & -2 & 0 \\ 0 & 0 & -3 \end{bmatrix}, \quad B = \begin{bmatrix} 0 & 0 & 0 \end{bmatrix}, \quad \text{and} \quad j = 5,$$

then the second-degree polynomial is $-x^2 - 2y^2 - 3z^2 = 5$, which has an empty solution set. That is, the surface represented has no points. However, if $j = -5$, then the second-degree polynomial is $-x^2 - 2y^2 - 3z^2 = -5$, which is identical to $x^2 + 2y^2 + 3z^2 = 5$. The assumptions $\lambda_1 > 0$ and $j \geq 0$ avoid such a redundant representation.

EXAMPLE 2

Consider a quadratic form in two variables with matrix A and assume that $\lambda_1 > 0$ and $f \geq 0$ in Equation (1) of Section 6.7. Then there are only three possible cases for the inertia of A, which we summarize as follows.

1. $\text{In}(A) = (2, 0, 0)$; then the quadratic form represents an ellipse.
2. $\text{In}(A) = (1, 1, 0)$; then the quadratic form represents a hyperbola.
3. $\text{In}(A) = (1, 0, 1)$; then the quadratic form represents a parabola.

This classification is identical to that given in Table 6.2 later in this section, taking the assumptions into account. ■

Note that the classification of the conic sections in Example 2 does not distinguish between special cases within a particular geometric class. For example, both $y = x^2$ and $x = y^2$ have inertia $(1, 0, 1)$.

Before classifying quadric surfaces, using inertia, we present the quadric surfaces in the standard forms met in analytic geometry and calculus. (In the following, a, b, and c are positive unless otherwise stated.)

ELLIPSOID (see Figure 6.11).

$$\frac{x^2}{a^2} + \frac{y^2}{b^2} + \frac{z^2}{c^2} = 1.$$

The special case $a = b = c$ is a sphere.

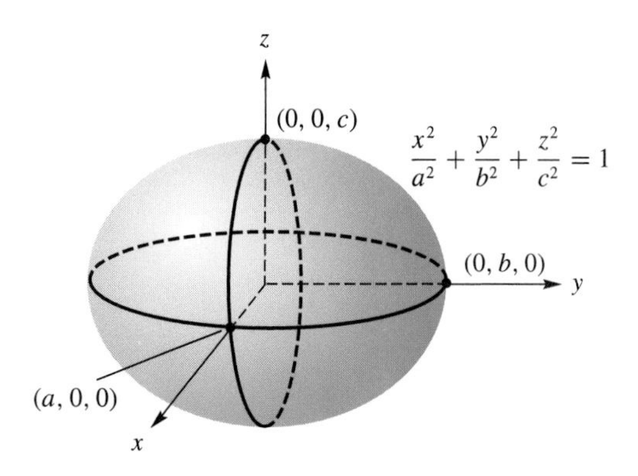

FIGURE 6.11

Ellipsoid

ELLIPTIC PARABOLOID (see Figure 6.12).

$$z = \frac{x^2}{a^2} + \frac{y^2}{b^2}, \qquad y = \frac{x^2}{a^2} + \frac{z^2}{c^2}, \qquad x = \frac{y^2}{b^2} + \frac{z^2}{c^2}.$$

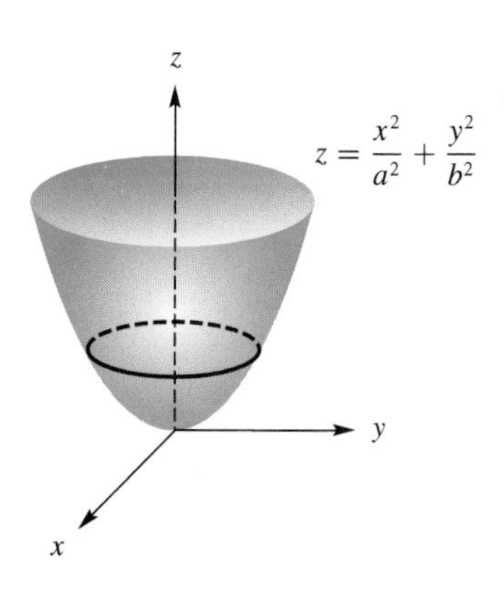

FIGURE 6.12

Elliptic paraboloid

A degenerate case of a parabola is a line, so a degenerate case of an elliptic paraboloid is an **elliptic cylinder** (see Figure 6.13), which is given by

$$\frac{x^2}{a^2} + \frac{y^2}{b^2} = 1, \qquad \frac{x^2}{a^2} + \frac{z^2}{c^2} = 1, \qquad \frac{y^2}{b^2} + \frac{z^2}{c^2} = 1.$$

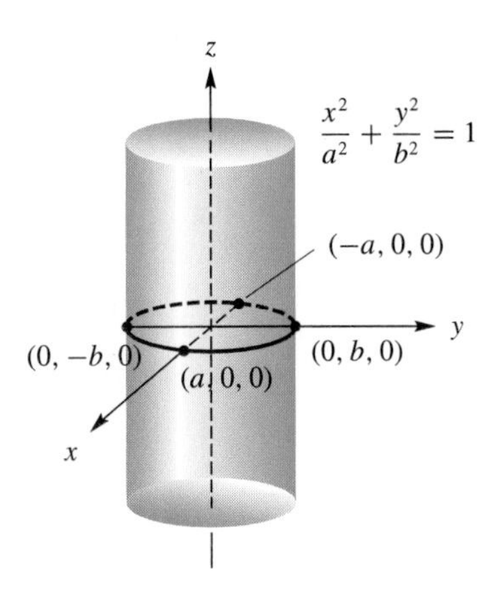

FIGURE 6.13

Elliptic cylinder

HYPERBOLOID OF ONE SHEET (see Figure 6.14).

$$\frac{x^2}{a^2} + \frac{y^2}{b^2} - \frac{z^2}{c^2} = 1, \qquad \frac{x^2}{a^2} - \frac{y^2}{b^2} + \frac{z^2}{c^2} = 1, \qquad -\frac{x^2}{a^2} + \frac{y^2}{b^2} + \frac{z^2}{c^2} = 1.$$

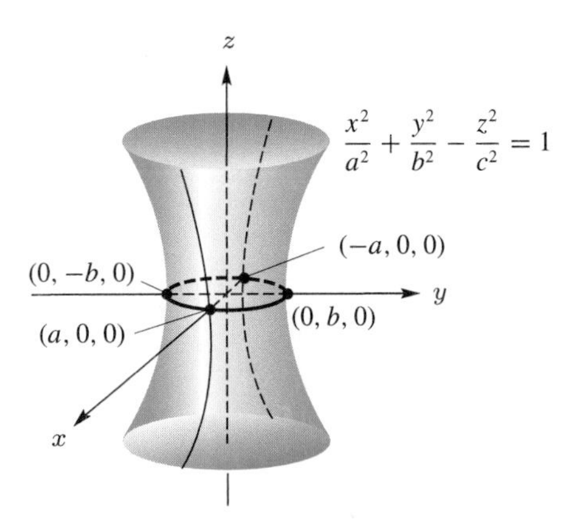

FIGURE 6.14

Hyperboloid of one sheet

A degenerate case of a hyperbola is a pair of lines through the origin; hence a degenerate case of a hyperboloid of one sheet is a **cone** (Figure 6.15), which is given by

$$\frac{x^2}{a^2} + \frac{y^2}{b^2} - \frac{z^2}{c^2} = 0, \qquad \frac{x^2}{a^2} - \frac{y^2}{b^2} + \frac{z^2}{c^2} = 0, \qquad -\frac{x^2}{a^2} + \frac{y^2}{b^2} + \frac{z^2}{c^2} = 0.$$

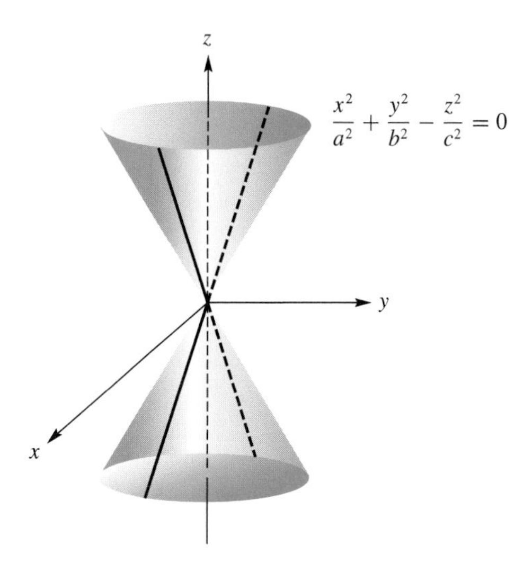

FIGURE 6.15

Cone

HYPERBOLOID OF TWO SHEETS (see Figure 6.16).

$$\frac{x^2}{a^2} - \frac{y^2}{b^2} - \frac{z^2}{c^2} = 1, \qquad -\frac{x^2}{a^2} - \frac{y^2}{b^2} + \frac{z^2}{c^2} = 1, \qquad -\frac{x^2}{a^2} + \frac{y^2}{b^2} - \frac{z^2}{c^2} = 1.$$

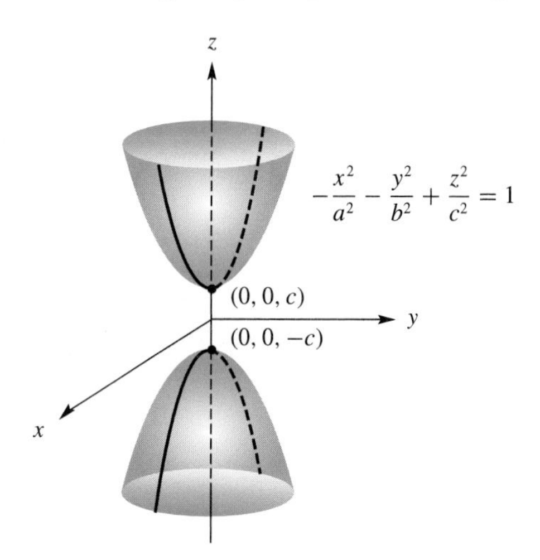

FIGURE 6.16

Hyperboloid of two sheets

HYPERBOLIC PARABOLOID (see Figure 6.17).

$$\pm z = \frac{x^2}{a^2} - \frac{y^2}{b^2}, \qquad \pm y = \frac{x^2}{a^2} - \frac{z^2}{b^2}, \qquad \pm x = \frac{y^2}{a^2} - \frac{z^2}{b^2}.$$

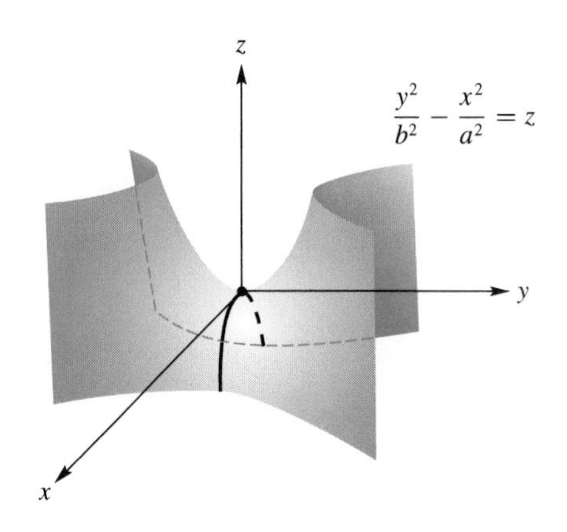

FIGURE 6.17

Hyperbolic paraboloid

A degenerate case of a parabola is a line, so a degenerate case of a hyperbolic paraboloid is a hyperbolic cylinder (see Figure 6.18), which is given by

$$\frac{x^2}{a^2} - \frac{y^2}{b^2} = \pm 1, \qquad \frac{x^2}{a^2} - \frac{z^2}{b^2} = \pm 1, \qquad \frac{y^2}{a^2} - \frac{z^2}{b^2} = \pm 1.$$

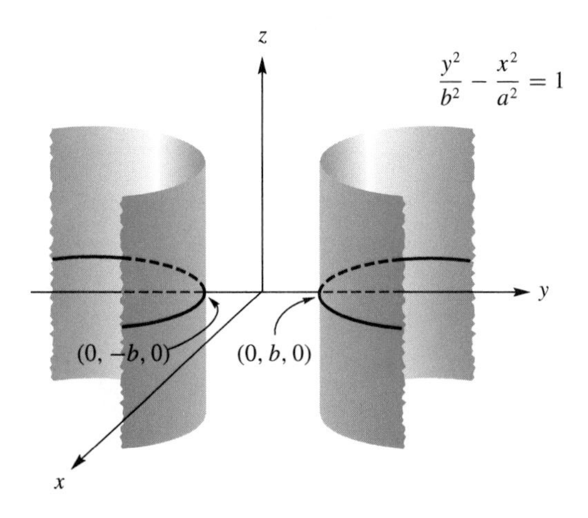

FIGURE 6.18

Hyperbolic cylinder

PARABOLIC CYLINDER (see Figure 6.19). One of a or b is not zero.

$$x^2 = ay + bz, \qquad y^2 = ax + bz, \qquad z^2 = ax + by.$$

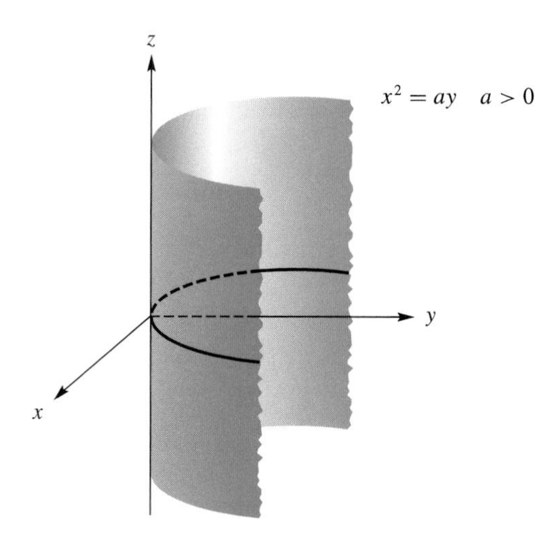

FIGURE 6.19

Parabolic cylinder

For a quadratic form in three variables with matrix A, under the assumptions $\lambda_1 > 0$ and $j \geq 0$ in (2), there are exactly six possibilities for the inertia of A. We present these in Table 6.2. As with the conic section classification in Example 2, the classification of quadric surfaces in Table 6.2 does not distinguish between special cases within a particular geometric class.

TABLE 6.2 Identification of the Quadric Surfaces

$\text{In}(A) = (3, 0, 0)$	Ellipsoid
$\text{In}(A) = (2, 0, 1)$	Elliptic paraboloid
$\text{In}(A) = (2, 1, 0)$	Hyperboloid of one sheet
$\text{In}(A) = (1, 2, 0)$	Hyperboloid of two sheets
$\text{In}(A) = (1, 1, 1)$	Hyperbolic paraboloid
$\text{In}(A) = (1, 0, 2)$	Parabolic cylinder

EXAMPLE 3

Classify the quadric surface represented by the quadratic form $\mathbf{x}^T A\mathbf{x} = 3$, where

$$A = \begin{bmatrix} 0 & 2 & 2 \\ 2 & 0 & 2 \\ 2 & 2 & 0 \end{bmatrix} \quad \text{and} \quad \mathbf{x} = \begin{bmatrix} x \\ y \\ z \end{bmatrix}.$$

Solution From Example 1 we have that $\text{In}(A) = (1, 2, 0)$ and hence the quadric surface is a hyperboloid of two sheets. ■

EXAMPLE 4

Classify the quadric surface given by

$$2x^2 + 4y^2 - 4z^2 + 6yz - 5x + 3y = 2.$$

Solution Rewrite the second-degree polynomial as a quadratic form in three variables to identify the matrix A of the quadratic form. We have

$$A = \begin{bmatrix} 2 & 0 & 0 \\ 0 & 4 & 3 \\ 0 & 3 & -4 \end{bmatrix}.$$

Its eigenvalues are $\lambda_1 = 5$, $\lambda_2 = 2$, and $\lambda_3 = -5$ (verify). Thus $\text{In}(A) = (2, 1, 0)$ and hence the quadric surface is a hyperboloid of one sheet. ■

The classification of a quadric surface is much easier than the problem of transforming it to the standard forms that are used in analytic geometry and calculus. The algebraic steps to obtain an equation in standard form from a second-degree polynomial equation (1) require, in general, a rotation and translation of axes, as mentioned earlier. The rotation requires both the eigenvalues and eigenvectors of the matrix A of the quadratic form. The eigenvectors of A are used to form an orthogonal matrix P so that $\det(P) = 1$, and hence the change of variables $\mathbf{x} = P\mathbf{y}$ represents a rotation. The resulting associated quadratic form is that obtained in the principal axes theorem, Theorem 6.12; that is, all cross-product terms are eliminated. We illustrate this with the next example.

EXAMPLE 5

For the quadric surface in Example 4,

$$\mathbf{x}^T A\mathbf{x} + \begin{bmatrix} -5 & 3 & 0 \end{bmatrix} \mathbf{x} = 2,$$

determine the rotation so that all cross-product terms are eliminated.

Solution The eigenvalues of

$$A = \begin{bmatrix} 2 & 3 & 0 \\ 0 & 4 & 3 \\ 0 & 3 & -4 \end{bmatrix}$$

are

$$\lambda_1 = 5, \quad \lambda_2 = 2, \quad \lambda_3 = -5$$

and associated eigenvectors are (verify),

$$\mathbf{v}_1 = \begin{bmatrix} 0 \\ 3 \\ 1 \end{bmatrix}, \quad \mathbf{v}_2 = \begin{bmatrix} 1 \\ 0 \\ 0 \end{bmatrix}, \quad \mathbf{v}_3 = \begin{bmatrix} 0 \\ 1 \\ -3 \end{bmatrix}.$$

The eigenvectors $\mathbf{v}_i$ are mutually orthogonal, since they correspond to distinct eigenvalues of a symmetric matrix (see Theorem 6.7). We normalize the eigenvectors as

$$\mathbf{u}_1 = \frac{1}{\sqrt{10}}\begin{bmatrix} 0 \\ 3 \\ 1 \end{bmatrix}, \qquad \mathbf{u}_2 = \mathbf{v}_2, \qquad \mathbf{u}_3 = \frac{1}{\sqrt{10}}\begin{bmatrix} 0 \\ 1 \\ -3 \end{bmatrix}$$

and define $P = \begin{bmatrix} \mathbf{u}_1 & \mathbf{u}_2 & \mathbf{u}_3 \end{bmatrix}$. Then $\det(P) = 1$ (verify), so we let $\mathbf{x} = P\mathbf{y}$ and obtain the representation

$$(P\mathbf{y})^T A(P\mathbf{y}) + \begin{bmatrix} -5 & 3 & 0 \end{bmatrix} P\mathbf{y} = 2$$
$$\mathbf{y}^T(P^T AP)\mathbf{y} + \begin{bmatrix} -5 & 3 & 0 \end{bmatrix} P\mathbf{y} = 2.$$

Since $P^T AP = D$, and letting $\mathbf{y} = \begin{bmatrix} x' \\ y' \\ z' \end{bmatrix}$, we have

$$\mathbf{y}^T D\mathbf{y} + \begin{bmatrix} -5 & 3 & 0 \end{bmatrix} P\mathbf{y} = 2,$$

$$\mathbf{y}^T \begin{bmatrix} 5 & 0 & 0 \\ 0 & 2 & 0 \\ 0 & 0 & -5 \end{bmatrix} \mathbf{y} + \begin{bmatrix} \frac{9}{\sqrt{10}} & -5 & \frac{3}{\sqrt{10}} \end{bmatrix} \mathbf{y} = 2$$

(if $\det(P) \neq 1$, we redefine P by reordering its columns until we get its determinant to be 1), or

$$5x'^2 + 2y'^2 - 5z'^2 + \frac{9}{\sqrt{10}}x' - 5y' + \frac{3}{\sqrt{10}}z' = 2. \qquad ■$$

To complete the transformation to standard form, we introduce a change of variable to perform a translation that eliminates any first-degree terms. Algebraically, we complete the square in each of the three variables.

EXAMPLE 6

Continue with Example 5 to eliminate the first-degree terms.

Solution The last expression for the quadric surface in Example 5 can be written as

$$5x'^2 + \frac{9}{\sqrt{10}}x' + 2y'^2 - 5y' - 5z'^2 + \frac{3}{\sqrt{10}}z' = 2.$$

Completing the square in each variable, we have

$$5\left(x'^2 + \frac{9}{5\sqrt{10}}x' + \frac{81}{1000}\right) + 2\left(y'^2 - \frac{5}{2}y' + \frac{25}{16}\right) - 5\left(z'^2 - \frac{3}{5\sqrt{10}}z' + \frac{9}{1000}\right)$$

$$= 5\left(x' + \frac{9}{10\sqrt{10}}\right)^2 + 2\left(y' - \frac{5}{4}\right)^2 - 5\left(z' - \frac{3}{10\sqrt{10}}\right)^2$$

$$= 2 + \frac{405}{1000} + \frac{50}{16} - \frac{45}{1000} = \frac{5485}{1000}.$$

Letting

$$x'' = x' + \frac{9}{10\sqrt{10}}, \qquad y'' = y' - \frac{5}{4}, \qquad z'' = z' - \frac{3}{10\sqrt{10}},$$

we can write the equation of the quadric surface as

$$5x''^2 + 2y''^2 - 5z''^2 = \frac{5485}{1000} = 5.485.$$

This can be written in standard form as

$$\frac{x''^2}{\frac{5.485}{5}} + \frac{y''^2}{\frac{5.485}{2}} - \frac{z''^2}{\frac{5.485}{5}} = 1. \qquad \blacksquare$$

6.8 Exercises

In Exercises 1 *through* 14, *use inertia to classify the quadric surface given by each equation.*

1. $x^2 + y^2 + 2z^2 - 2xy - 4xz - 4yz + 4x = 8$.

2. $x^2 + 3y^2 + 2z^2 - 6x - 6y + 4z - 2 = 0$.

3. $z = 4xy$.

4. $x^2 + y^2 + z^2 + 2xy = 4$.

5. $x^2 - y = 0$.

6. $2xy + z = 0$.

7. $5y^2 + 20y + z - 23 = 0$.

8. $x^2 + y^2 + 2z^2 - 2xy + 4xz + 4yz = 16$.

9. $4x^2 + 9y^2 + z^2 + 8x - 18y - 4z - 19 = 0$.

10. $y^2 - z^2 - 9x - 4y + 8z - 12 = 0$.

11. $x^2 + 4y^2 + 4x + 16y - 16z - 4 = 0$.

12. $4x^2 - y^2 + z^2 - 16x + 8y - 6z + 5 = 0$.

13. $x^2 - 4z^2 - 4x + 8z = 0$.

14. $2x^2 + 2y^2 + 4z^2 + 2xy - 2xz - 2yz + 3x - 5y + z = 7$.

In Exercises 15 *through* 28, *classify the quadric surface given by each equation and determine its standard form.*

15. $x^2 + 2y^2 + 2z^2 + 2yz = 1$.

16. $x^2 + y^2 + 2z^2 - 2xy + 4xz + 4yz = 16$.

17. $2xz - 2z - 4y - 4z + 8 = 0$.

18. $x^2 + 3y^2 + 3z^2 - 4yz = 9$.

19. $x^2 + y^2 + z^2 + 2xy = 8$.

20. $-x^2 - y^2 - z^2 + 4xy + 4xz + 4yz = 3$.

21. $2x^2 + 2y^2 + 4z^2 - 4xy - 8xz - 8yz + 8x = 15$.

22. $4x^2 + 4y^2 + 8z^2 + 4xy - 4xz - 4yz + 6x - 10y + 2z = \frac{9}{4}$.

23. $2y^2 + 2z^2 + 4yz + \frac{16}{\sqrt{2}}x + 4 = 0$.

24. $x^2 + y^2 - 2z^2 + 2xy + 8xz + 8yz + 3x + z = 0$.

25. $-x^2 - y^2 - z^2 + 4xy + 4xz + 4yz + \frac{3}{\sqrt{2}}x - \frac{3}{\sqrt{2}}y = 6$.

26. $2x^2 + 3y^2 + 3z^2 - 2yz + 2x + \frac{1}{\sqrt{2}}y + \frac{1}{\sqrt{2}}z = \frac{3}{8}$.

27. $x^2 + y^2 - z^2 - 2x - 4y - 4z + 1 = 0$.

28. $-8x^2 - 8y^2 + 10z^2 + 32xy - 4xz - 4yz = 24$.

Supplementary Exercises

1. Find the eigenvalues and corresponding eigenvectors for each of the matrices in Supplementary Exercise 2 in Chapter 5. Which of these matrices are similar to a diagonal matrix?

2. Let

$$A = \begin{bmatrix} 1 & 0 & -4 \\ 0 & 5 & 4 \\ -4 & 4 & 3 \end{bmatrix}.$$

(a) Find the eigenvalues and associated eigenvectors of A.

(b) Is A similar to a diagonal matrix? If so, find a nonsingular matrix P such that $P^{-1}AP$ is diagonal. Is P unique? Explain.

(c) Find the eigenvalues of A^{-1}.

(d) Find the eigenvalues and associated eigenvectors of A^2.

3. Let A be any $n \times n$ real matrix.

(a) Prove that the coefficient of λ^{n-1} in the characteristic polynomial of A is given by $-\operatorname{Tr}(A)$ (see Exercise 29 in Section 1.2).

(b) Prove that $\operatorname{Tr}(A)$ is the sum of the eigenvalues of A.

(c) Prove that the constant coefficient of the characteristic polynomial of A is $\pm$ times the product of the eigenvalues of A.

4. Prove or disprove: Every nonsingular matrix is similar to a diagonal matrix.

5. Let $p(x) = a_0 + a_1x + a_2x^2 + \cdots + a_kx^k$ be a polynomial in x. Show that the eigenvalues of matrix $p(A) = a_0I_n + a_1A + a_2A^2 + \cdots + a_kA^k$ are $p(\lambda_i)$, $i = 1, 2, \ldots, n$, where λ_i are the eigenvalues of A.

6. Let $p_1(\lambda)$ be the characteristic polynomial of A_{11} and $p_2(\lambda)$ the characteristic polynomial of A_{22}. What is the characteristic polynomial of each of the following partitioned matrices?

(a) $A = \begin{bmatrix} A_{11} & O \\ O & A_{22} \end{bmatrix}$. (b) $A = \begin{bmatrix} A_{11} & A_{21} \\ O & A_{22} \end{bmatrix}$.

(*Hint*: See Exercises 16 and 19 in Section 5.2.)

7. Let $L\colon P_1 \to P_1$ be the linear operator defined by $L(at + b) = \dfrac{a+b}{2}t$. Let $S = \{2 - t, 3 + t\}$ be a basis for P_1.

(a) Find $\left[L(2-t)\right]_S$ and $\left[L(3+t)\right]_S$.

(b) Find a matrix A representing L with respect to S.

(c) Find the eigenvalues and associated eigenvectors of A.

(d) Find the eigenvalues and associated eigenvectors of L.

(e) Describe the eigenspace for each eigenvalue of L.

8. Let $V = M_{22}$ and let $L\colon V \to V$ be the linear operator defined by $L(A) = A^T$, for A in V. Let $S = \{A_1, A_2, A_3, A_4\}$, where

$$A_1 = \begin{bmatrix} 1 & 0 \\ 0 & 0 \end{bmatrix}, \quad A_2 = \begin{bmatrix} 0 & 1 \\ 0 & 0 \end{bmatrix},$$

$$A_3 = \begin{bmatrix} 0 & 0 \\ 1 & 0 \end{bmatrix}, \quad \text{and} \quad A_4 = \begin{bmatrix} 0 & 0 \\ 0 & 1 \end{bmatrix},$$

be a basis for V.

(a) Find $\left[L(A_i)\right]_S$ for $i = 1, 2, 3, 4$.

(b) Find the matrix B representing L with respect to S.

(c) Find the eigenvalues and associated eigenvectors of B.

(d) Find the eigenvalues and associated eigenvectors of L.

(e) Show that one of the eigenspaces is the set of all 2×2 symmetric matrices and that the other is the set of all 2×2 skew symmetric matrices.

9. Let V be the *real* vector space of trigonometric polynomials of the form $a + b\sin x + c\cos x$. Let $L\colon V \to V$ be the linear operator defined by $L(\mathbf{v}) = \dfrac{d}{dx}\left[\mathbf{v}\right]$. Find the eigenvalues and associated eigenvectors of L. (*Hint*: Use the basis $S = \{1, \sin x, \cos x\}$ for V.)

10. Let V be the *complex* vector space (see Section B.2) of trigonometric polynomials

$$a + b\sin x + c\cos x.$$

For L as defined in Exercise 9, find the eigenvalues and associated eigenvectors.

11. Prove that if the matrix A is similar to a diagonal matrix, then A is similar to A^T.

Chapter 7

Differential Equations

7.1 DIFFERENTIAL EQUATIONS

A **differential equation** is an equation that involves an unknown function and its derivatives. An important, simple example of a differential equation is

$$\frac{d}{dt}x(t) = rx(t),$$

where r is a constant. The idea here is to find a function $x(t)$ that will satisfy the given differential equation. This differential equation is discussed further below. Differential equations occur often in all branches of science and engineering; linear algebra is helpful in the formulation and solution of differential equations. In this section we provide only a brief survey of the approach; books on differential equations deal with the subject in much greater detail, and several suggestions for further reading are given at the end of this chapter.

Homogeneous Linear Systems

We consider the **homogeneous linear system** of differential equations

$$\begin{aligned} x_1'(t) &= a_{11}x_1(t) + a_{12}x_2(t) + \cdots + a_{1n}x_n(t) \\ x_2'(t) &= a_{21}x_1(t) + a_{22}x_2(t) + \cdots + a_{2n}x_n(t) \\ &\vdots \\ x_n'(t) &= a_{n1}x_1(t) + a_{n2}x_2(t) + \cdots + a_{nn}x_n(t), \end{aligned} \tag{1}$$

where the a_{ij} are known constants. We seek functions $x_1(t), x_2(t), \ldots, x_n(t)$ defined and differentiable on the real line satisfying (1).

We can write (1) in matrix form by letting

$$\mathbf{x}(t) = \begin{bmatrix} x_1(t) \\ x_2(t) \\ \vdots \\ x_n(t) \end{bmatrix}, \qquad A = \begin{bmatrix} a_{11} & a_{12} & \cdots & a_{1n} \\ a_{21} & a_{22} & \cdots & a_{2n} \\ \vdots & \vdots & & \vdots \\ a_{n1} & a_{n2} & \cdots & a_{nn} \end{bmatrix},$$

and defining

$$\mathbf{x}'(t) = \begin{bmatrix} x_1'(t) \\ x_2'(t) \\ \vdots \\ x_n'(t) \end{bmatrix}.$$

Then (1) can be written as

$$\mathbf{x}'(t) = A\mathbf{x}(t). \tag{2}$$

We shall often write (2) more briefly as

$$\mathbf{x}' = A\mathbf{x}.$$

With this notation, a vector function

$$\mathbf{x}(t) = \begin{bmatrix} x_1(t) \\ x_2(t) \\ \vdots \\ x_n(t) \end{bmatrix}$$

satisfying (2) is called a **solution** to the given system.

We leave it to the reader to verify that if $\mathbf{x}^{(1)}(t), \mathbf{x}^{(2)}(t), \ldots, \mathbf{x}^{(n)}(t)$ are all solutions to (2), then any linear combination

$$\mathbf{x}(t) = b_1\mathbf{x}^{(1)}(t) + b_2\mathbf{x}^{(2)}(t) + \cdots + b_n\mathbf{x}^{(n)}(t) \tag{3}$$

is also a solution to (2).

It can be shown (Exercise 4) that the set of all solutions to the homogeneous linear system of differential equations (2) is a subspace of the vector space of differentiable real-valued functions.

A set of vector functions $\{\mathbf{x}^{(1)}(t), \mathbf{x}^{(2)}(t), \ldots, \mathbf{x}^{(n)}(t)\}$ is said to be a **fundamental system** for (1) if every solution to (1) can be written in the form (3). In this case, the right side of (3), where $b_1, b_2, \ldots, b_n$ are arbitrary constants, is said to be the **general solution** to (2).

It can be shown (see the book by Boyce and DiPrima or the book by Cullen cited in Further Readings) that any system of the form (2) has a fundamental system (in fact, infinitely many).

In general, differential equations arise in the course of solving physical problems. Typically, once a general solution to the differential equation has been obtained, the physical constraints of the problem impose certain definite values on the arbitrary constants in the general solution, giving rise to a **particular solution**. An important particular solution is obtained by finding a solution $\mathbf{x}(t)$ to Equation (2) such that $\mathbf{x}(0) = \mathbf{x}_0$, an **initial condition**, where $\mathbf{x}_0$ is a given vector. This problem is called an **initial value problem**. If the general solution (3) is known, then the initial value problem can be solved by setting $t = 0$ in (3) and determining the constants $b_1, b_2, \ldots, b_n$ so that

$$\mathbf{x}_0 = b_1\mathbf{x}^{(1)}(0) + b_2\mathbf{x}^{(2)}(0) + \cdots + b_n\mathbf{x}^{(n)}(0).$$

It is readily seen that this is actually an $n \times n$ linear system with unknowns $b_1, b_2, \ldots, b_n$. This linear system can also be written as

$$C\mathbf{b} = \mathbf{x}_0, \tag{4}$$

where

$$\mathbf{b} = \begin{bmatrix} b_1 \\ b_2 \\ \vdots \\ b_n \end{bmatrix}$$

and C is the $n \times n$ matrix whose columns are $\mathbf{x}^{(1)}(0), \mathbf{x}^{(2)}(0), \ldots, \mathbf{x}^{(n)}(0)$, respectively. It can be shown (see the book by Boyce and DiPrima or the book by Cullen cited in Further Readings) that if $\mathbf{x}^{(1)}(t), \mathbf{x}^{(2)}(t), \ldots, \mathbf{x}^{(n)}(t)$ form a fundamental system for (1), then C is nonsingular, so (4) always has a unique solution.

EXAMPLE 1

The simplest system of the form (1) is the single equation

$$\frac{dx}{dt} = ax, \tag{5}$$

where a is a constant. From calculus, the solutions to this equation are of the form

$$x = be^{at}; \tag{6}$$

that is, this is the general solution to (5). To solve the initial value problem

$$\frac{dx}{dt} = ax, \qquad x(0) = x_0,$$

we set $t = 0$ in (6) and obtain $b = x_0$. Thus the solution to the initial value problem is

$$x = x_0e^{at}. \qquad ■$$

The system (2) is said to be **diagonal** if the matrix A is diagonal. Then (1) can be rewritten as

$$\begin{aligned}
x_1'(t) &= a_{11}x_1(t) \\
x_2'(t) &= \qquad a_{22}x_2(t) \\
&\vdots \\
x_n'(t) &= \qquad\qquad a_{nn}x_n(t).
\end{aligned} \tag{7}$$

This system is easy to solve, since the equations can be solved separately. Applying the results of Example 1 to each equation in (7), we obtain

$$\begin{aligned} x_1(t) &= b_1 e^{a_{11}t} \\ x_2(t) &= b_2 e^{a_{22}t} \\ &\vdots \\ x_n(t) &= b_n e^{a_{nn}t}, \end{aligned} \tag{8}$$

where $b_1, b_2, \ldots, b_n$ are arbitrary constants. Writing (8) in vector form yields

$$\mathbf{x}(t) = \begin{bmatrix} b_1 e^{a_{11}t} \\ b_2 e^{a_{22}t} \\ \vdots \\ b_n e^{a_{nn}t} \end{bmatrix} = b_1 \begin{bmatrix} 1 \\ 0 \\ 0 \\ \vdots \\ 0 \end{bmatrix} e^{a_{11}t} + b_2 \begin{bmatrix} 0 \\ 1 \\ 0 \\ \vdots \\ 0 \end{bmatrix} e^{a_{22}t} + \cdots + b_n \begin{bmatrix} 0 \\ 0 \\ \vdots \\ 0 \\ 1 \end{bmatrix} e^{a_{nn}t}.$$

This implies that the vector functions

$$\mathbf{x}^{(1)}(t) = \begin{bmatrix} 1 \\ 0 \\ 0 \\ \vdots \\ 0 \end{bmatrix} e^{a_{11}t}, \quad \mathbf{x}^{(2)}(t) = \begin{bmatrix} 0 \\ 1 \\ 0 \\ \vdots \\ 0 \end{bmatrix} e^{a_{22}t}, \quad \ldots, \quad \mathbf{x}^{(n)}(t) = \begin{bmatrix} 0 \\ 0 \\ \vdots \\ 0 \\ 1 \end{bmatrix} e^{a_{nn}t}$$

form a fundamental system for the diagonal system (7).

EXAMPLE 2

The diagonal system

$$\begin{bmatrix} x_1' \\ x_2' \\ x_3' \end{bmatrix} = \begin{bmatrix} 3 & 0 & 0 \\ 0 & -2 & 0 \\ 0 & 0 & 4 \end{bmatrix} \begin{bmatrix} x_1 \\ x_2 \\ x_3 \end{bmatrix} \tag{9}$$

can be written as three equations:

$$\begin{aligned} x_1' &= 3x_1 \\ x_2' &= -2x_2 \\ x_3' &= 4x_3. \end{aligned}$$

Solving these equations, we obtain

$$x_1 = b_1 e^{3t}, \qquad x_2 = b_2 e^{-2t}, \qquad x_3 = b_3 e^{4t},$$

where b_1, b_2, and b_3 are arbitrary constants. Thus

$$\mathbf{x}(t) = \begin{bmatrix} b_1 e^{3t} \\ b_2 e^{-2t} \\ b_3 e^{4t} \end{bmatrix} = b_1 \begin{bmatrix} 1 \\ 0 \\ 0 \end{bmatrix} e^{3t} + b_2 \begin{bmatrix} 0 \\ 1 \\ 0 \end{bmatrix} e^{-2t} + b_3 \begin{bmatrix} 0 \\ 0 \\ 1 \end{bmatrix} e^{4t}$$

is the general solution to (9) and the functions

$$\mathbf{x}^{(1)}(t) = \begin{bmatrix} 1 \\ 0 \\ 0 \end{bmatrix} e^{3t}, \qquad \mathbf{x}^{(2)}(t) = \begin{bmatrix} 0 \\ 1 \\ 0 \end{bmatrix} e^{-2t}, \qquad \mathbf{x}^{(3)}(t) = \begin{bmatrix} 0 \\ 0 \\ 1 \end{bmatrix} e^{4t}$$

form a fundamental system for (9). ■

If the system (2) is not diagonal, then it cannot be solved as simply as the system in the preceding example. However, there is an extension of this method that yields the general solution in the case where A is diagonalizable. Suppose that A is diagonalizable and P is a nonsingular matrix such that

$$P^{-1}AP = D, \tag{10}$$

where D is diagonal. Then multiplying the given system

$$\mathbf{x}' = A\mathbf{x}$$

on the left by P^{-1}, we obtain

$$P^{-1}\mathbf{x}' = P^{-1}A\mathbf{x}.$$

Since $P^{-1}P = I_n$, we can rewrite the last equation as

$$P^{-1}\mathbf{x}' = (P^{-1}AP)(P^{-1}\mathbf{x}). \tag{11}$$

Temporarily, let

$$\mathbf{u} = P^{-1}\mathbf{x}. \tag{12}$$

Since P^{-1} is a constant matrix,

$$\mathbf{u}' = P^{-1}\mathbf{x}'. \tag{13}$$

Therefore, substituting (10), (12), and (13) into (11), we obtain

$$\mathbf{u}' = D\mathbf{u}. \tag{14}$$

Equation (14) is a diagonal system and can be solved by the methods just discussed. Before proceeding, however, let us recall from Theorem 6.4 that

$$D = \begin{bmatrix} \lambda_1 & 0 & \cdots & 0 \\ 0 & \lambda_2 & \cdots & 0 \\ \vdots & \vdots & & \vdots \\ 0 & 0 & \cdots & \lambda_n \end{bmatrix},$$

where $\lambda_1, \lambda_2, \ldots, \lambda_n$ are the eigenvalues of A, and that the columns of P are linearly independent eigenvectors of A associated, respectively, with $\lambda_1, \lambda_2, \ldots, \lambda_n$. From the discussion just given for diagonal systems, the general solution to (14) is

$$\mathbf{u}(t) = b_1\mathbf{u}^{(1)}(t) + b_2\mathbf{u}^{(2)}(t) + \cdots + b_n\mathbf{u}^{(n)}(t) = \begin{bmatrix} b_1 e^{\lambda_1 t} \\ b_2 e^{\lambda_2 t} \\ \vdots \\ b_n e^{\lambda_n t} \end{bmatrix},$$

where

$$\mathbf{u}^{(1)}(t) = \begin{bmatrix} 1 \\ 0 \\ 0 \\ \vdots \\ 0 \end{bmatrix} e^{\lambda_1 t}, \quad \mathbf{u}^{(2)}(t) = \begin{bmatrix} 0 \\ 1 \\ 0 \\ \vdots \\ 0 \end{bmatrix} e^{\lambda_2 t}, \quad \ldots, \quad \mathbf{u}^{(n)}(t) = \begin{bmatrix} 0 \\ 0 \\ \vdots \\ 0 \\ 1 \end{bmatrix} e^{\lambda_n t} \tag{15}$$

and $b_1, b_2, \ldots, b_n$ are arbitrary constants. From Equation (12), $\mathbf{x} = P\mathbf{u}$, so the general solution to the given system $\mathbf{x}' = A\mathbf{x}$ is

$$\mathbf{x}(t) = P\mathbf{u}(t) = b_1 P\mathbf{u}^{(1)}(t) + b_2 P\mathbf{u}^{(2)}(t) + \cdots + b_n P\mathbf{u}^{(n)}(t). \tag{16}$$

However, since the constant vectors in (15) are the columns of the identity matrix and $PI_n = P$, (16) can be rewritten as

$$\mathbf{x}(t) = b_1\mathbf{p}_1 e^{\lambda_1 t} + b_2\mathbf{p}_2 e^{\lambda_2 t} + \cdots + b_n\mathbf{p}_n e^{\lambda_n t}, \tag{17}$$

where $\mathbf{p}_1, \mathbf{p}_2, \ldots, \mathbf{p}_n$ are the columns of P, and therefore eigenvectors of A associated with $\lambda_1, \lambda_2, \ldots, \lambda_n$, respectively.

We summarize the discussion above in the following theorem.

Theorem 7.1

If the $n \times n$ matrix A has n linearly independent eigenvectors $\mathbf{p}_1, \mathbf{p}_2, \ldots, \mathbf{p}_n$ associated with the eigenvalues $\lambda_1, \lambda_2, \ldots, \lambda_n$, respectively, then the general solution to the homogeneous linear system of differential equations

$$\mathbf{x}' = A\mathbf{x}$$

is given by (17). ●

EXAMPLE 3

For the system

$$\mathbf{x}' = \begin{bmatrix} 1 & -1 \\ 2 & 4 \end{bmatrix} \mathbf{x},$$

the matrix

$$A = \begin{bmatrix} 1 & -1 \\ 2 & 4 \end{bmatrix}$$

has eigenvalues $\lambda_1 = 2$ and $\lambda_2 = 3$ with associated eigenvectors (verify)

$$\mathbf{p}_1 = \begin{bmatrix} 1 \\ -1 \end{bmatrix} \quad \text{and} \quad \mathbf{p}_2 = \begin{bmatrix} 1 \\ -2 \end{bmatrix}.$$

These eigenvectors are automatically linearly independent, since they are associated with distinct eigenvalues (proof of Theorem 6.5). Hence the general solution to the given system is

$$\mathbf{x}(t) = b_1 \begin{bmatrix} 1 \\ -1 \end{bmatrix} e^{2t} + b_2 \begin{bmatrix} 1 \\ -2 \end{bmatrix} e^{3t}.$$

In terms of components, this can be written as

$$\begin{aligned} x_1(t) &= b_1 e^{2t} + b_2 e^{3t} \\ x_2(t) &= -b_1 e^{2t} - 2b_2 e^{3t}. \end{aligned}$$

■

EXAMPLE 4

Consider the following homogeneous linear system of differential equations:

$$\mathbf{x}' = \begin{bmatrix} x_1' \\ x_2' \\ x_3' \end{bmatrix} = \begin{bmatrix} 0 & 1 & 0 \\ 0 & 0 & 1 \\ 8 & -14 & 7 \end{bmatrix} \begin{bmatrix} x_1 \\ x_2 \\ x_3 \end{bmatrix}.$$

The characteristic polynomial of A is (verify)

$$p(\lambda) = \lambda^3 - 7\lambda^2 + 14\lambda - 8$$

or

$$p(\lambda) = (\lambda - 1)(\lambda - 2)(\lambda - 4),$$

so the eigenvalues of A are $\lambda_1 = 1$, $\lambda_2 = 2$, and $\lambda_3 = 4$. Associated eigenvectors are (verify)

$$\begin{bmatrix} 1 \\ 1 \\ 1 \end{bmatrix}, \quad \begin{bmatrix} 1 \\ 2 \\ 4 \end{bmatrix}, \quad \begin{bmatrix} 1 \\ 4 \\ 16 \end{bmatrix},$$

respectively. The general solution is then given by

$$\mathbf{x}(t) = b_1 \begin{bmatrix} 1 \\ 1 \\ 1 \end{bmatrix} e^{t} + b_2 \begin{bmatrix} 1 \\ 2 \\ 4 \end{bmatrix} e^{2t} + b_3 \begin{bmatrix} 1 \\ 4 \\ 16 \end{bmatrix} e^{4t},$$

where b_1, b_2, and b_3 are arbitrary constants. ■

EXAMPLE 5

For the linear system of Example 4 solve the initial value problem determined by the **initial conditions** $x_1(0) = 4$, $x_2(0) = 6$, and $x_3(0) = 8$.

Solution We write our general solution in the form $\mathbf{x} = P\mathbf{u}$ as

$$\mathbf{x}(t) = \begin{bmatrix} 1 & 1 & 1 \\ 1 & 2 & 4 \\ 1 & 4 & 16 \end{bmatrix} \begin{bmatrix} b_1 e^t \\ b_2 e^{2t} \\ b_3 e^{4t} \end{bmatrix}.$$

Now

$$\mathbf{x}(0) = \begin{bmatrix} 4 \\ 6 \\ 8 \end{bmatrix} = \begin{bmatrix} 1 & 1 & 1 \\ 1 & 2 & 4 \\ 1 & 4 & 16 \end{bmatrix} \begin{bmatrix} b_1 e^0 \\ b_2 e^0 \\ b_3 e^0 \end{bmatrix}$$

or

$$\begin{bmatrix} 1 & 1 & 1 \\ 1 & 2 & 4 \\ 1 & 4 & 16 \end{bmatrix} \begin{bmatrix} b_1 \\ b_2 \\ b_3 \end{bmatrix} = \begin{bmatrix} 4 \\ 6 \\ 8 \end{bmatrix}. \tag{18}$$

Solving (18) by Gauss–Jordan reduction, we obtain (verify)

$$b_1 = \tfrac{4}{3}, \qquad b_2 = 3, \qquad b_3 = -\tfrac{1}{3}.$$

Therefore, the solution to the initial value problem is

$$\mathbf{x}(t) = \frac{4}{3}\begin{bmatrix} 1 \\ 1 \\ 1 \end{bmatrix} e^t + 3\begin{bmatrix} 1 \\ 2 \\ 4 \end{bmatrix} e^{2t} - \frac{1}{3}\begin{bmatrix} 1 \\ 4 \\ 16 \end{bmatrix} e^{4t}.$$

■

We now recall several facts from Chapter 6. If A does not have distinct eigenvalues, then we may or may not be able to diagonalize A. Let λ be an eigenvalue of A of multiplicity k. Then A can be diagonalized if and only if the dimension of the eigenspace associated with λ is k, that is, if and only if the rank of the matrix $(\lambda I_n - A)$ is $n - k$ (verify). If the rank of $(\lambda I_n - A)$ is $n - k$, then we can find k linearly independent eigenvectors of A associated with λ.

EXAMPLE 6

Consider the linear system

$$\mathbf{x}' = A\mathbf{x} = \begin{bmatrix} 1 & 0 & 0 \\ 0 & 3 & -2 \\ 0 & -2 & 3 \end{bmatrix} \mathbf{x}.$$

The eigenvalues of A are $\lambda_1 = \lambda_2 = 1$ and $\lambda_3 = 5$ (verify). The rank of the matrix

$$(1I_3 - A) = \begin{bmatrix} 0 & 0 & 0 \\ 0 & -2 & 2 \\ 0 & 2 & -2 \end{bmatrix}$$

is 1 and the linearly independent eigenvectors

$$\begin{bmatrix} 1 \\ 0 \\ 0 \end{bmatrix} \quad \text{and} \quad \begin{bmatrix} 0 \\ 1 \\ 1 \end{bmatrix}$$

are associated with the eigenvalue 1 (verify). The eigenvector

$$\begin{bmatrix} 0 \\ 1 \\ -1 \end{bmatrix}$$

is associated with the eigenvalue 5 (verify). The general solution to the given system is then

$$\mathbf{x}(t) = b_1 \begin{bmatrix} 1 \\ 0 \\ 0 \end{bmatrix} e^t + b_2 \begin{bmatrix} 0 \\ 1 \\ 1 \end{bmatrix} e^t + b_3 \begin{bmatrix} 0 \\ 1 \\ -1 \end{bmatrix} e^{5t},$$

where b_1, b_2, and b_3 are arbitrary constants. ■

If we cannot diagonalize A as in the examples, we have a considerably more difficult situation. Methods for dealing with such problems are discussed in more advanced books (see Further Readings).

Application—A Diffusion Process

The following example is a modification of an example presented by Derrick and Grossman in *Elementary Differential Equations with Applications* (see Further Readings).

EXAMPLE 7

Consider two adjoining cells separated by a permeable membrane and suppose that a fluid flows from the first cell to the second one at a rate (in milliliters per minute) that is numerically equal to three times the volume (in milliliters) of the fluid in the first cell. It then flows out of the second cell at a rate (in milliliters per minute) that is numerically equal to twice the volume in the second cell. Let $x_1(t)$ and $x_2(t)$ denote the volumes of the fluid in the first and second cells at time t, respectively. Assume that initially the first cell has 40 milliliters of fluid, while the second one has 5 milliliters of fluid. Find the volume of fluid in each cell at time t. See Figure 7.1.

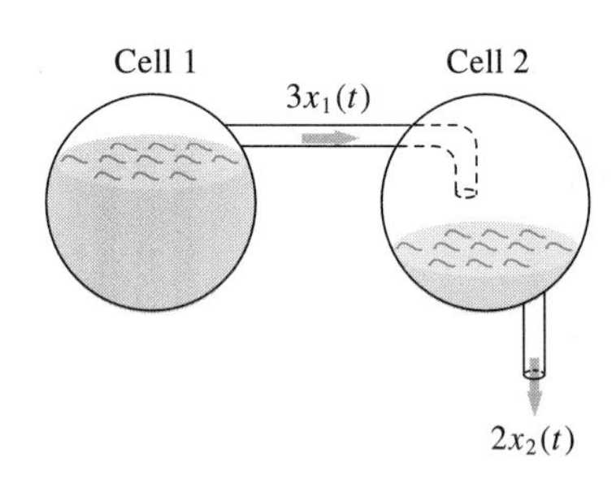

FIGURE 7.1

Solution The change in volume of the fluid in each cell is the difference between the amount flowing in and the amount flowing out. Since no fluid flows into the first cell, we have

$$\frac{dx_1(t)}{dt} = -3x_1(t),$$

where the minus sign indicates that the fluid is flowing out of the cell. The flow $3x_1(t)$ from the first cell flows into the second cell. The flow out of the second cell is $2x_2(t)$. Thus the change in volume of the fluid in the second cell is given by

$$\frac{dx_2(t)}{dt} = 3x_1(t) - 2x_2(t).$$

We have then obtained the linear system

$$\frac{dx_1(t)}{dt} = -3x_1(t)$$

$$\frac{dx_2(t)}{dt} = 3x_1(t) - 2x_2(t),$$

which can be written in matrix form as

$$\begin{bmatrix} x_1'(t) \\ x_2'(t) \end{bmatrix} = \begin{bmatrix} -3 & 0 \\ 3 & -2 \end{bmatrix} \begin{bmatrix} x_1(t) \\ x_2(t) \end{bmatrix}.$$

The eigenvalues of the matrix

$$A = \begin{bmatrix} -3 & 0 \\ 3 & -2 \end{bmatrix}$$

are (verify) $\lambda_1 = -3$, $\lambda_2 = -2$ and corresponding associated eigenvectors are (verify)

$$\begin{bmatrix} 1 \\ -3 \end{bmatrix}, \quad \begin{bmatrix} 0 \\ 1 \end{bmatrix}.$$

Hence the general solution is given by

$$\mathbf{x}(t) = \begin{bmatrix} x_1(t) \\ x_2(t) \end{bmatrix} = b_1 \begin{bmatrix} 1 \\ -3 \end{bmatrix} e^{-3t} + b_2 \begin{bmatrix} 0 \\ 1 \end{bmatrix} e^{-2t}.$$

Using the initial conditions, we find that (verify)

$$b_1 = 40, \qquad b_2 = 125.$$

Thus the volume of fluid in each cell at time t is given by

$$x_1(t) = 40e^{-3t}$$

$$x_2(t) = -120e^{-3t} + 125e^{-2t}.$$

■

It should be pointed out that many differential equations cannot be solved in the sense that we can write a formula for the solution. Numerical methods, some of which are studied in numerical analysis, exist for obtaining numerical solutions to differential equations; computer codes for some of these methods are widely available.

FURTHER READINGS

BOYCE, W.E., and R.C. DIPRIMA. *Elementary Differential Equations*, 6th ed. New York: John Wiley & Sons, Inc., 1996.

CULLEN, C.G. *Linear Algebra and Differential Equations*, 2nd ed. Boston: PWS-Kent, 1991.

DERRICK, W.R., and S.I. GROSSMAN. *Elementary Differential Equations with Applications*, 2nd ed. Reading, Mass.: Addison-Wesley, 1981.

DETTMAN, J.H. *Introduction to Linear Algebra and Differential Equations*. New York: Dover, 1986.

GOODE, S.W. *Differential Equations and Linear Algebra.* Upper Saddle River, N.J.: Prentice Hall, Inc., 1991.

RABENSTEIN, A.L. *Elementary Differential Equations with Linear Algebra*, 4th ed. Philadelphia: W.B. Saunders, 1992.

7.1 Exercises

1. Consider the linear system of differential equations

$$\begin{bmatrix} x_1' \\ x_2' \\ x_3' \end{bmatrix} = \begin{bmatrix} -3 & 0 & 0 \\ 0 & 4 & 0 \\ 0 & 0 & 2 \end{bmatrix} \begin{bmatrix} x_1 \\ x_2 \\ x_3 \end{bmatrix}.$$

(a) Find the general solution.

(b) Find the solution to the initial value problem determined by the initial conditions $x_1(0) = 3$, $x_2(0) = 4$, $x_3(0) = 5$.

2. Consider the linear system of differential equations

$$\begin{bmatrix} x_1' \\ x_2' \\ x_3' \end{bmatrix} = \begin{bmatrix} 1 & 0 & 0 \\ 0 & -2 & 1 \\ 0 & 0 & 3 \end{bmatrix} \begin{bmatrix} x_1 \\ x_2 \\ x_3 \end{bmatrix}.$$

(a) Find the general solution.

(b) Find the solution to the initial value problem determined by the initial conditions $x_1(0) = 2$, $x_2(0) = 7$, $x_3(0) = 20$.

3. Find the general solution to the linear system of differential equations

$$\begin{bmatrix} x_1' \\ x_2' \\ x_3' \end{bmatrix} = \begin{bmatrix} 4 & 0 & 0 \\ 3 & -5 & 0 \\ 2 & 1 & 2 \end{bmatrix} \begin{bmatrix} x_1 \\ x_2 \\ x_3 \end{bmatrix}.$$

4. Prove that the set of all solutions to the homogeneous linear system of differential equations $\mathbf{x}' = A\mathbf{x}$ is a subspace of the vector space of all differentiable real-valued functions. This subspace is called the **solution space** of the given linear system.

5. Find the general solution to the linear system of differential equations

$$\begin{bmatrix} x_1' \\ x_2' \\ x_3' \end{bmatrix} = \begin{bmatrix} 5 & 0 & 0 \\ 0 & -4 & 3 \\ 0 & 3 & 4 \end{bmatrix} \begin{bmatrix} x_1 \\ x_2 \\ x_3 \end{bmatrix}.$$

6. Find the general solution to the linear system of differential equations

$$\begin{bmatrix} x_1' \\ x_2' \end{bmatrix} = \begin{bmatrix} 3 & -2 \\ -2 & 3 \end{bmatrix} \begin{bmatrix} x_1 \\ x_2 \end{bmatrix}.$$

7. Find the general solution to the linear system of differential equations

$$\begin{bmatrix} x_1' \\ x_2' \\ x_3' \end{bmatrix} = \begin{bmatrix} -2 & -2 & 3 \\ 0 & -2 & 2 \\ 0 & 2 & 1 \end{bmatrix} \begin{bmatrix} x_1 \\ x_2 \\ x_3 \end{bmatrix}.$$

8. Find the general solution to the linear system of differential equations

$$\begin{bmatrix} x_1' \\ x_2' \\ x_3' \end{bmatrix} = \begin{bmatrix} 1 & 1 & 2 \\ 0 & 1 & 0 \\ 0 & 1 & 3 \end{bmatrix} \begin{bmatrix} x_1 \\ x_2 \\ x_3 \end{bmatrix}.$$

9. Consider two competing species that live in the same forest, and let $x_1(t)$ and $x_2(t)$ denote the respective populations of the species at time t. Suppose that the initial populations are $x_1(0) = 500$ and $x_2(0) = 200$. If the growth rates of the species are given by

$$\begin{aligned} x_1'(t) &= -3x_1(t) + 6x_2(t) \\ x_2'(t) &= x_1(t) - 2x_2(t), \end{aligned}$$

what is the population of each species at time t?

10. Suppose that we have a system consisting of two interconnected tanks, each containing a brine solution. Tank A contains $x(t)$ pounds of salt in 200 gallons of brine and tank B contains $y(t)$ pounds of salt in 300 gallons of brine. The mixture in each tank is kept uniform by constant stirring. When $t = 0$, brine is pumped from tank A to tank B at 20 gallons/minute and from tank B to tank A at 20 gallons/minute. Find the amount of salt in each tank at time t.

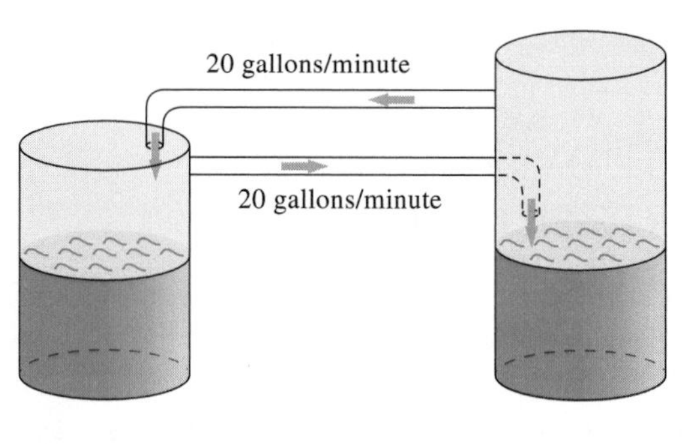

7.2 DYNAMICAL SYSTEMS

In Section 7.1 we studied how to solve homogeneous linear systems of differential equations for which an initial condition had been specified. We called such systems initial value problems and wrote then in the form

$$\mathbf{x}'(t) = A\mathbf{x}(t), \qquad \mathbf{x}(0) = \mathbf{x}_0, \tag{1}$$

where

$$\mathbf{x}(t) = \begin{bmatrix} x_1(t) \\ x_2(t) \\ \vdots \\ x_n(t) \end{bmatrix}, \qquad A = \begin{bmatrix} a_{11} & a_{12} & \cdots & \cdots & a_{1n} \\ a_{21} & a_{22} & \cdots & \cdots & a_{2n} \\ \vdots & \vdots & \cdots & \cdots & \vdots \\ \vdots & \vdots & \cdots & \cdots & \vdots \\ a_{n1} & a_{n2} & \cdots & \cdots & a_{nn} \end{bmatrix},$$

and $\mathbf{x}_0$ is a specified vector of constants. In the case that A was diagonalizable we used the eigenvalues and eigenvectors of A to construct a particular solution to (1).

In this section we focus our attention on the case $n = 2$, and for ease of reference we use x and y instead of x_1 and x_2. Such homogeneous linear systems of differential equations can be written in the form

$$\begin{aligned} \frac{dx}{dt} &= ax + by \\ \frac{dy}{dt} &= cx + dy, \end{aligned} \tag{2}$$

where a, b, c, and d are real constants, or

$$\mathbf{x}'(t) = \frac{d}{dt}\begin{bmatrix} x \\ y \end{bmatrix} = A\begin{bmatrix} x \\ y \end{bmatrix} = A\mathbf{x}(t), \tag{3}$$

where $A = \begin{bmatrix} a & b \\ c & d \end{bmatrix}$. For the systems (2) and (3) we try to describe properties of the solution based on the differential equation itself. This area is called the **qualitative theory** of differential equations and was studied extensively by J.H. Poincaré.†

†Jules Henri Poincaré (1854–1912) was born in Nancy, France, to a well-to-do family, many of whose members played key roles in the French government. As a youngster, he was clumsy and absent-minded, but showed great talent in mathematics. In 1873 he entered the École Polytechnique, from which he received his doctorate. He then began a university career, finally joining the University of Paris in 1881, where he remained until his death. Poincaré is considered the last of the universalists in mathematics, that is, someone who can work in all branches of mathematics, both pure and applied. His doctoral dissertation dealt with the existence of solutions to differential equations. In applied mathematics, he made contributions to the fields of optics, electricity, elasticity, thermodynamics, quantum mechanics, the theory of relativity, and cosmology. In pure mathematics, he was one of the principal creators of algebraic topology and made numerous contributions to algebraic geometry, analytic functions, and number theory. He was the first person to think of chaos in connection with his work in astronomy. In his later years, he wrote several books popularizing mathematics. In some of these books he dealt with the psychological processes involved in mathematical discovery and with the aesthetic aspects of mathematics.

The systems (2) or (3) are called **autonomous** differential equations since the rates of change $\frac{dx}{dt}$ and $\frac{dy}{dt}$ depend explicitly only on the values of x and y, not on the independent variable t. For our purposes we will call the independent variable t **time** and then (2) and (3) are said to be time-independent systems. Using this convention, the systems in (2) and (3) provide a model for the change of x and y as time goes by. Hence such systems are called **dynamical systems**. We will use this terminology throughout our discussion in this section.

A qualitative analysis of dynamical systems is interested in such questions as

- Are there any constant solutions?
- If there are constant solutions, do nearby solutions move toward or away from the constant solution?
- What is the behavior of solutions as $t \to \pm\infty$?
- Are there any solutions which oscillate?

Each of these questions has a geometric flavor. Hence we introduce a helpful device for studying the behavior of dynamical systems. If we regard t, time, as a parameter, then $x = x(t)$ and $y = y(t)$ will represent a curve in the xy-plane. Such a curve is called a **trajectory** or an **orbit** of the systems (2) and (3). The xy-plane is called the **phase plane** of the dynamical system.

EXAMPLE 1

The system

$$\mathbf{x}'(t) = A\mathbf{x}(t) = \begin{bmatrix} 0 & 1 \\ -1 & 0 \end{bmatrix} \mathbf{x}(t) \qquad \text{or} \qquad \begin{aligned} \frac{dx}{dt} &= y \\ \frac{dy}{dt} &= -x \end{aligned}$$

has the general solution[†]

$$\begin{aligned} x &= b_1 \sin(t) + b_2 \cos(t) \\ y &= b_1 \cos(t) - b_2 \sin(t). \end{aligned} \tag{4}$$

It follows that the trajectories satisfy[‡] $x^2 + y^2 = c^2$, where $c^2 = b_1^2 + b_2^2$. Hence the trajectories of this dynamical system are circles in the phase plane centered at the origin. We note that if an initial condition $\mathbf{x}(0) = \begin{bmatrix} k_1 \\ k_2 \end{bmatrix}$ is specified, then setting

[†]We verify this later in the section.

[‡]The trajectories can be obtained directly by noting that we can eliminate t to get $\frac{dy}{dx} = \frac{-x}{y}$. Then separating the variables gives $y\,dy = -x\,dx$ and upon integrating we get $\frac{y^2}{2} = -\frac{x^2}{2} + k^2$, or equivalently, $x^2 + y^2 = c^2$.

$t = 0$ in (4) gives the linear system

$$b_1 \sin(0) + b_2 \cos(0) = k_1$$
$$b_1 \cos(0) - b_2 \sin(0) = k_2.$$

It follows that the solution is $b_2 = k_1$ and $b_1 = k_2$ so that the corresponding particular solution to the initial value problem $\mathbf{x}'(t) = A\mathbf{x}(t)$, $\mathbf{x}_0 = \begin{bmatrix} k_1 \\ k_2 \end{bmatrix}$ determines the trajectory $x^2 + y^2 = k_1^2 + k_2^2$; that is, a circle centered at the origin with radius $\sqrt{k_1^2 + k_2^2}$. ■

A sketch of the trajectories of a dynamical system in the phase plane is called a **phase portrait**. A phase portrait usually contains the sketches of a few trajectories and an indication of the direction in which the curve is traversed. This is done by placing arrowheads on the trajectory to indicate the direction of motion of a point (x, y) as t increases. The direction is indicated by the **velocity vector**

$$\mathbf{v} = \begin{bmatrix} \dfrac{dx}{dt} \\ \dfrac{dy}{dt} \end{bmatrix}.$$

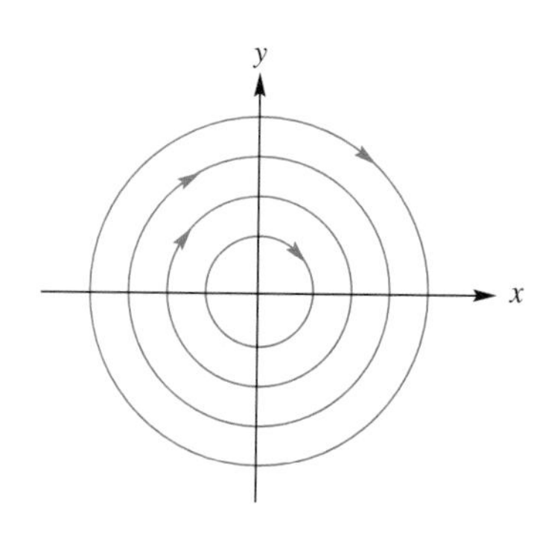

FIGURE 7.2

For the dynamical system in Example 1 we have $\mathbf{v} = \begin{bmatrix} y \\ -x \end{bmatrix}$. Thus in the phase plane for $x > 0$ and $y > 0$, the vector $\mathbf{v}$ is oriented downward to the right; hence these trajectories are traversed clockwise as indicated in Figure 7.2. (**Warning:** In other dynamical systems not all trajectories are traversed in the same direction.)

One of the questions posed earlier concerned the existence of constant solutions. For a dynamical system (which we regard as a model depicting the change of x and y as time goes by) to have a constant solution both $\dfrac{dx}{dt}$ and $\dfrac{dy}{dt}$ must be zero. That is, the system doesn't change. It follows that points in the phase plane that corresponds to solutions are determined by solving

$$\frac{dx}{dt} = ax + by = 0$$

$$\frac{dy}{dt} = cx + dy = 0,$$

which leads to the homogeneous linear system

$$A\mathbf{x} = \begin{bmatrix} a & b \\ c & d \end{bmatrix}\begin{bmatrix} x \\ y \end{bmatrix} = \begin{bmatrix} 0 \\ 0 \end{bmatrix}.$$

We know that one solution to this linear system is $x = 0$, $y = 0$ and that there exist other solutions if and only if $\det(A) = 0$. In Example 1, $A = \begin{bmatrix} 0 & 1 \\ -1 & 0 \end{bmatrix}$ and

$\det(A) = 1$. Thus for the dynamical system in Example 1, the only point in the phase plane that corresponds to a constant solution is $x = 0$ and $y = 0$, the origin.

Definition 7.1

A point in the phase plane at which both $\frac{dx}{dt}$ and $\frac{dy}{dt}$ are zero is called an **equilibrium point**, or **fixed point**, of the dynamical system. ▲

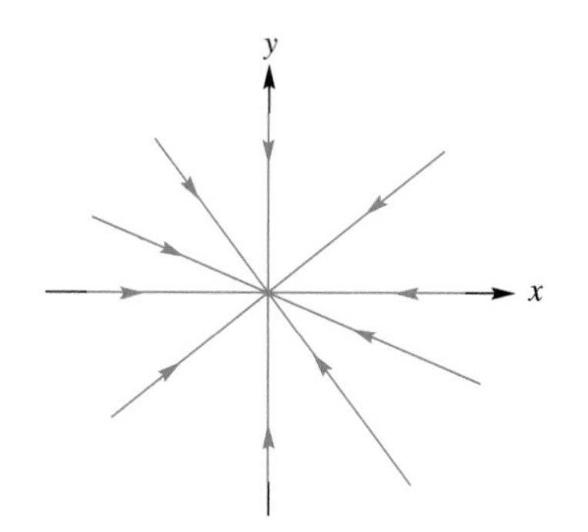

FIGURE 7.3

A stable equilibrium point at $(0, 0)$.

The behavior of trajectories near an equilibrium point is a way to characterize different types of equilibrium points. If trajectories through all points near an equilibrium point converge to the equilibrium point, then we say that the equilibrium point is **stable**, or an **attractor**. This is the case for the origin shown in Figure 7.3, where the trajectories are all straight lines heading into the origin. The dynamical system whose phase portrait is shown in Figure 7.3 is

$$\frac{dx}{dt} = -x$$
$$\frac{dy}{dt} = -y,$$

which we discuss later in Example 3. Another situation is shown in Figure 7.2, where again the only equilibrium point is the origin. In this case, trajectories through points near the equilibrium point stay a small distance away. In such a case, the equilibrium point is called **marginally stable**. At other times, nearby trajectories tend to move away from an equilibrium point. In such cases, we say that the equilibrium point is **unstable**, or a **repelling** point. (See Figure 7.4.) In addition we can have equilibrium points where nearby trajectories on one side move toward it and on the other side move away from it. Such an equilibrium point is called a **saddle point**. (See Figure 7.5.)

From the developments in Section 7.1, we expect that the eigenvalues and

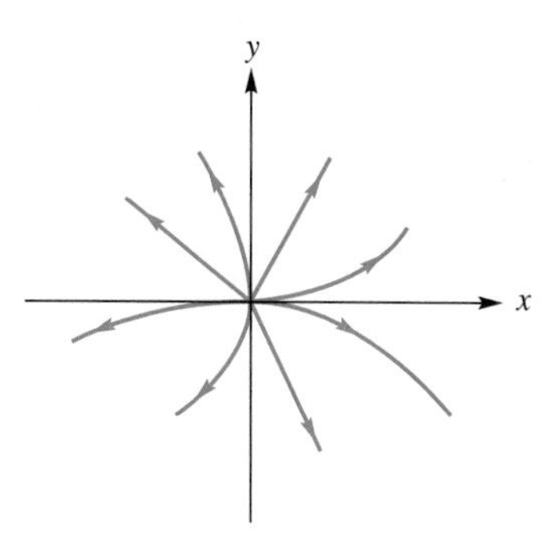

FIGURE 7.4

An unstable equilibrium point at $(0, 0)$.

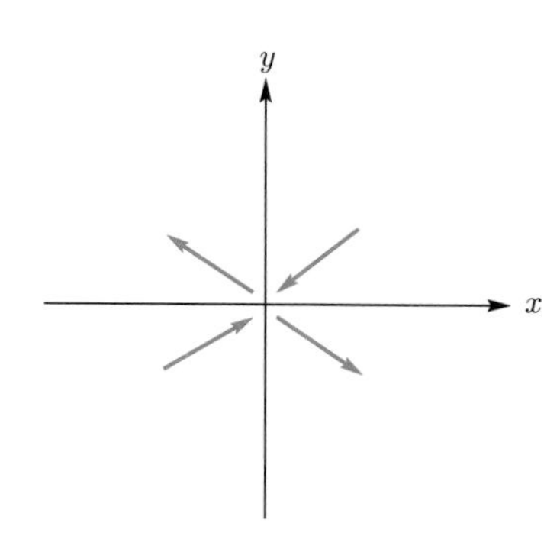

FIGURE 7.5

A saddle point at $(0, 0)$.

eigenvectors of the coefficient matrix $A = \begin{bmatrix} a & b \\ c & d \end{bmatrix}$ of the dynamical system will determine features of the phase portrait of the system. From Equation (17) in Section 7.1 we have

$$\mathbf{x}(t) = \begin{bmatrix} x \\ y \end{bmatrix} = b_1\mathbf{p}_1 e^{\lambda_1 t} + b_2\mathbf{p}_2 e^{\lambda_2 t}, \tag{5}$$

where λ_1 and λ_2 are the eigenvalues of A, and $\mathbf{p}_1$ and $\mathbf{p}_2$ are associated eigenvectors. In Sections 6.1 and 6.2, eigenvalues and eigenvectors were to be real numbers and real vectors, respectively. We no longer require them to be real; they can be complex. (See Section B.2 in Appendix B for a discussion of this more general case.) We also require that both λ_1 and λ_2 be nonzero.† Hence A is nonsingular. (Explain why.) And so the only equilibrium point is $x = 0$, $y = 0$, the origin.

To show how we use the eigen information from A to determine the phase portrait we treat the case of complex eigenvalues separately from the case of real eigenvalues.

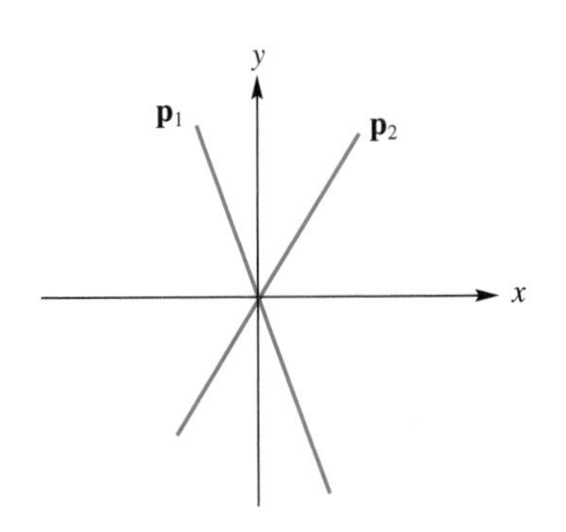

FIGURE 7.6

Case λ_1 and λ_2 Real

For real eigenvalues (and eigenvectors) the phase plane interpretation of Equation (5) is that $\mathbf{x}(t)$ is in span $\{\mathbf{p}_1, \mathbf{p}_2\}$. Hence $\mathbf{p}_1$ and $\mathbf{p}_2$ are trajectories. It follows that the eigenvectors $\mathbf{p}_1$ and $\mathbf{p}_2$ determine lines or rays through the origin in the phase plane, and a phase portrait for this case has the general form shown in Figure 7.6. To complete the portrait we need more than the special trajectories corresponding to the eigen directions. These other trajectories depend on the values of λ_1 and λ_2.

Eigenvalues negative and distinct: $\lambda_1 < \lambda_2 < 0$. From (5), as $t \to \infty$, $\mathbf{x}(t)$ gets small. Hence all the trajectories tend towards the equilibrium point at the origin as $t \to \infty$. See Example 2 and Figure 7.7.

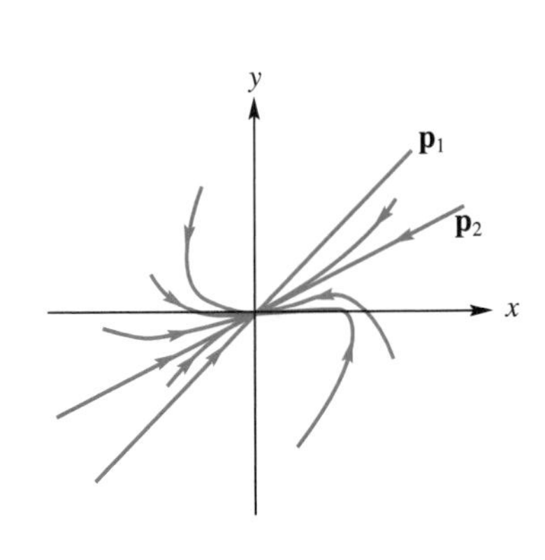

FIGURE 7.7

EXAMPLE 2

Determine the phase plane portrait of the dynamical system

$$\mathbf{x}'(t) = A\mathbf{x}(t) = \begin{bmatrix} -2 & -2 \\ 1 & -5 \end{bmatrix} \mathbf{x}(t).$$

We begin by finding the eigenvalues and associated eigenvectors of A. We find (verify)

$$\lambda_1 = -4, \quad \lambda_2 = -3, \quad \text{and} \quad \mathbf{p}_1 = \begin{bmatrix} 1 \\ 1 \end{bmatrix}, \quad \mathbf{p}_2 = \begin{bmatrix} 2 \\ 1 \end{bmatrix}.$$

It follows that

$$\mathbf{x}(t) = \begin{bmatrix} x \\ y \end{bmatrix} = b_1\mathbf{p}_1 e^{-4t} + b_2\mathbf{p}_2 e^{-3t}$$

†It can be shown that if both eigenvalues of A are zero, then all solutions to (2) as given in (5) are either constants or constants and straight lines. In addition, we can show that if one eigenvalue of A is zero and the other nonzero, then there is a line of equilibrium points. See Further Readings at the end of the chapter.

and as $t \to \infty$, $\mathbf{x}(t)$ gets small. It is helpful to rewrite this expression in the form

$$\mathbf{x}(t) = \begin{bmatrix} x \\ y \end{bmatrix} = b_1\mathbf{p}_1 e^{-4t} + b_2\mathbf{p}_2 e^{-3t} = e^{-3t}(b_1\mathbf{p}_1 e^{-t} + b_2\mathbf{p}_2).$$

As long as $b_2 \neq 0$, the term $b_1\mathbf{p}_1 e^{-t}$ is negligible in comparison to $b_2\mathbf{p}_2$. This implies that as $t \to \infty$ all trajectories, except those starting on $\mathbf{p}_1$, will align themselves in the direction of $\mathbf{p}_2$ as they get close to the origin. Hence the phase portrait appears like that given in Figure 7.7. The origin is an attractor. ■

Eigenvalues positive and distinct: $\lambda_1 > \lambda_2 > 0$. From (5), as $t \to \infty$, $\mathbf{x}(t)$ gets large. Hence all the trajectories tend to go away from the equilibrium point at the origin. The phase portrait for such dynamical systems is like that in Figure 7.7, except that all the arrowheads are reversed, indicating motion away from the origin. In this case (0, 0) is called an **unstable** equilibrium point.

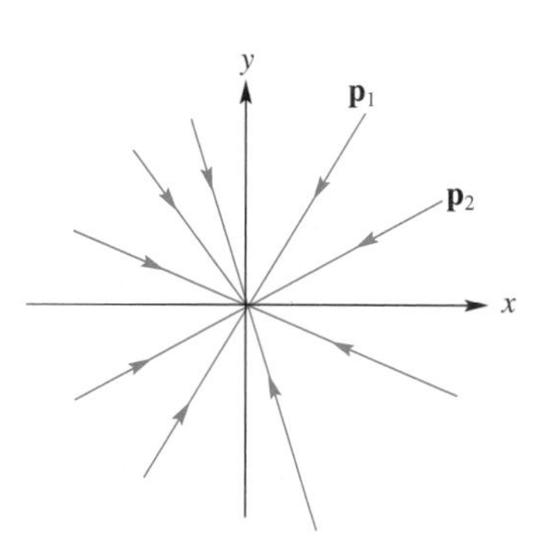

FIGURE 7.8

Both eigenvalues negative but equal: $\lambda_1 = \lambda_2 < 0$. All trajectories go to a stable equilibrium at the origin, but they may bend differently than the trajectories depicted in Figure 7.7. Their behavior depends upon the number of linearly independent eigenvectors of matrix A. If there are two linearly independent eigenvectors, then $\mathbf{x}(t) = e^{\lambda_1 t}(b_1\mathbf{p}_1 + b_2\mathbf{p}_2)$, which is a multiple of the constant vector $b_1\mathbf{p}_1 + b_2\mathbf{p}_2$. Thus it follows that all the trajectories are lines through the origin and, since $\lambda_1 < 0$, motion along them is towards the origin. See Figure 7.8. We illustrate this in Example 3.

EXAMPLE 3

The matrix A of the dynamical system

$$\mathbf{x}'(t) = A\mathbf{x}(t) = \begin{bmatrix} -1 & 0 \\ 0 & -1 \end{bmatrix}\mathbf{x}(t)$$

has eigenvalues $\lambda_1 = \lambda_2 = -1$ and corresponding eigenvectors

$$\mathbf{p}_1 = \begin{bmatrix} 1 \\ 0 \end{bmatrix} \quad \text{and} \quad \mathbf{p}_2 = \begin{bmatrix} 0 \\ 1 \end{bmatrix} \qquad \text{(verify.)}$$

It follows that

$$\mathbf{x}(t) = e^{-t}(b_1\mathbf{p}_1 + b_2\mathbf{p}_2) = e^{-t}\begin{bmatrix} b_1 \\ b_2 \end{bmatrix},$$

so $x = b_1 e^{-t}$ and $y = b_2 e^{-t}$. If $b_1 \neq 0$, then $y = \dfrac{b_2}{b_1}x$. If $b_1 = 0$, then we are on the trajectory in the direction of $\mathbf{p}_2$. It follows that all trajectories are straight lines through the origin as in Figures 7.3 and 7.8. ■

If there is only one linearly independent eigenvector, then it can be shown that all trajectories passing through points not on the eigenvector align themselves to be tangent to the eigenvector at the origin. We will not develop this case, but the phase

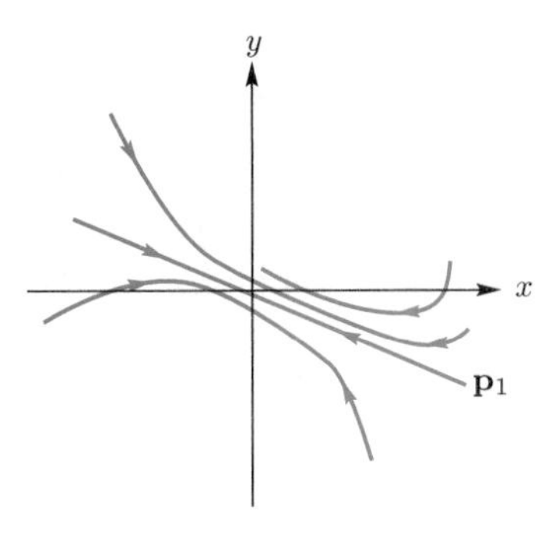

FIGURE 7.9

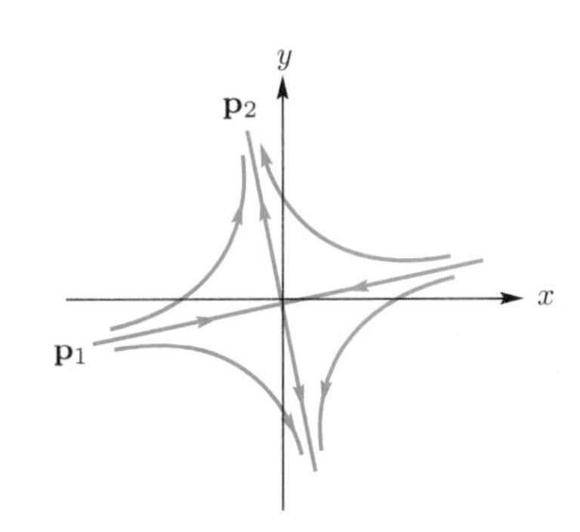

FIGURE 7.10

portrait is similar to Figure 7.9. In the case that the eigenvalues are positive and equal, the phase portraits for these two cases are like Figures 7.8 and 7.9 with the arrowheads reversed.

One positive eigenvalue and one negative eigenvalue: $\lambda_1 < 0 < \lambda_2$**.** From (5), as $t \to \infty$ one of the terms is increasing while the other term is decreasing. This causes a trajectory, that is not in the direction of an eigenvector, to head towards the origin but bend away as t gets larger. The origin in this case is called a **saddle point**. The phase portrait resembles Figure 7.10.

EXAMPLE 4

Determine the phase plane portrait of the dynamical system

$$\mathbf{x}'(t) = A\mathbf{x}(t) = \begin{bmatrix} 1 & -1 \\ -2 & 0 \end{bmatrix} \mathbf{x}(t).$$

We begin by finding the eigenvalues and associated eigenvectors of A. We find (verify)

$$\lambda_1 = -1, \quad \lambda_2 = 2, \quad \text{and} \quad \mathbf{p}_1 = \begin{bmatrix} 1 \\ 2 \end{bmatrix}, \quad \mathbf{p}_2 = \begin{bmatrix} 1 \\ -1 \end{bmatrix}.$$

It follows that the origin is a saddle point and that we have

$$\mathbf{x}(t) = \begin{bmatrix} x \\ y \end{bmatrix} = b_1\mathbf{p}_1 e^{-t} + b_2\mathbf{p}_2 e^{2t}.$$

We see that if $b_1 \neq 0$ and $b_2 = 0$, then the motion is in the direction of eigenvector $\mathbf{p}_1$ and towards the origin. Similarly, if $b_1 = 0$ and $b_2 \neq 0$, then the motion is in the direction of $\mathbf{p}_2$, but away from the origin. If we look at the components of the original system and eliminate t, we obtain (verify)

$$\frac{dy}{dx} = \frac{2x}{y - x}.$$

This expression tells us the slope along trajectories in the phase plane. Inspecting this expression we see the following: (Explain.)

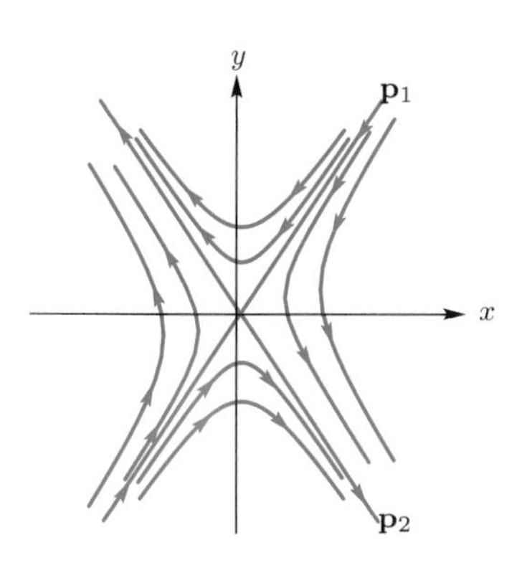

FIGURE 7.11

- All trajectories crossing the y-axis have horizontal tangents.
- As a trajectory crosses the line $y = x$ it has a vertical tangent.
- Whenever a trajectory crosses the x-axis it has slope -2.

Using the general form of a saddle point, shown in Figure 7.10, we can produce quite an accurate phase portrait for this dynamical system, shown in Figure 7.11. ■

Both Eigenvalues Complex Numbers

For real 2×2 matrices A, the characteristic equation $\det(\lambda I_2 - A) = 0$ is a quadratic polynomial. If the roots of this quadratic equation λ_1 and λ_2 are complex numbers, then they are conjugates of one another. (See Section B.2.) If $\lambda_1 = \alpha + \beta i$, where α and β are real numbers with $\beta \neq 0$, then $\lambda_2 = \overline{\lambda_1} = \alpha - \beta i$. Hence in Equation (5) we have the exponential of a complex number:

$$e^{\lambda_1 t} = e^{(\alpha+\beta i)t} = e^{\alpha t} e^{\beta i t}.$$

The term $e^{\alpha t}$ is a standard exponential function, but $e^{\beta i t}$ is quite different since $i = \sqrt{-1}$. Fortunately there is a simple way to express such an exponential function in terms of more manageable functions. We use Euler's identity,

$$e^{i\theta} = \cos(\theta) + i\,\sin(\theta),$$

which we state without proof. Using Euler's identity we have

$$e^{\lambda_1 t} = e^{(\alpha+\beta i)t} = e^{\alpha t} e^{\beta i t} = e^{\alpha t}(\cos(\beta t) + i\,\sin(\beta t))$$

and

$$\begin{aligned} e^{\lambda_2 t} = e^{(\alpha-\beta i)t} &= e^{\alpha t} e^{-\beta i t} \\ &= e^{\alpha t}(\cos(-\beta t) + i\,\sin(-\beta t)) = e^{\alpha t}(\cos(\beta t) - i\,\sin(\beta t)). \end{aligned}$$

It can be shown that the system given in (5) can be written so that the components $x(t)$ and $y(t)$ are linear combinations of $e^{\alpha t}\cos(\beta t)$ and $e^{\alpha t}\sin(\beta t)$ with real coefficients. The behavior of the trajectories can now be analyzed by considering the sign of α, since $\beta \neq 0$.

Complex eigenvalues: $\lambda_1 = \alpha + \beta i$ **and** $\lambda_2 = \alpha - \beta i$ **with** $\alpha = 0$**,** $\beta \neq 0$**.**
For this case $x(t)$ and $y(t)$ are linear combinations of $\cos(\beta t)$ and $\sin(\beta t)$. It can be shown that the trajectories are ellipses whose major and minor axes are determined by the eigenvectors. (See Example 1 for a particular case.) The motion is periodic since the trajectories are closed curves. The origin is a marginally stable equilibrium point, since the trajectories through points near the origin do not move very far away. (See Figure 7.2.)

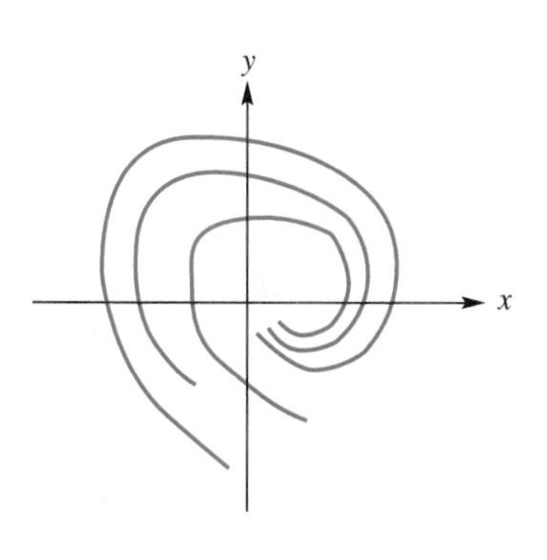

FIGURE 7.12

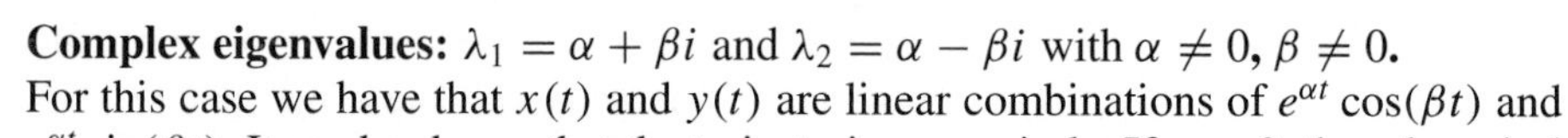

Complex eigenvalues: $\lambda_1 = \alpha + \beta i$ **and** $\lambda_2 = \alpha - \beta i$ **with** $\alpha \neq 0$**,** $\beta \neq 0$**.**
For this case we have that $x(t)$ and $y(t)$ are linear combinations of $e^{\alpha t}\cos(\beta t)$ and $e^{\alpha t}\sin(\beta t)$. It can be shown that the trajectories are spirals. If $\alpha > 0$, then the spiral goes outward, away from the origin. Thus the origin is an unstable equilibrium point. If $\alpha < 0$, then the spiral goes inward, towards the origin, and so the origin is a stable equilibrium point. The phase portrait is a collection of spirals as shown in Figure 7.12, with arrowheads appropriately affixed.

The dynamical system in (2) may appear quite special. However, the experience gained by the qualitative analysis we have presented is the key to understanding the behavior of nonlinear dynamical systems of the form

$$\frac{dx}{dt} = f(x, y)$$

$$\frac{dy}{dt} = g(x, y).$$

The extension of the phase plane and phase portraits to such nonlinear systems is beyond the scope of this brief introduction. Such topics are part of courses on differential equations and can be found in the books listed in Further Readings.

FURTHER READINGS

Boyce, W.E. and R.C. DiPrima, *Elementary Differential Equations*, 6th ed., New York: John Wiley & Sons, Inc., 1996.

Campbell, S.L., *An Introduction to Differential Equations and Their Applications*, 2nd ed., Belmont, California: Wadsworth Publishing Co., 1990.

Farlow, S.J., *An Introduction to Differential Equations and Their Applications*, New York: McGraw-Hill, Inc., 1994.

7.2 Exercises

For each of the dynamical systems in Exercises 1–10, *determine the nature of the equilibrium point at the origin, and describe the phase portrait.*

1. $\mathbf{x}'(t) = \begin{bmatrix} -1 & 0 \\ 0 & -3 \end{bmatrix} \mathbf{x}(t)$.

2. $\mathbf{x}'(t) = \begin{bmatrix} 2 & 0 \\ 0 & 1 \end{bmatrix} \mathbf{x}(t)$.

3. $\mathbf{x}'(t) = \begin{bmatrix} -1 & 2 \\ 0 & -1 \end{bmatrix} \mathbf{x}(t)$.

4. $\mathbf{x}'(t) = \begin{bmatrix} -2 & 0 \\ 3 & 1 \end{bmatrix} \mathbf{x}(t)$.

5. $\mathbf{x}'(t) = \begin{bmatrix} 1 & 1 \\ 3 & -1 \end{bmatrix} \mathbf{x}(t)$.

6. $\mathbf{x}'(t) = \begin{bmatrix} -1 & -1 \\ 1 & -1 \end{bmatrix} \mathbf{x}(t)$.

7. $\mathbf{x}'(t) = \begin{bmatrix} -2 & 1 \\ 2 & -3 \end{bmatrix} \mathbf{x}(t)$.

8. $\mathbf{x}'(t) = \begin{bmatrix} -2 & 1 \\ -1 & -2 \end{bmatrix} \mathbf{x}(t)$.

9. $\mathbf{x}'(t) = \begin{bmatrix} 3 & -13 \\ 1 & -3 \end{bmatrix} \mathbf{x}(t)$.

10. $\mathbf{x}'(t) = \begin{bmatrix} 3 & 2 \\ 2 & 3 \end{bmatrix} \mathbf{x}(t)$.

Supplementary Exercises

1. Let $A(t) = \begin{bmatrix} a_{ij}(t) \end{bmatrix}$ be an $n \times n$ matrix whose entries are all functions of t; $A(t)$ is called a **matrix function**. The derivative and integral of $A(t)$ is defined componentwise; that is,

$$\frac{d}{dt}[A(t)] = \left[\frac{d}{dt} a_{ij}(t) \right]$$

and

$$\int_a^t A(s)\, ds = \left[\int_a^t a_{ij}(s)\, ds \right].$$

For each of the following matrices $A(t)$, compute

$\frac{d}{dt}[A(t)]$ and $\int_0^t A(s)\,ds$.

(a) $A(t) = \begin{bmatrix} t^2 & \frac{1}{t+1} \\ 4 & e^{-t} \end{bmatrix}$.

(b) $A(t) = \begin{bmatrix} \sin 2t & 0 & 0 \\ 0 & 1 & -t \\ 0 & te^{t^2} & \frac{t}{t^2+1} \end{bmatrix}$.

2. The usual rules for differentiation and integration of functions introduced in calculus also apply to matrix functions. Let $A(t)$ and $B(t)$ be $n \times n$ matrix functions whose entries are differentiable and let c_1 and c_2 be real numbers. Prove the following properties.

(a) $\frac{d}{dt}[c_1A(t) + c_2B(t)] = c_1\frac{d}{dt}[A(t)] + c_2\frac{d}{dt}[B(t)]$.

(b) $\int_a^t (c_1A(s) + c_2B(s))\,ds =$

$$c_1\int_a^t A(s)\,ds + c_2\int_a^t B(s)\,ds.$$

(c) $\frac{d}{dt}[A(t)B(t)] = B(t)\frac{d}{dt}[A(t)] + A(t)\frac{d}{dt}[B(t)]$.

3. If A is an $n \times n$ matrix, then the matrix function

$$B(t) = I_n + At + A^2\frac{t^2}{2!} + A^3\frac{t^3}{3!} + \cdots$$

is called the **matrix exponential** function and we use the notation $B(t) = e^{At}$.

(a) Prove that $\frac{d}{dt}[e^{At}] = Ae^{At}$.

(b) Let

$$A = \begin{bmatrix} 0 & 1 \\ 0 & 0 \end{bmatrix} \quad \text{and} \quad B = \begin{bmatrix} 0 & 0 \\ 1 & 0 \end{bmatrix}.$$

Prove or disprove that $e^Ae^B = e^{A+B}$.

(c) Prove that $e^{iA} = \cos A + i \sin A$, where $i = \sqrt{-1}$ (see Section B.1).

4. Let A and B be $n \times n$ matrices that commute, that is, $AB = BA$. Prove that $e^Ae^B = e^{A+B}$.

5. Let $B(t) = \left[b_{ij}(t)\right]$ be a diagonal matrix function with $b_{ii}(t) = e^{\lambda_{ii}t}$, where λ_{ii} is a scalar, $i = 1, 2, \ldots, n$, and $b_{ij}(t) = 0$ if $i \neq j$. Let D be the diagonal matrix with diagonal entries λ_{ii}, $i = 1, 2, \ldots, n$. Prove that $B(t) = e^{Dt}$.

6. Let A be an $n \times n$ matrix that is diagonalizable with eigenvalues λ_i and associated eigenvectors $\mathbf{x}_i$, $i = 1, 2, \ldots, n$. Then we can choose the eigenvectors $\mathbf{x}_i$ so that they form a linearly independent set; the matrix P whose jth column is $\mathbf{x}_j$ is nonsingular, and $P^{-1}AP = D$, where D is the diagonal matrix whose diagonal entries are the eigenvalues of A. Prove that Equation (17) in Section 7.1 can be written as

$$\mathbf{x}(t) = Pe^{Dt}B,$$

where

$$B = \begin{bmatrix} b_1 \\ b_2 \\ \vdots \\ b_n \end{bmatrix}.$$

7. Let A be an $n \times n$ matrix and

$$\mathbf{x}(t) = \begin{bmatrix} x_1(t) \\ x_2(t) \\ \vdots \\ x_n(t) \end{bmatrix}.$$

Assume that A is diagonalizable, as in Exercise 6, and prove that the solution to the initial value problem

$$\mathbf{x}' = A\mathbf{x}$$
$$\mathbf{x}(0) = \mathbf{x}_0$$

can be written as

$$\mathbf{x}(t) = Pe^{Dt}P^{-1}\mathbf{x}_0 = e^{At}\mathbf{x}_0.$$

8. For $\mathbf{x}_0 = \begin{bmatrix} 1 \\ 1 \end{bmatrix}$ and each of the following matrices A, solve the initial value problem given in Exercise 7.

(a) $A = \begin{bmatrix} 2 & -1 \\ -1 & 2 \end{bmatrix}$.

(b) $A = \begin{bmatrix} -1 & 1 \\ 1 & -1 \end{bmatrix}$.

9. For $\mathbf{x}_0 = \begin{bmatrix} 1 \\ 0 \\ 1 \end{bmatrix}$ and each of the following matrices A, solve the initial value problem given in Exercise 7.

(a) $A = \begin{bmatrix} -1 & 1 & 0 \\ 0 & 3 & -12 \\ 1 & -1 & 0 \end{bmatrix}$.

(b) $A = \begin{bmatrix} 0 & 1 & 0 \\ 0 & 0 & 1 \\ 0 & 8 & -2 \end{bmatrix}$.

Chapter 8

MATLAB for Linear Algebra

INTRODUCTION

MATLAB is a versatile piece of computer software with linear algebra capabilities as its core. MATLAB stands for MATrix LABoratory. It incorporates portions of professionally developed projects of quality computer routines for linear algebra computation. The code employed by MATLAB is written in the C language however. Many of the routines/functions are written in the MATLAB language, and are upgraded as new versions of MATLAB are released. MATLAB is available for Microsoft Windows, Macintosh computers, and for Unix workstations.

MATLAB has a wide range of capabilities. In this book we will use only a small portion of its features. We will find that MATLAB's command structure is very close to the way we write algebraic expressions and linear algebra operations. The names of many MATLAB commands closely parallel those of the operations and concepts of linear algebra. We give descriptions of commands and features of MATLAB that relate directly to this course. A more detailed discussion of MATLAB commands can be found in *The MATLAB User's Guide* that accompanies the software and in the books *Experiments in Computational Matrix Algebra*, by David R. Hill (New York: Random House, 1988) and *Linear Algebra LABS with MATLAB*, 2nd ed., by David R. Hill and David E. Zitarelli (Upper Saddle River, N.J.: Prentice-Hall, Inc., 1996). Alternatively, the MATLAB software provides immediate on-screen descriptions, using the **help** command. Typing

help

displays a list of MATLAB subdirectories and alternative directories containing files corresponding to commands and data sets. Typing **help** *name*, where *name* is the name of a command, accesses information on the specific command named. In some cases the description displayed goes much further than we need for this course. Hence you may not fully understand all of the description displayed by **help**. We provide a list of the majority of MATLAB commands we use in this book in Section 8.9.

Once you initiate the MATLAB software, you will see the MATLAB logo appear and the MATLAB prompt ≫. The prompt ≫ indicates that MATLAB is awaiting a command. In Section 8.1 we describe how to enter matrices into MATLAB and give explanations of several commands. However, there are certain MATLAB features you should be aware of before you begin the material in Section 8.1.

☐ *Starting execution of a command.*
After you have typed a command name and any arguments or data required, you must press ENTER before it will begin to execute.

☐ *The command stack.*
As you enter commands, MATLAB saves a number of the most recent commands in a stack. Previous commands saved on the stack can be recalled using the **up arrow** key. The number of commands saved on the stack varies depending on the length of the commands and other factors.

☐ *Editing commands.*
If you make an error or mistype something in a command, you can use the **left arrow** or **right arrow** keys to position the cursor for corrections. The **home** key moves the cursor to the beginning of a command, and the **end** key moves the cursor to the end. The **backspace** and **delete** keys can be used to remove characters from a command line. The **insert** key is used to initiate the insertion of characters. Pressing the insert key a second time exits the insert mode. If MATLAB recognizes an error after you have pressed ENTER, then MATLAB responds with a beep and a message that helps define the error. You can recall the command line, using the up arrow key in order to edit the line.

☐ *Continuing commands.*
MATLAB commands that do not fit on a single line can be continued to the next line, using an ellipsis, which is three consecutive periods, followed by ENTER.

☐ *Stopping a command.*
To stop execution of a MATLAB command, press **Ctrl** and **C** simultaneously (on the Macintosh, press apple key and .), then press ENTER. Sometimes this sequence must be repeated.

☐ *Quitting.*
To quit MATLAB, type **exit** or **quit** followed by ENTER.

8.1 INPUT AND OUTPUT IN MATLAB

Matrix Input

To enter a matrix into MATLAB just type the entries enclosed in square brackets [...], with entries separated by a space and rows terminated with a semicolon. Thus, matrix

$$\begin{bmatrix} 9 & -8 & 7 \\ -6 & 5 & -4 \\ 11 & -12 & 0 \end{bmatrix}$$

is entered by typing

[9 −8 7;−6 5 −4;11 −12 0]

and the accompanying display is

```
ans =

     9    -8     7
    -6     5    -4
    11   -12     0
```

Notice that no brackets are displayed and that MATLAB has assigned this matrix the name **ans**. Every matrix in MATLAB must have a name. If you do not assign a matrix a name, then MATLAB assigns it **ans**, which is called the **default variable name**. To assign a matrix name we use the assignment operator =. For example,

A = [4 5 8;0 −1 6]

is displayed as

```
A =

     4     5     8
     0    -1     6
```

Warning

1. All rows must have the same number of entries.
2. MATLAB distinguishes between uppercase and lowercase letters. So matrix B is not the same as matrix **b**.
3. A matrix name can be reused. In such a case the "old" contents are lost.

To assign a matrix but *suppress the display of its entries*, follow the closing square bracket,], with a semicolon.

A = [4 5 8;0 −1 6];

assigns the same matrix to name A as above, but no display appears. To assign a currently defined matrix a new name, use the assignment operator =. Command **Z = A** assigns the contents of A to Z. Matrix **A** is still defined.

To determine the matrix names that are in use, use the **who** command. To delete a matrix, use the **clear** command followed by a space and then the matrix name. For example, the command

clear A

deletes name A and its contents from MATLAB. The command **clear** by itself deletes all currently defined matrices.

To determine the number of rows and columns in a matrix, use the **size** command, as in

size(A)

which, assuming that A has not been cleared, displays

```
ans =

     2     3
```

meaning that there are two rows and three columns in matrix A.

To see all of the components of a matrix, type its name. If the matrix is large, the display may be broken into subsets of columns that are shown successively. For example, use the command

hilb(9)

which displays the first seven columns followed by columns 8 and 9. (For information on command **hilb**, use **help hilb**.) If the matrix is quite large, the screen display will scroll too fast for you to see the matrix. To see a portion of a matrix, type command **more on** followed by ENTER, then type the matrix name or a command to generate it. Press the Space Bar to reveal more of the matrix. Continue pressing the Space Bar until the "--more--" no longer appears near the bottom of the screen. Try this with **hilb(20)**. To disable this paging feature, type command **more off**. If a scroll bar is available, you can use your mouse to move the scroll bar to reveal previous portions of displays.

We have the following conventions to see a portion of a matrix in MATLAB. For purposes of illustration, suppose matrix A has been entered into MATLAB as a 5×5 matrix.

- To see the (2, 3) entry of A, type

  A(2,3)

- To see the fourth row of A, type

  A(4,:)

- To see the first column of A, type

  A(:,1)

In the situations above, the : is interpreted to mean "all." The colon can also be used to represent a range of rows or columns. For example, typing

2:8

displays

```
ans =

     2     3     4     5     6     7     8
```

We can use this feature to display a subset of rows or columns of a matrix. As an illustration, to display rows 3 through 5 of matrix **A**, type

A(3:5,:)

Similarly, columns 1 through 3 are displayed by typing

A(:,1:3)

For more information on the use of the colon operator, type **help colon**. The colon operator in MATLAB is very versatile, but we will not need to use all of its features.

Display Formats

MATLAB stores matrices in decimal form and does its arithmetic computations using a decimal-type arithmetic. This decimal form retains about 16 digits, but not all digits must be shown. Between what goes on in the machine and what is shown on the screen are routines that convert or format the numbers into displays. Here we give an overview of the display formats that we will use. (For more information, see the *MATLAB User's Guide* or type **help format**.)

- If the matrix contains *all* integers, then the entire matrix is displayed as integer values; that is, no decimal points appear.
- If any entry in the matrix is not exactly represented as an integer, then the entire matrix is displayed in what is known as **format short**. Such a display shows four places behind the decimal point, and the last place may have been rounded. The exception to this is zero. If an entry is exactly zero, then it is displayed as an integer zero. Enter the matrix

  Q = [5 0 1/3 2/3 7.123456]

  into MATLAB. The display is

```
Q =

    5.0000    0    0.3333    0.6667    7.1235
```

Warning If a value is displayed as 0.0000, then it is not identically zero. You should change to **format long**, discussed below, and display the matrix again.

- To see more than four places, change the display format. One way to proceed is to use the command

  format long

  which shows 15 places. The matrix Q above in format long is

```
Q =
   Columns 1 through 4
   5.00000000000000  0  0.33333333333333  0.66666666666667
   Column 5
   7.12345600000000
```

Other display formats use an exponent of 10. They are **format short e** and **format long e**. The "e-formats" are often used in numerical analysis. Try these formats with matrix Q.

☐ MATLAB can display values in rational form. The command **format rat**, short for rational display, is used. Inspect the output from the following sequence of MATLAB commands.

```
format short
V = [1   1/2   1/6   1/12]
```

displays

```
V =

    1.0000    0.5000    0.1667    0.0833
```

and

```
format rat
V
```

displays

```
V =

    1    1/2    1/6    1/12
```

Finally, type **format short** to return to a decimal display form.

Rational output is displayed in what is called "string" form. Strings are not numeric data and hence cannot be used with arithmetic operators. Thus rational output is for "looks" only.

When MATLAB starts, the format in effect is **format short**. If you change the format, it remains in effect until another format command is executed. Some MATLAB routines change the format within the routine.

8.2 MATRIX OPERATIONS IN MATLAB

The operations of addition, subtraction, and multiplication of matrices in MATLAB follow the same definitions as in Sections 1.2 and 1.3. If A and B are $m \times n$ matrices that have been entered into MATLAB, then their sum in MATLAB is computed using command

A+B

and their difference by command

A−B

(spaces can be used on either side of $+$ or $-$). If A is $m \times n$ and C is $n \times k$, then the product of A and C in MATLAB must be written as

A∗C

In MATLAB, ∗ must be specifically placed between the names of matrices to be multiplied. In MATLAB, writing AC does not perform an implied multiplication. In fact, MATLAB considers AC as a new matrix name and, if it has not been previously defined, an error will result. If the matrices involved are not compatible for the

operation specified, then an error message will be displayed. Compatibility for addition and subtraction means that the matrices are the same size. Matrices are compatible for multiplication if the number of columns in the first matrix equals the number of rows in the second.

EXAMPLE 1

Enter the matrices

$$A = \begin{bmatrix} 1 & 2 \\ 2 & 4 \end{bmatrix}, \quad \mathbf{b} = \begin{bmatrix} -3 \\ 1 \end{bmatrix}, \quad \text{and} \quad C = \begin{bmatrix} 3 & -5 \\ 5 & 2 \end{bmatrix}$$

into MATLAB and compute the following expressions. We display the results from MATLAB.

Solution (a) **A+C** displays

```
ans =

     4    -3
     7     6
```

(b) **A*C** displays

```
ans =

    13    -1
    26    -2
```

(c) **b*A** displays

```
??? Error using ==> *
Inner matrix dimensions must agree.
```

■

Scalar multiplication in MATLAB requires the use of the multiplication symbol $*$. For the matrix A in Example 1, $5A$ designates scalar multiplication in the book, while **5*A** is required in MATLAB.

In MATLAB the transpose operator (or symbol) is the single quotation mark or prime, $'$. Using the matrices in Example 1, in MATLAB

Q = C′ displays

```
Q =

     3     5
    -5     2
```

and

p = b′ displays

```
p =

    -3     1
```

As a convenience, we can enter column matrices into MATLAB using $'$. To enter the matrix

$$\mathbf{x} = \begin{bmatrix} 1 \\ 3 \\ -5 \end{bmatrix},$$

we can use either the command

x = [1;3;−5]

or the command

x = [1 3 −5]′

Suppose that we are given the linear system $A\mathbf{x} = \mathbf{b}$, where coefficient matrix A and right side $\mathbf{b}$ have been entered into MATLAB. The augmented matrix $\left[A \mid \mathbf{b} \right]$ is formed in MATLAB by typing

[A b]

or, if we want to name it **aug**, by typing

aug = [A b]

No bar will be displayed separating the right side from the coefficient matrix. Using matrices A and $\mathbf{b}$ from Example 1, form the augmented matrix in MATLAB for the system $A\mathbf{x} = \mathbf{b}$.

Forming augmented matrices is a special case of building matrices in MATLAB. Essentially we can "paste together" matrices as long as sizes are appropriate. Using the matrices A, $\mathbf{b}$, and C in Example 1, we give some examples:

[A C] displays
```
ans =
     1     2     3    -5
     2     4     5     2
```

[A;C] displays
```
ans =
     1     2
     2     4
     3    -5
     5     2
```

[A b C] displays
```
ans =
     1     2    -3     3    -5
     2     4     1     5     2
```

[C A;A C] displays
```
ans =
     3    -5     1     2
     5     2     2     4
     1     2     3    -5
     2     4     5     2
```

MATLAB has a command to build diagonal matrices by entering only the diagonal entries. The command is **diag**, and

D = diag([1 2 3]) displays
```
D =
     1     0     0
     0     2     0
     0     0     3
```

Command **diag** also works to "extract" a set of diagonal entries. If

$$R = \begin{bmatrix} 5 & 2 & 1 \\ -3 & 7 & 0 \\ 6 & 4 & -8 \end{bmatrix}$$

has been entered into MATLAB, then

diag(R) displays

```
ans =

     5
     7
    -8
```

Note that

diag(diag(R)) displays

```
ans =

     5     0     0
     0     7     0
     0     0    -8
```

For more information on **diag**, use **help**. Commands related to **diag** are **tril** and **triu**.

8.3 MATRIX POWERS AND SOME SPECIAL MATRICES

In MATLAB, to raise a matrix to a power we must use the exponentiation operator ^. If A is square and k is a positive integer, then A^k is denoted in MATLAB by

A^k

which corresponds to a matrix product of A with itself k times. The rules for exponents, given in Section 1.4, apply in MATLAB. In particular,

A^0

displays an identity matrix having the same size as A.

EXAMPLE 1

Enter matrices

$$A = \begin{bmatrix} 1 & -1 \\ 1 & 1 \end{bmatrix} \quad \text{and} \quad B = \begin{bmatrix} 1 & -2 \\ 2 & 1 \end{bmatrix}$$

into MATLAB and compute the following expressions. We display the MATLAB results.

Solution

(a) **A^2** displays

```
ans =

     0    -2
     2     0
```

(b) **(A∗B)^2** displays

```
ans =
    -8     6
    -6    -8
```

(c) **(B−A)^3** displays

```
ans =
     0     1
    -1     0
```

■

The $n \times n$ identity matrix is denoted by I_n throughout this book. MATLAB has a command to generate I_n when it is needed. The command is **eye**, and it behaves as follows:

eye(2)	displays a 2×2 identity matrix.
eye(5)	displays a 5×5 identity matrix.
t = 10;eye(t)	displays a 10×10 identity matrix.
eye(size(A))	displays an identity matrix the same size as A.

Two other MATLAB commands, **zeros** and **ones**, behave in a similar manner. The command **zeros** produces a matrix of all zeros, and the command **ones** generates a matrix of all ones. Rectangular matrices of size $m \times n$ can be generated using

zeros(m,n), ones(m,n)

where m and n have been previously defined with positive integer values in MATLAB. Using this convention we can generate a column with four zeros, using the command

zeros(4,1)

From algebra you are familiar with polynomials in x such as

$$4x^3 - 5x^2 + x - 3 \quad \text{and} \quad x^4 - x - 6.$$

The evaluation of such polynomials at a value of x is readily handled in MATLAB, using the command **polyval**. Define the coefficients of the polynomial as a vector (a row or column matrix) with the coefficient of the largest power first, the coefficient of the next largest power second, and so on down to the constant term. If any power is explicitly missing, its coefficient must be set to zero in the corresponding position in the coefficient vector. In MATLAB, for the polynomials above we have coefficient vectors

v = [4 −5 1 −3] and **w = [1 0 0 −1 −6]**

respectively. The command

polyval(v,2)

evaluates the first polynomial at $x = 2$ and displays the computed value of 11. Similarly, the command

t = −1;polyval(w,t)

evaluates the second polynomial at $x = -1$ and displays the value -4.

Polynomials in a square matrix A have the form

$$5A^3 - A^2 + 4A - 7I.$$

Note that the constant term in a matrix polynomial is an identity matrix of the same size as A. This convention is a natural one if we recall that the constant term in an ordinary polynomial is the coefficient of x^0 and that $A^0 = I$. We often meet matrix polynomials when evaluating a standard polynomial such as $p(x) = x^4 - x - 6$ at an $n \times n$ matrix A. The resulting matrix polynomial is

$$p(A) = A^4 - A - 6I_n.$$

Matrix polynomials can be evaluated in MATLAB using the command **polyvalm**. Define the square matrix A and the coefficient vector

w = [1 0 0 −1 −6]

for $p(x)$ in MATLAB. Then the command

polyvalm(w,A)

produces the value of $p(A)$, which will be a matrix the same size as A.

EXAMPLE 2

Let

$$A = \begin{bmatrix} 1 & -1 & 2 \\ -1 & 0 & 1 \\ 0 & 3 & 1 \end{bmatrix} \quad \text{and} \quad p(x) = 2x^3 - 6x^2 + 2x + 3.$$

To compute $p(A)$ in MATLAB, use the following commands. We show the MATLAB display below the commands.

A = [1 −1 2;−1 0 1;0 3 1];
v = [2 −6 2 3];
Q = polyvalm(v,A)

```
Q =

   -13   -18    10
    -6   -25    10
     6    18   -17
```

■

At times you may want a matrix with integer entries to use in testing some matrix relationship. MATLAB commands can generate such matrices quite easily. Type

C = fix(10∗rand(4))

and you will see displayed a 4×4 matrix C with integer entries. To investigate what this command does, use **help** with the commands **fix** and **rand**.

EXAMPLE 3

In MATLAB, generate several $k \times k$ matrices A for $k = 3, 4, 5$ and display $B = A + A^T$. Look over the matrices displayed and try to determine a property that these matrices share. We show several such matrices below. Your results may not be the same because of the random number generator **rand**.

```
k = 3;
A = fix(10*rand(k));
B = A+A'
```

The display is

```
B =

     4     6    11
     6    18    11
    11    11     0
```

Using the **up arrow** key recall the previous commands one at a time, pressing ENTER after each command. This time the matrix displayed may be

```
B =

     0     5    10
     5     6     6
    10     6    10
```

See Exercise 16(a) at the end of Section 1.4. ■

8.4 ELEMENTARY ROW OPERATIONS IN MATLAB

The solution of linear systems of equations as discussed in Section 1.5 uses elementary row operations to obtain a sequence of linear systems whose augmented matrices are row equivalent. Row equivalent linear systems have the same solutions, hence we choose elementary row operations to produce row equivalent systems that are not difficult to solve. It is shown that linear systems in **reduced row echelon form** are readily solved using the Gauss–Jordan procedure, and systems in **row echelon form** are solved using Gaussian elimination with back substitution. Using either of these procedures requires that we perform row operations that introduce zeros into the augmented matrix of the linear system. We show how to perform such row operations using MATLAB. The arithmetic involved is done by the MATLAB software, and we are able to concentrate on the strategy to produce the reduced row echelon form or row echelon form.

Given a linear system $A\mathbf{x} = \mathbf{b}$, we enter the coefficient matrix A and the right side $\mathbf{b}$ into MATLAB. We form the augmented matrix (see Section 8.2) as

```
C = [A   b]
```

Now we are ready to begin applying row operations to the augmented matrix C. Each row operation replaces an existing row by a new row. Our strategy is to construct the row operation so that the resulting new row moves us closer to the goal of reduced row echelon form or row echelon form. There are many different choices that can be made for the sequence of row operations to transform $\left[A \mid \mathbf{b}\right]$ to one of these forms. Naturally we try to use the fewest number of row operations, but many times it is convenient to avoid introducing fractions (if possible), especially when doing calculations by hand. Since MATLAB will be doing the arithmetic for us, we need not be concerned about fractions, but it is visually pleasing to avoid them anyway.

As described in Section 1.5, there are three row operations. They are

- Interchange two rows.
- Multiply a row by a nonzero number.
- Add a multiple of one row to another row.

To perform these operations on an augmented matrix $C = \left[A \mid \mathbf{b}\right]$ in MATLAB, we employ the colon operator, which was discussed in Section 8.1. We illustrate the technique on the linear system in Example 7 of Section 1.5. When the augmented matrix is entered into MATLAB, we have

```
C =

     1     2     3     9
     2    -1     1     8
     3     0    -1     3
```

To produce the reduced row echelon form, we proceed as follows.

Description	*MATLAB Commands and Display*
add (−2) times row 1 to row 2	**C(2,:) = −2 * C(1,:) + C(2,:)**

[Explanation of MATLAB command: Row 2 is replaced by (or set equal to) the sum of −2 times row 1 and row 2.]	`C =`
	`1  2  3  9`
	`0  -5  -5  -10`
	`3  0  -1  3`

add (−3) times row 1 to row 3	**C(3,:) = −3 * C(1,:) + C(3,:)**

	`C =`
	`1  2  3  9`
	`0  -5  -5  -10`
	`0  -6  -10  -24`

multiply row 2 by (−1/5)	**C(2,:) = (− 1/5) * C(2,:)**

[Explanation of MATLAB command: Row 2 is replaced by (or set equal to) $\left(-\frac{1}{5}\right)$ times row 2.]	`C =`
	`1  2  3  9`
	`0  1  1  2`
	`0  -6  -10  -24`

add (−2) times row 2 to row 1 **C(1,:) = −2 * C(2,:) + C(1,:)**

```
C =
     1     0     1     5
     0     1     1     2
     0    -6   -10   -24
```

add 6 times row 2 to row 3 **C(3,:) = 6 * C(2,:) + C(3,:)**

```
C =
     1     0     1     5
     0     1     1     2
     0     0    -4   -12
```

multiply row 3 by (−1/4) **C(3,:) = (− 1/4) * C(3,:)**

```
C =
     1     0     1     5
     0     1     1     2
     0     0     1     3
```

add (−1) times row 3 to row 2 **C(2,:) = −1 * C(3,:) + C(2,:)**

```
C =
     1     0     1     5
     0     1     0    -1
     0     0     1     3
```

add (−1) times row 3 to row 1 **C(1,:) = −1 * C(3,:) + C(1,:)**

```
C =
     1     0     0     2
     0     1     0    -1
     0     0     1     3
```

This last augmented matrix implies that the solution of the linear system is $x = 2$, $y = -1$, $z = 3$.

In the preceding reduction of the augmented matrix to reduced row echelon form, no row interchanges were required. Suppose at some stage we had to interchange rows 2 and 3 of the augmented matrix C. To accomplish this we use a temporary storage area. (We choose to name this area **temp** here.) In MATLAB we proceed as follows.

Description	***MATLAB Commands***
Assign row 2 to temporary storage.	**temp = C(2,:);**
Assign the contents of row 3 to row 2.	**C(2,:) = C(3,:);**
Assign the contents of row 2 contained in temporary storage to row 3.	**C(3,:) = temp;**

(The semicolons at the end of each command just suppress the display of the contents.)

Using the colon operator and the assignment operator, =, as above, we can instruct MATLAB to perform row operations to obtain the reduced row echelon form or row echelon form of a matrix. MATLAB does the arithmetic and we concentrate on choosing the row operations to perform the reduction. We also must enter the appropriate MATLAB command. If we mistype a multiplier or row number, the error can be corrected, but the correction process requires a number of steps. To permit us to concentrate completely on choosing row operations for the reduction process, there is a routine called **reduce** in the set of auxiliary MATLAB routines available to users of this book. Once you have incorporated these routines into MATLAB, you can type **help reduce** and see the following display.

```
REDUCE  Perform row reduction on matrix A by
        explicitly choosing row operations to use.
        A row operation can be ''undone,'' but
        this feature cannot be used in succession.
        This routine is for small matrices, real or
        complex.

        Use the form ===> reduce <=== to select a demo
        or enter your own matrix A

        or in the form ===> reduce(A) <===
```

Routine **reduce** alleviates all the command typing and instructs MATLAB to perform the associated arithmetic. To use **reduce**, enter the augmented matrix C of your system as discussed previously and type

reduce(C)

We display the first three steps of **reduce** for Example 7 in Section 1.5. The matrices involved will be the same as those in the first three steps of the reduction process above, where we made direct use of the colon operator to perform the row operations in MATLAB. Screen displays are shown between rows of plus signs below, and all input appears in boxes.

```
++++++++++++++++++++++++++++++++++++++++++

     ***** "REDUCE" a Matrix by Row Reduction *****

The current matrix is:

A =
     1    2    3    9
     2   -1    1    8
     3    0   -1    3
```

```
                 OPTIONS
 <1>  Row(i) <===> Row(j)
 <2>  k * Row(i)    (k not zero)
 <3>  k * Row(i) + Row(j) ===> Row(j)
 <4>  Turn on rational display.
 <5>  Turn off rational display.
<-1>  "Undo" previous row operation.
 <0>  Quit reduce!
      ENTER your choice ===> [3]

Enter multiplier. [-2]

Enter first row number. [1]

Enter number of row that changes. [2]
```

Comment: Option 3 in the menu above means the same as

add a multiple of one row to another row

The input above performs the operation in the form

multiplier * (first row) + (second row)

```
++++++++++++++++++++++++++++++++++++++++++++

  ***** Replacement by Linear Combination Complete *****

The current matrix is:

A =
    1    2    3    9
    0   -5   -5  -10
    3    0   -1    3

                 OPTIONS
 <1>  Row(i) <===> Row(j)
 <2>  k * Row(i)    (k not zero)
 <3>  k * Row(i) + Row(j) ===> Row(j)
 <4>  Turn on rational display.
 <5>  Turn off rational display.
<-1>  "Undo" previous row operation.
 <0>  Quit reduce!
      ENTER your choice ===> [3]

Enter multiplier. [-3]
```

```
Enter first row number.  1

Enter number of row that changes.  3

++++++++++++++++++++++++++++++++++++++++

  ***** Replacement by Linear Combination Complete *****

The current matrix is:

A =
    1    2    3    9
    0   -5   -5  -10
    0   -6  -10  -24

                OPTIONS
 <1>  Row(i) <===> Row(j)
 <2>  k*Row(i)    (k not zero)
 <3>  k*Row(i)+Row(j) ===> Row(j)
 <4>  Turn on rational display.
 <5>  Turn off rational display.
<-1>  "Undo" previous row operation.
 <0>  Quit reduce!
      ENTER your choice ===>  2

Enter multiplier.  -1/5

Enter row number.  2
++++++++++++++++++++++++++++++++++++++++

         ***** Multiplication Complete *****

The current matrix is:

A =
    1    2    3    9
    0    1    1    2
    0   -6  -10  -24

                OPTIONS
 <1>  Row(i) <===> Row(j)
 <2>  k*Row(i)    (k not zero)
 <3>  k*Row(i)+Row(j) ===> Row(j)
 <4>  Turn on rational display.
 <5>  Turn off rational display.
<-1>  "Undo" previous row operation.
 <0>  Quit reduce!
      ENTER your choice ===>
++++++++++++++++++++++++++++++++++++++++
```

At this point you should complete the reduction of this matrix to reduced row echelon form using **reduce**.

1. Although options 1–3 in **reduce** appear in symbols, they have the same meaning as the phrases used to describe the row operations near the beginning of this section. Option <3> forms a *linear combination* of rows to replace a row. This terminology will be used later in this course and appears in certain displays of **reduce**. (See Sections 8.7 and 1.5.)
2. Within routine **reduce**, the matrix on which the row operations are performed is called A, regardless of the name of your input matrix.

EXAMPLE 1

Solve the following linear system using **reduce**.

$$\tfrac{1}{3}x + \tfrac{1}{4}y = \tfrac{13}{6}$$
$$\tfrac{1}{7}x + \tfrac{1}{9}y = \tfrac{59}{63}$$

Solution Enter the augmented matrix into MATLAB and name it C.

C = [1/3 1/4 13/6;1/7 1/9 59/63]

```
C =
      0.3333   0.2500   2.1667
      0.1429   0.1111   0.9365
```

Then type

reduce(C)

The steps from **reduce** are displayed below. The steps are shown with decimal displays, unless you choose the rational display option <4>. The corresponding rational displays are shown in braces in the following examples for illustrative purposes. Ordinarily, the decimal and rational displays are not shown simultaneously.

```
++++++++++++++++++++++++++++++++++++++++

      ***** "REDUCE" a Matrix by Row Reduction *****

The current matrix is:

A =
    0.3333    0.2500    2.1667               {1/3     1/4     13/6 }
    0.1429    0.1111    0.9365               {1/7     1/9     59/63}
```

```
                 OPTIONS
 <1>   Row(i) <===> Row(j)
 <2>   k*Row(i)     (k not zero)
 <3>   k*Row(i)+Row(j) ===> Row(j)
 <4>   Turn on rational display.
 <5>   Turn off rational display.
<-1>   "Undo" previous row operation.
 <0>   Quit reduce!
       ENTER your choice ===> [2]

Enter multiplier. [1/A(1,1)]

Enter row number. [1]

++++++++++++++++++++++++++++++++++++++++

          ***** Row Multiplication Complete *****

The current matrix is:

A =
    1.0000    0.7500    6.5000             {1      3/4     13/2 }
    0.1429    0.1111    0.9365             {1/7     1/9     59/63}

                 OPTIONS
 <1>   Row(i) <===> Row(j)
 <2>   k*Row(i)     (k not zero)
 <3>   k*Row(i)+Row(j) ===> Row(j)
 <4>   Turn on rational display.
 <5>   Turn off rational display.
<-1>   "Undo" previous row operation.
 <0>   Quit reduce!
       ENTER your choice ===> [3]

Enter multiplier. [-A(2,1)]

Enter first row number. [1]

Enter number of row that changes. [2]

++++++++++++++++++++++++++++++++++++++++

  ***** Replacement by Linear Combination Complete *****

The current matrix is:

A =
    1.0000    0.7500    6.5000             {1    3/4        13/2 }
         0    0.0040    0.0079             {0    1/252      1/126}
```

```
                    OPTIONS
 <1>   Row(i) <===> Row(j)
 <2>   k*Row(i)     (k not zero)
 <3>   k*Row(i) + Row(j) ===> Row(j)
 <4>   Turn on rational display.
 <5>   Turn off rational display.
<-1>   "Undo" previous row operation.
 <0>   Quit reduce!
       ENTER your choice ===> 2

Enter multiplier.  1/A(2,2)

Enter row number.  2

+++++++++++++++++++++++++++++++++++++++++++++

          ***** Row Multiplication Complete *****

The current matrix is:

A =
    1.0000    0.7500    6.5000             {1    3/4    13/2}
         0    1.0000    2.0000             {0    1      2   }

                    OPTIONS
 <1>   Row(i) <===> Row(j)
 <2>   k*Row(i)     (k not zero)
 <3>   k*Row(i) + Row(j) ===> Row(j)
 <4>   Turn on rational display.
 <5>   Turn off rational display.
<-1>   "Undo" previous row operation.
 <0>   Quit reduce!
       ENTER your choice ===> 3

Enter multiplier.  -A(1,2)

Enter first row number.  2

Enter number of row that changes.  1

+++++++++++++++++++++++++++++++++++++++++++++

  ***** Replacement by Linear Combination Complete *****

The current matrix is:

A =
    1.0000         0    5.0000             {1    0    5}
         0    1.0000    2.0000             {0    1    2}
```

```
            OPTIONS
 <1>   Row(i) <===> Row(j)
 <2>   k * Row(i)     (k not zero)
 <3>   k * Row(i) + Row(j) ===> Row(j)
 <4>   Turn on rational display.
 <5>   Turn off rational display.
<-1>   "Undo" previous row operation.
 <0>   Quit reduce!
       ENTER your choice ===> [0]

    **** ===> REDUCE is over.  Your final matrix is:

  A =

      1.0000         0    5.0000
           0    1.0000    2.0000

++++++++++++++++++++++++++++++++++++++++
```

It follows that the solution to the system is $x = 5$, $y = 2$. ■

The **reduce** routine forces you to concentrate on the strategy of the row reduction process. Once you have used **reduce** on a number of linear systems, the reduction process becomes a fairly systematic computation. The reduced row echelon form of a matrix is used in many places in linear algebra to provide information related to concepts. As such, the reduced row echelon form of a matrix becomes one step of more involved computational processes. Hence MATLAB provides an automatic way to obtain the **r**educed **r**ow **e**chelon **f**orm. The command is **rref**. Once you have entered the matrix A under consideration, where A could represent an augmented matrix, just type

rref(A)

and MATLAB responds by displaying the reduced row echelon form of A.

EXAMPLE 2

In Section 1.5, Example 14 asks for the solution of the homogeneous system

$$\begin{aligned} x_1 + x_2 + x_3 + x_4 &= 0 \\ x_1 \phantom{{}+ 2x_2 + x_3} + x_4 &= 0 \\ x_1 + 2x_2 + x_3 \phantom{{}+ x_4} &= 0. \end{aligned}$$

Form the augmented matrix C in MATLAB to obtain

```
C =

     1     1     1     1     0
     1     0     0     1     0
     1     2     1     0     0
```

Next type

rref(C)

and MATLAB displays

```
ans =

     1     0     0     1     0
     0     1     0    -1     0
     0     0     1     1     0
```

It follows that unknown x_4 can be chosen arbitrarily—say, $x_4 = r$, where r is any real number. Hence the solution is

$$x_1 = -r, \qquad x_2 = r, \qquad x_3 = -r, \qquad x_4 = r. \qquad \blacksquare$$

8.5 MATRIX INVERSES IN MATLAB

As discussed in Section 1.6, for a square matrix A to be nonsingular, the reduced row echelon form of A must be the identity matrix. Hence in MATLAB we can determine if A is singular or nonsingular by computing the reduced row echelon form of A using either **reduce** or **rref**. If the result is the identity matrix, then A is nonsingular. Such a computation determines whether an inverse exists, but does not explicitly compute the inverse when it does exist. To compute the inverse of A we can proceed as in Section 1.6 and find the reduced row echelon form of $\left[A \;\vdots\; I_n \right]$. If the resulting matrix is $\left[I_n \;\vdots\; Q \right]$, then $Q = A^{-1}$. In MATLAB, once a nonsingular matrix A has been entered, the inverse can be found step by step by using

reduce([A eye(size(A))])

or computed immediately by using

rref([A eye(size(A))])

For example, if we use the matrix A in Example 4 of Section 1.6, then

$$A = \begin{bmatrix} 1 & 1 & 1 \\ 0 & 2 & 3 \\ 5 & 5 & 1 \end{bmatrix}.$$

Entering matrix A into MATLAB and typing the command

rref([A eye(size(A))])

displays

```
ans =

    1.0000         0         0    1.6250   -0.5000   -0.1250
         0    1.0000         0   -1.8750    0.5000    0.3750
         0         0    1.0000    1.2500         0   -0.2500
```

To extract the inverse matrix we use

Ainv = ans(:,4:6)

and obtain

```
Ainv =

     1.6250   -0.5000   -0.1250
    -1.8750    0.5000    0.3750
     1.2500         0   -0.2500
```

To see the result in rational display, use

format rat

Ainv

which gives

```
Ainv =

      13/8    -1/2    -1/8
     -15/8     1/2     3/8
       5/4       0    -1/4
```

Type command

format short

to turn off the rational display. Thus our previous MATLAB commands can be used in a manner identical to the way the hand computations are described in Section 1.6.

For convenience, there is a routine that computes inverses directly. The command is **invert**. For the preceding matrix A we would type

invert(A)

and the result would be identical to that obtained in **Ainv** by using **rref**. If the matrix is not square or is singular, an error message will appear.

8.6 VECTORS IN MATLAB

An n-vector $\mathbf{x}$ (see Section 2.2) in MATLAB can be represented either as a column matrix with n elements,

$$\mathbf{x} = \begin{bmatrix} x_1 \\ x_2 \\ \vdots \\ x_n \end{bmatrix},$$

or as a row matrix with n elements,

$$\mathbf{x} = \begin{bmatrix} x_1 & x_2 & \cdots & x_n \end{bmatrix}.$$

In a particular problem or exercise, choose one way of representing the n-vectors and stay with that form.

The vector operations of Section 2.2 correspond to operations on $n \times 1$ matrices or columns. If the n-vector is represented by row matrices in MATLAB, then the vector operations correspond to operations on $1 \times n$ matrices. These are just special cases of addition, subtraction, and scalar multiplication of matrices, which were discussed in Section 8.2.

The norm or length of vector $\mathbf{x}$ in MATLAB is obtained by using the command

norm(x)

This command computes the square root of the sum of the squares of the components of $\mathbf{x}$, which is equal to $\|\mathbf{x}\|$, as discussed in Section 3.1.

The distance between vectors $\mathbf{x}$ and $\mathbf{y}$ in R^n in MATLAB is given by

norm(x − y)

EXAMPLE 1

Let

$$\mathbf{u} = \begin{bmatrix} 2 \\ 1 \\ 1 \\ -1 \end{bmatrix} \quad \text{and} \quad \mathbf{v} = \begin{bmatrix} 3 \\ 1 \\ 2 \\ 0 \end{bmatrix}.$$

Enter these vectors in R^4 into MATLAB as columns. Then

norm(u)

displays

```
ans =

    2.6458
```

while

norm(v)

gives

```
ans =

    3.7417
```

and

norm(u − v)

gives

```
ans =
    1.7321
```

■

The dot product of a pair of vectors $\mathbf{u}$ and $\mathbf{v}$ in R^n in MATLAB is computed by the command

dot(u,v)

For the vectors in Example 1, MATLAB gives the dot product as

```
ans =
     9
```

As discussed in Section 3.1, the notion of a dot product is useful to define the angle between n-vectors. Equation (8) in Section 3.1 tells us that the cosine of the angle θ between **u** and **v** is given by

$$\cos\theta = \frac{\mathbf{u}\cdot\mathbf{v}}{\|\mathbf{u}\|\,\|\mathbf{v}\|}.$$

In MATLAB the cosine of the angle between **u** and **v** is computed by the command

dot(u,v)/(norm(u) * norm(v))

The angle θ can be computed by taking the arccosine of the value of the previous expression. In MATLAB the arccosine function is denoted by **acos**. The result will be an angle in radians.

EXAMPLE 2

For the vectors **u** and **v** in Example 1, the angle between the vectors is computed as

c = dot(u,v)/(norm(u) * norm(v));
angle = acos(c)

which displays

```
angle =
     0.4296
```

and is approximately 24.61°. ■

8.7 APPLICATIONS OF LINEAR COMBINATIONS IN MATLAB

The notion of a linear combination, as discussed in Sections 1.2 and 2.3, is fundamental to a wide variety of topics in linear algebra. The ideas of span, linear independence, linear dependence, and basis are based on forming linear combinations of vectors. In addition, the elementary row operations discussed in Sections 1.5 and 8.4 are essentially of the form, "Replace an existing row by a linear combination of rows." This is clearly the case when we add a multiple of one row to another row. (See the menu for the routine **reduce** in Section 8.4.) From this point of view, it follows that the reduced row echelon form and the row echelon form are processes for implementing linear combinations of rows of a matrix. Hence the MATLAB routines **reduce** and **rref** should be useful in solving problems that involve linear combinations.

Here we discuss how to use MATLAB to solve problems dealing with linear combinations, span, linear independence, linear dependence, and basis. The basic strategy is to set up a linear system related to the problem and ask questions such as "Is there a solution?" or "Is the only solution the trivial solution?"

The Linear Combination Problem

Given a vector space V and a set of vectors $S = \{\mathbf{v}_1, \mathbf{v}_2, \ldots, \mathbf{v}_k\}$ in V, determine if $\mathbf{v}$, belonging to V, can be expressed as a linear combination of the members of S. That is, can we find some set of scalars $a_1, a_2, \ldots, a_k$ so that

$$a_1\mathbf{v}_1 + a_2\mathbf{v}_2 + \cdots + a_k\mathbf{v}_k = \mathbf{v}?$$

There are several common situations.

CASE 1 If the vectors in S are column matrices, then we construct (as shown in Example 6 of Section 2.3) a linear system whose coefficient matrix A is

$$A = \begin{bmatrix} \mathbf{v}_1 & \mathbf{v}_2 & \cdots & \mathbf{v}_k \end{bmatrix}$$

and whose right side is $\mathbf{v}$. Let $\mathbf{c} = \begin{bmatrix} c_1 & c_2 & \cdots & c_k \end{bmatrix}^T$ and $\mathbf{b} = \mathbf{v}$; then transform the linear system $A\mathbf{c} = \mathbf{b}$ using **reduce** or **rref** in MATLAB. If the system is shown to be consistent, so that no rows of the form $\begin{bmatrix} 0 & 0 & \cdots & 0 & \vdots & q \end{bmatrix}$, $q \neq 0$, occur, then the vector $\mathbf{v}$ can be written as a linear combination of the vectors in S. In that case the solution to the system gives the values of the coefficients.

***Caution*:** Many times we need only determine if the system is consistent to decide whether $\mathbf{v}$ is a linear combination of the members of S. Read the question carefully.

EXAMPLE 1

To apply MATLAB to Example 6 of Section 2.3, proceed as follows. Define

A = [1 2 1;1 0 2; 1 1 0]′
b = [2 1 5]′

Then use the command

rref([A b])

to give

```
ans =

     1     0     0     1
     0     1     0     2
     0     0     1    -1
```

Recall that this display represents the reduced row echelon form of an augmented matrix. It follows that the system is consistent, with solution

$$a_1 = 1, \qquad a_2 = 2, \qquad a_3 = -1.$$

Hence $\mathbf{v}$ is a linear combination of $\mathbf{v}_1$, $\mathbf{v}_2$, and $\mathbf{v}_3$. ■

CASE 2 If the vectors in S are row matrices, then we construct the coefficient matrix

$$A = \begin{bmatrix} \mathbf{v}_1^T & \mathbf{v}_2^T & \cdots & \mathbf{v}_k^T \end{bmatrix}$$

and set $\mathbf{b} = \mathbf{v}^T$. Proceed as described in Case 1.

CASE 3 If the vectors in S are polynomials, then associate a column of coefficients with each polynomial. Make sure any missing terms in the polynomial are associated with a zero coefficient. One way to proceed is to use the coefficient of the highest-power term as the first entry of the column, the coefficient of the next-highest-power term as the second entry, and so on. For example,

$$t^2 + 2t + 1 \longrightarrow \begin{bmatrix} 1 \\ 2 \\ 1 \end{bmatrix}, \qquad t^2 + 2 \longrightarrow \begin{bmatrix} 1 \\ 0 \\ 2 \end{bmatrix}, \qquad 3t - 2 \longrightarrow \begin{bmatrix} 0 \\ 3 \\ -2 \end{bmatrix}.$$

The linear combination problem is now solved as in Case 1.

CASE 4 If the vectors in S are $m \times n$ matrices, then associate with each such matrix A_j a column $\mathbf{v}_j$ formed by stringing together its columns one after the other. In MATLAB this transformation is done by using the **reshape** command. Then we proceed as in Case 1.

EXAMPLE 2

Given matrix

$$P = \begin{bmatrix} 1 & 2 & 3 \\ 4 & 5 & 6 \end{bmatrix}.$$

To associate a column matrix as described above within MATLAB, first enter P into MATLAB, then type the command

v = reshape(P,6,1)

which gives

```
v =

     1
     4
     2
     5
     3
     6
```

For more information, type **help reshape**. ■

The Span Problem

There are two common types of problems related to span. The first is:

> Given the set of vectors $S = \{\mathbf{v}_1, \mathbf{v}_2, \dots, \mathbf{v}_k\}$ and the vector $\mathbf{v}$ in a vector space V, is $\mathbf{v}$ in span S?

This is identical to the linear combination problem addressed above because we want to know if $\mathbf{v}$ is a linear combination of the members of S. As shown above, we can use MATLAB in many cases to solve this problem.

The second type of problem related to span is:

> Given the set of vectors $S = \{\mathbf{v}_1, \mathbf{v}_2, \ldots, \mathbf{v}_k\}$ in a vector space V, does span $S = V$?

Here we are asked if every vector in V can be written as a linear combination of the vectors in S. In this case the linear system constructed has a right side that contains arbitrary values that correspond to an arbitrary vector in V. (See Example 1 in Section 2.4.) Since MATLAB manipulates only numerical values in routines such as **reduce** and **rref**, we cannot use MATLAB here to (fully) answer this question.

For the second type of spanning question, there is a special case that arises frequently and can be handled in MATLAB. In Section 2.5 the concept of the dimension of a vector space is discussed. The dimension of a vector space V is the number of vectors in a basis (see Section 2.5), which is the smallest number of vectors that can span V. If we know that V has dimension k and the set S has k vectors, then we can proceed as follows to see if span $S = V$. Develop a linear system $A\mathbf{c} = \mathbf{b}$ associated with the span question. If the reduced row echelon form of the coefficient matrix A has the form

$$\begin{bmatrix} I_k \\ \mathbf{0} \end{bmatrix},$$

where $\mathbf{0}$ is a submatrix of all zeros, then any vector in V is expressible in terms of the members of S. In fact, S is a basis for V. In MATLAB we can use routine **reduce** or **rref** on matrix A. If A is square, we can also use **det**. Try this strategy on Example 1 in Section 2.4.

Another spanning question involves finding a set that spans the set of solutions of a homogeneous system of equations $A\mathbf{x} = \mathbf{0}$. The strategy in MATLAB is to find the reduced row echelon form of $\left[\, A \mid \mathbf{0} \,\right]$, using the command

rref(A)

(There is no need to include the augmented column since it is all zeros.) Then form the general solution of the system and express it as a linear combination of columns. The columns form a spanning set for the solution set of the system. See Example 3 in Section 2.4.

The Linear Independence/Dependence Problem

The linear independence or dependence of a set of vectors $S = \{\mathbf{v}_1, \mathbf{v}_2, \ldots, \mathbf{v}_k\}$ is a linear combination question. Set S is linearly independent if the *only* time the linear combination $c_1\mathbf{v}_1 + c_2\mathbf{v}_2 + \cdots + c_k\mathbf{v}_k$ gives the zero vector is when $c_1 = c_2 = \cdots = c_k = 0$. If we can produce the zero vector with any one of the coefficients $c_j \neq 0$, then S is linearly dependent. Following the discussion on linear combination problems, we produce the associated linear system

$$A\mathbf{c} = \mathbf{0}.$$

Note that the linear system is homogeneous. We have the following result:

S is linearly independent if and only if $A\mathbf{c} = \mathbf{0}$
has only the trivial solution.

Otherwise, S is linearly dependent. See Examples 5 and 6 in Section 2.4. Once we have the homogeneous system $A\mathbf{c} = \mathbf{0}$, we can use MATLAB routine **reduce** or **rref** to analyze whether or not the system has a nontrivial solution.

A special case arises if we have k vectors in a set S in a vector space V whose dimension is k (see Section 2.5). Let the linear system associated with the linear combination problem be $A\mathbf{c} = \mathbf{0}$. It can be shown that

S is linearly independent if and only if
the reduced row echelon form of A is $\begin{bmatrix} I_k \\ \mathbf{0} \end{bmatrix}$,

where $\mathbf{0}$ is a submatrix of all zeros. In fact we can extend this further to say S is a basis for V. (See Theorem 2.11.) In MATLAB we can use **reduce** or **rref** on A to aid in the analysis of such a situation.

8.8 LINEAR TRANSFORMATIONS IN MATLAB

We consider the special case of linear transformations $L\colon R^n \to R^m$. Such linear transformations can be represented by an $m \times n$ matrix A. (See Section 4.3.) Then, for $\mathbf{x}$ in R^n, $L(\mathbf{x}) = A\mathbf{x}$, which is in R^m. For example, suppose that $L\colon R^4 \to R^3$ is given by $L(\mathbf{x}) = A\mathbf{x}$, where matrix

$$A = \begin{bmatrix} 1 & -1 & -2 & -2 \\ 2 & -3 & -5 & -6 \\ 1 & -2 & -3 & -4 \end{bmatrix}.$$

The image of

$$\mathbf{x} = \begin{bmatrix} 1 \\ 2 \\ -1 \\ 0 \end{bmatrix}$$

under L is

$$L(\mathbf{x}) = A\mathbf{x} = \begin{bmatrix} 1 & -1 & -2 & -2 \\ 2 & -3 & -5 & -6 \\ 1 & -2 & -3 & -4 \end{bmatrix} \begin{bmatrix} 1 \\ 2 \\ -1 \\ 0 \end{bmatrix} = \begin{bmatrix} 1 \\ 1 \\ 0 \end{bmatrix}.$$

The **range of a linear transformation** L is the subspace of R^m consisting of the set of all images of vectors from R^n. It is easily shown that

$$\text{range } L = \text{column space of } A.$$

(See Example 7 in Section 4.2.) It follows that we "know the range of L" when we have a basis for the column space of A. There are two simple ways to find a basis for the column space of A:

1. The transposes of the nonzero rows of **rref(A′)** form a basis for the column space. (See Example 5 in Section 2.8.)
2. If the columns containing the leading 1's of **rref(A)** are $k_1 < k_2 < \cdots < k_r$, then columns $k_1, k_2, \ldots, k_r$ of A are a basis for the column space of A. (See Example 5 in Section 2.8.)

For the matrix A given above, we have

$$\mathbf{rref(A')} = \begin{bmatrix} 1 & 0 & -1 \\ 0 & 1 & 1 \\ 0 & 0 & 0 \\ 0 & 0 & 0 \end{bmatrix},$$

and hence $\left\{ \begin{bmatrix} 1 \\ 0 \\ -1 \end{bmatrix}, \begin{bmatrix} 0 \\ 1 \\ 1 \end{bmatrix} \right\}$ is a basis for the range of L. Using method 2,

$$\mathbf{rref(A)} = \begin{bmatrix} 1 & 0 & -1 & 0 \\ 0 & 1 & 1 & 2 \\ 0 & 0 & 0 & 0 \end{bmatrix}.$$

Thus it follows that columns 1 and 2 of A are a basis for the column space of A and hence a basis for the range of L. In addition, routine **lisub** can be used. Use **help** for directions.

The **kernel of a linear transformation** is the subspace of all vectors in R^n whose image is the zero vector in R^m. This corresponds to the set of all vectors $\mathbf{x}$ satisfying

$$L(\mathbf{x}) = A\mathbf{x} = \mathbf{0}.$$

Hence it follows that the kernel of L, denoted $\ker L$, is the set of all solutions to the homogeneous system

$$A\mathbf{x} = \mathbf{0},$$

which is the null space of A. Thus we "know the kernel of L" when we have a basis for the null space of A. To find a basis for the null space of A, we form the general solution to $A\mathbf{x} = \mathbf{0}$ and "separate it into a linear combination of columns using the arbitrary constants that are present." The columns employed form a basis for the null space of A. This procedure uses **rref(A)**. For the matrix A, we have

$$\mathbf{rref(A)} = \begin{bmatrix} 1 & 0 & -1 & 0 \\ 0 & 1 & 1 & 2 \\ 0 & 0 & 0 & 0 \end{bmatrix}.$$

If we choose the variables corresponding to columns without leading 1's to be arbitrary, we have $x_3 = r$ and $x_4 = t$. It follows that the general solution to $A\mathbf{x} = \mathbf{0}$ is given by

$$\mathbf{x} = \begin{bmatrix} x_1 \\ x_2 \\ x_3 \\ x_4 \end{bmatrix} = \begin{bmatrix} r \\ -r - 2t \\ r \\ t \end{bmatrix} = r \begin{bmatrix} 1 \\ -1 \\ 1 \\ 0 \end{bmatrix} + t \begin{bmatrix} 0 \\ -2 \\ 0 \\ 1 \end{bmatrix}.$$

Thus columns

$$\begin{bmatrix} 1 \\ -1 \\ 1 \\ 0 \end{bmatrix} \quad \text{and} \quad \begin{bmatrix} 0 \\ -2 \\ 0 \\ 1 \end{bmatrix}$$

form a basis for ker L. See also routine **homsoln**, which will display the general solution of a homogeneous linear system. In addition, the command **null** will produce an orthonormal basis for the null space of a matrix. Use **help** for further information on these commands.

In summary, appropriate use of the **rref** command in MATLAB will give bases for both the kernel and range of the linear transformation $L(\mathbf{x}) = A\mathbf{x}$.

8.9 MATLAB COMMAND SUMMARY

In this section we list the principal MATLAB commands and operators used in this book. The list is divided into two parts: commands that come with the MATLAB software, and special instructional routines available to users of this book. For ease of reference we have included a brief description of each instructional routine that is available to users of this book. These descriptions are also available using MATLAB's **help** command once the installation procedures are complete. A description of any MATLAB command can be obtained by using **help**. (See the introduction to this chapter.)

Built-in MATLAB Commands

ans	**inv**	**roots**
clear	**norm**	**rref**
conj	**null**	**size**
det	**ones**	**sqrt**
diag	**pi**	**sum**
dot	**poly**	**tril**
eig	**polyval**	**triu**
exit	**polyvalm**	**zeros**
eye	**quit**	\
fix	**rand**	;
format	**rank**	:
help	**rat**	′ (prime)
hilb	**real**	+, −, *, /, ^
image	**reshape**	

Supplemental Instructional Commands

adjoint	**forsub**	**planelt**
bksub	**gschmidt**	**reduce**
cofactor	**homsoln**	**vec2demo**
crossprd	**invert**	**vec3demo**
crossdemo	**lsqline**	
dotprod	**lupr**	

Both **rref** and **reduce** are used in many sections. Several utilities required by the instructional commands are **arrowh**, **mat2strh**, and **blkmat**. The description given below is that displayed in response to the **help** command. In the description of several commands, the notation differs slightly from that in the text.

Description of Instructional Commands

ADJOINT

Compute the classical adjoint of a square matrix A. If A is not square an empty matrix is returned.
*** This routine should only be used by students to check adjoint computations and should not be used as part of a routine to compute inverses. See invert or inv.

Use in the form ==> adjoint(A) <==

BKSUB

Perform back substitution on upper triangular system Ax=b. If A is not square, upper triangular, and nonsingular, an error message is displayed. In case of an error the solution returned is all zeros.

Use in the form ==> bksub(A,b) <==

COFACTOR

Computes the (i,j)-cofactor of matrix A. If A is not square, an error message is displayed.
*** This routine should only be used by students to check cofactor computations.

Use in the form ==> cofactor(i,j,A) <==

CROSSDEMO

Display a pair of three-dimensional vectors and their cross product.

The input vectors u and v are displayed in a three-dimensional perspective along with their cross product. For visualization purposes a set of coordinate 3-D axes are shown.

Use in the form ==> crossdemo(u,v) <==

CROSSPRD

Compute the cross product of vectors x and y in 3-space. The output is a vector orthogonal to both of the original vectors x and y. The output is returned as a row matrix with 3 components [v1 v2 v3] which is interpreted as v1*i + v2*j + v3*k where i, j, k are the unit vectors in the x, y, and z directions respectively.

Use in the form ==> v=crossprd(x,y) <==

DOTPROD

The dot product of two n-vectors x and y is computed. The vectors can be either rows, columns, or matrices of the same size. For complex vectors the dot product of x and y is computed as the conjugate transpose of the first times the second.

Use in the form ==> dotprod(x,y) <==

FORSUB

Perform forward substitution on a lower triangular system Ax=b. If A is not square, lower triangular, and nonsingular, an error message is displayed. In case of an error the solution returned is all zeros.

Use in the form ==> forsub(A,b) <==

GSCHMIDT

The Gram-Schmidt process on the columns in matrix x. The orthonormal basis appears in the columns of y unless there is a second argument, in which case y contains only an orthogonal basis. The second argument can have any value.

Use in the form ==> y=gschmidt(x) <==
or ==> y=gschmidt(x,v) <==

HOMSOLN

Find the general solution of a homogeneous system of equations. The routine returns a set of basis vectors for the null space of Ax=0.

Use in the form ==> ns=homsoln(A) <==

If there is a second argument, the general solution is displayed.

Use in the form ==> homsoln(A,1) <==

This option assumes that the general solution has at most 10 arbitrary constants.

INVERT

Compute the inverse of a matrix A by using the reduced row echelon form applied to [A I]. If A is singular, a warning is given.

Use in the form ==> B = invert(A) <==

LSQLINE

This routine will construct the equation of the least squares line to a data set of ordered pairs and then graph the line and the data set. A short menu of options is available, including evaluating the equation of the line at points.

Use in the form ==> c = lsqline(x,y) or lsqline(x,y) <==

Here x is a vector containing the x-coordinates and y is a vector containing the corresponding y-coordinates. On output, c contains the coefficients of the least squares line:

$$y = c(1) * x + c(2)$$

LUPR

Perform LU-factorization on matrix A by explicitly choosing row operations to use. No row interchanges are permitted, hence it is possible that the factorization cannot be found. It is recommended that the multipliers be constructed in terms of the elements of matrix U, like $-U(3,2)/U(2,2)$, since the displays of matrices L and U do not show all the decimal places available. A row operation can be "undone," but this feature cannot be used in succession.

This routine uses the utilities mat2strh and blkmat.

Use in the form ==> [L,U] = lupr(A) <==

PLANELT

Demonstration of plane linear transformations:

Rotations, Reflections, Expansions/Compressions, Shears

Or you may specify your own transformation.

Graphical results of successive plane linear transformations can be seen using a multiple window display. Standard figures can be chosen or you may choose to use your own figure.

Use in the form ==> planelt <==

REDUCE

Perform row reduction on matrix A by explicitly choosing row operations to use. A row operation can be "undone," but this feature cannot be used in succession. This routine is for small matrices, real or complex.

Use in the form ==> reduce <== to select a demo or enter your own matrix A or in the form ==> reduce(A) <==

VEC2DEMO

A graphical demonstration of vector operations for two-dimensional vectors.

Select vectors u=[x1 x2] and v=[y1 y2]. They will be displayed graphically along with their sum, difference, and a scalar multiple.

Use in the form ==> vec2demo(u,v) <==

or ==> vec2demo <==

In the latter case you will be prompted for input.

VEC3DEMO

Display a pair of three-dimensional vectors, their sum, difference and scalar multiples.

The input vectors u and v are displayed in a 3-dimensional perspective along with their sum, difference and selected scalar multiples. For visualization purposes a set of coordinate 3-D axes are shown.

Use in the form ==> vec3demo(u,v) <==

or in the form ==> vec3demo <== to choose a demo or be prompted for input.

Chapter 9

MATLAB Exercises

9.1 INTRODUCTION

In Chapter 8 we gave a brief survey of MATLAB and its functionality for use in an elementary linear algebra course. This chapter consists of exercises that are designed to be solved using MATLAB. However, we do not ask that users of this text write programs. The user is merely asked to use MATLAB to solve specific numerical problems.

The exercises in this chapter complement those given in Chapters 1–7 and exploit the computational capabilities of MATLAB. To extend the instructional capabilities of MATLAB we have developed a set of pedagogical routines, called scripts or M-files, to illustrate concepts, streamline step-by-step computational procedures, and demonstrate geometric aspects of topics, using graphical displays. We feel that MATLAB and our instructional M-files provide an opportunity for a working partnership between the student and the computer that in many ways forecasts situations that will occur once the student joins the technological workforce of the 21st century.

The exercises in this chapter are keyed to topics rather than to individual sections of the text. Short descriptive headings and references to MATLAB commands in Chapter 8 supply information about the sets of exercises.

Basic Matrix Properties

In order to use MATLAB in this section, you should first read Sections 8.1 and 8.2, which give basic information about MATLAB and about matrix operations in MATLAB. You are urged to do any examples or illustrations of MATLAB commands that appear in Sections 8.1 and 8.2 before trying these exercises.

ML.1. In MATLAB, enter the following matrices.

$$A = \begin{bmatrix} 5 & 1 & 2 \\ -3 & 0 & 1 \\ 2 & 4 & 1 \end{bmatrix},$$

$$B = \begin{bmatrix} 4*2 & 2/3 \\ 1/201 & 5-8.2 \\ 0.00001 & (9+4)/3 \end{bmatrix}.$$

Using MATLAB commands, display the following.

(a) a_{23}, b_{32}, b_{12}.

(b) $\text{row}_1(A), \text{col}_3(A), \text{row}_2(B)$.

(c) Type MATLAB command **format long** and display matrix B. Compare the elements of B from part (a) with the current display. Note that **format short** displays four decimal places rounded. Reset the format to **format short**.

ML.2. In MATLAB, type the command **H = hilb(5);** (Note that the last character is a semicolon, which suppresses the display of the contents of matrix H. See Section 8.1.) For more information on the **hilb** command, type **help hilb**. Using MATLAB commands, do the following:

(a) Determine the size of H.

(b) Display the contents of H.

(c) Extract as a matrix the first three columns.

(d) Extract as a matrix the last two rows.

ML.3. Sometimes it is convenient to see the contents of a matrix displayed as rational numbers.

(a) In MATLAB, type the commands:

format rat

H = hilb(5)

Note the fractions that appear as entries.

(b) Warning: format rat is for viewing purposes only. All MATLAB computations use decimal-style expressions. Besides, format rat may only display an approximation. In MATLAB, type the commands:

format rat

pi

format long

pi−355/113

Note that the value shown in format rat is only an approximation to π.

Matrix Operations

ML.1. In MATLAB, type the command **clear**, then enter the following matrices:

$$A = \begin{bmatrix} 1 & 1/2 \\ 1/3 & 1/4 \\ 1/5 & 1/6 \end{bmatrix}, \quad B = \begin{bmatrix} 5 & -2 \end{bmatrix},$$

$$C = \begin{bmatrix} 4 & 5/4 & 9/4 \\ 1 & 2 & 3 \end{bmatrix}.$$

Using MATLAB commands, compute each of the following, if possible. Recall that a prime in MATLAB indicates transpose.

(a) $A*C$

(b) $A*B$

(c) $A+C'$

(d) $B*A-C'*A$

(e) $(2*C-6*A')*B'$

(f) $A*C-C*A$

(g) $A*A'+C'*C$

ML.2. Enter the coefficient matrix of the system

$$\begin{aligned} 2x + 4y + 6z &= -12, \\ 2x - 3y - 4z &= \ \ 15, \\ 3x + 4y + 5z &= \ -8. \end{aligned}$$

into MATLAB and call it A. Enter the right side of the system and call it **b**. Form the augmented matrix associated with this linear system, using the MATLAB command **[A b]**. To give the augmented matrix a name, such as **aug**, use the command **aug = [A b]**. (Do not type the period!) Note that no bar appears between the coefficient matrix and the right side in the MATLAB display.

ML.3. Repeat the preceding exercise with the following linear system:

$$\begin{aligned} 4x - 3y + 2z - \ w &= -5, \\ 2x + \ y - 3z \qquad &= \ \ 7, \\ -x + 4y + \ z + 2w &= \ \ 8. \end{aligned}$$

ML.4. Enter matrices

$$A = \begin{bmatrix} 1 & -1 & 2 \\ 3 & 2 & 4 \\ 4 & -2 & 3 \\ 2 & 1 & 5 \end{bmatrix}$$

and

$$B = \begin{bmatrix} 1 & 0 & -1 & 2 \\ 3 & 3 & -3 & 4 \\ 4 & 2 & 5 & 1 \end{bmatrix}$$

into MATLAB.

(a) Using MATLAB commands, assign $\text{row}_2(A)$ to **R** and $\text{col}_3(B)$ to **C**. Let $\mathbf{V} = \mathbf{R} * \mathbf{C}$. What is **V** in terms of the entries of the product $\mathbf{A} * \mathbf{B}$?

(b) Using MATLAB commands, assign $\text{col}_2(B)$ to **C**; then compute $\mathbf{V} = \mathbf{A} * \mathbf{C}$. What is **V** in terms of the entries of the product $\mathbf{A} * \mathbf{B}$?

(c) Using MATLAB commands, assign $\text{row}_3(A)$ to **R**, then compute $\mathbf{V} = \mathbf{R} * \mathbf{B}$. What is **V** in terms of the entries of the product $\mathbf{A} * \mathbf{B}$?

ML.5. Use the MATLAB command **diag** to form each of the following diagonal matrices. Using **diag** we can form diagonal matrices without typing in all the entries. (To refresh your memory about command **diag**, use MATLAB's help feature.)

(a) The 4×4 diagonal matrix with main diagonal $\begin{bmatrix} 1 & 2 & 3 & 4 \end{bmatrix}$.

(b) The 5×5 diagonal matrix with main diagonal $\begin{bmatrix} 0 & 1 & \frac{1}{2} & \frac{1}{3} & \frac{1}{4} \end{bmatrix}$.

(c) The 5×5 scalar matrix with all 5's on the diagonal.

ML.6. MATLAB has some commands which behave quite differently from the standard definitions of +, −, and *. Enter the following matrices into MATLAB:

$$A = \begin{bmatrix} 4 & -2 & 1 \\ 0 & 5 & 8 \end{bmatrix}, \quad B = \begin{bmatrix} -2 & 9 & 4 \\ 7 & -3 & 8 \end{bmatrix}.$$

Execute each of the following commands and then write a description of the action taken.

(a) **A.*B**

(b) **A./B** and **B./A**

(c) **A.∧2**

Powers of a Matrix

In order to use MATLAB in this section, you should first have read Chapter 8 *through Section* 8.3.

ML.1. Use MATLAB to find the smallest positive integer k in each of the following cases.

(a) $A^k = I_3$ for $A = \begin{bmatrix} 0 & 0 & 1 \\ 1 & 0 & 0 \\ 0 & 1 & 0 \end{bmatrix}$.

(b) $A^k = A$ for $A = \begin{bmatrix} 0 & 1 & 0 & 0 \\ -1 & 0 & 0 & 0 \\ 0 & 0 & 0 & 1 \\ 0 & 0 & 1 & 0 \end{bmatrix}$.

ML.2. Use MATLAB to display the matrix A in each of the following cases. Find the smallest value of k such that A^k is a zero matrix. Here **tril**, **ones**, **triu**, **fix**, and **rand** are MATLAB commands. (To see a description, use **help**.)

(a) $A =$ **tril(ones(5), −1)**

(b) $A =$ **triu(fix(10 ∗ rand(7)), 2)**

ML.3. Let $A = \begin{bmatrix} 1 & -1 & 0 \\ 0 & 1 & -1 \\ -1 & 0 & 1 \end{bmatrix}$. Using command **polyvalm** in MATLAB, compute the following matrix polynomials:

(a) $A^4 - A^3 + A^2 + 2I_3$.

(b) $A^3 - 3A^2 + 3A$.

ML.4. Let $A = \begin{bmatrix} 0.1 & 0.3 & 0.6 \\ 0.2 & 0.2 & 0.6 \\ 0.3 & 0.3 & 0.4 \end{bmatrix}$. Using MATLAB, compute each of the following matrix expressions:

(a) $(A^2 - 7A)(A + 3I_3)$.

(b) $(A - I_3)^2 + (A^3 + A)$.

(c) Look at the sequence $A, A^2, A^3, \ldots, A^8, \ldots$. Does it appear to be converging to a matrix? If so, to what matrix?

ML.5. Let $A = \begin{bmatrix} 1 & \frac{1}{2} \\ 0 & \frac{1}{3} \end{bmatrix}$. Use MATLAB to compute members of the sequence $A, A^2, A^3, \ldots, A^k, \ldots$. Write a description of the behavior of this matrix sequence.

ML.6. Let $A = \begin{bmatrix} \frac{1}{2} & \frac{1}{3} \\ 0 & -\frac{1}{5} \end{bmatrix}$. Repeat Exercise ML.5.

ML.7. Let $A = \begin{bmatrix} 1 & -2 & 1 \\ -1 & 1 & 2 \\ 0 & 2 & 1 \end{bmatrix}$. Use MATLAB to do the following:

(a) Compute A^TA and AA^T. Are they equal?

(b) Compute $B = A + A^T$ and $C = A - A^T$. Show that B is symmetric and C is skew symmetric.

(c) Determine a relationship between $B + C$ and A.

Row Operations and Echelon Forms

In order to use MATLAB in this section, you should first have read Chapter 8 *through Section* 8.4.

ML.1. Let

$$A = \begin{bmatrix} 4 & 2 & 2 \\ -3 & 1 & 4 \\ 1 & 0 & 3 \\ 5 & -1 & 5 \end{bmatrix}.$$

Find the matrices obtained by performing the following row operations in succession on matrix A. Do the row operations directly, using the colon operator.

(a) Multiply row 1 by $\frac{1}{4}$.

(b) Add 3 times row 1 to row 2.

(c) Add (-1) times row 1 to row 3.

(d) Add (-5) times row 1 to row 4.

(e) Interchange rows 2 and 4.

ML.2. Let

$$A = \begin{bmatrix} \frac{1}{2} & \frac{1}{3} & \frac{1}{4} & \frac{1}{5} \\ \frac{1}{3} & \frac{1}{4} & \frac{1}{5} & \frac{1}{6} \\ 1 & \frac{1}{2} & \frac{1}{3} & \frac{1}{4} \end{bmatrix}.$$

Find the matrices obtained by performing the following row operations in succession on matrix A. Do the row operations directly, using the colon operator.

(a) Multiply row 1 by 2.

(b) Add $\left(-\frac{1}{3}\right)$ times row 1 to row 2.

(c) Add (-1) times row 1 to row 3.

(d) Interchange rows 2 and 3.

ML.3. Use **reduce** to find the reduced row echelon form of matrix A in Exercise ML.1.

ML.4. Use **reduce** to find the reduced row echelon form of matrix A in Exercise ML.2.

ML.5. Use **reduce** to find all solutions to the linear system in Exercise 6(a) in Section 1.5.

ML.6. Use **reduce** to find all solutions to the linear system in Exercise 7(b) in Section 1.5.

ML.7. Use **reduce** to find all solutions to the linear system in Exercise 8(b) in Section 1.5.

ML.8. Use **reduce** to find all solutions to the linear system in Exercise 9(a) in Section 1.5.

ML.9. Let

$$A = \begin{bmatrix} 1 & 2 \\ 2 & 4 \end{bmatrix}.$$

Use **reduce** to find a nontrivial solution to the homogeneous system

$$(5I_2 - A)\mathbf{x} = \mathbf{0}.$$

[*Hint*: In MATLAB, enter matrix A; then use the command **reduce(5 ∗ eye(size(A)) − A)**]

ML.10. Let

$$A = \begin{bmatrix} 1 & 5 \\ 5 & 1 \end{bmatrix}.$$

Use **reduce** to find a nontrivial solution to the homogeneous system

$$(-4I_2 - A)\mathbf{x} = \mathbf{0}.$$

[*Hint*: In MATLAB, enter matrix A; then use the command **reduce(− 4 ∗ eye(size(A)) − A)**]

ML.11. Use **rref** in MATLAB to solve the linear systems in Exercise 18 in Section 1.6.

ML.12. MATLAB has an immediate command for solving square linear systems $A\mathbf{x} = \mathbf{b}$. Once the coefficient matrix A and right side $\mathbf{b}$ are entered into MATLAB, command

$$\mathbf{x = A\backslash b}$$

displays the solution as long as A is considered nonsingular. The backslash command, \, does not use reduced row echelon form, but does initiate numerical methods that are discussed briefly in Section 1.8. For more details on the command see D. R. Hill, *Experiments in Computational Matrix Algebra*, New York: Random House, 1988.

(a) Use \ to solve Exercise 8(a) in Section 1.5

(b) Use \ to solve Exercise 6(b) in Section 1.5.

ML.13. The \ command behaves differently than **rref**. Use both \ and **rref** to solve $A\mathbf{x} = \mathbf{b}$, where

$$A = \begin{bmatrix} 1 & 2 & 3 \\ 4 & 5 & 6 \\ 7 & 8 & 9 \end{bmatrix}, \quad \mathbf{b} = \begin{bmatrix} 1 \\ 0 \\ 0 \end{bmatrix}.$$

LU-Factorization

Routine **lupr** *provides a step-by-step procedure in MATLAB for obtaining the LU-factorization discussed in Section* 1.8. *Once we have the LU-factorization, routines* **forsub** *and* **bksub** *can be used to perform the forward and back substitution, respectively. Use* **help** *for further information on these routines.*

ML.1. Use **lupr** in MATLAB to find an LU-factorization of

$$A = \begin{bmatrix} 2 & 8 & 0 \\ 2 & 2 & -3 \\ 1 & 2 & 7 \end{bmatrix}.$$

ML.2. Use **lupr** in MATLAB to find an LU-factorization of

$$A = \begin{bmatrix} 8 & -1 & 2 \\ 3 & 7 & 2 \\ 1 & 1 & 5 \end{bmatrix}.$$

ML.3. Solve the linear system in Example 2 in Section 1.8 using **lupr**, **forsub**, and **bksub** in MATLAB. Check your LU-factorization, using Example 3 in Section 1.8.

ML.4. Solve Exercises 7 and 8 in Section 1.8, using **lupr**, **forsub**, and **bksub** in MATLAB.

Matrix Inverses

In order to use MATLAB in this section, you should first have read Chapter 8 *through Section* 8.5.

ML.1. Using MATLAB, determine which of the following matrices are nonsingular. Use command **rref**.

(a) $A = \begin{bmatrix} 1 & 2 \\ -2 & 1 \end{bmatrix}$.

(b) $A = \begin{bmatrix} 1 & 2 & 3 \\ 4 & 5 & 6 \\ 7 & 8 & 9 \end{bmatrix}$.

(c) $A = \begin{bmatrix} 1 & 2 & 3 \\ 4 & 5 & 6 \\ 7 & 8 & 0 \end{bmatrix}$.

ML.2. Using MATLAB, determine which of the following matrices are nonsingular. Use command **rref**.

(a) $A = \begin{bmatrix} 1 & 2 \\ 2 & 4 \end{bmatrix}$. (b) $A = \begin{bmatrix} 1 & 0 & 0 \\ 0 & 1 & 0 \\ 1 & 1 & 1 \end{bmatrix}$.

(c) $A = \begin{bmatrix} 1 & 2 & 1 \\ 0 & 1 & 2 \\ 1 & 0 & 0 \end{bmatrix}$.

ML.3. Using MATLAB, determine the inverse of each of the following matrices. Use command **rref([A eye(size(A))])**.

(a) $A = \begin{bmatrix} 1 & 3 \\ 1 & 2 \end{bmatrix}$. (b) $A = \begin{bmatrix} 1 & 1 & 2 \\ 2 & 1 & 1 \\ 1 & 2 & 1 \end{bmatrix}$.

ML.4. Using MATLAB, determine the inverse of each of the following matrices. Use command **rref([A eye(size(A))])**.

(a) $A = \begin{bmatrix} 2 & 1 \\ 2 & 3 \end{bmatrix}$. (b) $A = \begin{bmatrix} 1 & -1 & 2 \\ 0 & 2 & 1 \\ 1 & 0 & 0 \end{bmatrix}$.

ML.5. Using MATLAB, determine a positive integer t so that $(tI - A)$ is singular.

(a) $A = \begin{bmatrix} 1 & 3 \\ 3 & 1 \end{bmatrix}$. (b) $A = \begin{bmatrix} 4 & 1 & 2 \\ 1 & 4 & 1 \\ 0 & 0 & -4 \end{bmatrix}$.

Vectors (Geometrically)

Exercises ML.1–ML.3 *use the routine* **vec2demo**, *which provides a graphical display of vectors in the plane. For a pair of vectors* $\mathbf{u} = (x_1, y_1)$ *and* $\mathbf{v} = (x_2, y_2)$, *routine* **vec2demo** *graphs* **u** *and* **v**, $\mathbf{u}+\mathbf{v}$, $\mathbf{u}-\mathbf{v}$, *and a scalar multiple. Once the vectors* **u** *and* **v** *are entered into MATLAB, type*

vec2demo(u, v)

For further information, use **help vec2demo**.

ML.1. Use the routine **vec2demo** with each of the following pairs of vectors. (Square brackets are used in MATLAB.)

(a) $\mathbf{u} = [2 \quad 0]$, $\mathbf{v} = [0 \quad 3]$

(b) $\mathbf{u} = [-3 \quad 1]$, $\mathbf{v} = [2 \quad 2]$

(c) $\mathbf{u} = [5 \quad 2]$, $\mathbf{v} = [-3 \quad 3]$

ML.2. Use the routine **vec2demo** with each of the following pairs of vectors. (Square brackets are used in MATLAB.)

(a) $\mathbf{u} = [2 \quad -2]$, $\mathbf{v} = [1 \quad 3]$

(b) $\mathbf{u} = [0 \quad 3]$, $\mathbf{v} = [-2 \quad 0]$

(c) $\mathbf{u} = [4 \quad -1]$, $\mathbf{v} = [-3 \quad 5]$

ML.3. Choose pairs of vectors **u** and **v** to use with **vec2demo**.

ML.4. As an aid for visualizing vector operations in R^3, we have **vec3demo**. This routine provides a graphical display of vectors in 3-space. For a pair of vectors **u** and **v**, routine **vec3demo** graphs **u** and **v**, **u** + **v**, **u** − **v**, and a scalar multiple. Once the pair of vectors from R^3 are entered into MATLAB, type

vec3demo(u, v)

Use **vec3demo** on each of the following pairs from R^3. (Square brackets are used in MATLAB.)

(a) $\mathbf{u} = [2, 6, 4]^T, \mathbf{v} = [6, 2, -5]'$

(b) $\mathbf{u} = [3, -5, 4]^T, \mathbf{v} = [7, -1, -2]'$

(c) $\mathbf{u} = [4, 0, -5]^T, \mathbf{v} = [0, 6, 3]'$

Vector Spaces

The concepts discussed in this section are not easily implemented in MATLAB routines. The items in Definition 2.4 *must hold for* all *vectors. Just because we demonstrate in MATLAB that a property of Definition* 2.4 *holds for a few vectors it does not imply that it holds for all such vectors. You must guard against such faulty reasoning. However, if, for a particular choice of vectors, we show that a property fails in MATLAB, then we have established that the property does not hold in all possible cases. Hence the property is considered to be false. In this way we might be able to show that a set is not a vector space.*

ML.1. Let V be the set of all 2×2 matrices with operations given by the following MATLAB commands:

$$A \oplus B \quad \text{is} \quad A.*B$$
$$k \odot A \quad \text{is} \quad k + A$$

Is V a vector space? (*Hint*: Enter some 2×2 matrices and experiment with the MATLAB commands to understand their behavior before checking the conditions in Definition 2.4.)

ML.2. Following Example 5 in Section 2.2, we discuss the vector space P_n of polynomials of degree n or less. Operations on polynomials of degree n can be performed in linear algebra software by associating a row matrix of size $n+1$ with polynomial $p(t)$ of P_n. The row matrix consists of the coefficients of $p(t)$, using the association

$$p(t) = a_n t^n + a_{n-1} t^{n-1} + \cdots + a_1 t + a_0 \rightarrow \begin{bmatrix} a_n & a_{n-1} & \cdots & a_1 & a_0 \end{bmatrix}.$$

If any term of $p(t)$ is explicitly missing, a zero is used for its coefficient. Then the addition of polynomials corresponds to matrix addition, and multiplication of a polynomial by a scalar corresponds to scalar multiplication of matrices. Use MATLAB to perform the following operations on polynomials, using the matrix association described above. Let $n = 3$ and

$$p(t) = 2t^3 + 5t^2 + t - 2,$$
$$q(t) = t^3 + 3t + 5.$$

(a) $p(t) + q(t)$. (b) $5p(t)$.

(c) $3p(t) - 4q(t)$.

Subspaces

ML.1. Let V be R_3 and let W be the subset of V of vectors of the form $\begin{bmatrix} 2 & a & b \end{bmatrix}$, where a and b are any real numbers. Is W a subspace of V? Use the following MATLAB commands to help you determine the answer.

```
a1 = fix(10 * randn);
a2 = fix(10 * randn);
b1 = fix(10 * randn);
b2 = fix(10 * randn);
v = [2 a1 b1]
w = [2 a2 b2]
v + w
3 * v
```

ML.2. Let V be P_2 and let W be the subset of V of vectors of the form $ax^2 + bx + 5$, where a and b are arbitrary real numbers. With each such polynomial in W we associate a vector $\begin{bmatrix} a & b & 5 \end{bmatrix}$ in R^3. Construct commands like those in Exercise ML.1 to show that W is not a subspace of V.

Before solving the following MATLAB exercises, you should have read Section 8.7.

ML.3. Use MATLAB to determine if vector **v** is a linear combination of the members of set S.

(a) $S = \{\mathbf{v}_1, \mathbf{v}_2, \mathbf{v}_3\}$
$= \{[1\ 0\ 0\ 1], [0\ 1\ 1\ 0], [1\ 1\ 1\ 1]\}$
$\mathbf{v} = [0\ 1\ 1\ 1]$

(b) $S = \{\mathbf{v}_1, \mathbf{v}_2, \mathbf{v}_3\}$

$$= \left\{ \begin{bmatrix} 1 \\ 2 \\ -1 \end{bmatrix}, \begin{bmatrix} 2 \\ -1 \\ 0 \end{bmatrix}, \begin{bmatrix} -1 \\ 8 \\ -3 \end{bmatrix} \right\}$$

$$\mathbf{v} = \begin{bmatrix} 0 \\ 5 \\ -2 \end{bmatrix}$$

ML.4. Use MATLAB to determine if $\mathbf{v}$ is a linear combination of the members of set S. If it is, express $\mathbf{v}$ in terms of the members of S.

(a) $S = \{\mathbf{v}_1, \mathbf{v}_2, \mathbf{v}_3\}$
$= \{[1\ 2\ 1], [3\ 0\ 1], [1\ 8\ 3]\}$
$\mathbf{v} = [-2\ 14\ 4]$

(b) $S = \{A_1, A_2, A_3\}$

$$= \left\{ \begin{bmatrix} 1 & 2 \\ 1 & 0 \end{bmatrix}, \begin{bmatrix} 2 & -1 \\ 1 & 2 \end{bmatrix}, \begin{bmatrix} -3 & 1 \\ 0 & 1 \end{bmatrix} \right\}$$

$\mathbf{v} = I_2$.

ML.5. Use MATLAB to determine if $\mathbf{v}$ is a linear combination of the members of set S. If it is, express $\mathbf{v}$ in terms of the members of S.

(a) $S = \{\mathbf{v}_1, \mathbf{v}_2, \mathbf{v}_3, \mathbf{v}_4\}$

$$= \left\{ \begin{bmatrix} 1 \\ 2 \\ 1 \\ 0 \\ 1 \end{bmatrix}, \begin{bmatrix} 0 \\ 1 \\ 2 \\ -1 \\ 1 \end{bmatrix}, \begin{bmatrix} 2 \\ 1 \\ 0 \\ 0 \\ -1 \end{bmatrix}, \begin{bmatrix} -2 \\ 1 \\ 1 \\ 1 \\ 1 \end{bmatrix} \right\}$$

$$\mathbf{v} = \begin{bmatrix} 0 \\ -1 \\ 1 \\ -2 \\ 1 \end{bmatrix}$$

(b) $S = \{p_1(t), p_2(t), p_3(t)\}$
$= \{2t^2 - t + 1, t^2 - 2, t - 1\}$
$\mathbf{v} = p(t) = 4t^2 + t - 5$.

ML.6. In each part, determine whether $\mathbf{v}$ belongs to span S, where

$$S = \{\mathbf{v}_1, \mathbf{v}_2, \mathbf{v}_3\}$$
$$= \{[1\ \ 1\ \ 0\ \ 1], [1\ \ -1\ \ 0\ \ 1], [0\ \ 1\ \ 2\ \ 1]\}.$$

(a) $\mathbf{v} = [2\ \ 3\ \ 2\ \ 3]$

(b) $\mathbf{v} = [2\ \ -3\ \ -2\ \ 3]$

(c) $\mathbf{v} = [0\ \ 1\ \ 2\ \ 3]$

ML.7. In each part, determine whether $p(t)$ belongs to span S, where

$$S = \{p_1(t), p_2(t), p_3(t)\}$$
$$= \{t - 1, t + 1, t^2 + t + 1\}.$$

(a) $p(t) = t^2 + 2t + 4$.

(b) $p(t) = 2t^2 + t - 2$.

(c) $p(t) = -2t^2 + 1$.

Linear Independence/Dependence

ML.1. Determine if S is linearly independent or linearly dependent.

(a) $S = \{[1\ \ 0\ \ 0\ \ 1], [0\ \ 1\ \ 1\ \ 0], [1\ \ 1\ \ 1\ \ 1]\}$.

(b) $S = \left\{ \begin{bmatrix} 1 & 2 \\ 1 & 0 \end{bmatrix}, \begin{bmatrix} 2 & -1 \\ 1 & 2 \end{bmatrix}, \begin{bmatrix} -3 & 1 \\ 0 & 1 \end{bmatrix} \right\}$.

(c) $S = \left\{ \begin{bmatrix} 1 \\ 2 \\ 1 \\ 0 \\ 1 \end{bmatrix}, \begin{bmatrix} 0 \\ 1 \\ 2 \\ -1 \\ 1 \end{bmatrix}, \begin{bmatrix} 2 \\ 1 \\ 0 \\ 0 \\ -1 \end{bmatrix}, \begin{bmatrix} -2 \\ 1 \\ 1 \\ 1 \\ 1 \end{bmatrix} \right\}$.

(d) $S = \{2t^2 - t + 3, t^2 + 2t - 1, 4t^2 - 7t + 11\}$.

ML.2. Find a spanning set of the solution space of $A\mathbf{x} = \mathbf{0}$, where

$$A = \begin{bmatrix} 1 & 2 & 0 & 1 \\ 1 & 1 & 1 & 2 \\ 2 & -1 & 5 & 7 \\ 0 & 2 & -2 & -2 \end{bmatrix}.$$

ML.3. Let

$$\mathbf{v}_1 = \begin{bmatrix} 2 \\ -1 \\ 3 \end{bmatrix}, \quad \mathbf{v}_2 = \begin{bmatrix} 1 \\ 2 \\ -1 \end{bmatrix}, \quad \mathbf{v}_3 = \begin{bmatrix} 4 \\ -7 \\ 1 \end{bmatrix}.$$

Determine if $\mathbf{v}$ is in span $\{\mathbf{v}_1, \mathbf{v}_2, \mathbf{v}_3\}$ for each of the following. If it is, find the coefficients that express $\mathbf{v}$ as a linear combination of $\mathbf{v}_1$, $\mathbf{v}_2$, and $\mathbf{v}_3$.

(a) $\mathbf{v} = \begin{bmatrix} 1 \\ 1 \\ 1 \end{bmatrix}$ (b) $\mathbf{v} = \begin{bmatrix} 3 \\ 1 \\ 2 \end{bmatrix}$

(c) $\mathbf{v} = \begin{bmatrix} 4 \\ -2 \\ 6 \end{bmatrix}$

Bases and Dimension

In order to use MATLAB in this section, you should have read Section 8.7. In the following exercises we relate the theory developed in that section to computational procedures in MATLAB that aid in analyzing the situation.

*To determine if a set $S = \{\mathbf{v}_1, \mathbf{v}_2, \ldots, \mathbf{v}_k\}$ is a basis for a vector space V, the definition requires that we show span $S = V$ and S is linearly independent. However, if we know that dim $V = k$, then Theorem 2.11 implies that we need only show that either span $S = V$ or S is linearly independent. The linear independence, in this special case, is easily analyzed using MATLAB's **rref** command. Construct the homogeneous system $A\mathbf{c} = \mathbf{0}$ associated with the linear independence/dependence question. Then S is linearly independent if and only if*

$$\mathbf{rref(A)} = \begin{bmatrix} \mathbf{I}_k \\ \mathbf{0} \end{bmatrix}.$$

In Exercises ML.1 through ML.6, if this special case can be applied, do so; otherwise, determine if S is a basis for V in the conventional manner.

ML.1. $S = \{\begin{bmatrix}1 & 2 & 1\end{bmatrix}, \begin{bmatrix}2 & 1 & 1\end{bmatrix}, \begin{bmatrix}2 & 2 & 1\end{bmatrix}\}$ in $V = R_3$.

ML.2. $S = \{2t-2, t^2-3t+1, 2t^2-8t+4\}$ in $V = P_2$.

ML.3. $S = \{\begin{bmatrix}1 & 1 & 0 & 0\end{bmatrix}, \begin{bmatrix}2 & 1 & 1 & -1\end{bmatrix}, \begin{bmatrix}0 & 0 & 1 & 1\end{bmatrix}, \begin{bmatrix}1 & 2 & 1 & 2\end{bmatrix}\}$ in $V = R_4$.

ML.4. $S = \{\begin{bmatrix}1 & 2 & 1 & 0\end{bmatrix}, \begin{bmatrix}2 & 1 & 3 & 1\end{bmatrix}, \begin{bmatrix}2 & -2 & 4 & 2\end{bmatrix}\}$ in $V =$ span S.

ML.5. $S = \{\begin{bmatrix}1 & 2 & 1 & 0\end{bmatrix}, \begin{bmatrix}2 & 1 & 3 & 1\end{bmatrix}, \begin{bmatrix}2 & 2 & 1 & 2\end{bmatrix}\}$ in $V =$ span S.

ML.6. $V =$ the subspace of R_3 of all vectors of the form $\begin{bmatrix}a & b & c\end{bmatrix}$, where $b = 2a - c$ and $S = \{\begin{bmatrix}0 & 1 & -1\end{bmatrix}, \begin{bmatrix}1 & 1 & 1\end{bmatrix}\}$.

*In Exercises ML.7 through ML.9, use MATLAB's **rref** command to determine a subset of S that is a basis for span S. See Example 4 in Section 2.5.*

ML.7. $S = \{\begin{bmatrix}1 & 1 & 0 & 0\end{bmatrix}, \begin{bmatrix}-2 & -2 & 0 & 0\end{bmatrix}, \begin{bmatrix}1 & 0 & 2 & 1\end{bmatrix}, \begin{bmatrix}2 & 1 & 2 & 1\end{bmatrix}, \begin{bmatrix}0 & 1 & 1 & 1\end{bmatrix}\}$.
What is dim span S? Does span $S = R_4$?

ML.8. $S = \left\{\begin{bmatrix}1 & 2 \\ 1 & 2\end{bmatrix}, \begin{bmatrix}1 & 0 \\ 1 & 1\end{bmatrix}, \begin{bmatrix}0 & 2 \\ 0 & 1\end{bmatrix}, \begin{bmatrix}2 & 4 \\ 2 & 4\end{bmatrix}, \begin{bmatrix}1 & 0 \\ 0 & 1\end{bmatrix}\right\}$.
What is dim span S? Does span $S = M_{22}$?

ML.9. $S = \{t-2, 2t-1, 4t-2, t^2-t+1, t^2+2t+1\}$.
What is dim span S? Does span $S = P_2$?

*An interpretation of Theorem 2.10 in Section 2.5 is that any linearly independent subset S of vector space V can be extended to a basis for V. Following the ideas in Example 9 in Section 2.5, use MATLAB's **rref** command to extend S to a basis for V in Exercises ML.10 through ML.12.*

ML.10. $S = \{\begin{bmatrix}1 & 1 & 0 & 0\end{bmatrix}, \begin{bmatrix}1 & 0 & 1 & 0\end{bmatrix}\}$, $V = R_4$.

ML.11. $S = \{t^3 - t + 1, t^3 + 2\}$, $V = P_3$.

ML.12. $S = \{\begin{bmatrix}0 & 3 & 0 & 2 & -1\end{bmatrix}\}$,
$V =$ the subspace of R_5 consisting of all vectors of the form $\begin{bmatrix}a & b & c & d & e\end{bmatrix}$, where $c = a$, $b = 2d + e$.

Coordinates and Change of Basis

*Finding the coordinates of a vector with respect to a basis is a linear combination problem. Hence, once the corresponding linear system is constructed, we can use MATLAB routine **reduce** or **rref** to find its solution. The solution gives us the desired coordinates. (The discussion in Section 8.7 will be helpful as an aid for constructing the necessary linear system.)*

ML.1. Let $V = R^3$ and

$$S = \left\{\begin{bmatrix}1 \\ 2 \\ 1\end{bmatrix}, \begin{bmatrix}2 \\ 1 \\ 0\end{bmatrix}, \begin{bmatrix}1 \\ 0 \\ 2\end{bmatrix}\right\}.$$

Show that S is a basis for V and find $\begin{bmatrix}\mathbf{v}\end{bmatrix}_S$ for each of the following vectors.

(a) $\mathbf{v} = \begin{bmatrix}8 \\ 4 \\ 7\end{bmatrix}$ (b) $\mathbf{v} = \begin{bmatrix}2 \\ 0 \\ -3\end{bmatrix}$

(c) $\mathbf{v} = \begin{bmatrix}4 \\ 3 \\ 3\end{bmatrix}$

ML.2. Let $V = R_4$ and $S = \{\begin{bmatrix}1 & 0 & 1 & 1\end{bmatrix}, \begin{bmatrix}1 & 2 & 1 & 3\end{bmatrix}, \begin{bmatrix}0 & 2 & 1 & 1\end{bmatrix}, \begin{bmatrix}0 & 1 & 0 & 0\end{bmatrix}\}$.
Show that S is a basis for V and find $\begin{bmatrix}\mathbf{v}\end{bmatrix}_S$ for each of the following vectors.

(a) $\mathbf{v} = \begin{bmatrix}4 & 12 & 8 & 14\end{bmatrix}$

(b) $\mathbf{v} = \begin{bmatrix}\frac{1}{2} & 0 & 0 & 0\end{bmatrix}$

(c) $\mathbf{v} = \begin{bmatrix}1 & 1 & 1 & \frac{7}{3}\end{bmatrix}$

ML.3. Let V be the vector space of all 2×2 matrices and

$$S = \left\{ \begin{bmatrix} 1 & 2 \\ 1 & 2 \end{bmatrix}, \begin{bmatrix} 0 & 2 \\ 1 & 0 \end{bmatrix}, \begin{bmatrix} 3 & 1 \\ -1 & 0 \end{bmatrix}, \begin{bmatrix} -1 & 0 \\ 0 & 0 \end{bmatrix} \right\}.$$

Show that S is a basis for V and find $\left[\mathbf{v}\right]_S$ for each of the following vectors.

(a) $\mathbf{v} = \begin{bmatrix} 1 & 0 \\ 0 & 1 \end{bmatrix}$ (b) $\mathbf{v} = \begin{bmatrix} 2 & \frac{10}{3} \\ \frac{7}{6} & 2 \end{bmatrix}$

(c) $\mathbf{v} = \begin{bmatrix} 1 & 1 \\ 1 & 1 \end{bmatrix}$

Finding the transition matrix $P_{S \leftarrow T}$ from the T-basis to the S-basis is also a linear combination problem. $P_{S \leftarrow T}$ is the matrix whose columns are the coordinates of the vectors in T with respect to the S-basis. Following the ideas developed in Example 3 *of Section* 2.7, *we can find matrix $P_{S \leftarrow T}$ using routine* **reduce** *or* **rref**. *The idea is to construct a matrix A whose columns correspond to the vectors in S (see Section* 8.7) *and a matrix B whose columns correspond to the vectors in T. Then MATLAB command* **rref([A B])** *gives* $[\mathbf{I} \;\; \mathbf{P_{S \leftarrow T}}]$.

In Exercises ML.4 *through* ML.6, *use the MATLAB techniques described above to find the transition matrix $P_{S \leftarrow T}$ from the T-basis to the S-basis.*

ML.4. $V = R^3$,

$$S = \left\{ \begin{bmatrix} 1 \\ 1 \\ 0 \end{bmatrix}, \begin{bmatrix} 0 \\ 1 \\ 1 \end{bmatrix}, \begin{bmatrix} 1 \\ 0 \\ 1 \end{bmatrix} \right\},$$

$$T = \left\{ \begin{bmatrix} 2 \\ 1 \\ 1 \end{bmatrix}, \begin{bmatrix} 1 \\ 2 \\ 1 \end{bmatrix}, \begin{bmatrix} 1 \\ 1 \\ 2 \end{bmatrix} \right\}.$$

ML.5. $V = P_3$, $S = \{t - 1, t + 1, t^2 + t, t^3 - t\}$, $T = \{t^2, 1 - t, 2 - t^2, t^3 + t^2\}$.

ML.6. $V = R_4$, $S = \{\begin{bmatrix} 1 & 2 & 3 & 0 \end{bmatrix}, \begin{bmatrix} 0 & 1 & 2 & 3 \end{bmatrix}, \begin{bmatrix} 3 & 0 & 1 & 2 \end{bmatrix}, \begin{bmatrix} 2 & 3 & 0 & 1 \end{bmatrix}\}$, T = natural basis.

ML.7. Let $V = R^3$ and suppose that we have bases

$$S = \left\{ \begin{bmatrix} 1 \\ 1 \\ 1 \end{bmatrix}, \begin{bmatrix} 1 \\ 2 \\ 1 \end{bmatrix}, \begin{bmatrix} 0 \\ 1 \\ 1 \end{bmatrix} \right\},$$

$$T = \left\{ \begin{bmatrix} 1 \\ 0 \\ 1 \end{bmatrix}, \begin{bmatrix} 1 \\ 1 \\ 0 \end{bmatrix}, \begin{bmatrix} 0 \\ 1 \\ 2 \end{bmatrix} \right\},$$

and

$$U = \left\{ \begin{bmatrix} 2 \\ 1 \\ 1 \end{bmatrix}, \begin{bmatrix} -1 \\ 2 \\ 1 \end{bmatrix}, \begin{bmatrix} 1 \\ -2 \\ 1 \end{bmatrix} \right\}.$$

(a) Find the transition matrix P from U to T.

(b) Find the transition matrix Q from T to S.

(c) Find the transition matrix Z from U to S.

(d) Does $Z = PQ$ or QP?

Homogeneous Linear Systems

In Exercises ML.1 *through* ML.3, *use MATLAB's* **rref** *command to aid in finding a basis for the null space of A. You may also use routine* **homsoln**. *For directions, use* **help**.

ML.1. $A = \begin{bmatrix} 1 & 1 & 2 & 2 & 1 \\ 2 & 0 & 4 & 2 & 4 \\ 1 & 1 & 2 & 2 & 1 \end{bmatrix}$.

ML.2. $A = \begin{bmatrix} 2 & 2 & 2 \\ 1 & 2 & 1 \\ 3 & 1 & 0 \\ 0 & 0 & 1 \\ 1 & 0 & 0 \end{bmatrix}$.

ML.3. $A = \begin{bmatrix} 1 & 4 & 7 & 0 \\ 2 & 5 & 8 & -1 \\ 3 & 6 & 9 & -2 \end{bmatrix}$.

ML.4. For the matrix

$$A = \begin{bmatrix} 1 & 2 \\ 2 & 1 \end{bmatrix}$$

and $\lambda = 3$, the homogeneous system $(\lambda I_2 - A)\mathbf{x} = \mathbf{0}$ has a nontrivial solution. Find such a solution using MATLAB commands.

ML.5. For the matrix

$$A = \begin{bmatrix} 1 & 2 & 3 \\ 3 & 2 & 1 \\ 2 & 1 & 3 \end{bmatrix}$$

and $\lambda = 6$, the homogeneous linear system $(\lambda I_3 - A)\mathbf{x} = \mathbf{0}$ has a nontrivial solution. Find such a solution using MATLAB commands.

Rank of a Matrix

Given a matrix A, the nonzero rows of **rref(A)** *form a basis for the row space of A and the nonzero rows of* **rref(A′)** *transformed to columns give a basis for the column space of A.*

ML.1. Solve Exercises 1 through 4 in Section 2.8 using MATLAB.

To find a basis for the row space of A that consists of rows of A, we compute **rref(A′)**. *The leading 1's point to the original rows of A that give us a basis for the row space. See Example 4 in Section 2.8.*

ML.2. Determine two bases for the row space of A that have no vectors in common.

(a) $A = \begin{bmatrix} 1 & 3 & 1 \\ 2 & 5 & 0 \\ 4 & 11 & 2 \\ 6 & 9 & 1 \end{bmatrix}$.

(b) $A = \begin{bmatrix} 2 & 1 & 2 & 0 \\ 0 & 0 & 0 & 0 \\ 1 & 2 & 2 & 1 \\ 4 & 5 & 6 & 2 \\ 3 & 3 & 4 & 1 \end{bmatrix}$.

ML.3. Repeat Exercise ML.2 for column spaces.

To compute the rank of a matrix A in MATLAB, use the command **rank(A)**.

ML.4. Compute the rank and nullity of each of the following matrices.

(a) $\begin{bmatrix} 3 & 2 & 1 \\ 1 & 2 & -1 \\ 2 & 1 & 3 \end{bmatrix}$.

(b) $\begin{bmatrix} 1 & 2 & 1 & 2 & 1 \\ 2 & 1 & 0 & 0 & 2 \\ 1 & -1 & -1 & -2 & 1 \\ 3 & 0 & -1 & -2 & 3 \end{bmatrix}$.

ML.5. Using only the rank command, determine which of the following linear systems is consistent.

(a) $\begin{bmatrix} 1 & 2 & 4 & -1 \\ 0 & 1 & 2 & 0 \\ 3 & 1 & 1 & -2 \end{bmatrix} \begin{bmatrix} x_1 \\ x_2 \\ x_3 \\ x_4 \end{bmatrix} = \begin{bmatrix} 21 \\ 8 \\ 16 \end{bmatrix}$.

(b) $\begin{bmatrix} 1 & 2 & 1 \\ 1 & 1 & 0 \\ 2 & 1 & -1 \end{bmatrix} \begin{bmatrix} x_1 \\ x_2 \\ x_3 \end{bmatrix} = \begin{bmatrix} 3 \\ 3 \\ 3 \end{bmatrix}$.

(c) $\begin{bmatrix} 1 & 2 \\ 2 & 0 \\ 2 & 1 \\ -1 & 2 \end{bmatrix} \begin{bmatrix} x_1 \\ x_2 \end{bmatrix} = \begin{bmatrix} 3 \\ 2 \\ 3 \\ 2 \end{bmatrix}$.

Standard Inner Product

In order to use MATLAB in this section, you should first have read Section 8.6.

ML.1. In MATLAB the dot product of a pair of vectors can be computed using the **dot** command. If the vectors **v** and **w** have been entered into MATLAB as either rows or columns, their dot product is computed from the MATLAB command **dot(v, w)**. If the vectors do not have the same number of elements, an error message is displayed.

(a) Use **dot** to compute the dot product of each of the following pairs of vectors.

(i) $\mathbf{v} = \begin{bmatrix} 1 & 4 & -1 \end{bmatrix}$, $\mathbf{w} = \begin{bmatrix} 7 & 2 & 0 \end{bmatrix}$

(ii) $\mathbf{v} = \begin{bmatrix} 2 \\ -1 \\ 0 \\ 6 \end{bmatrix}$, $\mathbf{w} = \begin{bmatrix} 4 \\ 2 \\ 3 \\ -1 \end{bmatrix}$

(b) Let $\mathbf{a} = \begin{bmatrix} 3 & -2 & 1 \end{bmatrix}$. Find a value for k so that the dot product of $\mathbf{a}$ with $\mathbf{b} = \begin{bmatrix} k & 1 & 4 \end{bmatrix}$ is zero. Verify your results in MATLAB.

(c) For each of the following vectors **v**, compute **dot(v,v)** in MATLAB.

(i) $\mathbf{v} = \begin{bmatrix} 4 & 2 & -3 \end{bmatrix}$

(ii) $\mathbf{v} = \begin{bmatrix} -9 & 3 & 1 & 0 & 6 \end{bmatrix}$

(iii) $\mathbf{v} = \begin{bmatrix} 1 \\ 2 \\ -5 \\ -3 \end{bmatrix}$

What sign is each of these dot products? Explain why this is true for almost all vectors **v**. When is it not true?

ML.2. Determine the norm, or length, of each of the following vectors using MATLAB.

(a) $\mathbf{u} = \begin{bmatrix} 2 \\ 2 \\ -1 \end{bmatrix}$ (b) $\mathbf{v} = \begin{bmatrix} 0 \\ 4 \\ -3 \\ 0 \end{bmatrix}$

(c) $\mathbf{w} = \begin{bmatrix} 1 \\ 0 \\ 1 \\ 0 \\ 3 \end{bmatrix}$

ML.3. Determine the distance between each of the following pairs of vectors using MATLAB.

(a) $\mathbf{u} = \begin{bmatrix} 2 \\ 0 \\ 3 \end{bmatrix}, \mathbf{v} = \begin{bmatrix} 2 \\ -1 \\ 1 \end{bmatrix}$

(b) $\mathbf{u} = \begin{bmatrix} 2 & 0 & 0 & 1 \end{bmatrix}, \mathbf{v} = \begin{bmatrix} 2 & 5 & -1 & 3 \end{bmatrix}$

(c) $\mathbf{u} = \begin{bmatrix} 1 & 0 & 4 & 3 \end{bmatrix}, \mathbf{v} = \begin{bmatrix} -1 & 1 & 2 & 2 \end{bmatrix}$

ML.4. Determine the lengths of the sides of the triangle ABC, which has vertices in R^3, given by $\mathbf{A}(1, 3, -2)$, $\mathbf{B}(4, -1, 0)$, $\mathbf{C}(1, 1, 2)$. (*Hint*: Determine a vector for each side and compute its length.)

ML.5. Determine the dot product of each one of the following pairs of vectors using MATLAB.

(a) $\mathbf{u} = \begin{bmatrix} 5 & 4 & -4 \end{bmatrix}, \mathbf{v} = \begin{bmatrix} 3 & 2 & 1 \end{bmatrix}$

(b) $\mathbf{u} = \begin{bmatrix} 3 & -1 & 0 & 2 \end{bmatrix}$
$\mathbf{v} = \begin{bmatrix} -1 & 2 & -5 & -3 \end{bmatrix}$

(c) $\mathbf{u} = \begin{bmatrix} 1 & 2 & 3 & 4 & 5 \end{bmatrix}, \mathbf{v} = -\mathbf{u}$.

ML.6. The norm, or length, of a vector can be computed using dot products as follows:

$$\|\mathbf{u}\| = \sqrt{\mathbf{u} \cdot \mathbf{u}}.$$

In MATLAB, the right side of the preceding expression is computed as

sqrt(dot(u, u))

Verify this alternative procedure on the vectors in Exercise ML.2.

ML.7. In MATLAB, if the n-vectors **u** and **v** are entered as columns, then

$$\mathbf{u}' * \mathbf{v} \quad \text{or} \quad \mathbf{v}' * \mathbf{u}$$

gives the dot product of vectors **u** and **v**. Verify this using the vectors in Exercise ML.5.

ML.8. Use MATLAB to find the angle between each of the following pairs of vectors. (To convert the angle from radians to degrees, multiply by 180/pi.)

(a) $\mathbf{u} = \begin{bmatrix} 3 & 2 & 4 & 0 \end{bmatrix}, \mathbf{v} = \begin{bmatrix} 0 & 2 & -1 & 0 \end{bmatrix}$

(b) $\mathbf{u} = \begin{bmatrix} 2 & 2 & -1 \end{bmatrix}, \mathbf{v} = \begin{bmatrix} 2 & 0 & 1 \end{bmatrix}$

(c) $\mathbf{u} = \begin{bmatrix} 1 & 0 & 0 & 2 \end{bmatrix}, \mathbf{v} = \begin{bmatrix} 0 & 3 & -4 & 0 \end{bmatrix}$

ML.9. Use MATLAB to find a unit vector in the direction of the vectors in Exercise ML.2.

Cross Product

There are two MATLAB routines that apply to the material in this section. They are **cross**, *which computes the cross product of a pair of* 3*-vectors; and* **crossdemo**, *which displays graphically a pair of vectors and their cross product. Using routine* **dot** *with* **cross**, *we can carry out the computations in Example* 6. (*For directions on the use of MATLAB routines, use* **help** *followed by a space and the name of the routine.*)

ML.1. Use **cross** in MATLAB to find the cross product of each of the following pairs of vectors.

(a) $\mathbf{u} = \mathbf{i} - 2\mathbf{j} + 3\mathbf{k}, \mathbf{v} = \mathbf{i} + 3\mathbf{j} + \mathbf{k}$.

(b) $\mathbf{u} = \begin{bmatrix} 1 & 0 & 3 \end{bmatrix}, \mathbf{v} = \begin{bmatrix} 1 & -1 & 2 \end{bmatrix}$

(c) $\mathbf{u} = \begin{bmatrix} 1 & 2 & -3 \end{bmatrix}, \mathbf{v} = \begin{bmatrix} 2 & -1 & 2 \end{bmatrix}$

ML.2. Use routine **cross** to find the cross product of each of the following pairs of vectors.

(a) $\mathbf{u} = \begin{bmatrix} 2 & 3 & -1 \end{bmatrix}, \mathbf{v} = \begin{bmatrix} 2 & 3 & 1 \end{bmatrix}$

(b) $\mathbf{u} = 3\mathbf{i} - \mathbf{j} + \mathbf{k}, \mathbf{v} = 2\mathbf{u}$.

(c) $\mathbf{u} = \begin{bmatrix} 1 & -2 & 1 \end{bmatrix}, \mathbf{v} = \begin{bmatrix} 3 & 1 & -1 \end{bmatrix}$

ML.3. Use **crossdemo** in MATLAB to display the vectors **u**, **v**, and their cross product.

(a) $\mathbf{u} = \mathbf{i} + 2\mathbf{j} + 4\mathbf{k}, \mathbf{v} = -2\mathbf{i} + 4\mathbf{j} + 3\mathbf{k}$.

(b) $\mathbf{u} = \begin{bmatrix} -2 & 4 & 5 \end{bmatrix}, \mathbf{v} = \begin{bmatrix} 0 & 1 & -3 \end{bmatrix}$

(c) $\mathbf{u} = \begin{bmatrix} 2 & 2 & 2 \end{bmatrix}, \mathbf{v} = \begin{bmatrix} 3 & -3 & 3 \end{bmatrix}$

ML.4. Use MATLAB to find the volume of the parallelepiped with vertex at the origin and edges $\mathbf{u} = \begin{bmatrix} 3 & -2 & 1 \end{bmatrix}, \mathbf{v} = \begin{bmatrix} 1 & 2 & 3 \end{bmatrix}$, and $\mathbf{w} = \begin{bmatrix} 2 & -1 & 2 \end{bmatrix}$

ML.5. The angle of intersection of two planes in 3-space is the same as the angle of intersection of perpendiculars to the planes. Find the angle of intersection of plane Π_1 determined by **x** and **y** and plane Π_2 determined by **v**, **w**, where

$$\mathbf{x} = \begin{bmatrix} 2 & -1 & 2 \end{bmatrix}, \quad \mathbf{y} = \begin{bmatrix} 3 & -2 & 1 \end{bmatrix}$$

$$\mathbf{v} = \begin{bmatrix} 1 & 3 & 1 \end{bmatrix}, \quad \mathbf{w} = \begin{bmatrix} 0 & 2 & -1 \end{bmatrix}$$

The Gram–Schmidt Process

The Gram–Schmidt process takes a basis S for a subspace W of V and produces an orthonormal basis T for W. The algorithm to produce the orthonormal basis T is implemented in MATLAB in routine **gschmidt**. *Type* **help gschmidt** *for directions.*

ML.1. Use **gschmidt** to produce an orthonormal basis for R^3 from the basis

$$S = \left\{ \begin{bmatrix} 1 \\ 1 \\ 0 \end{bmatrix}, \begin{bmatrix} 1 \\ 0 \\ 0 \end{bmatrix}, \begin{bmatrix} 0 \\ 1 \\ 1 \end{bmatrix} \right\}.$$

Your answer will be in decimal form; rewrite it in terms of $\sqrt{2}$.

ML.2. Use **gschmidt** to produce an orthonormal basis for R_4 from the basis $S = \{\begin{bmatrix} 1 & 0 & 1 & 1 \end{bmatrix}, \begin{bmatrix} 1 & 2 & 1 & 3 \end{bmatrix}, \begin{bmatrix} 0 & 2 & 1 & 1 \end{bmatrix}, \begin{bmatrix} 0 & 1 & 0 & 0 \end{bmatrix}\}$.

ML.3. In R_3, $S = \{\begin{bmatrix} 0 & -1 & 1 \end{bmatrix}, \begin{bmatrix} 0 & 1 & 1 \end{bmatrix}, \begin{bmatrix} 1 & 1 & 1 \end{bmatrix}\}$ is a basis. Use S to find an orthonormal basis T and then find $\left[\mathbf{v}\right]_T$ for each of the following vectors.

(a) $\mathbf{v} = \begin{bmatrix} 1 & 2 & 0 \end{bmatrix}$ (b) $\mathbf{v} = \begin{bmatrix} 1 & 1 & 1 \end{bmatrix}$

(c) $\mathbf{v} = \begin{bmatrix} -1 & 0 & 1 \end{bmatrix}$

ML.4. Find an orthonormal basis for the subspace of R_4 consisting of all vectors of the form

$$\begin{bmatrix} a & 0 & a+b & b+c \end{bmatrix},$$

where a, b, and c are any real numbers.

ML.5. Let $\mathbf{v} = \begin{bmatrix} 2 \\ 1 \end{bmatrix}$.

(a) Find a nonzero vector $\mathbf{w}$ orthogonal to $\mathbf{v}$.

(b) Compute

$$\mathbf{u}_1 = \mathbf{v}_1/\,\text{norm}(\mathbf{v})$$

$$\mathbf{u}_2 = \mathbf{w}/\,\text{norm}(\mathbf{w})$$

and form matrix

$$\mathbf{U} = \begin{bmatrix} \mathbf{u}_1 & \mathbf{u}_2 \end{bmatrix}.$$

Now compute $\mathbf{U}' * \mathbf{U}$. How are $\mathbf{U}$ and $\mathbf{U}'$ related?

(c) Choose $\mathbf{x}$ to be any nonzero vector in R^2. Compare the length of $\mathbf{x}$ and $\mathbf{U} * \mathbf{x}$.

(d) Choose a pair of nonzero vectors $\mathbf{x}$ and $\mathbf{y}$ in R^2. Compare **dot** $(\mathbf{x}, \mathbf{y})$ and **dot** $(\mathbf{U} * \mathbf{x}, \mathbf{U} * \mathbf{y})$.

Projections

ML.1. Find the projection of **v** onto **w**. (Recall that we have routines **dot** and **norm** available in MATLAB.)

(a) $\mathbf{v} = \begin{bmatrix} 1 \\ 5 \\ -1 \\ 2 \end{bmatrix}$, $\mathbf{w} = \begin{bmatrix} 0 \\ 1 \\ 2 \\ 1 \end{bmatrix}$

(b) $\mathbf{v} = \begin{bmatrix} 1 \\ -2 \\ 3 \\ 0 \\ 1 \end{bmatrix}$, $\mathbf{w} = \begin{bmatrix} 1 \\ 1 \\ 1 \\ 1 \\ 1 \end{bmatrix}$

ML.2. Let $S = \{\mathbf{w1}, \mathbf{w2}\}$, where

$$\mathbf{w1} = \begin{bmatrix} 1 \\ 0 \\ 1 \\ 1 \end{bmatrix} \quad \text{and} \quad \mathbf{w2} = \begin{bmatrix} 1 \\ 1 \\ -1 \\ 0 \end{bmatrix}$$

and let $W = \text{span } S$.

(a) Show that S is an orthogonal basis for W.

(b) Let

$$\mathbf{v} = \begin{bmatrix} 2 \\ 1 \\ 2 \\ 1 \end{bmatrix}$$

Compute $\text{proj}_{\mathbf{w1}}\mathbf{v}$.

(c) For vector **v** in part (b), compute $\text{proj}_W \mathbf{v}$.

ML.3. Plane Π in R^3 has orthogonal basis $\{\mathbf{w1}, \mathbf{w2}\}$, where

$$\mathbf{w1} = \begin{bmatrix} 1 \\ 2 \\ 3 \end{bmatrix} \quad \text{and} \quad \mathbf{w2} = \begin{bmatrix} 0 \\ -3 \\ 2 \end{bmatrix}$$

(a) Find the projection of

$$\mathbf{v} = \begin{bmatrix} 2 \\ 4 \\ 8 \end{bmatrix}$$

onto Π.

(b) Find the distance from **v** to Π.

ML.4. Let W be the subspace of R^4 with basis

$$S = \left\{ \begin{bmatrix} 1 \\ 1 \\ 0 \\ 1 \end{bmatrix}, \begin{bmatrix} 2 \\ -1 \\ 0 \\ 0 \end{bmatrix}, \begin{bmatrix} 0 \\ 1 \\ 0 \\ 1 \end{bmatrix} \right\} \quad \text{and} \quad \mathbf{v} = \begin{bmatrix} 0 \\ 0 \\ 1 \\ 1 \end{bmatrix}$$

(a) Find $\text{proj}_W \mathbf{v}$.

(b) Find the distance from $\mathbf{v}$ to W.

ML.5. Let

$$T = \begin{bmatrix} 1 & 0 \\ 0 & 1 \\ 1 & 1 \\ 1 & 0 \\ 1 & 0 \end{bmatrix} \quad \text{and} \quad \mathbf{b} = \begin{bmatrix} 1 \\ 1 \\ 1 \\ 1 \\ 1 \end{bmatrix}$$

(a) Show that the system $T\mathbf{x} = \mathbf{b}$ is inconsistent.

(b) Since $T\mathbf{x} = \mathbf{b}$ is inconsistent, $\mathbf{b}$ is not in the column space of T. One approach to find an approximate solution is to find a vector $\mathbf{y}$ in the column space of T so that $T\mathbf{y}$ is as close as possible to $\mathbf{b}$. We can do this by finding the projection $\mathbf{p}$ of $\mathbf{b}$ onto the column space of T. Find this projection $\mathbf{p}$ (which will be $T\mathbf{y}$).

Least Squares

Routine **lsqline** *in MATLAB will compute the least squares line for data you supply and graph both the line and the data points. To use* **lsqline**, *put the x-coordinates of your data into a vector* **x** *and the corresponding y-coordinates into a vector* **y** *and then type* **lsqline(x, y)**. *For more information, use* **help lsqline**.

ML.1. Solve Exercise 6 in Section 3.6 in MATLAB using **lsqline**.

ML.2. Use **lsqline** to determine the solution to Exercise 11 in Section 3.6.

ML.3. An experiment was conducted on the temperatures of a fluid in a newly designed container. The following data were obtained.

Time (minutes)	0	2	3	5	9
Temperature (°F)	185	170	166	152	110

(a) Determine the least squares line.

(b) Estimate the temperature at $x = 1, 6, 8$ minutes.

(c) Estimate the time at which the temperature of the fluid was 160°F.

ML.4. At time $t = 0$ an object is dropped from a height of 1 meter above a fluid. A recording device registers the height of the object above the surface of the fluid at $\frac{1}{2}$ second intervals, with a negative value indicating the object is below the surface of the fluid. The following table of data is the result.

Time (seconds)	Depth (meters)
0	1
0.5	0.88
1	0.54
1.5	0.07
2	−0.42
2.5	−0.80
3	−0.99
3.5	−0.94
4	−0.65
4.5	−0.21

(a) Determine the least squares quadratic polynomial.

(b) Estimate the depth at $t = 5$ and $t = 6$ seconds.

(c) Estimate the time the object breaks through the surface of the fluid the second time.

ML.5. Determine the least squares quadratic polynomial for the table of data given below. Use this data model to predict the value of y when $x = 7$.

x	y
−3	0.5
−2.5	0
−2	−1.125
−1.5	−1.875
−1	−1
0	0.9375
0.5	2.8750
1	4.75
1.5	8.25
2	11.5

Linear Transformations

MATLAB cannot be used to show that a function between vector spaces is a linear transformation. However, MATLAB can be used to construct an example that shows that a function is not a linear transformation. The following exercises illustrate this point.

ML.1. Let $L: R^n \to R^1$ be defined by $L(\mathbf{u}) = \|\mathbf{u}\|$.

(a) Find a pair of vectors $\mathbf{u}$ and $\mathbf{v}$ in R^2 such that

$$L(\mathbf{u}+\mathbf{v}) \neq L(\mathbf{u}) + L(\mathbf{v}).$$

Use MATLAB to do the computations. It follows that L is not a linear transformation.

(b) Find a pair of vectors $\mathbf{u}$ and $\mathbf{v}$ in R^3 such that

$$L(\mathbf{u}+\mathbf{v}) \neq L(\mathbf{u}) + L(\mathbf{v}).$$

Use MATLAB to do the computations.

ML.2. Let $L: M_{nn} \to R^1$ be defined by $L(A) = \det(A)$.

(a) Find a pair of 2×2 matrices A and B such that

$$L(A+B) \neq L(A) + L(B).$$

Use MATLAB to do the computations. It follows that L is not a linear transformation.

(b) Find a pair of 3×3 matrices A and B such that

$$L(A+B) \neq L(A) + L(B).$$

It follows that L is not a linear transformation. Use MATLAB to do the computations.

ML.3. Let $L: M_{nn} \to R^1$ be defined by $L(A) = \text{rank } A$.

(a) Find a pair of 2×2 matrices A and B such that

$$L(A+B) \neq L(A) + L(B).$$

It follows that L is not a linear transformation. Use MATLAB to do the computations.

(b) Find a pair of 3×3 matrices A and B such that

$$L(A+B) \neq L(A) + L(B).$$

It follows that L is not a linear transformation. Use MATLAB to do the computations.

Kernel and Range of Linear Transformations

In order to use MATLAB in this section, you should first read Section 8.8. Find a basis for the kernel and range of the linear transformation $L(\mathbf{x}) = A\mathbf{x}$ for each of the following matrices A.

ML.1. $A = \begin{bmatrix} 1 & 2 & 5 & 5 \\ -2 & -3 & -8 & -7 \end{bmatrix}.$

ML.2. $A = \begin{bmatrix} -3 & 2 & -7 \\ 2 & -1 & 4 \\ 2 & -2 & 6 \end{bmatrix}.$

ML.3. $A = \begin{bmatrix} 3 & 3 & -3 & 1 & 11 \\ -4 & -4 & 7 & -2 & -19 \\ 2 & 2 & -3 & 1 & 9 \end{bmatrix}.$

Matrix of a Linear Transformation

In MATLAB, follow the steps given in Section 4.3 to find the matrix of $L: R^n \to R^m$. The solution technique used in the MATLAB exercises dealing with coordinates and change of basis will be helpful here.

ML.1. Let $L: R^3 \to R^2$ be given by

$$L\left(\begin{bmatrix} x \\ y \\ z \end{bmatrix}\right) = \begin{bmatrix} 2x - y \\ x + y - 3z \end{bmatrix}.$$

Find the matrix A representing L with respect to the bases

$$S = \{\mathbf{v}_1, \mathbf{v}_2, \mathbf{v}_3\} = \left\{\begin{bmatrix} 1 \\ 1 \\ 1 \end{bmatrix}, \begin{bmatrix} 1 \\ 2 \\ 1 \end{bmatrix}, \begin{bmatrix} 0 \\ 1 \\ -1 \end{bmatrix}\right\}$$

and

$$T = \{\mathbf{w}_1, \mathbf{w}_2\} = \left\{\begin{bmatrix} 1 \\ 2 \end{bmatrix}, \begin{bmatrix} 2 \\ 1 \end{bmatrix}\right\}.$$

ML.2. Let $L: R^3 \to R^4$ be given by $L(\mathbf{v}) = C\mathbf{v}$, where

$$C = \begin{bmatrix} 1 & 2 & 0 \\ 2 & 1 & -1 \\ 3 & 1 & 0 \\ -1 & 0 & 2 \end{bmatrix}.$$

Find the matrix A representing L with respect to the bases

$$S = \{\mathbf{v}_1, \mathbf{v}_2, \mathbf{v}_3\} = \left\{\begin{bmatrix} 1 \\ 0 \\ 1 \end{bmatrix}, \begin{bmatrix} 2 \\ 0 \\ 1 \end{bmatrix}, \begin{bmatrix} 0 \\ 1 \\ 2 \end{bmatrix}\right\}$$

and

$$T = \{\mathbf{w}_1, \mathbf{w}_2, \mathbf{w}_3\}$$
$$= \left\{ \begin{bmatrix} 1 \\ 1 \\ 1 \\ 2 \end{bmatrix}, \begin{bmatrix} 1 \\ 1 \\ 1 \\ 0 \end{bmatrix}, \begin{bmatrix} 0 \\ 1 \\ 1 \\ -1 \end{bmatrix}, \begin{bmatrix} 0 \\ 0 \\ 1 \\ 0 \end{bmatrix} \right\}.$$

ML.3. Let $L: R^2 \to R^2$ be defined by

$$L\left(\begin{bmatrix} x \\ y \end{bmatrix}\right) = \begin{bmatrix} -x + 2y \\ 3x - y \end{bmatrix}$$

and let

$$S = \{\mathbf{v}_1, \mathbf{v}_2\} = \left\{ \begin{bmatrix} 1 \\ 2 \end{bmatrix}, \begin{bmatrix} -1 \\ 1 \end{bmatrix} \right\}$$

and

$$T = \{\mathbf{w}_1, \mathbf{w}_2\} = \left\{ \begin{bmatrix} -2 \\ 1 \end{bmatrix}, \begin{bmatrix} 1 \\ 1 \end{bmatrix} \right\}$$

be bases for R^2.

(a) Find the matrix A representing L with respect to S.

(b) Find the matrix B representing L with respect to T.

(c) Find the transition matrix P from T to S.

(d) Verify that $B = P^{-1}AP$.

Linear Transformations on Plane Geometric Figures

The routine **planelt** *in MATLAB provides a geometric visualization for the standard algebraic approach to linear transformations by illustrating* **plane linear transformations**, *which are linear transformations from R^2 to R^2. The name, of course, follows from the fact that we are mapping points in the plane into points in a corresponding plane.*
In MATLAB, type the command

planelt

A description of the routine will be displayed. Read it, then press ENTER. You will then see a set of General Directions. Read them carefully, then press ENTER. A set of Figure Choices will then be displayed. As you do the following exercises, record the figures requested on a separate sheet of paper. (After each menu choice, press ENTER. To return to a menu when the graphics are visible, press ENTER.)

ML.1. In **planelt**, choose figure #4 (the triangle). Then choose option #1 (see the triangle). Return to the menu and choose option #2 (use this figure). Next perform the following linear transformations.

(a) Reflect the triangle about the y-axis. Record the Current Figure.

(b) Now rotate the result from part (a) 60°. Record the figure.

(c) Return to the menu, restore the original figure, reflect it about the line $y = x$, then dilate it in the x-direction by a factor of 2. Record the figure.

(d) Repeat the experiment in part (c) but interchange the order of the linear transformations. Record the figure.

(e) Are the results from parts (c) and (d) the same? Compare the two figures you recorded.

(f) What does your answer in part (e) imply about the order of the linear transformations as applied to the triangle?

ML.2. Restore the original triangle selected in Exercise ML.1.

(a) Reflect it about the x-axis. Predict the result before pressing ENTER. (Call this linear transformation L_1.)

(b) Then reflect the figure resulting in part (a) about the y-axis. Predict the result before pressing ENTER. (Call this linear transformation L_2.)

(c) Record the figure that resulted from parts (a) and (b).

(d) Inspect the relationship between the Current Figure and the Original Figure. What (single) transformation do you think will accomplish the same result? (Use names that appear in the transformation menu. Call the linear transformation you select L_3.)

(e) Write a formula involving reflection L_1 about the x-axis, reflection L_2 about the y-axis, and the transformation L_3 that expresses the relationship you saw in part (d).

(f) Experiment with the formula in part (e) on several other figures until you can determine whether or not the formula in part (e) is correct in general. Write a brief summary of your experiments, observations, and conclusions.

ML.3. Choose the unit square as the figure.

(a) Reflect it about the x-axis, reflect the resulting figure about the y-axis, and then reflect that figure about the line $y = -x$. Record the figure.

(b) Compare the Current Figure with the Original Figure. Denote the reflection about the x-axis as L_1, the reflection about the y-axis as L_2, and the reflection about the line $y = -x$ as L_3. What formula relating these linear transformations does your comparison suggest when L_1 is followed by L_2, and then by L_3 on this figure?

(c) If M_i denotes the standard matrix representing the linear transformation L_i in part (b), to what matrix is $M_3 * M_2 * M_1$ equal? Does this result agree with your conclusion in part (b)?

(d) Experiment with the successive application of these three linear transformations on other figures.

ML.4. In routine **planelt** you can enter any figure you like and perform linear transformations on it. Follow the directions on entering your own figure and experiment with various linear transformations. It is recommended that you draw the figure on graph paper first and assign the coordinates of its vertices.

ML.5. On a figure, **planelt** allows you to select any 2×2 matrix to use as a linear transformation. Perform the following experiment. Choose a singular matrix and apply it to each of the stored figures. Write a brief summary of your experiments, observations, and conclusions about the behavior of "singular" linear transformations.

Determinants by Row Reduction

In order to use MATLAB in this section, you should first have read Chapter 8 *through Section* 8.5.

ML.1. Use the routine **reduce** to perform row operations and keep track by hand of the changes in the determinant as in Example 8 in Section 5.2.

(a) $A = \begin{bmatrix} 2 & 1 & 3 \\ 1 & 3 & 2 \\ 3 & 2 & 1 \end{bmatrix}.$

(b) $A = \begin{bmatrix} 0 & 1 & 3 & -2 \\ -2 & 1 & 1 & 1 \\ 2 & 0 & 1 & 2 \\ 1 & 0 & 0 & 1 \end{bmatrix}.$

ML.2. Use routine **reduce** to perform row operations and keep track by hand of the changes in the determinant as in Example 8 in Section 5.2.

(a) $A = \begin{bmatrix} 1 & 0 & 2 \\ 0 & 2 & 1 \\ 2 & 1 & 0 \end{bmatrix}.$

(b) $A = \begin{bmatrix} 1 & 2 & 0 & 0 \\ 2 & 1 & 2 & 0 \\ 0 & 2 & 1 & 2 \\ 0 & 0 & 2 & 1 \end{bmatrix}.$

ML.3. MATLAB has the command **det**, which returns the value of the determinant of a matrix. Just type **det(A)**. Find the determinant of each of the following matrices using **det**.

(a) $A = \begin{bmatrix} 1 & -1 & 1 \\ 1 & 1 & -1 \\ -1 & 1 & 1 \end{bmatrix}.$

(b) $A = \begin{bmatrix} 1 & 2 & 3 & 4 \\ 2 & 3 & 4 & 5 \\ 3 & 4 & 5 & 6 \\ 4 & 5 & 6 & 7 \end{bmatrix}.$

ML.4. Use **det** (see Exercise ML.3) to compute the determinant of each of the following.

(a) **5∗eye(size(A)) − A**, where

$$A = \begin{bmatrix} 2 & 3 & 0 \\ 4 & 1 & 0 \\ 0 & 0 & 5 \end{bmatrix}.$$

(b) **(3∗eye(size(A)) − A)^2**, where

$$A = \begin{bmatrix} 1 & 1 \\ 5 & 2 \end{bmatrix}.$$

(c) **invert(A)∗A**, where

$$A = \begin{bmatrix} 1 & 1 & 0 \\ 0 & 1 & 0 \\ 1 & 0 & 1 \end{bmatrix}.$$

ML.5. Determine a positive integer t so that **det(t∗eye(size(A)) − A) = 0**, where

$$A = \begin{bmatrix} 5 & 2 \\ -1 & 2 \end{bmatrix}.$$

Determinants by Cofactor Expansion

ML.1. In MATLAB there is a routine **cofactor** that computes the (i, j) cofactor of a matrix. For directions on using this routine, type **help cofactor**. Use **cofactor** to check your hand computations for the matrix A in Exercise 1 in Section 5.4.

ML.2. Use the **cofactor** routine (see Exercise ML.1) to compute the cofactor of the elements in the second row of

$$A = \begin{bmatrix} 1 & 5 & 0 \\ 2 & -1 & 3 \\ 3 & 2 & 1 \end{bmatrix}.$$

ML.3. Use the **cofactor** routine to evaluate the determinant of A using Theorem 5.10.

$$A = \begin{bmatrix} 4 & 0 & -1 \\ -2 & 2 & -1 \\ 0 & 4 & -3 \end{bmatrix}.$$

ML.4. Use the **cofactor** routine to evaluate the determinant of A using Theorem 5.10.

$$A = \begin{bmatrix} -1 & 2 & 0 & 0 \\ 2 & -1 & 2 & 0 \\ 0 & 2 & -1 & 2 \\ 0 & 0 & 2 & -1 \end{bmatrix}.$$

ML.5. In MATLAB there is a routine **adjoint**, which computes the adjoint of a matrix. For directions on using this routine, type **help adjoint**. Use **adjoint** to aid in computing the inverses of the matrices in Exercise 7 in Section 5.4.

Eigenvalues and Eigenvectors

MATLAB has a pair of commands that can be used to find the characteristic polynomial and eigenvalues of a matrix. Command **poly(A)** *gives the coefficients of the characteristic polynomial of matrix A, starting with the highest-degree term. If we set* $\mathbf{v}$ = **poly(A)** *and then use the command* **roots(v)**, *we obtain the roots of the characteristic polynomial of A. This process can also find complex eigenvalues, which are discussed in Appendix* B.2.

Once we have an eigenvalue λ of A, we can use **rref** *or* **homsoln** *to find a corresponding eigenvector from the linear system* $(\lambda I - A)\mathbf{x} = \mathbf{0}$.

ML.1. Find the characteristic polynomial of each of the following matrices using MATLAB.

(a) $A = \begin{bmatrix} 1 & 2 \\ 2 & -1 \end{bmatrix}$.

(b) $A = \begin{bmatrix} 2 & 4 & 0 \\ 1 & 2 & 1 \\ 0 & 4 & 2 \end{bmatrix}$.

(c) $A = \begin{bmatrix} 1 & 0 & 0 & 0 \\ 2 & -2 & 0 & 0 \\ 0 & 0 & 2 & -1 \\ 0 & 0 & -1 & 2 \end{bmatrix}$.

ML.2. Use the **poly** and **roots** commands in MATLAB to find the eigenvalues of the following matrices:

(a) $A = \begin{bmatrix} 1 & -3 \\ 3 & -5 \end{bmatrix}$.

(b) $A = \begin{bmatrix} 3 & -1 & 4 \\ -1 & 0 & 1 \\ 4 & 1 & 2 \end{bmatrix}$.

(c) $A = \begin{bmatrix} 2 & -2 & 0 \\ 1 & -1 & 0 \\ 1 & -1 & 0 \end{bmatrix}$.

(d) $A = \begin{bmatrix} 2 & 4 \\ 3 & 6 \end{bmatrix}$.

ML.3. In each of the following cases, λ is an eigenvalue of A. Use MATLAB to find an associated eigenvector.

(a) $\lambda = 3$, $A = \begin{bmatrix} 1 & 2 \\ -1 & 4 \end{bmatrix}$.

(b) $\lambda = -1$, $A = \begin{bmatrix} 4 & 0 & 0 \\ 1 & 3 & 0 \\ 2 & 1 & -1 \end{bmatrix}$.

(c) $\lambda = 2$, $A = \begin{bmatrix} 2 & 1 & 2 \\ 2 & 2 & -2 \\ 3 & 1 & 1 \end{bmatrix}$.

ML.4. Use MATLAB to determine if A is diagonalizable. If it is, find a nonsingular matrix P so that $P^{-1}AP$ is diagonal.

(a) $A = \begin{bmatrix} 0 & 2 \\ -1 & 3 \end{bmatrix}$.

(b) $A = \begin{bmatrix} 1 & -3 \\ 3 & -5 \end{bmatrix}$.

(c) $A = \begin{bmatrix} 0 & 0 & 4 \\ 5 & 3 & 6 \\ 6 & 0 & 5 \end{bmatrix}$.

ML.5. Use MATLAB and the hint in Exercise 19 in Section 6.2 to compute A^{30}, where

$$A = \begin{bmatrix} -1 & 1 & -1 \\ -2 & 2 & -1 \\ -2 & 2 & -1 \end{bmatrix}.$$

ML.6. Repeat Exercise ML.5 for

$$A = \begin{bmatrix} -1 & 1.5 & -1.5 \\ -2 & 2.5 & -1.5 \\ -2 & 2.0 & -1.0 \end{bmatrix}.$$

Display your answer in both **format short** and **format long**.

ML.7. Use MATLAB to investigate the sequences

$$A, A^3, A^5, \ldots \quad \text{and} \quad A^2, A^4, A^6, \ldots$$

for matrix A in Exercise ML.5. Write a brief description of the behavior of these sequences. Describe $\lim_{n\to\infty} A^n$.

Diagonalization

The MATLAB command **eig** *will produce the eigenvalues and a set of orthonormal eigenvectors for a symmetric matrix A. Use the command in the form*

$$[\mathbf{V}, \mathbf{D}] = \mathbf{eig}(\mathbf{A}).$$

The matrix V will contain the orthonormal eigenvectors, and matrix D will be diagonal containing the corresponding eigenvalues.

ML.1. Use **eig** to find the eigenvalues of A and an orthogonal matrix P so that $P^{-1}AP$ is diagonal.

(a) $A = \begin{bmatrix} 6 & 6 \\ 6 & 6 \end{bmatrix}$.

(b) $A = \begin{bmatrix} 1 & 2 & 2 \\ 2 & 1 & 2 \\ 2 & 2 & 1 \end{bmatrix}$.

(c) $A = \begin{bmatrix} 4 & 1 & 0 \\ 1 & 4 & 1 \\ 0 & 1 & 4 \end{bmatrix}$.

ML.2. Command **eig** can be applied to any matrix, but the matrix V of eigenvectors need not be orthogonal. For each of the following, use **eig** to determine which matrices A are such that V is orthogonal. If V is not orthogonal, then discuss briefly whether it can or cannot be replaced by an orthogonal matrix of eigenvectors.

(a) $A = \begin{bmatrix} 1 & 2 \\ -1 & 4 \end{bmatrix}$.

(b) $A = \begin{bmatrix} 2 & 1 & 2 \\ 2 & 2 & -2 \\ 3 & 1 & 1 \end{bmatrix}$.

(c) $A = \begin{bmatrix} 1 & -3 \\ 3 & -5 \end{bmatrix}$.

(d) $A = \begin{bmatrix} 1 & 0 & 0 \\ 0 & 1 & 1 \\ 0 & 1 & 1 \end{bmatrix}$.

Appendix A

Preliminaries

In this appendix, which can be consulted as the need arises, we present the basic ideas of sets and functions that are used in Chapters 2, 3, 4, and 6.

A.1 SETS

A **set** is a collection, class, aggregate, or family of objects, which are called **elements**, or **members**, of the set. A set will be denoted by a capital letter, and an element of a set by a lowercase letter. A set S is specified either by describing all the elements of S, or by stating a property that determines, unequivocally, whether an element is or is not an element of S. Let $S = \{1, 2, 3\}$ be the set of all positive integers < 4. Then a real number belongs to S if it is a positive integer < 4. Thus S has been described in both ways. Sets A and B are said to be **equal** if each element of A belongs to B and if each element of B belongs to A. We write $A = B$. Thus $\{1, 2, 3\} = \{3, 2, 1\} = \{2, 1, 3\}$, and so on. If A and B are sets such that every element of A belongs to B, then A is said to be a **subset** of B. The set of all rational numbers is a subset of the set of all real numbers; the set $\{1, 3\}$ is a subset of $\{1, 2, 3\}$; the set of all isosceles triangles is a subset of the set of all triangles. We can see that every set is a subset of itself. The **empty set** is the set that has no elements in it. The set of all real numbers whose squares equal -1 is empty because the square of a real number is never negative.

A.2 FUNCTIONS

A **function** f from a set S into a set T is a rule that assigns to each element s of S a unique element t of T. We denote the function f by $f\colon S \to T$ and write $t = f(s)$. Functions constitute the basic ingredient of calculus and other branches of mathematics, and the reader has dealt extensively with them. The set S is called the **domain** of f; the set T is called the **codomain** of f; the subset $f(S)$ of T consisting of all the elements $f(s)$, for s in S, is called the **range** of f or the **image** of S under f. As examples of functions we consider the following:

1. Let $S = T =$ the set of all real numbers. Let $f: S \to T$ be defined by the rule $f(s) = s^2$, for s in S.
2. Let $S =$ the set of all real numbers and let $T =$ the set of all nonnegative real numbers. Let $f: S \to T$ be defined by the rule $f(s) = s^2$, for s in S.
3. Let $S =$ three-dimensional space, where each point is described by x-, y-, and z-coordinates (x_1, x_2, x_3). Let $T =$ the (x, y)-plane as a subset of S. Let $f: S \to T$ be defined by the rule $f((x_1, x_2, x_3)) = (x_1, x_2, 0)$. To see what f does, we take a point (x_1, x_2, x_3) in S, draw a line from (x_1, x_2, x_3) perpendicular to T, the (x, y)-plane, and find the point of intersection $(x_1, x_2, 0)$ of this line with the (x, y)-plane. This point is the image of (x_1, x_2, x_3) under f; f is called a **projection function** (Figure A.1).

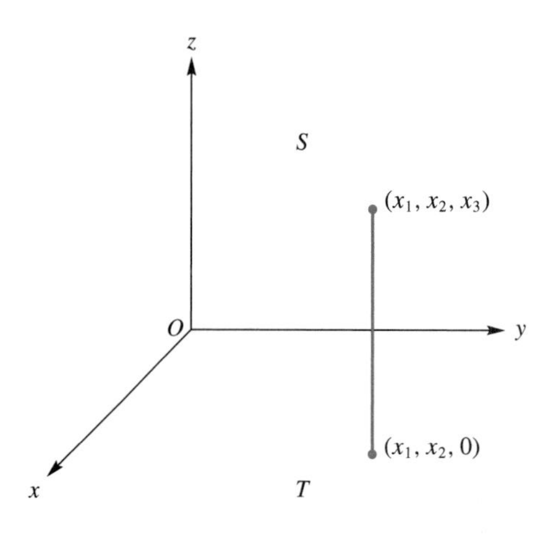

FIGURE A.1

4. Let $S = T =$ the set of all real numbers. Let $f: S \to T$ be defined by the rule $f(s) = 2s + 1$, for s in S.
5. Let $S =$ the x-axis in the (x, y)-plane and let $T =$ the (x, y)-plane. Let $f: S \to T$ be defined by the rule $f((s, 0)) = (s, 1)$, for s in S.
6. Let $S =$ the set of all real numbers. Let $T =$ the set of all positive real numbers. Let $f: S \to T$ be defined by the rule $f(s) = e^s$, for s in S.

There are two properties of functions that we need to distinguish. A function $f: S \to T$ is called **one-to-one** if $f(s_1) \neq f(s_2)$ whenever s_1 and s_2 are distinct elements of S. An equivalent statement is that if $f(s_1) = f(s_2)$, then we must have $s_1 = s_2$ (see Figure A.2). A function $f: S \to T$ is called **onto** if the range of f is all of T, that is, if for any given t in T there is at least one s in S such that $f(s) = t$ (see Figure A.3).

We now examine the listed functions:

1. f is not one-to-one, for if $f(s_1) = f(s_2)$, it need not follow that

$$s_1 = s_2 \qquad [f(2) = f(-2) = 4].$$

Since the range of f is the set of nonnegative real numbers, f is not onto.

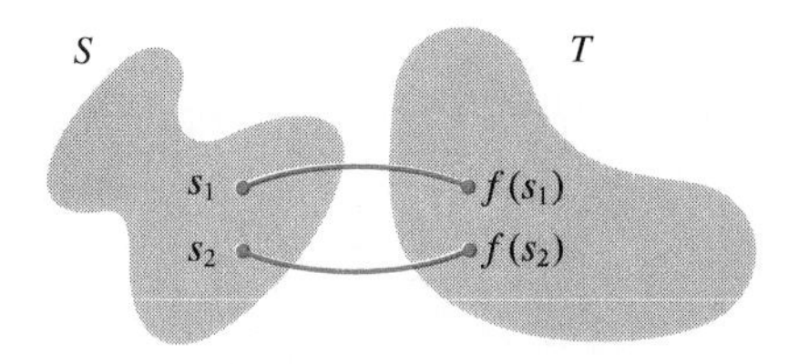

(a) f is one-to-one.

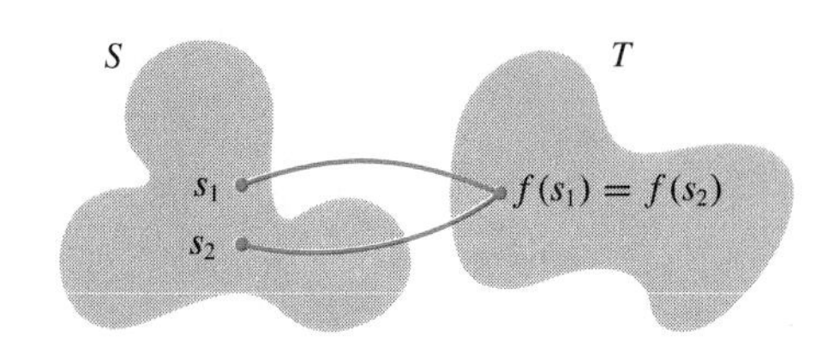

(b) f is not one-to-one.

FIGURE A.2

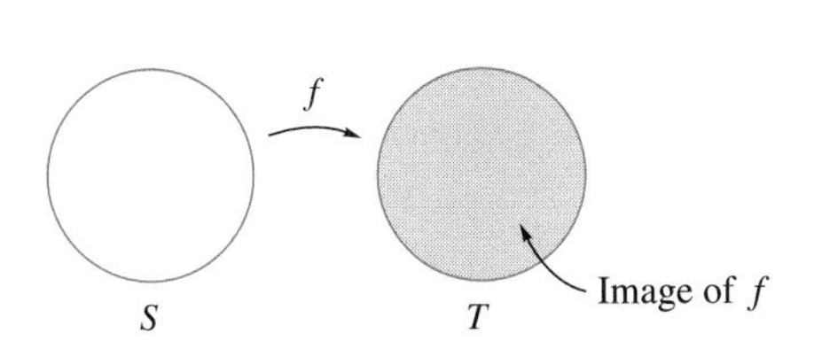

(a) f is onto.

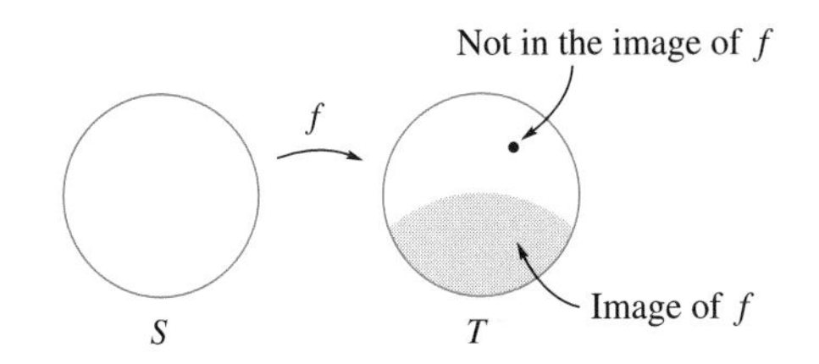

(b) f is not onto.

FIGURE A.3

Thus if $t = -4$, then there is no s such that $f(s) = -4$.

2. f is not one-to-one, but is onto. For if t is a given nonnegative real number, then $s = \sqrt{t}$ is in S and $f(s) = t$. Note that the codomain makes a difference here. The *formulas* are the same in 1 and 2, but the *functions* are not.
3. f is not one-to-one, for if $f((a_1, a_2, a_3)) = f((b_1, b_2, b_3))$, then $(a_1, a_2, 0) = (b_1, b_2, 0)$, so $a_1 = b_1$ and $a_2 = b_2$. However, b_3 need not equal a_3. The range of f is T; that is, f is onto. For let $(x_1, x_2, 0)$ be any element of T. Can we find an element (a_1, a_2, a_3) of S such that $f((a_1, a_2, a_3)) = (x_1, x_2, 0)$? We merely let $a_1 = x_1$, $a_2 = x_2$, and let $a_3 =$ any real number we wish, say $a_3 = 5$.
4. f is one-to-one, for if $f(s_1) = f(s_2)$, then $2s_1 + 1 = 2s_2 + 1$, which means that $s_1 = s_2$. Also, f is onto, for given a real number t we seek a real number s so that $f(s) = t$; that is, we need to solve $2s + 1 = t$ for s, which we can do, obtaining $s = \frac{1}{2}(t - 1)$.
5. f is one-to-one but f is not onto because not every element in T has 1 for its y-coordinate.
6. f is one-to-one and onto because $e^{s_1} \neq e^{s_2}$ if $s_1 \neq s_2$, and for any positive t we can always solve $t = e^s$, obtaining $s = \ln t$.

If $f: S \to T$ and $g: T \to U$ are functions, then we can define a new function $g \circ f$, by $(g \circ f)(s) = g(f(s))$, for s in S. The function $g \circ f: S \to U$ is called the **composite** of f and g. Thus, if f and g are the functions 4 and 6 above, then $g \circ f$ is defined by $(g \circ f)(s) = g(f(s)) = e^{2s+1}$, and $f \circ g$ is defined by $(f \circ g)(s) = f(g(s)) = 2e^s + 1$. The function $i: S \to S$ defined by $i(s) = s$, for s in S, is called the **identity function** on S. A function $f: S \to T$ for which there is a function $g: T \to S$ such that $g \circ f = i_S =$ identity function on S and $f \circ g = i_T =$ identity function on T is called an **invertible function**, and g is called an **inverse** of f. It can be shown that a function can have at most one inverse. It is not difficult to show, and we do so in Chapter 4 for a special case, that a function $f: S \to T$ is invertible if and only if it is one-to-one and onto. The inverse of f, if it exists, is denoted by f^{-1}. If f is invertible, then f^{-1} is also one-to-one and onto. We recall that "if and only if" means that both the statement and its converse are true. That is, if $f: S \to T$ is invertible, then f is one-to-one and onto; if $f: S \to T$ is one-to-one and onto, then f is invertible. Functions 4 and 6 are invertible; the inverse of function 4 is $g: T \to S$ defined by $g(t) = \frac{1}{2}(t - 1)$ for t in T; the inverse of function 6 is $g: T \to S$, defined by $g(t) = \ln t$.

Appendix B

Complex Numbers

B. 1 COMPLEX NUMBERS

Complex numbers are usually introduced in an algebra course to "complete" the solution to the quadratic equation

$$ax^2 + bx + c = 0.$$

In using the quadratic formula

$$x = \frac{-b \pm \sqrt{b^2 - 4ac}}{2a},$$

the case in which $b^2 - 4ac < 0$ is not resolved unless we can cope with the square roots of negative numbers. In the sixteenth century mathematicians and scientists justified this "completion" of the solution of quadratic equations by intuition. Naturally, a controversy arose, with some mathematicians denying the existence of these numbers and others using them along with real numbers. The use of complex numbers did not lead to any contradictions, and the idea proved to be an important milestone in the development of mathematics.

A **complex number** c is of the form $c = a + bi$, where a and b are real numbers and where $i = \sqrt{-1}$; a is called the **real part** of c, and b is called the **imaginary part** of c. The term "imaginary part" arose from the mysticism surrounding the beginnings of complex numbers; however, these numbers are as "real" as the real numbers.

EXAMPLE 1

(a) $5 - 3i$ has real part 5 and imaginary part -3;

(b) $-6 + \sqrt{2}\,i$ has real part -6 and imaginary part $\sqrt{2}$. ■

The symbol $i = \sqrt{-1}$ has the property that $i^2 = -1$, and we can deduce the following relationships:

$$i^3 = -i, \quad i^4 = 1, \quad i^5 = i, \quad i^6 = -1, \quad i^7 = -i, \quad \ldots.$$

These results will be handy for simplifying operations involving complex numbers.

We say that two complex numbers $c_1 = a_1 + b_1 i$ and $c_2 = a_2 + b_2 i$ are **equal** if their real and imaginary parts are equal, that is, if $a_1 = a_2$ and $b_1 = b_2$. Of course, every real number a is a complex number with its imaginary part zero: $a = a + 0i$.

Operations on Complex Numbers

If $c_1 = a_1 + b_1 i$ and $c_2 = a_2 + b_2 i$ are complex numbers, then their **sum** is

$$c_1 + c_2 = (a_1 + a_2) + (b_1 + b_2)i,$$

and their **difference** is

$$c_1 - c_2 = (a_1 - a_2) + (b_1 - b_2)i.$$

In words, to form the sum of two complex numbers, add the real parts and add the imaginary parts. The **product** of c_1 and c_2 is

$$\begin{aligned} c_1 c_2 &= (a_1 + b_1 i) \cdot (a_2 + b_2 i) = a_1 a_2 + (a_1 b_2 + b_1 a_2)i + b_1 b_2 i^2 \\ &= (a_1 a_2 - b_1 b_2) + (a_1 b_2 + b_1 a_2)i. \end{aligned}$$

A special case of multiplication of complex numbers occurs when c_1 is real. In this case we obtain the simple result

$$c_1 c_2 = c_1 \cdot (a_2 + b_2 i) = c_1 a_2 + c_1 b_2 i.$$

If $c = a + bi$ is a complex number, then the **conjugate** of c is the complex number $\overline{c} = a - bi$. It is not difficult to show that if c and d are complex numbers, then the following basic properties of complex arithmetic hold:

1. $\overline{\overline{c}} = c$.
2. $\overline{c + d} = \overline{c} + \overline{d}$.
3. $\overline{cd} = \overline{c}\,\overline{d}$.
4. c is a real number if and only if $c = \overline{c}$.
5. $c\overline{c}$ is a nonnegative real number and $c\overline{c} = 0$ if and only if $c = 0$.

We prove property 4 here and leave the others as exercises. Let $c = a + bi$ so that $\overline{c} = a - bi$. If $c = \overline{c}$, then $a + bi = a - bi$, so $b = 0$ and c is real. On the other hand, if c is real, then $c = a$ and $\overline{c} = a$, so $c = \overline{c}$.

EXAMPLE 2

Let $c_1 = 5 - 3i$, $c_2 = 4 + 2i$, and $c_3 = -3 + i$.

(a) $c_1 + c_2 = (5 - 3i) + (4 + 2i) = 9 - i$.

(b) $c_2 - c_3 = (4 + 2i) - (-3 + i) = (4 - (-3)) + (2 - 1)i = 7 + i$.

(c) $c_1c_2 = (5-3i)\cdot(4+2i) = 20+10i-12i-6i^2 = 26-2i.$

(d) $$\begin{aligned} c_1\overline{c}_3 &= (5-3i)\cdot\overline{(-3+i)} = (5-3i)\cdot(-3-i) \\ &= -15-5i+9i+3i^2 \\ &= -18+4i. \end{aligned}$$

(e) $$\begin{aligned} 3c_1+2\overline{c}_2 &= 3(5-3i)+2\overline{(4+2i)} = (15-9i)+2(4-2i) \\ &= (15-9i)+(8-4i) = 23-13i. \end{aligned}$$

(f) $c_1\overline{c}_1 = (5-3i)\overline{(5-3i)} = (5-3i)(5+3i) = 34.$ ■

When we consider systems of linear equations with complex coefficients, we will need to divide complex numbers to complete the solution process and obtain a reasonable form for the solution. Let $c_1 = a_1+b_1i$ and $c_2 = a_2+b_2i$. If $c_2 \neq 0$, that is, if $a_2 \neq 0$ or $b_2 \neq 0$, then we can **divide** c_1 by c_2:

$$\frac{c_1}{c_2} = \frac{a_1+b_1i}{a_2+b_2i}.$$

To conform to our practice of expressing a complex number in the form real part + imaginary part $\cdot\, i$, we must simplify the foregoing expression for c_1/c_2. To simplify this complex fraction, we multiply the numerator and the denominator by the conjugate of the denominator. Thus, dividing c_1 by c_2 gives the complex number

$$\frac{c_1}{c_2} = \frac{a_1+b_1i}{a_2+b_2i} = \frac{(a_1+b_1i)(a_2-b_2i)}{(a_2+b_2i)(a_2-b_2i)} = \frac{a_1a_2+b_1b_2}{a_2^2+b_2^2} - \frac{a_1b_2+a_2b_1}{a_2^2+b_2^2}\,i.$$

EXAMPLE 3

Let $c_1 = 2-5i$ and $c_2 = -3+4i$. Then

$$\frac{c_1}{c_2} = \frac{2-5i}{-3+4i} = \frac{(2-5i)(-3-4i)}{(-3+4i)(-3-4i)} = \frac{-26+7i}{(-3)^2+(4)^2} = -\frac{26}{25}+\frac{7}{25}i.$$ ■

Finding the reciprocal of a complex number is a special case of division of complex numbers. If $c = a+bi$, $c \neq 0$, then

$$\begin{aligned} \frac{1}{c} = \frac{1}{a+bi} &= \frac{a-bi}{(a+bi)(a-bi)} = \frac{a-bi}{a^2+b^2} \\ &= \frac{a}{a^2+b^2} - \frac{b}{a^2+b^2}\,i. \end{aligned}$$

EXAMPLE 4

(a) $$\frac{1}{2+3i} = \frac{2-3i}{(2+3i)(2-3i)} = \frac{2-3i}{2^2+3^2} = \frac{2}{13}-\frac{3}{13}i.$$

(b) $$\frac{1}{i} = \frac{-i}{i(-i)} = \frac{-i}{-i^2} = \frac{-i}{-(-1)} = -i.$$ ■

Summarizing, we can say that complex numbers are mathematical objects for which addition, subtraction, multiplication, and division are defined in such a way that these operations on real numbers can be derived as special cases. In fact, it can be shown that complex numbers form a mathematical system that is called a field.

Geometric Representation of Complex Numbers

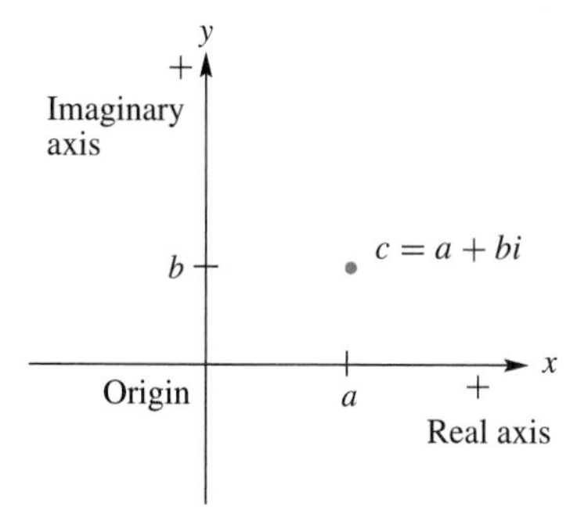

FIGURE B.1

Complex plane.

A complex number $c = a + bi$ may be regarded as an ordered pair (a, b) of real numbers. This ordered pair of real numbers corresponds to a point in the plane. Such a correspondence naturally suggests that we represent $a + bi$ as a point in the **complex plane**, where the horizontal axis is used to represent the real part of c and the vertical axis is used to represent the imaginary part of c. To simplify matters, we call these the **real axis** and the **imaginary axis**, respectively (see Figure B.1).

EXAMPLE 5

Plot the complex numbers $c = 2 - 3i$, $d = 1 + 4i$, $e = -3$, and $f = 2i$ in the complex plane.

Solution See Figure B.2. ■

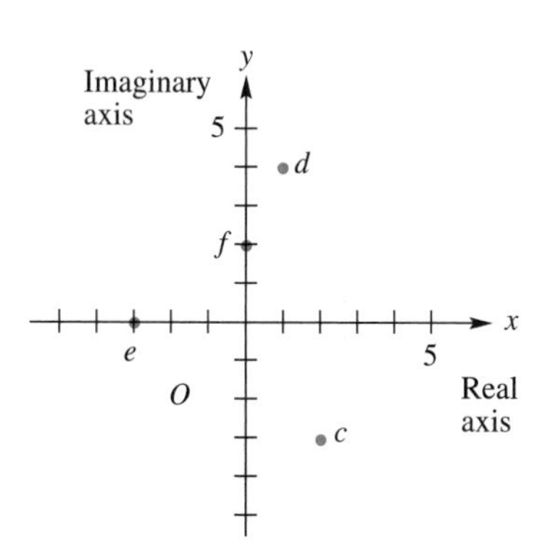

FIGURE B.2

The rules concerning inequality of real numbers, such as less than and greater than, *do not apply to complex numbers*. There is no way to arrange the complex numbers according to size. However, using the geometric representation from the complex plane, we can attach a notion of size to a complex number by measuring its distance from the origin. The distance from the origin to $c = a + bi$ is called the **absolute value**, or **modulus**, of the complex number, and is denoted by $|c| = |a + bi|$. Using the formula for the distance between ordered pairs of real numbers, we obtain

$$|c| = |a + bi| = \sqrt{a^2 + b^2}.$$

It follows that $c\bar{c} = |c|^2$ (verify).

EXAMPLE 6

Referring to Example 5: $|c| = \sqrt{13}$; $|d| = \sqrt{17}$; $|e| = 3$; $|f| = 2$. ■

A different, but related, interpretation of a complex number is obtained if we associate with $c = a + bi$ the vector OP, where O is the origin $(0, 0)$ and P is the point (a, b). There is an obvious correspondence between this representation and vectors in the plane discussed in calculus, which we reviewed in Section 2.1. Using a vector representation, addition and subtraction of complex numbers can be viewed as the corresponding vector operations. These are represented in Figures 2.5, 2.6, and 2.9. We will not pursue the manipulation of complex numbers by vector operations here, but such a point of view is important for the development and study of complex variables.

Matrices with Complex Entries

If the entries of a matrix are complex numbers, we can perform the matrix operations of addition, subtraction, multiplication, and scalar multiplication in a manner completely analogous to that for real matrices. The validity of these operations can be verified using properties of complex arithmetic and just imitating the proofs for real matrices presented in the text. We illustrate these concepts in the following example.

EXAMPLE 7

Let

$$A = \begin{bmatrix} 4+i & -2+3i \\ 6+4i & -3i \end{bmatrix}, \quad B = \begin{bmatrix} 2-i & 3-4i \\ 5+2i & -7+5i \end{bmatrix},$$

$$C = \begin{bmatrix} 1+2i & i \\ 3-i & 8 \\ 4+2i & 1-i \end{bmatrix}.$$

(a) $$A+B = \begin{bmatrix} (4+i)+(2-i) & (-2+3i)+(3-4i) \\ (6+4i)+(5+2i) & (-3i)+(-7+5i) \end{bmatrix}$$
$$= \begin{bmatrix} 6 & 1-i \\ 11+6i & -7+2i \end{bmatrix}.$$

(b) $$B-A = \begin{bmatrix} (2-i)-(4+i) & (3-4i)-(-2+3i) \\ (5+2i)-(6+4i) & (-7+5i)-(-3i) \end{bmatrix}$$
$$= \begin{bmatrix} -2-2i & 5-7i \\ -1-2i & -7+8i \end{bmatrix}.$$

(c) $$CA = \begin{bmatrix} 1+2i & i \\ 3-i & 8 \\ 4+2i & 1-i \end{bmatrix} \begin{bmatrix} 4+i & -2+3i \\ 6+4i & -3i \end{bmatrix}$$

$$= \begin{bmatrix} (1+2i)(4+i)+(i)(6+4i) & (1+2i)(-2+3i)+(i)(-3i) \\ (3-i)(4+i)+(8)(6+4i) & (3-i)(-2+3i)+(8)(-3i) \\ (4+2i)(4+i)+(1-i)(6+4i) & (4+2i)(-2+3i)+(1-i)(-3i) \end{bmatrix}$$

$$= \begin{bmatrix} -2+15i & -5-i \\ 61+31i & -3-13i \\ 24+10i & -17+5i \end{bmatrix}.$$

(d) $$(2+i)B = \begin{bmatrix} (2+i)(2-i) & (2+i)(3-4i) \\ (2+i)(5+2i) & (2+i)(-7+5i) \end{bmatrix}$$
$$= \begin{bmatrix} 5 & 10-5i \\ 8+9i & -19+3i \end{bmatrix}.$$

■

Just as we can compute the conjugate of a complex number, we can compute the **conjugate of a matrix** by computing the conjugate of each entry of the matrix. We denote the conjugate of a matrix A by $\overline{A}$, and write

$$\overline{A} = \left[\overline{a_{ij}}\right].$$

EXAMPLE 8

Referring to Example 7, we find that

$$\overline{A} = \begin{bmatrix} 4-i & -2-3i \\ 6-4i & 3i \end{bmatrix} \quad \text{and} \quad \overline{B} = \begin{bmatrix} 2+i & 3+4i \\ 5-2i & -7-5i \end{bmatrix}.$$

■

The following properties of the conjugate of a matrix hold:

1. $\overline{\overline{A}} = A$.
2. $\overline{A+B} = \overline{A} + \overline{B}$.
3. $\overline{AB} = \overline{A}\,\overline{B}$.
4. For any real number k, $\overline{kA} = k\,\overline{A}$.
5. For any complex number c, $\overline{cA} = \overline{c}\,\overline{A}$.
6. $(\overline{A})^T = \overline{A^T}$.
7. If A is nonsingular, then $(\overline{A})^{-1} = \overline{A^{-1}}$.

We prove properties 5 and 6 here and leave the others as exercises. First, property 5: If c is complex, the (i, j) entry of $\overline{cA}$ is

$$\overline{ca_{ij}} = \overline{c}\,\overline{a_{ij}},$$

which is the (i, j) entry of $\overline{c}\,\overline{A}$. Next, property 6: The (i, j) entry of $(\overline{A})^T$ is $\overline{a_{ji}}$, which is the (i, j) entry of $\overline{A^T}$.

Special Types of Complex Matrices

As we have already seen, certain types of real matrices satisfy some important properties. The same situation applies to complex matrices and we now discuss several of these types of matrices.

An $n \times n$ complex matrix A is called **Hermitian** if

$$\overline{A^T} = A.$$

This is equivalent to saying that $\overline{a_{ji}} = a_{ij}$ for all i and j. Every real symmetric matrix is Hermitian [Exercise 11(c)], so we may consider Hermitian matrices as the analogs of real symmetric matrices.

EXAMPLE 9

The matrix

$$A = \begin{bmatrix} 2 & 3+i \\ 3-i & 5 \end{bmatrix}$$

is Hermitian, since

$$\overline{A^T} = \overline{\begin{bmatrix} 2 & 3-i \\ 3+i & 5 \end{bmatrix}} = \begin{bmatrix} 2 & 3+i \\ 3-i & 5 \end{bmatrix} = A.$$

■

An $n \times n$ complex matrix A is called **unitary** if

$$(\overline{A^T})A = A(\overline{A^T}) = I_n.$$

This is equivalent to saying that $\overline{A^T} = A^{-1}$. Every real orthogonal matrix is unitary [Exercise 12(a)], so we may consider unitary matrices as the analogs of real orthogonal matrices.

EXAMPLE 10

The matrix

$$A = \begin{bmatrix} \frac{1}{\sqrt{3}} & \frac{1+i}{\sqrt{3}} \\ \frac{1-i}{\sqrt{3}} & -\frac{1}{\sqrt{3}} \end{bmatrix}$$

is unitary, since (verify)

$$(\overline{A^T})A = \begin{bmatrix} \frac{1}{\sqrt{3}} & \frac{1+i}{\sqrt{3}} \\ \frac{1-i}{\sqrt{3}} & -\frac{1}{\sqrt{3}} \end{bmatrix} \begin{bmatrix} \frac{1}{\sqrt{3}} & \frac{1+i}{\sqrt{3}} \\ \frac{1-i}{\sqrt{3}} & -\frac{1}{\sqrt{3}} \end{bmatrix} = I_2$$

and similarly, $A(\overline{A^T}) = I_2$. ■

There is one more type of complex matrix that is important. An $n \times n$ complex matrix is called **normal** if

$$(\overline{A^T})\,A = A\,(\overline{A^T}).$$

EXAMPLE 11

The matrix

$$A = \begin{bmatrix} 5-i & -1+i \\ -1-i & 3-i \end{bmatrix}$$

is normal, since (verify)

$$(\overline{A^T})\,A = A\,(\overline{A^T}) = \begin{bmatrix} 28 & -8+8i \\ -8-8i & 12 \end{bmatrix}.$$

Moreover, A is not Hermitian, since $\overline{A^T} \neq A$ (verify). ■

Complex Numbers and Roots of Polynomials

A polynomial of degree n with real coefficients has n complex roots; some, all, or none of which may be real numbers. Thus the polynomial $f_1(x) = x^4 - 1$ has the roots i, $-i$, 1, and -1; the polynomial $f_2(x) = x^2 - 1$ has the roots 1 and -1; and the polynomial $f_3(x) = x^2 + 1$ has the roots i and $-i$. We make use of this result later, when we extend the definition of an eigenvalue (Section 6.1) to allow it to be a complex number.

B.1 EXERCISES

1. Let $c_1 = 3+4i$, $c_2 = 1-2i$, and $c_3 = -1+i$. Compute each of the following and simplify as much as possible.

(a) $c_1 + c_2$. (b) $c_3 - c_1$.

(c) $c_1 c_2$. (d) $c_2 \overline{c_3}$.

(e) $4c_3 + \overline{c_2}$. (f) $(-i)\cdot c_2$.

(g) $\overline{3c_1 - ic_2}$. (h) $c_1c_2c_3$.

2. Write in the form $a+bi$.

(a) $\dfrac{1+2i}{3-4i}$. (b) $\dfrac{2-3i}{3-i}$.

(c) $\dfrac{(2+i)^2}{i}$. (d) $\dfrac{1}{(3+2i)(1+i)}$.

3. Represent each complex number as a point and as a vector in the complex plane.

(a) $4+2i$. (b) $-3+i$.

(c) $3-2i$. (d) $i(4+i)$.

4. Find the modulus of each complex number in Exercise 3.

5. If $c = a+bi$, then we can denote the real part of c by $\operatorname{Re}(c)$ and the imaginary part of c by $\operatorname{Im}(c)$.

(a) For any complex numbers $c_1 = a_1 + b_1 i$, $c_2 = a_2 + b_2 i$, prove that $\operatorname{Re}(c_1 + c_2) = \operatorname{Re}(c_1) + \operatorname{Re}(c_2)$ and $\operatorname{Im}(c_1 + c_2) = \operatorname{Im}(c_1) + \operatorname{Im}(c_2)$.

(b) For any real number k, prove that $\operatorname{Re}(kc) = k\operatorname{Re}(c)$ and $\operatorname{Im}(kc) = k\operatorname{Im}(c)$.

(c) Is part (b) true if k is a complex number?

(d) Prove or disprove:

$$\operatorname{Re}(c_1c_2) = \operatorname{Re}(c_1)\cdot\operatorname{Re}(c_2).$$

6. In the complex plane sketch the vectors corresponding to c and $\overline{c}$ for $c = 2+3i$ and $c = -1+4i$. Geometrically, we can say that $\overline{c}$ is the reflection of c with respect to the real axis. (See also Example 8 in Section 4.1.)

7. Let

$$A = \begin{bmatrix} 2+2i & -1+3i \\ -2 & 1-i \end{bmatrix},$$

$$B = \begin{bmatrix} 2i & 1+2i \\ 0 & 3-i \end{bmatrix}, \qquad C = \begin{bmatrix} 2+i \\ -i \end{bmatrix}.$$

Compute each of the following and simplify each entry as $a+bi$.

(a) $A+B$. (b) $(1-2i)C$. (c) AB.

(d) BC. (e) $A - 2I_2$. (f) $\overline{B}$.

(g) $A\overline{C}$. (h) $(\overline{A+B})C$.

8. Let A and B be $m \times n$ complex matrices, and let C be an $n \times n$ nonsingular matrix.

(a) Prove that $\overline{A+B} = \overline{A} + \overline{B}$.

(b) Prove that for any real number k, $\overline{kA} = k\overline{A}$.

(c) Prove that $(\overline{C})^{-1} = \overline{C^{-1}}$.

9. If $A = \begin{bmatrix} 0 & i \\ i & 0 \end{bmatrix}$, compute A^2, A^3, and A^4. Give a general rule for A^n, n a positive integer.

10. Which of the following matrices are Hermitian, unitary, or normal?

(a) $\begin{bmatrix} 3 & 2+i \\ 2-i & 4 \end{bmatrix}$. (b) $\begin{bmatrix} 2 & 1-i \\ 3+i & -2 \end{bmatrix}$.

(c) $\begin{bmatrix} \frac{1-i}{2} & \frac{1+i}{2} \\ \frac{1+i}{2} & \frac{1-i}{2} \end{bmatrix}$. (d) $\begin{bmatrix} 1 & -1 \\ 1 & 1 \end{bmatrix}$.

(e) $\begin{bmatrix} 1 & 3-i & 4-i \\ 3+i & -2 & 2+i \\ 4+i & 2-i & 3 \end{bmatrix}$.

(f) $\begin{bmatrix} 3 & \frac{3-i}{2} & \frac{4-i}{2} \\ \frac{3-i}{2} & -2 & 2+i \\ \frac{4-i}{2} & 2-i & 5 \end{bmatrix}$.

(g) $\begin{bmatrix} 3+2i & -1 \\ -i & 2+i \end{bmatrix}$. (h) $\begin{bmatrix} i & i \\ -i & 1 \end{bmatrix}$.

(i) $\begin{bmatrix} 1 & 0 & 0 \\ 0 & \frac{1+i}{\sqrt{3}} & \frac{1}{\sqrt{3}} \\ 0 & -\frac{1}{\sqrt{3}} & \frac{1-i}{\sqrt{3}} \end{bmatrix}$.

(j) $\begin{bmatrix} 4+7i & -2-i \\ 1-2i & 3+4i \end{bmatrix}$.

11. (a) Prove that the diagonal entries of a Hermitian matrix must be real.

(b) Prove that every Hermitian matrix A can be written as $A = B + iC$, where B is real and symmetric and C is real and skew symmetric (see Definition 1.8). [*Hint*: Consider $B = (A + \overline{A})/2$ and $C = (A - \overline{A})/2i$.]

(c) Prove that every real symmetric matrix is Hermitian.

12. (a) Show that every real orthogonal matrix is unitary.

(b) Show that if A is a unitary matrix, then A^T is unitary.

(c) Show that if A is a unitary matrix, then A^{-1} is unitary.

13. Let A be an $n \times n$ complex matrix.

(a) Show that A can be written as $B + iC$, where B and C are Hermitian.

(b) Show that A is normal if and only if

$$BC = CB.$$

[*Hint*: Consider $B = (A + \overline{A^T})/2$ and $C = (A - \overline{A^T})/2i$.]

14. (a) Prove that every Hermitian matrix is normal.

(b) Prove that every unitary matrix is normal.

(c) Find a 2×2 normal matrix that is neither Hermitian nor unitary.

15. An $n \times n$ complex matrix A is called **skew Hermitian** if

$$\overline{A^T} = -A.$$

Show that a matrix $A = B + iC$, where B and C are real matrices, is skew Hermitian if and only if B is skew symmetric and C is symmetric.

16. Find all the roots.

(a) $x^2 + x + 1 = 0$.

(b) $x^3 + 2x^2 + x + 2 = 0$.

(c) $x^5 + x^4 - x - 1 = 0$.

17. Let $p(x)$ denote a polynomial and let A be a square matrix. Then $p(A)$ is called a **matrix polynomial**, or a **polynomial in the matrix** A. For $p(x) = 2x^2 + 5x - 3$, compute $p(A) = 2A^2 + 5A - 3I_n$ for each of the following.

(a) $A = \begin{bmatrix} -3 & 0 \\ 0 & -3 \end{bmatrix}$. (b) $A = \begin{bmatrix} 1 & 2 \\ 0 & 1 \end{bmatrix}$.

(c) $A = \begin{bmatrix} 0 & i \\ i & 0 \end{bmatrix}$. (d) $A = \begin{bmatrix} 1 & i \\ 0 & 0 \end{bmatrix}$.

18. Let $p(x) = x^2 + 1$.

(a) Determine two different 2×2 matrices A of the form kI_2 that satisfy $p(A) = O_2$.

(b) Verify that $p(A) = O_2$, for $A = \begin{bmatrix} 1 & 2 \\ -1 & -1 \end{bmatrix}$.

19. Find all the 2×2 matrices A of the form kI_2 that satisfy $p(A) = O_2$ for $p(x) = x^2 - x - 2$.

20. In Supplementary Exercise 4 in Chapter 1, we introduced the concept of a square root of a matrix with real entries. We can generalize the notion of a square root of a matrix if we permit complex entries.

(a) Compute a complex square root of

$$A = \begin{bmatrix} -1 & 0 \\ 0 & 0 \end{bmatrix}.$$

(b) Compute a complex square root of

$$A = \begin{bmatrix} -2 & 2 \\ 2 & -2 \end{bmatrix}.$$

B.2 COMPLEX NUMBERS IN LINEAR ALGEBRA

Almost everything in the first six chapters of this book remains true if we cross out the word *real* and replace it by the word *complex*. In Section B.1 we introduced matrices with complex entries. Here we consider linear systems of equations with complex entries, determinants of complex matrices, complex vector spaces, complex inner products, and eigenvalues and eigenvectors of complex matrices. The theory we have developed for these topics, using real numbers, is immediately applicable to the case with complex numbers. We could have developed Chapters 1 through 6 with complex numbers, but we limited ourselves to the real numbers for the following reasons:

1. The results would not look markedly different.
2. For many applications the real number situation is adequate; the transition to complex numbers is, as we have just remarked, not too difficult.
3. The computational effort of complex arithmetic is, in most cases, more than double that resulting with only real numbers.

As a matter of fact, we can do most of the first six chapters not only over the real numbers and the complex numbers, but also over any *field* (the real numbers and the complex numbers are familiar examples of a field).

The primary goal of this appendix is to provide an easy transition to complex numbers in linear algebra. This is of particular importance in Chapter 6, where complex eigenvalues and eigenvectors arise naturally for matrices with real entries. Hence we only restate the main theorems in the complex case and provide a discussion and examples of the major ideas needed to accomplish this transition. It will soon be evident that the increased computational effort of complex arithmetic becomes quite tedious if done by hand.

Solving Linear Systems with Complex Entries

The results and techniques dealing with the solution of linear systems that we developed in Chapter 1 carry over directly to linear systems with complex coefficients. We shall illustrate row operations and echelon forms for such systems with Gauss–Jordan reduction, using complex arithmetic.

EXAMPLE 1

Solve the following linear system by Gauss–Jordan reduction:

$$\begin{aligned}(1+i)x_1 + (2+i)x_2 &= 5\\ (2-2i)x_1 + \qquad ix_2 &= 1+2i.\end{aligned}$$

Solution We form the augmented matrix and use elementary row operations to transform it to reduced row echelon form. For the augmented matrix $\left[A \mid B\right]$,

$$\left[\begin{array}{cc|c} 1+i & 2+i & 5\\ 2-2i & i & 1+2i\end{array}\right],$$

multiply the first row by $1/(1+i)$ to obtain

$$\left[\begin{array}{cc|c} 1 & \frac{3}{2}-\frac{1}{2}i & \frac{5}{2}-\frac{5}{2}i \\ 2-2i & i & 1+2i \end{array}\right].$$

We now add $-(2-2i)$ times the first row to the second row to get

$$\left[\begin{array}{cc|c} 1 & \frac{3}{2}-\frac{1}{2}i & \frac{5}{2}-\frac{5}{2}i \\ 0 & -2+5i & 1+12i \end{array}\right].$$

Multiply the second row by $1/(-2+5i)$ to obtain

$$\left[\begin{array}{cc|c} 1 & \frac{3}{2}-\frac{1}{2}i & \frac{5}{2}-\frac{5}{2}i \\ 0 & 1 & 2-i \end{array}\right],$$

which is in row echelon form. To get to reduced row echelon form, we add $-\left(\frac{3}{2}-\frac{1}{2}i\right)$ times the second row to the first row to obtain

$$\left[\begin{array}{cc|c} 1 & 0 & 0 \\ 0 & 1 & 2-i \end{array}\right].$$

Hence the solution is $x_1 = 0$ and $x_2 = 2-i$. ■

If you carry out the arithmetic for the row operations in the preceding example, you will feel the burden of the complex arithmetic even though there were just two equations in two unknowns. Gaussian elimination with back substitution can also be used on linear systems with complex coefficients.

EXAMPLE 2

Suppose that the augmented matrix of a linear system has been transformed to the following matrix in row echelon form:

$$\left[\begin{array}{ccc|c} 1 & 0 & 1+i & -1 \\ 0 & 1 & 3i & 2+i \\ 0 & 0 & 1 & 2i \end{array}\right].$$

The back substitution procedure gives us

$$\begin{aligned} x_3 &= 2i \\ x_2 &= 2+i-3i(2i) = 2+i+6 = 8+i \\ x_1 &= -1-(1+i)(2i) = -1-2i+2 = 3-2i. \end{aligned}$$ ■

We can alleviate the tediousness of complex arithmetic by using computers to solve linear systems with complex entries. However, we must still pay a high price because the execution time will be approximately twice as long as that for the same size linear system with all real entries. We can illustrate this by showing how to transform an $n \times n$ linear system with complex coefficients to a $2n \times 2n$ linear system with only real coefficients.

EXAMPLE 3

Consider the linear system

$$\begin{aligned}(2+i)x_1 + \ (1+i)x_2 &= 3+6i\\ (3-i)x_1 + (2-2i)x_2 &= 7-i.\end{aligned}$$

If we let $x_1 = a_1 + b_1 i$ and $x_2 = a_2 + b_2 i$, with a_1, b_1, a_2, and b_2 real numbers, then we can write this system in matrix form as

$$\begin{bmatrix} 2+i & 1+i \\ 3-i & 2-2i \end{bmatrix}\begin{bmatrix} a_1 + b_1 i \\ a_2 + b_2 i \end{bmatrix} = \begin{bmatrix} 3+6i \\ 7-i \end{bmatrix}.$$

We first rewrite the given linear system as

$$\left(\begin{bmatrix} 2 & 1 \\ 3 & 2 \end{bmatrix} + i\begin{bmatrix} 1 & 1 \\ -1 & -2 \end{bmatrix}\right)\left(\begin{bmatrix} a_1 \\ a_2 \end{bmatrix} + i\begin{bmatrix} b_1 \\ b_2 \end{bmatrix}\right) = \begin{bmatrix} 3 \\ 7 \end{bmatrix} + i\begin{bmatrix} 6 \\ -1 \end{bmatrix}.$$

Multiplying, we have

$$\left(\begin{bmatrix} 2 & 1 \\ 3 & 2 \end{bmatrix}\begin{bmatrix} a_1 \\ a_2 \end{bmatrix} - \begin{bmatrix} 1 & 1 \\ -1 & -2 \end{bmatrix}\begin{bmatrix} b_1 \\ b_2 \end{bmatrix}\right)$$
$$+ i\left(\begin{bmatrix} 2 & 1 \\ 3 & 2 \end{bmatrix}\begin{bmatrix} b_1 \\ b_2 \end{bmatrix} + \begin{bmatrix} 1 & 1 \\ -1 & -2 \end{bmatrix}\begin{bmatrix} a_1 \\ a_2 \end{bmatrix}\right) = \begin{bmatrix} 3 \\ 7 \end{bmatrix} + i\begin{bmatrix} 6 \\ -1 \end{bmatrix}.$$

The real and imaginary parts on both sides of the equation must agree, respectively, and so we have

$$\begin{bmatrix} 2 & 1 \\ 3 & 2 \end{bmatrix}\begin{bmatrix} a_1 \\ a_2 \end{bmatrix} - \begin{bmatrix} 1 & 1 \\ -1 & -2 \end{bmatrix}\begin{bmatrix} b_1 \\ b_2 \end{bmatrix} = \begin{bmatrix} 3 \\ 7 \end{bmatrix}$$

and

$$\begin{bmatrix} 2 & 1 \\ 3 & 2 \end{bmatrix}\begin{bmatrix} b_1 \\ b_2 \end{bmatrix} + \begin{bmatrix} 1 & 1 \\ -1 & -2 \end{bmatrix}\begin{bmatrix} a_1 \\ a_2 \end{bmatrix} = \begin{bmatrix} 6 \\ -1 \end{bmatrix}.$$

This leads to the linear system

$$\begin{aligned} 2a_1 + a_2 - b_1 - b_2 &= 3\\ 3a_1 + 2a_2 + b_1 + 2b_2 &= 7\\ a_1 + a_2 + 2b_1 + b_2 &= 6\\ -a_1 - 2a_2 + 3b_1 + 2b_2 &= -1,\end{aligned}$$

which can be written as

$$\begin{bmatrix} 2 & 1 & -1 & -1 \\ 3 & 2 & 1 & 2 \\ 1 & 1 & 2 & 1 \\ -1 & -2 & 3 & 2 \end{bmatrix}\begin{bmatrix} a_1 \\ a_2 \\ b_1 \\ b_2 \end{bmatrix} = \begin{bmatrix} 3 \\ 7 \\ 6 \\ -1 \end{bmatrix}.$$

This linear system of four equations in four unknowns is now solved as in Chapter 1. The solution is (verify) $a_1 = 1$, $a_2 = 2$, $b_1 = 2$, and $b_2 = -1$. Thus $x_1 = 1 + 2i$ and $x_2 = 2 - i$ is the solution to the given linear system. ■

Determinants of Complex Matrices

The definition of a determinant and all the properties derived in Chapter 5 apply to matrices with complex entries. The following example is an illustration.

EXAMPLE 4

Let A be the coefficient matrix of Example 3. Compute $|A|$.

Solution

$$\begin{aligned}\begin{vmatrix} 2+i & 1+i \\ 3-i & 2-2i \end{vmatrix} &= (2+i)(2-2i) - (3-i)(1+i) \\ &= (6-2i) - (4+2i) \\ &= 2-4i.\end{aligned}$$

■

Complex Vector Spaces

A **complex vector space** is defined exactly as was a real vector space in Definition 2.4, except that the scalars in properties 5 through 8 are permitted to be complex numbers. The terms *complex* vector space and *real* vector space emphasize the set from which the scalars are chosen. It happens that, in order to satisfy the closure property of scalar multiplication [Definition 2.4(b)] in a complex vector space, we must, in most examples, consider vectors that involve complex numbers.

Most of the real vector spaces of Chapter 2 have complex vector space analogs.

EXAMPLE 5

(a) Consider C^n, the set of all $n \times 1$ matrices

$$\begin{bmatrix} a_1 \\ a_2 \\ \vdots \\ a_n \end{bmatrix}$$

with complex entries. Let the operation $\oplus$ be matrix addition and let the operation $\odot$ be multiplication of a matrix by a complex number. We can verify that C^n is a complex vector space by using the properties of matrices established in Section 1.3 and the properties of complex arithmetic established in Section B.1. (Note that if the operation $\odot$ is taken as multiplication of a matrix by a real number, then C^n is a real vector space whose vectors have complex components.)

(b) The set of all $m \times n$ matrices, with complex entries with matrix addition as $\oplus$ and multiplication of a matrix by a complex number as $\odot$, is a complex vector space (verify). We denote this vector space by C_{mn}.

(c) The set of polynomials, with complex coefficients with polynomial addition as $\oplus$ and multiplication of a polynomial by a complex constant as $\odot$, forms a complex vector space. Verification follows the pattern of Example 5 in Section 2.2.

(d) The set of complex-valued continuous functions defined on the interval $[a, b]$ (i.e., all functions of the form $f(t) = f_1(t) + if_2(t)$, where f_1 and f_2 are real-valued continuous functions on $[a, b]$), with $\oplus$ defined by $(f \oplus g)(t) = f(t) + g(t)$ and $\odot$ defined by $(c \odot f)(t) = cf(t)$ for a complex scalar c, forms a complex vector space. The corresponding real vector space is given in Example 6 in Section 2.2 for the interval $(-\infty, \infty)$. ■

A **complex vector subspace** W of a complex vector space V is defined as in Definition 2.5, but with real scalars replaced by complex ones. The analog of Theorem 2.3 can be proved to show that a nonempty subset W of a complex vector space V is a complex vector subspace if and only if the following conditions hold:

(a) If $\mathbf{u}$ and $\mathbf{v}$ are any vectors in W, then $\mathbf{u} \oplus \mathbf{v}$ is in W.

(b) If c is any complex number and $\mathbf{u}$ is any vector in W, then $c \odot \mathbf{u}$ is in W.

EXAMPLE 6

(a) Let W be the set of all vectors in C_{31} of the form $\begin{bmatrix} a \\ 0 \\ b \end{bmatrix}$, where a and b are complex numbers. It follows that

$$\begin{bmatrix} a \\ 0 \\ b \end{bmatrix} \oplus \begin{bmatrix} d \\ 0 \\ e \end{bmatrix} = \begin{bmatrix} a+d \\ 0 \\ b+e \end{bmatrix}$$

belongs to W and, for any complex scalar c,

$$c \odot \begin{bmatrix} a \\ 0 \\ b \end{bmatrix} = \begin{bmatrix} ca \\ 0 \\ cb \end{bmatrix}$$

belongs to W. Hence W is a complex vector subspace of C_{31}.

(b) Let W be the set of all vectors in C_{mn} having only real entries. If $A = \begin{bmatrix} a_{ij} \end{bmatrix}$ and $B = \begin{bmatrix} b_{ij} \end{bmatrix}$ belong to W, then so will $A \oplus B$, because if a_{ij} and b_{ij} are real, then so is their sum. However, if c is any complex scalar and A belongs to W, then $c \odot A = cA$ can have entries ca_{ij} that need not be real numbers. It follows that $c \odot A$ need not belong to W, so W is not a complex vector subspace. ■

Linear Independence and Basis in Complex Vector Spaces

The notions of linear combinations, spanning sets, linear dependence, linear independence, and basis are unchanged for complex vector spaces, except that we use complex scalars (see Sections 2.4 and 2.5).

EXAMPLE 7

Let V be the complex vector space C^3. Let

$$\mathbf{v}_1 = \begin{bmatrix} 1 \\ i \\ 0 \end{bmatrix}, \quad \mathbf{v}_2 = \begin{bmatrix} i \\ 0 \\ 1+i \end{bmatrix}, \quad \text{and} \quad \mathbf{v}_3 = \begin{bmatrix} 1 \\ 1 \\ 1 \end{bmatrix}.$$

(a) Determine whether $\mathbf{v} = \begin{bmatrix} -1 \\ -3+3i \\ -4+i \end{bmatrix}$ is a linear combination of $\mathbf{v}_1$, $\mathbf{v}_2$, and $\mathbf{v}_3$.

(b) Determine whether $\{\mathbf{v}_1, \mathbf{v}_2, \mathbf{v}_3\}$ spans C^3.

(c) Determine whether $\{\mathbf{v}_1, \mathbf{v}_2, \mathbf{v}_3\}$ is a linearly independent subset of C^3.

(d) Is $\{\mathbf{v}_1, \mathbf{v}_2, \mathbf{v}_3\}$ a basis for C^3?

Solution

(a) We proceed as in Example 1 of Section 2.4. We form a linear combination of $\mathbf{v}_1$, $\mathbf{v}_2$, and $\mathbf{v}_3$, with unknown coefficients a_1, a_2, and a_3, respectively, and set it equal to $\mathbf{v}$:

$$a_1\mathbf{v}_1 + a_2\mathbf{v}_2 + a_3\mathbf{v}_3 = \mathbf{v}.$$

If we substitute the vectors $\mathbf{v}_1$, $\mathbf{v}_2$, $\mathbf{v}_3$, and $\mathbf{v}$ into this expression, we obtain (verify) the linear system

$$\begin{aligned} a_1 + \quad\quad ia_2 + a_3 &= -1 \\ ia_1 \quad\quad\quad\quad + a_3 &= -3+3i \\ (1+i)a_2 + a_3 &= -4+i. \end{aligned}$$

We next investigate the consistency of this linear system by using elementary row operations to transform its augmented matrix to either row echelon or reduced row echelon form. A row echelon form is (verify)

$$\left[\begin{array}{ccc|c} 1 & i & 1 & -1 \\ 0 & 1 & 1-i & -3+4i \\ 0 & 0 & 1 & -3 \end{array}\right],$$

which implies that the system is consistent; hence $\mathbf{v}$ is a linear combination of $\mathbf{v}_1$, $\mathbf{v}_2$, and $\mathbf{v}_3$. In fact, back substitution gives (verify) $a_1 = 3$, $a_2 = i$, and $a_3 = -3$.

(b) Let $\mathbf{v} = \begin{bmatrix} c_1 \\ c_2 \\ c_3 \end{bmatrix}$ be an arbitrary vector of C^3. We form the linear combination

$$a_1\mathbf{v}_1 + a_2\mathbf{v}_2 + a_3\mathbf{v}_3 = \mathbf{v}$$

and solve for a_1, a_2, and a_3. The resulting linear system is

$$\begin{aligned} a_1 + \quad\quad ia_2 + a_3 &= c_1 \\ ia_1 \quad\quad\quad\quad + a_3 &= c_2 \\ (1+i)a_2 + a_3 &= c_3. \end{aligned}$$

Transforming the augmented matrix to row echelon form, we obtain (verify)

$$\left[\begin{array}{ccc|c} 1 & i & 1 & c_1 \\ 0 & 1 & 1-i & c_2 - ic_1 \\ 0 & 0 & 1 & -c_3 + (1+i)(c_2 - ic_1) \end{array}\right].$$

Hence we can solve for a_1, a_2, a_3 for any choice of complex numbers c_1, c_2, c_3, which implies that $\{\mathbf{v}_1, \mathbf{v}_2, \mathbf{v}_3\}$ spans C^3.

(c) Proceeding as in Example 6 of Section 2.4, we form the equation

$$a_1\mathbf{v}_1 + a_2\mathbf{v}_2 + a_3\mathbf{v}_3 = \mathbf{0}$$

and solve for a_1, a_2, and a_3. The resulting homogeneous system is

$$\begin{aligned} a_1 + \quad\quad ia_2 + a_3 &= 0 \\ ia_1 \quad\quad\quad\quad + a_3 &= 0 \\ (1+i)a_2 + a_3 &= 0. \end{aligned}$$

Transforming the augmented matrix to row echelon form, we obtain (verify)

$$\left[\begin{array}{ccc|c} 1 & i & 1 & 0 \\ 0 & 1 & 1-i & 0 \\ 0 & 0 & 1 & 0 \end{array}\right],$$

and hence the only solution is $a_1 = a_2 = a_3 = 0$, showing that $\{\mathbf{v}_1, \mathbf{v}_2, \mathbf{v}_3\}$ is linearly independent.

(d) Yes, because $\mathbf{v}_1$, $\mathbf{v}_2$, and $\mathbf{v}_3$ span C^3 [part (b)] and they are linearly independent [part (c)]. ■

Just as for real vector spaces, the questions of spanning sets, linearly independent or linearly dependent sets, and basis in a complex vector space are resolved by using an appropriate linear system. The definition of the dimension of a complex vector space is the same as that given in Definition 2.11. In discussing the dimension of a complex vector space like C^n, we must adjust our intuitive picture. For example, C^1 consists of all complex multiples of a single nonzero vector. This collection can be put into one-to-one correspondence with the complex numbers themselves, that is, with all the points in the complex plane (see Figure B.1). Since the elements of a two-dimensional real vector space can be put into a one-to-one correspondence with the points of R^2 (see Section 2.1), we see that a complex vector space of dimension one has a geometric model that is in one-to-one correspondence with a geometric model of a two-dimensional real vector space. Similarly, a complex vector space of dimension two is the same, geometrically, as a four-dimensional real vector space.

Complex Inner Products

Let V be a complex vector space. An **inner product** on V is a function that assigns to each ordered pair of vectors $\mathbf{u}$, $\mathbf{v}$ in V a complex number $(\mathbf{u}, \mathbf{v})$ satisfying:

(a) $(\mathbf{u}, \mathbf{v}) \geq 0$; $(\mathbf{u}, \mathbf{u}) = 0$ if and only if $\mathbf{u} = \mathbf{0}_V$.

(b) $\overline{(\mathbf{v}, \mathbf{u})} = (\mathbf{u}, \mathbf{v})$ for any $\mathbf{u}$, $\mathbf{v}$ in V.
(c) $(\mathbf{u} + \mathbf{v}, \mathbf{w}) = (\mathbf{u}, \mathbf{w}) + (\mathbf{v}, \mathbf{w})$ for any $\mathbf{u}$, $\mathbf{v}$, $\mathbf{w}$ in V.
(d) $(c\mathbf{u}, \mathbf{v}) = c(\mathbf{u}, \mathbf{v})$ for any $\mathbf{u}$, $\mathbf{v}$ in V and c a complex scalar.

Remark Observe how similar this definition is to Definition 3.2 of a real inner product.

EXAMPLE 8

We can define the **standard inner product** on C^n be defining $(\mathbf{u}, \mathbf{v})$ for

$$\mathbf{u} = \begin{bmatrix} u_1 \\ u_2 \\ \vdots \\ u_n \end{bmatrix} \quad \text{and} \quad \mathbf{v} = \begin{bmatrix} v_1 \\ v_2 \\ \vdots \\ v_n \end{bmatrix} \quad \text{in } C^n \text{ as}$$

$$(\mathbf{u}, \mathbf{v}) = u_1\overline{v}_1 + u_2\overline{v}_2 + \cdots + u_n\overline{v}_n,$$

which can also be expressed as

$$(\mathbf{u}, \mathbf{v}) = \mathbf{u}^T\,\overline{\mathbf{v}}.$$

Thus, if

$$\mathbf{u} = \begin{bmatrix} 1-i \\ 2 \\ -3+2i \end{bmatrix} \quad \text{and} \quad \mathbf{v} = \begin{bmatrix} 3+2i \\ 3-4i \\ -3i \end{bmatrix}$$

are vectors in C^3, then

$$\begin{aligned}(\mathbf{u}, \mathbf{v}) &= (1-i)\overline{(3+2i)} + 2\overline{(3-4i)} + (-3+2i)\overline{(-3i)} \\ &= (1-5i) + (6+8i) + (-6-9i) \\ &= 1-6i.\end{aligned}$$

■

A complex vector space that has a complex inner product defined on it is called a **complex inner product space**. If V is a complex inner product space, then we can define the length of a vector $\mathbf{u}$ in V exactly as in the real case:

$$\|\mathbf{u}\| = \sqrt{(\mathbf{u}, \mathbf{u})}.$$

Moreover, the vectors $\mathbf{u}$ and $\mathbf{v}$ in V are said to be **orthogonal** if $(\mathbf{u}, \mathbf{v}) = 0$.

Complex Eigenvalues and Eigenvectors

Let V be an n-dimensional complex vector space and let $L\colon V \to V$ be a linear operator. If $S = \{\mathbf{x}_1, \mathbf{x}_2, \ldots, \mathbf{x}_n\}$ is a basis for V such that L is represented by the $n \times n$ complex matrix A, then for $\mathbf{x}$ in V,

$$L(\mathbf{x}) = A\mathbf{x}. \tag{1}$$

Imitating Definition 6.1, we say that the complex number λ is an **eigenvalue** of L if there exists a nonzero complex vector $\mathbf{x}$ in V such that

$$L(\mathbf{x}) = \lambda\mathbf{x}. \tag{2}$$

Every nonzero complex vector $\mathbf{x}$ satisfying this equation is called an **eigenvector** of L **associated with the eigenvalue** λ. Substituting Equation (1) into Equation (2), we have the relationship

$$A\mathbf{x} = \lambda\mathbf{x}. \tag{3}$$

If λ is a complex scalar and $\mathbf{x} \neq \mathbf{0}_V$ is a vector in the complex vector space V such that Equation (3) is satisfied, then we say that λ is an **eigenvalue** of the complex matrix A and $\mathbf{x}$ is an **eigenvector** of A **associated with** λ. Equation (3) can be rewritten as the homogeneous system

$$(\lambda I_n - A)\mathbf{x} = \mathbf{0}. \tag{4}$$

This homogeneous system has a nonzero solution $\mathbf{x}$ if and only if

$$\det(\lambda I_n - A) = 0$$

has a solution. Following the development in Section 6.1, we call $\det(\lambda I_n - A)$ the **characteristic polynomial** of the matrix A, which is a complex polynomial of degree n in λ. As in Section 6.1, the eigenvalues of the complex matrix A are the complex roots of the characteristic polynomial. According to the Fundamental Theorem of Algebra, an nth-degree polynomial has exactly n roots, if multiple roots are counted. Even if $\det(\lambda I_n - A)$ has only one root of multiplicity n, we then know that the characteristic polynomial of any complex matrix A, or equivalently of any linear operator L mapping a complex vector space into itself, always has at least one root (possibly complex). This is quite different from the case of real vector spaces discussed in Section 6.2, because there are real matrices that do not have real eigenvalues (see Example 5 in Section 6.2).

EXAMPLE 9

Let

$$A = \begin{bmatrix} 0 & 1 \\ -1 & 0 \end{bmatrix}.$$

Then the characteristic polynomial of A is

$$\det(\lambda I_2 - A) = \lambda^2 + 1.$$

If we interpret A as representing a linear transformation between real vector spaces, then A has no (real) eigenvalues because the roots of $f(\lambda) = \lambda^2 + 1$ are i and $-i$. However, if we interpret A as representing a linear transformation between complex vector spaces, then we know that there is always at least one (complex) eigenvalue, and in this case the eigenvalues are i and $-i$. Associated (complex) eigenvectors are obtained by finding a nontrivial solution of the homogeneous systems

$$(iI_2 - A)\mathbf{x} = \mathbf{0} \quad \text{and} \quad (-iI_2 - A)\mathbf{x} = \mathbf{0},$$

respectively. An eigenvector associated with eigenvalue i is $\begin{bmatrix} 1 \\ i \end{bmatrix}$ and an eigenvector associated with eigenvalue $-i$ is $\begin{bmatrix} -1 \\ i \end{bmatrix}$ (verify). ■

In the case of complex matrices, we have the following analogs of the theorems presented in Section 6.4, which show the role played by the special matrices discussed in Section 6.4.

Theorem B.1

If A is a Hermitian matrix, then the eigenvalues of A are all real. Moreover, eigenvectors belonging to distinct eigenvalues are orthogonal (complex analog of Theorems 6.6 *and* 6.7). ●

Theorem B.2

If A is a Hermitian matrix, then there exists a unitary matrix U such that $U^{-1}AU = D$, a diagonal matrix. The eigenvalues of A lie on the main diagonal of D (complex analog of Theorem 6.9). ●

In Section 6.2 we proved that if A is a real symmetric matrix, then there exists an orthogonal matrix P such that $P^{-1}AP = D$, a diagonal matrix, and conversely, if there is an orthogonal matrix P such that $P^{-1}AP$ is a diagonal matrix, then A is a symmetric matrix. For complex matrices, the situation is more complicated. The converse of Theorem B.2 is not true. That is, if A is a matrix for which there exists a unitary matrix U such that $U^{-1}AU = D$, a diagonal matrix, then A need not be a Hermitian matrix. The correct statement involves normal matrices. The following result can be established.

Theorem B.3

If A is a normal matrix, then there exists a unitary matrix U such that $U^{-1}AU = D$, a diagonal matrix. Conversely, if A is a matrix for which there exists a unitary matrix U such that $U^{-1}AU = D$, a diagonal matrix, then A is a normal matrix. ●

Every real $n \times n$ matrix can be considered as an $n \times n$ complex matrix. Hence the eigenvalue and eigenvector computations for real matrices carried out in Chapter 6 can "be fixed up" by permitting complex roots of the characteristic polynomial to be called eigenvalues. Of course, this means that we would need to use complex arithmetic to solve for an associated eigenvector. However, this "fix" does not apply when we consider a linear operator $L: V \to V$ where V is a real vector space, because a complex number with nonzero imaginary part does not belong to the field of scalars (the real numbers) in this case. Thus, if $\lambda = a + bi$, $b \neq 0$, is a root of the characteristic polynomial of L, the equation

$$L(\mathbf{x}) = \lambda \mathbf{x}$$

cannot hold. However, the information derived from such a procedure can reveal fundamental properties of L.

So, if as in Chapter 6, we restrict ourselves to real vector spaces, we must permit only real numbers to be eigenvalues. The eigen-problem for symmetric matrices, Section 6.4, needs no "fix" because the eigenvalues of any real symmetric matrix are all real numbers (see Theorem 6.6).

B.2 EXERCISES

1. Solve by using Gauss–Jordan reduction.

(a) $\begin{aligned}(1+2i)x_1 + (-2+i)x_2 &= 1-3i\\ (2+i)x_1 + (-1+2i)x_2 &= -1-i.\end{aligned}$

(b) $\begin{aligned}2ix_1 - (1-i)x_2 &= 1+i\\ (1-i)x_1 + x_2 &= 1-i.\end{aligned}$

(c) $\begin{aligned}(1+i)x_1 - x_2 &= -2+i\\ 2ix_1 + (1-i)x_2 &= i.\end{aligned}$

2. Transform the given augmented matrix of a linear system to row echelon form and solve by back substitution.

(a) $\left[\begin{array}{ccc|c} 2 & i & 0 & 1-i \\ 0 & 3i & -2+i & 4 \\ 0 & 0 & 2+i & 2-i \end{array}\right]$.

(b) $\left[\begin{array}{ccc|c} i & 2 & 1+i & 3i \\ 0 & 1-i & 0 & 2+i \\ 0 & 0 & 3 & 6-3i \end{array}\right]$.

3. Solve by Gaussian elimination with back substitution.

(a) $\begin{aligned}ix_1 + (1+i)x_2 \qquad &= i\\ (1-i)x_1 + x_2 - ix_3 &= 1\\ ix_2 + x_3 &= 1.\end{aligned}$

(b) $\begin{aligned}x_1 + ix_2 + (1-i)x_3 &= 2+i\\ ix_1 \qquad + (1+i)x_3 &= -1+i\\ 2ix_2 - x_3 &= 2-i.\end{aligned}$

4. Compute the determinant and simplify as much as possible.

(a) $\begin{vmatrix} 1+i & -1 \\ 2i & 1+i \end{vmatrix}$.

(b) $\begin{vmatrix} 2-i & 1+i \\ 1+2i & -(1-i) \end{vmatrix}$.

(c) $\begin{vmatrix} 1+i & 2 & 2-i \\ i & 0 & 3+i \\ -2 & 1 & 1+2i \end{vmatrix}$.

(d) $\begin{vmatrix} 2 & 1-i & 0 \\ 1+i & -1 & i \\ 0 & -i & 2 \end{vmatrix}$.

5. Find the inverse of each of the following matrices if possible.

(a) $\begin{bmatrix} i & 2 \\ 1+i & -i \end{bmatrix}$. (b) $\begin{bmatrix} 2 & i & 3 \\ 1+i & 0 & 1-i \\ 2 & 1 & 2+i \end{bmatrix}$.

6. Determine whether the following subsets W of C_{22} are complex vector subspaces.

(a) W is the set of all 2×2 complex matrices with zeros on the main diagonal.

(b) W is the set of all 2×2 complex matrices that have diagonal entries with real part equal to zero.

(c) W is the set of all symmetric 2×2 complex matrices.

7. (a) Prove or disprove: The set W of all $n \times n$ Hermitian matrices is a complex vector subspace of C_{nn}.

(b) Prove or disprove: The set W of all $n \times n$ Hermitian matrices is a real vector subspace of the real vector space of all $n \times n$ complex matrices.

8. Prove or disprove: The set W of all $n \times n$ unitary matrices is a complex vector subspace of C_{nn}.

9. Let $W = \text{span}\{\mathbf{v}_1, \mathbf{v}_2, \mathbf{v}_3\}$, where

$$\mathbf{v}_1 = \begin{bmatrix} -1+i \\ 2 \\ 1 \end{bmatrix}, \quad \mathbf{v}_2 = \begin{bmatrix} 1 \\ 1+i \\ i \end{bmatrix},$$

$$\mathbf{v}_3 = \begin{bmatrix} -5+2i \\ -1-3i \\ 2-3i \end{bmatrix}.$$

(a) Does $\mathbf{v} = \begin{bmatrix} i \\ 0 \\ 0 \end{bmatrix}$ belong to W?

(b) Is the set $\{\mathbf{v}_1, \mathbf{v}_2, \mathbf{v}_3\}$ linearly independent or linearly dependent?

10. Let $\{\mathbf{v}_1, \mathbf{v}_2, \mathbf{v}_3\}$ be a basis for a complex vector space V. Determine whether or not $\mathbf{w}$ is in span $\{\mathbf{w}_1, \mathbf{w}_2\}$.

(a) $\mathbf{w}_1 = i\mathbf{v}_1 + (1 - i)\mathbf{v}_2 + 2\mathbf{v}_3$,
$\mathbf{w}_2 = (2 + i)\mathbf{v}_1 + 2i\mathbf{v}_2 + (3 - i)\mathbf{v}_3$,
$\mathbf{w} = (-2 - 3i)\mathbf{v}_1 + (3 - i)\mathbf{v}_2 + (-2 - 2i)\mathbf{v}_3$.

(b) $\mathbf{w}_1 = 2i\mathbf{v}_1 + \mathbf{v}_2 + (1 - i)\mathbf{v}_3$,
$\mathbf{w}_2 = 3i\mathbf{v}_1 + (1 + i)\mathbf{v}_2 + 3\mathbf{v}_3$,
$\mathbf{w} = (2 + 3i)\mathbf{v}_1 + (2 + i)\mathbf{v}_2 + (4 - 2i)\mathbf{v}_3$.

11. Find the eigenvalues and associated eigenvectors of the following complex matrices.

(a) $A = \begin{bmatrix} 1 & 1 \\ -1 & 1 \end{bmatrix}$. (b) $A = \begin{bmatrix} 1 & i \\ -i & 1 \end{bmatrix}$.

(c) $A = \begin{bmatrix} 2 & 0 & 0 \\ 0 & 2 & i \\ 0 & -i & 2 \end{bmatrix}$.

12. For each of the parts in Exercise 11, find a matrix P such that $P^{-1}AP = D$, a diagonal matrix. For part (c), find three different matrices P that diagonalize A.

13. (a) Prove that if A is Hermitian, then the eigenvalues of A are real.

(b) Verify that A in Exercise 11(c) is Hermitian.

(c) Are the eigenvectors associated with an eigenvalue of a Hermitian matrix guaranteed to be real vectors? Explain.

14. Prove that an $n \times n$ complex matrix A is unitary if and only if the columns (rows) of A form an orthonormal set with respect to the standard inner product on C^n. (*Hint*: See Theorem 6.8.)

15. Show that if A is a skew Hermitian matrix (see Exercise 15 in Section B.1) and λ is an eigenvalue of A, then the real part of λ is zero.

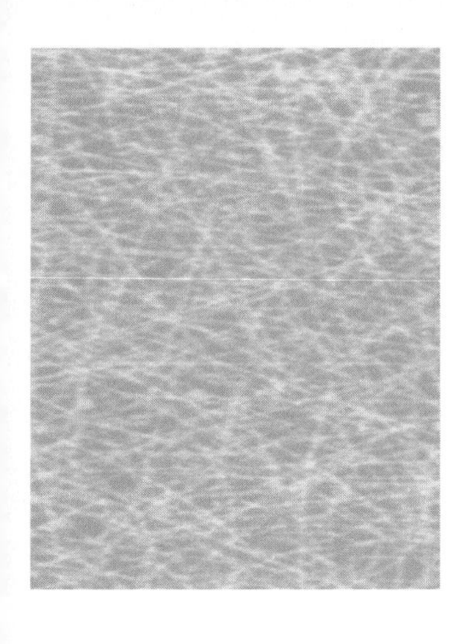

ANSWERS TO ODD-NUMBERED EXERCISES

CHAPTER 1

Section 1.1, p. 7

1. $x_1 = 4, x_2 = 2.$

3. $x_1 = -4, x_2 = 2, x_3 = 10.$

5. $x_1 = 2, x_2 = -1, x_3 = -2.$

7. $x_1 = -20, x_2 = \left(\frac{1}{4}\right) x_3 + 8, x_3 =$ any real number.

9. This linear system has no solution. It is inconsistent.

11. $x_1 = 5, x_2 = 1.$

13. This linear system has no solution. It is inconsistent.

15. (a) $t = 10$. (b) One value is $t = 3$.

(c) The choice $t = 3$ in part (b) was arbitrary. Any choice for t, other than $t = 10$, makes the system inconsistent. Hence there are infinitely many ways to choose a value for t in part (b).

19. $x_1 = 2, x_2 = 1, x_3 = 0.$

21. There is no such value of r.

Section 1.2, p. 18

1. (a) $-3, -5, 4$. (b) $4, 5$. (c) $2, 6, -4$.

3. (a) $\begin{bmatrix} 5 & -5 & 8 \\ 4 & 2 & 9 \\ 5 & 3 & 4 \end{bmatrix}$.

(b) $AB = \begin{bmatrix} 14 & 8 \\ 16 & 9 \end{bmatrix}$, $BA = \begin{bmatrix} 1 & 2 & 3 \\ 4 & 5 & 10 \\ 7 & 8 & 17 \end{bmatrix}$.

(c) $\begin{bmatrix} 0 & 10 & -9 \\ 8 & -1 & -2 \\ -5 & -4 & 3 \end{bmatrix}$.

(d) Impossible. (e) $\begin{bmatrix} 19 & -8 \\ 32 & 30 \end{bmatrix}$.

5. (a) $\begin{bmatrix} -2 & 12 \\ 2 & 17 \\ 10 & 13 \end{bmatrix}$. (b) $\begin{bmatrix} 8 & 12 \\ 12 & -1 \end{bmatrix}$.

(c) $3(2A) = \begin{bmatrix} 6 & 12 & 18 \\ 12 & 6 & 24 \end{bmatrix} = 6A$.

(d) Impossible. (e) Impossible.

7. (a) $\begin{bmatrix} 1 & 2 \\ 2 & 1 \\ 3 & 4 \end{bmatrix}$. (b) $(A^T)^T = A$.

(c) $\begin{bmatrix} 14 & 16 \\ 8 & 9 \end{bmatrix}$. (d) Same as (c).

(e) $(C + E)^T = C^T + E^T = \begin{bmatrix} 5 & 4 & 5 \\ -5 & 2 & 3 \\ 8 & 9 & 4 \end{bmatrix}$.

(f) $A(2B) = 2(AB) = \begin{bmatrix} 28 & 16 \\ 32 & 18 \end{bmatrix}$.

11. There are infinitely many choices. For example, $r = 1$, $s = 0$; or $r = 0$, $s = 2$; or $r = 10$, $s = -18$.

17.
$$\begin{aligned} -2x_1 - x_2 \qquad\quad + 4x_4 &= 5 \\ -3x_1 + 2x_2 + 7x_3 + 8x_4 &= 3 \\ x_1 \qquad\qquad\quad + 2x_4 &= 4 \\ 3x_1 \qquad + x_3 + 3x_4 &= 6. \end{aligned}$$

19. $\begin{bmatrix} 2 & 3 & 0 \\ 0 & 3 & 1 \\ 2 & -1 & 0 \end{bmatrix} \begin{bmatrix} x_1 \\ x_2 \\ x_3 \end{bmatrix} = \begin{bmatrix} 0 \\ 0 \\ 0 \end{bmatrix}$.

21. The linear systems are equivalent. That is, they have the same solutions.

23. (a) $\begin{aligned} 2x_1 + x_2 &= 4 \\ 3x_2 &= 2. \end{aligned}$

(b) $\begin{aligned} x_1 + 3x_3 + x_4 &= 2 \\ 2x_1 + x_2 + 4x_3 + 3x_4 &= 5 \\ -x_1 + 2x_2 + 5x_3 + 4x_4 &= 8. \end{aligned}$

25. (a) $\begin{aligned} x_1 + 3x_2 &= 3 \\ 2x_1 &= 1 \\ x_1 - x_2 &= 4. \end{aligned}$

(b) $\begin{aligned} 2x_1 + 3x_2 &= 0 \\ x_1 - x_2 + x_3 &= 0 \\ 2x_2 - x_3 &= 0 \\ x_1 + 2x_2 + 3x_3 &= 0. \end{aligned}$

27. No.

Section 1.3, p. 27

9. One such pair is $A = \begin{bmatrix} 1 & 1 \\ 2 & 2 \end{bmatrix}$ and $B = \begin{bmatrix} 1 & 1 \\ -1 & -1 \end{bmatrix}$.
Another pair is $A = \begin{bmatrix} 2 & 6 \\ 4 & 12 \end{bmatrix}$ and $B = \begin{bmatrix} -3 & -3 \\ 1 & 1 \end{bmatrix}$.
Yet another pair is $A = O_2$ and $B =$ any 2×2 matrix with at least one nonzero element. There are infinitely many such pairs.

11. There are many such pairs of matrices. For example, $A = \begin{bmatrix} 1 & 1 \\ 0 & 1 \end{bmatrix}$ and $B = \begin{bmatrix} 1 & -1 \\ 0 & 1 \end{bmatrix}$ or $A = \begin{bmatrix} 1 & 1 \\ 1 & 2 \end{bmatrix}$ and $B = \begin{bmatrix} 2 & -1 \\ -1 & 1 \end{bmatrix}$. Note also that for $A = k\begin{bmatrix} 1 & 0 \\ 0 & 1 \end{bmatrix}$ and $B = \left(\frac{1}{k}\right)\begin{bmatrix} 1 & 0 \\ 0 & 1 \end{bmatrix}$, $k \neq 0, \pm 1$, we have $A \neq B$ and $AB = \begin{bmatrix} 1 & 0 \\ 0 & 1 \end{bmatrix}$.

21. $r = 2$.

23. $s = r^2$.

29. $k = \pm\sqrt{\frac{1}{6}}$.

Section 1.4, p. 38

9. $B = \begin{bmatrix} 1 & 3 \\ 3 & 1 \end{bmatrix}$ is such that $AB = BA$. There are infinitely many such matrices B.

25. (a) $A^{-1} = \begin{bmatrix} -\frac{2}{13} & \frac{3}{13} \\ \frac{5}{13} & -\frac{1}{13} \end{bmatrix}$.

(b) $A^{-1} = \begin{bmatrix} -\frac{1}{3} & \frac{2}{3} \\ \frac{2}{3} & -\frac{1}{3} \end{bmatrix}$

27. $\begin{bmatrix} 11 & 19 \\ 7 & 0 \end{bmatrix}$.

29. (a) $\begin{bmatrix} \frac{6}{13} \\ \frac{11}{13} \end{bmatrix}$. (b) $\begin{bmatrix} \frac{8}{13} \\ \frac{19}{13} \end{bmatrix}$.

31. Possible answer: $\begin{bmatrix} 1 & 2 \\ -3 & 4 \end{bmatrix}$ and $\begin{bmatrix} -1 & -2 \\ 3 & -4 \end{bmatrix}$.

Section 1.5, p. 57

1. (a) Possible answer:

$$B = \begin{bmatrix} 1 & 2 & -3 & 1 \\ 0 & 1 & 2 & -1 \\ 0 & 0 & 1 & -\frac{7}{4} \\ 0 & 0 & 0 & 1 \end{bmatrix},$$

$$C = \begin{bmatrix} 1 & 2 & -3 & 1 \\ 0 & 1 & 0 & \frac{5}{2} \\ 0 & 0 & 1 & -\frac{7}{4} \\ 0 & 0 & 0 & 1 \end{bmatrix}.$$

(b) $D = I_4$.

5. (a) $x_1 = 1, x_2 = 2, x_3 = -2$.

(b) $x_1 = 1, x_2 = 2, x_3 = -2$.

7. (a) $x_1 = -1, x_2 = 4, x_3 = -3$.

(b) $x_1 = x_2 = x_3 = 0$.

(c) $x_1 = r, x_2 = -2r, x_3 = r$, where $r =$ any real number.

(d) $x_1 = -2r, x_2 = r, x_3 = 0$, where $r =$ any real number.

9. (a) $x_1 = 1, x_2 = 2, x_3 = 2$.

(b) $x_1 = x_2 = x_3 = 0$.

11. $X = \begin{bmatrix} r \\ r \end{bmatrix}, r \neq 0$.

13. $X = \begin{bmatrix} -\frac{1}{2}r \\ \frac{1}{2}r \\ r \end{bmatrix}, r \neq 0$.

15. (a) $a = \pm\sqrt{3}$. (b) $a \neq \pm\sqrt{3}$.

(c) There is no value of a such that this system has infinitely many solutions.

17. (a) $a = -3$. (b) $a \neq \pm 3$. (c) $a = 3$.

23. (a) $\begin{bmatrix} 1 & 0 & 0 & 0 & 0 \\ 2 & 1 & 0 & 0 & 0 \\ 3 & \frac{5}{3} & 1 & 0 & 0 \end{bmatrix}$.

(b) $\begin{bmatrix} 1 & 0 & 0 & 0 & 0 \\ 0 & 1 & 0 & 0 & 0 \\ 0 & 0 & 1 & 0 & 0 \end{bmatrix}$.

25. $-3a - b + c = 0$.

Section 1.6, p. 67

3. (a) $\begin{bmatrix} 1 & -4 & 0 & 0 \\ 0 & 1 & 0 & 0 \\ 0 & 0 & 1 & 0 \\ 0 & 0 & 0 & 1 \end{bmatrix}$.

(b) $\begin{bmatrix} 1 & 0 & 0 & 0 \\ 0 & 0 & 1 & 0 \\ 0 & 1 & 0 & 0 \\ 0 & 0 & 0 & 1 \end{bmatrix}$.

(c) $\begin{bmatrix} 1 & 0 & 0 & 0 \\ 0 & 1 & 0 & 0 \\ 0 & 0 & 4 & 0 \\ 0 & 0 & 0 & 1 \end{bmatrix}$.

5. (a) $C = \begin{bmatrix} 1 & 0 & 0 \\ 0 & 1 & 0 \\ 0 & 0 & 1 \end{bmatrix}$.

(b) $B = \begin{bmatrix} 1 & 0 & 0 \\ 0 & 1 & 0 \\ 2 & 0 & 1 \end{bmatrix}$.

(c) A and B are inverses of each other.

7. $A^{-1} = \begin{bmatrix} -2 & \frac{3}{2} \\ 1 & -\frac{1}{2} \end{bmatrix}$.

9. (a) Singular. (b) $A^{-1} = \begin{bmatrix} \frac{1}{2} & -\frac{1}{4} \\ \frac{1}{6} & \frac{1}{12} \end{bmatrix}$.

(c) $A^{-1} = \begin{bmatrix} 0 & 1 & -1 \\ 2 & -2 & -1 \\ -1 & 1 & 1 \end{bmatrix}$. (d) Singular.

11. (a) $A^{-1} = \begin{bmatrix} 1 & 0 & -1 \\ 1 & -1 & 2 \\ -1 & 1 & -1 \end{bmatrix}$.

(b) $A^{-1} = \begin{bmatrix} \frac{7}{3} & -\frac{1}{3} & -\frac{1}{3} & -\frac{2}{3} \\ \frac{4}{9} & -\frac{1}{9} & -\frac{4}{9} & \frac{1}{9} \\ -\frac{1}{9} & -\frac{2}{9} & \frac{1}{9} & \frac{2}{9} \\ -\frac{5}{3} & \frac{2}{3} & \frac{2}{3} & \frac{1}{3} \end{bmatrix}$.

(c) Singular.

(d) $A^{-1} = \begin{bmatrix} \frac{3}{2} & -1 & \frac{1}{2} \\ \frac{1}{2} & 0 & -\frac{1}{2} \\ -\frac{3}{2} & 1 & \frac{1}{2} \end{bmatrix}$.

(e) Singular.

13. $A = \begin{bmatrix} 1 & 0 \\ 3 & 1 \end{bmatrix} \begin{bmatrix} 1 & 0 \\ 0 & -2 \end{bmatrix} \begin{bmatrix} 1 & 2 \\ 0 & 1 \end{bmatrix}$.

15. $A = \begin{bmatrix} \frac{1}{2} & -1 \\ -\frac{1}{2} & 2 \end{bmatrix}$.

17. (a) and (b).

19. A^{-1} exists for $a \neq 0$. Then

$$A^{-1} = \begin{bmatrix} 0 & 1 & 0 \\ 1 & -1 & 0 \\ -\dfrac{2}{a} & \dfrac{1}{a} & \dfrac{1}{a} \end{bmatrix}.$$

Section 1.7, p. 73

3. (a) $\begin{bmatrix} 1 & 0 & 0 & 0 \\ 0 & 1 & 0 & 0 \\ 0 & 0 & 0 & 0 \end{bmatrix}$. (b) $\begin{bmatrix} 1 & 0 & 0 \\ 0 & 1 & 0 \\ 0 & 0 & 1 \\ 0 & 0 & 0 \end{bmatrix}$.

(c) $\begin{bmatrix} 1 & 0 & 0 & 0 \\ 0 & 1 & 0 & 0 \\ 0 & 0 & 1 & 0 \\ 0 & 0 & 0 & 0 \\ 0 & 0 & 0 & 0 \end{bmatrix}$. (d) I_3.

7. $B = \begin{bmatrix} 1 & 0 & 0 & 0 \\ 0 & 1 & 0 & 0 \\ 0 & 0 & 0 & 0 \\ 0 & 0 & 0 & 0 \end{bmatrix}$, $P = \begin{bmatrix} -1 & 1 & 0 & 0 \\ 2 & -1 & 0 & 0 \\ -4 & 0 & 1 & 0 \\ 0 & 0 & 0 & 1 \end{bmatrix}$,

$Q = \begin{bmatrix} 1 & 0 & -1 & 2 \\ 0 & 1 & 1 & 5 \\ 0 & 0 & 1 & 0 \\ 0 & 0 & 0 & 1 \end{bmatrix}$.

Section 1.8, p. 80

1. $X = \begin{bmatrix} 1 \\ 2 \\ 1 \end{bmatrix}$.

3. $X = \begin{bmatrix} 1 \\ 0 \\ 2 \\ -4 \end{bmatrix}$.

5. $L = \begin{bmatrix} 1 & 0 & 0 \\ 2 & 1 & 0 \\ 2 & -2 & 1 \end{bmatrix}, U = \begin{bmatrix} 2 & 3 & 4 \\ 0 & -1 & 2 \\ 0 & 0 & -2 \end{bmatrix},$
$X = \begin{bmatrix} 4 \\ -2 \\ 1 \end{bmatrix}$.

7. $L = \begin{bmatrix} 1 & 0 & 0 \\ 0.5 & 1 & 0 \\ 0.25 & -1.5 & 1 \end{bmatrix}, U = \begin{bmatrix} 4 & 2 & 3 \\ 0 & -1 & 3.5 \\ 0 & 0 & 5.5 \end{bmatrix},$
$X = \begin{bmatrix} 2 \\ -2 \\ -1 \end{bmatrix}$.

9. $L = \begin{bmatrix} 1 & 0 & 0 & 0 \\ 0.5 & 1 & 0 & 0 \\ -1 & 0.2 & 1 & 0 \\ 2 & -0.4 & 2 & 1 \end{bmatrix},$
$U = \begin{bmatrix} 2 & 1 & 0 & -4 \\ 0 & -0.5 & 0.25 & 1 \\ 0 & 0 & 0.2 & 2 \\ 0 & 0 & 0 & 2 \end{bmatrix}, X = \begin{bmatrix} 0.5 \\ 2 \\ -2 \\ 1.5 \end{bmatrix}$.

Supplementary Exercises, p. 81

1. (a) 3. (b) 6. (c) 10. (d) $\frac{n}{2}(n+1)$.

3. $\begin{bmatrix} 1 & 0 \\ 0 & 1 \end{bmatrix}, \begin{bmatrix} -1 & 0 \\ 0 & -1 \end{bmatrix}, \begin{bmatrix} 1 & b \\ 0 & -1 \end{bmatrix}, \begin{bmatrix} -1 & b \\ 0 & 1 \end{bmatrix},$
where b is any real number.

5. (a) $a = -4$ or $a = 2$.

(b) Any real number.

CHAPTER 2

Section 2.1, p. 94

1.

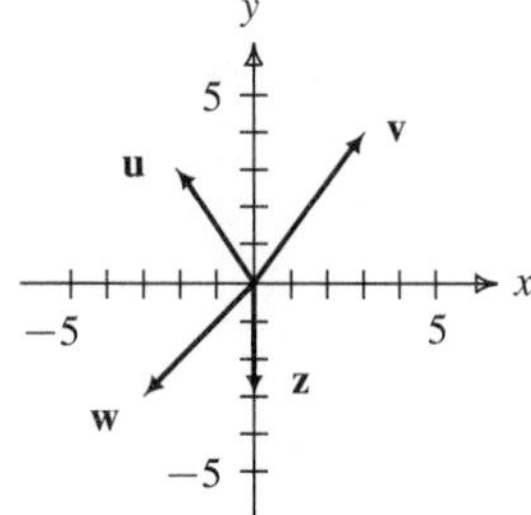

3. Tail $(-1, -4)$.

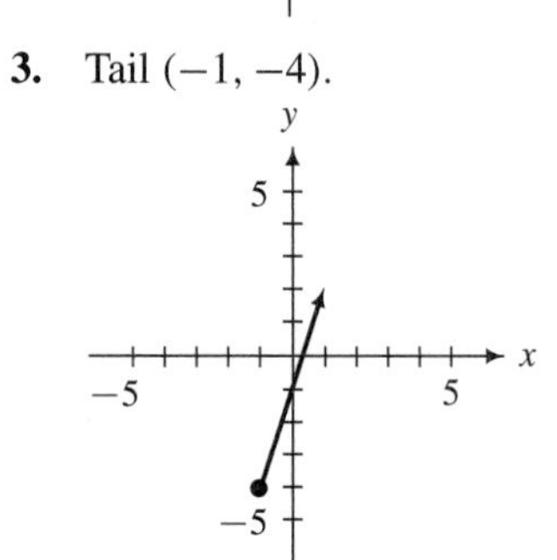

5. $a = 3, b = -1$.

7. (a) $\begin{bmatrix} 2 \\ 3 \end{bmatrix}$. (b) $\begin{bmatrix} -1 \\ 3 \\ -1 \end{bmatrix}$.

9. (a) $\begin{bmatrix} 4 \\ 3 \end{bmatrix}$. (b) $\begin{bmatrix} 2 \\ 3 \\ 7 \end{bmatrix}$.

11. (a) $\mathbf{u}+\mathbf{v} = \begin{bmatrix} 0 \\ 8 \end{bmatrix}, \mathbf{u}-\mathbf{v} = \begin{bmatrix} 4 \\ -2 \end{bmatrix},$
$2\mathbf{u} = \begin{bmatrix} 4 \\ 6 \end{bmatrix}, 3\mathbf{u}-2\mathbf{v} = \begin{bmatrix} 10 \\ -1 \end{bmatrix}$.

(b) $\mathbf{u}+\mathbf{v} = \begin{bmatrix} 3 \\ 5 \end{bmatrix}, \mathbf{u}-\mathbf{v} = \begin{bmatrix} -3 \\ 1 \end{bmatrix},$
$2\mathbf{u} = \begin{bmatrix} 0 \\ 6 \end{bmatrix}, 3\mathbf{u}-2\mathbf{v} = \begin{bmatrix} -6 \\ 5 \end{bmatrix}$.

(c) $\mathbf{u}+\mathbf{v} = \begin{bmatrix} 5 \\ 8 \end{bmatrix}, \mathbf{u}-\mathbf{v} = \begin{bmatrix} -1 \\ 4 \end{bmatrix},$
$2\mathbf{u} = \begin{bmatrix} 4 \\ 12 \end{bmatrix}, 3\mathbf{u}-2\mathbf{v} = \begin{bmatrix} 0 \\ 14 \end{bmatrix}$.

13. (a) $\begin{bmatrix} 1 \\ 5 \\ 3 \end{bmatrix}$. (b) $\begin{bmatrix} -4 \\ -3 \\ -1 \end{bmatrix}$.

(c) $\begin{bmatrix} 1 \\ 6 \\ 2 \end{bmatrix}$. (d) $\begin{bmatrix} -7 \\ 1 \\ 13 \end{bmatrix}$.

15. (a) $r = \frac{1}{2}, s = \frac{3}{2}$.

(b) $r = -2, s = 1, t = -1$.

(c) $r = -2, s = 6$.

17. Impossible.

19. Possible answer: $c_1 = -2, c_2 = -1, c_3 = 1$.

Section 2.2, p. 102

3. Properties (3), (4), (b), (5), (6), and (7).

5. Properties (5), (6), and (8).

7. Property (8).

13. No.

Section 2.3, p. 111

1. (b), (c), and (d).

3. (b) and (c).

5. (a) and (c).

7. (a) and (c).

9. (a) and (b).

11. (b), (c), and (e).

23. (b) and (c).

25. (a) Yes. (b) No. (c) Yes. (d) No.

29. (a) and (c).

31. (a) Possible answer: $x = 2 + 2t, y = -3 + 5t, z = 1 + 4t$.

(b) Possible answer: $x = -3 + 8t, y = -2 + 7t, z = -2 + 6t$.

Section 2.4, p. 122

1. (a), (c), and (d).

3. (a) and (d).

5. No.

7. Possible answer: $\left\{\begin{bmatrix} -1 \\ -1 \\ 1 \\ 0 \end{bmatrix}\right\}$.

9. No.

11. (a) and (c) are linearly dependent; (b) is linearly independent.

(a) $\begin{bmatrix} 3 & 6 & 6 \end{bmatrix} = 2\begin{bmatrix} 1 & 1 & 0 \end{bmatrix} + 1\begin{bmatrix} 0 & 2 & 3 \end{bmatrix} + 1\begin{bmatrix} 1 & 2 & 3 \end{bmatrix}$.

(c) $\begin{bmatrix} 0 & 0 & 0 \end{bmatrix} = 0\begin{bmatrix} 1 & 1 & 0 \end{bmatrix} + 0\begin{bmatrix} 0 & 2 & 3 \end{bmatrix} + 0\begin{bmatrix} 1 & 2 & 3 \end{bmatrix}$.

13. (a) and (b) are linearly independent; (c) is linearly dependent: $t + 13 = 3(2t^2 + t + 1) - 2(3t^2 + t - 5)$.

15. (b) is linearly dependent:

$$\begin{bmatrix} 1 \\ 2 \\ -2 \end{bmatrix} = 2\begin{bmatrix} 1 \\ 1 \\ -1 \end{bmatrix} + \begin{bmatrix} 0 \\ 1 \\ 1 \end{bmatrix} - \begin{bmatrix} 1 \\ 1 \\ 1 \end{bmatrix}.$$

17. For $c \neq \pm 2$.

Section 2.5, p. 137

1. (a) and (d).

3. (a) and (d).

5. (c).

7. (a) is a basis for R^3 and

$$\begin{bmatrix} 2 \\ 1 \\ 3 \end{bmatrix} = \frac{3}{2}\begin{bmatrix} 1 \\ 1 \\ 1 \end{bmatrix} + \frac{1}{2}\begin{bmatrix} 1 \\ 2 \\ 3 \end{bmatrix} - \frac{3}{2}\begin{bmatrix} 0 \\ 1 \\ 0 \end{bmatrix}.$$

9. (a) forms a basis.

$$5t^2 - 3t + 8 = 5(t^2 + t) - 8(t - 1).$$

11. $\left\{\begin{bmatrix} 1 \\ 2 \\ 2 \end{bmatrix}, \begin{bmatrix} 3 \\ 2 \\ 1 \end{bmatrix}\right\}$; $\dim W = 2$.

13. $\{t^3 + t^2 - 2t + 1, t^2 + 1\}$.

15. For $a \neq -1, 0, 1$.

17. $S = \left\{\begin{bmatrix} 1 & 0 & 0 \\ 0 & 0 & 0 \\ 0 & 0 & 0 \end{bmatrix}, \begin{bmatrix} 0 & 0 & 0 \\ 0 & 1 & 0 \\ 0 & 0 & 0 \end{bmatrix}, \begin{bmatrix} 0 & 0 & 0 \\ 0 & 0 & 0 \\ 0 & 0 & 1 \end{bmatrix}\right\}$.

19. (a) $\left\{\begin{bmatrix} 1 \\ 1 \\ 0 \end{bmatrix}, \begin{bmatrix} 0 \\ 1 \\ 1 \end{bmatrix}\right\}$. (b) $\left\{\begin{bmatrix} 1 \\ 1 \\ 0 \end{bmatrix}, \begin{bmatrix} 0 \\ 0 \\ 1 \end{bmatrix}\right\}$.

(c) $\left\{\begin{bmatrix} 1 \\ 0 \\ 2 \end{bmatrix}, \begin{bmatrix} 0 \\ 1 \\ 1 \end{bmatrix}\right\}$.

21. $\{t^2 + 2, t - 3\}$.

23. (a) 3. (b) 2.

25. (a) 2. (b) 1. (c) 2. (d) 2.

27. (a) 4. (b) 3. (c) 3. (d) 4.

29. $\{t^3+t, t^2-t, t^3, 1\}$.

31. 2.

33. The set of all vectors of the form $\begin{bmatrix} a \\ a+2b \\ -2a+b \\ a-2b \end{bmatrix}$, where a, b, c, and d are real numbers.

Section 2.6, p. 147

1. (a) $\mathbf{x} = \begin{bmatrix} \frac{1}{2}r+s \\ r \\ s \end{bmatrix}$, where r and s are any real numbers.

(b) $\mathbf{x} = r\begin{bmatrix} \frac{1}{2} \\ 1 \\ 0 \end{bmatrix} + s\begin{bmatrix} 1 \\ 0 \\ 1 \end{bmatrix}$.

(c)

3. $\left\{\begin{bmatrix} 0 \\ -1 \\ 0 \\ 1 \end{bmatrix}, \begin{bmatrix} 2 \\ -3 \\ 1 \\ 0 \end{bmatrix}\right\}$; dimension = 2.

5. $\left\{\begin{bmatrix} 1 \\ -\frac{8}{3} \\ -\frac{4}{3} \\ 1 \end{bmatrix}\right\}$; dimension = 1.

7. $\left\{\begin{bmatrix} -2 \\ 1 \\ 0 \\ 0 \\ 0 \end{bmatrix}, \begin{bmatrix} -3 \\ 0 \\ 1 \\ 1 \\ 0 \end{bmatrix}\right\}$; dimension = 2.

9. $\left\{\begin{bmatrix} -2 \\ \frac{1}{2} \\ 0 \\ 0 \\ 1 \end{bmatrix}, \begin{bmatrix} -1 \\ 1 \\ 0 \\ 1 \\ 0 \end{bmatrix}\right\}$; dimension = 2.

11. $\left\{\begin{bmatrix} -3 \\ 2 \\ 0 \\ 1 \end{bmatrix}, \begin{bmatrix} 5 \\ -4 \\ 1 \\ 0 \end{bmatrix}\right\}$.

13. $\left\{\begin{bmatrix} -1 \\ 1 \end{bmatrix}\right\}$.

15. $\left\{\begin{bmatrix} 1 \\ -2 \\ 1 \end{bmatrix}\right\}$.

17. $\lambda = 3$ or -4.

19. $\lambda = 0$ or 1.

Section 2.7, p. 161

1. $\begin{bmatrix} 3 \\ -2 \end{bmatrix}$. **3.** $\begin{bmatrix} 2 \\ -1 \end{bmatrix}$. **5.** $\begin{bmatrix} 1 \\ -1 \\ 0 \\ 2 \end{bmatrix}$.

7. $\begin{bmatrix} 0 \\ 3 \end{bmatrix}$. **9.** $4t-3$. **11.** $\begin{bmatrix} -1 & 0 \\ 9 & 7 \end{bmatrix}$.

13. (a) $[\mathbf{v}]_T = \begin{bmatrix} -7 \\ 4 \end{bmatrix}$; $[\mathbf{w}]_T = \begin{bmatrix} 7 \\ -1 \end{bmatrix}$.

(b) $P_{S \leftarrow T} = \begin{bmatrix} 1 & 2 \\ -1 & -1 \end{bmatrix}$.

(c) $[\mathbf{v}]_S = \begin{bmatrix} 1 \\ 3 \end{bmatrix}$; $[\mathbf{w}]_S = \begin{bmatrix} 5 \\ -6 \end{bmatrix}$.

(d) Same as (c).

(e) $Q_{T \leftarrow S} = \begin{bmatrix} -1 & -2 \\ 1 & 1 \end{bmatrix}$.

(f) Same as (a).

15. (a) $[\mathbf{v}]_T = \begin{bmatrix} 3 \\ 2 \\ -7 \end{bmatrix}$; $[\mathbf{w}]_T = \begin{bmatrix} 2 \\ 3 \\ -3 \end{bmatrix}$.

(b) $P_{S \leftarrow T} = \begin{bmatrix} 2 & 1 & 0 \\ 1 & -\frac{2}{5} & \frac{3}{5} \\ 0 & \frac{2}{5} & \frac{2}{5} \end{bmatrix}$.

(c) $\left[\mathbf{v}\right]_S = \begin{bmatrix} 8 \\ -2 \\ -2 \end{bmatrix}; \left[\mathbf{w}\right]_S = \begin{bmatrix} 7 \\ -1 \\ 0 \end{bmatrix}$.

(d) Same as (c).

(e) $Q_{T\leftarrow S} = \begin{bmatrix} \frac{1}{3} & \frac{1}{3} & -\frac{1}{2} \\ \frac{1}{3} & -\frac{2}{3} & 1 \\ -\frac{1}{3} & \frac{2}{3} & \frac{3}{2} \end{bmatrix}$.

(f) $\left[\mathbf{v}\right]_T = Q_{T\leftarrow S}\left[\mathbf{v}\right]_S = \begin{bmatrix} 3 \\ 2 \\ -7 \end{bmatrix}$.

$\left[\mathbf{w}\right]_T = Q_{T\leftarrow S}\left[\mathbf{w}\right]_S = \begin{bmatrix} 2 \\ 3 \\ -3 \end{bmatrix}$; same as (a).

17. (a) $\left[\mathbf{v}\right]_T = \begin{bmatrix} 1 \\ 1 \\ 1 \\ 0 \end{bmatrix}; \left[\mathbf{w}\right]_T = \begin{bmatrix} 2 \\ -2 \\ 1 \\ -1 \end{bmatrix}$.

(b) $P_{S\leftarrow T} = \begin{bmatrix} 1 & 0 & 0 & 1 \\ \frac{1}{3} & \frac{2}{3} & -\frac{2}{3} & 0 \\ \frac{1}{3} & -\frac{1}{3} & \frac{1}{3} & 0 \\ -\frac{1}{3} & \frac{1}{3} & \frac{2}{3} & 0 \end{bmatrix}$.

(c) $\left[\mathbf{v}\right]_S = \begin{bmatrix} 1 \\ \frac{1}{3} \\ \frac{1}{3} \\ \frac{2}{3} \end{bmatrix}; \left[\mathbf{w}\right]_S = \begin{bmatrix} 1 \\ -\frac{4}{3} \\ \frac{5}{3} \\ -\frac{2}{3} \end{bmatrix}$.

(d) Same as (c).

(e) $Q_{T\leftarrow S} = \begin{bmatrix} 0 & 1 & 2 & 0 \\ 0 & 1 & 0 & 1 \\ 0 & 0 & 1 & 1 \\ 1 & -1 & -2 & 0 \end{bmatrix}$.

(f) Same as (a).

19. $\begin{bmatrix} 1 \\ 1 \end{bmatrix}$.

21. $\begin{bmatrix} 1 \\ 1 \\ -1 \end{bmatrix}$.

23. $S = \{t+1, 5t-2\}$.

25. $S = \{-t+5, t-3\}$.

Section 2.8, p. 175

1. A possible basis is $\{\mathbf{v}_1, \mathbf{v}_2, \mathbf{v}_3\}$, where

$\mathbf{v}_1 = \begin{bmatrix} 1 \\ 0 \\ 0 \end{bmatrix}, \mathbf{v}_2 = \begin{bmatrix} 0 \\ 1 \\ 0 \end{bmatrix}$, and $\mathbf{v}_3 = \begin{bmatrix} 0 \\ 0 \\ 1 \end{bmatrix}$.

(a) $\begin{bmatrix} 3 \\ 4 \\ 12 \end{bmatrix} = 3\mathbf{v}_1 + 4\mathbf{v}_2 + 12\mathbf{v}_3$.

(b) $\begin{bmatrix} 3 \\ 2 \\ 2 \end{bmatrix} = 3\mathbf{v}_1 + 2\mathbf{v}_2 + 2\mathbf{v}_3$.

(c) $\begin{bmatrix} 1 \\ 2 \\ 6 \end{bmatrix} = \mathbf{v}_1 + 2\mathbf{v}_2 + 6\mathbf{v}_3$.

3. A possible answer is $\left\{\begin{bmatrix} 1 & 0 \\ 0 & 0 \end{bmatrix}, \begin{bmatrix} 0 & 1 \\ 0 & 0 \end{bmatrix}, \begin{bmatrix} 0 & 0 \\ 1 & 0 \end{bmatrix}, \begin{bmatrix} 0 & 0 \\ 0 & 1 \end{bmatrix}\right\}$.

5. (a) $\{(1, 0, -1), (0, 1, 0)\}$.

(b) $\{(1, 2, -1), (1, 9, -1)\}$.

7. (a) $\left\{\begin{bmatrix} 1 \\ 0 \\ 0 \\ 0 \end{bmatrix}, \begin{bmatrix} 0 \\ 1 \\ 0 \\ \frac{1}{5} \end{bmatrix}, \begin{bmatrix} 0 \\ 0 \\ 1 \\ \frac{3}{5} \end{bmatrix}\right\}$.

(b) $\left\{\begin{bmatrix} 1 \\ 1 \\ 3 \\ 2 \end{bmatrix}, \begin{bmatrix} -2 \\ -1 \\ 2 \\ 1 \end{bmatrix}, \begin{bmatrix} 0 \\ 0 \\ 5 \\ 3 \end{bmatrix}\right\}$.

9. (a) 3. (b) 5.

13. (a) 2. (b) 3.

15. B and C are equivalent; A, D, and E are equivalent.

17. Neither.

19. (b).

21. (b).

23. (a) 2. (b) 3.

25. Has a unique solution.

27. Linearly dependent.

29. (a) Linearly dependent.

(b) Linearly independent.

Supplementary Exercises, p. 178

1. (b) $k = 0$.

3. (a) No. (b) Yes. (c) Yes.

11. $a = 1$ or $a = 2$.

13. $k \neq 1, -1$.

17. (a) $\mathbf{b} = \begin{bmatrix} a \\ b \\ c \end{bmatrix}$, where $b + c - 3a = 0$.

(b) Any $\mathbf{b}$.

25. (a) $T = \left\{ \begin{bmatrix} 2 \\ 2 \\ 1 \end{bmatrix}, \begin{bmatrix} 3 \\ 1 \\ 2 \end{bmatrix}, \begin{bmatrix} 5 \\ 4 \\ 4 \end{bmatrix} \right\}$.

(b) $S = \left\{ \begin{bmatrix} \frac{6}{5} \\ -\frac{2}{5} \\ \frac{13}{5} \end{bmatrix}, \begin{bmatrix} \frac{2}{5} \\ \frac{1}{5} \\ -\frac{4}{5} \end{bmatrix}, \begin{bmatrix} -\frac{1}{5} \\ \frac{2}{5} \\ -\frac{3}{5} \end{bmatrix} \right\}$.

CHAPTER 3

Section 3.1, p. 189

1. (a) 1. (b) 0. (c) $\sqrt{5}$.

3. (a) 1. (b) $\sqrt{2}$.

5. (a) $\sqrt{74}$. (b) $\sqrt{58}$.

7. (a) 0. (b) 3. (c) 8.

9. (a) $\dfrac{-14}{\sqrt{5}\sqrt{41}}$ (b) $\dfrac{-6}{\sqrt{5}\sqrt{41}}$.

11. (a) 1, 0, 0.

(b) $\dfrac{1}{\sqrt{14}}, \dfrac{3}{\sqrt{14}}, \dfrac{2}{\sqrt{14}}$.

(c) $\dfrac{-1}{\sqrt{14}}, \dfrac{-2}{\sqrt{14}}, \dfrac{-3}{\sqrt{14}}$.

(d) $\dfrac{4}{\sqrt{29}}, \dfrac{-3}{\sqrt{29}}, \dfrac{2}{\sqrt{29}}$.

17. (a) $\mathbf{v}_1$ and $\mathbf{v}_4$, $\mathbf{v}_1$ and $\mathbf{v}_6$, $\mathbf{v}_3$ and $\mathbf{v}_4$, $\mathbf{v}_3$ and $\mathbf{v}_6$, $\mathbf{v}_4$ and $\mathbf{v}_5$, $\mathbf{v}_5$ and $\mathbf{v}_6$.

(b) $\mathbf{v}_1$ and $\mathbf{v}_5$, $\mathbf{v}_4$ and $\mathbf{v}_6$.

(c) $\mathbf{v}_1$ and $\mathbf{v}_3$, $\mathbf{v}_3$ and $\mathbf{v}_5$.

19. (b).

21.

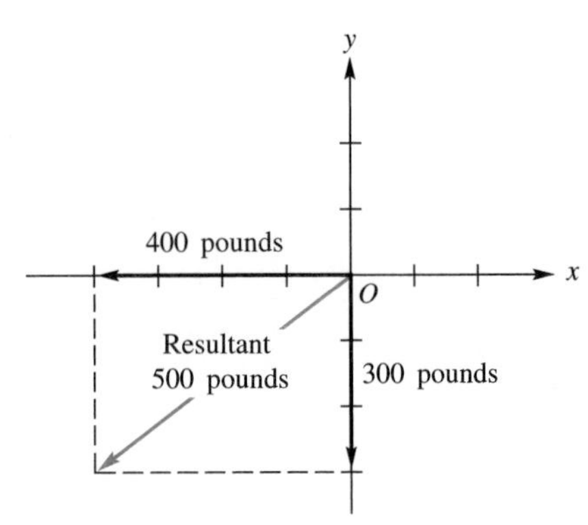

25. $\frac{5}{2}$.

27. $a = -2, b = 2$.

Section 3.2, p. 198

1. (a) $-15\mathbf{i} - 2\mathbf{j} + 9\mathbf{k}$. (b) $-3\mathbf{i} + 3\mathbf{j} + 3\mathbf{k}$.
(c) $7\mathbf{i} + 5\mathbf{j} - \mathbf{k}$. (d) $0\mathbf{i} + 0\mathbf{j} + 0\mathbf{k}$.

5. $(\mathbf{u} \times \mathbf{v}) \cdot \mathbf{w} = 24$.

13. $\frac{3}{2}\sqrt{10}$.

15. 1.

17. (a).

19. (a) $x - z + 2 = 0$. (b) $3x + y - 14z + 47 = 0$.

21. $4x - 4y + z + 16 = 0$.

23. $x = -2 + 2t,\ y = 5 - 3t,\ z = -3 + 4t$.

25. $\left\{ \begin{bmatrix} \frac{2}{3} \\ 1 \\ 0 \end{bmatrix}, \begin{bmatrix} \frac{5}{3} \\ 0 \\ 1 \end{bmatrix} \right\}$.

Section 3.3, p. 211

9. (a) -8. (b) 0. (c) 1.

11. (a) $\frac{3}{2}$. (b) 1. (c) $\frac{1}{2}\sin^2 1$.

13. (a) 2. (b) $2\sqrt{5}$. (c) $2\sqrt{2}$.

15. (a) 1. (b) 0. (c) $\dfrac{2\sin^2 1}{\sqrt{4 - \sin^2 2}}$.

25. If $\mathbf{u} = \begin{bmatrix} u_1 & u_2 \end{bmatrix}$ and $\mathbf{v} = \begin{bmatrix} v_1 & v_2 \end{bmatrix}$, then $(\mathbf{u}, \mathbf{v}) = 3u_1v_1 - 2u_1v_2 - 2u_2v_1 + 3u_2v_2$.

27. (a) $\sqrt{1 - \sin^2 1}$. (b) $\frac{1}{30}$.

29. (a) Orthonormal. (b) Neither.
(c) Neither.

31. $a = 0$.

33. $a = 5$.

35. $B = \begin{bmatrix} b_{11} & b_{12} \\ b_{21} & b_{22} \end{bmatrix}$ with $b_{11} + 3b_{21} + 2b_{12} + 4b_{22} = 0$.

Section 3.4, p. 224

1. (a) $\left\{ \begin{bmatrix} 1 \\ 2 \end{bmatrix}, \begin{bmatrix} -4 \\ 2 \end{bmatrix} \right\}$.

(b) $\left\{ \frac{1}{\sqrt{5}} \begin{bmatrix} 1 \\ 2 \end{bmatrix}, \frac{1}{\sqrt{5}} \begin{bmatrix} -2 \\ 1 \end{bmatrix} \right\}$.

3. $\left\{ \frac{1}{\sqrt{3}} \begin{bmatrix} 1 & 1 & -1 & 0 \end{bmatrix}, \frac{1}{\sqrt{33}} \begin{bmatrix} -2 & 4 & 2 & 3 \end{bmatrix} \right\}$.

5. $\left\{ \sqrt{3}\, t, 2 - 3t \right\}$.

7. $\left\{ \sqrt{3}\, t, \dfrac{\sin 2\pi t + \left(\frac{3}{2\pi}\right) t}{\sqrt{\frac{1}{2} - \frac{3}{4\pi^2}}} \right\}$.

9. $\left\{ \begin{bmatrix} \frac{2}{3} \\ -\frac{2}{3} \\ \frac{1}{3} \end{bmatrix}, \begin{bmatrix} \frac{2}{3} \\ \frac{1}{3} \\ -\frac{2}{3} \end{bmatrix}, \begin{bmatrix} \frac{1}{3} \\ \frac{2}{3} \\ \frac{2}{3} \end{bmatrix} \right\}$.

11. Possible answer:

$$\left\{ \frac{1}{\sqrt{3}} \begin{bmatrix} 1 \\ 1 \\ 1 \end{bmatrix}, \frac{1}{\sqrt{6}} \begin{bmatrix} -1 \\ -1 \\ 2 \end{bmatrix}, \frac{1}{\sqrt{2}} \begin{bmatrix} -1 \\ 1 \\ 0 \end{bmatrix} \right\}.$$

13. $\left\{ \begin{bmatrix} \frac{1}{\sqrt{2}} \\ \frac{1}{\sqrt{2}} \\ 0 \end{bmatrix}, \begin{bmatrix} -\frac{1}{\sqrt{6}} \\ \frac{1}{\sqrt{6}} \\ \frac{2}{\sqrt{6}} \end{bmatrix} \right\}$.

15. $\left\{ \begin{bmatrix} \frac{1}{\sqrt{2}} \\ 0 \\ -\frac{1}{\sqrt{2}} \end{bmatrix}, \begin{bmatrix} -\frac{1}{\sqrt{6}} \\ \frac{2}{\sqrt{6}} \\ -\frac{1}{\sqrt{6}} \end{bmatrix} \right\}$.

17. $\left\{ \frac{1}{\sqrt{26}} \begin{bmatrix} -3 \\ 4 \\ 1 \end{bmatrix} \right\}$.

23. (b) $\begin{bmatrix} 9 \\ 9 \\ 9 \end{bmatrix}$. (c) $\|\mathbf{v}\| = 9\sqrt{3}$.

25. (b) $\left[\mathbf{v}\right]_S = \begin{bmatrix} 1 \\ 2 \\ 3 \\ 4 \end{bmatrix}$.

27. Possible answer:

$$\left\{ \begin{bmatrix} 1 & 0 \\ 0 & 0 \end{bmatrix}, \frac{1}{\sqrt{2}} \begin{bmatrix} 0 & 1 \\ 1 & 0 \end{bmatrix}, \frac{1}{\sqrt{6}} \begin{bmatrix} 0 & -1 \\ 1 & 2 \end{bmatrix} \right\}.$$

29. (a) $Q = \begin{bmatrix} \frac{1}{\sqrt{2}} & \frac{1}{\sqrt{2}} \\ -\frac{1}{\sqrt{2}} & \frac{1}{\sqrt{2}} \end{bmatrix} \approx \begin{bmatrix} 0.7071 & 0.7071 \\ -0.7071 & 0.7071 \end{bmatrix}$,

$$R = \begin{bmatrix} \sqrt{2} & -\frac{1}{\sqrt{2}} \\ 0 & \frac{5}{\sqrt{2}} \end{bmatrix} \approx \begin{bmatrix} 1.4142 & 0.7071 \\ 0 & 3.5355 \end{bmatrix}.$$

(b) $Q = \begin{bmatrix} \frac{\sqrt{3}}{3} & \frac{1}{\sqrt{6}} \\ -\frac{\sqrt{3}}{3} & -\frac{1}{\sqrt{6}} \\ \frac{\sqrt{3}}{3} & -\frac{2}{\sqrt{6}} \end{bmatrix} \approx \begin{bmatrix} 0.5774 & 0.4082 \\ -0.5774 & -0.4082 \\ 0.5774 & -0.8165 \end{bmatrix}$,

$$R = \begin{bmatrix} \sqrt{3} & \frac{5}{\sqrt{3}} \\ 0 & \frac{2}{\sqrt{6}} \end{bmatrix} \approx \begin{bmatrix} 1.7321 & 2.8868 \\ 0 & 0.8165 \end{bmatrix}.$$

(c) $Q = \begin{bmatrix} \frac{1}{\sqrt{6}} & \frac{4}{\sqrt{21}} & -\frac{1}{\sqrt{14}} \\ \frac{2}{\sqrt{6}} & -\frac{1}{\sqrt{21}} & \frac{2}{\sqrt{14}} \\ -\frac{1}{\sqrt{6}} & \frac{2}{\sqrt{21}} & \frac{3}{\sqrt{14}} \end{bmatrix} \approx$

$$\begin{bmatrix} 0.4082 & 0.8729 & -0.2673 \\ 0.8165 & -0.2182 & 0.5345 \\ -0.4082 & 0.4364 & 0.8018 \end{bmatrix},$$

$$R = \begin{bmatrix} \frac{6}{\sqrt{6}} & -\frac{8}{\sqrt{6}} & \frac{1}{\sqrt{6}} \\ 0 & \frac{7}{\sqrt{21}} & \frac{1}{\sqrt{21}} \\ 0 & 0 & \frac{19}{\sqrt{14}} \end{bmatrix} \approx$$

$$\begin{bmatrix} 2.4495 & -3.2660 & 0.4082 \\ 0 & 1.5275 & 0.2182 \\ 0 & 0 & 5.0780 \end{bmatrix}.$$

Section 3.5, p. 241

1. (a) $\left\{\begin{bmatrix} \frac{3}{2} \\ 1 \\ 0 \end{bmatrix}, \begin{bmatrix} -\frac{1}{2} \\ 0 \\ 1 \end{bmatrix}\right\}$.

(b) The set of all points $P(x, y, z)$ such that $2x - 3y + z = 0$. $W^{\perp}$ is the plane whose normal is $\mathbf{w}$.

3. $\left\{\begin{bmatrix} -\frac{17}{5} & \frac{6}{5} & 5 & 1 & 0 \end{bmatrix}, \begin{bmatrix} \frac{8}{5} & \frac{1}{5} & -3 & 0 & 1 \end{bmatrix}\right\}$.

5. $\left\{\frac{45}{14}t^3 - \frac{55}{14}t^2 + t, \frac{130}{7}t^3 - \frac{130}{7}t^2 + 1\right\}$.

7. $\begin{bmatrix} -3 \\ -2 \\ 1 \end{bmatrix}$.

9. Null space of A has basis $\left\{\begin{bmatrix} 2 \\ -1 \\ 1 \\ 0 \end{bmatrix}, \begin{bmatrix} 3 \\ -2 \\ 0 \\ 1 \end{bmatrix}\right\}$.

Basis for row space of A is $\left\{\begin{bmatrix} 1 & 0 & -2 & -3 \end{bmatrix}, \begin{bmatrix} 0 & 1 & 1 & 2 \end{bmatrix}\right\}$.

Null space of A^T has basis $\left\{\begin{bmatrix} -\frac{7}{5} \\ -\frac{13}{10} \\ 1 \end{bmatrix}\right\}$.

Basis for column space of A^T is $\left\{\begin{bmatrix} 1 \\ 0 \\ \frac{7}{5} \end{bmatrix}, \begin{bmatrix} 0 \\ 1 \\ \frac{13}{10} \end{bmatrix}\right\}$.

11. (a) $\begin{bmatrix} 3 \\ 0 \\ -1 \end{bmatrix}$. (b) $\begin{bmatrix} 2 \\ 0 \\ 3 \end{bmatrix}$. (c) $\begin{bmatrix} -5 \\ 0 \\ 1 \end{bmatrix}$.

13. (a) $2 \sin t$.

(b) $\dfrac{\pi^2}{3} - 4 \cos t$.

(c) $\left(\dfrac{e^{\pi} - e^{-\pi}}{2\pi}\right) + \left(\dfrac{e^{-\pi} - e^{\pi}}{2\pi}\right) \cos t + \left(\dfrac{e^{\pi} - e^{-\pi}}{2\pi}\right) \sin t$.

15. $\mathbf{w} = \begin{bmatrix} -\frac{1}{5} \\ 2 \\ -\frac{2}{5} \end{bmatrix}, \mathbf{u} = \begin{bmatrix} \frac{6}{5} \\ 0 \\ -\frac{3}{5} \end{bmatrix}$.

17. $\mathbf{w} = 2 \sin t - 1, \mathbf{u} = t - 1 - [2 \sin t - 1] = t - 2 \sin t$.

19. $\frac{3}{5}\sqrt{5}$.

21. $\sqrt{\dfrac{2\pi^3}{3} - 4\pi} \approx 2.847$.

23. $$\operatorname{proj}_W e^t = \frac{1}{2\pi}\left(e^{\pi} - e^{-\pi}\right) + \frac{1}{\pi}\left(-\frac{1}{2}e^{\pi} + \frac{1}{2}e^{-\pi}\right)\cos t + \frac{1}{\pi}\left(\frac{1}{2}e^{\pi} - \frac{1}{2}e^{-\pi}\right)\sin t + \frac{1}{\pi}\left(\frac{1}{5}e^{\pi} - \frac{1}{5}e^{-\pi}\right)\cos 2t + \frac{1}{\pi}\left(-\frac{2}{5}e^{\pi} + \frac{2}{5}e^{-\pi}\right)\sin 2t.$$

Section 3.6, p. 248

3. $\widehat{\mathbf{x}} \approx \begin{bmatrix} -1.5333 \\ -1.8667 \\ 4.2667 \end{bmatrix}$.

9. $\widehat{\mathbf{x}} = \begin{bmatrix} -\frac{5}{11} \\ \frac{4}{11} \\ 0 \end{bmatrix}$.

11. (a) $y = 690.1029x - 659.2857$, where $x =$ the year ($1970 = 0$) and $y =$ debt per capita.

(b) 20,044.

13. (b) $\ln y = -0.0272x + 3.2709$

(c) $r = 26.3350$, $s = -0.0272$

(d) 16.14 mm.

Supplementary Exercises, p. 250

1. $\left\{\begin{bmatrix} 2 \\ 1 \\ 0 \end{bmatrix}, \begin{bmatrix} -\frac{1}{5} \\ \frac{2}{5} \\ 1 \end{bmatrix}\right\}$.

3. $\mathbf{v} = -\sqrt{2}\begin{bmatrix} \frac{1}{\sqrt{2}} \\ 0 \\ -\frac{1}{\sqrt{2}} \end{bmatrix} + 2\begin{bmatrix} 0 \\ 1 \\ 0 \end{bmatrix} + 2\sqrt{2}\begin{bmatrix} \frac{1}{\sqrt{2}} \\ 0 \\ \frac{1}{\sqrt{2}} \end{bmatrix}$.

5. Vector in P closest to $\mathbf{v}$ is $\dfrac{1}{122}\begin{bmatrix} 59 \\ 271 \\ 50 \end{bmatrix}$; distance is $\dfrac{9}{\sqrt{122}}$.

9. (a) Possible answer: $\left\{\dfrac{1}{\sqrt{30}}\begin{bmatrix} -5 \\ 2 \\ 1 \\ 0 \end{bmatrix}, \dfrac{1}{\sqrt{30}}\begin{bmatrix} 2 \\ 5 \\ 0 \\ 1 \end{bmatrix}\right\}$.

(b) Possible answer:

$$\left\{ \frac{1}{\sqrt{30}} \begin{bmatrix} -5 \\ 2 \\ 1 \\ 0 \end{bmatrix}, \frac{1}{\sqrt{255}} \begin{bmatrix} -5 \\ -14 \\ 3 \\ 5 \end{bmatrix} \right\}.$$

11. (a) $\left\{ \begin{bmatrix} -1 \\ 0 \\ 1 \end{bmatrix} \right\}$.

(c) (i) $\mathbf{w} = \frac{1}{2} \begin{bmatrix} 1 \\ 0 \\ 1 \end{bmatrix}, \mathbf{u} = -\frac{1}{2} \begin{bmatrix} -1 \\ 0 \\ 1 \end{bmatrix}$;

(ii) $\mathbf{w} = \begin{bmatrix} 2 \\ 2 \\ 2 \end{bmatrix}, \mathbf{u} = \begin{bmatrix} -1 \\ 0 \\ 1 \end{bmatrix}$.

13. $\left\{ \frac{1}{\sqrt{2}}, \sqrt{\frac{3}{2}}\,t, \sqrt{\frac{5}{8}}\,(3t^2 - 1) \right\}$.

15. $\sqrt{\frac{8}{175}} \approx 0.214$.

27. (a) 7, 4.1231, 3.

(b) 5, 3.6055, 3.

(c) 2, 2, 2.

CHAPTER 4

Section 4.1, p. 263

1. (b).

3. (a) and (b).

7. (a) $\begin{bmatrix} 15 & 5 & 4 & 8 \\ -5 & -1 & 10 & 2 \end{bmatrix}$.

9. (a) $A = \begin{bmatrix} 0 & 1 \\ 1 & 0 \end{bmatrix}$.

(b) $A = \begin{bmatrix} 1 & -3 \\ 2 & -1 \\ 0 & 2 \end{bmatrix}$.

(c) $A = \begin{bmatrix} 1 & 4 & 0 \\ 0 & 0 & -1 \\ 0 & 1 & 1 \end{bmatrix}$.

11. (a) $\begin{bmatrix} 2 \\ 15 \end{bmatrix}$. (b) $\begin{bmatrix} 2a_1 + 3a_2 + 2a_3 \\ -4a_1 - 5a_2 + 3a_3 \end{bmatrix}$.

13. (a) $2t^3 - 5t^2 + 2t + 3$.

(b) $at^3 + bt^2 + at + c$.

15. a = any real number, $b = 0$.

17. (a) Reflection about the y-axis.

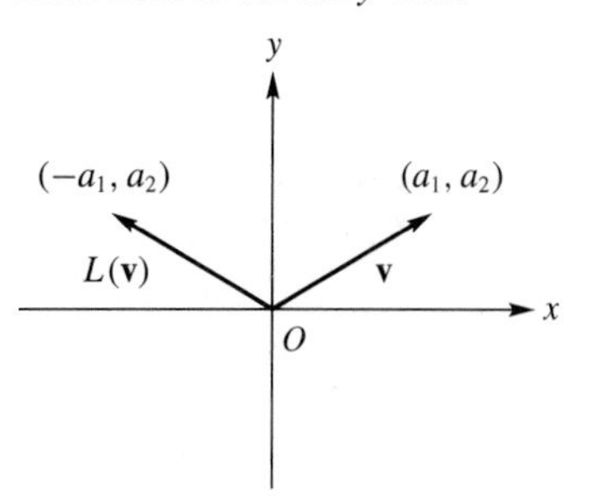

(b) Reflection about the origin.

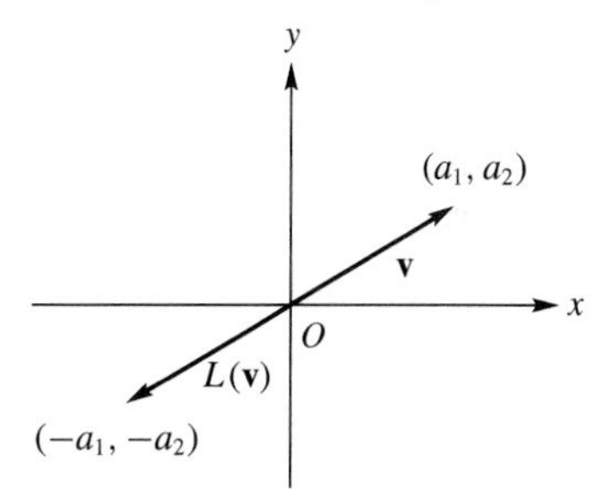

(c) Counterclockwise rotation through $\pi/2$ radians.

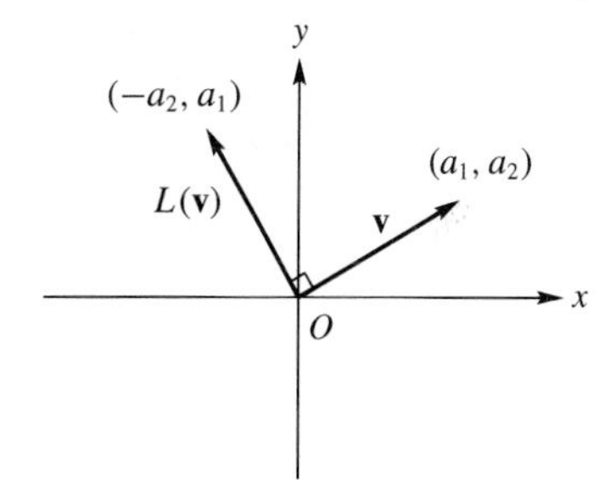

Section 4.2, p. 279

1. (a) Yes. (b) No. (c) Yes. (d) No.

(e) All vectors $\begin{bmatrix} 0 \\ a \end{bmatrix}$, where a is any real number; that is, the y-axis.

(f) All vectors $\begin{bmatrix} a \\ 0 \end{bmatrix}$, where a is any real number; that is, the x-axis.

3. (a) No. (b) Yes. (c) Yes. (d) Yes.

(e) All vectors of the form $\begin{bmatrix} -r & -s & r & s \end{bmatrix}$, where r and s are real numbers.

(f) $\{\begin{bmatrix} 1 & 0 \end{bmatrix}, \begin{bmatrix} 0, 1 \end{bmatrix}\}$.

5. (a) $\{\begin{bmatrix} 1 & -1 & -1 & 1 \end{bmatrix}\}$. (b) 1.

(c) $\{[1\ \ 0\ \ 1],[1\ \ 0\ \ 0],[0\ \ 1\ \ 1]\}$. (d) 3.

7. (a) 0. (b) 6.

9. (a) $\{1\}$. (b) $\{[1\ \ 0],[0\ \ 1]\}$.

11. (a) $\ker L = \left\{\begin{bmatrix} 0 & 0 \\ 0 & 0 \end{bmatrix}\right\}$, so $\ker L$ has no basis.

(b) $\left\{\begin{bmatrix} 1 & 0 \\ 1 & 0 \end{bmatrix}, \begin{bmatrix} 1 & 1 \\ 0 & 1 \end{bmatrix}, \begin{bmatrix} 0 & 1 \\ 0 & 0 \end{bmatrix}, \begin{bmatrix} 0 & 0 \\ 1 & 1 \end{bmatrix}\right\}$

13. (a) $\dim \ker L = 1$, $\dim \operatorname{range} L = 2$.

(b) $\dim \ker L = 1$, $\dim \operatorname{range} L = 2$.

(c) $\dim \ker L = 0$, $\dim \operatorname{range} L = 4$.

17. No.

19. (b) $L^{-1}\left(\begin{bmatrix} 2 \\ 3 \\ 4 \end{bmatrix}\right) = \begin{bmatrix} \frac{3}{2} \\ -\frac{1}{2} \\ \frac{1}{2} \end{bmatrix}$.

21. 1.

23. (b) $\begin{bmatrix} -2a_1 - a_3 \\ -2a_1 - a_2 + 2a_3 \\ -2a_1 + a_2 - a_3 \end{bmatrix}$.

25. (a) 2. (b) 1.

Section 4.3, p. 289

1. (a) $\begin{bmatrix} 1 & 2 \\ 2 & -1 \end{bmatrix}$. (b) $\begin{bmatrix} 1 & -\frac{1}{2} \\ 1 & \frac{3}{4} \end{bmatrix}$.

(c) $\begin{bmatrix} 3 & 2 \\ -4 & 4 \end{bmatrix}$. (d) $\begin{bmatrix} -2 & 2 \\ \frac{1}{2} & 2 \end{bmatrix}$.

(e) $\begin{bmatrix} 5 \\ 0 \end{bmatrix}$.

3. (a) $\begin{bmatrix} 1 & 0 & 1 & 1 \\ 0 & 1 & 2 & 1 \\ -1 & -2 & 1 & 0 \end{bmatrix}$.

(b) $\begin{bmatrix} 1 & 0 & 2 & 1 \\ 1 & 1 & 3 & 3 \\ -5 & -3 & -4 & -5 \end{bmatrix}$.

5. (a) $\begin{bmatrix} 1 & 2 & 1 \\ 1 & 0 & 0 \\ 0 & 1 & 1 \end{bmatrix}$. (b) $\begin{bmatrix} 8 \\ 1 \\ 5 \end{bmatrix}$.

7. (a) $\begin{bmatrix} 10 \\ 5 \\ 5 \end{bmatrix}$. (b) $\begin{bmatrix} 4 \\ 2 \\ 2 \end{bmatrix}$.

9. $\begin{bmatrix} 0 & 1 & 0 & 0 \\ 0 & 0 & 0 & 0 \\ 0 & 0 & 1 & 1 \\ 0 & 0 & 0 & 1 \end{bmatrix}$.

11. (a) $\begin{bmatrix} 0 & -3 & 2 & 0 \\ -2 & -3 & 0 & 2 \\ 3 & 0 & 3 & -3 \\ 0 & 3 & -2 & 0 \end{bmatrix}$.

(b) $\begin{bmatrix} 0 & 3 & -2 & 3 \\ 0 & -9 & -2 & -6 \\ 0 & 3 & 6 & 0 \\ 0 & 4 & 0 & 3 \end{bmatrix}$.

(c) $\begin{bmatrix} 0 & 3 & -2 & 0 \\ -3 & -6 & 1 & 3 \\ 3 & 0 & 3 & -3 \\ 1 & 3 & -1 & -1 \end{bmatrix}$.

(d) $\begin{bmatrix} 0 & -3 & 2 & -3 \\ 0 & -5 & -2 & -3 \\ 0 & 3 & 6 & 0 \\ 0 & 3 & -2 & 3 \end{bmatrix}$.

13. (a) $\begin{bmatrix} 1 & 0 \\ 0 & -1 \end{bmatrix}$. (b) $\begin{bmatrix} 0 & -1 \\ -1 & 0 \end{bmatrix}$.

(c) $\begin{bmatrix} \frac{1}{2} & -\frac{1}{2} \\ -\frac{1}{2} & -\frac{1}{2} \end{bmatrix}$. (d) $\begin{bmatrix} 1 & -1 \\ -1 & -1 \end{bmatrix}$.

17. (a) $\begin{bmatrix} 1 & 0 \\ 0 & 1 \end{bmatrix}$. (b) $\begin{bmatrix} 1 & 0 \\ 0 & 1 \end{bmatrix}$.

(c) $\begin{bmatrix} \frac{3}{5} & -\frac{2}{5} \\ \frac{1}{5} & \frac{1}{5} \end{bmatrix}$. (d) $\begin{bmatrix} 1 & 2 \\ -1 & 3 \end{bmatrix}$.

19. $\begin{bmatrix} 0 & -1 \\ 1 & 0 \end{bmatrix}$.

Section 4.4, p. 297

3. (a) $\begin{bmatrix}-a_1+4a_2-a_3 & 3a_1-a_2+3a_3 & 4a_1+3a_2+5a_3\end{bmatrix}$.

(b) $\begin{bmatrix}5 & -4 & -4\end{bmatrix}$.

(c) $\begin{bmatrix}-1 & 4 & -1\\ 3 & -1 & 3\\ 4 & 3 & 5\end{bmatrix}$.

(d) $\begin{bmatrix}2a_1+2a_3 & -4a_1-2a_2-2a_3 & -2a_1-4a_2-6a_3\end{bmatrix}$.

(e) $\begin{bmatrix}14 & -28 & 14\end{bmatrix}$.

(f) $\begin{bmatrix}2 & 16 & 2\\ 0 & -10 & 6\\ 10 & 0 & 2\end{bmatrix}$.

5. (a) $\begin{bmatrix}5\\ 10\\ 0\end{bmatrix}$.

(b) $\ker L_1 =$ all vectors of the form $\begin{bmatrix}-r-s\\ r\\ s\end{bmatrix}$,

$\ker L_2 =$ all vectors of the form $\begin{bmatrix}-r-2s\\ r\\ s\end{bmatrix}$,

$\ker L_1 \cap \ker L_2 =$ all vectors of the form $\begin{bmatrix}-r\\ r\\ 0\end{bmatrix}$.

(c) All vectors of the form $\begin{bmatrix}-r\\ r\\ 0\end{bmatrix}$.

(d) They are the same.

7. (a) $C = \begin{bmatrix}2 & 2\\ 1 & -1\\ -1 & 1\\ 1 & -1\end{bmatrix}$.

(b) $A = \begin{bmatrix}1 & 1\\ \frac{2}{3} & -\frac{2}{3}\\ \frac{1}{3} & -\frac{1}{3}\end{bmatrix}$, $B = \begin{bmatrix}2 & 0 & 0\\ 0 & 1 & 1\\ 0 & -1 & -1\\ 0 & 1 & 1\end{bmatrix}$.

9. (a) $\begin{bmatrix}2 & 8 & -2\\ 4 & 2 & 6\\ 2 & -2 & 4\end{bmatrix}$. (b) $\begin{bmatrix}10 & 17 & 7\\ 11 & 8 & 13\\ 3 & -1 & 4\end{bmatrix}$.

11. (a) 6. (b) 6. (c) 12. (d) 12.

15. (a) $L(t^2) = t+3$, $L(t) = 2t+4$, $L(1) = -2t-1$.

(b) $(a+2b-2c)t + (3a+4b-c)$.

(c) $-16t-18$.

17. Possible answers: $L\left(\begin{bmatrix}a_1\\ a_2\end{bmatrix}\right) = \begin{bmatrix}a_2\\ a_1\end{bmatrix}$; $L\left(\begin{bmatrix}a_1\\ a_2\end{bmatrix}\right) = \begin{bmatrix}-a_1\\ -a_2\end{bmatrix}$.

19. Possible answers: $L\left(\begin{bmatrix}a_1\\ a_2\end{bmatrix}\right) = \begin{bmatrix}a_1\\ 0\end{bmatrix}$;

$$L\left(\begin{bmatrix}a_1\\ a_2\end{bmatrix}\right) = \begin{bmatrix}\dfrac{a_1+a_2}{2}\\ \dfrac{a_1+a_2}{2}\end{bmatrix}.$$

21. $\begin{bmatrix}2 & 0 & -1\\ -2 & -1 & 2\\ 1 & 1 & -1\end{bmatrix}$.

Section 4.5, p. 305

3. $\begin{bmatrix}-2 & 2\\ \frac{1}{2} & 2\end{bmatrix}$.

5. (a) $A = \begin{bmatrix}1 & 0\\ 0 & -1\end{bmatrix}$. (b) $B = \begin{bmatrix}\frac{1}{3} & \frac{4}{3}\\ \frac{2}{3} & -\frac{1}{3}\end{bmatrix}$.

(c) $P = \begin{bmatrix}1 & 1\\ -1 & 2\end{bmatrix}$.

9. $\begin{bmatrix}\frac{13}{2} & \frac{1}{2} & -\frac{3}{2}\\ -\frac{5}{2} & \frac{1}{2} & -\frac{1}{2}\end{bmatrix}$.

13. $\begin{bmatrix}1 & 0\\ 0 & 1\end{bmatrix}$.

15. $\begin{bmatrix}0 & -1\\ 1 & 0\end{bmatrix}$.

Section 4.6, p. 310

1. (a) $\begin{bmatrix}-1 & 0\\ 0 & 1\end{bmatrix}$. (b) $\begin{bmatrix}0 & 1\\ 1 & 0\end{bmatrix}$.

(c) $\begin{bmatrix}0 & -1\\ 1 & 0\end{bmatrix}$. (d) $\begin{bmatrix}\frac{\sqrt{3}}{2} & \frac{1}{2}\\ -\frac{1}{2} & \frac{\sqrt{3}}{2}\end{bmatrix}$.

3. (a) $\begin{bmatrix}1 & 0\\ k & 1\end{bmatrix}$.

(b)

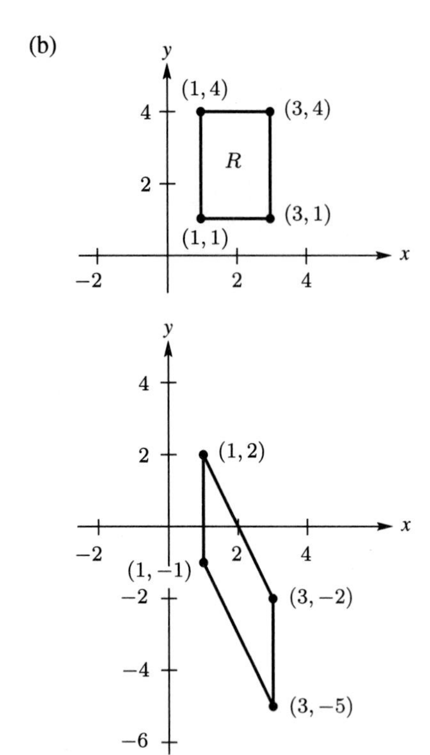

5. (a) $\begin{bmatrix} k & 0 \\ 0 & 1 \end{bmatrix}$.

(b)

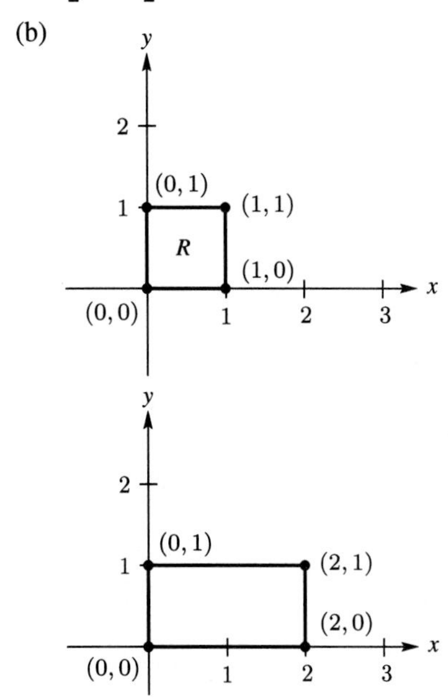

7. $(-10, 15)$, $(3, 12)$, $(-5, 2)$.

9.

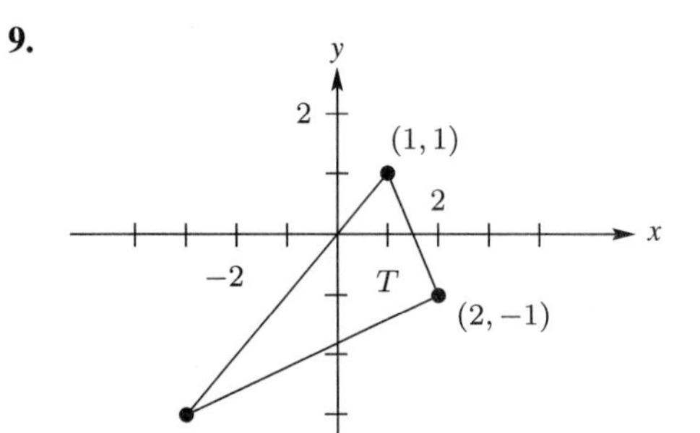

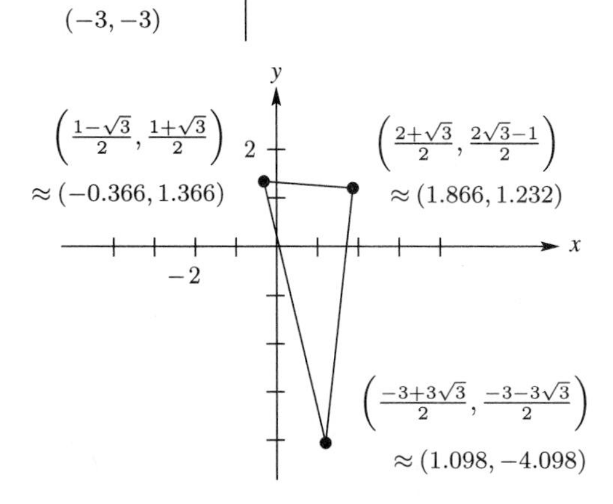

11. The image of the vertices of T under L consists of the points $(-9, -18)$, $(0, 0)$, and $(3, 6)$. Thus the image of T under L is a line segment.

Supplementary Exercises, p. 312

3. No.

5. (a) $8t + 7$. (b) $\frac{1}{2}(3a+b)t+\frac{1}{2}(3a-b)$.

7. (a) No. (b) No. (c) Yes. (d) No.
(e) $\{t^3 + t^2, t + 1\}$. (f) $\{t^3, t\}$.

9. (a) $\begin{bmatrix} 1 & 0 & 0 & 0 \\ 0 & 0 & 1 & 0 \\ 0 & 1 & 0 & 0 \\ 0 & 0 & 0 & 1 \end{bmatrix}$.

(b) $\begin{bmatrix} 1 & 1 & 0 & -1 \\ -1 & -1 & 1 & 1 \\ 0 & 1 & 0 & 0 \\ 0 & -1 & 0 & 1 \end{bmatrix}$.

(c) $\begin{bmatrix} 1 & 0 & 0 & 1 \\ 0 & 0 & 1 & 0 \\ 1 & 1 & 0 & 0 \\ 0 & 0 & 1 & 1 \end{bmatrix}$.

(d) $\begin{bmatrix} 2 & 1 & -1 & 0 \\ -2 & -1 & 2 & 0 \\ 1 & 1 & 0 & 0 \\ -1 & -1 & 1 & 1 \end{bmatrix}$.

11. (b) $\ker L$ = the set of all continuous functions f on $[0, 1]$ such that $f(0) = 0$.

(c) Yes.

13. (a) $\begin{bmatrix} 2 & 0 & 1 \\ 0 & 1 & 2 \\ -1 & 0 & -2 \end{bmatrix}$. (b) $4t^2 - 4t + 1$.

15. $\begin{bmatrix} 0 & 0 & 0 & 0 \\ 3 & 0 & 0 & 0 \\ 0 & 2 & 0 & 0 \\ 0 & 0 & 1 & 0 \end{bmatrix}$.

CHAPTER 5

Section 5.1, p. 319

1. (a) 5. (b) 7. (c) 4.

3. (a) Even (b) Odd (c) Even

5. (a) −. (b) +. (c) −.

7. (a) 9.

(b) Number of inversions in 416235 is 6; number of inversions in 436215 is 9.

9. (a) 0. (b) 0.

11. (a) 9. (b) 0. (c) 144.

13. (a) $t^2 - 3t - 4$. (b) $t^3 - 4t^2 + 3t$.

15. (a) $t = 4, t = -1$. (b) $t = 1, t = 0, t = 3$.

Section 5.2, p. 328

1. (a) 3. (b) 2. (c) 24. (d) 29.

(e) 4. (f) −30.

3. 3.

5. 8.

7. (a) 2. (b) −120.

(c) $(t-1)(t-2)(t-3) = t^3 - 6t^2 + 11t - 6$.

(d) $t^2 - 2t - 11$.

23. 32.

25. (b) is nonsingular.

27. The system has a nontrivial solution.

Section 5.3, p. 337

1. (a) 1. (b) 7. (c) 2. (d) 10.

3. (a) −2. (b) 9. (c) −2. (d) −1.

5. (a) 3. (d) 29. (e) 4.

7. (a) 4. (c) −30. (f) 0.

11. (a) $t = 0, t = 5$. (b) $t = 1, t = 4$.

15. (a) 6. (b) $(3, -6), (-1, 2), (13, -14)$.

(c) 24.

17. $\frac{41}{2}$.

Section 5.4, p. 343

3. (a) $\begin{bmatrix} 24 & -42 & -30 \\ 19 & -2 & -30 \\ -4 & 32 & 30 \end{bmatrix}$. (b) 150.

5. (a) $A^{-1} = \begin{bmatrix} 1 & 0 & -1 \\ 1 & -1 & 2 \\ -1 & 1 & -1 \end{bmatrix}$.

(b) $A^{-1} = \begin{bmatrix} \frac{7}{3} & -\frac{1}{3} & -\frac{1}{3} & -\frac{2}{3} \\ \frac{4}{9} & -\frac{1}{9} & -\frac{4}{9} & \frac{1}{9} \\ -\frac{1}{9} & -\frac{2}{9} & \frac{1}{9} & \frac{2}{9} \\ -\frac{5}{3} & \frac{2}{3} & \frac{2}{3} & \frac{1}{3} \end{bmatrix}$.

(c) Singular. (d) $A^{-1} = \begin{bmatrix} \frac{3}{2} & -1 & \frac{1}{2} \\ \frac{1}{2} & 0 & -\frac{1}{2} \\ -\frac{3}{2} & 1 & \frac{1}{2} \end{bmatrix}$.

(e) Singular.

7. (a) $-\frac{1}{28}\begin{bmatrix} -30 & -5 & 9 & 46 \\ -32 & 4 & 4 & 36 \\ -12 & -2 & -2 & 24 \\ 16 & -2 & -2 & -32 \end{bmatrix}$.

(b) $\begin{bmatrix} \frac{3}{14} & -\frac{3}{7} & \frac{1}{7} \\ \frac{1}{7} & \frac{5}{7} & -\frac{4}{7} \\ -\frac{1}{14} & \frac{1}{7} & \frac{2}{7} \end{bmatrix}$. (c) $\begin{bmatrix} \frac{2}{9} & -\frac{1}{9} \\ \frac{1}{6} & \frac{1}{6} \end{bmatrix}$.

9. $\begin{bmatrix} \dfrac{d}{ad-bc} & \dfrac{-b}{ad-bc} \\ \dfrac{-c}{ad-bc} & \dfrac{a}{ad-bd} \end{bmatrix}$.

11. $\begin{bmatrix} \frac{1}{4} & 0 & 0 \\ 0 & -\frac{1}{3} & 0 \\ 0 & 0 & \frac{1}{2} \end{bmatrix}$.

Section 5.5, p. 348

1. $x_1 = -2, x_2 = 0, x_3 = 1$.

3. Yes.

5. $x_1 = 1, x_2 = 2, x_3 = -2$.

7. $x_1 = x_2 = x_3 = 0$.

9. $x_1 = 1, x_2 = \frac{2}{3}, x_3 = -\frac{2}{3}$.

11. Yes.

13. For $c \neq \pm 2$.

15. (a) $-15\mathbf{i} - 2\mathbf{j} + 9\mathbf{k}$. (b) $-3\mathbf{i} + 3\mathbf{j} + 3\mathbf{k}$.
(c) $7\mathbf{i} + 5\mathbf{j} - \mathbf{k}$. (d) $3\mathbf{i} - 8\mathbf{j} - \mathbf{k}$.

Supplementary Exercises, p. 350

1. (a) 5. (b) 4. (c) 36. (d) 5.

CHAPTER 6

Section 6.1, p. 362

1. The only eigenvalue of L is $\lambda = -1$. Every nonzero vector in R^2 is an eigenvector of L associated with λ.

3. The eigenvalues are $\lambda_1 = 1, \lambda_2 = -1, \lambda_3 = 0$. Associated eigenvectors are $1, t^2$, and t, respectively.

5. (a) $p(\lambda) = \lambda^2 - 5\lambda + 7$.
(b) $p(\lambda) = \lambda^3 - 4\lambda^2 + 7$.
(c) $p(\lambda) = (\lambda - 4)(\lambda - 2)(\lambda - 3) = \lambda^3 - 9\lambda^2 + 26\lambda - 24$.
(d) $p(\lambda) = \lambda^2 - 7\lambda + 6$.

7. (a) $p(\lambda) = \lambda^2 - 5\lambda + 6$. The eigenvalues are $\lambda_1 = 2$ and $\lambda_2 = 3$. Associated eigenvectors are

$$\mathbf{x}_1 = \begin{bmatrix} 1 \\ -1 \end{bmatrix} \text{ and } \mathbf{x}_2 = \begin{bmatrix} 1 \\ -2 \end{bmatrix}.$$

(b) $p(\lambda) = \lambda^3 - 7\lambda^2 + 14\lambda - 8$. The eigenvalues are $\lambda_1 = 1, \lambda_2 = 2$, and $\lambda_3 = 4$. Associated eigenvectors are

$$\mathbf{x}_1 = \begin{bmatrix} -1 \\ 1 \\ 1 \end{bmatrix}, \quad \mathbf{x}_2 = \begin{bmatrix} 1 \\ 0 \\ 0 \end{bmatrix}, \quad \text{and}$$

$$\mathbf{x}_3 = \begin{bmatrix} 7 \\ -4 \\ 2 \end{bmatrix}.$$

(c) $p(\lambda) = \lambda^3 - 5\lambda^2 + 2\lambda + 8$. The eigenvalues are $\lambda_1 = -1, \lambda_2 = 2$, and $\lambda_3 = 4$. Associated eigenvectors are

$$\mathbf{x}_1 = \begin{bmatrix} 1 \\ 0 \\ -1 \end{bmatrix}, \quad \mathbf{x}_2 = \begin{bmatrix} -2 \\ -3 \\ 2 \end{bmatrix}, \quad \text{and}$$

$$\mathbf{x}_3 = \begin{bmatrix} 8 \\ 5 \\ 2 \end{bmatrix}.$$

(d) $p(\lambda) = \lambda^3 - 7\lambda^2 + 14\lambda - 8$. The eigenvalues are $\lambda_1 = 2, \lambda_2 = 4$, and $\lambda_3 = 1$. Associated eigenvectors are

$$\mathbf{x}_1 = \begin{bmatrix} 1 \\ 0 \\ 0 \end{bmatrix}, \quad \mathbf{x}_2 = \begin{bmatrix} \frac{7}{2} \\ -2 \\ 1 \end{bmatrix}, \quad \text{and}$$

$$\mathbf{x}_3 = \begin{bmatrix} -1 \\ 1 \\ 1 \end{bmatrix}.$$

11. The characteristic polynomial of A is

$$p(\lambda) = (\lambda - 1)(\lambda + 1)(\lambda - 3)(\lambda - 2).$$

The eigenvalues of L are $\lambda_1 = 1, \lambda_2 = -1, \lambda_3 = 3$ and $\lambda_4 = 2$. Associated eigenvectors are

$$\begin{bmatrix} 1 & 0 \\ 0 & 0 \end{bmatrix}, \quad \begin{bmatrix} 1 & -1 \\ 0 & 0 \end{bmatrix},$$

$$\begin{bmatrix} 9 & 3 \\ 4 & 0 \end{bmatrix}, \quad \text{and} \quad \begin{bmatrix} -29 & -7 \\ -9 & 3 \end{bmatrix}.$$

15. (a) $\left\{ \begin{bmatrix} 1 \\ 0 \\ 1 \end{bmatrix}, \begin{bmatrix} 0 \\ 1 \\ 0 \end{bmatrix} \right\}$.

(b) $\left\{ \begin{bmatrix} -1 \\ 0 \\ 1 \end{bmatrix} \right\}$.

(c) $\left\{ \begin{bmatrix} -1 \\ 0 \\ 1 \end{bmatrix}, \begin{bmatrix} 0 \\ 1 \\ 0 \end{bmatrix} \right\}$.

(d) $\left\{ \begin{bmatrix} 0 \\ 0 \\ 1 \\ 0 \end{bmatrix} \right\}$.

Section 6.2, p. 373

1. L is not diagonalizable. The eigenvalues of L are $\lambda_1 = \lambda_2 = \lambda_3 = 0$. The set of associated eigenvectors does not form a basis for P_2.

3. $\{t^2, t, 1\}$.

5. The eigenvalues of L are $\lambda_1 = 2$, $\lambda_2 = -3$, and $\lambda_3 = 4$. Associated eigenvectors are t^2, $t^2 - 5t$, and $9t^2 + 4t + 14$. L is diagonalizable.

7. (a) Not diagonalizable.
(b) Diagonalizable.
(c) Not diagonalizable.
(d) Not diagonalizable.

9. $\begin{bmatrix} 3 & 5 & -5 \\ 5 & 3 & -5 \\ 5 & 5 & -7 \end{bmatrix}$.

11. (a) $P = \begin{bmatrix} 1 & 2 & 1 \\ 0 & 1 & 0 \\ 0 & 0 & -3 \end{bmatrix}$.
(b) $P = \begin{bmatrix} -1 & -1 & 1 \\ 1 & 0 & 1 \\ 0 & 1 & 1 \end{bmatrix}$.
(c) Not possible.
(d) Not possible.

13. $D = \begin{bmatrix} -3 & 0 & 0 \\ 0 & 4 & 0 \\ 0 & 0 & 4 \end{bmatrix}$, $P = \begin{bmatrix} -1 & 0 & 0 \\ 0 & 0 & 1 \\ 1 & 1 & 1 \end{bmatrix}$.

15. (a) $D = \begin{bmatrix} 6 & 0 \\ 0 & 1 \end{bmatrix}$. (b) $D = \begin{bmatrix} 0 & 0 \\ 0 & 7 \end{bmatrix}$.
(c) $D = \begin{bmatrix} 2 & 0 & 0 \\ 0 & 4 & 0 \\ 0 & 0 & 1 \end{bmatrix}$.
(d) $D = \begin{bmatrix} 1 & 0 & 0 \\ 0 & 2 & 0 \\ 0 & 0 & 1 \end{bmatrix}$.

17. (a) Defective.
(b) Not defective.
(c) Not defective.
(d) Not defective.

19. $\begin{bmatrix} 768 & -1280 \\ 256 & -768 \end{bmatrix}$.

Section 6.3, p. 380

1. $\begin{bmatrix} 8 \\ 2 \\ 1 \end{bmatrix}$.

3. (b) and (c).

5. (a) $\mathbf{x}^{(1)} = \begin{bmatrix} 0.7 \\ 0.3 \end{bmatrix}$, $\mathbf{x}^{(2)} = \begin{bmatrix} 0.61 \\ 0.39 \end{bmatrix}$, $\mathbf{x}^{(3)} = \begin{bmatrix} 0.583 \\ 0.417 \end{bmatrix}$.
(b) Since all entries of T are positive, it is regular; $\mathbf{u} = \begin{bmatrix} 0.571 \\ 0.429 \end{bmatrix}$.

7. (a) and (d).

9. (a) $\begin{bmatrix} \frac{3}{7} \\ \frac{4}{7} \end{bmatrix}$. (b) $\begin{bmatrix} \frac{1}{8} \\ \frac{7}{8} \end{bmatrix}$.
(c) $\begin{bmatrix} \frac{4}{11} \\ \frac{4}{11} \\ \frac{3}{11} \end{bmatrix}$. (d) $\begin{bmatrix} \frac{1}{11} \\ \frac{4}{11} \\ \frac{6}{11} \end{bmatrix}$.

11. (a) 0.69.
(b) 20.7 percent of the population will be farmers.

13. (a) 35 percent, 37.5 percent.
(b) 40 percent.

Section 6.4, p. 393

7. $P^T P = I_3$.

9. (a) If B is the given matrix, verify that $B^T B = I_2$.

15. A is similar to $D = \begin{bmatrix} 0 & 0 \\ 0 & 4 \end{bmatrix}$ and $P = \begin{bmatrix} \frac{1}{\sqrt{2}} & \frac{1}{\sqrt{2}} \\ -\frac{1}{\sqrt{2}} & \frac{1}{\sqrt{2}} \end{bmatrix}$.

17. A is similar to $D = \begin{bmatrix} 0 & 0 & 0 \\ 0 & 0 & 0 \\ 0 & 0 & 4 \end{bmatrix}$ and

$$P = \begin{bmatrix} 1 & 0 & 0 \\ 0 & -\frac{1}{\sqrt{2}} & \frac{1}{\sqrt{2}} \\ 0 & \frac{1}{\sqrt{2}} & \frac{1}{\sqrt{2}} \end{bmatrix}.$$

19. A is similar to $D = \begin{bmatrix} -2 & 0 & 0 \\ 0 & 1 & 0 \\ 0 & 0 & 1 \end{bmatrix}$ and

$$P = \begin{bmatrix} \frac{1}{\sqrt{3}} & -\frac{1}{\sqrt{2}} & -\frac{1}{\sqrt{6}} \\ \frac{1}{\sqrt{3}} & \frac{1}{\sqrt{2}} & -\frac{1}{\sqrt{6}} \\ \frac{1}{\sqrt{3}} & 0 & \frac{2}{\sqrt{6}} \end{bmatrix}.$$

21. A is similar to $D = \begin{bmatrix} 3 & 0 \\ 0 & 1 \end{bmatrix}$.

23. A is similar to $D = \begin{bmatrix} 1 & 0 & 0 \\ 0 & 2 & 0 \\ 0 & 0 & 0 \end{bmatrix}$.

25. A is similar to $D = \begin{bmatrix} 1 & 0 & 0 \\ 0 & 0 & 0 \\ 0 & 0 & 2 \end{bmatrix}$.

27. A is similar to $D = \begin{bmatrix} 2 & 0 & 0 \\ 0 & -2 & 0 \\ 0 & 0 & 4 \end{bmatrix}$.

Section 6.5, p. 404

1. (a) 5, 1. (b) 2, 0.
(c) $\sqrt{5}, \sqrt{6}$. (d) $0, 0, 1, \sqrt{5}$.

3. $A = USV^T = \begin{bmatrix} \frac{1}{\sqrt{2}} & \frac{1}{\sqrt{2}} \\ -\frac{1}{\sqrt{2}} & \frac{1}{\sqrt{2}} \end{bmatrix} \begin{bmatrix} \sqrt{2} & 0 \\ 0 & \sqrt{2} \end{bmatrix} \begin{bmatrix} 1 & 0 \\ 0 & 1 \end{bmatrix}^T$.

5. (a) 10.8310, 0.8310.
(b) 18.9245, 3.8400, 0.3440.
(c) 25.4368, 1.7226, 0.

Section 6.6, p. 414

1. (a) $\begin{bmatrix} x & y \end{bmatrix} \begin{bmatrix} -3 & \frac{5}{2} \\ \frac{5}{2} & -2 \end{bmatrix} \begin{bmatrix} x \\ y \end{bmatrix}$.

(b) $\begin{bmatrix} x_1 & x_2 & x_3 \end{bmatrix} \begin{bmatrix} 2 & \frac{3}{2} & -\frac{5}{2} \\ \frac{3}{2} & 0 & \frac{7}{2} \\ -\frac{5}{2} & \frac{7}{2} & 0 \end{bmatrix} \begin{bmatrix} x_1 \\ x_2 \\ x_3 \end{bmatrix}$.

(c) $\begin{bmatrix} x_1 & x_2 & x_3 \end{bmatrix} \begin{bmatrix} 3 & \frac{1}{2} & -\frac{1}{2} \\ \frac{1}{2} & 1 & -2 \\ -\frac{1}{2} & -2 & -2 \end{bmatrix} \begin{bmatrix} x_1 \\ x_2 \\ x_3 \end{bmatrix}$.

3. (a) $\begin{bmatrix} -1 & 0 & 0 \\ 0 & 2 & 0 \\ 0 & 0 & 0 \end{bmatrix}$.

(b) $\begin{bmatrix} 3 & 0 & 0 \\ 0 & 0 & 0 \\ 0 & 0 & 0 \end{bmatrix}$.

(c) $\begin{bmatrix} 4 & 0 & 0 \\ 0 & -2 & 0 \\ 0 & 0 & -2 \end{bmatrix}$.

5. $3x'^2 - 2y'^2$.

7. $y_1^2 - y_3^2$

9. $-2y_1^2 + 5y_2^2 - 5y_3^2$.

11. y'^2.

13. $y_1^2 + y_2^2 - y_3^2$.

15. $y_1^2 - y_2^2$.

17. $y_1^2 - y_2^2$, rank $= 2$, signature $= 0$.

21. g_1, g_2, and g_4.

23. (a), (b), and (c).

Section 6.7, p. 423

1. Ellipse.

3. Hyperbola.

5. Two intersecting lines.

7. Circle.

9. Point.

11. Ellipse; $\frac{x'^2}{2} + y'^2 = 1$.

13. Circle; $\frac{x'^2}{5^2} + \frac{y'^2}{5^2} = 1$.

15. Pair of parallel lines; $y' = 2$, $y' = -2$; $y'^2 = 4$.

17. Point $(1, 3)$; $x'^2 + y'^2 = 0$.

19. Possible answer: ellipse; $\frac{x'^2}{12} + \frac{y'^2}{4} = 1$.

21. Possible answer: pair of parallel lines $y' = \frac{2}{\sqrt{10}}$ and $y' = -\frac{2}{\sqrt{10}}$.

23. Possible answer: two intersecting lines $y' = 3x'$ and $y' = -3x'$; $9x'^2 - y'^2 = 0$.

25. Possible answer: parabola; $y''^2 = -4x''$.

27. Possible answer: hyperbola; $\frac{x''^2}{4} - \frac{y''^2}{9} = 1$.

29. Possible answer: hyperbola; $\frac{x''^2}{\frac{9}{8}} - \frac{y''^2}{\frac{9}{8}} = 1$.

Section 6.8, p.434

1. Hyperboloid of one sheet.

3. Hyperbolic paraboloid.

5. Parabolic cylinder.

7. Parabolic cylinder.

9. Ellipsoid.

11. Elliptic paraboloid.

13. Hyperbolic paraboloid.

15. Ellipsoid; $x'^2 + y'^2 + \frac{z'^2}{\frac{1}{3}} = 1$.

17. Hyperbolic paraboloid; $\frac{x''^2}{4} - \frac{y''^2}{4} = z''$.

19. Elliptic paraboloid; $\frac{x'^2}{4} + \frac{y'^2}{8} = 1$.

21. Hyperboloid of one sheet; $\frac{x''^2}{2} + \frac{y''^2}{4} - \frac{z''}{4} = 1$.

23. Parabolic cylinder; $x''^2 = \frac{4}{\sqrt{2}} y''$.

25. Hyperboloid of two sheets; $\frac{x''^2}{\frac{7}{4}} - \frac{y''^2}{\frac{7}{4}} - \frac{z''^2}{\frac{7}{4}} = 1$.

27. Cone; $x''^2 + y''^2 - z''^2 = 0$.

Supplementary Exercises, p. 435

1. (a) $\lambda_1 = 1, \lambda_2 = 1, \lambda_3 = 4$;

associated eigenvectors: $\begin{bmatrix} 1 \\ 0 \\ 0 \end{bmatrix}, \begin{bmatrix} 0 \\ 0 \\ 1 \end{bmatrix}, \begin{bmatrix} 4 \\ 3 \\ 0 \end{bmatrix}$.

(b) $\lambda_1 = 3, \lambda_2 = 4, \lambda_3 = -1$;

associated eigenvectors: $\begin{bmatrix} 0 \\ 0 \\ 1 \end{bmatrix}, \begin{bmatrix} 1 \\ -1 \\ 0 \end{bmatrix}, \begin{bmatrix} 2 \\ 3 \\ 0 \end{bmatrix}$.

(c) $\lambda_1 = 1, \lambda_2 = 2, \lambda_3 = 3$;

associated eigenvectors: $\begin{bmatrix} 1 \\ 1 \\ 1 \end{bmatrix}, \begin{bmatrix} 1 \\ 2 \\ 4 \end{bmatrix}, \begin{bmatrix} 1 \\ 3 \\ 9 \end{bmatrix}$.

(d) $\lambda_1 = -3, \lambda_2 = 1, \lambda_3 = -1$;

associated eigenvectors: $\begin{bmatrix} 1 \\ -3 \\ 9 \end{bmatrix}, \begin{bmatrix} 1 \\ 1 \\ 1 \end{bmatrix}, \begin{bmatrix} -1 \\ 1 \\ -1 \end{bmatrix}$.

7. (a) $\begin{bmatrix} -\frac{3}{10} \\ \frac{1}{5} \end{bmatrix}, \begin{bmatrix} -\frac{6}{5} \\ \frac{4}{5} \end{bmatrix}$.

(b) $A = \frac{1}{10}\begin{bmatrix} -3 & -12 \\ 2 & 8 \end{bmatrix}$.

(c) The eigenvalues are $\lambda_1 = 0, \lambda_2 = \frac{1}{2}$. Associated eigenvectors are

$$\mathbf{x}_1 = \begin{bmatrix} -4 \\ 1 \end{bmatrix} \quad \text{and} \quad \mathbf{x}_2 = \begin{bmatrix} -3 \\ 2 \end{bmatrix}.$$

(d) The eigenvalues are $\lambda_1 = 0$ and $\lambda_2 = \frac{1}{2}$. Associated eigenvectors are $p_1(t) = 5t - 5$ and $p_2(t) = 5t$.

(e) The eigenspace for $\lambda_1 = 0$ is the subspace of P_1 with basis $\{5t - 5\}$. The eigenspace for $\lambda_2 = \frac{1}{2}$ is the subspace of P_1 with basis $\{5t\}$.

9. The only eigenvalue is $\lambda_1 = 0$ and an associated eigenvector is $p_1(t) = 1$.

CHAPTER 7

Section 7.1, p. 446

1. (a) $\mathbf{x}(t) = \begin{bmatrix} x_1(t) \\ x_2(t) \\ x_3(t) \end{bmatrix} = \begin{bmatrix} b_1 e^{-3t} \\ b_2 e^{4t} \\ b_3 e^{2t} \end{bmatrix}$

$$= b_1 \begin{bmatrix} 1 \\ 0 \\ 0 \end{bmatrix} e^{-3t} + b_2 \begin{bmatrix} 0 \\ 1 \\ 0 \end{bmatrix} e^{4t} + b_3 \begin{bmatrix} 0 \\ 0 \\ 1 \end{bmatrix} e^{2t}.$$

(b) $\begin{bmatrix} 3e^{-3t} \\ 4e^{4t} \\ 5e^{2t} \end{bmatrix} = 3\begin{bmatrix} 1 \\ 0 \\ 0 \end{bmatrix} e^{-3t} + 4\begin{bmatrix} 0 \\ 1 \\ 0 \end{bmatrix} e^{4t} + 5\begin{bmatrix} 0 \\ 0 \\ 1 \end{bmatrix} e^{2t}$.

3. $\mathbf{x}(t) = b_1 \begin{bmatrix} 6 \\ 2 \\ 7 \end{bmatrix} e^{4t} + b_2 \begin{bmatrix} 0 \\ 7 \\ -1 \end{bmatrix} e^{-5t} + b_3 \begin{bmatrix} 0 \\ 0 \\ 1 \end{bmatrix} e^{2t}$.

5. $\mathbf{x}(t) = b_1 \begin{bmatrix} 1 \\ 0 \\ 0 \end{bmatrix} e^{5t} + b_2 \begin{bmatrix} 0 \\ 1 \\ 3 \end{bmatrix} e^{5t} + b_3 \begin{bmatrix} 0 \\ -3 \\ 1 \end{bmatrix} e^{-5t}$.

7. $\mathbf{x}(t) = b_1 \begin{bmatrix} 1 \\ 0 \\ 0 \end{bmatrix} e^{-2t} + b_2 \begin{bmatrix} 1 \\ 1 \\ 2 \end{bmatrix} e^{2t} + b_3 \begin{bmatrix} -7 \\ -2 \\ 1 \end{bmatrix} e^{-3t}$.

9. $\mathbf{x}(t) = \begin{bmatrix} 440 + 60e^{-5t} \\ 220 - 20e^{-5t} \end{bmatrix}$.

Section 7.2, p. 455

1. The origin is a stable equilibrium. The phase portrait shows all trajectories tending toward the origin.

3. The origin is a stable equilibrium. The phase portrait shows all trajectories tending towards the origin with those passing through points not on the eigenvector aligning themselves to be tangent to the eigenvector at the origin.

5. The origin is a saddle point. The phase portrait shows trajectories not in the direction of an eigenvector heading towards the origin, but bending away as $t \to \infty$.

7. The origin is a stable equilibrium. The phase portrait shows all trajectories tending towards the origin.

9. The origin is called marginally stable.

Supplementary Exercises, p. 455

1. (a) $\dfrac{d}{dt}[A(t)] = \begin{bmatrix} 2t & \dfrac{-1}{(t+1)^2} \\ 0 & -e^{-t} \end{bmatrix}$

$$\int_0^t A(s)\,ds = \begin{bmatrix} \dfrac{t^3}{3} & \ln(1+t) \\ 4t & -e^{-t}+1 \end{bmatrix}.$$

(b) $\dfrac{d}{dt}[A(t)] = \begin{bmatrix} 2\cos 2t & 0 & 0 \\ 0 & 0 & -1 \\ 0 & e^{t^2} + 2t^2e^{t^2} & \dfrac{1-t^2}{(t^2+1)^2} \end{bmatrix}$.

$\displaystyle\int_0^t A(s)\,ds =$

$$\begin{bmatrix} -\dfrac{\cos 2t}{2} + \dfrac{1}{2} & 0 & 0 \\ 0 & t & -\dfrac{t^2}{2} \\ 0 & \dfrac{e^{t^2}}{2} - \dfrac{1}{2} & \dfrac{1}{2}\ln(t^2+1) \end{bmatrix}.$$

9. (a) $\mathbf{x}(t) = \frac{2}{5}\begin{bmatrix} 4 \\ 4 \\ 1 \end{bmatrix} + \frac{7}{20}\begin{bmatrix} -1 \\ -6 \\ 1 \end{bmatrix} e^{5t} + \frac{1}{4}\begin{bmatrix} -1 \\ 2 \\ 1 \end{bmatrix} e^{-3t}$.

(b) $\mathbf{x}(t) = \frac{7}{8}\begin{bmatrix} 1 \\ 0 \\ 0 \end{bmatrix} + \frac{1}{12}\begin{bmatrix} 1 \\ 2 \\ 4 \end{bmatrix} e^{2t} + \frac{1}{24}\begin{bmatrix} 1 \\ -4 \\ 16 \end{bmatrix} e^{-4t}$.

CHAPTER 9

Basic Matrix Properties, p. 492

ML.1. (a) Commands: **A(2,3)**, **B(3,2)**, **B(1,2)**.

(b) For $\text{row}_1(\mathbf{A})$, use command **A(1,:)**.
For $\text{col}_3(\mathbf{A})$, use command **A(:,3)**.
For $\text{row}_2(\mathbf{B})$, use command **B(2,:)**.
(In this context the colon means "all.")

(c) Matrix B in **format long** is

$$\begin{bmatrix} 8.00000000000000 & 0.666666666666667 \\ 0.00497512437811 & -3.200000000000000 \\ 0.00001000000000 & 4.333333333333333 \end{bmatrix}.$$

Matrix Operations, p. 492

ML.1. (a) $\begin{bmatrix} 4.5000 & 2.2500 & 3.7500 \\ 1.5833 & 0.9167 & 1.5000 \\ 0.9667 & 0.5833 & 0.9500 \end{bmatrix}$.

(b)
```
??? Error using ==> *
Inner matrix dimensions must
agree.
```

(c) $\begin{bmatrix} 5.0000 & 1.5000 \\ 1.5833 & 2.2500 \\ 2.4500 & 3.1667 \end{bmatrix}$.

(d)
```
??? Error using ==> *
Inner matrix dimensions must
agree.
```

(e)
```
??? Error using ==> *
Inner matrix dimensions must
agree.
```

(f)
```
??? Error using ==> -
Inner matrix dimensions must
agree.
```

(g) $\begin{bmatrix} 18.2500 & 7.4583 & 12.2833 \\ 7.4583 & 5.7361 & 8.9208 \\ 12.2833 & 8.9208 & 14.1303 \end{bmatrix}$.

ML.3. $\begin{bmatrix} 4 & -3 & 2 & -1 & -5 \\ 2 & 1 & -3 & 0 & 7 \\ -1 & 4 & 1 & 2 & 8 \end{bmatrix}$.

ML.5. (a) $\begin{bmatrix} 1 & 0 & 0 & 0 \\ 0 & 2 & 0 & 0 \\ 0 & 0 & 3 & 0 \\ 0 & 0 & 0 & 4 \end{bmatrix}$.

(b) $\begin{bmatrix} 0 & 0 & 0 & 0 & 0 \\ 0 & 1.0000 & 0 & 0 & 0 \\ 0 & 0 & 0.5000 & 0 & 0 \\ 0 & 0 & 0 & 0.3333 & 0 \\ 0 & 0 & 0 & 0 & 0.2500 \end{bmatrix}$.

(c) $\begin{bmatrix} 5 & 0 & 0 & 0 & 0 & 0 \\ 0 & 5 & 0 & 0 & 0 & 0 \\ 0 & 0 & 5 & 0 & 0 & 0 \\ 0 & 0 & 0 & 5 & 0 & 0 \\ 0 & 0 & 0 & 0 & 5 & 0 \\ 0 & 0 & 0 & 0 & 0 & 5 \end{bmatrix}$.

Powers of a Matrix, p. 493

ML.1. (a) $k = 3$. (b) $k = 5$.

ML.3. (a) $\begin{bmatrix} 0 & -2 & 4 \\ 4 & 0 & -2 \\ -2 & 4 & 0 \end{bmatrix}$. (b) $\begin{bmatrix} 0 & 0 & 0 \\ 0 & 0 & 0 \\ 0 & 0 & 0 \end{bmatrix}$.

ML.5. The sequence seems to be converging to

$$\begin{bmatrix} 1.0000 & 0.7500 \\ 0 & 0 \end{bmatrix}.$$

ML.7. (a) $A^T A = \begin{bmatrix} 2 & -3 & -1 \\ -3 & 9 & 2 \\ -1 & 2 & 6 \end{bmatrix}$,

$AA^T = \begin{bmatrix} 6 & -1 & -3 \\ -1 & 6 & 4 \\ -3 & 4 & 5 \end{bmatrix}$.

(b) $B = \begin{bmatrix} 2 & -3 & 1 \\ -3 & 2 & 4 \\ 1 & 4 & 2 \end{bmatrix}$,

$C = \begin{bmatrix} 0 & -1 & 1 \\ 1 & 0 & 0 \\ -1 & 0 & 0 \end{bmatrix}$.

(c) $B + C = \begin{bmatrix} 2 & -4 & 2 \\ -2 & 2 & 4 \\ 0 & 4 & 2 \end{bmatrix}$,

$B + C = 2A$.

Row Operations and Echelon Forms, p. 494

ML.1. (a) $\begin{bmatrix} 1.0000 & 0.5000 & 0.5000 \\ -3.0000 & 1.0000 & 4.0000 \\ 1.0000 & 0 & 3.0000 \\ 5.0000 & -1.0000 & 5.0000 \end{bmatrix}$.

(b) $\begin{bmatrix} 1.0000 & 0.5000 & 0.5000 \\ 0 & 2.5000 & 5.5000 \\ 1.0000 & 0 & 3.0000 \\ 5.0000 & -1.0000 & 5.0000 \end{bmatrix}$.

(c) $\begin{bmatrix} 1.0000 & 0.5000 & 0.5000 \\ 0 & 2.5000 & 5.5000 \\ 0 & -0.5000 & 2.5000 \\ 5.0000 & -1.0000 & 5.0000 \end{bmatrix}$.

(d) $\begin{bmatrix} 1.0000 & 0.5000 & 0.5000 \\ 0 & 2.5000 & 5.5000 \\ 0 & -0.5000 & 2.5000 \\ 0 & -3.5000 & 2.5000 \end{bmatrix}$.

(e) $\begin{bmatrix} 1.0000 & 0.5000 & 0.5000 \\ 0 & -3.5000 & 2.5000 \\ 0 & -0.5000 & 2.5000 \\ 0 & 2.5000 & 5.5000 \end{bmatrix}$.

ML.3. $\begin{bmatrix} 1 & 0 & 0 \\ 0 & 1 & 0 \\ 0 & 0 & 1 \\ 0 & 0 & 0 \end{bmatrix}$.

ML.5. $x = -2 + r$, $y = -1$, $z = 8 - 2r$, $w = r$, $r =$ any real number.

ML.7. $x_1 = -r + 1$, $x_2 = r + 2$, $x_3 = r - 1$, $x_4 = r$, $r =$ any real number.

ML.9. $\mathbf{x} = \begin{bmatrix} 0.5r \\ r \end{bmatrix}$.

ML.11. Exercise 15:

(a) Unique solution: $x = -1$, $y = 4$, $z = -3$.

(b) The only solution is the trivial one.

Exercise 16:

(a) $x = r$, $y = -2r$, $z = r$, where r is any real number.

(b) Unique solution: $x = 1$, $y = 2$, $z = 2$.

ML.13. The \ command yields a matrix showing that the system is inconsistent. The **rref** command leads to the display of a warning that the result may contain large roundoff errors.

LU-Factorization, p. 495

ML.1. $L = \begin{bmatrix} 1 & 0 & 0 \\ 1 & 1 & 0 \\ 0.5 & 0.3333 & 1 \end{bmatrix}$,

$U = \begin{bmatrix} 2 & 8 & 0 \\ 0 & -6 & -3 \\ 0 & 0 & 8 \end{bmatrix}$.

ML.3. $L = \begin{bmatrix} 1.0000 & 0 & 0 & 0 \\ 0.5000 & 1.0000 & 0 & 0 \\ -2.0000 & -2.0000 & 1.0000 & 0 \\ -1.0000 & 1.0000 & -2.0000 & 1.0000 \end{bmatrix}$,

$U = \begin{bmatrix} 6 & -2 & -4 & 4 \\ 0 & -2 & -4 & -1 \\ 0 & 0 & 5 & -2 \\ 0 & 0 & 0 & 8 \end{bmatrix}$,

$\mathbf{z} = \begin{bmatrix} 2 \\ -5 \\ 2 \\ -32 \end{bmatrix}$, $\mathbf{x} = \begin{bmatrix} 4.5000 \\ 6.9000 \\ -1.2000 \\ -4.0000 \end{bmatrix}$.

Matrix Inverses, p. 495

ML.1. (a) and (c).

ML.3. (a) $\begin{bmatrix} -2 & 3 \\ 1 & -1 \end{bmatrix}$.

(b) $\begin{bmatrix} -\frac{1}{4} & \frac{3}{4} & -\frac{1}{4} \\ -\frac{1}{4} & -\frac{1}{4} & \frac{3}{4} \\ \frac{3}{4} & -\frac{1}{4} & -\frac{1}{4} \end{bmatrix}$.

ML.5. (a) $t = 4$. (b) $t = 3$.

Subspaces, p. 496

ML.3. (a) No. (b) Yes.

ML.5. (a) $0\mathbf{v}_1 + \mathbf{v}_2 - \mathbf{v}_3 - \mathbf{v}_4 = \mathbf{v}$.

(b) $p_1(t) + 2p_2(t) + 2p_3(t) = p(t)$.

ML.7. (a) Yes. (b) Yes. (c) Yes.

Linear Independence/Dependence, p. 497

ML.1. (a) Linearly dependent.

(b) Linearly independent.

(c) Linearly independent.

Bases and Dimension, p. 498

ML.1. Basis.

ML.3. Basis.

ML.5. Basis.

ML.7. dim span $S = 3$, span $S \neq R^4$.

ML.9. dim span $S = 3$, span $S = P_2$.

ML.11. $\{t^3 - t + 1, t^3 + 2, t, 1\}$.

Coordinates and Change of Basis, p. 498

ML.1. (a) $\begin{bmatrix} 1 \\ 2 \\ 3 \end{bmatrix}$. (b) $\begin{bmatrix} -1 \\ 2 \\ -1 \end{bmatrix}$. (c) $\begin{bmatrix} 1 \\ 1 \\ 1 \end{bmatrix}$.

ML.3. (a) $\begin{bmatrix} 0.5000 \\ -0.5000 \\ 0 \\ -0.5000 \end{bmatrix}$. (b) $\begin{bmatrix} 1.0000 \\ 0.5000 \\ 0.3333 \\ 0 \end{bmatrix}$.

(c) $\begin{bmatrix} 0.5000 \\ 0.1667 \\ -0.3333 \\ -1.5000 \end{bmatrix}$.

ML.5. $\begin{bmatrix} -0.5000 & -1.0000 & -0.5000 & 0 \\ -0.5000 & 0 & 1.5000 & 0 \\ 1.0000 & 0 & -1.0000 & 1.0000 \\ 0 & 0 & 0 & 1.0000 \end{bmatrix}$.

ML.7. (a) $\begin{bmatrix} 1.0000 & -1.6667 & 2.3333 \\ 1.0000 & 0.6667 & -1.3333 \\ 0 & 1.3333 & -0.6667 \end{bmatrix}$.

(b) $\begin{bmatrix} 2 & 0 & 1 \\ -1 & 1 & -1 \\ 0 & -1 & 2 \end{bmatrix}$.

(c) $\begin{bmatrix} 2 & -2 & 4 \\ 0 & 1 & -3 \\ -1 & 2 & 0 \end{bmatrix}$. (d) QP.

Homogeneous Linear Systems, p. 499

ML.1. $\left\{ \begin{bmatrix} -2 \\ 0 \\ 1 \\ 0 \\ 0 \end{bmatrix}, \begin{bmatrix} -1 \\ -1 \\ 0 \\ 1 \\ 0 \end{bmatrix}, \begin{bmatrix} -2 \\ 1 \\ 0 \\ 0 \\ 1 \end{bmatrix} \right\}$.

ML.3. $\left\{ \begin{bmatrix} 1 \\ -2 \\ 1 \\ 0 \end{bmatrix}, \begin{bmatrix} \frac{4}{3} \\ -\frac{1}{3} \\ 0 \\ 1 \end{bmatrix} \right\}$.

ML.5. $\mathbf{x} = \begin{bmatrix} t \\ t \\ t \\ t \end{bmatrix}$, where t is any nonzero real number.

Rank of a Matrix, p. 500

ML.3. (a) The original columns of A and

$$\left\{\begin{bmatrix}1\\0\\2\\0\end{bmatrix}, \begin{bmatrix}0\\1\\1\\0\end{bmatrix}, \begin{bmatrix}0\\0\\0\\1\end{bmatrix}\right\}.$$

(b) The first two columns of A and

$$\left\{\begin{bmatrix}1\\0\\0\\1\\1\end{bmatrix}, \begin{bmatrix}0\\0\\1\\2\\1\end{bmatrix}\right\}.$$

ML.5. (a) Consistent. (b) Inconsistent.

(c) Inconsistent.

Standard Inner Product, p. 500

ML.3. (a) 2.2361. (b) 5.4772. (c) 3.1623.

ML.5. (a) 19. (b) −11. (c) −55.

ML.9. (a) $\begin{bmatrix}0.6667\\0.6667\\-0.3333\end{bmatrix}$ or in rational form $\begin{bmatrix}\frac{2}{3}\\\frac{2}{3}\\-\frac{1}{3}\end{bmatrix}$.

(b) $\begin{bmatrix}0\\0.8000\\-0.6000\\0\end{bmatrix}$ or in rational form $\begin{bmatrix}0\\\frac{4}{5}\\-\frac{3}{5}\\0\end{bmatrix}$.

(c) $\begin{bmatrix}0.3015\\0\\0.3015\\0\end{bmatrix}$.

Cross Product, p. 501

ML.1. (a) $\begin{bmatrix}-11 & 2 & 5\end{bmatrix}$. (b) $\begin{bmatrix}3 & 1 & -1\end{bmatrix}$.

(c) $\begin{bmatrix}1 & -8 & -5\end{bmatrix}$.

ML.5. 8.

The Gram–Schmidt Process, p. 502

ML.1. $\left\{\begin{bmatrix}0.7071\\0.7071\\0\end{bmatrix}, \begin{bmatrix}0.7071\\-0.7071\\0\end{bmatrix}, \begin{bmatrix}0\\0\\1.0000\end{bmatrix}\right\}$

$$= \left\{\begin{bmatrix}\frac{\sqrt{2}}{2}\\\frac{\sqrt{2}}{2}\\0\end{bmatrix}, \begin{bmatrix}\frac{\sqrt{2}}{2}\\-\frac{\sqrt{2}}{2}\\0\end{bmatrix}, \begin{bmatrix}0\\0\\1\end{bmatrix}\right\}.$$

ML.3. (a) $\begin{bmatrix}-1.4142\\1.4142\\1.0000\end{bmatrix}$. (b) $\begin{bmatrix}0\\1.4142\\1.0000\end{bmatrix}$.

(c) $\begin{bmatrix}0.7071\\0.7071\\-1.0000\end{bmatrix}$.

Projections, p. 502

ML.1. (a) $\begin{bmatrix}0\\\frac{5}{6}\\\frac{5}{3}\\\frac{5}{6}\end{bmatrix}$. (b) $\begin{bmatrix}\frac{3}{5}\\\frac{3}{5}\\\frac{3}{5}\\\frac{3}{5}\\\frac{3}{5}\end{bmatrix}$.

ML.3. (a) $\begin{bmatrix}2.4286\\3.9341\\7.9011\end{bmatrix}$.

(b) $\sqrt{(2.4286-2)^2 + (3.9341-4)^2 + (7.9011-8)^2}$

≈ 0.4448.

ML.5. $\mathbf{p} = \begin{bmatrix}0.8571\\0.5714\\1.4286\\0.8571\\0.8571\end{bmatrix}$.

Least Squares, p. 503

ML.1. $y = 1.87 + 1.345t$.

ML.3. (a) $T = -8.278t + 188.1$, where t = time.

(b) $T(1) = 179.7778°$ F.
$T(6) = 138.3889°$ F.
$T(8) = 121.8333°$ F.

(c) 3.3893 minutes.

ML.5. $y = 1.0204x^2 + 3.1238x + 1.0507$,
when $x = 7$, $y = 72.9169$.

Kernel and Range of Linear Transformations, p. 504

ML.1. Basis for ker L: $\left\{\begin{bmatrix}-1\\-2\\1\\0\end{bmatrix}, \begin{bmatrix}1\\-3\\0\\1\end{bmatrix}\right\}$.

Basis for range L: $\left\{\begin{bmatrix}1\\0\end{bmatrix}, \begin{bmatrix}0\\1\end{bmatrix}\right\}$.

ML.3. Basis for ker L: $\left\{\begin{bmatrix}-2\\0\\1\\-2\\1\end{bmatrix}, \begin{bmatrix}-1\\1\\0\\0\\0\end{bmatrix}\right\}$.

Basis for range L: $\left\{\begin{bmatrix}1\\0\\0\end{bmatrix}, \begin{bmatrix}0\\1\\0\end{bmatrix}, \begin{bmatrix}0\\0\\1\end{bmatrix}\right\}$.

Matrix of a Linear Transformation, p. 504

ML.1. $A = \begin{bmatrix}-1 & 0 & 3\\1 & 0 & -2\end{bmatrix}$.

ML.3. (a) $A = \begin{bmatrix}1.3333 & -0.3333\\-1.6667 & -3.3333\end{bmatrix}$.

(b) $B = \begin{bmatrix}-3.6667 & 0.3333\\-3.3333 & 1.6667\end{bmatrix}$.

(c) $P = \begin{bmatrix}-0.3333 & 0.6667\\1.6667 & -0.3333\end{bmatrix}$.

Determinants by Row Reduction, p. 506

ML.1. (a) -18. (b) 5.

ML.3. (a) 4. (b) 0.

ML.5. $t = 3, t = 4$.

Determinants by Cofactor Expansion, p. 507

ML.1. $A_{11} = -11, A_{23} = -2, A_{31} = 2$.

ML.3. 0.

ML.5. (a) $\dfrac{1}{28}\begin{bmatrix}30 & 5 & -9 & -46\\32 & -4 & -4 & -36\\12 & 2 & 2 & -24\\-16 & 2 & 2 & 32\end{bmatrix}$.

(b) $\dfrac{1}{14}\begin{bmatrix}3 & -6 & 2\\2 & 10 & -8\\-1 & 2 & 4\end{bmatrix}$.

(c) $\dfrac{1}{18}\begin{bmatrix}4 & -2\\3 & 3\end{bmatrix}$.

Eigenvalues and Eigenvectors, p. 507

ML.1. (a) $\lambda^2 - 5$. (b) $\lambda^3 - 6\lambda^2 + 4\lambda + 8$.
(c) $\lambda^4 - 3\lambda^3 - 3\lambda^2 + 11\lambda - 6$.

ML.3. (a) $\begin{bmatrix}1\\1\end{bmatrix}$. (b) $\begin{bmatrix}0\\0\\1\end{bmatrix}$. (c) $\begin{bmatrix}1\\-2\\1\end{bmatrix}$.

ML.5. $\begin{bmatrix}1 & -1 & 1\\0 & 0 & 1\\0 & 0 & 1\end{bmatrix}$.

ML.7. The sequence $A, A^3, A^5, \ldots$ converges to
$\begin{bmatrix}-1 & 1 & -1\\-2 & 2 & -1\\-2 & 2 & -1\end{bmatrix}$.

The sequence $A^2, A^4, A^6, \ldots$ converges to
$\begin{bmatrix}1 & -1 & 1\\0 & 0 & 1\\0 & 0 & 1\end{bmatrix}$.

Diagonalization, p. 508

ML.1. (a) $\lambda_1 = 0, \lambda_2 = 12$; $P = \begin{bmatrix}0.7071 & 0.7071\\-0.7071 & 0.7071\end{bmatrix}$.

(b) $\lambda_1 = -1, \lambda_2 = -1, \lambda_3 = 5$;

$$P = \begin{bmatrix}0.7743 & -0.2590 & 0.5774\\-0.6115 & -0.5411 & 0.5774\\-0.1629 & 0.8001 & 0.5774\end{bmatrix}.$$

(c) $\lambda_1 = 5.4142, \lambda_2 = 4.0000, \lambda_3 = 2.5858$.

$$P = \begin{bmatrix}0.5000 & -0.7071 & -0.5000\\0.7071 & -0.0000 & 0.7071\\0.5000 & 0.7071 & -0.5000\end{bmatrix}.$$

APPENDIX B

Section B.1, p. 520

1. (a) $4 + 2i$. (b) $-4 - 3i$. (c) $11 - 2i$.

(d) $-3 + i$. (e) $-3 + 6i$. (f) $-2 - i$.

(g) $7 - 11i$. (h) $-9 + 13i$.

3. (a)

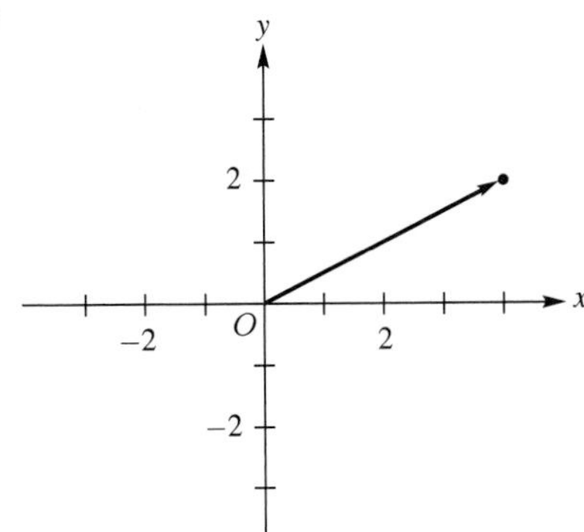

(b)

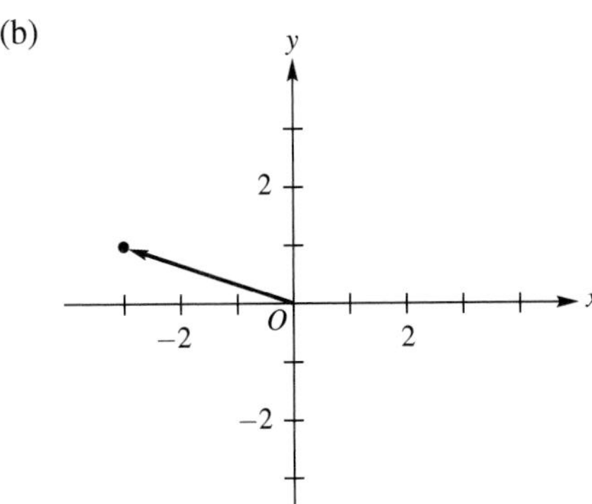

(c)

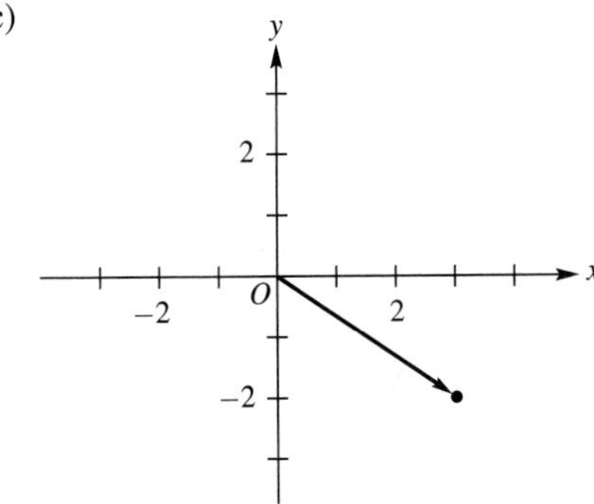

(d)

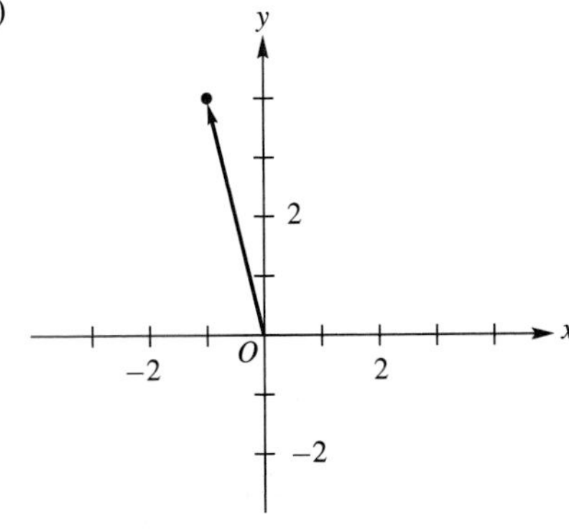

7. (a) $\begin{bmatrix} 2+4i & 5i \\ -2 & 4-2i \end{bmatrix}$. (b) $\begin{bmatrix} 4-3i \\ -2-i \end{bmatrix}$.

(c) $\begin{bmatrix} -4+4i & -2+16i \\ -4i & -8i \end{bmatrix}$.

(d) $\begin{bmatrix} 3i \\ -1-3i \end{bmatrix}$. (e) $\begin{bmatrix} 2i & -1+3i \\ -2 & -1-i \end{bmatrix}$.

(f) $\begin{bmatrix} -2i & 1-2i \\ 0 & 3+i \end{bmatrix}$. (g) $\begin{bmatrix} 3+i \\ -3+3i \end{bmatrix}$.

(h) $\begin{bmatrix} 3-6i \\ -2-6i \end{bmatrix}$.

9. $A^2 = \begin{bmatrix} -1 & 0 \\ 0 & -1 \end{bmatrix}$, $A^3 = \begin{bmatrix} 0 & -i \\ -i & 0 \end{bmatrix}$,

$A^4 = \begin{bmatrix} 1 & 0 \\ 0 & 1 \end{bmatrix}$, $A^{4n} = I_2$, $A^{4n+1} = A$,

$A^{4n+2} = A^2 = -I_2$, $A^{4n+3} = A^3 = -A$.

17. (a) $\begin{bmatrix} 0 & 0 \\ 0 & 0 \end{bmatrix}$. (b) $\begin{bmatrix} 4 & 18 \\ 0 & 4 \end{bmatrix}$.

(c) $\begin{bmatrix} -5 & 5i \\ 5i & -5 \end{bmatrix}$. (d) $\begin{bmatrix} 4 & 7i \\ 0 & -3 \end{bmatrix}$.

19. $\begin{bmatrix} 2 & 0 \\ 0 & 2 \end{bmatrix}$, $\begin{bmatrix} -1 & 0 \\ 0 & -1 \end{bmatrix}$.

Section B.2, p. 532

1. (a) No solution. (b) No solution.

(c) $x_1 = \frac{3}{4} + \frac{5}{4}i$, $x_2 = \frac{3}{2} + i$.

3. (a) $x_1 = i$, $x_2 = 1$, $x_3 = 1 - i$.

(b) $x_1 = 0$, $x_2 = -i$, $x_3 = i$.

5. (a) $\frac{1}{5}\begin{bmatrix} 2+i & 2-4i \\ 3-i & -2-i \end{bmatrix}$.

(b) $\frac{1}{6}\begin{bmatrix} i & 1-3i & 1 \\ -2-3i & 2i & 3+2i \\ 1 & 2i & -i \end{bmatrix}$.

9. (a) Yes. (b) Linearly independent.

11. (a) The eigenvalues are $\lambda_1 = 1+i$, $\lambda_2 = 1-i$. Associated eigenvectors are

$$\mathbf{x}_1 = \begin{bmatrix} -i \\ 1 \end{bmatrix} \quad \text{and} \quad \mathbf{x}_2 = \begin{bmatrix} i \\ 1 \end{bmatrix}.$$

(b) The eigenvalues are $\lambda_1 = 0$, $\lambda_2 = 2$. Associated eigenvectors are

$$\mathbf{x}_1 = \begin{bmatrix} -i \\ 1 \end{bmatrix} \quad \text{and} \quad \mathbf{x}_2 = \begin{bmatrix} i \\ 1 \end{bmatrix}.$$

(c) The eigenvalues are $\lambda_1 = 1, \lambda_2 = 2, \lambda_3 = 3$.
Associated eigenvectors are

$$\mathbf{x}_1 = \begin{bmatrix} 0 \\ -i \\ 1 \end{bmatrix}, \quad \mathbf{x}_2 = \begin{bmatrix} 1 \\ 0 \\ 0 \end{bmatrix}, \quad \text{and} \quad \mathbf{x}_3 = \begin{bmatrix} 0 \\ i \\ 1 \end{bmatrix}.$$

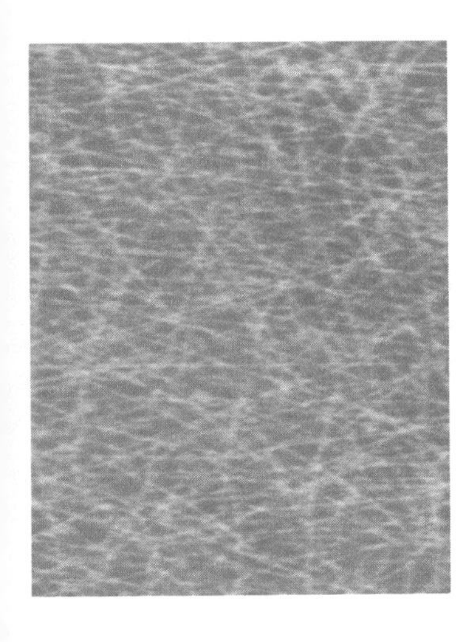

INDEX

Page Index to Lemmas, Theorems, and Corollaries

[*For Index to Lemmas, Theorems, and Corollaries, see preceeding two pages.*]

Built-in MATLAB Commands

ans	**inv**	**roots**
clear	**norm**	**rref**
conj	**null**	**size**
det	**ones**	**sqrt**
diag	**pi**	**sum**
dot	**poly**	**tril**
eig	**polyval**	**triu**
exit	**polyvalm**	**zeros**
eye	**quit**	\
fix	**rand**	;
format	**rank**	:
help	**rat**	′ (prime)
hilb	**real**	+, −, *, /, ^
image	**reshape**	

Supplemental Instructional Commands

adjoint	**forsub**	**planelt**
bksub	**gschmidt**	**reduce**
cofactor	**homsoln**	**vec2demo**
crossprd	**invert**	**vec3demo**
crossdemo	**lsqline**	
dotprod	**lupr**	